NEC Exam Preparation

2008 Edition

Mike Holt Enterprises, Inc.
1.888.NEC.CODE • www.MikeHolt.com • Info@MikeHolt.com

NOTICE TO THE READER

The publisher does not warrant or guarantee any of the products described herein or perform any independent analysis in connection with any of the product information contained herein. The publisher does not assume, and expressly disclaims, any obligation to obtain and include information other than that provided to it by the manufacturer.

The reader is expressly warned to consider and adopt all safety precautions that might be indicated by the activities herein and to avoid all potential hazards. By following the instructions contained herein, the reader willingly assumes all risks in connection with such instructions.

The publisher makes no representation or warranties of any kind, including but not limited to, the warranties of fitness for particular purpose or merchantability, nor are any such representations implied with respect to the material set forth herein, and the publisher takes no responsibility with respect to such material. The publisher shall not be liable for any special, consequential, or exemplary damages resulting, in whole or part, from the reader's use of, or reliance upon, this material.

Mike Holt's Illustrated Guide to
NEC Exam Preparation
2008 Edition

First Printing: April 2008

Technical Illustrator: Mike Culbreath
Cover Design: Tracy Jette
Layout Design and Typesetting: Cathleen Kwas and Tara Moffitt

COPYRIGHT © 2008 Charles Michael Holt, Sr.
ISBN 978-978-1-932685-36-7

For more information, call 1.888.NEC.CODE (888-632-2633), or E-mail Info@MikeHolt.com.

All rights reserved. No part of this work covered by the copyright hereon may be reproduced or used in any form or by any means graphic, electronic, or mechanical, including photocopying, recording, taping, or information storage and retrieval systems without the written permission of the publisher. You can request permission to use material from this text, phone 1.888.NEC.CODE (888-632-2633), Info@MikeHolt.com, or www.MikeHolt.com.

NEC, NFPA, and National Electrical Code are registered trademarks of the National Fire Protection Association.

 This logo is a registered trademark of Mike Holt Enterprises, Inc.

To request examination copies of this or other Mike Holt Publications:
Phone: 1.888.NEC.CODE (888-632-2633) • Fax: 1.352.360.0983
E-mail: Info@MikeHolt.com
or visit Mike Holt online: www.MikeHolt.com
You can download a sample PDF of all our publications by visiting www.MikeHolt.com

I dedicate this book to the
Lord Jesus Christ,
my mentor and teacher.
Proverbs 16:3

One Team

To Our Instructors and Students:

We are committed to providing you the finest product with the fewest errors, but we are realistic and know that there will be errors found and reported after the printing of this book. The last thing we want is for you to have problems finding, communicating, or accessing this information. It is unacceptable to us for there to be even one error in our textbooks or answer keys. For this reason, we are asking you to work together with us as One Team.

Students: Please report any errors that you may find to your instructor.
Instructors: Please communicate these errors to us.

Our Commitment:

We will continue to list all of the corrections that come through for all of our textbooks and answer keys on our Website. We will always have the most up-to-date answer keys available to instructors to download from our instructor Website. We do not want you to have problems finding this updated information, so we're outlining where to go for all of this below:

To view textbook and answer key corrections: Students and instructors go to our Website, www.MikeHolt.com, click on "Books" in the sidebar of links, and then click on "Corrections."

To download the most up-to-date answer keys: Instructors go to our Website, www.MikeHolt.com, click on "Instructors" in the sidebar of links and then click on "Answer Keys." On this page you will find instructions for accessing and downloading these answer keys.

If you are not registered as an instructor you will need to register. Your registration will be sent to our educational director who in turn reviews and approves your registration. In your approval E-mail will be the login and password so you can have access to all of the answer keys. If you have a situation that needs immediate attention, please contact the office directly at 1.888.NEC.CODE.

Call 1.888.NEC.CODE or visit us online at www.MikeHolt.com

Table of Contents

CHAPTER 1—ELECTRICAL THEORY 1

UNIT 1—ELECTRICIAN'S MATH AND BASIC ELECTRICAL FORMULAS 3
Introduction to Unit 1 3
Part A—Electrician's Math 3
Introduction 3
1.1 Whole Numbers 3
1.2 Decimals 3
1.3 Fractions 3
1.4 Percentages 4
1.5 Multiplier 4
1.6 Percent Increase 5
1.7 Reciprocals 5
1.8 Squaring a Number 6
1.9 Parentheses 7
1.10 Square Root 7
1.11 Volume 8
1.12 Kilo 8
1.13 Rounding Off 9
1.14 Testing Your Answer for Reasonableness 9

Part B—Basic Electrical Formulas 10
Introduction 10
1.15 Electrical Circuit 10
1.16 Power Source 10
1.17 Conductance 11
1.18 Circuit Resistance 11
1.19 Ohm's Law 12
1.20 Ohm's Law and Alternating Current 12
1.21 Ohm's Law Formula Circle 12
1.22 PIE Formula Circle 14
1.23 Formula Wheel 15
1.24 Using the Formula Wheel 16
1.25 Power Losses of Conductors 16
1.26 Cost of Power 17
1.27 Power Changes with the Square of the Voltage 18

Conclusion to Unit 1 20
Unit 1 Practice Questions 21
Unit 1 Challenge Questions 26

UNIT 2—ELECTRICAL CIRCUITS 28
Introduction to Unit 2 28
Part A—Series Circuits 28
2.1 Practical Uses of the Series Circuit 28
2.2 Understanding Series Calculations 30
2.3 Series Circuit Calculations 33
2.4 Power Calculations 33
2.5 Variations 33
2.6 Series Circuit Notes 33
2.7 Series-Connected Power Supplies 34

Part B—Parallel Circuits 34
Introduction 34
2.8 Practical Uses of the Parallel Circuit 35
2.9 Understanding Parallel Calculations 36
2.10 Circuit Resistance 37
2.11 Parallel Circuit Notes 39
2.12 Parallel-Connected Power Supplies 40

Part C—Series-Parallel Circuits 40
Introduction 40
2.13 Review of Series and Parallel Circuits 40
2.14 Working With Series-Parallel Circuits 41
2.15 Voltage 42

Part D—Multiwire Branch Circuits 42
Introduction 42
2.16 Neutral Conductor 43
2.17 Grounded Conductor 43
2.18 Current Flow on the Neutral Conductor 43
2.19 Balanced Systems 44
2.20 Unbalanced Current 44
2.21 Multiwire Branch Circuits 46
2.22 Dangers of Multiwire Branch Circuits 47
2.23 *NEC* Requirements 48

Conclusion to Unit 2 49
Unit 2 Practice Questions 50
Unit 2 Challenge Questions 55

UNIT 3—UNDERSTANDING ALTERNATING CURRENT 59
Introduction to Unit 3 59
Part A—Understanding Alternating Current 59
Introduction 59
3.1 Current Flow 59
3.2 Why Alternating Current is Used 60
3.3 How Alternating Current is Produced 60
3.4 AC Generator 60
3.5 Waveform 61
3.6 Sine Wave 62
3.7 Frequency 62
3.8 Phase 62
3.9 Degrees 63
3.10 Lead or Lag 63
3.11 Values of Alternating Current 64

Part B—Capacitance 64
Introduction 64
3.12 Charged Capacitor 65
3.13 Electrical Field 65
3.14 Discharging a Capacitor 66

Table of Contents

3.15	Determining Capacitance	66
3.16	Uses of Capacitors	67
3.17	Phase Relationship	67

Part C—Induction 68
Introduction 68

3.18	Self-Induction	69
3.19	Induced Voltage and Applied Current	69
3.20	Conductor AC Resistance	70
3.21	Conductor Shape	70
3.22	Magnetic Cores	71
3.23	Self-Induced and Applied Voltage	72
3.24	Inductive Reactance	72
3.25	Phase Relationship	73
3.26	Uses of Induction	73

Part D—Power Factor 73
Introduction 73

3.27	Apparent Power (Volt-Amperes)	74
3.28	True Power (Watts)	74
3.29	Power Factor	75
3.30	Unity Power Factor	75
3.31	Power Factor Formulas	75
3.32	Cost of True Power	76
3.33	Effects of Power Factor	77

Part E—Efficiency 78
Introduction 78

3.34	Efficiency Formulas	79

Conclusion to Unit 3 80

Unit 3 Practice Questions 81

Unit 3 Challenge Questions 88

UNIT 4—MOTORS AND TRANSFORMERS 92

Introduction to Unit 4 92

Part A—Motor Basics 92
Introduction 92

4.1	Motor Principles	93
4.2	Dual-Voltage AC Motors	93
4.3	Motor Horsepower Ratings	93
4.4	Motor Current Ratings	94
4.5	Calculating Motor FLA	94
4.6	Motor-Starting Current	95
4.7	Motor-Running Current	95
4.8	Motor Locked-Rotor Current (LRC)	96
4.9	Motor Overload Protection	96
4.10	Direct-Current Motor Principles	97
4.11	Direct-Current Motor Types	97
4.12	Reversing the Rotation of a DC Motor	98
4.13	AC Induction Motor	98
4.14	Alternating-Current Motor Types	98
4.15	Reversing the Rotation of an AC Motor	99

Part B—Transformers 99
Introduction 99

4.16	Transformer Basics	99
4.17	Secondary Induced Voltage	100
4.18	Efficiency	101
4.19	Transformer Turns Ratio	101
4.20	Autotransformers	103
4.21	Power Losses	103
4.22	Transformer kVA Rating	105
4.23	Current Flow	105
4.24	Current Rating	105

Conclusion to Unit 4 107

Unit 4 Practice Questions 108

Unit 4 Challenge Questions 112

CHAPTER 2—NEC CALCULATIONS 117

UNIT 5—RACEWAY AND BOX CALCULATIONS 119

Introduction to Unit 5 119

Part A—Raceway Fill 119
Introduction 119

5.1	Understanding the NEC, Chapter 9 Tables	119
5.2	Raceway Calculations	126
5.3	Wireways	128
5.4	Tips for Raceway Calculations	130

Part B—Outlet Box Fill Calculations [314.16] 130
Introduction 130

5.5	Sizing Box—Conductors All the Same Size [Table 314.16(A)]	130
5.6	Conductor Equivalents	131
5.7	Outlet Box Sizing [314.16(B)]	134

Part C—Pull Boxes, Junction Boxes, and Conduit Bodies 136
Introduction 136

5.8	Pull/Junction Box Sizing Requirements	137
5.9	Pull/Junction Box Sizing Tips	139
5.10	Pull Box Examples	139

Summary 141

Conclusion to Unit 5 142

Unit 5 Practice Questions 143

Unit 5 Challenge Questions 147

UNIT 6—CONDUCTOR SIZING AND PROTECTION CALCULATIONS 149

Introduction to Unit 6 149

Part A—General Conductor Requirements 149
Introduction 149

6.1	Conductor Insulation [Table 310.13]	149
6.2	Conductor Sizing [110.6]	152
6.3	Smallest Conductor Size [310.5]	152
6.4	Conductor Size—Terminal Temperature Rating [110.14(C)]	152
6.5	Conductors in Parallel	154
6.6	NEC Requirements for Conductors in Parallel [310.4]	155
6.7	Overcurrent Protection [Article 240]	157
6.8	Overcurrent Protection of Conductors—General Requirements [240.4]	159
6.9	Overcurrent Protection of Conductors—Specific Requirements	163

Part B—Conductor Ampacity	174
Introduction	174
6.10 Conductor Ampacity	174
6.11 Ambient Temperature Correction Factor [Table 310.16]	174
6.12 Conductor Bundling Ampacity Adjustment Factor [Table 310.15(B)(2)(a)]	177
6.13 Ambient and Conductor Bundling Adjustment	180
6.14 Current-Carrying Conductors	182
6.15 Conductor Sizing Summary	184
Conclusion to Unit 6	185
Unit 6—Practice Questions	186
Unit 6—Challenge Questions	191

UNIT 7—MOTOR AND AIR-CONDITIONING CALCULATIONS ... 193

Introduction to Unit 7	193
Part A—Motor Calculations	193
7.1 Scope of Article 430	193
7.2 FLC Versus Motor Nameplate	194
7.3 Highest Rated Motor [430.17]	195
7.4 Branch-Circuit Conductor Size	195
7.5 Feeder Conductor Size [430.24]	196
7.6 Overload Protection [430.6(A)(2) and 430.32(A)]	197
7.7 Branch-Circuit Short-Circuit and Ground-Fault Protection [430.51]	200
7.8 Branch-Circuit Summary	203
7.9 Feeder Protection [430.61]	205
7.10 Motor VA Calculations	206
Part B—Air-Conditioning Calculations	207
7.11 Scope of Article 440	207
7.12 Other Articles	207
7.13 Short-Circuit and Ground-Fault Protection	207
7.14 Conductor Sizing for Single Motor-Compressor	209
Conclusion to Unit 7	210
Unit 7—Practice Questions	211
Unit 7—Challenge Questions	215

UNIT 8—VOLTAGE-DROP CALCULATIONS ... 217

Introduction to Unit 8	217
PART A—CONDUCTOR RESISTANCE CALCULATIONS	217
Introduction	217
8.1 Conductor Resistance	217
8.2 Conductor Resistance—Direct-Current Circuits [Chapter 9, Table 8]	219
8.3 Conductor Resistance—Alternating-Current Circuits	220
8.4 Alternating-Current Resistance	220
8.5 AC Resistance as Compared to DC Resistance	222
PART B—Voltage-Drop Considerations	223
Introduction	223
8.6 NEC Voltage-Drop Recommendations	223
8.7 Determining Circuit Conductors' Voltage Drop—Ohm's Law Method	224
8.8 Determining Circuit Conductors' Voltage Drop—Formula Method	226
8.9 Sizing Conductors to Prevent Excessive Voltage Drop	227
8.10 Limiting Conductor Length to Minimize Voltage Drop	228
8.11 Limiting Current to Limit Voltage Drop	229
8.12 Fire Pump Motor Circuits	230
8.13 Sizing Conductors with Loads at Different Distances	233
Conclusion to Unit 8	234
Unit 8 Practice Questions	235
Unit 8 Challenge questions	240

UNIT 9—DWELLING UNIT CALCULATIONS ... 243

Introduction to Unit 9	243
Part A—General Requirements	243
Introduction	243
9.1 General Requirements	243
9.2 Voltages [220.5(A)]	244
9.3 Fraction of an Ampere [220.5(B)]	244
9.4 Small-Appliance Circuits [210.11(C)(1)]	245
9.5 Cooking Equipment—Branch Circuit [Table 220.55, Note 4]	250
9.6 Laundry Circuit [210.11(C)(2)]	253
9.7 Lighting and Receptacles	253
PART B—Standard Method—Feeder/Service Load Calculations	255
Introduction	255
9.8 Dwelling Unit Feeder/Service Load Calculations (Article 220, Part III)	255
9.9 Dwelling Unit Example	257
Part C—Optional Method—Feeder/Service Load Calculations	264
Introduction	264
9.10 Dwelling Unit Optional Calculations [220.82]	264
9.11 Optional Calculation Example	265
Part D—Other Topics of Interest	266
9.12 Neutral Calculations [220.61]	266
9.13 Grounding and Bonding of Service Equipment	268
Conclusion to Unit 9	273
Unit 9 Practice Questions	274
Unit 9 Challenge Questions	280

CHAPTER 3—ADVANCED NEC CALCULATIONS ... 283

UNIT 10—MULTIFAMILY DWELLING CALCULATIONS ... 285

Introduction to Unit 10	285
10.1 Multifamily Dwelling Calculations—General	286
Part A—Standard Method—Feeder/Service Load Calculations	287
Introduction	287
10.2 Multifamily Dwelling Unit Calculation Examples—Standard Method	287
10.3 Multifamily Dwelling Calculations—Standard Method Example	295

Table of Contents

Part B—Optional Method—Feeder/Service Load Calculations ...297
Introduction..297
10.4 Multifamily Dwelling Unit Calculations [220.84]—
Optional Method 297
10.5 Multifamily—Optional Method Example 1 [220.84].........297
10.6 Multifamily—Optional Method Example 2 [220.84].........300

Conclusion to Unit 10—Multifamily Dwelling Calculations..........302

Unit 10 Practice Questions 303

Unit 10 Challenge Questions 307

UNIT 11—COMMERCIAL CALCULATIONS 309
Introduction to Unit 11 309

Part A—General 309
Introduction...309
11.1 General Requirements..309
11.2 Conductor Ampacity [Article 100] 309
11.3 Conductor Overcurrent Protection [240.4] 311
11.4 Voltages [220.5(A)]...311
11.5 Rounding an Ampere [220.5(B)]....................................312
11.6 Lighting—Demand Factors [Tables 220.12 and 220.42] 313
11.7 Lighting Without Demand Factors
[215.2(A)(1), 230.42(A)(1), and Table 220.12]....................313
11.8 Lighting—Miscellaneous 313
11.9 Multioutlet Receptacle Assembly [220.14(H)] 314
11.10 Receptacle VA Load 315
11.11 Banks and Offices—General Lighting and Receptacles
[220.14(K)] 316
11.12 Sign Circuit [220.14(F) and 600.5] 317

Part B—Examples 318
11.13 Bank/Office Building Example 318
11.14 Marina [555.12] 321
11.15 Mobile/Manufactured Home Park [550.31] 322
11.16 Recreational Vehicle Park [551.73] 322

Part C—Optional Method—Feeder/Service Load Calculations ...323
Introduction...323
11.17 New Restaurant—Optional Method [220.88].................323

Part D—Welders 325
Introduction...325
11.18 Arc Welders..325
11.19 Resistance Welders..326
11.20 Light Industrial Calculation..327

Conclusion to Unit 11 329

Unit 11 Practice Questions 330

Unit 11 Challenge Questions 333

UNIT 12—TRANSFORMER CALCULATIONS.............................335
Introduction to Unit 12 335

Part A—General 335
Introduction...335
12.1 Transformer Basics 335
12.2 Secondary Induced Voltage..336
12.3 Autotransformers...337

12.4 Power Losses..337
12.5 Efficiency 339
12.6 Delta/Delta Connected Transformers...........................339
12.7 Delta/Wye Connected Transformers 343
12.8 Transformer Turns Ratio...344
12.9 Transformer kVA Rating..345
12.10 Current Flow...346
12.11 Line Currents 346

Part B—NEC Requirements...350
12.12 Transformer Overcurrent Protection.............................350
12.13 Primary Conductor Sizing 351
12.14 Secondary Conductor Sizing..352
12.15 Grounding and Bonding 355
12.16 Available Short-Circuit Current 358

Conclusion to Unit 12 361

Unit 12 Practice Questions 362

Unit 12 Challenge Questions 365

CHAPTER 4—NEC PRACTICE QUIZZES, CHALLENGE QUIZZES, AND FINAL NEC EXAMS

Practice Quizzes

Unit 1 Practice Questions in Straight Order—
Article 90.1 through 225.6 369

Unit 1 Practice Questions in Random Order—
Article 100 through 215.10 378

Unit 2 Practice Questions in Straight Order—
Article 225.16 through 250.102...383

Unit 2 Practice Questions in Random Order—
Article 90.2 through 250.104..392

Unit 3 Practice Questions in Straight Order—
Article 250.118 through 334.2 397

Unit 3 Practice Questions in Random Order—
Article 90.2 through 320.30..406

Unit 4 Practice Questions in Straight Order—
Article 100 through 378.10 411

Unit 4 Practice Questions in Random Order—
Article 90.1 through 358.12..420

Unit 5 Practice Questions in Straight Order—
Article 378.10 through 410.151 425

Unit 5 Practice Questions in Random Order—
Article 90.4 through 410.90..434

Unit 6 Practice Questions in Straight Order—
Article 410.151 through 502.1 438

Unit 6 Practice Questions in Random Order—
Article 90.2 through 502.10..448

Unit 7 Practice Questions in Straight Order—
Article 502.10 through 550.2..452

Table of Contents

Unit 7 Practice Questions in Random Order—
Article 90.2 through 547.10 461

Unit 8 Practice Questions in Straight Order—
Article 550.13 through 680.22 465

Unit 8 Practice Questions in Random Order—
Article 90.2 through 680.22 474

Unit 9 Practice Questions in Straight Order—
Article 680.23 through 760.1 478

Unit 9 Practice Questions in Random Order—
Article 90.3 through 760.2 487

Unit 10 Practice Questions in Straight Order—
Article 760.24 through Chapter 9 491

Unit 10 Practice Questions in Random Order—
Article 90.3 through Annex C 500

Challenge Quizzes

Challenge Quiz 1—
Article 90 through Chapter 9 509

Challenge Quiz 2—
Article 90 through Chapter 9 514

Challenge Quiz 3—
Article 90 through Chapter 9 519

Challenge Quiz 4—
Article 90 through Chapter 9 524

Challenge Quiz 5—
Article 90 through Chapter 9 529

Challenge Quiz 6—
Article 90 through Chapter 9 534

Challenge Quiz 7—
Article 90 through Chapter 9 539

Final Exams

Final Exam 1 Practice Questions in Random Order—
Article 90 through Annex C 543

Final Exam 2 Practice Questions in Random Order—
Article 90 through Annex C 552

Introduction

Mike Holt's Illustrated Guide to NEC Exam Preparation

This edition of Mike Holt's *Illustrated Guide to NEC Exam Preparation* is intended to provide you with the knowledge to pass your exam the first time. The writing style of this textbook, as with all of Mike Holt's products, is meant to be informative, practical, informal, easy to read, and applicable for today's electrical professional. Also, just like all of Mike Holt's textbooks, it contains hundreds of detailed color-coded illustrations that apply to today's electrical installations.

Passing an important electrical exam is the dream of all those who care about improving themselves and their families. Unfortunately, many don't pass the exam at all, and few pass it the first time. The primary reason people fail their exam is because they're not prepared on the technical material and/or they don't know how to take the exam.

Most electrical exams contain questions on electrical theory, basic electrical calculations, the *Code*, and important and difficult *National Electrical Code* calculations. This textbook contains hundreds of illustrations, examples, and almost 3,200 practice questions covering all of these subjects and 36 practice quizzes. Mike Holt's *Illustrated Guide to NEC Exam Preparation* is intended to be used with the 2008 *National Electrical Code*.

Scope of NEC Exam Preparation

This textbook covers the general installation requirements contained in the *NEC* Articles 90 through 830 that Mike considers to be of critical importance. This textbook contains the following stipulations:

- **Power Systems and Voltage.** All power-supply systems are assumed to be solidly grounded and any of the following voltages: 120V single-phase, 120/240V single-phase, 120/208V three-phase, 120/240V three-phase, or 277/480V three-phase, unless identified otherwise.

- **Electrical Calculations.** Unless the question or example specifies three-phase, the questions and examples are based on a single-phase power supply.

- **Rounding.** All calculations are rounded to the nearest ampere in accordance with 220.5(B).

- **Conductor Material.** Conductors are considered copper, unless aluminum is identified or specified.

- **Conductor Sizing.** Conductors are sized based on a THHN copper conductor terminating on a 75°C terminal in accordance with 110.14(C), unless the question or example identifies otherwise.

- **Overcurrent Device.** The term "overcurrent device" in this textbook refers to a molded case circuit breaker, unless identified otherwise. Where a fuse is identified, it's to be of the single-element type, also known as a "one-time fuse," unless identified otherwise.

Author's Comment: Because the neutral conductor of a solidly grounded system is always grounded to the earth, it's both a "grounded conductor" and a "neutral" conductor. To make it easier for the reader of this textbook, we will refer to the "grounded" conductor of a solidly grounded system as the "neutral" conductor.

How to Use This Textbook

This textbook is divided into four chapters, which are further divided into units. Chapters 1 through 3 are the calculations section of the textbook.

- Chapter 1. Electrical Theory
- Chapter 2. *NEC* Calculations
- Chapter 3. Advanced *NEC* Calculations

Each unit of this part of the textbook contains objectives, explanations with graphics, examples, steps for calculations, formulas, and practice calculations. As you read through the textbook, be sure to take the time to review the author's comments, graphics, and examples with your 2008 *Code* book.

The writing style of this textbook is intended to be informal and relaxed. The color graphics are included to help make the learning easy and fun.

Introduction

After you've read each unit, you need to take time to complete the unit questions. Each of the 12 units in chapters 1 through 3 include the following:

- Unit Practice Questions
- Unit Challenge Questions

The calculations study of the first three chapters will help you learn to use the *NEC* tables and perform calculations based on the *Code*. The material in Chapter 4 will help you in answering *NEC* questions.

Chapter 4 includes three types of *Code* quizzes. You may want to intersperse the *NEC* review questions with your study of the first three chapters of the textbook, or you may want to complete the first three chapters first, and then work on the *NEC* review questions. Be sure to use your 2008 *Code* book to answer the review questions (quizzes). These quizzes are:

- **Review Quiz.** Each quiz contains 100 questions that are in *Code* order and take you step by step through each of the nine chapters and Annex C of the *National Electrical Code*.

- **Practice Quiz.** Practice quizzes each contain 50 questions that are presented in random *Code* order and contain questions different than those in the review quizzes.

- **Challenge Quiz.** Each quiz contains 50 questions that cover all nine chapters and Annex C of the *NEC*. The questions in the challenge quizzes don't follow the chapters of the *Code* book (as the other quizzes do); they're organized in a random manner, and you might find them harder to answer. Scores for the first few challenge quizzes might be somewhat lower than you would like to see. But, as you go through this textbook and take the review and practice quizzes, you'll learn a great deal and gain a better understanding of the material. This improved knowledge and understanding will help you improve on the challenge quizzes as you proceed.

If you have difficulty with a question or section in the textbook, skip it for the moment and get back to it later. Don't frustrate yourself! The answer key is intended to be very clear, so be sure to review those questions that you miss.

If there are *NEC* questions that you don't understand after reading the answer key, don't hesitate to ask your co-workers or fellow students about them. Additional study of the *Code* questions on a chapter-by-chapter basis is available using our *Understanding the National Electrical Code* book which includes many illustrations and clear explanations.

Journeyman or Master/Contractor Units

If you're using this textbook to pass a Journeyman Exam, complete Units 1 through 9 in Chapters 1 and 2 of the textbook (including the unit calculation practice and challenge questions), as well as the *NEC* quizzes in Chapter 4.

If you're using this textbook to pass a Master/Contractor's Exam, complete Units 1 through 12 in its entirety (including the unit calculation practice and challenge questions) as well as the *NEC* quizzes in Chapter 4.

Cross-References and Author's Comments

This textbook contains several *NEC* cross-references to other related *Code* requirements to help you develop a better understanding of how the *NEC* rules relate to one another. These cross-references are identified by *Code* section numbers in brackets, an example of which is "[90.4]."

Author's Comments were written by Mike to help you (the reader) better understand the *NEC* by bringing to your attention items you should be aware of.

Special Sections and Examples

Additional information to better help you understand a concept is identified with light green shading. In addition, examples are highlighted with a yellow background.

 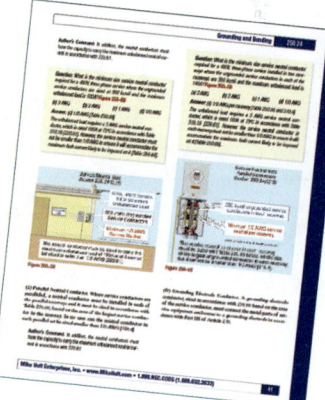

Difficult Concepts

As you progress through this textbook, you might find you don't understand every explanation, example, calculation, or comment. Don't get frustrated, and don't get down on yourself. Remember, this is the *National Electrical Code*, and

Introduction

sometimes the best attempt to explain a concept isn't enough to make it perfectly clear. If you're still confused, visit www.MikeHolt.com, and post your question on the *Code* Forum for help.

Different Interpretations

Some electricians, contractors, instructors, inspectors, engineers, and others enjoy the challenge of discussing the *Code* requirements, hopefully in a positive and a productive manner. This give-and-take is important to the process of better understanding *NEC* requirements and applications. However, if you're going to get into an *NEC* discussion, please don't spout out what you think without having the actual *Code* in your hand. The professional way of discussing an *NEC* requirement is by referring to a specific section, rather than talking in vague generalities.

Textbook Errors and Corrections

If you believe there's an error of any kind in this textbook (typographical, grammatical, or technical), no matter how insignificant, please let us know.

Any errors found after printing are listed on our Website, so if you find an error, first check to see if it has already been corrected. Go to www.MikeHolt.com, click on the "Books" link, and then the "Corrections" link (www.MikeHolt.com/bookcorrections.htm).

If you don't find the error listed on the Website, contact us by E-mailing us at Corrections@MikeHolt.com. Be sure to include the book title, page number, and any other pertinent information.

Journeyman and Master/Contractor Comprehensive Library

The Comprehensive Exam Preparation Library is based on a full term of live classes. You'll learn everything you need to know to pass your exam the first time. Mike's dynamic style and detailed graphics easily explain the most difficult subjects. The Journeyman Comprehensive Library includes four textbooks, a simulated exam, and eighteen DVDs. The Master/Contractor Comprehensive Library includes everything in the Journeyman Library, plus three additional Calculation DVDs. For more information, visit www.MikeHolt.com.

Journeyman and Master/Contractor Intermediate Library

The Journeyman Intermediate Library includes two textbooks and seven DVDs. The Master/Contractor Intermediate Library includes everything in the Journeyman Library, plus three additional Calculation DVDs. For more information, visit www.MikeHolt.com.

Passing Your Exam

This textbook was designed to help you prepare for your *National Electrical Code* exam.

Understanding the Emotional Aspects of Learning

To learn effectively, you must develop an attitude that learning is a process that will help you grow both personally and professionally. The learning process has an emotional as well as an intellectual component that we must recognize. To understand what affects our learning, consider the following:

Positive Image. Many feel disturbed by the expectations of being treated like children and we often feel threatened with the learning experience.

Uniqueness. Each of us will understand the subject matter from different perspectives and we all have some unique learning challenges and needs.

Resistance to Change. People tend to resist change and resist information that appears to threaten their comfort level of knowledge. However, we often support new ideas that support our existing beliefs.

Dependence and Independence. The dependent person is afraid of disapproval and often will not participate in class discussion and will tend to wrestle alone. The independent person spends too much time asserting differences and too little time trying to understand others' views.

Fearful. Most of us feel insecure and afraid of learning, until we understand the process. We fear that our performance will not match the standard set by us or by others.

Egocentric. Our ego tendency is to prove someone is wrong, with a victorious surge of pride. Learning together without a win-lose attitude can be an exhilarating learning experience.

Emotional. It's difficult to discard our cherished ideas in the face of contrary facts when overpowered by the logic of others.

Getting the Best Grade

Studies have concluded that for students to get their best grades, they must learn to get the most from their natural abilities. It's not how long you study or how high your IQ is, it's what you do and how you study that counts the most. To get your best grade, you must make a decision to do your best and follow as many of the following techniques and suggestions as possible.

Reality. These instructions are a basic guide to help you get the maximum grade. It's unreasonable to think that all of the instructions can be followed to the letter all of the time. Day-to-day events and unexpected situations must be taken into consideration.

Support. You need encouragement in your studies and you need support from your loved ones and employer. To properly prepare for your exam, you need to study 10 to 15 hours per week for about 3 to 6 months.

Communication with Your Family. Good communication with your family is very important because studying every night and on weekends can cause much tension and stress. Try to get their support, cooperation, and encouragement during this difficult time. Let them know the benefits to the family and what passing the exam means. Be sure to plan some special time with them during this preparation period; don't go overboard and neglect them.

Stress. Stress can really take the wind out of you. It takes practice, but get into the habit of relaxing before you begin your studies. Stretch, do a few sit-ups and push-ups, take a 20-minute walk, or a few slow, deep breaths. Close your eyes for a couple of minutes, and deliberately relax the muscle

Passing Your Exam

groups that are associated with tension, such as the shoulders, back, neck, and jaw.

Attitude. Maintaining a positive attitude is important. It helps keep you going and helps keep you from becoming discouraged.

Training. Preparing for the exam is the same as training for any other event. Get plenty of rest and avoid intoxicating drugs, including alcohol. Stretching or exercising each day for at least 10 minutes helps you get in a better mood. Eat light meals such as pasta, chicken, fish, vegetables, fruit, and so forth. Try to avoid red meat, butter, sugar, salt, and high-fat content foods. They slow you down and make you tired and sleepy.

Eye Care. It's very important to have your eyes checked! Human eyes weren't designed to constantly focus on something less than an arm's length away. Our eyes were designed for survival, spotting food and enemies at a distance. Your eyes  will be under tremendous stress because of prolonged reading, which can result in headaches, fatigue, nausea, squinting, or eyes that burn, ache, water, or tire easily. Be sure to tell your eye doctor that you are studying to pass an exam (bring this textbook and the *Code* book with you) and that you expect to do a tremendous amount of reading and writing. Reading glasses can reduce eye discomfort.

Reducing Eye Strain. Be sure to look up occasionally, away from near tasks to distant objects. Your work area should be three times brighter than the rest of the room. Don't read under a single lamp in a dark room. Try to eliminate glare. Mixing of fluorescent and incandescent lighting can be helpful.

Posture. Sit up straight, with your chest up and your shoulders back, so both eyes are an equal distance from what you're viewing.

Getting Organized. Our lives are so busy that simply making time for homework and exam preparation is almost impossible. You can't waste time looking for a pencil or missing paper. Keep everything you 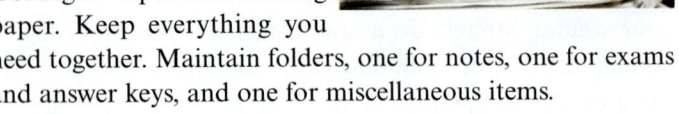 need together. Maintain folders, one for notes, one for exams and answer keys, and one for miscellaneous items.

Study Location. It's very important that you have a private study area available at all times. Keep your materials there. The dining room table is not a good spot.

Time Management. Time management and planning are very important. There simply are not enough hours in the day to get everything done. Make a schedule that allows time for work, rest, study, meals, family, and recreation. Establish a schedule that's consistent from day to day. Have a calendar and immediately plan your exam preparation schedule. Try to follow the same routine each week and try not to become overtired. Learn to pace yourself to accomplish as much as you can without the need for cramming.

Speak Up in Class. If you're in a classroom setting, the most important part of the learning process is participation. If you don't understand the instructor's point, ask for clarification. Don't try to get attention by asking 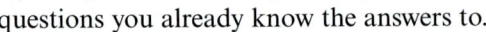 questions you already know the answers to.

Study With a Friend. Studying with a friend can make learning more enjoyable. You can push and encourage each other. You're more likely to study if someone else is depending on you. Students who study together perform above average because they try different approaches and explain their solutions to each other. This kind of interaction is a significant aid to learning retention. Those who study alone may spend much of their time reading and rereading the text and trying the same approach time after time with less success.

Study Anywhere/Anytime. To make the most of your limited time, always keep a copy of the book(s) with you. Any time you get a free minute, study! Continue to study any chance you get. You can study at the supply house when waiting for your material; you can study during your coffee break, or even while you're at the doctor's office. Become creative!

You need to find your best study time. For some, it's late at night when the house is quiet. For others, it's the first thing in the morning before things get going.

Set Priorities. Once you begin your study, stop all phone calls, TV shows, radio, snacks, and other interruptions. You can always take care of it later.

Passing Your Exam

Preparing to Take the Exam

Get Advance Information. Check with the testing authorities in advance to be certain you know what reference books and what kind of calculator you're allowed to take with you to the exam site. Find out if you're allowed to make notes in your *Code* book, or if the testing site will provide you with a *Code* book. If your state is using computerized testing, find the Website of the testing company and review any information that they make available. Take any sample online test(s) on the testing company's Website so you're familiar with the testing process.

Have the Proper Supplies. First of all, make sure you have everything needed several days before the exam. The night before the exam is not the time to be out buying pencils, calculators, and batteries. You should have a checklist (prepared in advance) of everything you can possibly need the night before the exam. The following is a sample checklist to get you started:

- Six sharpened #2H pencils or two mechanical pencils with extra #2H leads. The type with the larger lead is better for filling in the answer key.

- Two calculators. Most examining boards require quiet, paperless calculators. Solar calculators are great, but there may not be enough light to operate them. Find out what type of calculators will be allowed for the exam. Some exams don't allow scientific calculators, so be sure that you don't find yourself with a calculator you can't use!

- Spare batteries. Two sets of extra batteries should be taken. It's very unlikely you'll need them, but...

- Extra glasses if you use them.

- A wristwatch to ensure you'll stay on track.

- All your reference materials, even the ones not on the list. Let the proctors tell you which ones aren't permitted.

- Bring something to drink. Coffee is excellent.

- Some fruit, nuts, aspirin, etc.

- Know where the exam is going to take place and how long it takes to get there. Arrive at least 30 minutes early.

- Bring a jacket.

Meals. It's also a good idea to pack a lunch rather than going out. It can give you a little extra time to review the material for the afternoon portion of the exam, and it reduces the chance of coming back late.

Taking the Exam

Being prepared for an exam means more than just knowing electrical concepts, the *Code*, and the calculations. Have you felt prepared for an exam, and then choked when actually taking it? Many good and knowledgeable people didn't pass their exam because they didn't know "how to take an exam."

Taking exams is a learned process that takes practice and involves strategies. The following suggestions are designed to help you develop your strategies:

Relax. This is easier said than done, but it's one of the most important factors in passing your exam. Stress and tension cause us to choke or forget. Everyone has had experiences where they became tense and couldn't think straight. The first step is becoming aware of the tension, and the second step is to make a deliberate effort to relax. Make sure you're comfortable; remove clothes if you're hot, or put on a jacket if you're cold.

There are many ways to relax and you have to find a method that works for you. Two of the easiest methods that work very well for many people follow:

- **Breathing Technique:** Take a few slow deep breaths every few minutes. Don't confuse this with hyperventilation, which is abnormally fast breathing.

- **Single-Muscle Relaxation:** When we're tense or stressful, many of us do things like clench our jaw, squint our eyes, or tense our shoulders without even being aware of it. If you find a muscle group that does this, deliberately relax that one group. The rest of the muscles will automatically relax also. Try to repeat this every few minutes, and it will help you stay more relaxed during the exam.

Understand the Question. To answer a question correctly, you must first understand the question. One word in a question can totally change the meaning of it. Carefully read every word of every question. Underlining key words in the question will help you focus.

Passing Your Exam

Skip the Difficult Questions. Contrary to popular belief, you don't have to answer one question before going on to the next one. The irony is that the question you get stuck on is one that you'll probably get wrong anyway. This will result in not having enough time to answer the easy questions. You'll get all stressed-out worrying that you won't complete the exam on time, and a chain reaction is started. More people fail their exams this way than for any other reason.

Many states are using computerized testing, but even in the computerized format you're usually able to mark a question for later review and return to that question after you've answered the easier ones.

The following strategy should be used to avoid getting into this situation:

- **First Pass:** Answer the questions you know. Give yourself about 30 seconds for each question. If you can't find the answer in your reference book within the 30 seconds, go on to the next question. Chances are that you'll come across the answers while looking up another question. The total time for the first pass should be 25 percent of the exam time.

- **Second Pass:** This pass is done the same as the first pass except that you allow a little more time for each question, about 60 seconds. If you still can't find the answer, go on to the next one. Don't get stuck. The total time for the second pass should be about 30 percent of the exam time.

- **Third Pass:** See how much time is left and subtract 30 minutes. Spend the remaining time equally on each question. If you still haven't answered the question, it's time to make an educated guess. Never leave a question unanswered.

- **Fourth Pass:** Use the last 30 minutes of the exam to transfer your answers from the exam booklet to the answer key. Read each question and verify that you selected the correct answer on the test book. Transfer the answers carefully to the answer key. With the remaining time, see if you can find the answer to those questions you guessed at.

Guessing. When time is running out and you still have unanswered questions, GUESS! Never leave a question unanswered.

You can improve your chances of getting a question correct by the process of elimination. Many times, there are a couple of choices that can be easily eliminated as illogical or outside the range of expected answers.

How do you pick one of the remaining answers? Some people toss a coin, others count up how many of the answers were As, Bs, Cs, and Ds and use the one with the most as the basis for their guess.

Checking Your Work. The first thing to check (and you should be watching out for this during the whole exam) is to make sure you mark the answer in the correct spot. People have failed the exam by one-half of a point. When they reviewed their exam, they found they correctly answered several questions on the test booklet, but marked the wrong spot on the exam answer sheet. They knew the answer was "(b) False" but marked in "(d)" in error. Another thing to be very careful of, is marking the answer for, let's say question 7, in the spot reserved for question 8.

Changing Answers.  When re-reading the question and checking the answers during the fourth pass, resist the urge to change an answer. In most cases, your first choice is best and if you aren't sure, stick with the first choice. Only change answers if you are positive that you made a mistake. Multiple choice exams are graded electronically so be sure to thoroughly erase any answer that you changed. Also erase any stray pencil marks from the answer sheet.

Rounding Off. You should always round your answers to the same number of places as the exam's answers.

Example: If an exam has multiple choice of:

(a) 2.10 (b) 2.20
(c) 2.30 (d) none of these

And your calculation comes out to 2.16, don't choose the answer (d) none of these. The correct answer is (b) 2.20, because the answers in this case are rounded off to the nearest tenth.

Example: It could be rounded to tens, such as:

(a) 50
(b) 60
(c) 70
(d) none of these.

For this group, an answer such as 67 will be (c) 70, while an answer of 63 will be (b) 60. The general rule is to check the question's choice of answers and then round off your answer to match it.

Things to be Careful of:

- Don't get stuck on any one question.
- Read each question carefully.
- Be sure you mark the answer in the correct spot on the answer sheet.
- Don't get flustered or extremely tense.

Summary:

- Make sure everything is ready and packed the night before the exam.
- Don't try to cram the night before the exam—if you don't know it by then, it's too late!
- Have a good breakfast. Get the thermos and energy snacks ready.
- Take all your reference books. Let the proctors tell you what you can't use.
- Know where the exam is to be held and arrive early.
- Bring identification and your confirmation papers from the license board if this is required.
- Review your *NEC* while you wait for your exam to begin.
- Try to stay relaxed.
- Determine the time per question for each pass and don't forget to save 30 minutes for transferring your answers to the answer key.
- Remember—in the first pass answer only the easy questions. In the second pass, spend a little more time with each question, but don't get stuck. In the third pass, use the remainder of the time minus 30 minutes to answer the remaining questions. In the fourth pass, check your work and transfer the answers to the answer key.

How to Use the *National Electrical Code*

This textbook is to be used with the *NEC*, not as a replacement for the *NEC*, so be sure to have a copy of the 2008 *National Electrical Code* handy. Compare what Mike explains in the text to your *Code* book, and discuss those topics that you find difficult to understand with others. As you read through this textbook, be sure to take the time to review the text with the outstanding graphics and examples.

The *National Electrical Code* is written for persons who understand electrical terms, theory, safety procedures, and electrical trade practices. These individuals include electricians, electrical contractors, electrical inspectors, electrical engineers, designers, and other qualified persons. The *Code* is not written to serve as an instructive or teaching manual for untrained individuals [90.1(C)].

Learning to use the *NEC* is somewhat like learning to play the game of chess; it's a great game if you enjoy mental warfare. When learning to play chess, you must first learn the names of the game pieces, how the pieces are placed on the board, and how each piece moves.

Once you understand the fundamentals of the game of chess, you're ready to start playing the game. Unfortunately, at this point all you can do is make crude moves, because you really don't understand how all the information works together. To play chess well, you'll need to learn how to use your knowledge by working on subtle strategies before you can work your way up to the more intriguing and complicated moves.

Not a Game

Electrical work isn't a game, and it must be taken very seriously. Learning the basics of electricity, important terms and concepts, as well as the basic layout of the *NEC* gives you just enough knowledge to be dangerous. There are thousands of specific and unique applications of electrical installations, and the *Code* doesn't cover every one of them. To safely apply the *NEC*, you must understand the purpose of a rule and how it affects the safety aspects of the installation.

NEC Terms and Concepts

The *NEC* contains many technical terms, so it's crucial for *Code* users to understand their meanings and their applications. If you don't understand a term used in a *Code* rule, it will be impossible to properly apply the *NEC* requirement. Be sure you understand that Article 100 defines the terms that apply to two or more articles. For example, the term "Dwelling Unit" applies to many articles. If you don't know what a dwelling unit is, how can you apply the *Code* requirements for it?

In addition, many articles have terms unique for that specific article. This means that the definitions of those terms are only applicable for that given article. For example, Section 250.2 contains the definitions of terms that only apply to Article 250, Grounding and Bonding.

Small Words, Grammar, and Punctuation

It's not only the technical words that require close attention, because even the simplest of words can make a big difference to the intent of a rule. The word "or" can imply alternate choices for equipment wiring methods, while "and" can mean an additional requirement. Let's not forget about grammar and punctuation. The location of a comma "," can dramatically change the requirement of a rule.

Slang Terms or Technical Jargon

Electricians, engineers, and other trade-related professionals use slang terms or technical jargon that isn't shared by all. This makes it very difficult to communicate because not everybody understands the intent or application of those slang terms. So where possible, be sure you use the proper word, and don't use a word if you don't understand its definition and application. For example, lots of electricians use the term "pigtail" when describing the short conductor for the connection of a receptacle, switch, luminaire, or equipment. Although they may understand this, not everyone does.

NEC Style and Layout

Before we get into the details of the *NEC*, we need to take a few moments to understand its style and layout. Understanding the structure and writing style of the *Code* is very important before it can be used effectively. If you think about it, how can you use something if you don't know how it works? The *National Electrical Code* is organized into ten components.

1. Table of Contents
2. Article 90 (Introduction to the *Code*)
3. Chapters 1 through 9 (major categories)
4. Articles 90 through 830 (individual subjects)
5. Parts (divisions of an article)
6. Sections and Tables (*Code* requirements)
7. Exceptions (*Code* permissions)
8. Fine Print Notes (explanatory material)
9. Annexes (information)
10. Index

1. Table of Contents. The Table of Contents displays the layout of the Chapters, Articles, and Parts as well as the page numbers. It's an excellent resource and should be referred to periodically to observe the interrelationship of the various *NEC* components. When attempting to locate the rules for a particular situation, knowledgeable *Code* users often go first to the Table of Contents to quickly find the specific *NEC* part that applies.

2. Introduction. The *NEC* begins with Article 90, the introduction to the *Code*. It contains the purpose of the *NEC*, what is covered and what is not covered along with how the *Code* is arranged. It also gives information on enforcement and how mandatory and permissive rules are written as well as how explanatory material is included. Article 90 also includes information on formal interpretations, examination of equipment for safety, wiring planning, and information about formatting units of measurement.

3. Chapters. There are nine chapters, each of which is divided into articles. The articles fall into one of four groupings: General Requirements (Chapters 1 through 4), Specific Requirements (Chapters 5 through 7), Communications Systems (Chapter 8), and Tables (Chapter 9).

- Chapter 1 General
- Chapter 2 Wiring and Protection
- Chapter 3 Wiring Methods and Materials
- Chapter 4 Equipment for General Use
- Chapter 5 Special Occupancies
- Chapter 6 Special Equipment
- Chapter 7 Special Conditions
- Chapter 8 Communications Systems (Telephone, Data, Satellite, and Cable TV)
- Chapter 9 Tables–Conductor and Raceway Specifications

4. Articles. The *NEC* contains approximately 140 articles, each of which covers a specific subject. For example:

- Article 110 General Requirements
- Article 250 Grounding and Bonding
- Article 300 Wiring Methods
- Article 430 Motors and Motor Controllers
- Article 500 Hazardous (Classified) Locations
- Article 680 Swimming Pools, Fountains, and Similar Installations
- Article 725 Remote-Control, Signaling, and Power-Limited Circuits
- Article 800 Communications Systems

5. Parts. Larger articles are subdivided into parts.

Author's Comment: Because the parts of a *Code* article aren't included in the section numbers, we have a tendency to forget what "Part" the *NEC* rule is relating to. For example, Table 110.34(A) contains the working space clearances for electrical equipment. If we aren't careful, we might think this table applies to all electrical installations, but Table 110.34(A) is located in Part III, which contains the requirements for Over 600 Volts, Nominal installations. The rules for working clearances for electrical equipment for systems 600V, nominal, or less are contained in Table 110.26(A)(1), which is located in Part II—600 Volts, Nominal, or Less.

6. Sections and Tables.

Sections. Each *NEC* rule is called a *Code* section. A *Code* section may be broken down into subsections by letters in parentheses "(A), (B)," etc. Numbers in parentheses (1), (2), etc., may further break down a subsection, and lowercase letters (a), (b), etc., further break the rule down to the third level. For example, the rule requiring all receptacles in a dwelling unit bathroom to be GFCI protected is contained in Section 210.8(A)(1). Section 210.8(A)(1) is located in Chapter 2, Article 210, Section 8, subsection (A), sub-subsection (1).

Many in the industry incorrectly use the term "Article" when referring to a *Code* section. For example, they say "Article 210.8," when they should say "Section 210.8."

Tables. Many *Code* requirements are contained within tables, which are lists of *NEC* requirements placed in a systematic arrangement. The titles of the tables are extremely important; you must read them carefully in order to understand the contents, applications, limitations, etc., of each table in the *Code*. Many times notes are provided in or below a table; be sure to read them as well since they are also part of the requirement. For example, Note 1 for Table 300.5 explains how to measure the cover when burying cables and raceways, and Note 5 explains what to do if solid rock is encountered.

How to Use the *National Electrical Code*

7. Exceptions. Exceptions are *Code* requirements or allowances that provide an alternative method to a specific requirement. There are two types of exceptions—mandatory and permissive. When a rule has several exceptions, those exceptions with mandatory requirements are listed before the permissive exceptions.

Mandatory Exception. A mandatory exception uses the words "shall" or "shall not." The word "shall" in an exception means that if you're using the exception, you're required to do it in a particular way. The phrase "shall not" means it isn't permitted.

Permissive Exception. A permissive exception uses words such as "shall be permitted," which means it's acceptable (but not mandatory) to do it in this way.

8. Fine Print Note (FPN). A fine print note contains explanatory material intended to clarify a rule or give assistance, but it isn't a *Code* requirement [90.5(C)].

9. Annexes. Annexes aren't a part of the *NEC* requirements, and are included in the *Code* for informational purposes only.

10. Index. The Index at the back of the *NEC* is helpful in locating a specific rule.

> **Author's Comment:** Changes to the *NEC* since the previous edition(s), are identified by shading, but rules that have been relocated aren't identified as a change. A bullet symbol "•" is located on the margin to indicate the location of a rule that was deleted from a previous edition.

How to Locate a Specific Requirement

How to go about finding what you're looking for in the *Code* depends, to some degree, on your experience with the *NEC*. *Code* experts typically know the requirements so well they just go to the correct rule without any outside assistance. The Table of Contents might be the only thing very experienced *NEC* users need to locate the requirement they're looking for. On the other hand, average *Code* users should use all of the tools at their disposal, and that includes the Table of Contents and the Index.

Table of Contents. Let's work out a simple example: What *NEC* rule specifies the maximum number of disconnects permitted for a service? If you're an experienced *Code* user, you'll know Article 230 applies to "Services," and because this article is so large, it's divided up into multiple parts (actually eight parts). With this knowledge, you can quickly go to the Table of Contents and see that it lists Service Equipment Disconnecting Means requirements in Part VI.

> **Author's Comment:** The number 70 precedes all page numbers because the *NEC* is NFPA standard number 70.

Index. If you use the Index, which lists subjects in alphabetical order, to look up the term "service disconnect," you'll see there's no listing. If you try "disconnecting means," then "services," you'll find the Index specifies the rule is located in Article 230, Part VI. Because the *NEC* doesn't give a page number in the Index, you'll need to use the Table of Contents to find the page number, or flip through the *Code* to Article 230, then continue to flip through pages until you find Part VI.

Many people complain that the *NEC* only confuses them by taking them in circles. As you gain experience in using the *Code* and deepen your understanding of words, terms, principles, and practices, you will find the *NEC* much easier to understand and use than you originally thought.

Customizing Your *Code* Book

One way to increase your comfort level with the *Code* is to customize it to meet your needs. You can do this by highlighting and underlining important *NEC* requirements, and by attaching tabs to important pages.

Highlighting. As you read through this textbook, be sure you highlight those requirements in the *Code* that are the most important or relevant to you. Use yellow for general interest and orange for important requirements you want to find quickly. Be sure to highlight terms in the Index and Table of Contents as you use them.

Underlining. Underline or circle key words and phrases in the *NEC* with a red pen (not a lead pencil) and use a six-inch ruler to keep lines straight and neat. This is a very handy way to make important requirements stand out. A small six-inch ruler also comes in handy for locating specific information in the many *Code* tables.

Tabbing the *NEC*. By placing tabs on *Code* articles, sections, and tables, it will make it easier for you to use the *NEC*. However, too many tabs will defeat the purpose. You can order a custom set of *Code* tabs online at www.MikeHolt.com, or by calling 1.888.NEC.CODE.

Acknowledgments

About the Author

Mike Holt worked his way up through the electrical trade from an apprentice electrician to become one of the most recognized experts in the world as it relates to electrical power installations. He was a Journeyman Electrician, Master Electrician, and Electrical Contractor. Mike came from the real world, and he has a unique understanding of how the *NEC* relates to electrical installations from a practical standpoint. You'll find his writing style to be simple, nontechnical, and practical.

Did you know that he didn't finish high school? So if you struggled in high school or if you didn't finish it at all, don't let this get you down, you're in good company. As a matter of fact, Mike Culbreath, Master Electrician, who produces the finest electrical graphics in the history of the electrical industry, didn't finish high school either. So two high school dropouts produced the text and graphics in this textbook! However, realizing success depends on one's continuing pursuit of education, Mike immediately attained his GED (as did Mike Culbreath) and ultimately attended the University of Miami's Graduate School for a Master's degree in Business Administration (MBA).

Mike Holt resides in Central Florida, is the father of seven children, and has many outside interests and activities. He is a five-time National Barefoot Water-Ski Champion (1988, 1999, 2005, 2006, and 2007); he has set many national records and continues to train year-round at a World competition level [www.barefootwaterskier.com].

What sets him apart from some is his commitment to living a balanced lifestyle; he places God first, then family, career, and self.

Special Acknowledgments

First, I want to thank God for my godly wife who is always by my side and my children, Belynda, Melissa, Autumn, Steven, Michael, Meghan, and Brittney.

A special thank you must be sent to the staff at the National Fire Protection Association (NFPA), publishers of the NEC—in particular Jeff Sargent for his assistance in answering my many Code questions over the years. Jeff, you're a "first class" guy, and I admire your dedication and commitment to helping others understand the NEC. Other former NFPA staff members I would like to thank include John Caloggero, Joe Ross, and Dick Murray for their help in the past.

A personal thank you goes to Sarina, my long-time friend and office manager. It has been wonderful working side-by-side with you for over 25 years nurturing this company's growth from its small beginnings.

Mike Holt Enterprises Team

Graphic Illustrator

Mike Culbreath devoted his career to the electrical industry and worked his way up from an apprentice electrician to master electrician. While working as a journeyman electrician, he suffered a serious on-the-job knee injury. With a keen interest in continuing education for electricians, he completed courses at Mike Holt Enterprises, Inc. and then passed the exam to receive his Master Electrician's license. In 1986, after attending classes at Mike Holt Enterprises, Inc. he joined the staff to update material and later studied computer graphics and began illustrating Mike Holt's textbooks and magazine articles. He's worked with the company for over 20 years and, as Mike Holt has proudly acknowledged, has helped to transform his words and visions into lifelike graphics.

Acknowledgments

Technical Editorial Director

Steve Arne has been involved in the electrical industry since 1974 working in various positions from electrician to full-time instructor and department chair in technical postsecondary education. Steve has developed curriculum for many electrical training courses and has developed university business and leadership courses. Currently, Steve offers occasional exam prep and continuing education *Code* classes. Steve believes that as a teacher he understands the joy of helping others as they learn and experience new insights. His goal is to help others understand more of the technological marvels that surround us. Steve thanks God for the wonders of His creation and for the opportunity to share it with others.

Steve and his lovely wife Deb live in Rapid City, South Dakota where they are both active in their church and community. They have two grown children and five grandchildren.

Technical *Code* Consultant

Ryan Jackson is a combination inspector for Draper City, Utah. He is certified as a building, electrical, mechanical, and plumbing inspector. He's also certified as a building plans examiner and electrical plans examiner. Ryan is the senior electrical inspector for Draper City, and also teaches seminars on the *NEC*. Ryan is very active in the Utah Chapter of IAEI, where he is currently president. He also enjoys staying active in the *NEC* change process, and loves to help people with their *Code* problems. On Mike Holt's *Code* Forum, he has been involved in nearly 5,000 topics.

Ryan enjoys reading, going to college football games, and spending time with his wife Sharie and their two children, Kaitlynn and Aaron.

Editorial Team

I would like to thank Toni Culbreath and Barbara Parks who worked tirelessly to proofread and edit the final stages of this publication. Their attention to detail and dedication to this project is greatly appreciated.

Production Team

I would like to thank Tara Moffitt and Cathleen Kwas who worked as a team to do the layout and production of this book. Their desire to create the best possible product for our customers is appreciated.

Video Team Members

Steve Arne and **Mike Culbreath** (members of the Mike Holt Enterprises Team) were video team members, along with the following highly qualified professionals:

Paul Abernathy
NEC Consultant/Electrical Inspector
Virginia Continuing Education Provider
Harrisonburg/Richmond, Virginia
www.ElectricalSeminars.net

Paul W. Abernathy is an Electrical Inspector for the City of Richmond, Virginia, ICC and IAEI Certified Electrical Inspector, Eaton Certified Contractors Network training provider and travels around the country doing electrical seminars training electricians and inspectors. He has 20 years in the electrical field as a Master Electrician, *NEC* Consultant, Electrical Educator and Electrical Contractor as well as being a member of the NFPA Electrical Section Committee.

In 1988 he graduated first in his electrical class at the Charlottesville-Albemarle Technical Education Center and went on to take first place in the regional VICA competition among other electrical students. He also took "Top Gun" honors at the 2007 Mike Holt Enterprises Instructors Conference held in Orlando, Florida as well as voted 2007 "interNACHI" member of the year.

Paul lives in Richmond, Virginia and Harrisonburg, Virginia, with his wife Darlene and their son Zachary..

Dennis Carlson
Master Electrician/Electrical Instructor
The Industrial Company (T.I.C.)
Steamboat Springs, Colorado

Dennis Carlson started in the electrical trade in 1980 with Simmons Electric based in Greeley, Colorado and did residential and commercial work with them until 1985. He then went to work for

Acknowledgments

Atkinson Electric based out of Ft. Collins, Colorado until 1987, also doing residential and commercial work.

Dennis has since worked on various industrial projects for TIC (The Industrial Company) as a journeyman, electrical foreman, and project coordinator. He is presently the Manager of the Electrical Craft Training Program for TIC.

Dennis instructs levels first, second, third, and fourth year trainees. With the use of Mike Holt's *NEC Exam Preparation* book, he has had a success rate of 90 percent with his fourth year trainees on the Colorado State electrical exam.

Dennis lives in Steamboat Springs, with his wife Cindy. Their son Josh will be attending college in the Fall of 2008, studying for an engineering degree, and specializing in robotics. Their daughter is enrolled at CSU in Ft. Collins studying for a degree in Art.

Wes Gerrans
Electrical Technology Instructor
Northwest Kansas Technical College
Goodland, Kansas

Wes Gerrans, upon obtaining a Bachelor's degree from Andrews University in 1965, began his career as an Industrial Arts instructor at Eau Claire Michigan High School. After teaching for four years, he worked in the construction trades for the next 23 years. In 1992 he completed an Electrical Technology course at Northwest Kansas Technical School. It was at this time that he first had contact with Mike Holt when he helped with the editing of one of Mike's earlier books. He then worked for Archer Daniels Midland as an industrial electrician until 2002, when he went back to NWKTC, now a college, as an instructor in the electrical program.

Wes remains happily married to his wife of 43 years, Jacque, who continues to be a source of strength and encouragement. She retired from being a school psychologist this year and is now enjoying a more leisurely pace at home. Their children are both married. Their son, Neil, is a teacher in Seattle, WA; their daughter, Lisa, teaches in Lincoln, NE.

Other areas of interest for Wes have included a commercial fishing business in Alaska, rebuilding wrecked vehicles, and restoring antique furniture. He is an active leader and teacher in his church. He enjoys participating in community projects, has a passion for teaching, and for sharing the Good News of Jesus.

John Mills
Master Electrician/Electrical Instructor
Premier Electrical Company/Dade County School Board
Miami Springs, Florida

John Mills is an instructor for Mike Holt Enterprises, Inc. and is a Certified Electrical Contractor, Certified Inspector, Plan Reviewer, and CBO. John started in the electrical trade in 1980 as a helper and attended an ABC Apprenticeship Program, then worked his way up to Master Electrician. John has been an instructor for 20 years, teaching master and journeyman exam preparation courses and Florida State certification courses.

John resides in Miami, Florida with his wife Corina and their four sons. John's hobbies are working with wood and metal. He makes wooden bowls and furniture from trees that have been cut down. He enjoys making things for his children as well as fishing, family picnics, and vacations with his family.

Lynn E. Palmer
Electrical Inspector
Construction Supervisor
State of California

Lynn Palmer started his career as an electrician in 1979. He later went to work as an Electrical Inspector for the County of Sacramento. Lynn is a member of the International Association of Electrical Inspectors, Sacramento Valley Chapter, where he also served as Chairman, and was a member of IBEW Local 340 Sacramento. He became a licensed Electrical Contractor in 1985, and is the owner of L & B Palmer Electric, www.lbpalmerelectric.com.

Lynn has taught a variety of apprenticeship, journeyman and *Code* training classes. He currently teaches in the Building Inspection Technology Program at Cosumnes River College in Sacramento. After working for the County of Sacramento for 10 years, he went to work for the State of California as a Construction Supervisor.

Lynn currently resides in Elk Grove, California, with his wife Brenda and two children, Christopher and Jennifer.

Acknowledgments

Eric Stromberg
Electrical Engineer/Instructor
Dow Chemical
Lake Jackson, TX

Eric Stromberg enrolled in the University of Houston in 1976, with Electrical Engineering as his major. During the first part of his college years, Eric worked for a company that specialized in installing professional sound systems. Later, he worked for a small electrical company and eventually became a Journeyman Electrician. After graduation from college in 1982, Eric went to work for an electronics company that specialized in fire alarm systems for high-rise buildings. He became a state licensed Fire Alarm Installation Superintendent and was also a member of IBEW Local Union 716. In 1989, Eric began a career with the Dow Chemical Company as an Electrical Engineer designing power distribution systems for large industrial facilities. In 1997, he began teaching *National Electrical Code* classes.

Eric currently resides in Lake Jackson, Texas, with his wife Jane and three children: Ainsley, Austin, and Brieanna.

James Thomas
Master Electrician/Electrical Instructor
James Sprunt Community College
Kenansville, North Carolina

James Thomas has been working in the electrical field since 1985, and is licensed in the State of North Carolina. He has worked in residential, commercial, and industrial electrical jobs. James has been an instructor teaching electrical and electronic courses (including the *National Electrical Code*) full time at James Sprunt Community College in Kenansville, NC for over 16 years and has taught thousands of individuals. He has taught at Pitt Community College in Greenville, NC; Cape Fear Community College in Wilmington, NC; and Sampson Community College in Clinton, NC.

James earned a diploma in Electrical Installation and Maintenance; a Bachelor's degree in Business Management from Mount Olive College in Mount Olive, NC; and a Master's degree in Industrial Technology from East Carolina University in Greenville NC. James also served in the Air Force Reserve in Goldsboro NC for over 19 years.

CHAPTER 1: ELECTRICAL THEORY

Unit 1   **Electrician's Math and Basic Electrical Formulas**

Unit 2   **Electrical Circuits**

Unit 3   **Understanding Alternating Current**

Unit 4   **Motors and Transformers**

CHAPTER 1 Notes

UNIT 1: Electrician's Math and Basic Electrical Formulas

INTRODUCTION TO UNIT—ELECTRICIAN'S MATH AND BASIC ELECTRICAL FORMULAS

In order to construct a building that will last into the future, a strong foundation is a prerequisite. The foundation is a part of the building that isn't visible in the finished structure, but is essential in erecting a building that will have the necessary strength to endure.

The math and basic electrical concepts of this unit are very similar to the foundation of a building. The concepts in this unit are the essential basics that you must understand, because you'll build upon them as you study electrical circuits and systems. As your studies continue, you'll find that a good foundation in electrical theory and math will help you understand why the *NEC* contains certain provisions.

This unit includes math, and electrical fundamentals. You'll be amazed at how often your electrical studies return to the basics of this unit. Ohm's law and the electrical formulas related to it are the foundation of all electrical circuits.

Every student begins at a different level of understanding, and you may find this unit an easy review, or you may find it requires a high level of concentration. In any case, be certain that you fully understand the concepts of this unit and are able to successfully complete the questions at the end of the unit before going on. A solid foundation will help in your successful study of the rest of this book.

PART A—ELECTRICIAN'S MATH

Introduction

Numbers can take different forms:

Whole numbers: 1, 20, 300, 4,000, 5,000
Decimals: 0.80, 1.25, 0.75, 1.15
Fractions: ½, ¼, ⅝, ⁴⁄₃
Percentages: 80%, 125%, 250%, 500%

You'll need to be able to convert these numbers from one form to another and back again, because all of these number forms are part of electrical work and electrical calculations.

You'll also need to be able to do some basic algebra. Many people have a fear of algebra, but as you work through the material here you'll see there is nothing to fear.

1.1 Whole Numbers

Whole numbers are exactly what the term implies. These numbers don't contain any fractions, decimals, or percentages. Another name for whole numbers is "integers."

1.2 Decimals

The decimal method is used to display numbers other than whole numbers, fractions, or percentages such as, 0.80, 1.25, 1.732, and so on.

1.3 Fractions

A fraction represents part of a whole number. If you use a calculator for adding, subtracting, multiplying, or dividing, you need to convert the fraction to a decimal or whole number. To change a fraction to a decimal or whole number, divide the numerator (the top number) by the denominator (the bottom number).

▶ **Examples**

⅙ = *one divided by six = 0.166*
⅖ = *two divided by five = 0.40*
³⁄₆ = *three divided by six = 0.50*
⁵⁄₄ = *five divided by four = 1.25*
⁷⁄₂ = *seven divided by two = 3.50*

Unit 1 Electrician's Math and Basic Electrical Formulas

1.4 Percentages

A percentage is another method used to display a value. One hundred percent (100%) means all of a value; fifty percent (50%) means one-half of a value, and twenty-five percent (25%) means one-fourth of a value.

For convenience in multiplying or dividing by a percentage, convert the percentage value to a whole number or decimal, and then use the result for the calculation. When changing a percent value to a decimal or whole number, drop the percentage symbol and move the decimal point two places to the left. **Figure 1–1**

▶ **Example 1**

Question: An overcurrent device (circuit breaker or fuse) must be sized no less than 125 percent of the continuous load. If the load is 80A, the overcurrent device will have to be sized no smaller than _____. **Figure 1–2**

(a) 75A   (b) 80A   (c) 100A   (d) 125A

Answer: (c) 100A

Step 1: Convert 125 percent to a decimal: 1.25

Step 2: Multiply the value of the load 80A by 1.25 = 100A

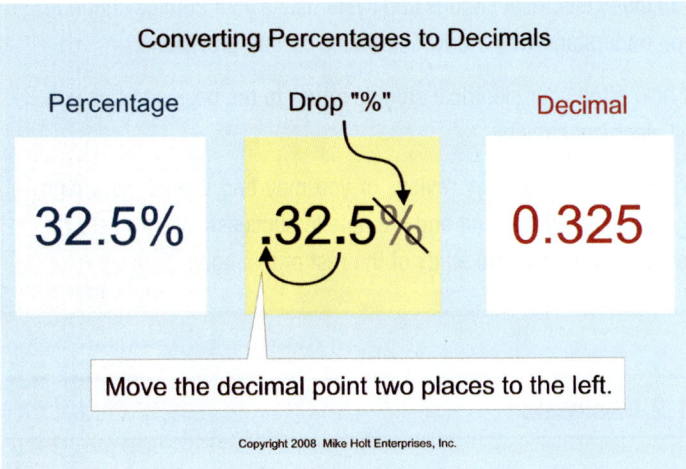

Figure 1–1

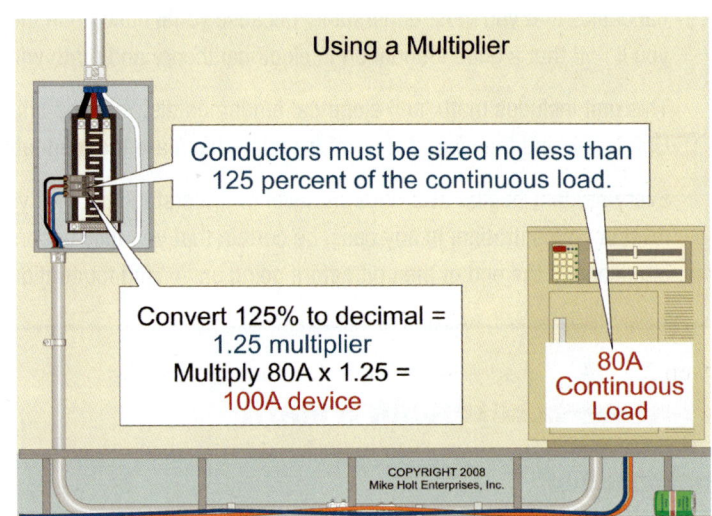

Figure 1–2

▶ **Examples**

Percentage	Number
32.50%	0.325
80%	0.80
125%	1.25
250%	2.50

1.5 Multiplier

When a number needs to be changed by multiplying it by a percentage, the percentage is called a multiplier. The first step is to convert the percentage to a decimal, then multiply the original number by the decimal value.

▶ **Example 2**

Question: The maximum continuous load on an overcurrent device is limited to 80 percent of the device rating. If the overcurrent device is rated 50A, what is the maximum continuous load permitted on the overcurrent device? **Figure 1–3**

(a) 40A   (b) 50A   (c) 75A   (d) 100A

Answer: (a) 40A

Step 1: Convert 80 percent to a decimal: 0.80

Step 2: Multiply the value of the device rating 50A by 0.80 = 40A

Mike Holt's Illustrated Guide to NEC Exam Preparation

Electrician's Math and Basic Electrical Formulas — Unit 1

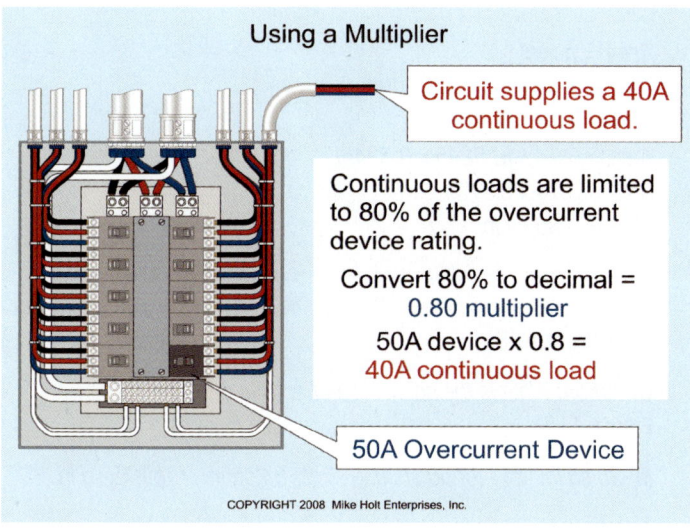

Figure 1–3

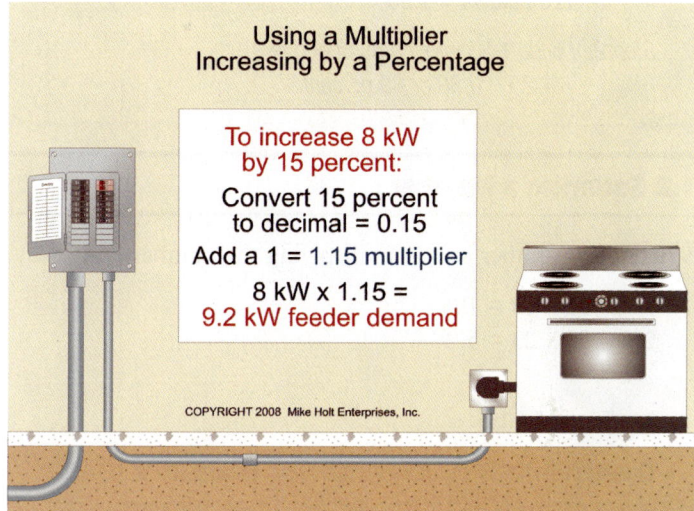

Figure 1–4

1.6 Percent Increase

Use the following steps to increase a number by a specific percentage:

Step 1: Convert the percent to a decimal value.

Step 2: Add one to the decimal value to create the multiplier.

Step 3: Multiply the original number by the multiplier found in Step 2.

▶ **Example 1**

Question: How do you increase the whole number 45 by 35 percent?

Step 1: Convert 35 percent to decimal form: 0.35

Step 2: Add one to the decimal value: 1 + 0.35 = 1.35

Step 3: Multiply 45 by the multiplier: 1.35: 45 x 1.35 = 60.75

▶ **Example 2**

Question: If the feeder demand load for a range is 8 kW and it is required to be increased by 15 percent, the total calculated load will be _____. Figure 1–4

(a) 6.8 kW   (b) 8 kW   (c) 9.20 kW   (d) 15 kW

Answer: (c) 9.20 kW

(continued in next column)

Step 1: Convert the percentage increase required to decimal form: 15 percent = 0.15

Step 2: Add one to the decimal: 1 + 0.15 = 1.15

Step 3: Multiply 8 by the multiplier 1.15: 8 kW x 1.15 = 9.20 kW

1.7 Reciprocals

To obtain the reciprocal of a number, convert the number into a fraction with the number one as the numerator (the top number). It's also possible to calculate the reciprocal of a decimal number. Determine the reciprocal of a decimal number by following these steps:

Step 1: Convert the number to a decimal value.

Step 2: Divide the value into the number one.

▶ **Example 1**

Question: What is the reciprocal of 80 percent?

(a) 0.80   (b) 100%   (c) 125%   (d) 150%

Answer: (c) 125%

Step 1: Convert 80 percent into a decimal (move the decimal two places to the left): 80 percent = 0.80

Step 2: Divide 0.80 into the number one: 1/0.80 = 1.25 or 125 percent

Unit 1 Electrician's Math and Basic Electrical Formulas

▶ Example 2

Question: What is the reciprocal of 125 percent?

(a) 75%   (b) 0.80   (c) 100%   (d) 125%

Answer: (b) 0.80

Step 1: Convert 125 percent into a decimal:
125 percent = 1.25

Step 2: Divide 1.25 into the number one:
1/1.25 = 0.80 or 80 percent

1.8 Squaring a Number

Squaring a number means multiplying the number by itself.

$10^2 = 10 \times 10 = 100$
$23^2 = 23 \times 23 = 529$

▶ Example 1

Question: What is the power consumed in watts by a 12 AWG conductor that is 200 ft long, and has a total resistance of 0.40 ohms, if the current (I) in the circuit conductors is 16A?

Formula: Power = $I^2 \times R$
(Answers are rounded to the nearest 50).

(a) 50W   (b) 100W   (c) 150W   (d) 200W

Answer: (b) 100W

$P = I^2 \times R$
I = 16A
R = 0.40 ohms
$P = 16A^2 \times 0.40$ ohms
$P = 16A \times 16A \times 0.40$ ohms
$P = 102.40W$

▶ Example 2

Question: What is the area in square inches (sq in.) of a trade size 1 raceway with an inside diameter of 1.049 in.?

Formula: Area = $\pi \times r^2$
$\pi = 3.14$
r = radius (equal to 0.50 of the diameter)

(a) 0.34 sq in.   (b) 0.50 sq in.   (c) 0.86 sq in.   (d) 1 sq in.

Answer: (c) 0.86 sq in.

(continued in next column)

Area = $\pi \times r^2$
Area = $3.14 \times (0.50 \times 1.049)^2$
Area = 3.14×0.5245^2
Area = $3.14 \times (0.5245 \times 0.5245)$
Area = 3.14×0.2751
Area = 0.86 sq in.

▶ Example 3

Question: What is the sq in. area of an 8 in. pizza?
Figure 1–5A

(a) 25 sq in.   (b) 50 sq in.   (c) 64 sq in.   (d) 75 sq in.

Answer: (b) 50 sq in.

Area = $\pi \times r^2$
Area = $3.14 \times (0.50 \times 8)^2$
Area = 3.14×4^2
Area = $3.14 \times 4 \times 4$
Area = 3.14×16
Area = 50 sq in.

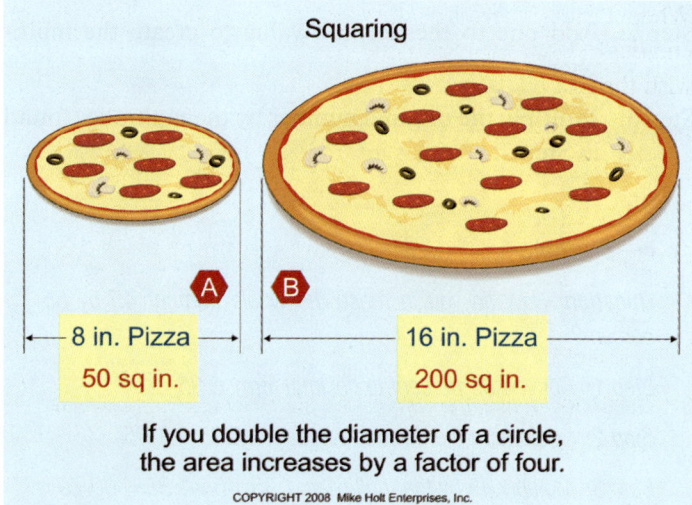

Figure 1–5

Mike Holt's Illustrated Guide to NEC Exam Preparation

▶ **Example 4**

Question: What is the sq in. area of a 16 in. pizza? **Figure 1–5B**

(a) 100 sq in.   (b) 150 sq in.
(c) 200 sq in.   (d) 256 sq in.

Answer: (c) 200 sq in.

Area = π × r²
Area = 3.14 × (0.50 × 16)²
Area = 3.14 × 8²
Area = 3.14 × 8 × 8
Area = 3.14 × 64
Area = 200 sq in.

Author's Comment: As you see in Examples C and D, if you double the diameter of the circle, the area contained in the circle is increased by a factor of four! By the way, a large pizza is always cheaper per sq in. than a small pizza.

1.9 Parentheses

Whenever numbers are in parentheses, complete the mathematical function within the parentheses before proceeding with the rest of the problem.

Parentheses are used to group steps of a process in the correct order. For instance, the sum of 3 and 15 added to the product of 4 and 2.

(3 + 15) + (4 x 2) = 18 + 8 = 26

▶ **Example**

Question: What is the current of a 36,000W, 208V, three-phase load? **Figure 1–6**

Ampere (I) = Watts/(E x 1.732)

(a) 50A   (b) 100A   (c) 150A   (d) 360A

Answer: (b) 100A

Step 1: Perform the operation inside the parentheses first—determine the product of:
208V x 1.732 = 360V

Step 2: Divide 36,000W by 360V = 100A

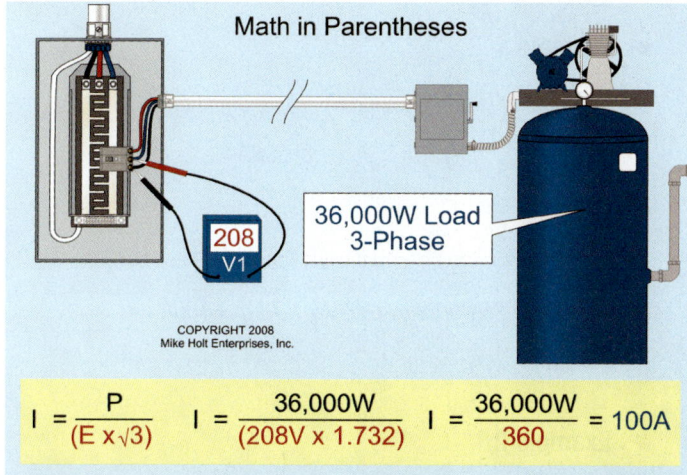

Whenever numbers are in parentheses, we must complete the mathematical function within the parentheses before proceeding with the rest of the problem.

Figure 1–6

1.10 Square Root

Deriving the square root of a number ($\sqrt{n}$) is the opposite of squaring a number. The square root of 36 is a number that, when multiplied by itself, gives the product 36. The $\sqrt{36}$ equals six (6), because six, multiplied by itself (6^2) equals the number 36.

Because it's difficult to do this manually, just use the square root key of your calculator.

$\sqrt{3}$: Following your calculator's instructions, enter the number 3, then press the square root key = 1.732.

$\sqrt{1,000}$: enter the number 1,000, then press the square root key = 31.62.

If your calculator doesn't have a square root key, don't worry about it. For all practical purposes in using this textbook, the only number you need to know the square root of is 3. The square root of 3 equals approximately 1.732.

To add, subtract, multiply, or divide a number by a square root value, determine the decimal value, then perform the math function.

Unit 1   Electrician's Math and Basic Electrical Formulas

▶ **Example 1**

Question: What is 36,000W/(208V x √3) equal to?

(a) 100A   (b) 120A   (c) 208A   (d) 360A

Answer: (a) 100A

Step 1: Determine the decimal value for the √3 = 1.732

Step 2: Divide 36,000W by (208V x 1.732) = 100A

▶ **Example 2**

Question: The phase voltage of a 120/208V system is equal to 208V/√3 which is _____.

(a) 120V   (b) 208V   (c) 360V   (d) 480V

Answer: (a) 120V

Step 1: Determine the decimal value for the √3 = 1.732

Step 2: Divide 208V by 1.732 = 120V

1.11 Volume

The volume of an enclosure is expressed in cubic inches (cu in.). It's determined by multiplying the length, by the width, by the depth of the enclosure.

▶ **Example**

Question: What is the volume of a box that has the dimensions of 4 x 4 x 1½ in.? **Figure 1–7**

(a) 12 cu in.   (b) 20 cu in.   (c) 24 cu in.   (d) 30 cu in.

Answer: (c) 24 cu in.

1½ = 1.50

4 x 4 x 1.50 = 24 cu in.

Author's Comment: The actual volume of a 4 in. square electrical box is less than 24 cu in. because the interior dimensions may be less than the nominal size and often corners are rounded, so the allowable volume is given in the *NEC* Table 314.16(A).

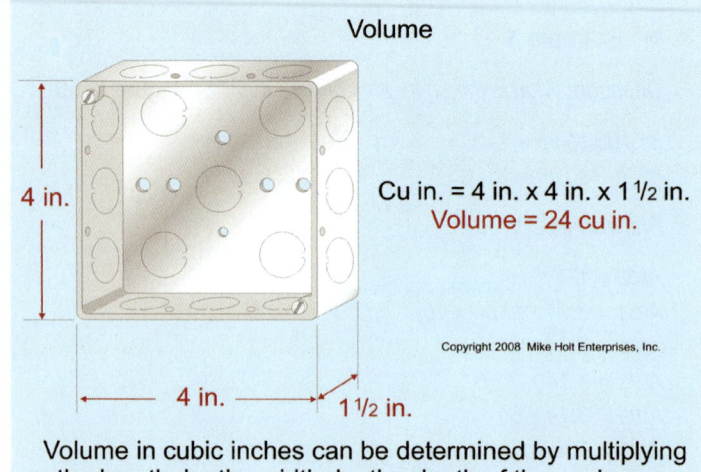

Volume in cubic inches can be determined by multiplying the length, by the width, by the depth of the enclosure.

Figure 1–7

1.12 Kilo

The letter "k" is used in the electrical trade to abbreviate the metric prefix "kilo", which represents a value of 1,000.

To convert a number which includes the "k" prefix to units, multiply the number preceding the "k" by 1,000.

▶ **Example 1**

Question: What is the wattage value for an 8 kW rated range?

(a) 8W   (b) 800W   (c) 4,000W   (d) 8,000W

Answer: (d) 8,000W

To convert a unit value to a "k" value, divide the number by 1,000 and add the "k" suffix.

▶ **Example 2**

Question: What is the kW rating of a 300W load? **Figure 1–8**

(a) 0.30 kW   (b) 30 kW   (c) 300 kW   (d) 3,000 kW

Answer: (a) 0.30 kW

kW = Watts/1,000

kW = 300W/1,000 = 0.30 kW

Mike Holt's Illustrated Guide to NEC Exam Preparation

Electrician's Math and Basic Electrical Formulas — Unit 1

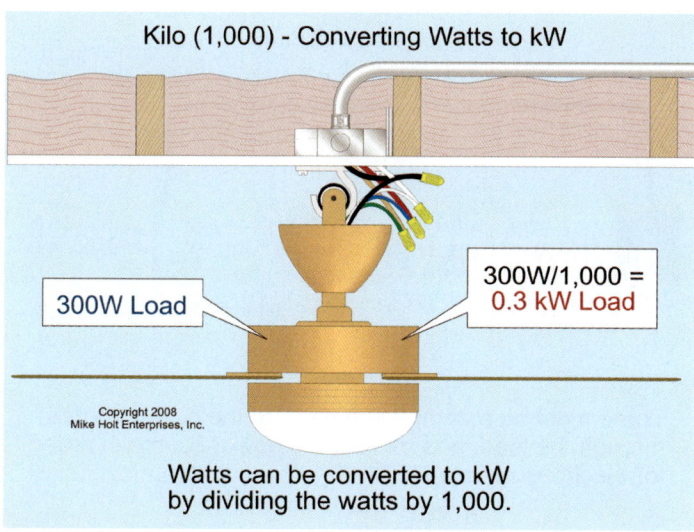

Figure 1–8

Author's Comment: The use of the letter "k" isn't limited to "kW." It is also used for kVA (1,000 volt-amperes), kcmil (1,000 circular mils) and other units such as kft (1,000 feet).

1.13 Rounding Off

There is no specific rule for rounding off, but rounding to two or three "significant digits" should be sufficient for most electrical calculations. Numbers below five are rounded down, while numbers five and above are rounded up.

0.1245—fourth number is five or above = 0.125 rounded up

1.674—fourth number is below five = 1.67 rounded down

21.99—fourth number is five or above = 22 rounded up

367.20—fourth number is below five = 367 rounded down

Rounding Answers for Multiple Choice Questions

You should round your answers in the same manner as the multiple choice selections given in the question.

▶ **Example**

Question: The sum* of 12, 17, 28, and 40 is equal to _____.

(a) 70 (b) 80 (c) 90 (d) 100

Answer: (d) 100

*A sum is the result of adding numbers.

The sum of these values equals 97, but this isn't listed as one of the choices. The multiple choice selections in this case are rounded off to the closest "tens."

1.14 Testing Your Answer for Reasonableness

When working with any mathematical calculation, don't just blindly do the calculation and assume it's correct. When you perform a mathematical calculation, you need to know if the answer is greater than or less than the values given in the problem. Always do a "reality check" to be certain that your answer is not nonsense. Even the best of us make mistakes at times, so always examine your answer to make sure it makes sense!

▶ **Example**

Question: The input of a transformer is 300W; the transformer efficiency is 90 percent. What is the transformer output? Figure 1–9

(a) 270W (b) 300W (c) 333W (d) 500W

Answer: (a) 270W

Since the output has to be less than the input (300W), you won't have to perform any mathematical calculation; the only multiple choice selection that's less than 300W is (a) 270W.

The math to work out the answer is:
300W x 0.90 = 270W

To check your multiplication, use division:
270W/0.90 = 300W

Author's Comment: One of the nice things about mathematical equations is that you can usually test to see if your answer is correct. To do this test, substitute the answer you arrived at back into the equation you are working with, and verify that it's indeed an equality. This method of checking your math will become easier once you know more of the formulas and how they relate to each other.

Unit 1 — Electrician's Math and Basic Electrical Formulas

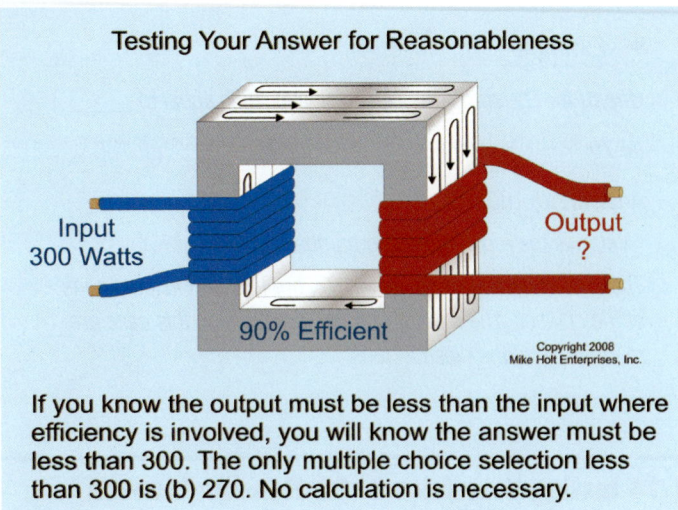

Figure 1–9

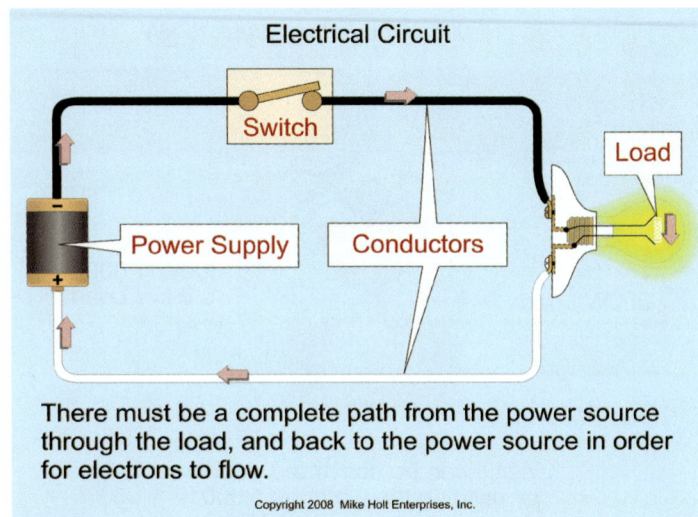

Figure 1–10

PART B—BASIC ELECTRICAL FORMULAS

Introduction

Now that you've mastered the math and understand some basics about electrical circuits, you're ready to take your knowledge of electrical formulas to the next level. One of the things we're going to do here is strengthen your proficiency with Ohm's Law.

Many false notions about the application of the *NEC* Article 250 and the *NEC* Chapter 3 wiring methods arise when people use Ohm's Law only to solve practice problems on paper but lack a real understanding of how it works and how to apply it. After completing this unit, you'll have that understanding, and won't be subject to those false notions—or the unsafe conditions they lead to.

But we won't stop with Ohm's Law. You're also going to have a high level of proficiency with the power equation. One of the tools for handling the power equation—and Ohm's Law—with ease is the power wheel. You'll be able to use it to solve all kinds of problems.

1.15 Electrical Circuit

A basic electrical circuit consists of the power source, the conductors, and the load. A switch can be placed in series with the circuit conductors to control the operation of the load (on or off). Figure 1–10

Author's Comment: According to the "electron current flow theory," current always flows from the negative terminal of the source, through the circuit and load, to the positive terminal of the source.

1.16 Power Source

The power necessary to move electrons out of their orbit around the nucleus of an atom can be produced by chemical, magnetic, photovoltaic, and other means. The two categories of power sources are direct current (dc) and alternating current (ac).

Direct Current

The polarity and the output voltage from a dc power source never change direction. One terminal is negative and the other is positive, relative to each other. Direct-current power is often produced by batteries, dc generators, and electronic power supplies. Figure 1–11

Direct current is used for electroplating, street trolley and railway systems, or where a smooth and wide range of speed control is required for a motor-driven application. Direct current is also used for control circuits and electronic instruments.

Electrician's Math and Basic Electrical Formulas — Unit 1

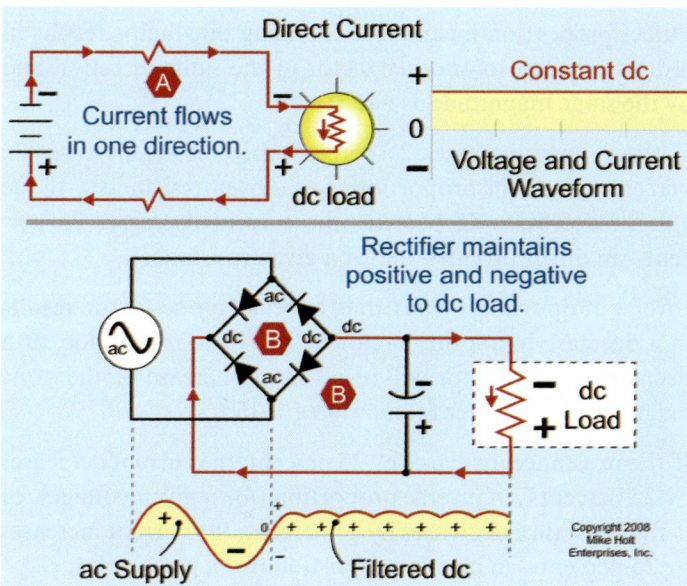

Figure 1–11

Alternating Current

Alternating-current power sources produce a voltage that changes polarity and magnitude. Alternating current is produced by an ac power source such as an ac generator. The major advantage of ac over dc is that voltage can be changed through the use of a transformer. **Figure 1–12**

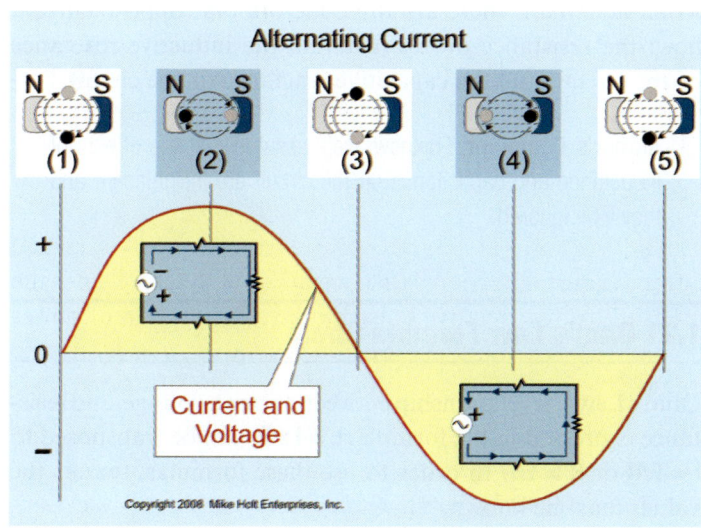

Figure 1–12

Author's Comment: Alternating current accounts for more than 90 percent of all electric power used throughout the world.

1.17 Conductance

Conductance or conductivity is the property of a metal that permits current to flow. The best conductors in order of their conductivity are silver, copper, gold, and aluminum. Copper is most often used for electrical applications. **Figure 1–13**

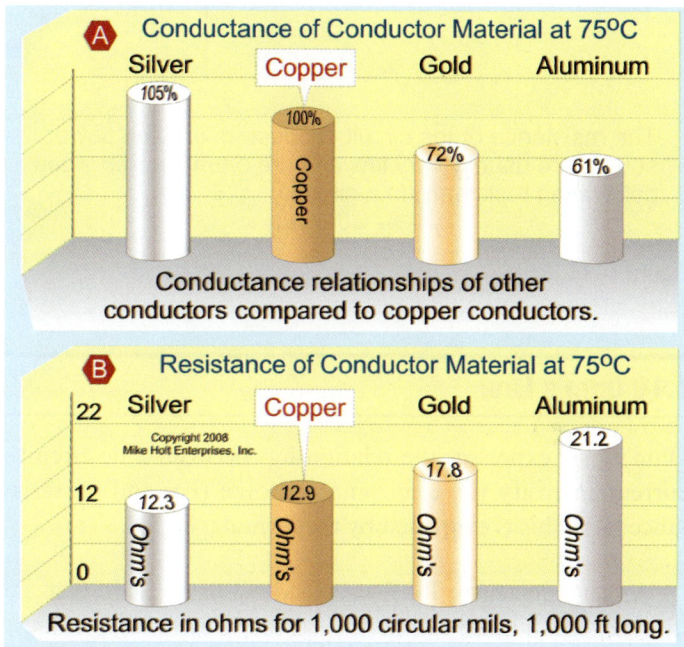

Figure 1–13

1.18 Circuit Resistance

The total resistance of a circuit includes the resistance of the power supply, the circuit wiring, and the load. Appliances such as heaters and toasters use high-resistance conductors to produce the heat needed for the application. Because the resistance of the power source and conductor are so much smaller than that of the load, they're generally ignored in circuit calculations. **Figure 1–14**

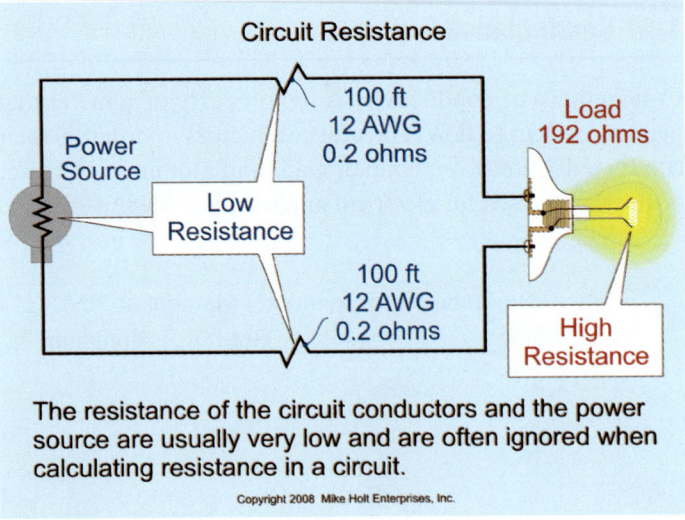

Figure 1–14

1.19 Ohm's Law

Ohm's Law expresses the relationship between a dc circuit's current intensity (I), electromotive force (E), and its resistance (R). This is expressed by the formula: **I = E/R**.

Author's Comment: The German physicist Georg Simon Ohm (1787-1854) stated that current is directly proportional to voltage, and inversely proportional to resistance. **Figure 1–15**

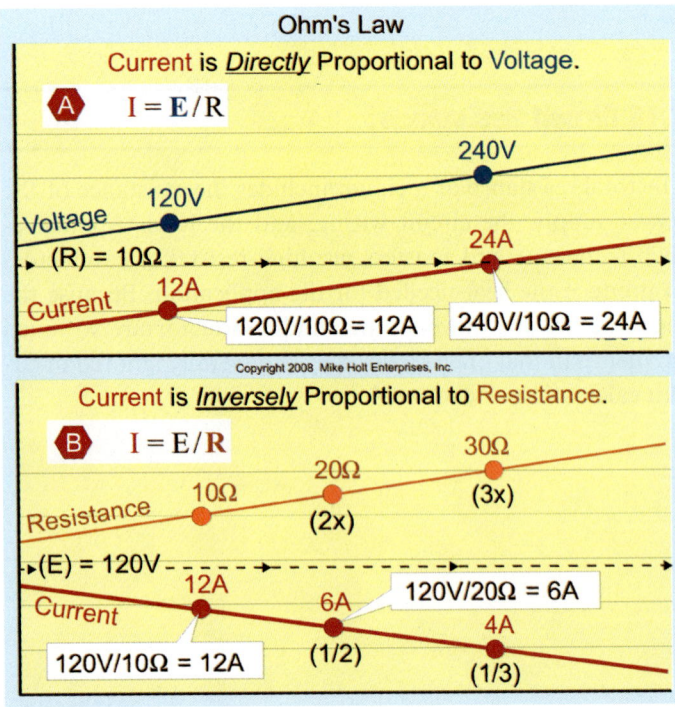

Figure 1–15

Direct proportion means that changing one factor results in a direct change to another factor in the same direction and by the same magnitude. **Figure 1–15A**

If the voltage increases 25 percent, the current increases 25 percent—in direct proportion (for a given resistance). If the voltage decreases 25 percent, the current decreases 25 percent—in direct proportion (for a given resistance).

Inverse proportion means that increasing one factor results in a decrease in another factor by the same magnitude, or a decrease in one factor will result in an increase of the same magnitude in another factor. **Figure 1–15B**

If the resistance increases by 25 percent, the current decreases by 25 percent—in inverse proportion (for a given voltage), or if the resistance decreases by 25 percent, the current increases by 25 percent—in inverse proportion (for a given voltage).

1.20 Ohm's Law and Alternating Current

Direct Current

In a dc circuit, the only opposition to current flow is the physical resistance of the material that the current flows through. This opposition is called resistance and is measured in ohms.

Alternating Current

In an ac circuit, there are three factors that oppose current flow: the resistance of the material, the inductive reactance of the circuit, and the capacitive reactance of the circuit.

Author's Comment: For now, we'll assume that the effects of inductance and capacitance on the circuit are insignificant and they'll be ignored.

1.21 Ohm's Law Formula Circle

Ohm's Law, the relationship between current, voltage, and resistance expressed in the formula, **E = I x R**, can be transposed to **I = E/R** or **R = E/I**. In order to use these formulas, two of the values must be known.

Author's Comment: Place your thumb on the unknown value in **Figure 1–16**, and the two remaining variables will "show" you the correct formula.

Electrician's Math and Basic Electrical Formulas — Unit 1

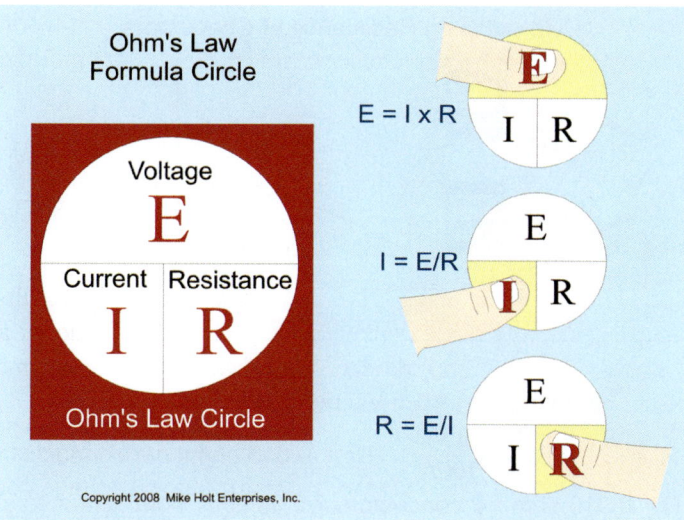

Figure 1–16

▶ Voltage-Drop Example

Question: What is the voltage drop over two 12 AWG conductors (resistance of 0.20 ohms per 100 ft) supplying a 16A load located 50 ft from the power supply? **Figure 1–18**

(a) 1.60V (b) 3.20V (c) 16V (d) 32V

Answer: (b) 3.20V

Step 1: What is the question? What is "E?"

Step 2: What do you know about the conductors?

I = 16A, R = 0.20 ohms. The NEC lists the ac resistance of 1,000 ft of 12 AWG as 2 ohms [Chapter 9, Table 8]. The resistance of 100 ft is equal to 0.20 ohms. **Figure 1–19**

Step 3: The formula is **E = I x R**

Step 4: The answer is E = 16A x 0.20 ohms

Step 5: The answer is E = 3.20V

▶ Current Example

Question: 120V supplies a lamp that has a resistance of 192 ohms. What is the current flow in the circuit? **Figure 1–17**

(a) 0.50A (b) 0.60A (c) 1.30A (d) 2.50A

Answer: (b) 0.60A

Step 1: What is the question? What is "I?"

Step 2: What do you know? E = 120V, R = 192 ohms

Step 3: The formula is **I = E/R**

Step 4: The answer is I = 120V/192 ohms

Step 5: The answer is I = 0.625A

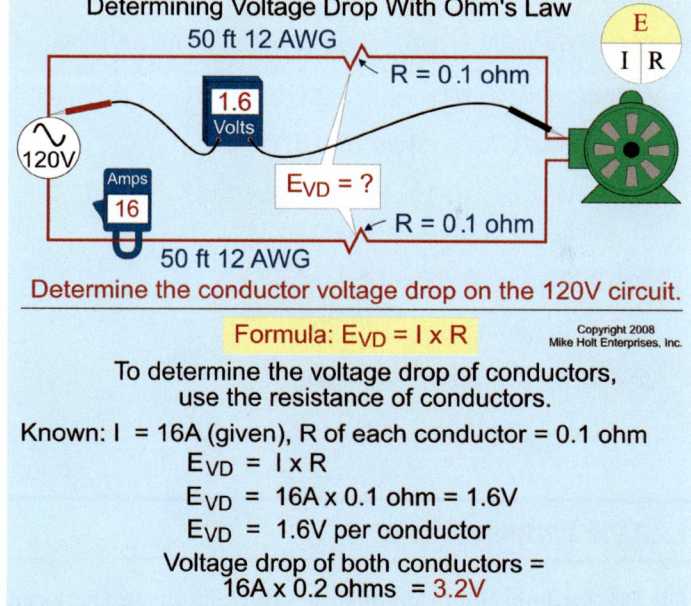

Figure 1–18

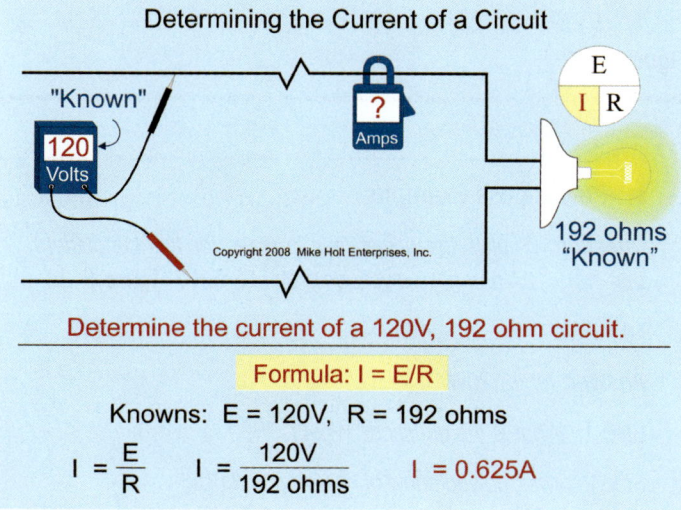

Figure 1–17

Unit 1 — Electrician's Math and Basic Electrical Formulas

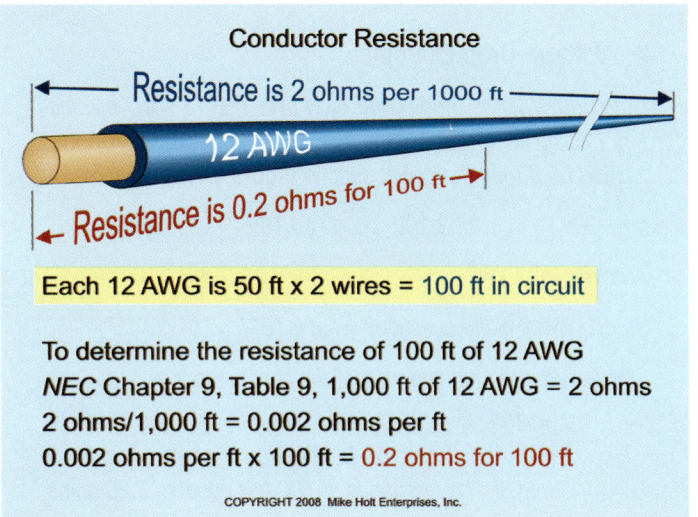

Figure 1–19

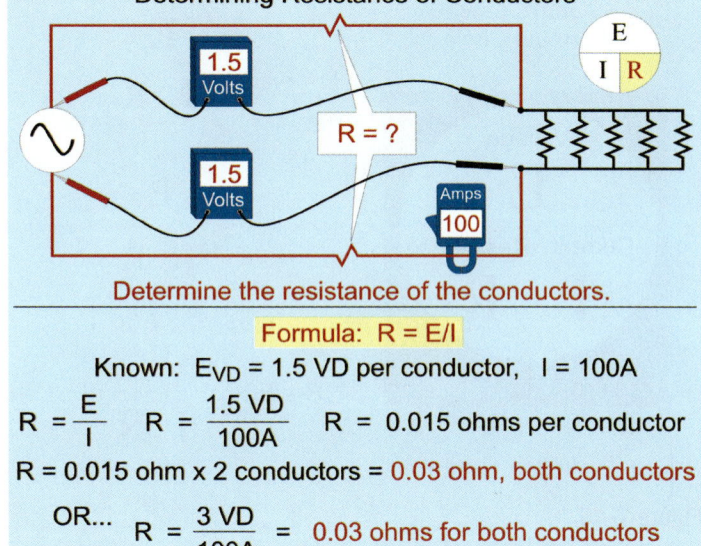

Figure 1–20

▶ **Resistance Example**

Question: What is the resistance of the circuit conductors when the conductor voltage drop is 3V and the current flowing in the circuit is 100A? **Figure 1–20**

(a) 0.03 ohms (b) 2 ohms (c) 30 ohms (d) 300 ohms

Answer: (a) 0.03 ohms

Step 1: What is the question? What is "R?"

Step 2: What do you know about the conductors?
E = 3V dropped, I = 100A

Step 3: The formula is **R = E/I**

Step 4: The answer is R = 3V/100A

Step 5: The answer is R = 0.03 ohms

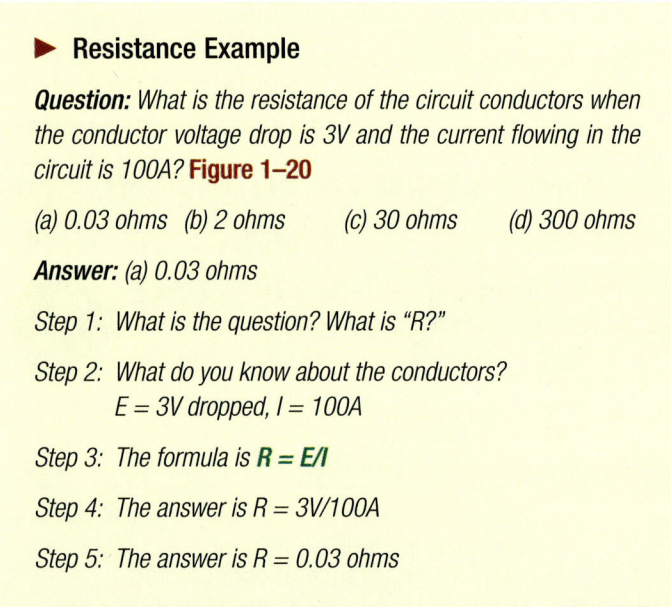

Figure 1–21

1.22 PIE Formula Circle

The PIE formula circle demonstrates the relationship between power, current, and voltage, and is expressed in the formula **P = I x E**. This formula can be transposed to **I = P/E** or **E = P/I**. In order to use these formulas, two of the values must be known.

Author's Comment: Place your thumb on the unknown value in **Figure 1–21** and the two remaining variables will "show" you the correct formula.

▶ **Power Loss Example**

Question: What is the power loss in watts for two conductors that carry 12A and have a voltage drop of 3.60V? **Figure 1–22**

(a) 4.30W (b) 24W (c) 43.20W (d) 432W

Answer: (c) 43.20W

Step 1: What is the question? What is "P?"

Step 2: What do you know? I = 12A, E = 3.60V

(continued in next column)

Step 3: The formula is **P = I x E**

Step 4: The answer is P = 12A x 3.60V

Step 5: The answer is 43.20W

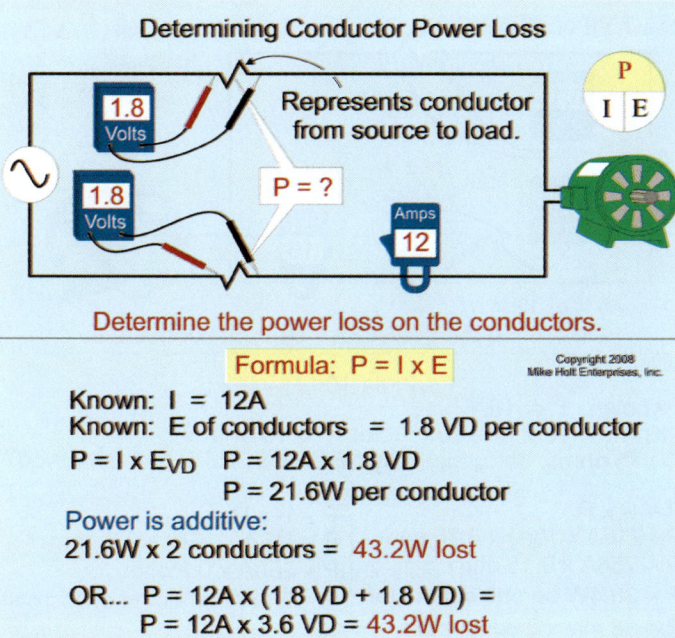

Figure 1–22

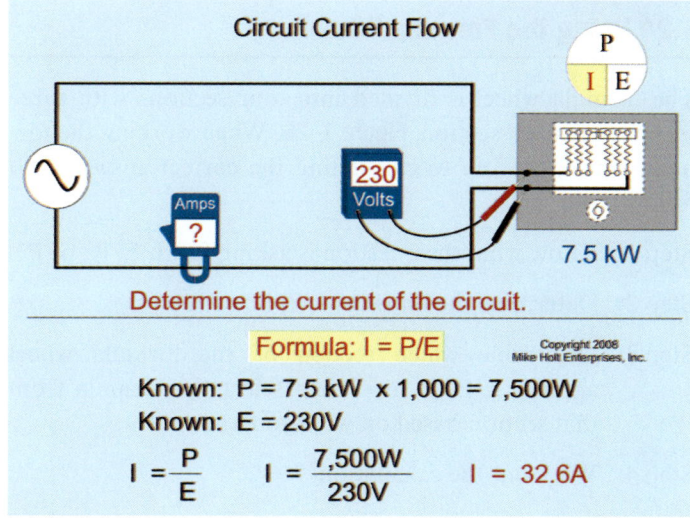

Figure 1–23

1.23 Formula Wheel

The formula wheel is a combination of the Ohm's Law and the PIE formula wheels. The formulas in the formula wheel can be used for dc circuits or ac circuits with unity power factor. **Figure 1–24**

Author's Comment: Unity power factor is explained in Unit 3. For the purpose of this unit, we'll assume a power factor of 1.0 for all ac circuits.

▶ **Current Example**

Question: What is the current flow in amperes through a 7.50 kW heat strip rated 230V when connected to a 230V power supply? **Figure 1–23**

(a) 25A   (b) 33A   (c) 39A   (d) 230A

Answer: (b) 33A

Step 1: What is the question? What is "I?"

Step 2: What do you know? P = 7,500W, E = 230V

Step 3: The formula is **I = P/E**

Step 4: The answer is I = 7,500W/230V

Step 5: The answer is 32.60A

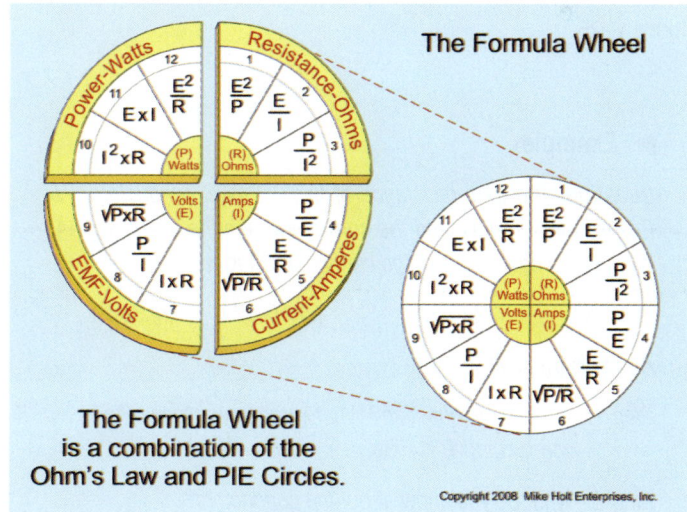

Figure 1–24

Unit 1   **Electrician's Math and Basic Electrical Formulas**

1.24 Using the Formula Wheel

The formula wheel is divided into four sections with three formulas in each section. **Figure 1–25**. When working the formula wheel, the key to calculating the correct answer is to follow these steps:

Step 1: Know what the question is asking for: I, E, R, or P.

Step 2: Determine the knowns: I, E, R, or P.

Step 3: Determine which section of the formula wheel applies: I, E, R, or P and select the formula from that section based on what you know.

Step 4: Work out the calculation.

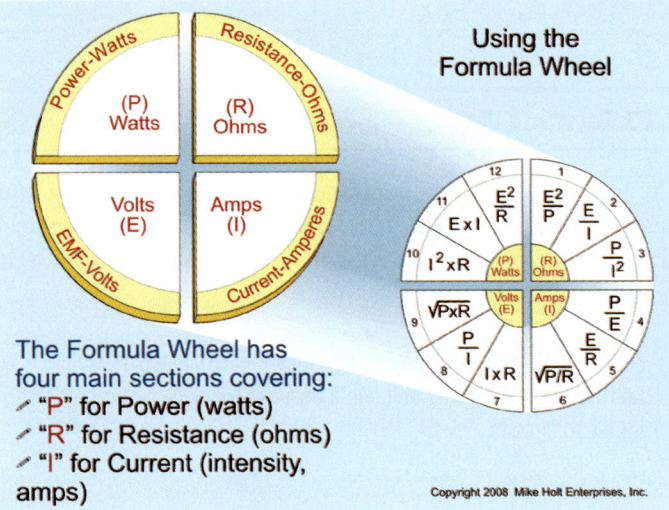

Figure 1–25

> **► Example**
>
> **Question:** The total resistance of two 12 AWG conductors, 75 ft long is 0.30 ohms, and the current through the circuit is 16A. What is the power loss of the conductors? **Figure 1–26**
>
> (a) 20W   (b) 75W   (c) 150W   (d) 300W
>
> **Answer:** (b) 75W
>
> Step 1: What is the question? What is the power loss of the conductors "P?"
>
> Step 2: What do you know about the conductors?
> I = 16A, R = 0.30 ohms
>
> Step 3: What is the formula? $P = I^2 \times R$
>
> *(continued in next column)*

> Step 4: Calculate the answer: $P = 16A^2 \times 0.30$ ohms = 76.80W
> The answer is 76.80W

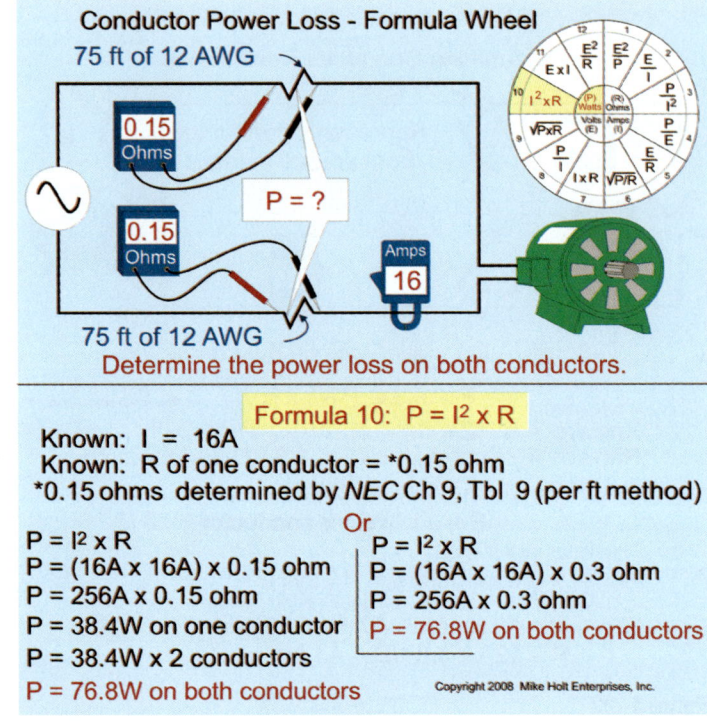

Figure 1–26

1.25 Power Losses of Conductors

Power in a circuit can be either "useful" or "wasted." Most of the power used by loads such as fluorescent lighting, motors, or stove elements is consumed in useful work. However, the heating of conductors, transformers, and motor windings is wasted work. Wasted work is still energy used; therefore it must be paid for, so we call these power losses.

> **► Example**
>
> **Question:** What is the conductor power loss in watts for a 10 AWG conductor that has a voltage drop of 3 percent and carries a current flow of 24A? **Figure 1–27**
>
> (a) 17W   (b) 173W   (c) 350W   (d) 450W
>
> **Answer:** (b) 173W
>
> *(continued in next column)*

Step 1: What is the problem asking you to find? What is wasted "P?"

Step 2: What do you know about the conductors?

I = 24A
E = 240V x 3%
E = 240V x 0.03
E = 7.20V

Step 3: The formula is **P = I x E**

Step 4: Calculate the answer: P = 24A x 7.20V = 172.80W
The answer is 172.80W

Step 1: Determine the power consumed:

P = I² x R
P = 24A² x 0.30 ohms
P = 172.80W

Step 2: Convert answer in Step 1 to kW:
P = 172.80W/1,000W
P = 0.1728 kW

Step 3: Determine cost per hour:
(0.086 dollars per kWh) x 0.17280 kW =
0.01486 dollars per hr

Step 4: Determine dollars per day:
0.01486 dollars per hr x (24 hrs per day) =
0.3567 dollars per day

Step 5: Determine dollars per year:
0.3567 dollars per day x (365 days per year) =
$130.20 per year

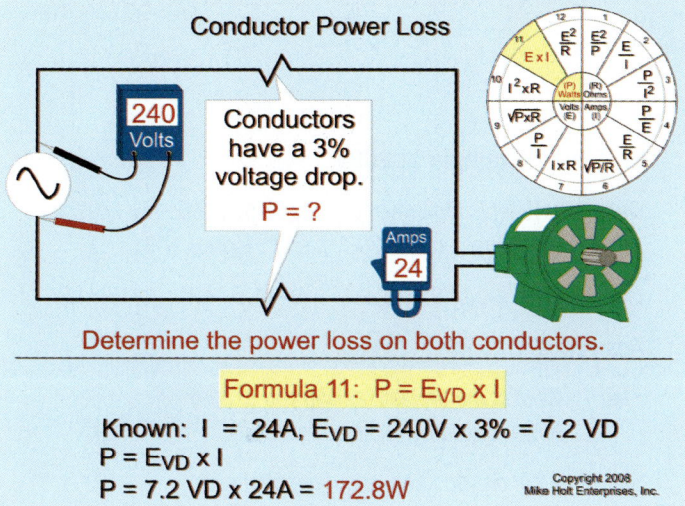

Figure 1–27

1.26 Cost of Power

Since electric bills are based on power consumed in watts, we should understand how to determine the cost of power.

▶ **Example**

Question: What does it cost per year (at 8.60 cents per kWh) for the power loss of two 10 AWG circuit conductors that have a total resistance of 0.30 ohms with a current flow of 24A? Figure 1–28

(a) $1.30   (b) $13.00
(c) $130.00   (d) $1,300.00

Answer: (c) $130.00

(continued in next column)

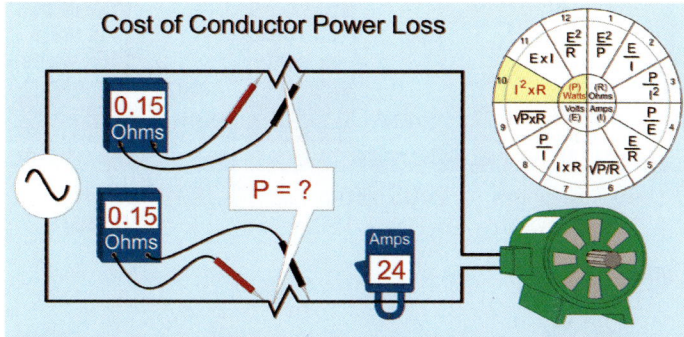

Figure 1–28

Author's Comment: That's a lot of money just to heat up two 10 AWG conductors for one circuit. Imagine how much it costs to heat up the conductors for an entire building!

Unit 1 Electrician's Math and Basic Electrical Formulas

1.27 Power Changes with the Square of the Voltage

The voltage applied to a resistor dramatically affects the power consumed by that resistor. Power is determined by the square of the voltage. This means that if the voltage is doubled, the power will increase four times. If the voltage is decreased 50 percent, the power will decrease to 25 percent of its original value. **Figure 1–29**

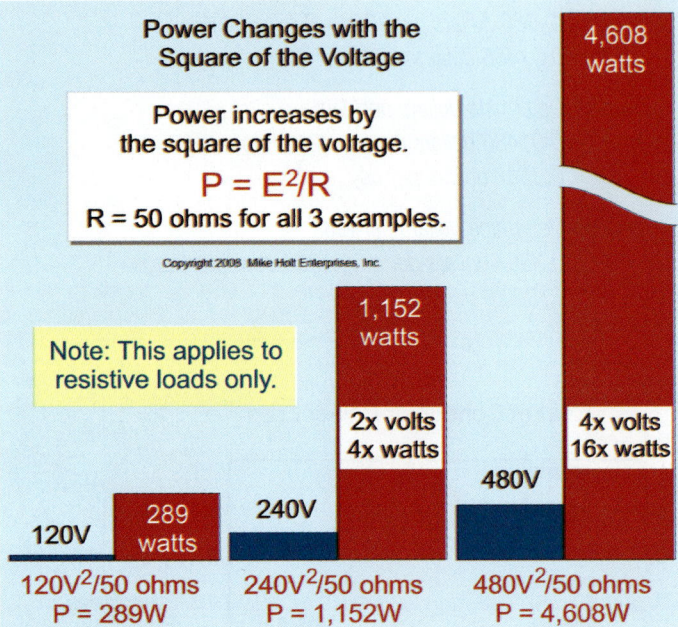

Figure 1–29

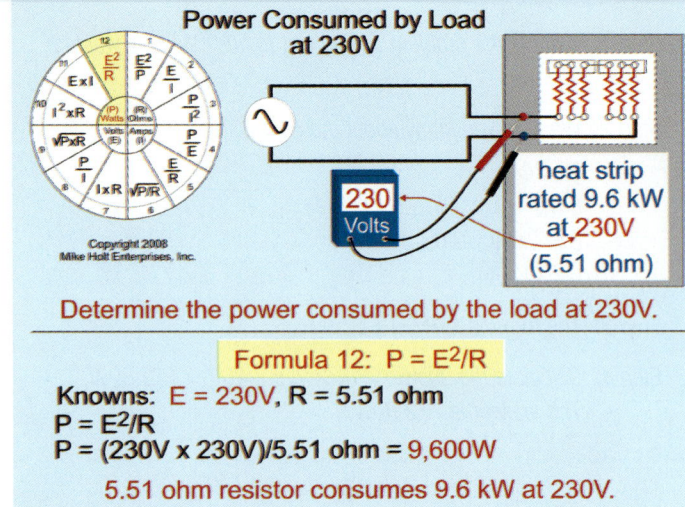

Figure 1–30

▶ **Power Example at 230V**

Question: What is the power consumed by a 9.60 kW heat strip rated 230V connected to a 230V circuit? **Figure 1–30**

(a) 7.85 kW (b) 9.60 kW (c) 11.57 kW (d) 25 kW

Answer: (b) 9.60 kW

Step 1: What is the problem asking you to find?
Power consumed by the resistance.

Step 2: What do you know about the heat strip?
You were given P = 9.60 kW in the statement of the problem.

▶ **Power Example at 208V**

Question: What is the power consumed by a 9.60 kW heat strip rated 230V connected to a 208V circuit? **Figure 1–31**

(a) 7.85 kW (b) 9.60 kW (c) 11.57 kW (d) 208 kW

Answer: (a) 7.85 kW

Step 1: What is the problem asking you to find?
Power consumed by the resistance.

Step 2: What do you know about the heat strip?
$E = 208V$, $R = E^2/P$
$R = 230V \times 230V/9,600W$
$R = 5.51$ ohms

Step 3: The formula to determine power is: $P = E^2/R$

Step 4: The answer is:
$P = 208V^2/5.51$ ohms
$P = 7,851W$ or 7.85 kW

Author's Comment: It's important to realize that the resistance of the heater unit doesn't change—it's a property of the material that the current flows through and isn't dependent on the voltage applied.

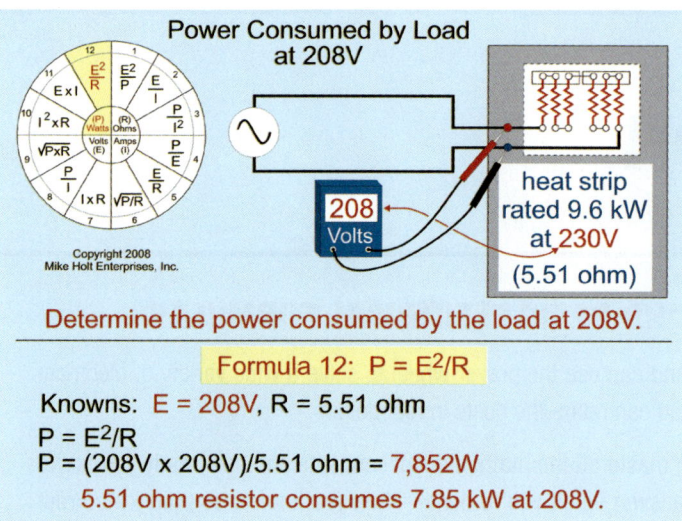

Determine the power consumed by the load at 208V.

Formula 12: $P = E^2/R$

Knowns: E = 208V, R = 5.51 ohm
$P = E^2/R$
P = (208V x 208V)/5.51 ohm = 7,852W

5.51 ohm resistor consumes 7.85 kW at 208V.

Figure 1–31

Thus, for a small change in voltage, there's a considerable change in power consumption by this heater.

Author's Comment: The current flow for this heat strip is I = P/E.

P = 7,851W
E = 208V
I = 7,851W/208V
I = 38A

▶ **Power Example at 240V**

Question: What is the power consumed by a 9.60 kW heat strip rated 230V connected to a 240V circuit? **Figure 1–32**

(a) 7.85 kW (b) 9.60 kW (c) 10.45 kW (d) 11.57 kW

Answer: (c) 10.45 kW

Step 1: What is the problem asking you to find?
Power consumed by the resistance.

Step 2: What do you know about the resistance?
R = 5.51 ohms*

*The resistance of the heat strip is determined by the formula $R = E^2/P$.

E = Nameplate voltage rating of the resistance, 230V
P = Nameplate power rating of the resistance, 9,600W
$R = E^2/P$
R = 230V²/9,600W
R = 5.51 ohms

(continued in next column)

Step 3: The formula to determine power is: $P = E^2/R$

Step 4: The answer is:
P = 240V x 240V/5.51 ohms
P = 10,454W
P = 10.45 kW

Author's Comment: The current flow for this heat strip is I = P/E.

P = 10,454W
E = 240V
I = 10,454W/240V
I = 44A

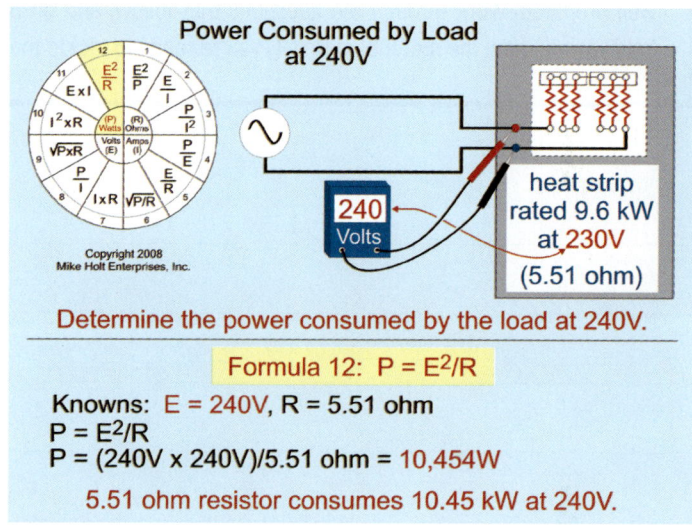

Determine the power consumed by the load at 240V.

Formula 12: $P = E^2/R$

Knowns: E = 240V, R = 5.51 ohm
$P = E^2/R$
P = (240V x 240V)/5.51 ohm = 10,454W

5.51 ohm resistor consumes 10.45 kW at 240V.

Figure 1–32

As you can see when the voltage changes, the power changes by the square of the change in the voltage, but the current changes in direct proportion to the change in the voltage.

Unit 1 — Conclusion

CONCLUSION TO UNIT 1—ELECTRICIAN'S MATH AND BASIC ELECTRICAL FORMULAS

You've gained skill in working with Ohm's Law and the power equation, and can use the power wheel to solve a wide variety of electrical problems. You also know how to calculate voltage drop and power loss, and can relate the costs in real dollars.

As you work through the practice questions, you'll see how well you have mastered the mathematical concepts and how ready you are to put them to use in electrical formulas. Always remember to check your answer when you're done—then you'll know you have a correct answer every time. As useful as these skills are, there is still more to learn. But, your mastery of the basic electrical formulas means you're well prepared. Work through the questions that follow, and go back over the instructional material if you have any difficulty. When you believe you know the material in Unit 1, you're ready to tackle the electrical circuits of Unit 2.

Practice Questions

UNIT 1—ELECTRICIAN'S MATH AND BASIC ELECTRICAL FORMULAS PRACTICE QUESTIONS

PART A—ELECTRICIAN'S MATH

1.3 Fractions

1. The decimal equivalent for the fraction "½" is _____.

    (a) 0.50
    (b) 0.70
    (c) 2
    (d) 5

2. The approximate decimal equivalent for the fraction "4/18" is _____.

    (a) 0.20
    (b) 2.50
    (c) 3.50
    (d) 4.50

1.4 Percentages

3. To change a percent value to a decimal or whole number, drop the percentage sign and move the decimal point two places to the _____.

    (a) right
    (b) left
    (c) depends
    (d) none of these

4. The decimal equivalent for "75 percent" is _____.

    (a) 0.075
    (b) 0.75
    (c) 7.50
    (d) 75

5. The decimal equivalent for "225 percent" is _____.

    (a) 0.225
    (b) 2.25
    (c) 22.5
    (d) 225

6. The decimal equivalent for "300 percent" is _____.

    (a) 0.03
    (b) 0.30
    (c) 3
    (d) 30.0

1.5 Multiplier

7. The method of increasing a number by another number is done by using a _____.

    (a) percentage
    (b) decimal
    (c) fraction
    (d) multiplier

8. An overcurrent device (circuit breaker or fuse) must be sized no less than 125 percent of the continuous load. If the load is 16A, the overcurrent device will have to be sized at no less than _____.

    (a) 17A
    (b) 20A
    (c) 23A
    (d) 30A

9. The maximum continuous load on an overcurrent device is limited to 80 percent of the device rating. If the overcurrent device is rated 100A, the maximum continuous load is _____.

    (a) 72A
    (b) 80A
    (c) 90A
    (d) 125A

Unit 1 — Practice Questions

1.6 Percent Increase

10. The feeder calculated load for an 8 kW load, increased by 20 percent is _____.

    (a) 8 kW
    (b) 9.60 kW
    (c) 10 kW
    (d) 12 kW

1.7 Reciprocals

11. What is the reciprocal of 1.25?

    (a) 0.80
    (b) 1.10
    (c) 1.25
    (d) 1.50

12. A continuous load requires an overcurrent device sized no smaller than 125 percent of the load. What is the maximum continuous load permitted on a 100A overcurrent device?

    (a) 75A
    (b) 80A
    (c) 100A
    (d) 125A

1.8 Squaring a Number

13. Squaring a number means multiplying the number by itself.

    (a) True
    (b) False

14. What is the power consumed in watts by a 12 AWG conductor that is 100 ft long and has a resistance (R) of 0.20 ohms, when the current (I) in the circuit is 16A?

    Formula: Power = $I^2 \times R$

    (a) 50W
    (b) 75W
    (c) 100W
    (d) 200W

15. What is the approximate area in sq in. of a trade size 2 raceway?

    Formula: Area = $\pi \times r^2$, $\pi = 3.14$, r = radius (½ of diameter)

    (a) 1 sq in.
    (b) 2 sq in.
    (c) 3 sq in.
    (d) 4 sq in.

16. The numeric equivalent of 4^2 is _____.

    (a) 2
    (b) 8
    (c) 16
    (d) 32

17. The numeric equivalent of 12^2 is _____.

    (a) 3.46
    (b) 24
    (c) 144
    (d) 1,728

1.9 Parentheses

18. What is the maximum distance that two 14 AWG conductors can be run if they carry 16A and the maximum allowable voltage drop is 10V?

    D = (Cmil x E_{vd})/(2 x K x I)
    Cmil area of 14 AWG = 4,110 cmil [Chapter 9, Table 8]
    D = (4,110 Cmil x 10V)/(2 x 12.90 ohms x 16A)

    (a) 50 ft
    (b) 75 ft
    (c) 100 ft
    (d) 150 ft

19. What is the current in amperes of an 18 kW, 208V, three-phase load?

    Current: I = VA/(E x $\sqrt{3}$)
    Current: I = 18,000W/(208V x 1.732)

    (a) 25A
    (b) 50A
    (c) 100A
    (d) 150A

1.10 Square Root

20. Deriving the square root of a number is almost the same as squaring a number.

    (a) True
    (b) False

21. What is the approximate square root of 1,000 ($\sqrt{1,000}$)?

    (a) 3
    (b) 32
    (c) 100
    (d) 500

22. The square root of 3 ($\sqrt{3}$) is _____.

    (a) 1.5
    (b) 1.732
    (c) 9
    (d) 729

1.11 Volume

23. The volume of an enclosure is expressed in _____, and is calculated by multiplying the length, by the width, by the depth of the enclosure.

    (a) cubic inches
    (b) weight
    (c) inch-pounds
    (d) none of these

24. What is the volume (in cubic inches) of a 4 x 4 x 1.50 in. box?

    (a) 20 cu in.
    (b) 24 cu in.
    (c) 30 cu in.
    (d) 33 cu in.

1.12 Kilo

25. What is the kW of a 75W load?

    (a) 0.075 kW
    (b) 0.75 kW
    (c) 7.50 kW
    (d) 75 kW

1.13 Rounding Off

26. The approximate sum of 2, 7, 8, and 9 is equal to _____.

    (a) 20
    (b) 25
    (c) 30
    (d) 35

1.14 Testing Your Answer for Reasonableness

27. The output power of a transformer is 100W and the transformer efficiency is 90 percent. What is the transformer input if the output is lower than the input?

    Formula: Input = Output/Efficiency

    (a) 90W
    (b) 100W
    (c) 110W
    (d) 125W

PART B—BASIC ELECTRICAL FORMULAS

1.15 Electrical Circuit

28. An electrical circuit consists of the _____.

    (a) power source
    (b) conductors
    (c) load
    (d) all of these

29. According to the electron flow theory, electrons leave the _____ terminal of the source, flow through the conductors and load(s), and return to the _____ terminal of the source.

    (a) positive, negative
    (b) negative, positive
    (c) negative, negative
    (d) positive, positive

1.16 Power Source

30. The polarity and the output voltage from a dc power source changes direction. One terminal will be negative and the other will be positive at one moment, then the terminals switch polarity.

    (a) True
    (b) False

31. Direct current is used for electroplating, street trolley and railway systems, or where a smooth and wide range of speed control is required for a motor-driven application.

    (a) True
    (b) False

32. The polarity and the output voltage from an ac power source never change direction.

    (a) True
    (b) False

33. The major advantage of ac over dc is the ease of voltage regulation by the use of a transformer.

    (a) True
    (b) False

1.17 Conductance

34. Conductance is the property that permits current to flow.

    (a) True
    (b) False

Unit 1 — Practice Questions

35. The best conductors, in order of their conductivity, are gold, silver, copper, and aluminum.

    (a) True
    (b) False

36. Conductance or conductivity is the property of metal that permits current to flow. The best conductors in order of their conductivity are _____.

    (a) gold, silver, copper, aluminum
    (b) gold, copper, silver, aluminum
    (c) silver, gold, copper, aluminum
    (d) silver, copper, gold, aluminum

1.18 Circuit Resistance

37. The circuit resistance includes the resistance of the _____.

    (a) power source
    (b) conductors
    (c) load
    (d) all of these

38. Often the resistance of the power source and conductor are ignored in circuit calculations.

    (a) True
    (b) False

1.19 Ohm's Law

39. The Ohm's Law formula, I = E/R, states that current is _____ proportional to the voltage, and _____ proportional to the resistance.

    (a) indirectly, inversely
    (b) inversely, directly
    (c) inversely, indirectly
    (d) directly, inversely

40. Ohm's Law demonstrates the relationship between circuit _____.

    (a) intensity
    (b) EMF
    (c) resistance
    (d) all of these

1.20 Ohm's Law and Alternating Current

41. In a dc circuit, the only opposition to current flow is the physical resistance of the material. This opposition is called reactance and is measured in ohms.

    (a) True
    (b) False

42. In an ac circuit, the factors that oppose current flow are _____.

    (a) resistance
    (b) inductive reactance
    (c) capacitive reactance
    (d) all of these

1.21 Ohm's Law Formula Circle

43. What is the voltage drop of two 12 AWG conductors (0.40 ohms) supplying a 16A load, located 100 ft from the power supply?

    Formula: $E_{vd} = I \times R$

    (a) 1.60V
    (b) 3.20V
    (c) 6.40V
    (d) 12.80V

44. What is the resistance of the circuit conductors when the conductor voltage drop is 7.20V and the current flow is 50A?

    Formula: $R = E/I$

    (a) 0.14 ohms
    (b) 0.30 ohms
    (c) 3 ohms
    (d) 14 ohms

1.22 PIE Formula Circle

45. What is the power loss in watts of a conductor that carries 24A and has a voltage drop of 7.20V?

    Formula: $P = I \times E$

    (a) 175W
    (b) 350W
    (c) 700W
    (d) 2,400W

46. What is the approximate power consumed by a 10 kW heat strip rated 230V, when connected to a 208V circuit?

    Formula: $P = E^2/R$

    (a) 8.20 kW
    (b) 9.30 kW
    (c) 10.90 kW
    (d) 11.20 kW

1.23 Formula Wheel

47. The formulas in the power wheel apply to _____.

    (a) dc
    (b) ac with unity power factor
    (c) dc or ac circuits
    (d) a and b

1.24 Using the Formula Wheel

48. When working any formula, the key to calculating the correct answer is following these four steps:

    Step 1: Know what the question is asking you to find.
    Step 2: Determine the knowns of the circuit.
    Step 3: Select the formula.
    Step 4: Work out the formula calculation.

    (a) True
    (b) False

1.25 Power Losses of Conductors

49. Power in a circuit can be either "useful" or "wasted." Wasted work is still energy used; therefore it must be paid for, so we call this _____.

    (a) resistance
    (b) inductive reactance
    (c) capacitive reactance
    (d) power loss

50. The total circuit resistance of two 12 AWG conductors (each 100 ft long) is 0.40 ohms. If the current of the circuit is 16A, what is the power loss of both conductors?

    Formula: $P = I^2 \times R$

    (a) 75W
    (b) 100W
    (c) 300W
    (d) 600W

51. What is the conductor power loss for a 120V circuit that has a 3 percent voltage drop and carries a current flow of 12A?

    Formula: $P = I \times E$

    (a) 43W
    (b) 86W
    (c) 172W
    (d) 1,440W

1.26 Cost of Power

52. What does it cost per year (at 8 cents per kWh) for the power loss of a 12 AWG conductor (100 ft long) that has a total resistance of 0.40 ohms and a current flow of 16A?

    Formula: Cost per Year = Power for the Year in kWh x $0.08

    (a) $33
    (b) $54
    (c) $72
    (d) $89

1.27 Power Changes with the Square of the Voltage

53. The voltage applied to a resistor dramatically affects the power consumed by that resistor because power is affected in direct proportion to the voltage.

    (a) True
    (b) False

54. What is the power consumed by a 10 kW heat strip that's rated 230V, if it's connected to a 115V circuit?

    Formula: $P = E^2/R$

    (a) 2.50 kW
    (b) 5 kW
    (c) 7.50 kW
    (d) 15 kW

Unit 1 — Challenge Questions

UNIT 1—ELECTRICIAN'S MATH AND BASIC ELECTRICAL FORMULAS CHALLENGE QUESTIONS

(•Indicates that 75% or fewer of those who took this exam answered the question correctly.)

PART A—ELECTRICIAN'S MATH

1.12 Kilo

1. •One kVA is equal to _____.

   (a) 100 VA
   (b) 1,000V
   (c) 1,000W
   (d) 1,000 VA

PART B—BASIC ELECTRICAL FORMULAS

1.17 Conductance

2. •Which of the following is the most conductive?

   (a) Bakelite
   (b) Oil
   (c) Air
   (d) Salt water

1.19 Ohm's Law

3. •If the contact resistance of a connection increases and the current of the circuit (load) remains the same, the voltage dropped across the connection will _____.

   (a) increase
   (b) decrease
   (c) remain the same
   (d) cannot be determined

4. •To double the current of a circuit when the voltage remains constant, the R (resistance) must be _____.

   (a) doubled
   (b) reduced by half
   (c) increased
   (d) none of these

5. •An ohmmeter is being used to test a relay coil. The equipment instructions indicate that the resistance of the coil should be between 30 and 33 ohms. The ohmmeter indicates that the actual resistance is less than 22 ohms. This reading would most likely indicate _____.

   (a) the coil is okay
   (b) an open coil
   (c) a shorted coil
   (d) a meter problem

1.23 Formula Wheel

6. •To calculate the energy consumed in watts by a resistive appliance, you need to know _____ of the circuit.

   (a) the voltage and current
   (b) the current and resistance
   (c) the voltage and resistance
   (d) any of these pairs of variables

7. •The power consumed of a resistor can be expressed by the formula $I^2 \times R$. If 120V is applied to a 10 ohm resistor, the power consumed will be _____.

   (a) 510W
   (b) 1,050W
   (c) 1,230W
   (d) 1,440W

8. •Power loss in a circuit because of heat can be determined by the formula _____.

   (a) $P = R \times I$
   (b) $P = I \times R$
   (c) $P = I^2 \times R$
   (d) none of these

9. •The energy consumed by a 5 ohm resistor is _____ than the energy consumed by a 10 ohm resistor, assuming the current in both cases remains the same.

    (a) more
    (b) less

10. When a load is rated 500W at 115V is connected to a 120V power supply, the current of the circuit will be _____. Tip: At 120V, the load consumes more than 500 ohms, but the resistance of the load remains constant.

    (a) 2.70A
    (b) 3.80A
    (c) 4.50A
    (d) 5.50A

1.27 Power Changes with the Square of the Voltage

11. A 120V rated toaster will produce _____ heat when supplied by 115V.

    (a) more
    (b) less
    (c) the same
    (d) none of these

12. •When a 100W, 115V lamp operates at 230V, the lamp will consume approximately _____.

    (a) 150W
    (b) 300W
    (c) 400W
    (d) 450W

13. •A 1,500W resistive heater is rated 230V and is connected to a 208V supply. The power consumed for this load at 208V will be approximately _____.

    (a) 1,225W
    (b) 1,625W
    (c) 1,750W
    (d) 1,850W

14. •The total resistance of a circuit is 10.20 ohms. The load has a resistance of 10 ohms and the wire has a resistance of 0.20 ohms. If the current of the circuit is 12A, then the power consumed by the circuit conductors (0.20 ohms) is approximately _____.

    (a) 8W
    (b) 29W
    (c) 39W
    (d) 45W

UNIT 2 Electrical Circuits

INTRODUCTION TO UNIT 2—ELECTRICAL CIRCUITS

Unit 1 laid a foundation for your continuing study of the topic of electricity. The math, basic electrical concepts, and electrical formulas you learned are essential to understanding the topics covered in Unit 2.

The basic requirements of a practical electrical circuit are the presence of an electrical source, an electrical load, and a complete path of current flow from the source to the load and returning again to the source. The path of current flow can be configured in many ways, and the way this path is designed is one factor that determines the amount of current flow at any specific point of the circuit, and how many volts are measured across any particular component. The configuration of the current-flow path can be expressed as a series path, a parallel path, or a combination series-parallel path.

In this unit, you'll learn about series circuits, parallel circuits, combinations of series and parallel circuits, and a special application called a multiwire branch circuit. Multiwire branch circuits have unique safety requirements, and the *Code* makes special provisions for using them. Be sure to understand the roles of current and voltage in each of these different types of circuits and how to find the total resistance of a given circuit. The questions at the end of the unit will help solidify your understanding of these circuits.

This unit makes use of some direct-current (dc) circuits and some alternating-current (ac) circuits, and their similarities in series and parallel resistive circuits will be explained. In Unit 3, more about the differences between direct-current and alternating-current circuits will be explained as you continue to learn about how electricity works.

PART A—SERIES CIRCUITS

Introduction

A series circuit is a circuit in which a specific amount of current leaves the voltage source and flows through every electrical device in a single path before it returns to the voltage source. **Figure 2–1**

It's important to understand that in a series circuit, current is identical through ALL circuit elements of the circuit.

2.1 Practical Uses of the Series Circuit

In a series circuit, if any part of the circuit is open, the current in the circuit will stop.

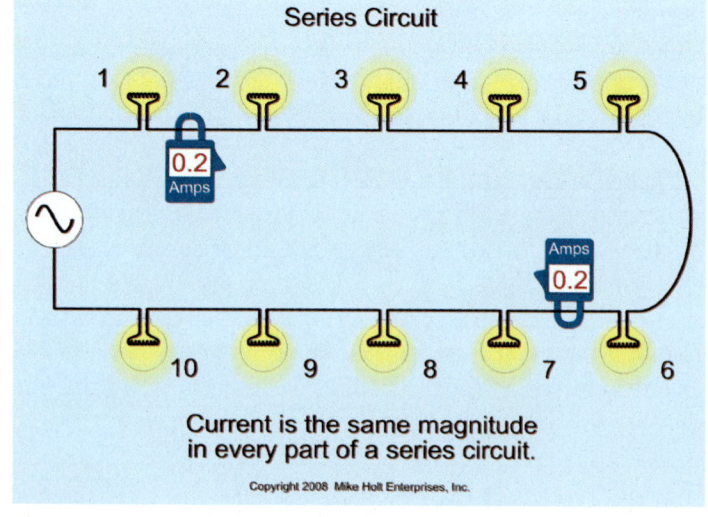

Figure 2–1

Electrical Circuits — Unit 2

Lighting

A series circuit is not generally useful for lighting because if one lamp burns out, all the other lamps on the circuit will go out. **Figure 2–2**

Control Circuit

Closed-loop (series) circuits are often used for the purpose of controlling (starting and stopping) electrical equipment. **Figure 2–3**

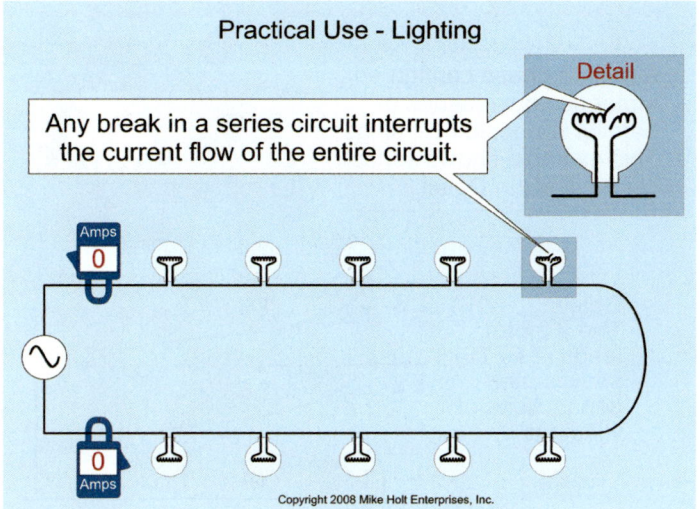

Figure 2–2

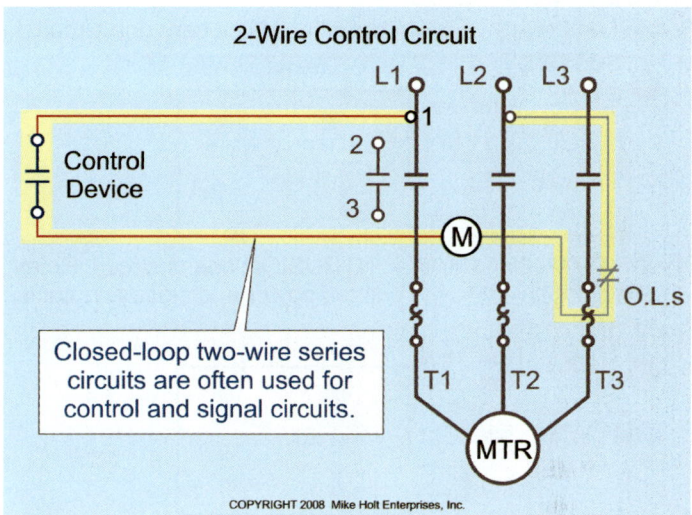

Figure 2–3

Author's Comment: There are some special applications such as airport runway lighting that make use of series circuits.

Older strings of Christmas lights are made in series circuits—if one lamp burns out, it requires examining EVERY lamp in order to find the bad one because the entire string stops burning when this occurs! Most of the newer Christmas light lamps used in series string lighting have a shorting bar. When one lamp burns out, the lamp leads are connected together by the shorting bar and the rest of the string stays lit.

Author's Comment: If, however, there are ten lamps in the string, and one burns out, the rest of the lamps will operate at 10 percent higher voltage, causing the other nine lamps to burn out more quickly.

For most practical purposes, series circuits, also called closed-loop systems, aren't used for building wiring; however, they are often used for control and signal circuits.

Signaling Circuit

Two-wire closed-loop (series) circuits are often used to give a signal that something has occurred. They can indicate that a door is open, a process is operating, or that there is fire or smoke. For example, the discontinuation of current flow when part of the circuit opens is important for the operation of burglar alarm circuits. **Figure 2–4**

Internal Equipment Wiring

The internal wiring of many types of equipment is connected in series. For example, a 115/230V rated motor connected to a 230V circuit must have the internal windings connected in series so that each winding will receive at least 115V. **Figure 2–5**

Unit 2 Electrical Circuits

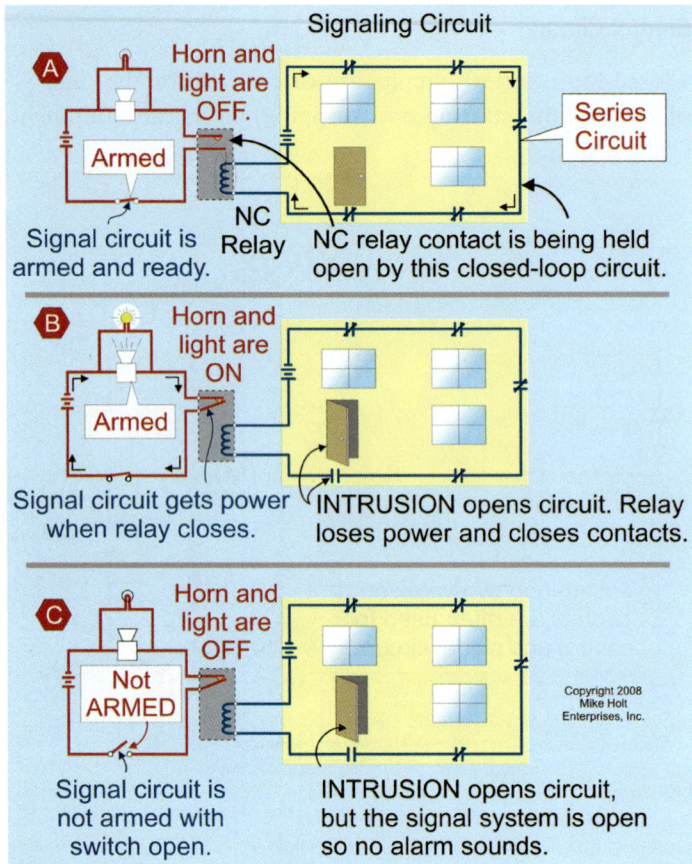

Figure 2–4

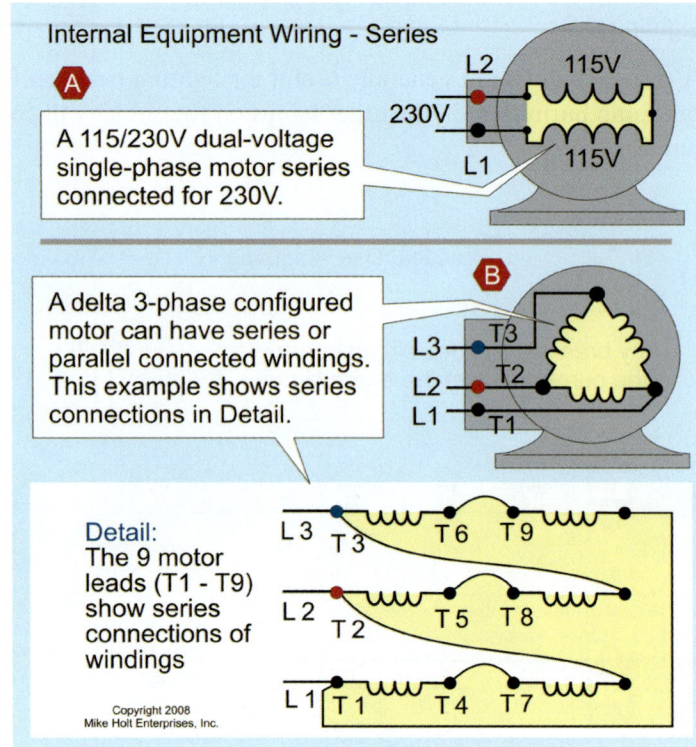

Figure 2–5

2.2 Understanding Series Calculations

It's important to understand the relationship between current, voltage, resistance, and power in series circuits. **Figure 2–6**

Resistance

Resistance opposes the flow of electrons. In a series circuit, the total circuit resistance is equal to the sum of all the series resistances. Resistance is additive: $R_T = R_1 + R_2 + R_3 + R_4$. **Figure 2–7**

R_1 Power Supply	0.05 ohms
R_2 Conductor No. 1	0.15 ohms
R_3 Appliance	7.15 ohms
R_4 Conductor No. 2	+ 0.15 ohms
R_T Total Resistance	7.50 ohms

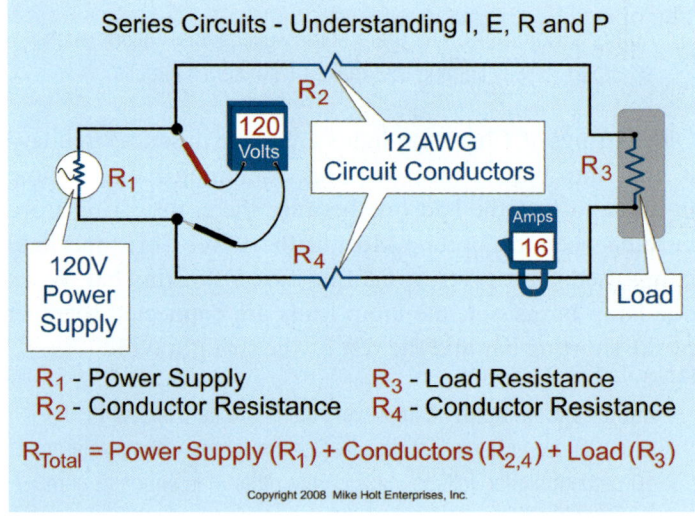

Figure 2–6

30   Mike Holt's Illustrated Guide to NEC Exam Preparation

Electrical Circuits — Unit 2

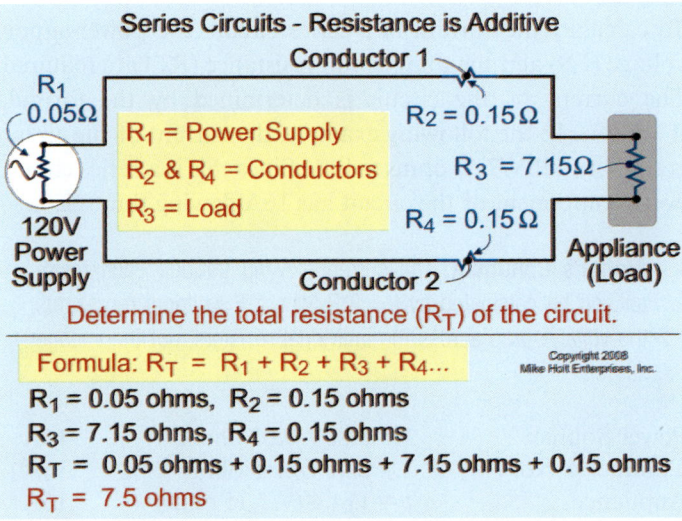

Figure 2–7

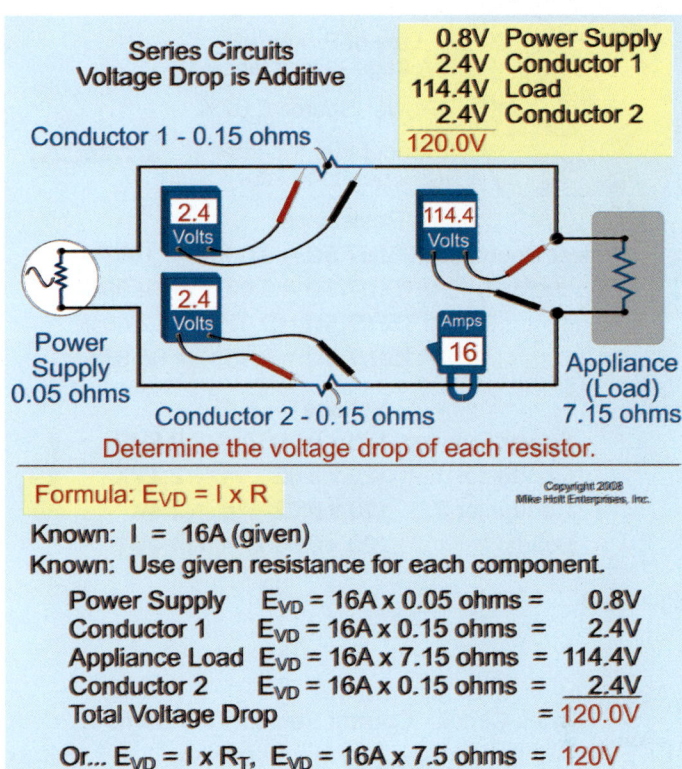

Figure 2–8

Voltage

The voltage, also called electromotive force (EMF), provides the pressure necessary to move electrons through the circuit. However, the power supply, the conductors, and the appliance all have resistance that opposes the current flow (although the power supply's resistance is usually ignored). The opposition (resistance) to current flow (amperes) results in a drop of the circuit voltage (voltage drop), and is calculated by the formula $E_{vd} = I \times R$.

Kirchoff's Voltage Law

Kirchoff's Voltage Law states that in a series circuit, the sum of the voltages across all of the resistors (or "voltage drops") in the series circuit is equal to the applied voltage.

The voltage drop across each resistor is determined by the formula: **Figure 2–8**

$E_{vd} = I \times R$

I = Current of circuit
R = Resistance of resistor

Power Supply	16A x 0.05 ohms	0.80V
Conductor 1	16A x 0.15 ohms	2.40V
Appliance	16A x 7.15 ohms	114.40V
Conductor 2	16A x 0.15 ohms	+ 2.40V
Total	16A x 7.50 ohms	120.00V

In addition, the voltage of the power supply is distributed or divided among the circuit resistors according to the Resistance Law of Proportion. The Resistance Law of Proportion means that the supply voltage is distributed among all the resistors, according to the proportion of resistance each resistor has relative to the total resistance. **Figure 2–9**

	Resistance	Percentage	Voltage
Power Source	0.05 ohms	0.67%	0.80V
Conductor No. 1	0.15 ohms	2.00%	2.40V
Appliance	7.15 ohms	95.33%	114.40V
Conductor No. 2	+ 0.15 ohms	2.00%	2.40V
Total	7.50 ohms	100%	120.00V

Kirchoff's Current Law

Kirchoff's Current Law states that the sum of currents flowing into a junction equals the sum of currents flowing away from the junction. Another way to say it is, current flowing through each resistor of a series circuit is the same. **Figure 2–10**

Unit 2 — Electrical Circuits

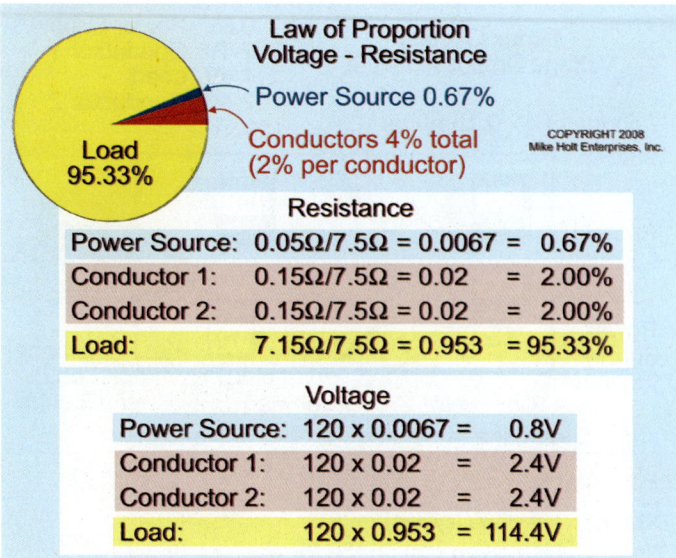

Figure 2–9

To calculate the current in a series circuit, the power-supply voltage (ES) and the total circuit resistance (RT) are required. The current of the circuit is determined by the formula **IT = ES/RT**. In the following example, the current of the circuit is equal to 120V/7.50 ohms = 16A. Since this is a series circuit, every component of the circuit has 16A flowing through it.

> **Author's Comment:** The current flowing through each resistor can be calculated by I = E/R, where E (voltage) represents the voltage drop across the individual resistors, not the voltage source.

Power Source	I = 0.80V/0.05 ohms	16A
Conductor No. 1	I = 2.40V/0.15 ohms	16A
Appliance	I = 114.40V/7.15 ohms	16A
Conductor No. 2	I = 2.40V/0.15 ohms	16A
Total Resistance	I = 120V/7.50	16A

Power

The power consumed in a series circuit equals the sum of the power consumed by all of the resistors in the series circuit. The Law of Conservation of Energy states that the power supply (battery, etc.) will only produce as much power as that consumed by the circuit elements. Power is a result of current flowing through a resistance and is calculated by the formula **P = I² × R**.

> ▶ **Example**
>
> | | | |
> |---|---|---|
> | Power Source | P = 16A² × 0.05 ohms | 12.80W |
> | Conductor No. 1 | P = 16A² × 0.15 ohms | 38.40W |
> | Appliance | P = 16A² × 7.15 ohms | 1,830.40W |
> | Conductor No. 2 | P = 16A² × 0.15 ohms | 38.40W |

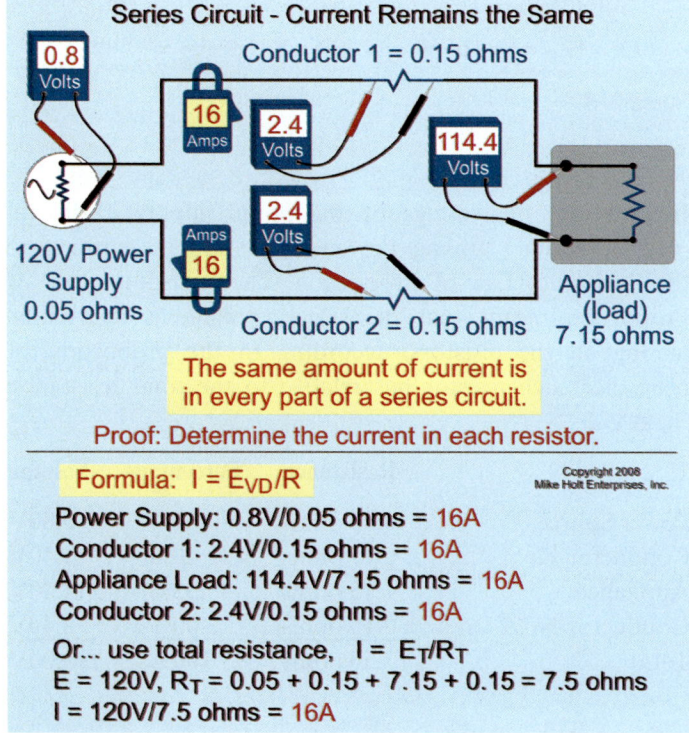

Figure 2–10

Power can also be calculated according to the Resistance Law of Proportion.

	Resistance	Percentage	Power
Power Source	0.05 ohms	0.67%	12.80W
Conductor No. 1	0.15 ohms	2.00%	38.40W
Appliance	7.15 ohms	95.33%	1,830.00W
Conductor No. 2	+ 0.15 ohms	2.00%	38.40W
Total Resistance	7.50 ohms	100%	1,920.00W

Electrical Circuits   Unit 2

2.3 Series Circuit Calculations

When performing series circuit calculations, follow these steps: **Figure 2–11**

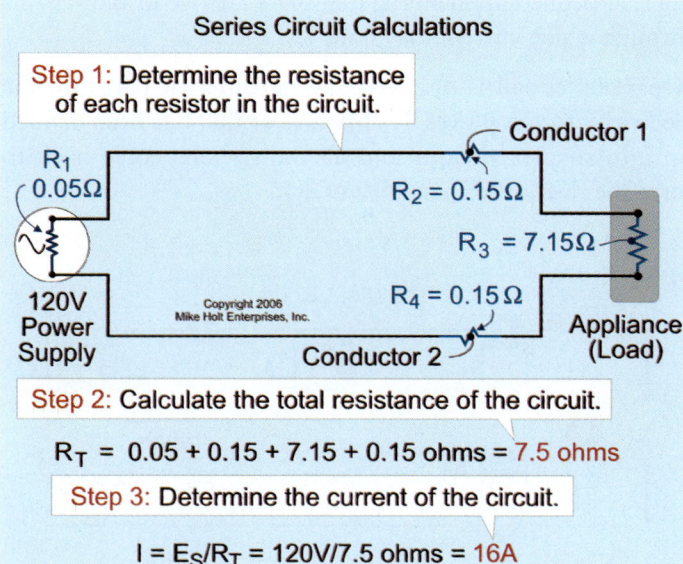

Figure 2–11

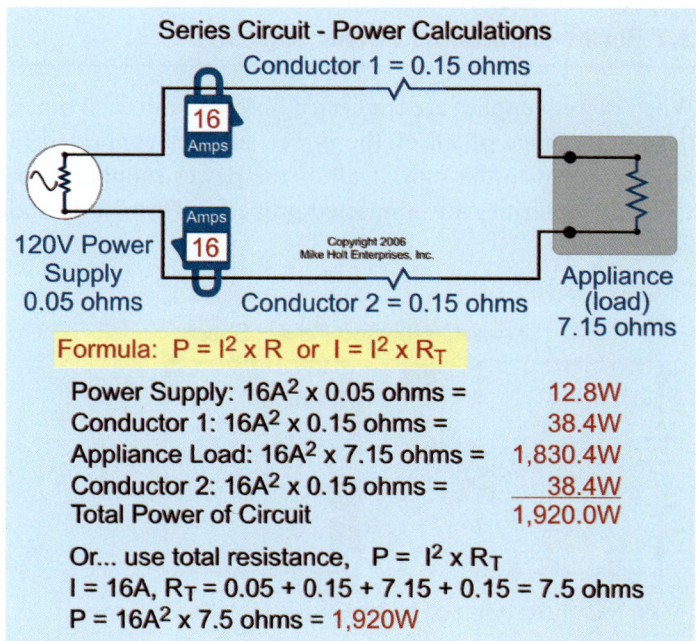

Figure 2–12

Step 1: Determine the resistance of each resistive element in the circuit. Often, the resistance of each element is given in the problem. If you know the nameplate voltage and power (wattage) rating of the appliance or equipment, determine its resistance by the formula $R = E^2/P$.

E = Nameplate voltage rating (squared)
P = Nameplate power rating

Step 2: Calculate the total resistance (RT) of the circuit, $R_T = R_1 + R_2 + R_3 + R_4$.

Step 3: The current of the circuit is determined by the formula $I = E_S/R_T$.
E_S = Voltage Source
R_T = Total circuit resistance (Step 2)

2.4 Power Calculations

If you know the current of the circuit and the resistance of each resistor, the power of each resistor is determined by the formula $P = I^2 \times R$. **Figure 2–12**

I^2 = Current of circuit (squared) (Step 3)
R = Resistance of the resistor (Step 1)

The power of the circuit is determined by adding up the power of all of the resistors, or by the formula $P = I^2 \times R_T$.

I^2 = Current of the circuit (squared)
R_T = Resistance total of the circuit

2.5 Variations

There are often many different ways to solve an electrical circuit problem involving voltage, current, resistance, and power. It's also often possible to verify or check one's work by solving the problem different ways.

2.6 Series Circuit Notes

Note 1: *The total resistance of a series circuit is equal to the sum of all of the resistances of the circuit.*

Note 2: *Current is the same value through all of the resistances.*

Note 3: *The sum of the voltage drops across all resistances equals the voltage of the source.*

Note 4: *The sum of the power consumed by all resistors equals the total power consumed by the circuit.*

Unit 2 | Electrical Circuits

2.7 Series-Connected Power Supplies

When power supplies are connected in series with each other, the total voltage of all of the power supplies connected in series is equal to the sum of all of the power supplies (provided the polarities are connected properly). **Figure 2–13** and **Figure 2–14**

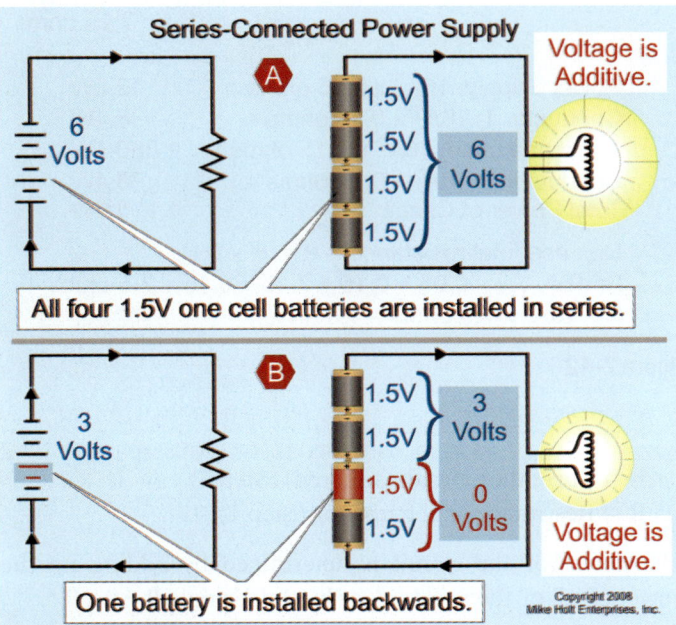

Figure 2–13

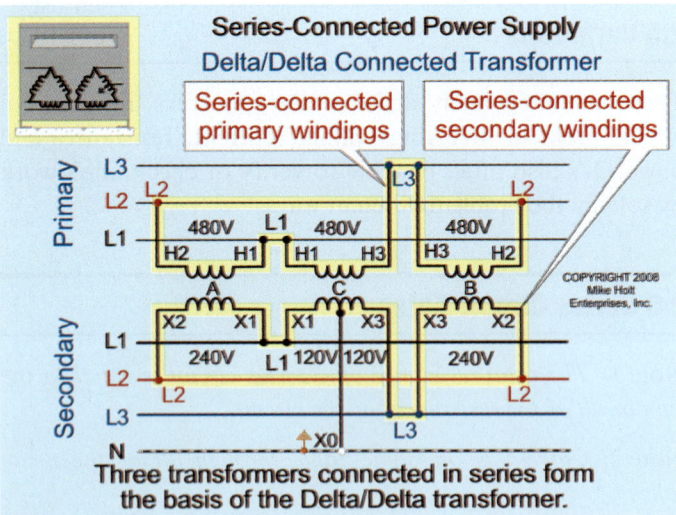

Figure 2–14

PART B—PARALLEL CIRCUITS

Introduction

"Parallel" is a term used to describe a method of connecting electrical components so that there are two or more paths through which current may flow.

A parallel circuit is one with several different paths for the electricity to travel over. It's like a river that has been divided up into smaller streams and all the streams come back to form the river once again. **Figure 2–15**

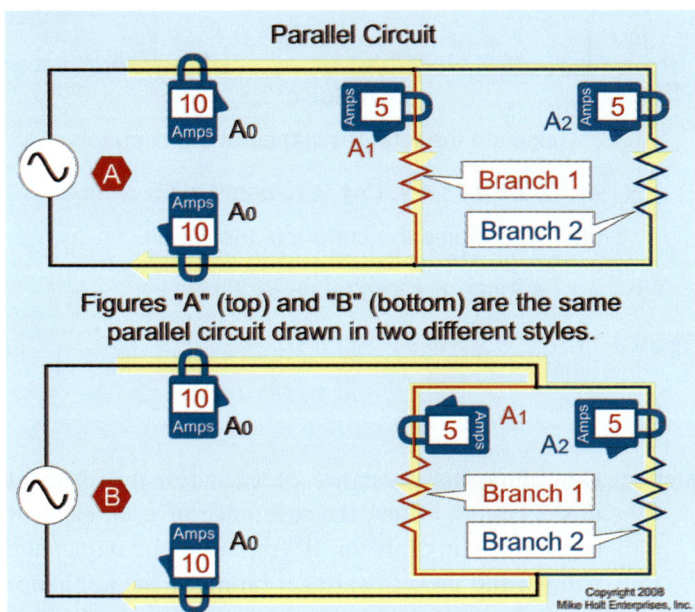

Figure 2–15

A parallel circuit has different characteristics than a series circuit. For one, the total resistance of a parallel circuit is not equal to the sum of the resistors. The total resistance in a parallel circuit is always less than any of the branch resistances. Adding more parallel resistances to the paths causes the total resistance in the circuit to decrease.

Mike Holt's Illustrated Guide to NEC Exam Preparation

Electrical Circuits | Unit 2

2.8 Practical Uses of the Parallel Circuit

Parallel circuits are used for most building wiring.

Receptacles

When wiring receptacles on a circuit, they are connected in parallel to each other. Figure 2–16

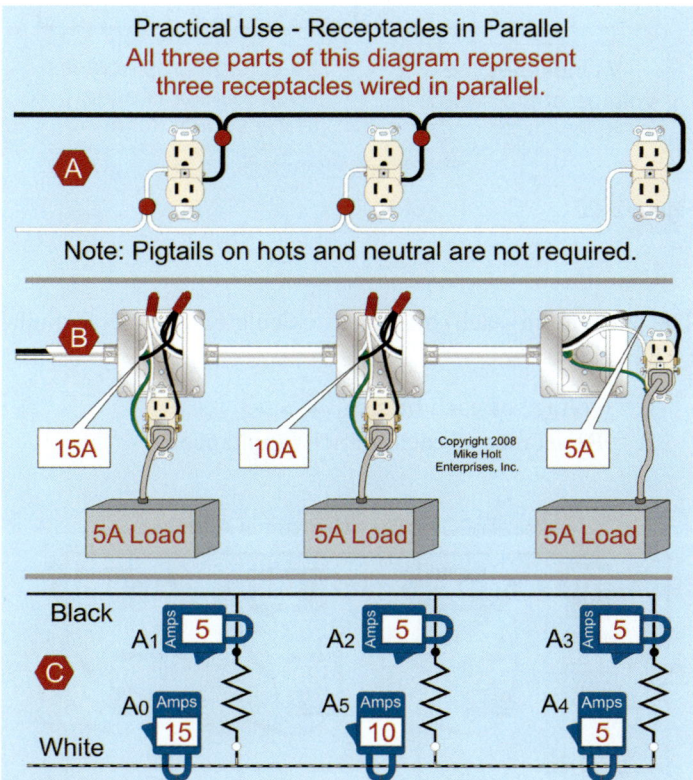

Figure 2–16

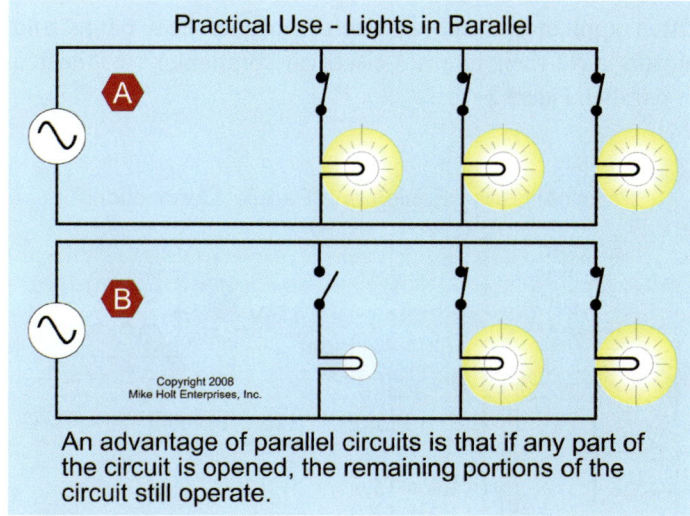

Figure 2–17

Lighting

Another example is lights connected in parallel to each other. Figure 2–17A

The major advantage of a parallel circuit is that if any branch of the circuit is opened or turned off, the power supply continues to provide voltage to the remaining parts of the circuit. Figure 2–17B

Other Uses

Parallel circuits, also called open-loop systems, are used for fire alarm pull stations and smoke detectors. If any initiating device (pull station or smoke detector) closes, the signal circuit is complete and the alarm sounds. Figure 2–18

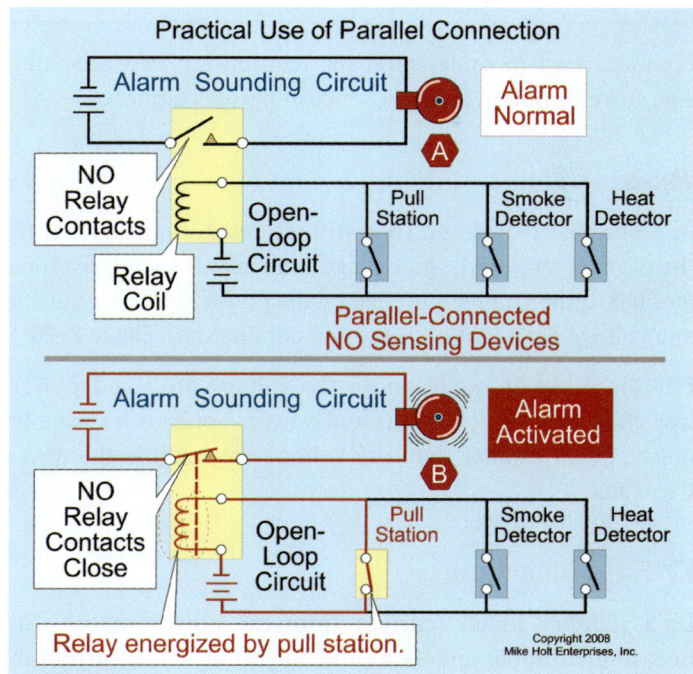

Figure 2–18

Mike Holt Enterprises, Inc. • www.MikeHolt.com • 1.888.NEC.CODE (1.888.632.2633)   35

Unit 2 — Electrical Circuits

Often appliances, such as electric furnace heat banks and motors have their internal electrical components connected in parallel. **Figure 2–19**

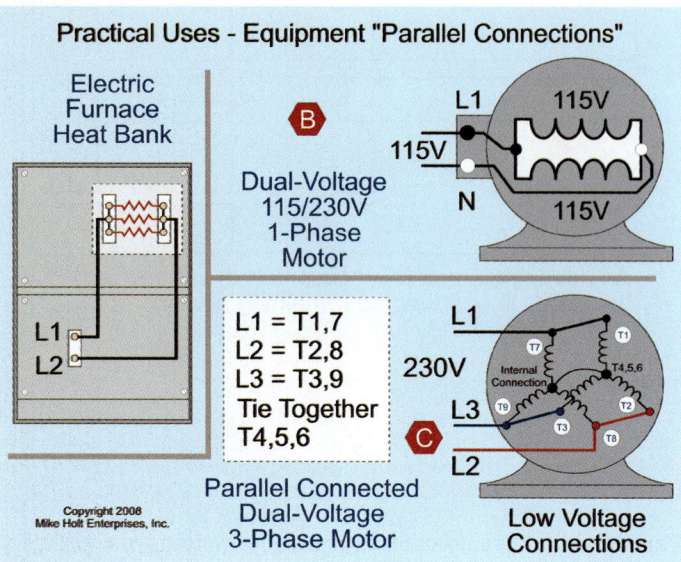

Figure 2–19

2.9 Understanding Parallel Calculations

It's important to understand the relationship between voltage, current, power, and resistance in parallel circuits.

Voltage

In a pure parallel circuit (one with no resistors in series with the parallel resistors), the voltage drop across each resistance is equal to the voltage supplied by the power source (ignoring any voltage drop in the source and conductors). **Figure 2–20**

For the moment, we'll ignore the voltage drop and power loss effects of the conductor and power supply as it's usually much, much smaller than the voltage drop across the resistive elements.

Kirchoff's Current Law

In a parallel circuit, current from the power source will branch in different directions and magnitudes. The current in each branch is dependent on the resistance of each branch. Kirchoff's Current Law states that the total current provided by the source to a parallel circuit is equal to the sum of the currents of all of the branches.

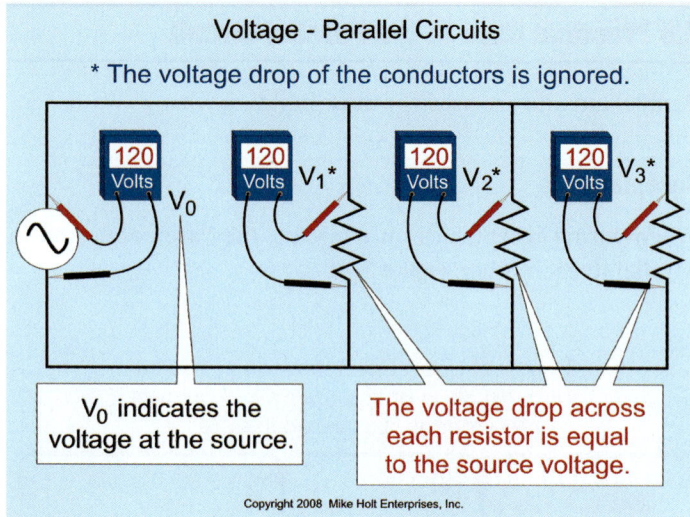

Figure 2–20

The current in each branch is calculated by the formula $I = E/R$. **Figure 2–21**

E = Voltage of each branch.
R = Resistance of each branch (appliance).

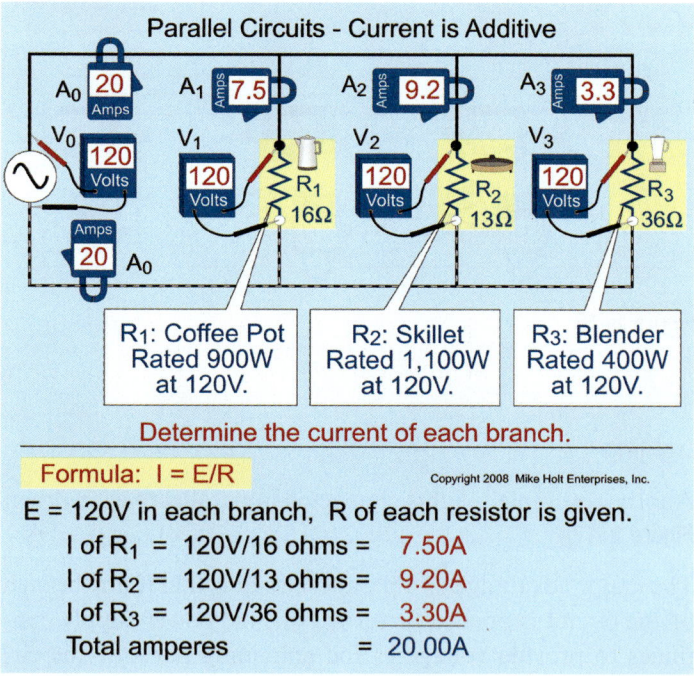

Figure 2–21

36 — Mike Holt's Illustrated Guide to NEC Exam Preparation

The current of each branch is as follows: I = E/R

Coffee Pot (R_1)	P = 120V/16 ohms	7.50A
Skillet (R_2)	P = 120V/13 ohms	9.20A
Blender (R_3)	P = 120V/36 ohms	3.30A
Total Current (R_T)		20.00A

Author's Comment: The resistance and currents have been rounded off.

Power

When current flows through a resistor, power is consumed. The power consumed by each branch of the parallel circuit is determined by the formulas:

$P = I^2 \times R$ or $P = E \times I$ or $P = E^2/R$

The total power consumed in a parallel circuit equals the sum of the power in each branch. **Figure 2–22**

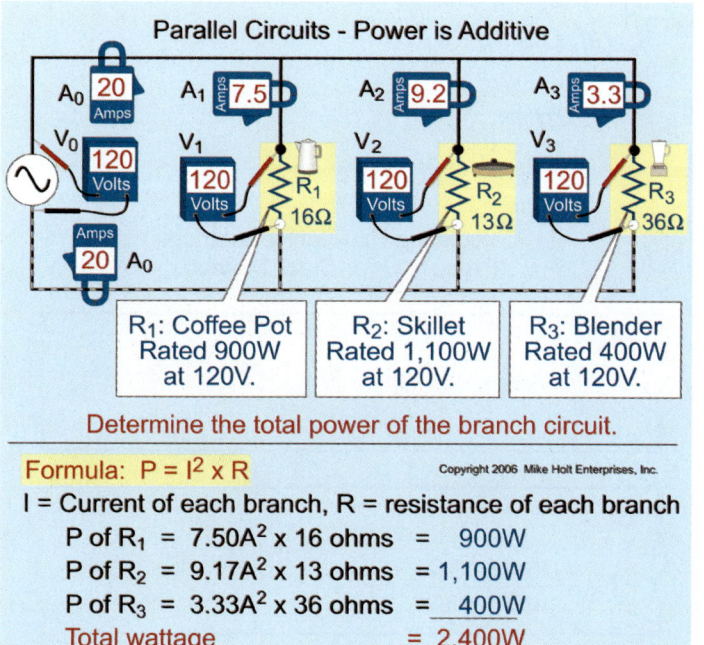

Figure 2–22

Using the formula $P = I^2 \times R$, determine the power of each resistor:

Coffee Pot	P = 7.50A² x 16 ohms	900W
Skillet	P = 9.20A² x 13 ohms	1,100W
Blender	P = 3.30A² x 36 ohms	+ 400W
Total Circuit Power		2,400W

2.10 Circuit Resistance

Calculating total circuit resistance is different in parallel and series circuits. In a series circuit, the total circuit resistance is equal to the sum of resistance. **Figure 2–23A**.

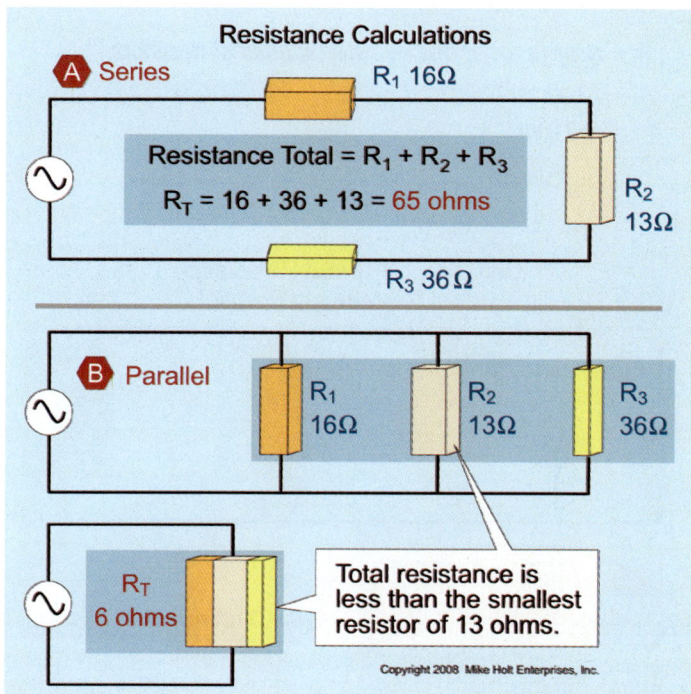

Figure 2–23

In a parallel circuit, the total circuit resistance is always less than the smallest individual resistance. **Figure 2–23B**

There are three basic methods of calculating the total resistance of a parallel circuit: the Equal Resistance method, the Product-Over-Sum method, and the Reciprocal method.

Equal Resistance Method

When all of the resistors of the parallel circuit have the same resistance, total circuit resistance is found by dividing the resistance of one resistive element by the total number of resistors in parallel.

Unit 2 — Electrical Circuits

▶ **Example 1**

Question: The total resistance of three 10 ohm resistors in parallel is _____. **Figure 2–24**

(a) 3.33 ohms (b) 10 ohms (c) 20 ohms (d) 30 ohms

Answer: (a) 3.33 ohms

R_T = Resistance of One Resistor/Number of Resistors
R_T = 10 ohms/3
R_T = 3.33 ohms

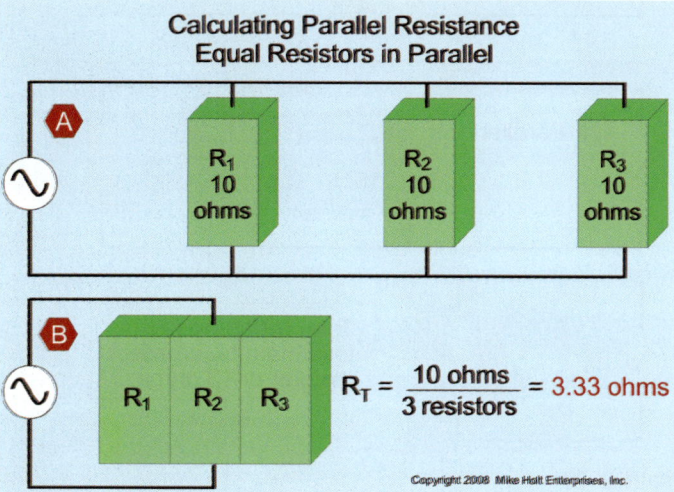

Figure 2–24

▶ **Example 2**

Question: The total resistance of ten 10 ohm resistors in parallel is _____.

(a) 1 ohm (b) 10 ohms
(c) 50 ohms (d) 100 ohms

Answer: (a) 1 ohm

R_T = Resistance of One Resistor/Number of Resistors
R_T = 10 ohms/10
R_T = 1 ohm

Product-Over-Sum Method

This method is used to calculate the resistance of two resistors at a time:

$R_T = (R_1 \times R_2)/(R_1 + R_2)$

Author's Comment: The term "product" means the answer obtained when numbers are multiplied. The term "sum" means the answer obtained by adding a group of numbers.

▶ **Example 1**

Question: The resistance of a 900W coffee pot is 16 ohms and the resistance of a 1,100W skillet is approximately 13 ohms; the appliances are connected in parallel. What is the total resistance of the two appliances? **Figure 2–25A** and **Figure 2–25B**

(a) 7.20 ohms (b) 13 ohms (c) 16 ohms (d) 29 ohms

Answer: (a) 7.20 ohms

$R_T = (R_1 \times R_2)/(R_1 + R_2)$
$R_T = (16 \times 13)/(16 + 13)$
$R_T = 7.20$ ohms

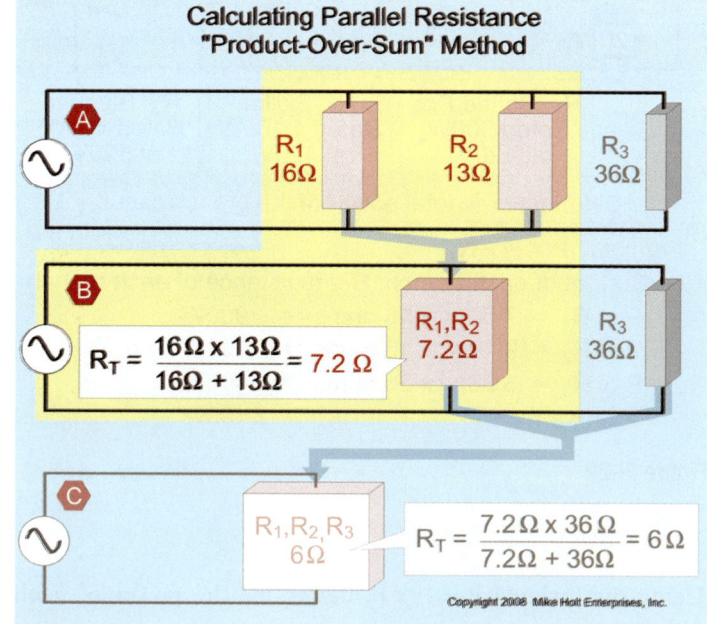

Figure 2–25

Electrical Circuits | Unit 2

The total resistance of a parallel circuit is always less than the smallest resistance.

The Product-Over-Sum method can be used to determine the resistance total for more than two resistors in a parallel circuit, but only two resistors can be dealt with at a time. If more than two resistors are in parallel, the "Product-Over-Sum" method must be applied several times, each time considering the equivalent resistance of the last two resistors looked at as a "new" resistance for the equation.

▶ **Example 2**

Question: *What is the total resistance of a 16 ohm resistor, a 13 ohm resistor, and a 36 ohm resistor connected in parallel?* **Figure 2–26**

(a) 6 ohms   (b) 26 ohms   (c) 43 ohms   (d) 65 ohms

Answer: (a) 6 ohms

The 16 and 13 ohm resistors are treated as an "equivalent" single resistor of 7.20 ohms (previous example). The resistance of the circuit will be calculated as follows: **Figure 2–26B** and **Figure 2–26C**

$R_T = (R_{1,2} \times R_3)/(R_{1,2} + R_3)$

$R_T = (7.20 \times 36)/(7.20 + 36)$

$R_T = 6$ ohms

Author's Comment: The answer must be less than the smallest resistor of the circuit (13 ohms).

Reciprocal Method

The advantage of the Reciprocal method in determining the total resistance of a parallel circuit is that this formula can be used for as many resistors as the parallel circuit contains.

$R_T = 1/(1/R_1 + 1/R_2 + 1/R_3 \ldots)$

▶ **Example**

Question: *What is the resistance total of 16 ohm, 13 ohm, and 36 ohm resistors connected in parallel?*

(a) 6 ohms   (b) 13 ohms   (c) 16 ohms   (d) 36 ohms

Answer: (a) 6 ohms

$R_T = 1/(1/16 \text{ ohms} + 1/13 \text{ ohms} + 1/36 \text{ ohms})$
$R_T = 1/(0.0625 \text{ ohms} + 0.0769 \text{ ohms} + 0.0278 \text{ ohms})$
$R_T = 1/(0.1672 \text{ ohms})$
$R_T = 6$ ohms

2.11 Parallel Circuit Notes

A parallel circuit has the following characteristics:

Note 1: *A parallel circuit has two or more paths for current to flow through.*

Note 2: *Total resistance is less than the smallest individual resistor. Total resistance in a parallel circuit is calculated with the following formula:*

$R_T = 1/(1/R_1 + 1/R_2 + 1/R_3 \ldots)$

Note 3: *The sum of the currents through each path is equal to the total current that flows from the source.*

Note 4: *Total power is equal to the sum of the branches' powers.*

Note 5: *Voltage is the same across each component of the parallel circuit.*

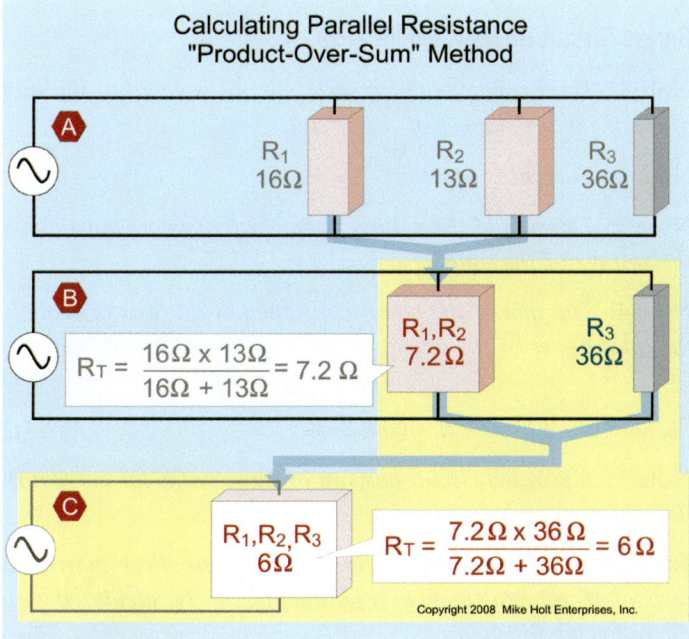

Figure 2–26

2.12 Parallel-Connected Power Supplies

When power supplies are connected in parallel, the voltage remains the same, but the current (or in the case of batteries the amp-hour capacity) increases. To place batteries in parallel to each other, connect them with the proper polarity, which is (+) to (+) and (−) to (−). **Figure 2–27A**

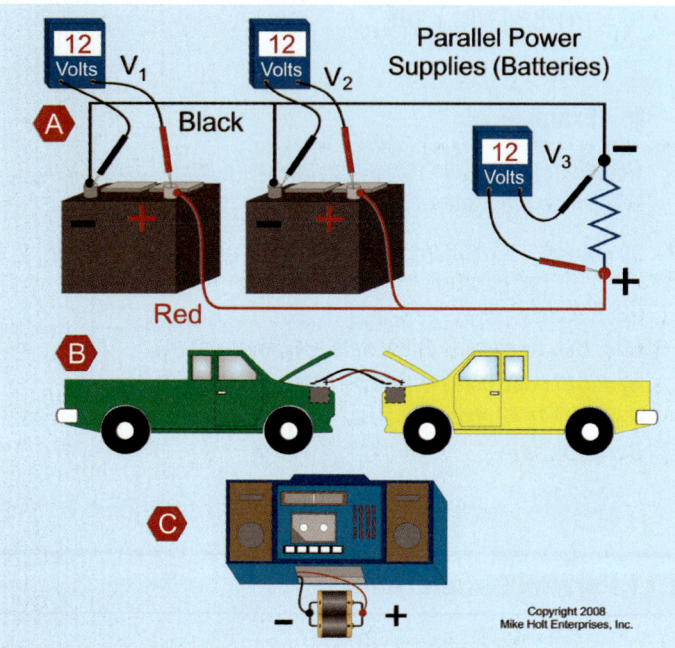

Figure 2–27

Author's Comment: When jumping a car battery, place the red cables on the positive (+) terminals and the black cables on the negative (−) terminals. **Figure 2–27B**

Batteries are often connected in parallel in radios, toys, and other appliances that operate on dc power. **Figure 2–27C**

PART C—SERIES-PARALLEL CIRCUITS

Introduction

A series-parallel circuit is a circuit that contains some resistors in series and some in parallel to each other. That portion of the series-parallel circuit that contains resistors in series must comply with the rules for series circuits. That portion of the series-parallel circuit that contains resistors in parallel complies with the rules for parallel circuits. However, it's good to remember that Ohm's Law always prevails. **Figure 2–28**

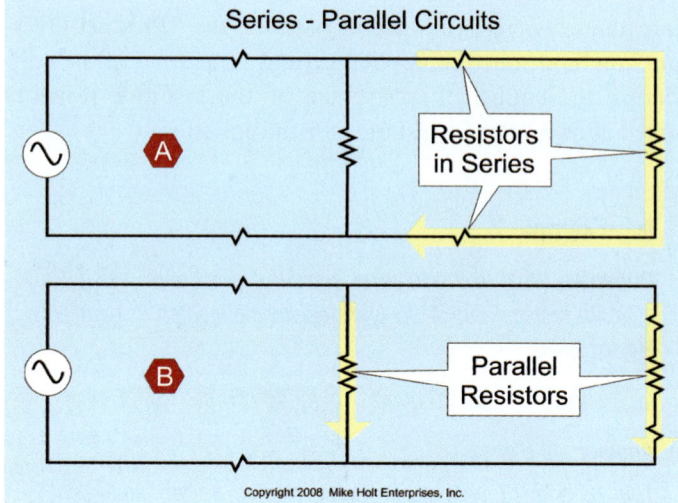

Figure 2–28

2.13 Review of Series and Parallel Circuits

To understand series-parallel circuits, we must review the rules for series and parallel circuits.

Series Circuit Review, Figure 2–29

Note 1: *Total resistance of a series circuit is equal to the sum of all of the resistances of the circuit.*

Note 2: *Current is constant.*

Note 3: *The sum of the voltage drop of all resistors must equal the voltage of the source.*

Note 4: *The sum of the power consumed by all resistors equals the total power of the circuit.*

Parallel Circuit Review, Figure 2–30

Note 1: *A parallel circuit has two or more paths for current to flow through.*

Note 2: *Total resistance is less than the smallest individual resistor. Total resistance in a parallel circuit is calculated with the following formula:*

$$R_T = 1/(1/R_1 + 1/R_2 + 1/R_3 ...)$$

Electrical Circuits | Unit 2

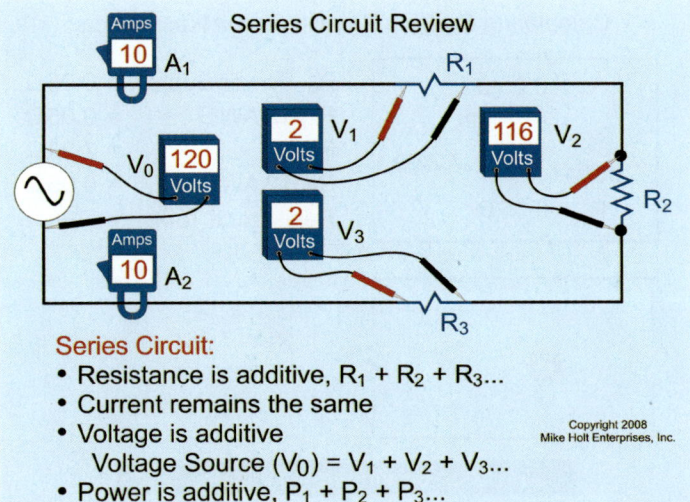

Figure 2–29

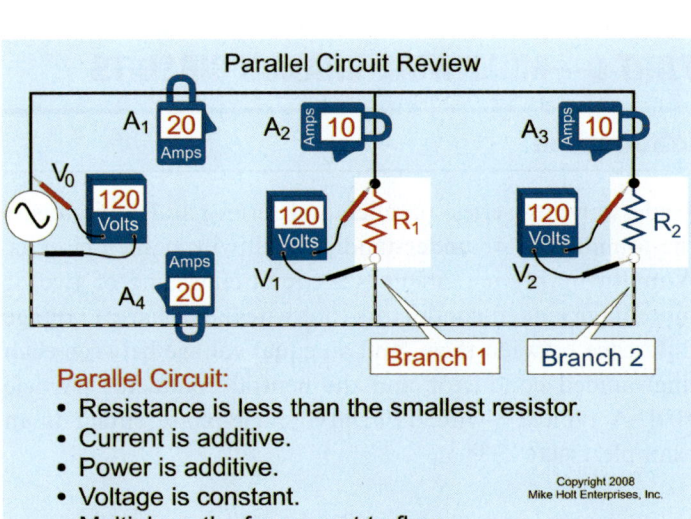

Figure 2–30

Note 3: *The sum of the currents through each path is equal to the total current that flows from the source.*

Note 4: *Total power is equal to the sum of the branches' powers.*

Note 5: *Voltage is the same across each component of the parallel circuit.*

2.14 Working With Series-Parallel Circuits

When working with series-parallel circuits, keep breaking the circuit down from series to parallel to series to parallel, etc., until you have only one resistance. It's best to redraw the circuit so you can see the series components and the parallel branches. Each circuit should be examined to determine the best plan of attack. Some turn out to be easier to analyze if you tackle the parallel elements first and then combine them with the series elements. **Figure 2–31**

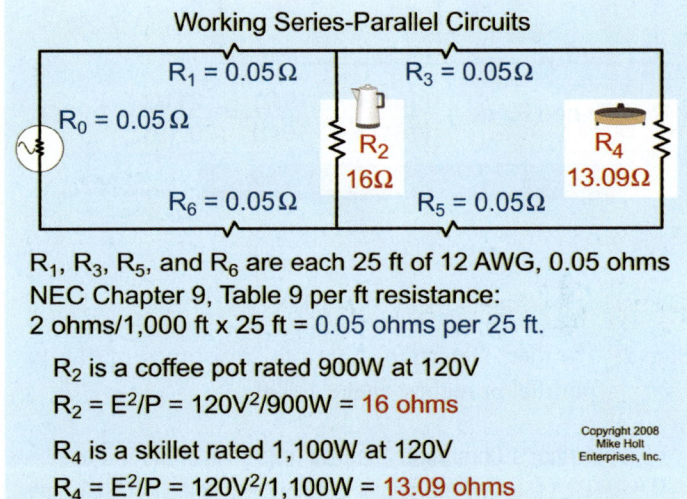

R_1, R_3, R_5, and R_6 are each 25 ft of 12 AWG, 0.05 ohms NEC Chapter 9, Table 9 per ft resistance:
2 ohms/1,000 ft x 25 ft = 0.05 ohms per 25 ft.

R_2 is a coffee pot rated 900W at 120V
$R_2 = E^2/P = 120V^2/900W = 16$ ohms

R_4 is a skillet rated 1,100W at 120V
$R_4 = E^2/P = 120V^2/1,100W = 13.09$ ohms

Figure 2–31

Other circuits are best worked by combining series elements first and then combining the result with the parallel resistances. In **Figure 2–31**, the series combination of R_3, R_4, and R_5 is the first step.

Step 1: Series: Determine the total resistance of each series branch using the formula: **Figure 2–32A**

$R_T = R_3 + R_4 + R_5$

R_3 Conductor (25 ft of 12 AWG)	0.05 ohms
R_4 Skillet (1,100W)	13.09 ohms
R_5 Conductor (25 ft of 12 AWG)	+ 0.05 ohms
	13.19 ohms

The circuit can now be redrawn showing the relationship between the two conductors and the two parallel branches. **Figure 2–32B**

Unit 2 — Electrical Circuits

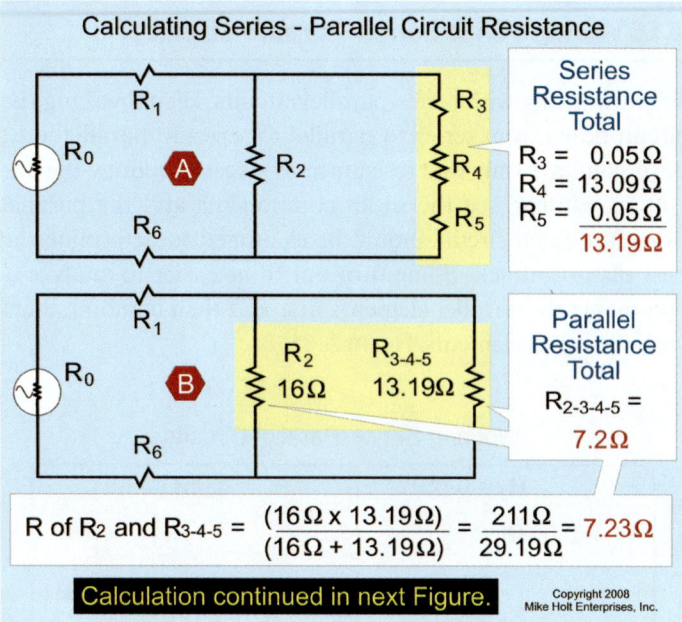

Figure 2–32

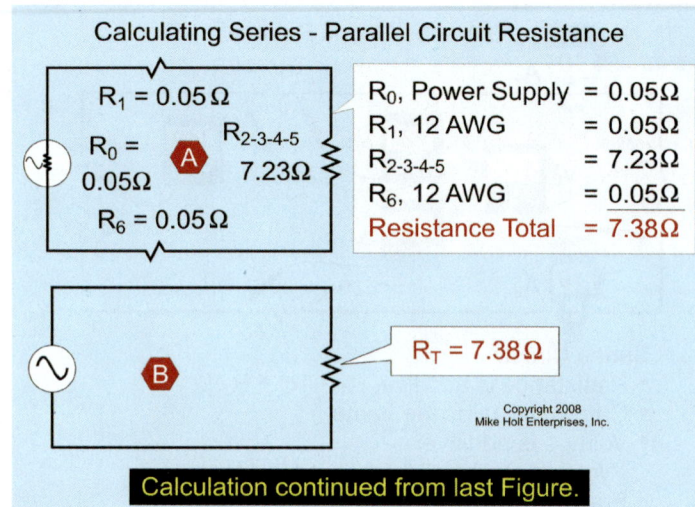

Figure 2–33

Step 2: Parallel: Determine the total resistance of the two parallel branches. **Figure 2–32B**

> **Author's Comment:** The total resistance of the two branches will be less than that of the smallest branch (13.19 ohms). Since we're only trying to determine the total resistance of two parallel branches, the Product-Over-Sum method can be used to determine the resistance.

$R_T = (R_2 \times R_{3,4,5})/(R_2 + R_{3,4,5})$
$R_T = (16\ \text{ohms} \times 13.19\ \text{ohms})/(16\ \text{ohms} + 13.19\ \text{ohms})$
$R_T = 7.23\ \text{ohms}$

> **Author's Comment:** When working with series-parallel circuits, keep breaking the circuit down from series to parallel to series to parallel, etc., until you have only one resistor. **Figure 2–33**

2.15 Voltage

Even though the current is different in the different resistors, remember that Ohm's Law always works. Every complicated problem is really just a series of easy problems waiting to be worked out. To calculate the voltage of each resistor, consider each one on a case-by-case basis and multiply its value by the current flowing through it.

PART D—MULTIWIRE BRANCH CIRCUITS

Introduction

Understanding series, parallel, and series-parallel circuits is the foundation for understanding multiwire branch circuits. A multiwire branch circuit is a circuit consisting of two or more ungrounded conductors (hot wires) that have a voltage difference between them, and an equal voltage between each ungrounded conductor and the neutral conductor [Article 100]. A typical 3-wire, 120/240V, single-phase circuit is an example. **Figure 2–34**

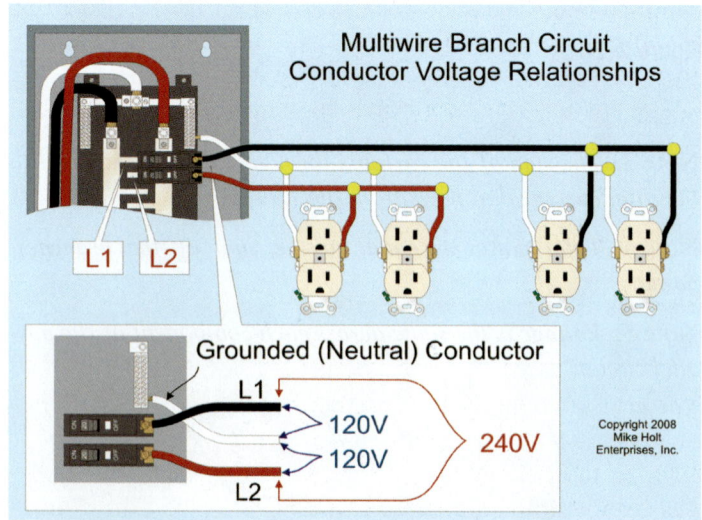

Figure 2–34

Electrical Circuits | Unit 2

2.16 Neutral Conductor

The *NEC* defines a neutral conductor as the conductor connected to the neutral point of a system that is intended to carry current under normal conditions [Article 100]. **Figure 2–35**

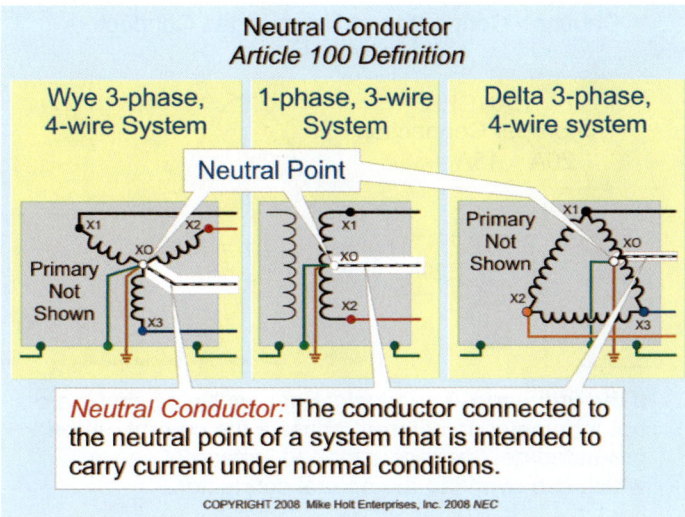

Figure 2–35

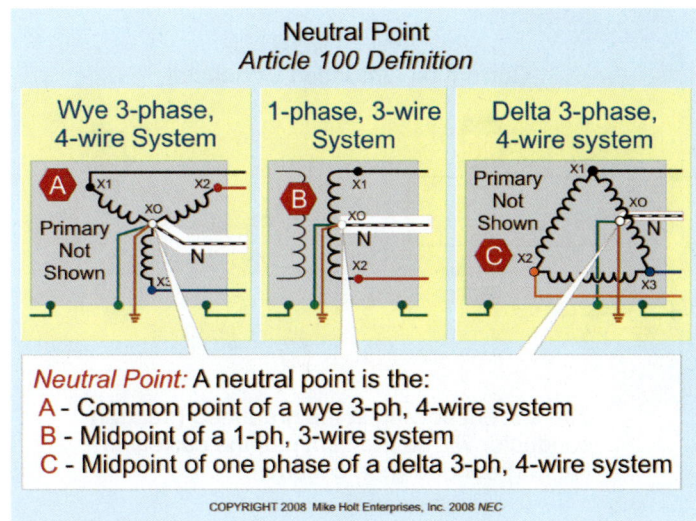

Figure 2–36

Author's Comment: The neutral conductor of a solidly grounded system is required to be grounded to the earth, therefore this conductor is also called a "grounded conductor."

The term "Neutral Point" in the *Code* is defined as the common point on a 4-wire, three-phase, 120/208V or 277/480V wye-connected system, the midpoint of a 3-wire, 120/240V single-phase system, or the midpoint of the single-phase portion of a 120/240V three-phase delta-connected system [Article 100]. **Figure 2–36**

The definition of "Neutral Point" describes the point to which the neutral conductor is connected.

2.17 Grounded Conductor

The grounded conductor, according to the *NEC*, is a circuit or system conductor that is intentionally grounded [Article 100]. In the case of home wiring (3-wire, 120/240V, single-phase), the grounded conductor is often called the neutral conductor, and it will be either white or gray in color in accordance with the *National Electrical Code* [200.6].

Author's Comment: When a grounded conductor is also used as a neutral conductor, this textbook will refer to it as the "neutral conductor" except in the *Code* questions at the end of each unit, which will follow the *NEC* wording.

2.18 Current Flow on the Neutral Conductor

To understand the current flow on the neutral conductor, review the following circuits.

2-Wire Circuit

The current flowing in the neutral conductor of a 2-wire circuit is the same as the current flowing in the ungrounded (hot) conductor. **Figure 2–37**

3-Wire, 120/240V, Single-Phase Circuit

The current flowing in the neutral conductor of a 3-wire, 120/240V, single-phase circuit equals the difference in current flowing in the ungrounded conductors (I_N = L1–L2). **Figure 2–38**

The current on the neutral conductor is equal to the difference in ungrounded conductor current because at any instant the currents on the two ungrounded conductors oppose each other. **Figure 2–39**

Unit 2 — Electrical Circuits

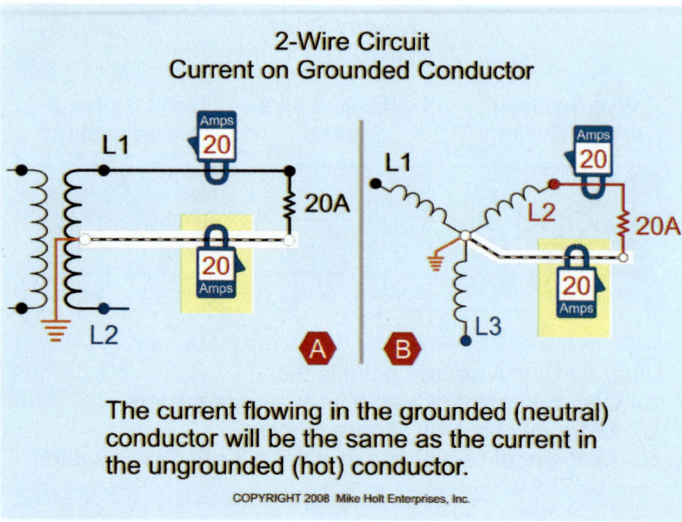

Figure 2–37

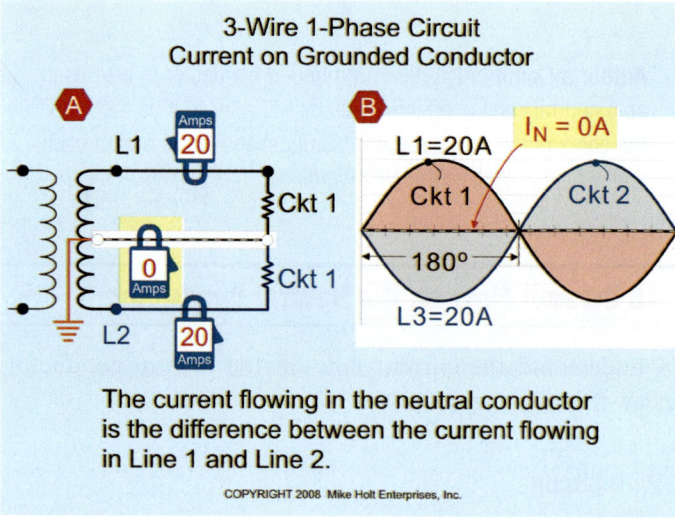

Figure 2–38

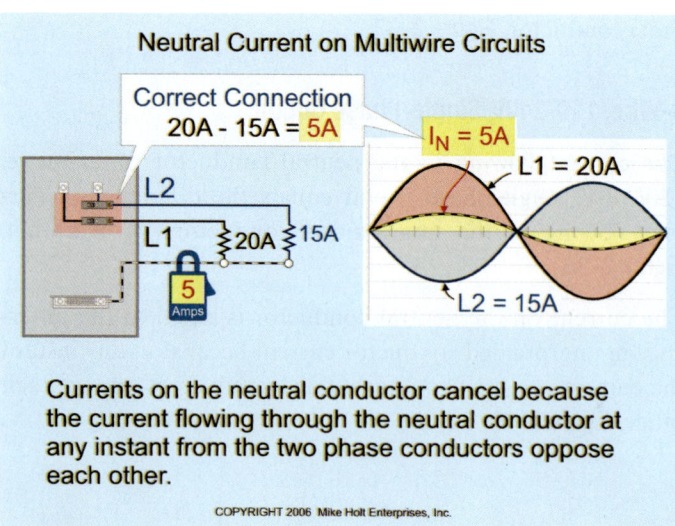

Figure 2–39

CAUTION: *If the ungrounded conductors of a multiwire branch circuit are incorrectly terminated so they are not on different lines, the currents on the ungrounded conductors will not cancel, but will add on the neutral conductor. This can cause the neutral current to be in excess of the neutral conductor rating.* Figure 2–40

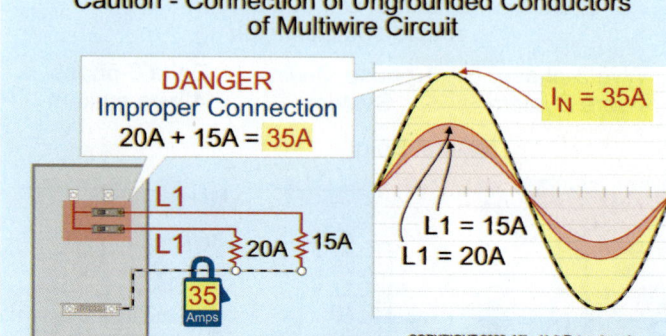

If the ungrounded conductors of a multiwire circuit are not terminated to different phases, the current on the grounded/neutral conductor add instead of cancel, which can overload the neutral conductor.

Figure 2–40

Author's Comment: This is one reason white neutral conductors sometimes turn brown or black.

2.19 Balanced Systems

If the current in each ungrounded conductor of a multiwire branch circuit is the same, the neutral conductor carries 0A. This applies to 3-wire, 120/240V, single-phase and all three-phase circuits, regardless of configuration or voltage. Figure 2–41

2.20 Unbalanced Current

The current flowing in the neutral conductor of a multiwire branch circuit is called "unbalanced current."

Electrical Circuits — Unit 2

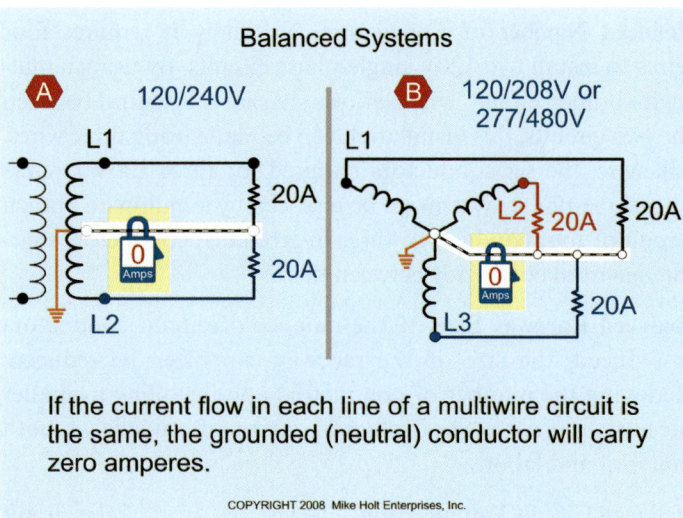

Figure 2–41

3-Wire, 120/240V, Single-Phase Circuit

The neutral conductor of a 3-wire, 120/240V, single-phase circuit only carries current when the current on the ungrounded conductors is not identical. The unbalanced current is determined by the following formula: **Figure 2–42**

$I_N = L1 - L2$

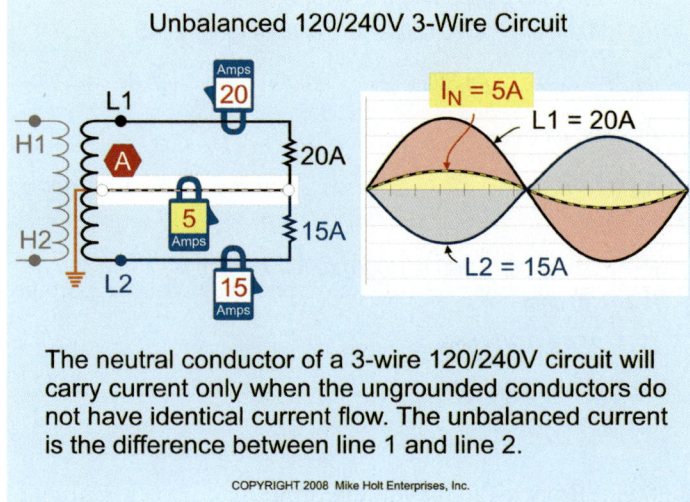

Figure 2–42

3-Wire Circuit from a 4-Wire, Three-Phase System

The neutral conductor of a 3-wire, 120/208V or 277/480V, three-phase circuit derived from a 4-wire, three-phase system always carries neutral current. The current on the neutral conductor of a 3-wire circuit supplied from a 4-wire, three-phase system is determined by the following formula:

$I_N = \sqrt{[(L1^2 + L2^2 + L3^2) - [(L1 \times L2) + (L2 \times L3) + (L1 \times L3)]]}$

> ▶ **Example**
>
> **Question:** What is the neutral current for a 3-wire, 120/208V, single-phase circuit, if each ungrounded conductor carries 20A, and the circuit is supplied from a 4-wire, 120/208V, three-phase system? **Figure 2–43**
>
> (a) 0A (b) 20A (c) 80A (d) 100A
>
> **Answer:** (b) 20A
>
> $I_N = \sqrt{[(20^2 + 20^2 + 0) - (20^2 + 0 + 0)]}$
> $I_N = \sqrt{400}$
> $I_N = 20A$

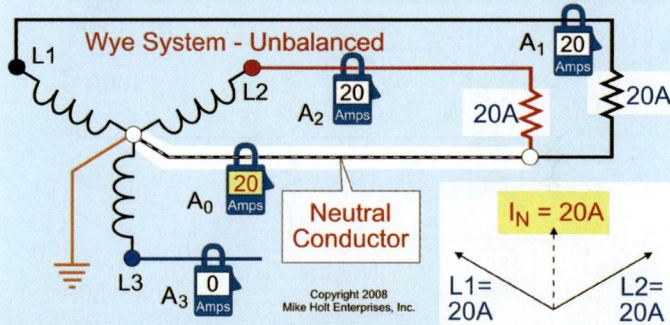

The neutral of a 3-wire circuit from a 4-wire wye system carries about the same current as the phase conductors.

$I_N = \sqrt{(L1^2 + L2^2 + L3^2) - [(L1 \times L2) + (L2 \times L3) + (L1 \times L3)]}$
$I_N = \sqrt{(20^2 + 20^2 + 0^2) - [(20 \times 20) + (20 \times 0) + (20 \times 0)]}$
$I_N = \sqrt{(400 + 400 + 0) - (400 + 0 + 0)}$
$I_N = \sqrt{800 - 400} \quad I_N = \sqrt{400} \quad I_N = 20A$

Figure 2–43

Unit 2 — Electrical Circuits

4-Wire, Three-Phase Circuit

The neutral conductor of a 4-wire, 120/208V or 277/480V, three-phase system carries neutral current when the ungrounded conductors are not identically loaded. The current on the neutral conductor of a 4-wire circuit supplied from a 4-wire system is determined by the following formula:

$$I_N = \sqrt{[(L1^2 + L2^2 + L3^2) - [(L1 \times L2) + (L2 \times L3) + (L1 \times L3)]]}$$

▶ **Example**

Question: What is the neutral current for a 4-wire, 120/208V, three-phase circuit, if Line 1 = 100A, Line 2 = 100A and Line 3 = 50A? **Figure 2–44**

(a) 0A   (b) 50A   (c) 100A   (d) 125A

Answer: (b) 50A

$$I_N = \sqrt{[(100^2 + 100^2 + 50^2) - [(100 \times 100) + (100 \times 50) + (100 \times 50)]]}$$

$$I_N = \sqrt{2,500A}$$

$$I_N = 50A$$

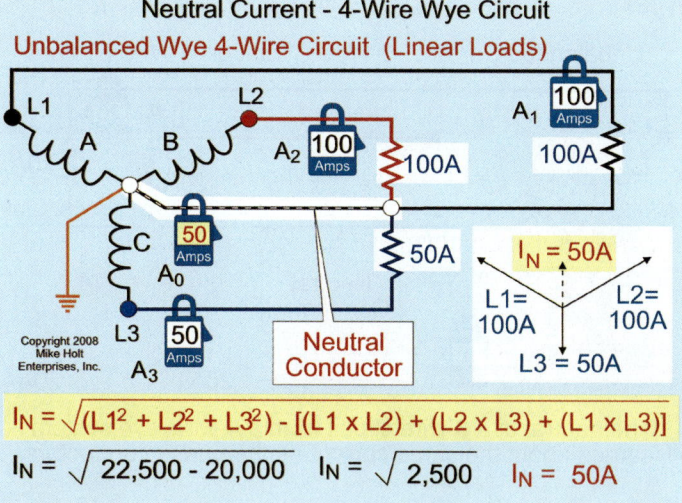

Figure 2–44

2.21 Multiwire Branch Circuits

Multiwire branch circuits are more cost-effective than 2-wire circuits in that they have fewer conductors for a given number of circuits, which enables the use of a smaller raceway. In addition, multiwire branch circuits result in lower circuit voltage drop.

Reduced Number of Conductors. Normally it requires four wires to install two 120V single-phase circuits. By using a multiwire branch circuit, which allows sharing the neutral between the two circuits, this installation can be made using three wires. Likewise, the six conductors required for three 120V circuits on a three-phase system can be replaced by a multiwire branch circuit of four conductors; three ungrounded conductors sharing a neutral conductor between them.

Reduced Raceway Size. If the number of circuit conductors is reduced, the size of the raceway can often be reduced. Reducing the number of conductors and installing a smaller raceway is very cost-effective in terms of savings in both material and labor.

Reduced Circuit Voltage Drop. The voltage drop of the circuit conductors is dependent upon the magnitude of current and conductor resistance: $E_{vd} = I \times R$.

2-Wire Circuit Voltage Drop

A typical 2-wire circuit has current flow over both the ungrounded and neutral conductors. Therefore, the circuit voltage drop includes the voltage drop of both conductors.

▶ **Example**

Question: What is the voltage drop of two 12 AWG conductors, each 75 ft long, supplying a 2-wire, 20A load? The resistance for 1,000 ft of 12 AWG copper is 2 ohms. **Figure 2–45**

(a) 2V   (b) 3V   (c) 4V   (d) 6V

Answer: (d) 6V

$E_{vd} = I \times R$

$I = 20A$

$R = (2 \text{ ohms per } 1,000 \text{ ft}/1,000) \times 75 \text{ ft} \times 2 \text{ wires}$

$R = 0.30$ ohms

$E_{vd} = 20A \times 0.30$ ohms

$E_{vd} = 6V$

Multiwire Branch-Circuit Voltage Drop

A balanced 3-wire, single-phase or 4-wire, three-phase multiwire branch circuit has current flowing only on the ungrounded circuit conductors. Therefore, the circuit voltage drop for the line-to-neutral only includes the voltage drop of one conductor. The line-to-line voltage drop is found by multiplying the line-to-neutral voltage drop by 1.732.

Electrical Circuits | Unit 2

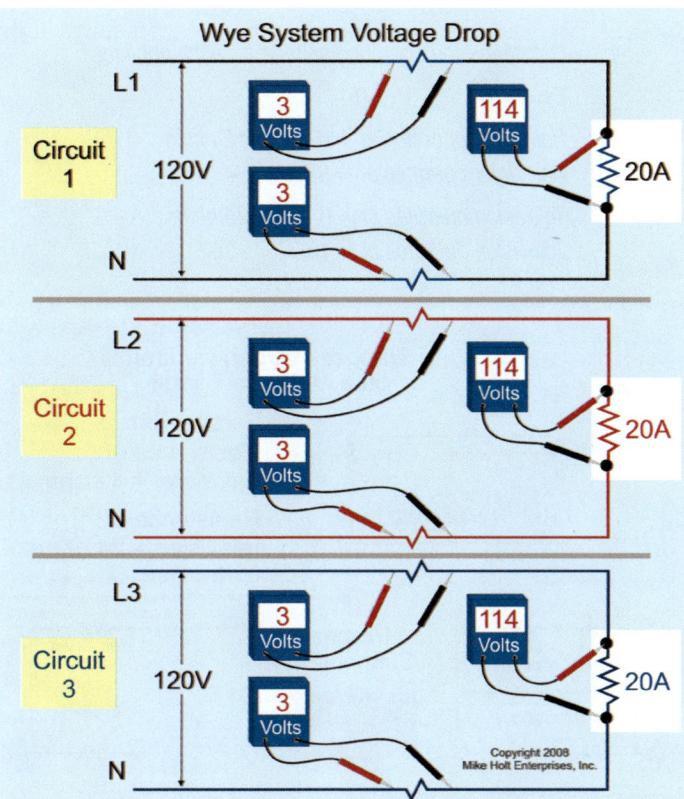

Figure 2–45

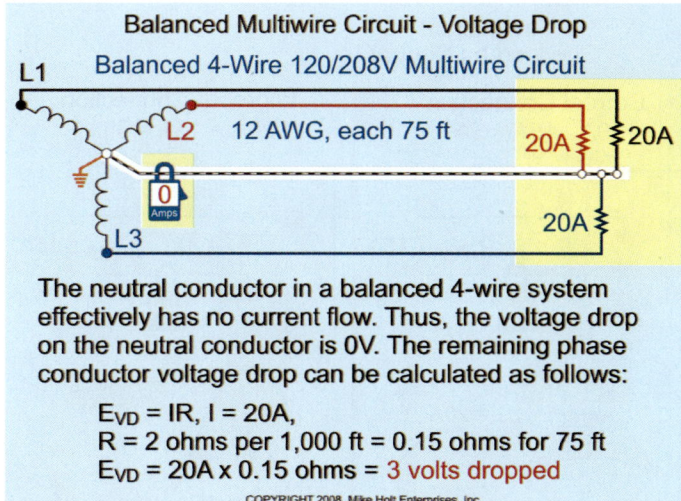

The neutral conductor in a balanced 4-wire system effectively has no current flow. Thus, the voltage drop on the neutral conductor is 0V. The remaining phase conductor voltage drop can be calculated as follows:

$E_{VD} = IR$, $I = 20A$,
$R = 2$ ohms per 1,000 ft = 0.15 ohms for 75 ft
$E_{VD} = 20A \times 0.15$ ohms = 3 volts dropped

COPYRIGHT 2008 Mike Holt Enterprises, Inc.

Figure 2–46

▶ **Example**

Question: What is the line-to neutral voltage drop over each line conductor of a balanced 4-wire multiwire branch circuit? Each conductor is 12 AWG, 75 ft long, supplying a 20A load. Figure 2–46

(a) 2V (b) 3V (c) 4V (d) 6V

Answer: (b) 3V

The neutral conductor in a balanced 4-wire system effectively has 0A of current flow (the three return currents cancel each other out because of their phase relationship). Thus, by Ohm's Law, the voltage drop over this conductor is 0V. The remaining phase conductor voltage drop can be calculated as follows:

$E_{vd} = I \times R$
$I = 20A$
$R = (2$ ohms per 1,000 ft/1,000) x 75 ft x 1 wire
$R = 0.15$ ohms
$E_{vd} = 20A \times 0.15$ ohms
$E_{vd} = 3V$

2.22 Dangers of Multiwire Branch Circuits

Multiwire branch circuits offer fewer conductors, reduced raceway size, and reduced voltage drop. However, improper wiring or mishandling of multiwire branch circuits can cause a fire hazard because of conductor overloading and/or the destruction of connected equipment.

Fire Hazard

Failure to terminate the ungrounded conductors to separate phases or lines may cause the neutral conductor to become overloaded from excessive neutral current, and the insulation may be damaged or destroyed. Conductor overheating is known to decrease insulating material service life, potentially resulting in a fire from arcing faults in hidden locations. It's difficult to predict just how long conductor insulation will last under normal operating conditions, but heat does decrease its life span. **Figure 2–47**

Destruction of Equipment as Well as Fire Hazard

The opening of the ungrounded or neutral conductor of a 2-wire circuit during the replacement of a device doesn't cause a safety hazard, so pigtailing of these conductors isn't required.

If the continuity of the neutral conductor of a multiwire branch circuit is interrupted (open), there can be a fire and/or destruction of electrical equipment resulting from overvoltage or undervoltage.

Unit 2 Electrical Circuits

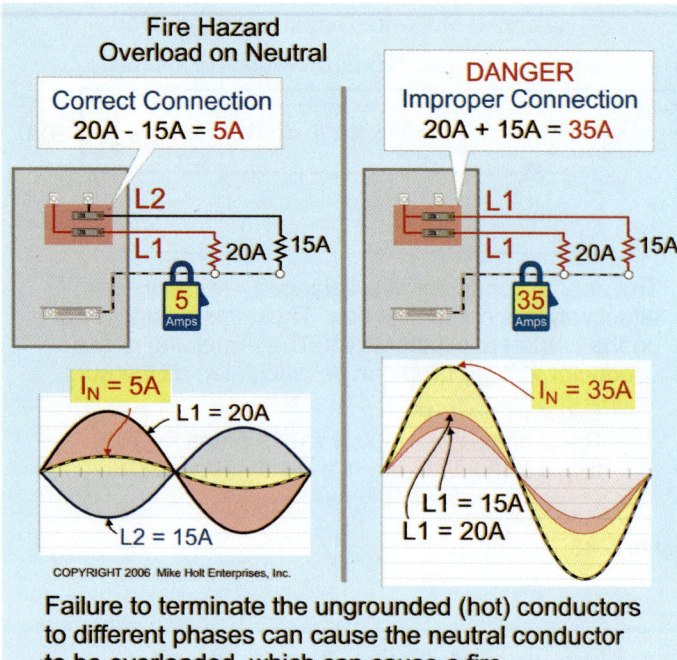

Failure to terminate the ungrounded (hot) conductors to different phases can cause the neutral conductor to be overloaded, which can cause a fire.

Figure 2–47

▶ **Example**

A 3-wire, 120/240V circuit supplies a 1,200W, 120V hair dryer and a 600W, 120V television. If the neutral conductor is interrupted, it will cause the 120V television to operate at 160V and consume 1,067W of power (instead of 600W) for only a few seconds before it burns up. **Figure 2–48**

Step 1: Determine the resistance of each appliance.

$R = E^2/P$

Hair Dryer
$R = 120V^2/1,200W$
$R = 12$ ohms

Television
$R = 120V^2/600W$
$R = 24$ ohms

Step 2: Determine the current of the circuit.

$I = E/R$
$I = 240V/(12$ ohms $+ 24$ ohms$)$
$I = 6.70A$

(continued in next column)

Step 3: Determine the operating voltage for each appliance.

$E = I \times R$

Hair dryer operates at $= 6.70A \times 12$ ohms
Hair dryer operates at $= 80V$

Television operates at $= 6.70A \times 24$ ohms
Television operates at $= 160V$

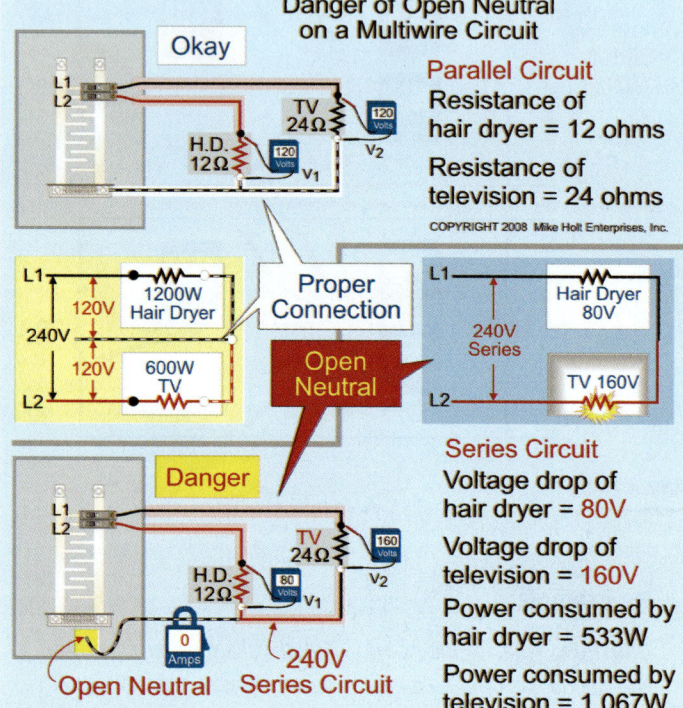

Figure 2–48

2.23 NEC Requirements

Because of the dangers associated with an open neutral conductor, the *NEC* specifies that the continuity of the neutral conductor of a multiwire branch circuit cannot be dependent upon any wiring device [300.13(B)]. In other words, the neutral conductors of a multiwire branch circuit must be spliced together, and a wire (pigtail) brought out to the device. This way, if the receptacle is removed, it doesn't result in an open neutral conductor. **Figure 2–49**

Electrical Circuits | Unit 2

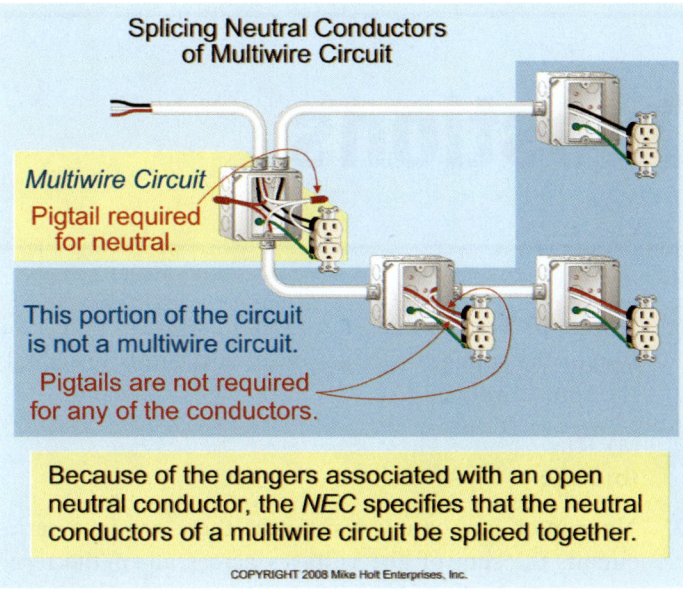

Figure 2–49

Unit 2 Conclusion

CONCLUSION TO UNIT 2—ELECTRICAL CIRCUITS

In completing this unit, you've laid another block in the foundation of your understanding. You've learned about series circuits, parallel circuits, and combinations of series and parallel circuits. It's important to be able to look at a circuit and visualize the relationship between the components and be able to understand why there are different levels of voltage or current in various locations of the circuit. This ability will greatly enhance your troubleshooting success on real branch circuits.

In this unit, you learned about a special circuit called the multiwire branch circuit. There are special *Code* considerations for multiwire branch circuits. You now know how important it is to correctly install multiwire branch circuits with neutral connections that are very securely made. A neutral must never be opened on an energized multiwire branch circuit, as this can result in placing both undervoltages and overvoltages on circuit components, which can result in damage.

You learned about Kirchoff's Current Law in this unit. If you understand the application of Kirchoff's law, you're less likely to be misled by some of the incorrect "myths" that are commonly accepted in the electrical industry concerning fault current flow, particularly as it relates to the practices of grounding and bonding. For instance, does current always take only the path of least resistance? Is current always seeking a path to the earth? Ponder these questions as you continue your study in the intriguing field of electricity.

Unit 2 Practice Questions

UNIT 2—ELECTRICAL CIRCUITS PRACTICE QUESTIONS

PART A—SERIES CIRCUITS

Introduction

1. A series circuit is a circuit in which a specific amount of current leaves the voltage source and flows through every electrical device in a single path before it returns to the voltage source.

   (a) True
   (b) False

2.1 Practical Uses of the Series Circuit

2. Series circuits are often used for _____ applications.

   (a) signal
   (b) control
   (c) a and b
   (d) none of these

3. A 115/230V rated motor connected to a 230V circuit must have the windings connected in series so that each winding will receive at least 230V.

   (a) True
   (b) False

2.2 Understanding Series Calculations

4. Resistance opposes the flow of electrons. In a series circuit, the total circuit resistance is equal to the sum of all of the resistances in series.

   (a) True
   (b) False

5. The opposition to current flow results in a voltage drop of the circuit voltage.

   (a) True
   (b) False

6. Kirchoff's Voltage Law states that in a series circuit, the sum of the voltage drops across all of the resistors equals the applied voltage.

   (a) True
   (b) False

7. No matter how many resistances there are in a series circuit, the sum of the voltages across all of the resistances equals the voltage of the source according to the Resistance Law of Proportion.

   (a) True
   (b) False

8. Kirchoff's Current Law states that in a series circuit, the current is _____ through the transformer, the conductors, and the appliance.

   (a) proportional
   (b) distributed
   (c) additive
   (d) the same

9. The power consumed in a series circuit is equal to the power consumed by the largest resistance in the series circuit.

   (a) True
   (b) False

2.3 Series Circuit Calculations

10. To determine the resistance of each resistive element in the circuit, use the formula $R = E^2/VA$. E is the rated voltage of the resistance, and VA is the rated power of the resistor.

    (a) True
    (b) False

11. To calculate the total resistance of the circuit, use the formula:

    $R_T = R_1 + R_2 + R_3 + ...$

    (a) True
    (b) False

12. The current of the circuit can be determined by the formula $I = E_S/R_T$.

    (a) True
    (b) False

2.4 Power Calculations

13. If you know the current of the circuit and the resistance of each resistor, the power of each resistor can be determined by the formula $P = I^2 \times R$.

    (a) True
    (b) False

2.5 Variations

14. There can never be variations in the formulas used or the order in which they are used for series circuits.

    (a) True
    (b) False

2.7 Series-Connected Power Supplies

15. When power supplies are connected in series, the circuit voltage remains the same as when only one power supply is connected to it, provided that all the polarities are connected properly.

    (a) True
    (b) False

PART B—PARALLEL CIRCUITS

Introduction

16. A parallel circuit has two or more paths in which current can flow.

    (a) True
    (b) False

2.9 Understanding Parallel Calculations

17. In a parallel circuit, the voltage drop across each resistance is equal to the sum of the voltage drops of each of the resistors in parallel.

    (a) True
    (b) False

18. According to Kirchoff's Current Law, the total current provided by the source to a parallel circuit equals the sum of the currents of all of the branches.

    (a) True
    (b) False

19. The total power consumed in a parallel circuit equals the sum of the power of all branches.

    (a) True
    (b) False

2.10 Circuit Resistance

20. In a parallel circuit, the total circuit resistance is always greater than the smallest resistance.

    (a) True
    (b) False

21. The total resistance of a parallel circuit can be calculated by the _____ method.

    (a) equal resistance
    (b) product-over-sum
    (c) reciprocal
    (d) any of these

22. According to the equal resistance method, when all the resistances of the parallel circuit have the same resistance, divide the resistance of one element by the largest resistor in parallel.

    (a) True
    (b) False

23. The product-over-sum method is used to calculate the resistance of _____ resistance(s) at a time.

    (a) one
    (b) two
    (c) three
    (d) four

24. The advantage of the reciprocal method is that the formula can be used for as many resistances as the parallel circuit contains.

    (a) True
    (b) False

Unit 2 — Practice Questions

2.12 Parallel-Connected Power Supplies

25. When power supplies are connected in parallel, the voltage remains the same, but the current or ampere-hour capacity is increased.

    (a) True
    (b) False

PART C—SERIES-PARALLEL CIRCUITS

Introduction

26. A _____ is a circuit that contains some resistances in series and some resistances in parallel with each other.

    (a) parallel circuit
    (b) series circuit
    (c) series-parallel circuit
    (d) none of these

27. That portion of the series-parallel circuit that contains resistances in series must comply with the rules for series circuits.

    (a) True
    (b) False

28. That portion of the series-parallel circuit that contains resistances in parallel must comply with the rules for parallel circuits.

    (a) True
    (b) False

2.14 Working With Series-Parallel Circuits

29. When working with series-parallel circuits, it is best to redraw the circuit so you can see the series components and the parallel branches.

    (a) True
    (b) False

PART D—MULTIWIRE BRANCH CIRCUITS

Introduction

30. A multiwire branch circuit has two or more ungrounded conductors having a potential difference between them, and having an equal difference of potential between each ungrounded conductor and the grounded conductor.

    (a) True
    (b) False

31. A neutral conductor is the conductor connected to the _____ of a system that is intended to carry neutral current under normal conditions.

    (a) grounding electrode
    (b) neutral point
    (c) intersystem bonding termination
    (d) earth

2.16 Neutral Conductor

32. An ungrounded delta system contains a neutral point.

    (a) True
    (b) False

33. The common point on a 4-wire, three-phase, 120/208V wye-connected system is defined by the *Code* as the _____.

    (a) wye point
    (b) neutral point
    (c) bridge point
    (d) ungrounded point

2.17 Grounded Conductor

34. The grounded conductor is a conductor that is intentionally grounded to the earth.

    (a) True
    (b) False

2.18 Current Flow on the Neutral Conductor

35. The current on the neutral conductor of a 2-wire circuit is _____ of the current on the ungrounded conductor.

    (a) 0 percent
    (b) 70 percent
    (c) 80 percent
    (d) 100 percent

36. A balanced 3-wire, 120/240V, single-phase circuit is connected so that the ungrounded conductors are from different transformer phases (Line 1 and Line 2). The current on the neutral conductor is _____ of the ungrounded conductor current.

    (a) 0 percent
    (b) 70 percent
    (c) 80 percent
    (d) 100 percent

37. The neutral conductor of a 3-wire, 120/240V, single-phase circuit only carries the unbalanced current when the circuit is not balanced.

    (a) True
    (b) False

38. If the ungrounded conductors of a multiwire branch circuit are not terminated to different phases, this can cause the neutral current to exceed the neutral conductor rating.

    (a) True
    (b) False

2.19 Balanced Systems

39. If the current in each ungrounded conductor of a multiwire branch circuit is the same, the neutral conductor carries 0A.

    (a) True
    (b) False

40. What is the neutral current for a 4-wire, 120/208V circuit, where L_1 = 20A, L_2 = 20A, and L_3 = 20A?

    (a) 0A
    (b) 10A
    (c) 20A
    (d) 60A

2.20 Unbalanced Current

41. The current flowing on the neutral conductor of a multiwire branch circuit is called unbalanced current.

    (a) True
    (b) False

42. The neutral conductor of a 3-wire, 120/240V, single-phase circuit only carries current when the current on the ungrounded conductors is not identical.

    (a) True
    (b) False

43. The neutral conductor of a 3-wire, 120/208V or 277/480V circuit supplied from a 4-wire, three-phase system never carries neutral current.

    (a) True
    (b) False

44. The neutral conductor of a 4-wire, 120/208V or 277/480V, three-phase system has neutral current flow when the ungrounded conductors are equally loaded.

    (a) True
    (b) False

2.21 Multiwire Branch Circuits

45. Multiwire branch circuits have more conductors for a given number of circuits, which requires the use of a larger raceway.

    (a) True
    (b) False

46. A balanced multiwire branch circuit has current flow only on the ungrounded conductors.

    (a) True
    (b) False

47. What is the voltage drop of two 12 AWG conductors, each 100 ft in length, supplying a 2-wire, 16A load? The resistance of 12 AWG conductors is 2 ohms per 1,000 ft.

    (a) 3.20V
    (b) 6.40V
    (c) 7.20V
    (d) 9.60V

48. What is the voltage drop of each ungrounded conductor of a 4-wire multiwire branch circuit? Each conductor is 12 AWG, 100 ft in length, supplying a 16A load. The resistance of 12 AWG conductors is 2 ohms per 1,000 ft.

    (a) 3.20V
    (b) 6.40V
    (c) 7.20V
    (d) 9.60V

2.22 Dangers of Multiwire Branch Circuits

49. Improper wiring or mishandling of multiwire branch circuits can cause _____ connected to the circuit.

    (a) overloading of the ungrounded conductors
    (b) overloading of the grounded conductors
    (c) destruction of equipment
    (d) b and c

50. The opening of the ungrounded or neutral conductor of a _____ circuit during the replacement of a device does not cause a safety hazard, so pigtailing of these conductors is not required.

    (a) 2-wire
    (b) 3-wire
    (c) 4-wire
    (d) all of these

2.23 NEC Requirements

51. Because of the dangers associated with an open neutral conductor, the continuity of the _____ conductor in a multiwire branch circuit cannot depend upon the receptacle.

    (a) ungrounded
    (b) grounded
    (c) a and b
    (d) none of these

Unit 2 Challenge Questions

UNIT 2—ELECTRICAL CIRCUITS CHALLENGE QUESTIONS

(•Indicates that 75% or fewer of those who took this exam answered the question correctly.)

PART A—SERIES CIRCUITS

2.2 Understanding Series Calculations

1. •A series circuit contains two resistors, one rated 4 ohms and the other rated 8 ohms. If the total voltage drop across both resistors equals 12V, then the current that passes through either resistor will be _____.

   (a) 1A
   (b) 2A
   (c) 4A
   (d) 8A

2. •A series circuit has four 40 ohm resistors and the power supply is 120V. The voltage drop of each resistor will be _____.

   (a) one-quarter of the source voltage
   (b) 30V
   (c) the same across each resistor
   (d) all of these

3. •The power consumed in a series circuit is _____.

   (a) the sum of the power consumed of each load
   (b) determined by the formula $P_T = I^2 \times R_T$
   (c) determined by the formula $P_T = E \times I$
   (d) all of these

4. •The reading on Voltmeter 2 (V2) is _____. See **Figure 2–50**.

   (a) 5V
   (b) 6V
   (c) 7V
   (d) 10V

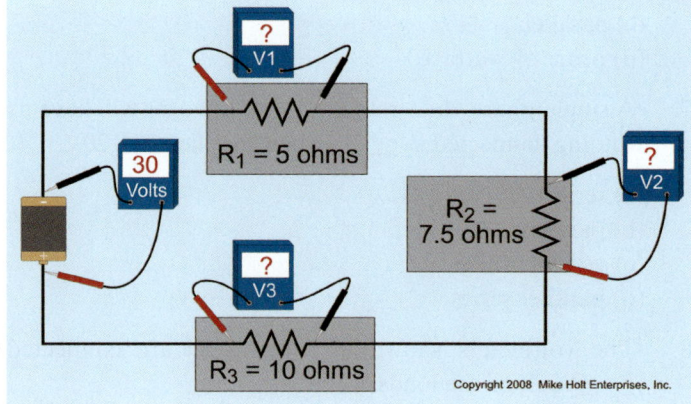

Figure 2–50

5. The voltmeter connected across the switch reads _____. See **Figure 2–51**.

   (a) 3V
   (b) 6V
   (c) 12V
   (d) 18V

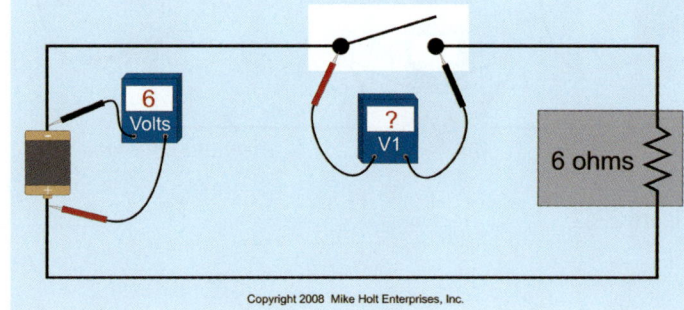

Figure 2–51

Unit 2 | Challenge Questions

PART B—PARALLEL CIRCUITS

2.9 Understanding Parallel Calculations

6. •In general, when multiple light bulbs are wired in a single luminaire, they are connected in _____ to each other.

   (a) series
   (b) series-parallel
   (c) parallel
   (d) order of wattage

7. •A single-phase, dual-rated, 120/240V motor will have its winding connected in _____ when supplied by 120V.

   (a) series
   (b) parallel
   (c) series-parallel
   (d) parallel-series

8. •The voltmeters shown in **Figure 2–52** are connected _____ each of the loads.

   (a) in series to
   (b) across
   (c) in parallel to
   (d) b and c

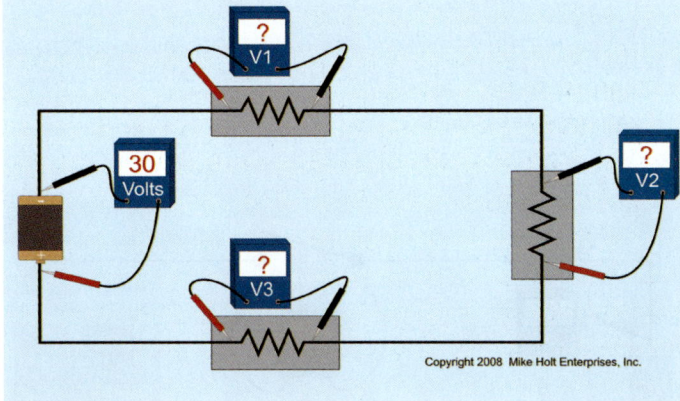

Figure 2–52

9. •If the supply voltage is 120V, the total energy consumed for four 10 ohm resistors will be more if all of the resistors are connected _____.

   (a) in series
   (b) in series-parallel
   (c) in parallel
   (d) any of these

2.10 Circuit Resistance

10. •A parallel circuit has three resistors. One resistor is rated 2 ohms, one is rated 3 ohms, and the other is rated 5 ohms. The total resistance of the parallel circuit is _____. Remember, the total resistance of any parallel circuit is always less than the smallest resistor.

    (a) 0.50 ohms
    (b) 1 ohm
    (c) 2 ohms
    (d) 3 ohms

Figure 2–53 applies to the next three questions:

11. •The total current of the circuit can be measured by ammeter _____. See **Figure 2–53**.

    (a) A1
    (b) A2
    (c) A3
    (d) none of these

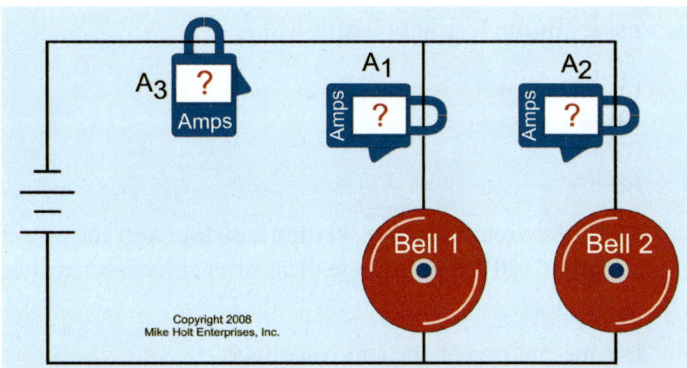

Figure 2–53

12. If Bell 2 consumes 12W of power when supplied by two 12V batteries (connected in series), the resistance of this bell is _____ . See **Figure 2–53**.

    (a) 12 ohms
    (b) 24 ohms
    (c) 36 ohms
    (d) 48 ohms

Challenge Questions — Unit 2

13. Determine the total circuit resistance of the parallel circuit based on the following facts. See **Figure 2–53**:

    1. The current on ammeter 1 reads 0.75A.
    2. The voltage of the circuit is 30V.
    3. Bell 2 has a resistance of 48 ohms.

    Tip: The total resistance of a parallel circuit is always less than the smallest resistor.

    (a) 22 ohms
    (b) 32 ohms
    (c) 42 ohms
    (d) 60 ohms

PART C—SERIES-PARALLEL CIRCUITS

2.14 Working With Series-Parallel Circuits

Figure 2–54 applies to the next three questions:

14. •The total current of this circuit can be read on _____. See **Figure 2–54**.

    (a) Ammeter 1
    (b) Ammeter 2
    (c) Ammeter 3
    (d) all of these

15. •The reading of Voltmeter 2 (V2) is _____. See **Figure 2–54**.

    (a) 1.50V
    (b) 4V
    (c) 5V
    (d) 8V

16. •The reading of V4 is _____. See **Figure 2–54**.

    (a) 1.50V
    (b) 3V
    (c) 5V
    (d) 8V

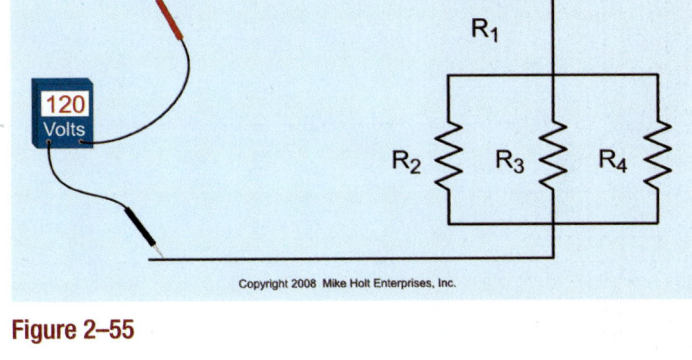

Figure 2–55

Figure 2–55 applies to the next two questions:

17. Resistor R1 has a resistance of 5 ohms and resistors R2, R3, and R4 have a resistance of 15 ohms each. The total resistance of this series-parallel circuit is _____. See **Figure 2–55**.

    (a) 10 ohms
    (b) 25 ohms
    (c) 35 ohms
    (d) 50 ohms

18. •What is the voltage drop across R_1, if R_1 is equal to 5 ohms and the total resistance of R_2, R_3, and R_4 is 5 ohms? See **Figure 2–55**.

    (a) 33V
    (b) 40V
    (c) 60V
    (d) 120V

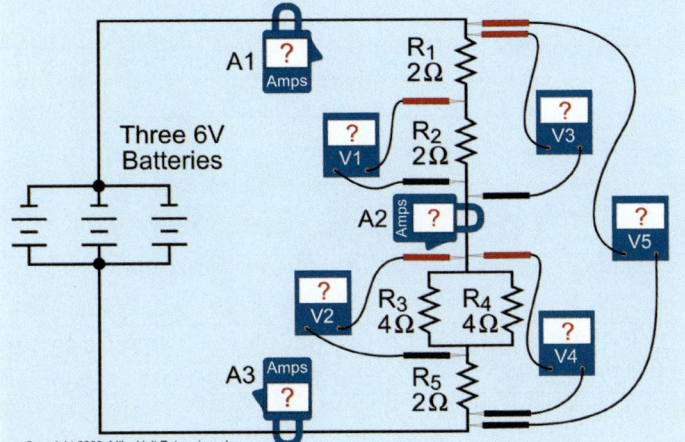

Figure 2–54

| Unit 2 | Challenge Questions |

PART D—MULTIWIRE BRANCH CIRCUITS

2.22 Dangers of Multiwire Branch Circuits

19. •If the neutral of the circuit in the diagram is opened, the circuit becomes one series circuit of 240V. Under this condition, the current of the circuit is _____. Tip: Determine the total resistance. See **Figure 2–56**.

(a) 0.25A
(b) 0.58A
(c) 0.67A
(d) 2.25A

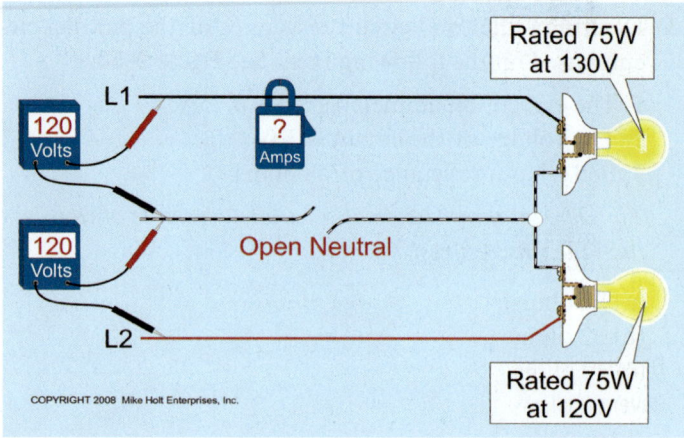

Figure 2–56

UNIT 3 — Understanding Alternating Current

INTRODUCTION TO UNIT 3—UNDERSTANDING ALTERNATING CURRENT

Direct-current (dc) circuits and alternating-current (ac) circuits have similarities as well as differences. In resistive circuits, dc and ac circuits have many similarities. The calculations of resistors used in series, parallel, or series-parallel combination circuits discussed in the preceding unit applies to both dc and ac circuits. In contrast, this unit focuses on the special characteristics of alternating current and how it's generated, distributed, and utilized.

The practical use of electric power depends on the efficient generation and distribution of power to the point of utilization. The principles of magnetism and induction are important in the generation and transformation of ac power. The flow of direct current in a circuit is opposed by resistance, but because of the changing current flow of an ac circuit, ac is opposed by a property called impedance. Impedance is a combination of resistance, inductive reactance, and capacitive reactance.

Be sure to study this unit carefully so you'll understand the differences between ac and dc circuits. There are some types of equipment that rely on the characteristics of the changing magnetic field developed by ac power to function, such as the ac motors and transformers that will be covered in Unit 4.

PART A—UNDERSTANDING ALTERNATING CURRENT

Introduction

Because ac current is inexpensive to transmit compared to dc current, ac current has become the dominant form of electricity in our modern infrastructure. In the early days of commercially available electric power, dc current was more common. But, economics won out. Applying ac current safely or effectively, however, requires an understanding of certain concepts that border on the complex. All of those concepts build on what you've already learned.

3.1 Current Flow

In order for current to flow in a circuit, the power supply must apply sufficient electromotive force (voltage) to cause the electrons to move. The movement of the electrons themselves doesn't produce any useful work. It's the effects that the moving electrons have on the loads they flow through that are important. The effects of electron movement are the same regardless of the direction of the current flow. **Figure 3–1**

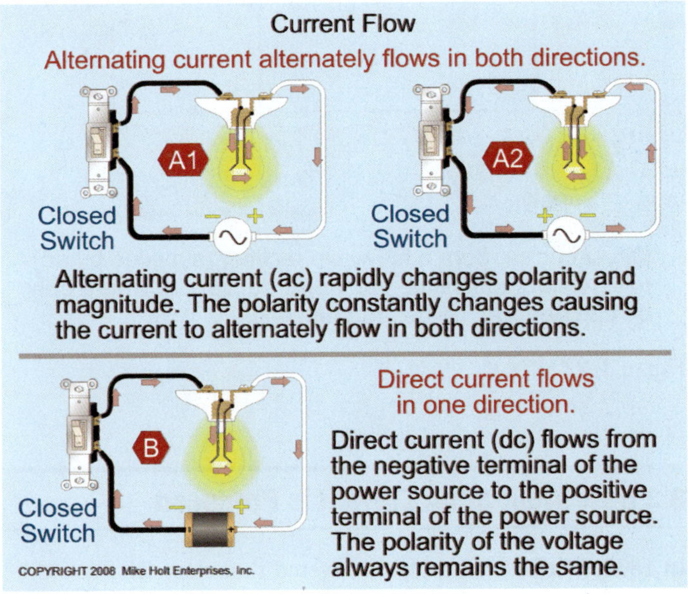

Figure 3–1

Unit 3 — Understanding Alternating Current

3.2 Why Alternating Current is Used

Alternating current is primarily used because it can be transmitted inexpensively by transforming to high-transmission voltage and then transforming this voltage back to lower distribution voltage levels. High-voltage transmission lines allow for the transmission of ac at high voltage, and relatively low current. High-voltage transmission with the resultant lower current results in reduced voltage drop in the transmission lines. Conductors and electrical equipment are sized based on the current in the lines, so the lower current levels also allow the use of smaller wire and smaller electrical equipment.

Direct-Current Use. There are other applications however, particularly in electronic equipment, where only direct current can perform the desired function. This is accomplished by rectifying ac to dc to power these electronic loads. **Figure 3–2**

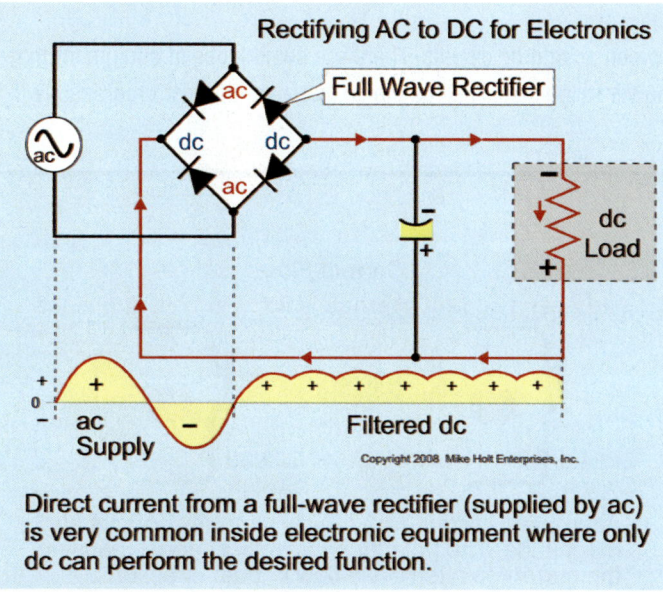

Figure 3–2

3.3 How Alternating Current is Produced

In 1831, Michael Faraday discovered that electricity could be produced from a source other than a battery. He knew that electricity produced magnetism, and wondered why magnetism couldn't produce electricity. Faraday discovered that when he moved a magnet inside a coil of wire, he got a pulse of electricity. When he pulled the magnet out, he got another pulse. He also got the same reaction when he moved the coil toward and away from the magnet.

Faraday's experiments revealed that when a magnetic field moves through a coil of wire, the lines of force of the magnetic field cause the electrons in the wire to flow in a specific direction. When the magnetic field moves in the opposite direction, electrons in the wire flow in the opposite direction. Electrons flow only when there is motion of the conductors relative to the magnetic field. **Figure 3–3**

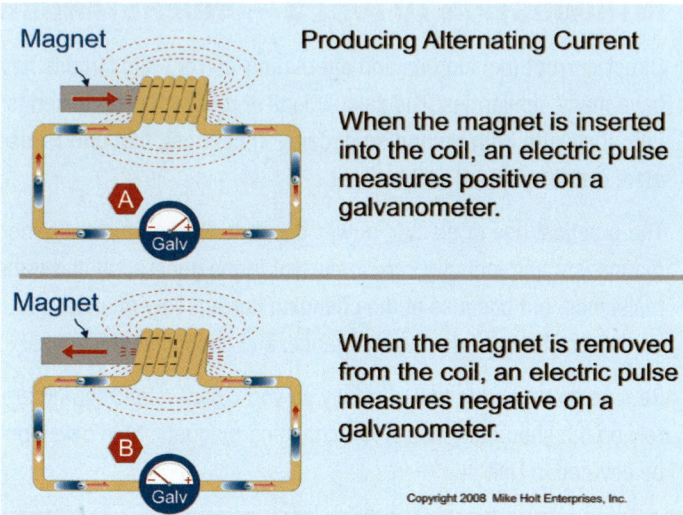

Figure 3–3

3.4 AC Generator

A simple ac generator consists of a loop of wire rotating between the lines of force between the opposite poles of a magnet. The halves of each conductor loop travel through the magnetic lines of force in opposite directions, causing the electrons within the conductor to move in a given direction. The magnitude of the voltage produced depends upon the number of turns of wire, the strength of the magnetic field, and the speed at which the coil rotates. **Figure 3–4**

The rotating conductor loop is called a rotor or armature. Slip or collector rings and carbon brushes are used to connect the output voltage from the generator to an external circuit.

In generators that produce large quantities of electricity, the conductor coils are stationary and the magnetic field revolves within the coils. The magnetic field is produced by an electromagnet, instead of a permanent magnet. Using electromagnets permits the strength of the magnetic field (and thus the lines of force) to be modified, thereby controlling the output voltage. **Figure 3–5**

Understanding Alternating Current   Unit 3

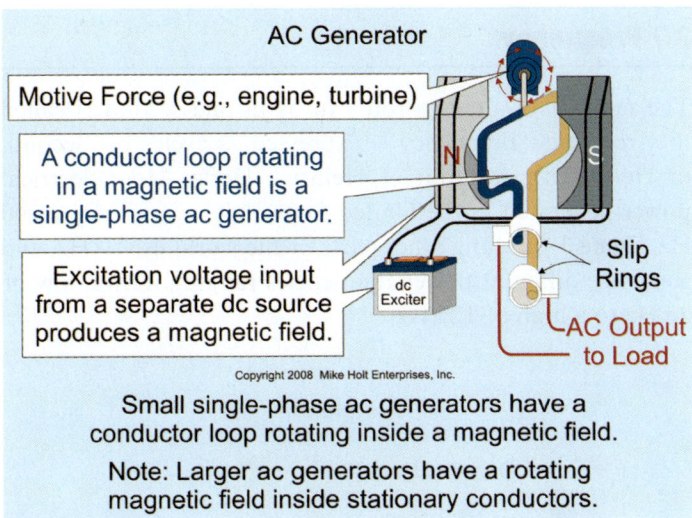

Figure 3–4

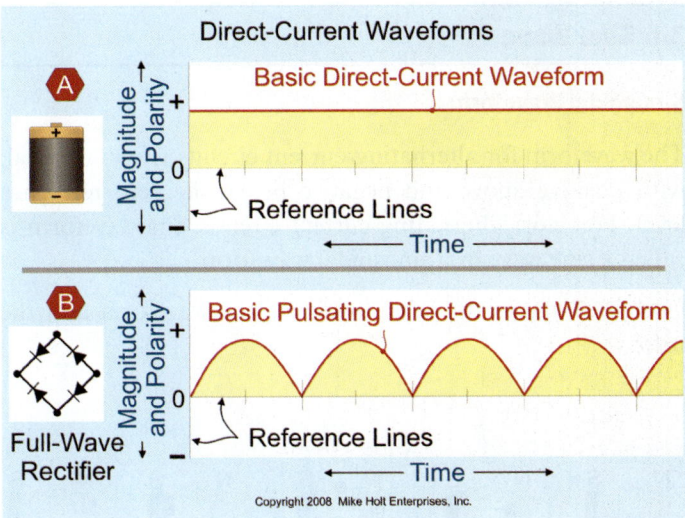

Figure 3–6

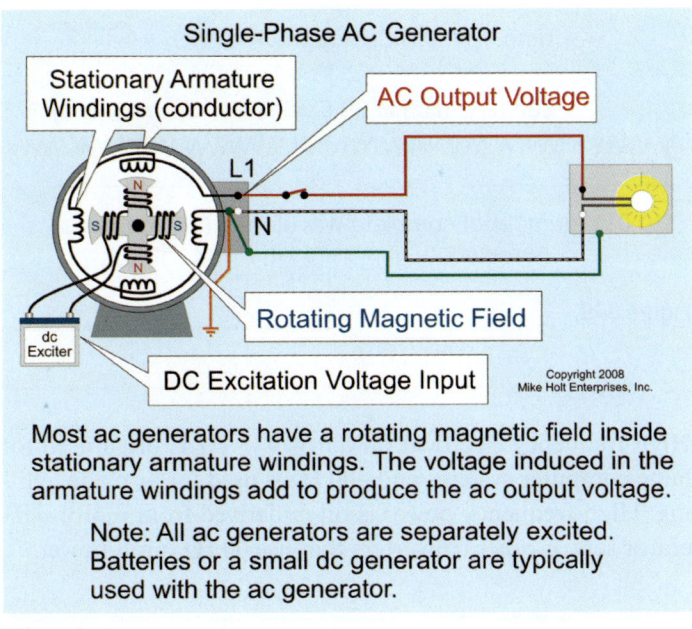

Figure 3–5

3.5 Waveform

A waveform image is used to display the level and direction of current and voltage.

Direct-Current Waveform

A direct-current waveform displays the direction (polarity) and magnitude of the current or voltage. Figure 3–6

Alternating-Current Waveform

An alternating-current waveform displays the level and direction of the current and voltage for every instant of time for one full revolution of the rotor. Figure 3–7

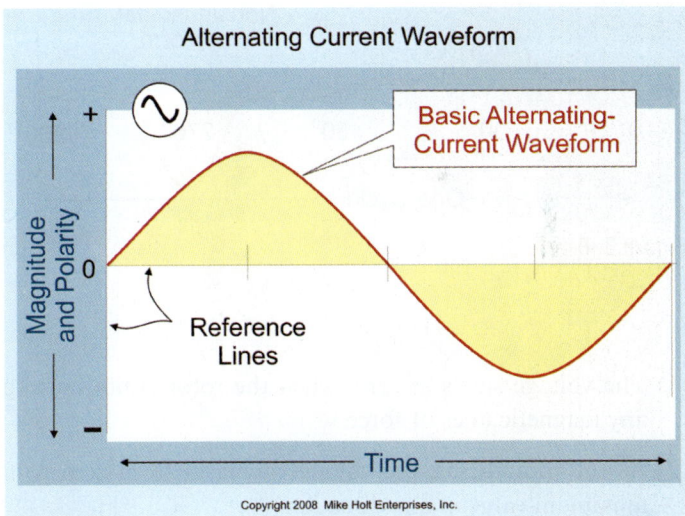

Figure 3–7

Mike Holt Enterprises, Inc. • www.MikeHolt.com • 1.888.NEC.CODE (1.888.632.2633)   61

Unit 3 — Understanding Alternating Current

3.6 Sine Wave

Sinusoidal Waveform

The waveform for alternating-current circuits is symmetrical, with positive above and negative below the zero reference level. For most alternating-current circuits, the waveform is called a sine wave or a sinusoidal waveform.

Figure 3–8 shows the relationship of the waveform and the rotor.

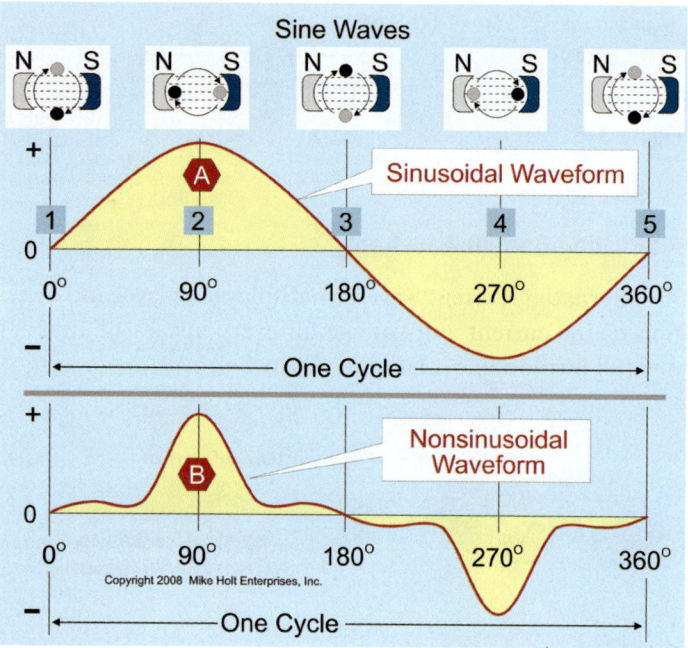

Figure 3–8

(1) The voltage starts at zero, when the rotor is not cutting any magnetic lines of force.

(2) As the rotor turns, the voltage increases from zero to a maximum value in one direction.

(3) It then decreases until it reaches zero.

(4) At zero, the voltage reverses polarity and increases until it reaches a maximum value at this opposite polarity.

(5) It decreases until it reaches zero again.

3.7 Frequency

The number of times a rotor turns in one second is called the frequency. Frequency is expressed as cycles per second, or Hertz (Hz) in honor of Heinrich Hertz. Most electrical power generated in the United States has a frequency of 60 Hz, Figure 3–9. Many other parts of the world use 50 Hz, and some use different power frequencies ranging from a low of 25 Hz to a high of 125 Hz.

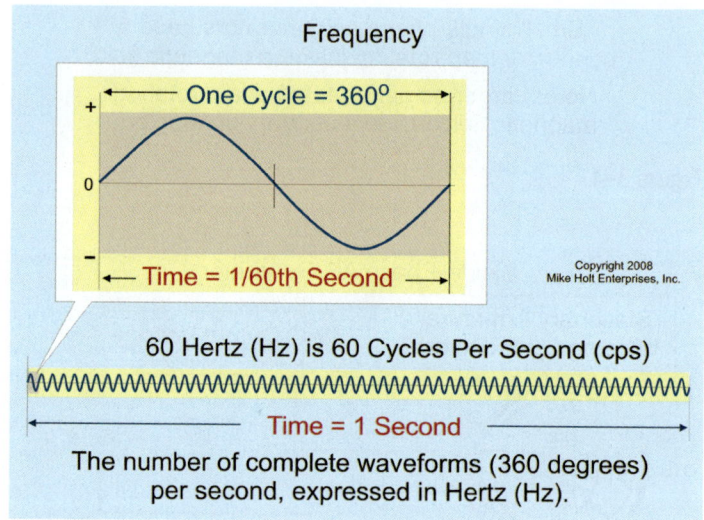

Figure 3–9

High-frequency electrical power of 415 Hz is often used for large computer systems and 400 Hz is used for airplane lighting. High-frequency power is often derived from motor-generator sets or other converters that use 60 Hz input power.

3.8 Phase

"Phase" is a term that indicates the time or degree relationship between two waveforms, such as voltage-to-current or voltage-to-voltage. When two waveforms are in step with each other, they are said to be "in phase." In a purely resistive ac circuit, the current and voltage are in phase. This means that, at every instant, the current is exactly in step with the applied voltage. They both reach their zero and peak values at the same time. Figure 3–10

Understanding Alternating Current — Unit 3

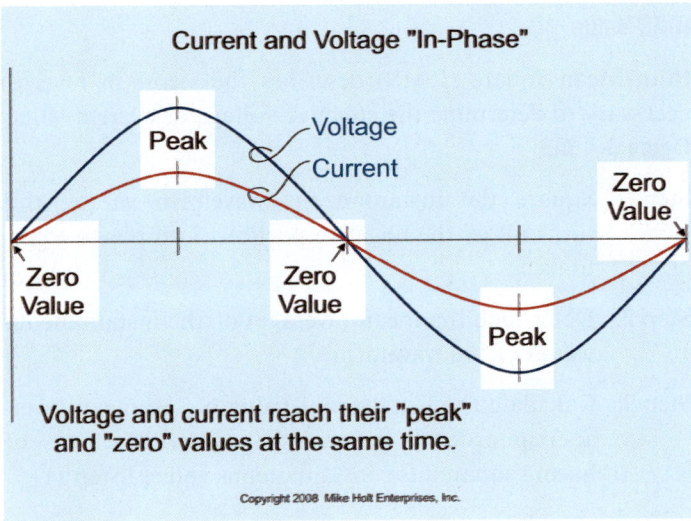

Figure 3–10

3.10 Lead or Lag

When describing the relationship between voltage and current, the reference waveform is always voltage—thus, a "lagging" waveform means that the voltage lags behind the current; a "leading" waveform means that the voltage leads the current.

Leading

It's easy to get confused as to which waveform leads and which one lags behind. The best way to remember this is to look at which waveform finishes its cycle first. In part A of **Figure 3–12**, the voltage waveform finishes its waveform cycle before the current waveform (designated by E2 in **Figure 3–12**), so the voltage waveform "leads" the current waveform.

3.9 Degrees

Phase differences are often expressed in degrees; one full waveform equals 360 degrees. For example, a three-phase generator has each of its windings out-of-phase with each other by 120 degrees. **Figure 3–11**

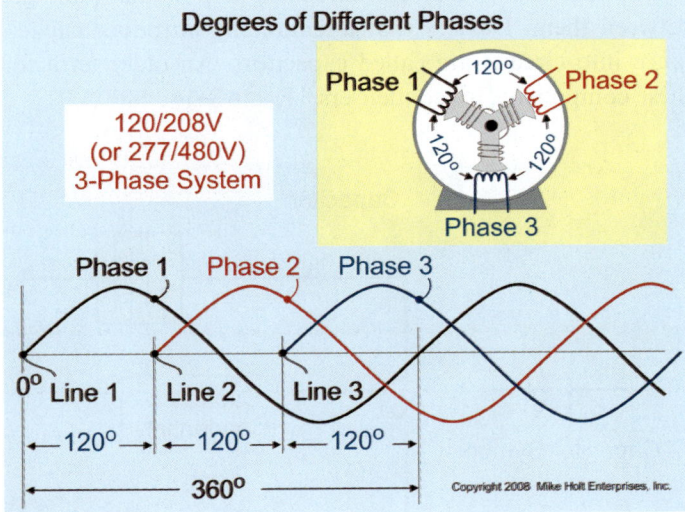

Figure 3–11

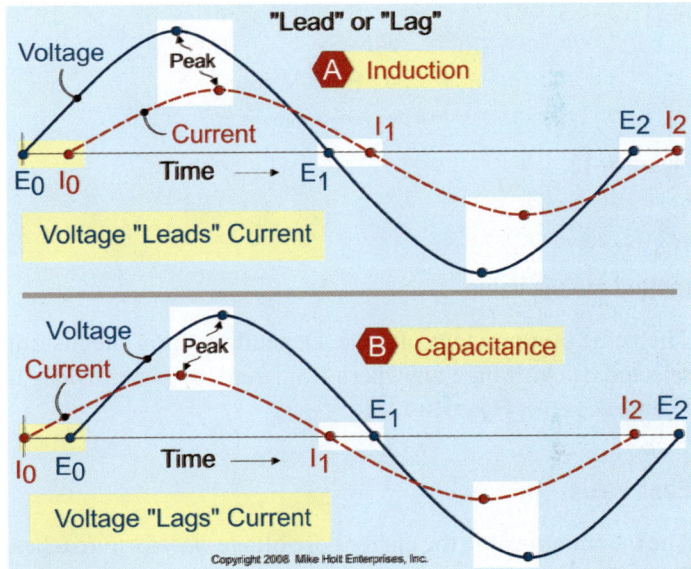

Figure 3–12

Lagging

In **Figure 3–12B**, the voltage waveform in Part B finishes its waveform cycle after the current waveform (designated by I2 in **Figure 3–12**). In this case the voltage "lags" the current.

Unit 3 — Understanding Alternating Current

3.11 Values of Alternating Current

There are many important values in alternating-current waveforms. Some of the most important include instantaneous, peak, and effective. **Figure 3–13**

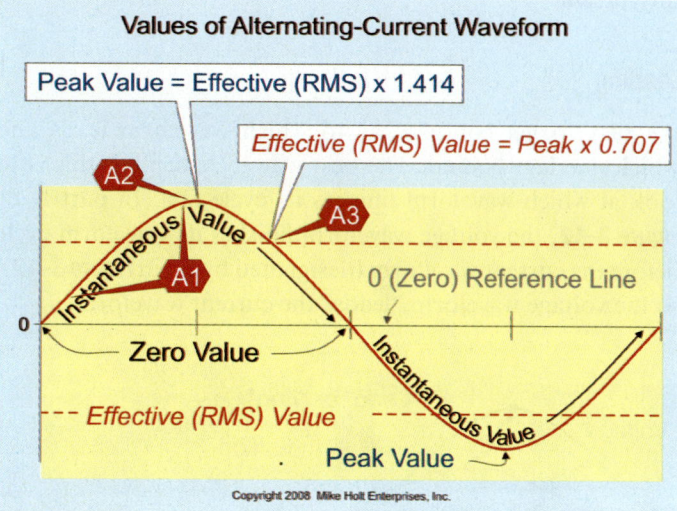

Figure 3–13

Instantaneous Value

The value at a moment of time. Depending upon the instant selected, it can range anywhere from zero, to peak, to negative peak value. **Figure 3–13A1**

Peak Value

The maximum value the current or voltage waveform reaches. **Figure 3–13A2**

For a pure sine wave:

Peak Value = Effective Value/0.707
Peak Value = Effective Value x 1.414
Effective Value = Peak Value x 0.707

Effective Value

Effective ac voltage or effective ac current is the equivalent value of dc voltage or dc current that will produce the same amount of heat in a resistor. **Figure 3–13A3** For a pure sine wave:

Effective Value = RMS Value

RMS Value

Root-Mean-Square (RMS) describes the steps (in reverse) necessary to determine the effective voltage or current value. **Figure 3–13A3**

Step 1: Square the instantaneous waveform values; this turns all of the negative portions into positive portions.

Step 2: Determine the mean (average) of the instantaneous values of the waveform.

Step 3: Calculate the square root value of the mean (average) in order to reverse the numerical effects of having squared the instantaneous values (Step 1).

PART B—CAPACITANCE

Introduction

Capacitance is the property of an electrical circuit that enables it to store electrical energy by means of an electrical field and to release that energy at a later time. Capacitance exists whenever an insulating material (dielectric) separates two conductors that have a difference of potential (voltage) between them. Devices that intentionally introduce capacitance into circuits are called capacitors. An older term for these components is "condensers." **Figure 3–14**

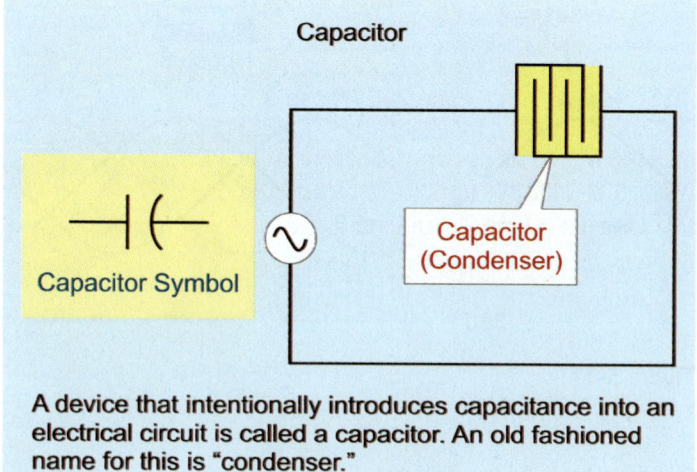

Figure 3–14

Capacitor Current Flow

Direct current does not flow through a capacitor. In an ac circuit, the electrons in the circuit move back and forth to alternately charge the capacitor, first in one direction, and then in the other. A capacitor permits current to flow because of its ability to store energy and then discharge the energy as the ac current flows in the opposite direction.

Capacitors as Filters

Figure 3–15 shows a diagram of a full-wave rectifier, which makes use of four rectifiers. A rectifier is a device that allows current to flow in only one direction, thus converting ac current to dc current. Each rectifier only provides current flow for one-half of the ac sine wave, which isn't a very practical dc current. Connecting four rectifiers into the full-wave rectifier circuit shown provides a bridge circuit which allows different rectifiers to alternately provide current to the load during each half of the sine wave alternation, while always keeping the same dc polarity on the load. This provides a partial solution, however the current produced is still not a pure dc current, but what is called "pulsating dc." A capacitor can be used to level out the pulsating dc into a filtered dc current.

> **Author's Comment:** As a "full-wave rectifier" converts the ac voltage to filtered dc voltage, the capacitor will be continuously charged. **Figure 3–15**

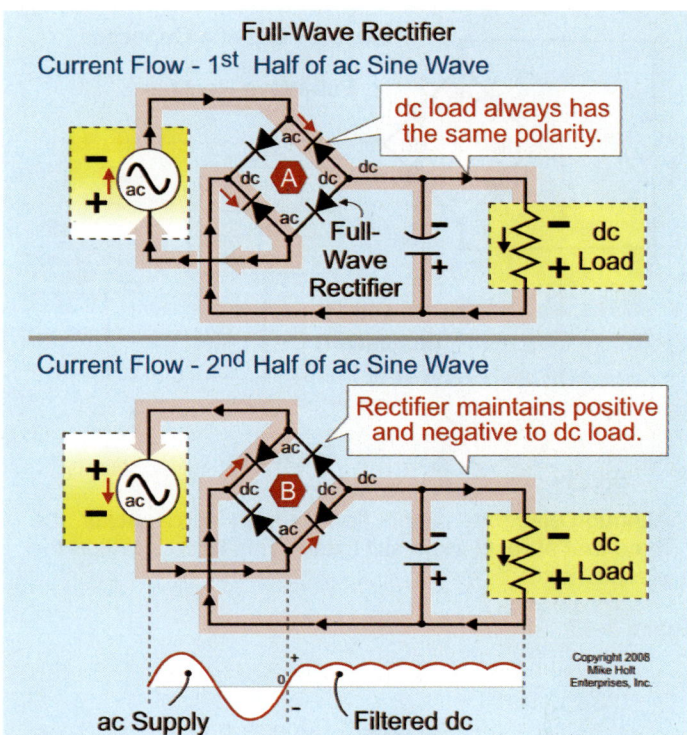

Figure 3–15

3.12 Charged Capacitor

When a capacitor has a potential difference between the conductors (plates), the capacitor is charged. One plate has an excess of free electrons and the other plate has a lack of them. The plate with the excess electrons has an overall negative charge (–), while the plate from which electrons were removed has an overall positive charge (+). A difference of potential (voltage) exists between the plates. **Figure 3–16**

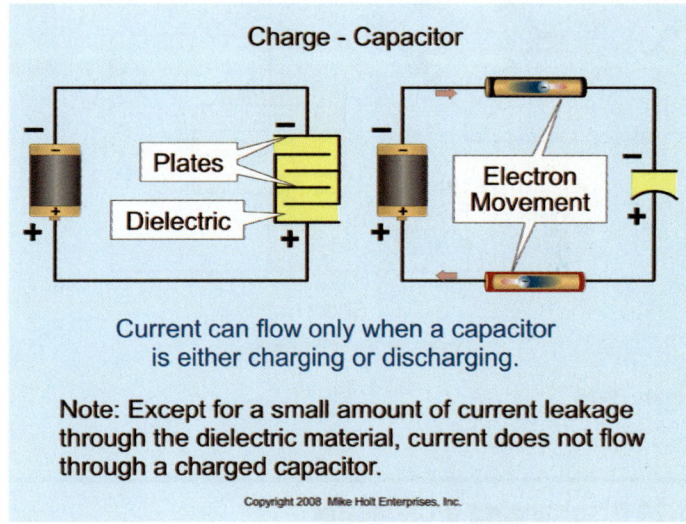

Figure 3–16

3.13 Electrical Field

Although the electrons can't flow, the force that attracts them still exists; this force is called the electrical field. The electrical field can be thought of as lines of electrical force that exist between the capacitor plates. **Figure 3–17**

The more the capacitor is charged, the stronger the electrical field. If the capacitor is overcharged, the electrons from the negative plate could be pulled through the insulation to the positive plate. If this happens, the capacitor is said to have broken down (shorted). **Figure 3–18**

Unit 3 — Understanding Alternating Current

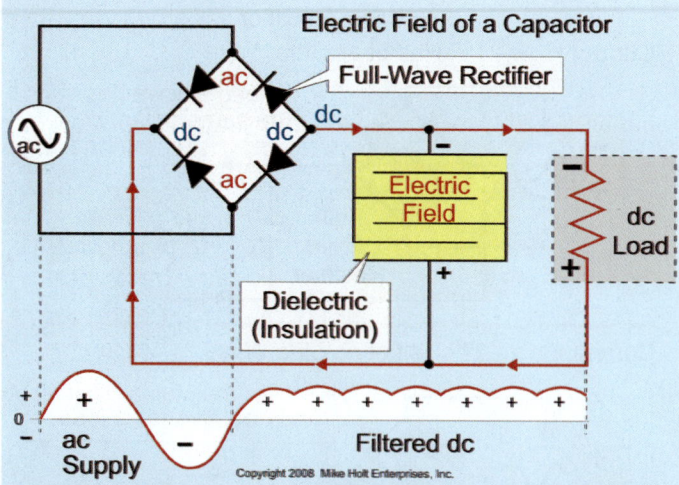

Figure 3–17

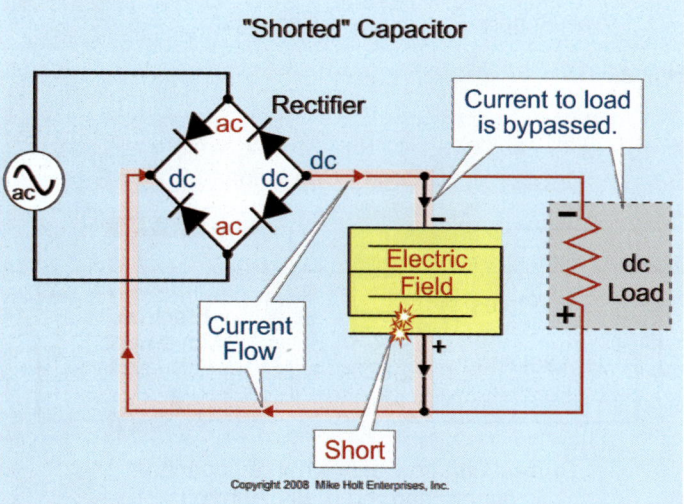

Figure 3–18

3.14 Discharging a Capacitor

To discharge a capacitor, all that is required is a conducting path connected across the terminals of the capacitor. The free electrons on the negative plate will then flow through the external circuit to the positive plate.

CAUTION: *Be careful when working on a circuit that contains capacitors (such as those found in variable speed drives). Capacitors can store large amounts of energy for a long period of time. They can present a shock hazard even after the power has been removed.*

3.15 Determining Capacitance

Factors that determine the capacitance of a capacitor are the surface area of the plates, the distance between the plates, and the insulating material or dielectric between the plates.

Plate Distance

Capacitance is inversely proportional to the distance between the capacitor plates. The closer the plates of the capacitor are to each other, the greater the capacitance and, conversely, the greater the distance is between the plates, the lower the capacitance. **Figure 3–19**

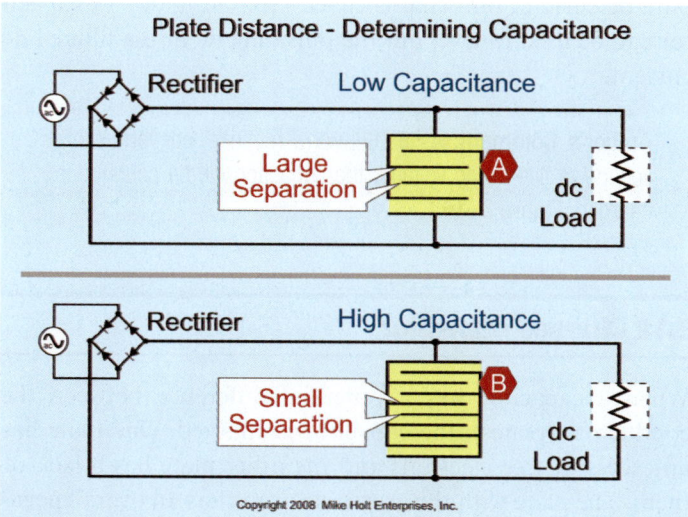

Figure 3–19

Surface Area

Capacitance is directly proportional to surface area. The greater the surface area of the plates is, the greater the capacitance. Connecting capacitors in parallel has the effect of increasing the plate surface area, and increasing the capacitance by the sum of the capacitors. Connecting capacitors in series has the effect of increasing the dielectric, and decreasing the capacitance. **Figure 3–20**

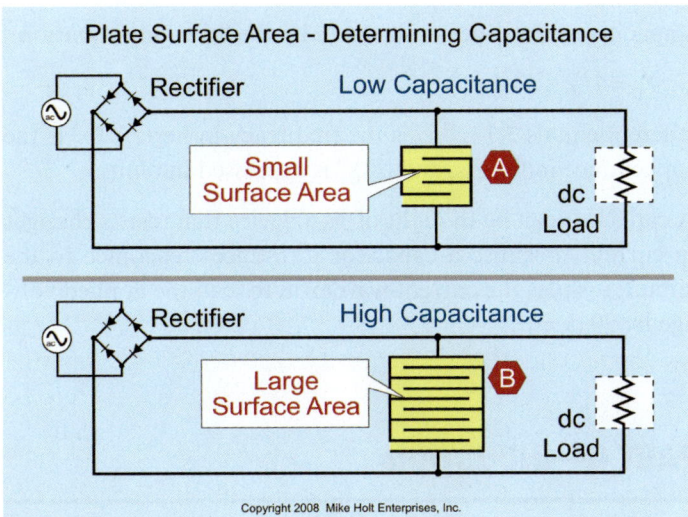

Figure 3–20

Dielectric Strength

Dielectric strength indicates the maximum voltage that can be applied across the dielectric safely. **Figure 3–21**

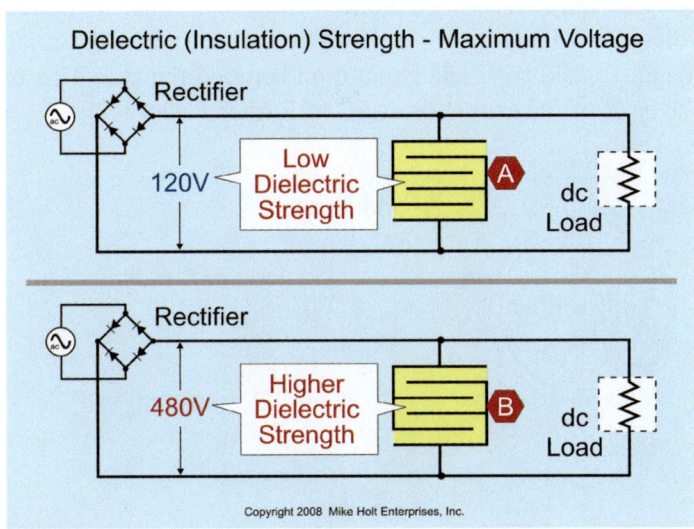

Figure 3–21

3.16 Uses of Capacitors

Capacitors have many uses in electronic circuits, as well as applications that are more familiar to electricians. Capacitors aid in starting single-phase ac motors by creating a phase displacement in the motor windings. Later in this unit, we'll discuss power factor and the role that capacitors play in correcting low power factor. Capacitors can be used to prevent arcing across the contacts of electric switches. A capacitor connected across the switch contacts provides a path for current flow until the switch is fully open and the danger of arcing has passed. **Figure 3–22**

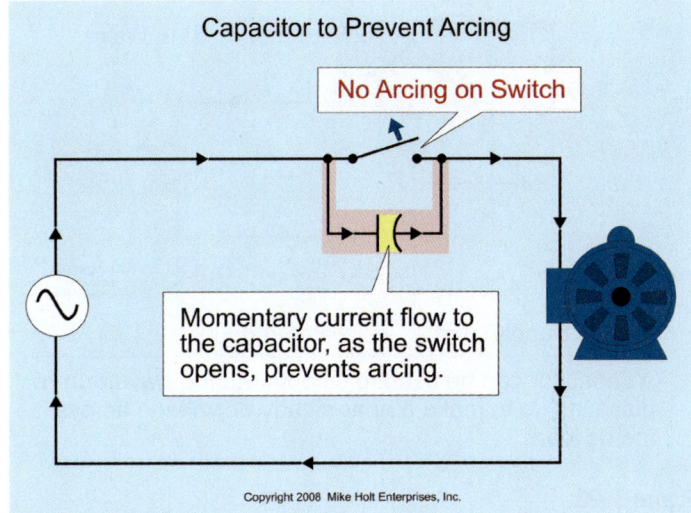

Figure 3–22

Electronic Power Supplies

One use of capacitors is to smooth out pulsating dc waveforms, such as those which would be present through a dc load if the capacitor wasn't present. A full-wave bridge rectifier transforms the ac waveform from the source into a pulsating dc waveform—the capacitor then smoothes out the waveform and makes a near-steady dc voltage across the dc load. **Figure 3–23**

3.17 Phase Relationship

A capacitor is a device that resists changes in voltage. As the ac sine wave reaches a positive peak, the capacitor fully charges to the same polarity. Then, as the ac current passes peak and decreases, the capacitor begins to discharge which has the effect of resisting the change of the ac circuit voltage value. This results in what is called capacitive reactance and a shifting of the current waveform out-of-phase to the voltage waveform. **Figure 3–24**

Unit 3 — Understanding Alternating Current

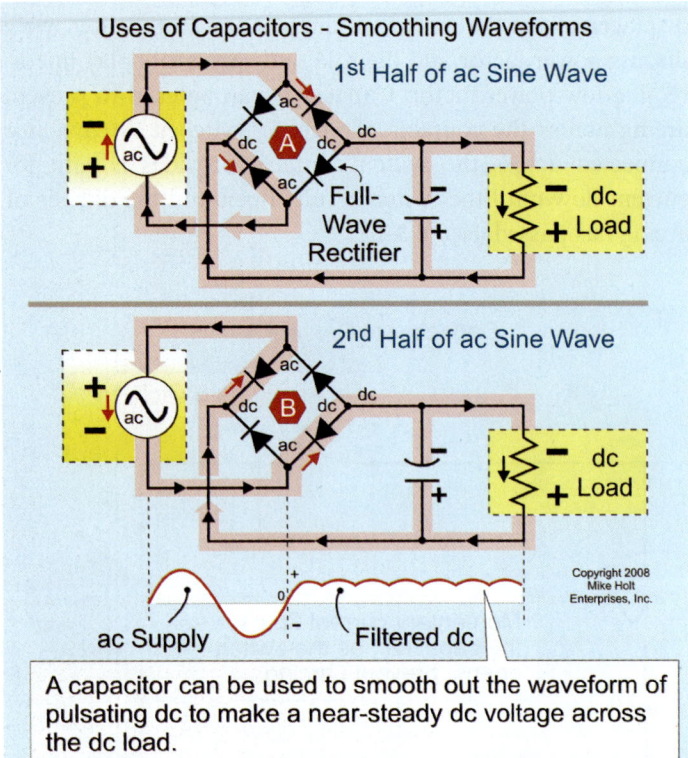

Figure 3–23

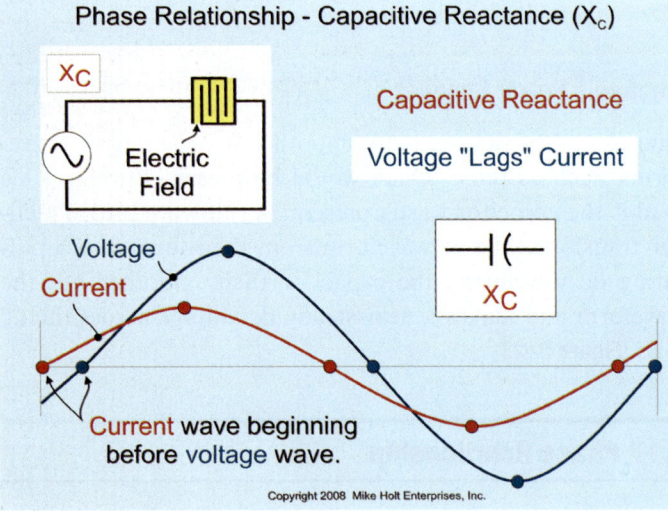

Figure 3–24

Capacitive reactance is calculated by the following equation:

$$X_C = 1/(2 \times \pi \times f \times C)$$

where π equals 3.14, "f" is the frequency in hertz, "C" is the capacitance in farads, and "X_C" is expressed in ohms.

A capacitor can be thought of as a device that resists changes in current. Because a capacitor introduces reactance to the circuit, it shifts the current waveform to lead the applied voltage by 90 degrees.

PART C—INDUCTION

Introduction

Because electrons spin, they have their own magnetic fields. When electrons move, the magnetic fields of the individual electrons combine to produce an overall magnetic field. The overall magnetic field extends outside the conductor. The greater the current flow, the greater the overall magnetic field. The movement of electrons caused by an external magnetic field is called induced current, and the associated potential that's established is called induced voltage. In order to induce voltage, all that's required is a conductor and an external magnetic field with relative motion between the two. This is the basis of the generator and transformer. **Figure 3–25**

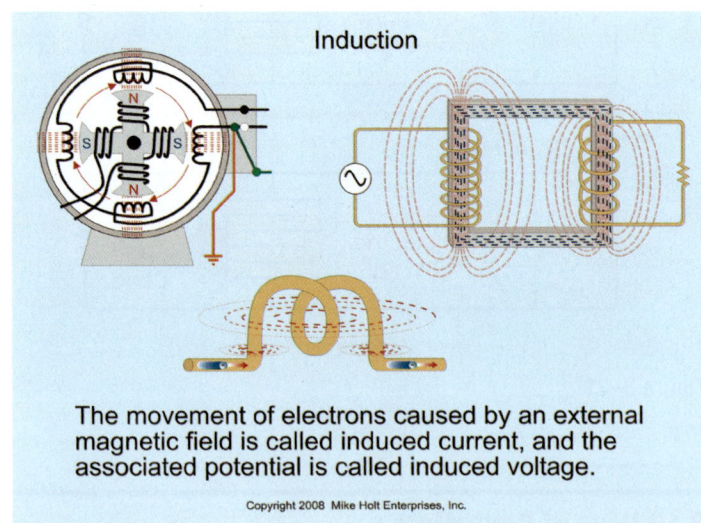

Figure 3–25

One way to think of this is that the capacitor responds to changes in current by increasing or decreasing its own amount of charge. Therefore, the voltage waveform change lags the current waveform change. The opposition offered to the flow of ac current by a capacitor is called capacitive reactance. This is expressed in ohms and abbreviated X_C.

Inductance is the property of an electrical circuit that enables it to store electrical energy by means of an electromagnetic field and to release this energy at a later time.

3.18 Self-Induction

As the ac current through a conductor increases, an expanding electromagnetic field is created through the conductor. The expanding magnetic flux lines cut through the conductor itself (which, in effect, is in motion relative to the field), thus inducing a voltage within the conductor.

When the current within the conductor decreases, the electromagnetic field collapses, and again the magnetic flux lines through the conductor cut through the conductor itself. The voltage induced within the conductor caused by its own expanding and collapsing magnetic field is known as self-induced voltage. **Figure 3–26**

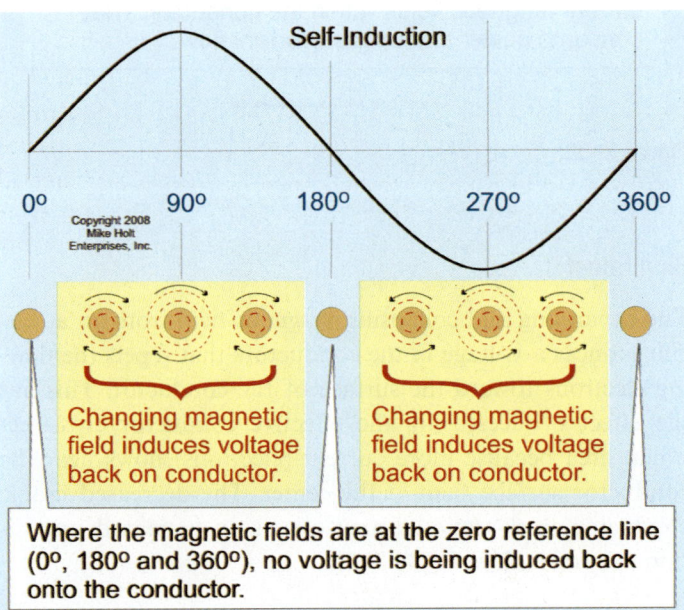

Figure 3–26

3.19 Induced Voltage and Applied Current

The induced voltage in a conductor carrying alternating current always opposes the change in current flowing through the conductor. The induced voltage that opposes the current flow is called "counter-electromotive force" (CEMF), or back-EMF.

The waveform of the induced voltage in the conductor (CEMF) is 90 degrees out-of-phase with the circuit current and it's 180 degrees out-of-phase with the applied voltage waveform. CEMF either opposes or aids the conductor current flow. **Figure 3–27**

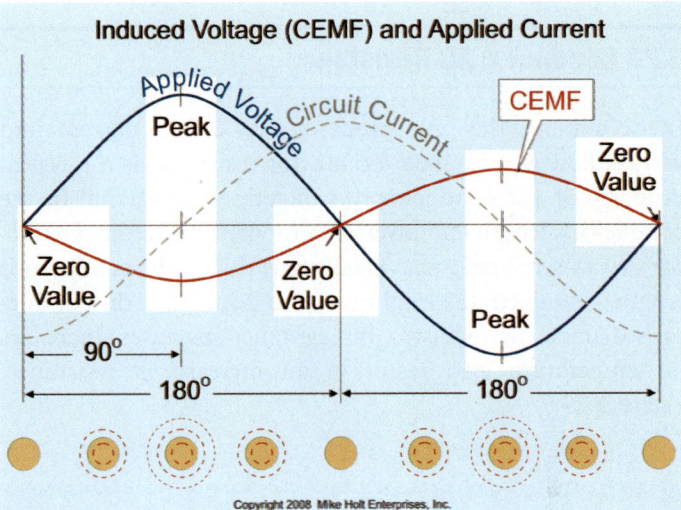

Figure 3–27

Opposes Current Flow

When alternating current increases, the polarity of the induced voltage (CEMF) within the conductor opposes the conductor's current and tries to prevent the current from increasing. **Figure 3–28A**

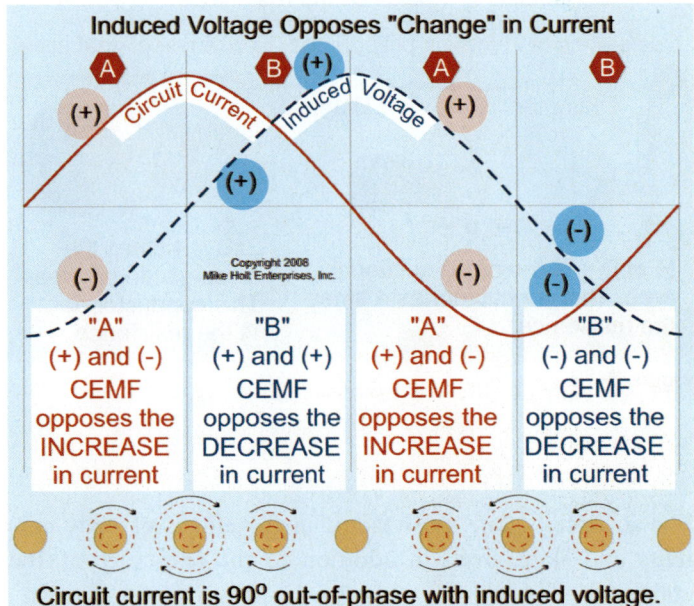

Figure 3–28

Unit 3 | Understanding Alternating Current

Aids Current Flow

When alternating current decreases, the polarity of the induced voltage (CEMF) within the conductor aids the conductor's current and tries to prevent the current from decreasing. **Figure 3–28B**

3.20 Conductor AC Resistance

In dc circuits, the only property that affects current and voltage flow is resistance. Conductor resistance is a physical property of the conductor. It's directly proportional to the conductor's length and inversely proportional to the conductor's cross-sectional area. This means that if the conductor's length is doubled, the total resistance doubles; if the conductor's diameter is reduced, the resistance increases. Increases in temperature also result in an increase in resistance. **Figure 3–29**

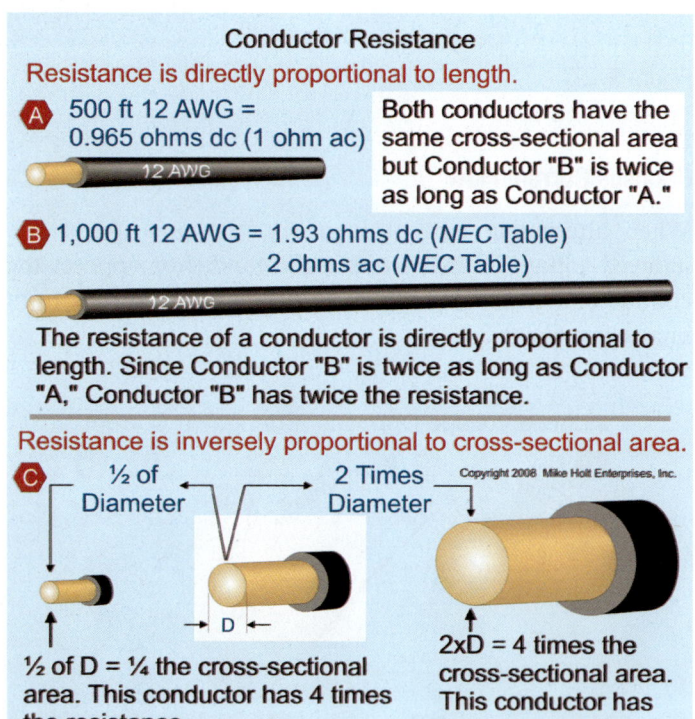

Figure 3–29

For ac circuits, one must factor in the effects of eddy currents and skin effect, in addition to the resistance of the conductor.

Eddy Currents

Eddy currents are small independent currents that are produced as a result of the expanding and collapsing magnetic field from an ac circuit. Eddy currents flow erratically through a conductor, consume power, and increase the opposition to current flow. **Figure 3–30**

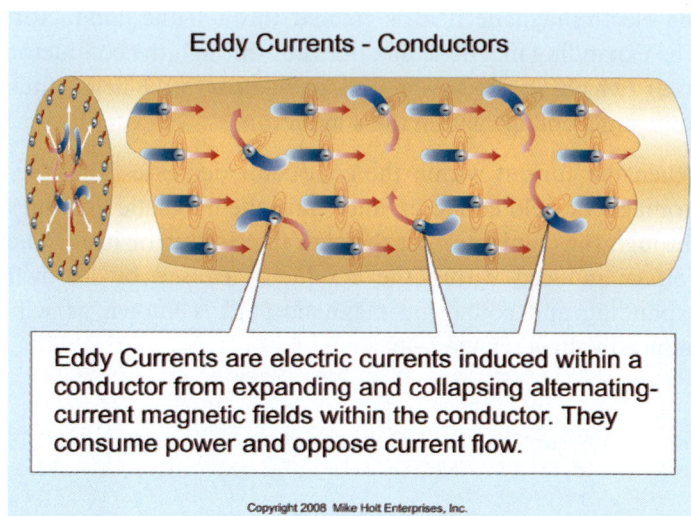

Figure 3–30

Skin Effect

The expanding and collapsing magnetic field from an ac circuit induces a voltage in the conductors that repels the flowing electrons toward the surface of the conductor. This has the effect of decreasing the effective conductor cross-sectional area because more current (electrons) flows near the conductor surface than at the center. The decreased usable conductor cross-sectional area causes an increased opposition to current flow. **Figure 3–31**

3.21 Conductor Shape

The physical shape of a conductor affects the amount of self-induced voltage within the conductor itself. When a conductor is coiled into adjacent loops (helically wound), it's called a winding. The expanding and collapsing magnetic flux lines of ac current flowing through the conductor loops interact and add together to create a strong overall magnetic field. As the combined flux lines expand and collapse they cut additional conductor loops, creating greater self-inductance in each conductor loop. **Figure 3–32**

Understanding Alternating Current — Unit 3

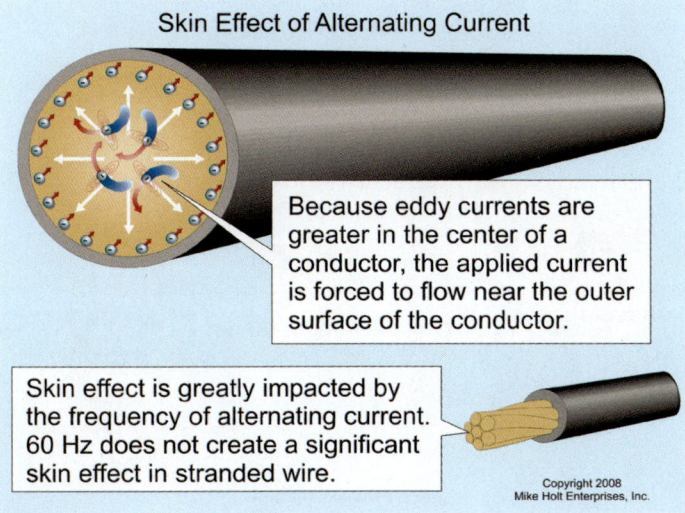

Figure 3–31

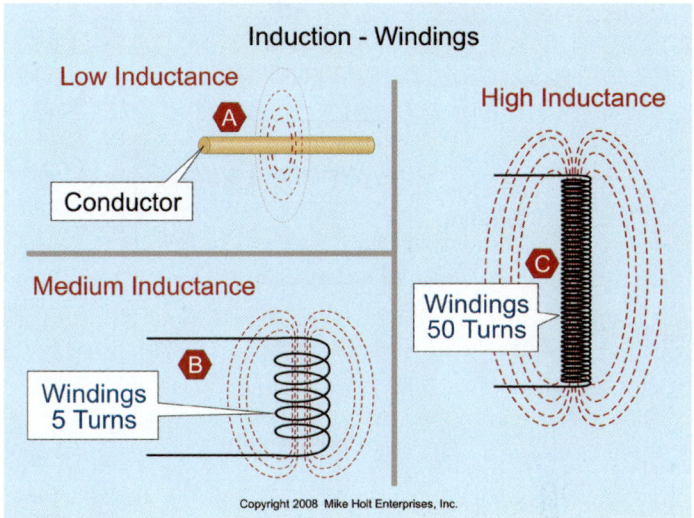

Figure 3–33

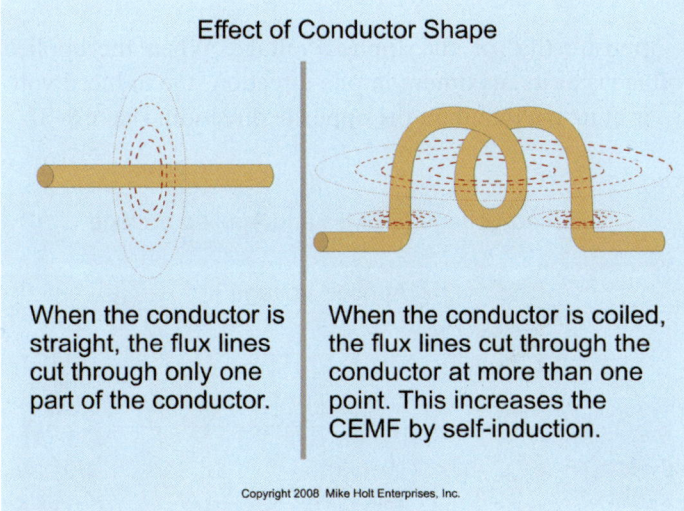

Figure 3–32

The amount of the self-induced voltage created within the winding is directly proportional to the current flow, the winding (conductor length and the number of turns), and the frequency at which the expanding and collapsing magnetic fields cut through the conductors of the winding.

Current

The greater the winding current is, and the greater the alternating magnetic field, the greater the CEMF will be within the winding.

Winding

The greater the number of winding conductor loops (turns) and the closer the windings, the greater the CEMF produced within the winding will be. **Figure 3–33**

Frequency

Self-induced voltage depends upon the frequency at which the magnetic field expands or collapses. Therefore, the greater the frequency is, the greater the CEMF induced within the winding will be.

3.22 Magnetic Cores

The core material also affects self-inductance in a winding (coil).

Core Material

Because an iron core provides an easy path for magnetic flux, windings with soft iron cores produce greater self-inductance than windings with an air core. **Figure 3–34**

Core Length

Longer cores result in fewer flux lines; this results in reduced self-inductance. If the core length is doubled, the CEMF decreases by 50 percent. **Figure 3–35**

Unit 3 — Understanding Alternating Current

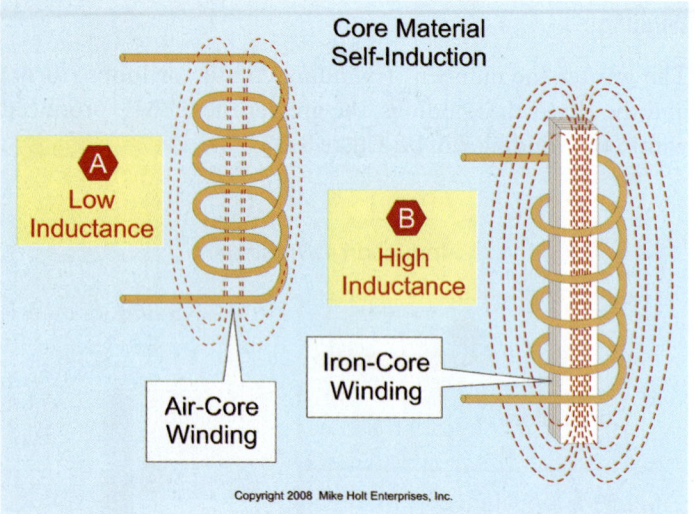

Figure 3–34

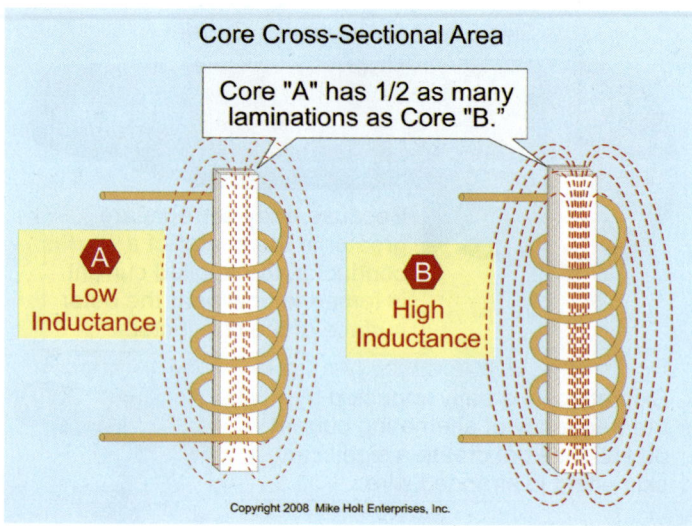

Figure 3–36

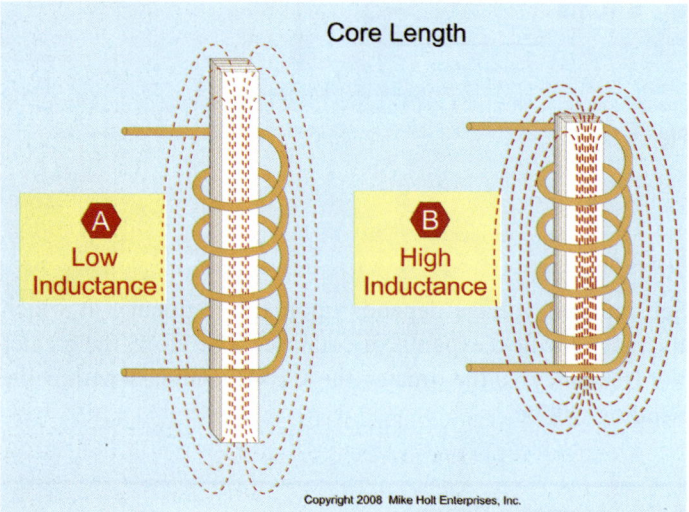

Figure 3–35

Core Area

Self-inductance (CEMF) is directly proportional to the cross-sectional area of the core, and inversely proportional to its length. This means that if the core area is doubled, the CEMF increases 200 percent. **Figure 3–36**

3.23 Self-Induced and Applied Voltage

A self-induced voltage waveform is 180 degrees out-of-phase with the applied voltage waveform. When the applied voltage increases or decreases, the polarity of the self-induced voltage is opposite that of the applied voltage. When the applied voltage is at its maximum in one direction, the induced voltage is at its maximum in the opposite direction. **Figure 3–37**

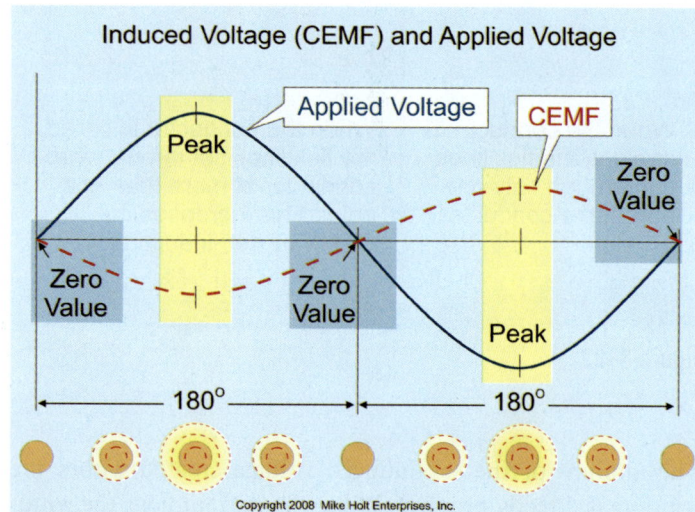

Figure 3–37

3.24 Inductive Reactance

Alternating-current flow in a conductor is limited by the conductor's resistance and self-induced voltage (CEMF). Self-induced voltage (CEMF) acts to oppose the change in current flowing in the conductor. This property is called "inductive reactance" and it's measured in ohms.

Understanding Alternating Current — Unit 3

Inductive Reactance

Inductive reactance is abbreviated X_L and can be calculated by the following equation, where "f" is frequency with units of hertz, "L" is inductance with units of Henrys, and $\pi = 3.14$.

$X_L = 2 \times \pi \times f \times L$

Author's Comment: Just remember that ac current flow contains an additional element (reactance) that opposes the flow of electrons, besides conductor resistance.

3.25 Phase Relationship

In a purely inductive circuit, the CEMF waveform is 90 degrees out-of-phase with the circuit current waveform and 180 degrees out-of-phase with the applied voltage waveform. As a result, the applied voltage waveform leads the current waveform by 90 degrees. **Figure 3–38**

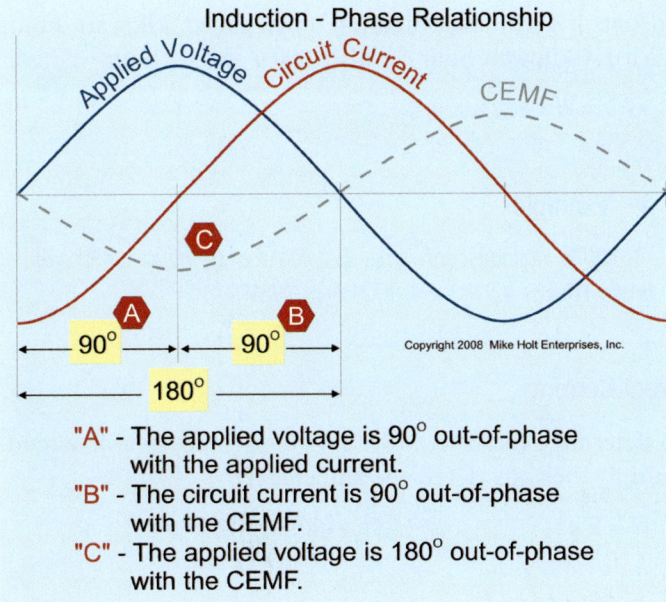

Figure 3–38

Author's Comment: Just remember that an ac circuit contains inductive reactance because the voltage and current are not in phase with each other.

3.26 Uses of Induction

The major use of induction is in transformers, motors, and generators. **Figure 3–39**

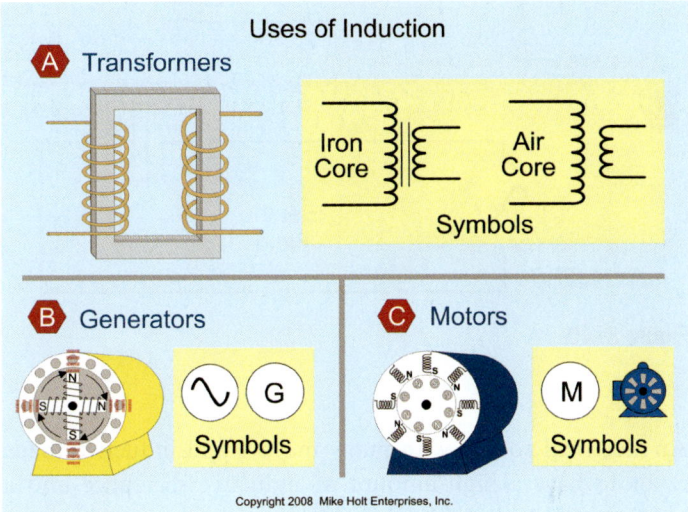

Figure 3–39

PART D—POWER FACTOR

Introduction

The out-of-phase relationship between voltage and current in ac circuits that serve reactive loads such as motors often leads to a situation where more power is being delivered to the load than is actually used. This situation is very common in industrial plants where a large number of high horsepower motors are installed, resulting in a highly inductive load and a low power factor.

Electric utilities typically impose penalties in their rate structure for customers with low power factor, so electricians should understand power factor and how to correct it in order to procure utility bill savings for their customers or their employer.

Inductors and Capacitors

Inductors and capacitors are energy storage devices in ac and dc circuits. Energy is stored in the electromagnetic field of an inductor and the electric field of a capacitor. **Figure 3–40**

Unit 3 — Understanding Alternating Current

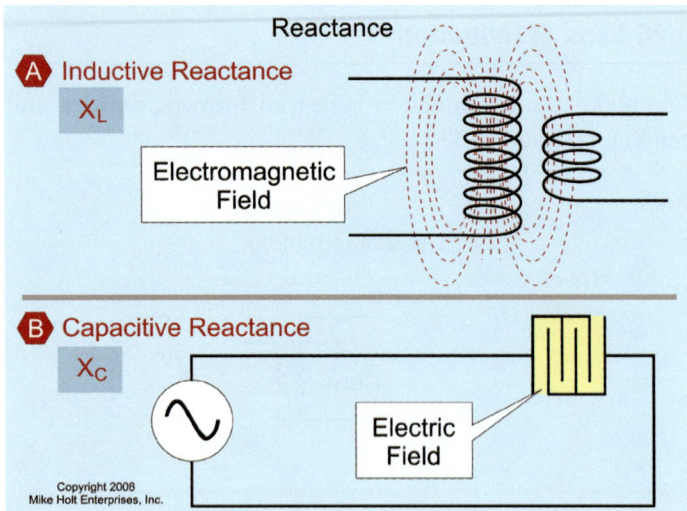

Figure 3–40

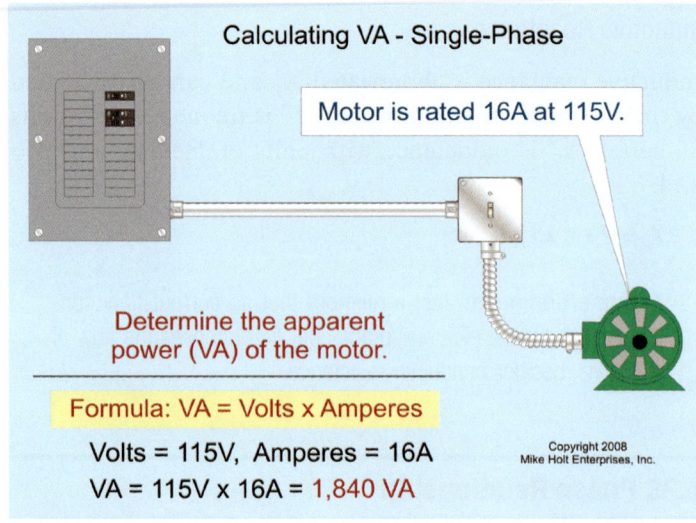

Figure 3–41

An electrical source with many motors and inductive loads tends to have a high amount of inductive reactance and a low power factor. Adding capacitors will help to counteract a high inductance and increase the power factor.

3.27 Apparent Power (Volt-Amperes)

If you measure voltage and current in an inductive or capacitive circuit, and multiply them together, the product is the apparent power supplied to the circuit by the source. Apparent power is expressed in volt-amperes (VA). Circuits and equipment must be sized to the circuit VA load.

Apparent Power = Volts x Amperes

▶ **Example**

Question: What is the apparent power in VA of a 1 hp, 115V motor that has a full-load current rating of 16A? **Figure 3–41**

(a) 1,320 VA   (b) 1,840 VA   (c) 1,920 VA   (d) 2,400 VA

Answer: (b) 1,840 VA

VA = E x I
VA = 115V x 16A
VA = 1,840 VA

3.28 True Power (Watts)

True power is the energy consumed, expressed in watts (W). Utility companies charge based on the total power consumed for one month, measured in units called kilowatt hours (kWh). A kilowatt hour equals 1,000 watts x 1 hour.

kWh = W x hrs/1,000

▶ **Example**

A 100W incandescent lamp, burning for 10 hours uses one kWh: 100W x 10 hrs = 1,000 Wh = 1 kWh

Direct Current

To determine the true power (W) consumed by a dc circuit, multiply the volts (E) by the amperes (I):

W = E x I

Alternating Current

In an ac circuit, true power (W) is determined by multiplying the circuit volts (E), by the amperes (I), times the power factor (PF):

W = E x I x PF

Understanding Alternating Current — Unit 3

3.29 Power Factor

AC inductive or capacitive reactive loads cause the voltage and current sine waves to be out-of-phase with each other. Power factor measures how far the current waveform is out-of-phase with the voltage waveform. **Figure 3–42**

Power factor is defined as a ratio of true power (W) to apparent power, which is expressed in volt-amperes (VA):

Power Factor (PF) = True Power (W)/Apparent Power (VA)

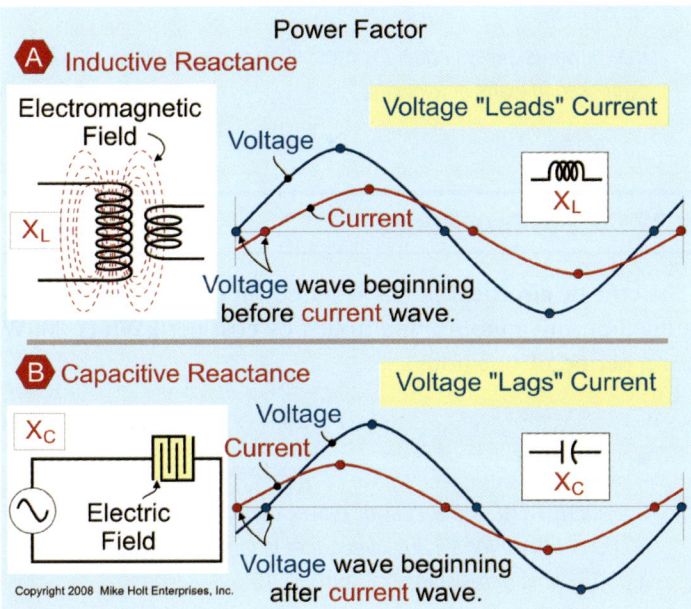

Figure 3–42

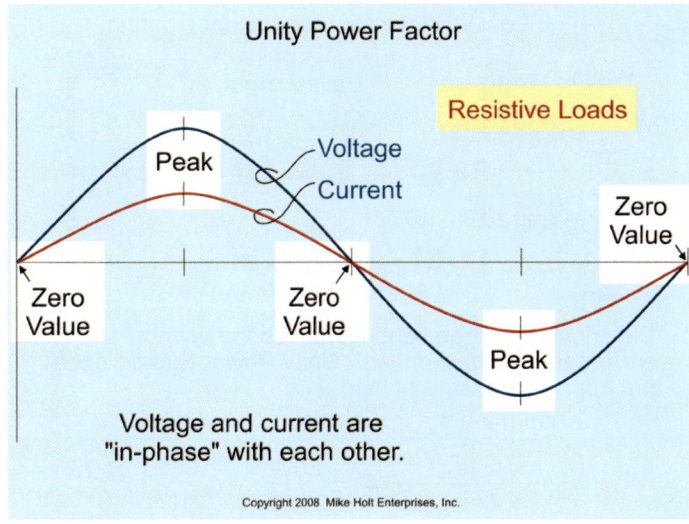

Figure 3–43

3.30 Unity Power Factor

When an ac circuit supplies power to a purely resistive load, such as incandescent lighting or heating elements, the circuit voltage and current are in-phase with each other.

Because the voltage and current reach their zero and peak values at the same time, there is no leading or lagging of the voltage to the current. Therefore, the power factor of the load is 100 percent (or 1.0), and this condition is called "unity power factor." **Figure 3–43**

3.31 Power Factor Formulas

The relationship between true power (Watts), apparent power (VA), and power factor can be shown as:

Power Factor (PF) = True Power (Watts)/Apparent Power (VA)
Apparent Power (VA) = True Power (W)/Power Factor (PF)
True Power (Watts) = Apparent Power (VA) x Power Factor (PF)

▶ **Power Factor (PF) Example**

Question: Assuming 100 percent efficiency, what is the power factor for each ballast rated 0.75A at 120V for a 2 x 4 fixture containing four 40W lamps (two lamps per ballast)? **Figure 3–44**

(a) 69%   (b) 75%   (c) 89%   (d) 95%

Answer: (c) 89%

PF = W/VA
PF = 80W/(0.75A x 120V)
PF = 80W/(90VA)
PF = 0.888 or 89%

Unit 3 — Understanding Alternating Current

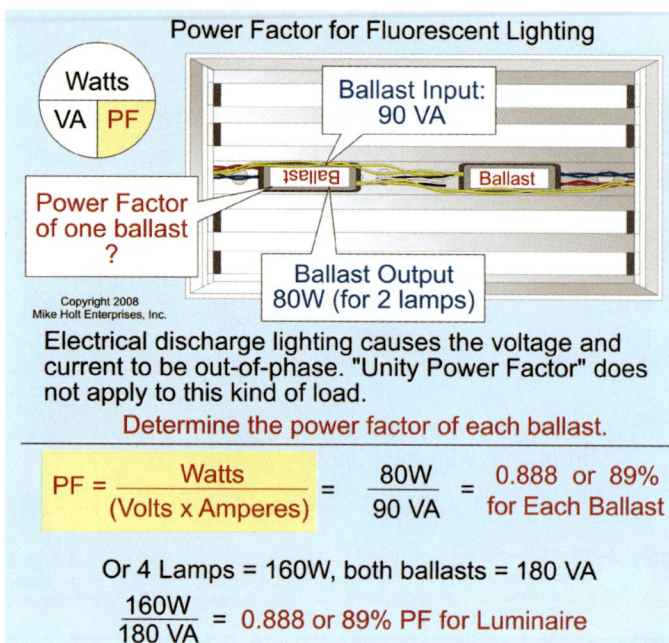

Figure 3–44

▶ Apparent Power (VA) Example

Question: What is the apparent power in VA of a fluorescent ballast that has a power factor of 89 percent when connected to two 40W lamps?

(a) 70W   (b) 80 VA   (c) 90 VA   (d) 100W

Answer: (c) 90 VA

VA = W/PF
VA = 80W/0.89 PF
VA = 89.90 VA

Apparent power is expressed in VA; therefore, neither (a) nor (d) can be the answer.

▶ True Power (Watts) Example

Question: What is the true power of a 16A load rated 120V, with a power factor of 85 percent?

(a) 1,632W   (b) 1,800W   (c) 1,920W   (d) 2,400W

Answer: (a) 1,632W

Watts = VA x PF
Watts = (120V x 16A) x 0.85 PF
Watts = 1,632W

True power (watts) is equal to volts times amperes times power factor.

True power is always equal to or less than apparent power and is expressed in watts or kilowatts.

3.32 Cost of True Power

The cost of electrical power is based on the true power consumed during a month, multiplied by cost per kWh (1,000W for a period of one hour).

▶ Example

Question: What is the cost of power consumed per month (at $0.09 per kWh) for a 120V circuit that is faulted to a ground rod with 25 ohms of resistance? **Figure 3–45**

(a) $10   (b) $25   (c) $37   (d) $55

Answer: (c) $37

Step 1: **Power per hour = E^2/R**
    E = 120V
    R = 25 ohms
    P = (120V x 120V)/25 ohms
    P = 576W

Step 2: Power consumed per day:
    576W x 24 = 13,824Wh or 13.824 kWh

Step 3: Power consumed in 30 days:
    13.824 kWh x 30 = 415 kWh

Step 4: Cost of power at $0.09 per kWh:
    415 kWh x $0.09 = $37.33

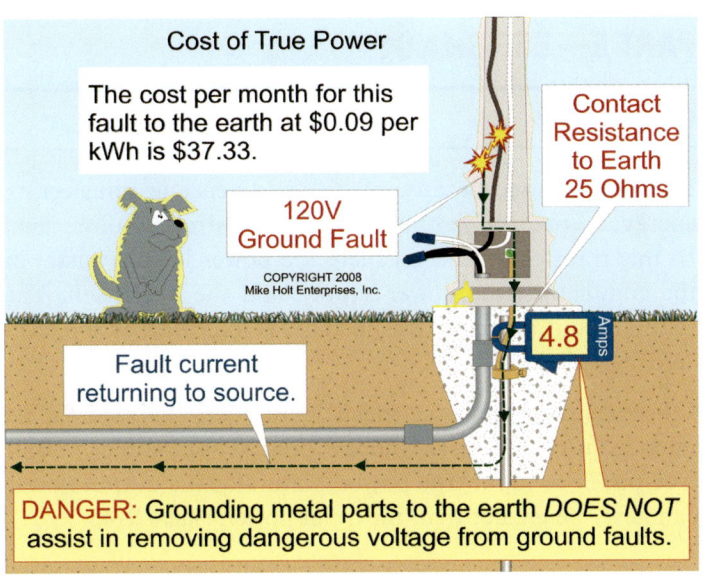

Figure 3–45

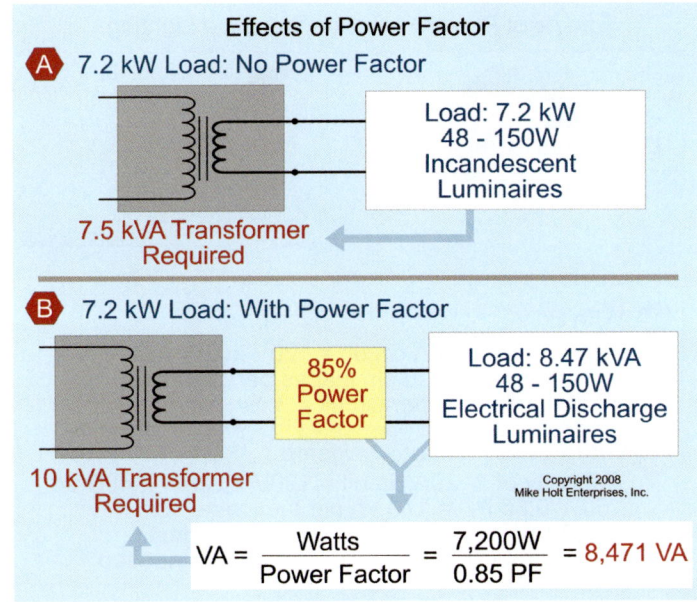

Figure 3–46

3.33 Effects of Power Factor

Apparent power (VA) is used for sizing circuits and equipment. Because the VA of the load is greater than the watts of the load, fewer loads can be supplied by each branch circuit. More circuits and panels, and larger transformers may be required.

▶ **Example 1**

Question: What size transformer is required for forty-eight, 150W incandescent luminaires (noncontinuous load)? Figure 3–46A

(a) 3 kVA (b) 5 kVA (c) 7.50 kVA (d) 10 kVA

Answer: (c) 7.50 kVA

kVA = (48 Fixtures x 150W)/1,000
kVA = 7,200W/1,000
kVA = 7.20 kVA

Author's Comment: Transformers are always sized based on kVA, not kW because transformers are an inductive load which results in apparent power. Apparent power is measured in kVA, while kW is used only for true power such as resistive loads.

▶ **Example 2**

Question: What size transformer is required for forty-eight, 150W electric discharge luminaires that have a power factor of 85 percent (noncontinuous load)? Figure 3–46B

(a) 3 kVA (b) 5 kVA (c) 7.50 kVA (d) 10 kVA

Answer: (d) 10 kVA

kW = (48 x 150W)/1,000
kW = 7,200W/1,000
kW = 7.20 kW

Apparent Power = kW/PF
Apparent Power = 7.20 kW/0.85 PF
Apparent Power = 8.47 kVA

Transformers are sized for apparent power (kVA), not true power.

▶ **Example 3**

Question: How many 20A, 120V circuits are required for forty-eight, 150W incandescent luminaires (noncontinuous load)? Figure 3–47A

(a) 2 circuits (b) 3 circuits (c) 4 circuits (d) 5 circuits

Answer: (b) 3 circuits

Each circuit has a capacity of: 120V x 20A = 2,400 VA
Each circuit can have: 2,400W/150W = 16 luminaires
The number of circuits required is:
48 luminaires/16 luminaires = 3 circuits

Unit 3 — Understanding Alternating Current

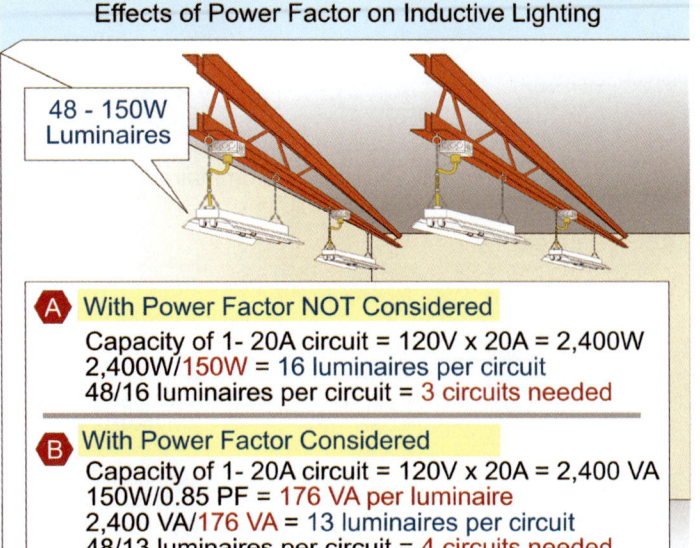

Figure 3–47

PART E—EFFICIENCY

Introduction

Electricity is used because of its convenience in transferring energy to operate lighting, heating, controls, motors, etc. In this transfer of energy, there are power losses (waste) in the conductors, the power supply, and the load itself. The total amount of power loss in watts is indicated by the term "efficiency."

Efficiency describes how much input energy is used for its intended useful purpose, and is expressed as a ratio of output true power to input true power. Naturally, the output power can never be greater than the input power. **Figure 3–48**

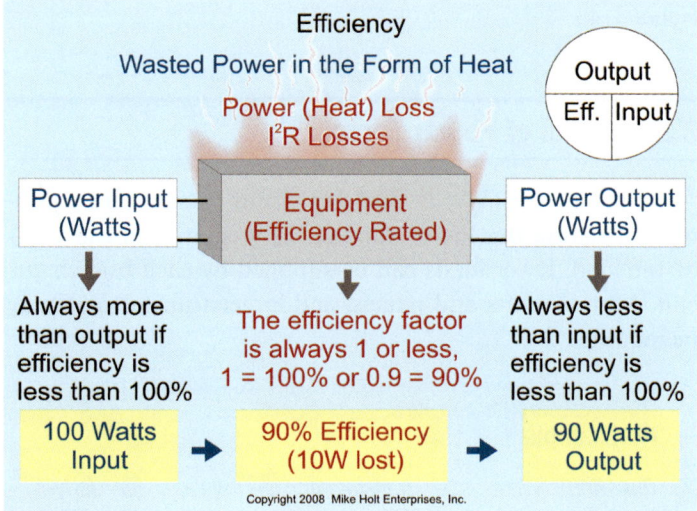

Figure 3–48

▶ **Example 4**

Question: How many 20A, 120V circuits are required for forty-eight, 150W electric discharge luminaires (noncontinuous load) that have a power factor of 85 percent? **Figure 3–47B**

(a) 4 circuits   (b) 6 circuits   (c) 8 circuits   (d) 12 circuits

Answer: (a) 4 circuits

Circuits are loaded according to VA, not watts!

VA of each luminaire equals:

VA = Watts/PF
VA = 150W/0.85 PF
VA = 176 VA

Each circuit has a capacity of: 120V x 20A = 2,400 VA
Each circuit can have: 2,400 VA/176 VA = 13 luminaires
The number of circuits required is:
48 luminaires/13 luminaires per circuit = 4 circuits

If the building had 480 luminaires, we would need 40 circuits instead of 30 and the transformer would need to be rated at least 100 kVA, instead of 75 kVA.

Author's Comment: The VA and/or amperage rating is commonly marked on the nameplate or ballast of inductive luminaires for use in branch circuit sizing. The circuit is not based on the lamp wattage rating alone. [220.14(D)]

If equipment were rated at 100 percent efficiency (there is none), that means that 100 percent of the input energy would be consumed for its intended useful purpose.

When equipment is rated at 90 percent efficiency, only 90 percent of the input power is used for its intended useful purpose. Another way of saying this is that 10 percent of the input power is wasted.

Power Loss

When energy is not used for its intended purpose, this condition is called power loss. Conductor resistance, mechanical friction, and other factors can contribute to increased power losses, or reduced efficiency rating.

Understanding Alternating Current — Unit 3

3.34 Efficiency Formulas

The formulas that are often used with efficiency calculations include:

Efficiency = Output Watts/Input Watts
Input Watts = Output Watts/Efficiency
Output Watts = Input Watts x Efficiency

▶ **Efficiency Example 1**

Question: If the output of a load is 640W and the input is 800W, what is the efficiency of the equipment? **Figure 3–49**

(a) 60%   (b) 70%   (c) 80%   (d) 100%

Answer: (c) 80%
Efficiency is always less than 100%.

Efficiency = Output Watts/Input Watts
Efficiency = 640W/800W
Efficiency = 0.80 or 80%

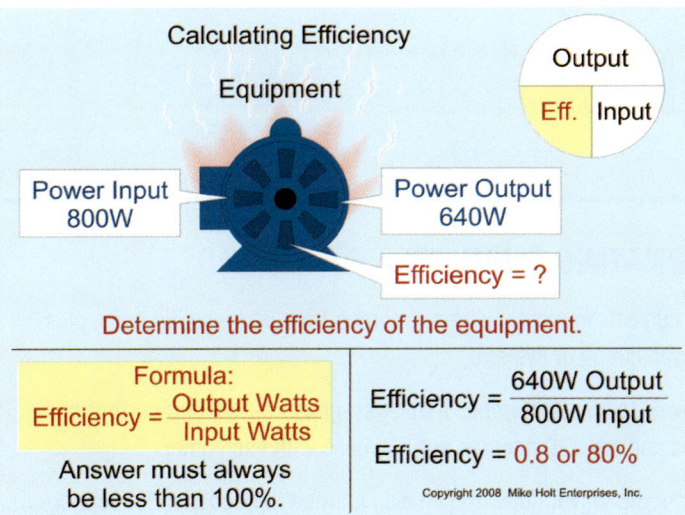

Figure 3–49

▶ **Efficiency Example 2**

Question: If the output of a 5 hp dc motor is 3,730W (746W x 5) and the input is 4,800W (40A x 120V), what is the efficiency of the motor? **Figure 3–50**

(a) 60%   (b) 78%   (c) 80%   (d) 100%

Answer: (b) 78%

(continued in next column)

Efficiency = Output Watts/Input Watts
Efficiency = 3,730W/4,800W
Efficiency = 0.777 or 77.70%

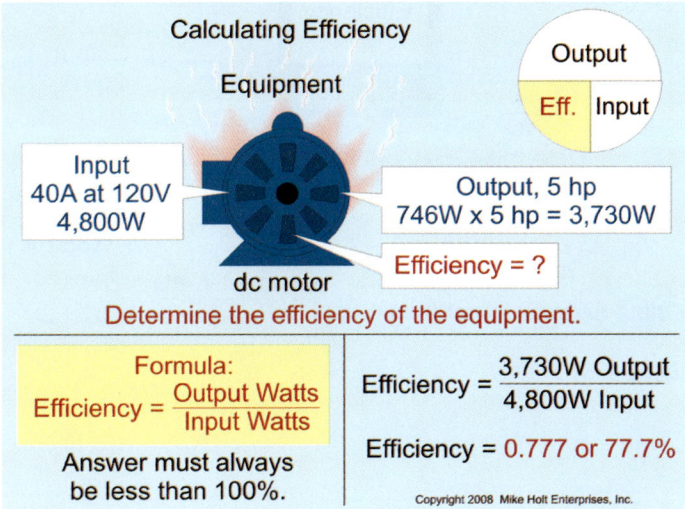

Figure 3–50

▶ **Input Example**

Question: If the output is 250W and the equipment is 88 percent efficient, what is the input power rating in watts? **Figure 3–51**

(a) 200W   (b) 250W   (c) 285W   (d) 325W

Answer: (c) 285W
Input is always greater than the output.

Input = Output Watts/Efficiency
Input = 250W/0.88 Efficiency
Input = 284W

▶ **Output Example**

Question: If the input power to a load is 479W and the equipment is rated at 87.60 percent efficiency, what is the output power rating in watts? **Figure 3–52**

(a) 350W   (b) 420W   (c) 500W   (d) 550W

Answer: (b) 420W
Output is always less than input.

Output Watts = Input Watts x Efficiency
Output Watts = 479W x 0.876 Efficiency
Output Watts = 419.60W

Unit 3 | **Understanding Alternating Current**

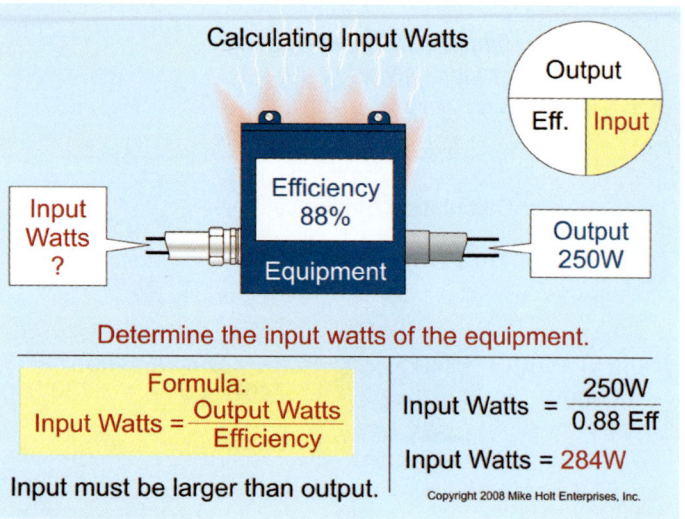

Figure 3–51

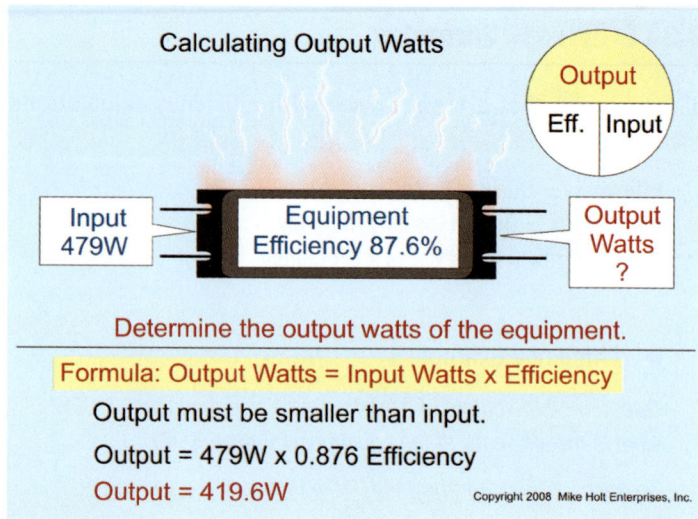

Figure 3–52

UNIT 3 Conclusion

CONCLUSION TO UNIT 3—UNDERSTANDING ALTERNATING CURRENT

You now have a solid understanding of the basic concepts of alternating current. You know where it comes from and why it's used, plus you understand the various properties that characterize the current in a given circuit or system.

Your knowledge of what lead and lag mean will help you understand power factor. Remember that three-phase alternating current has three waveforms that, when conditions are perfect, have sinusoidal shapes and are 120 degrees out-of-phase with each other.

You now understand what a capacitor is, some uses for it, and why it works the way it does. You've also learned the equation for capacitive reactance. Capacitors have many uses, not all of which have been discussed here. For example, capacitors are often combined in circuits with inductors to form "tank circuits" in devices called power conditioners.

Induction, like capacitance and impedance, is one of the fundamental properties of ac circuits. Now that you understand inductance, you're ready to learn the basic theory behind three important types of inductive devices—motors, generators, and transformers.

You understand apparent power, true power, power factor, and unity power factor, and you know that it's possible to use capacitors to correct power factor.

You know that no circuit is 100 percent efficient—all circuits and equipment have some power loss. The ability to calculate efficiency and power loss is now part of your skill set. While you don't normally have to use this skill, you may find it useful when troubleshooting or when deciding whether to replace old equipment.

Practice Questions

UNIT 3—UNDERSTANDING ALTERNATING CURRENT PRACTICE QUESTIONS

PART A—UNDERSTANDING ALTERNATING CURRENT

3.1 Current Flow

1. The movement of electrons themselves doesn't produce any useful work. It's the effects that the moving electrons have on the loads they flow through that are important.

    (a) True
    (b) False

3.2 Why Alternating Current is Used

2. Alternating current is primarily used because it can be transmitted inexpensively due to the ability to easily transform ac for high-voltage transmission.

    (a) True
    (b) False

3.3 How Alternating Current is Produced

3. When a magnetic field moves through a coil of wire, the lines of force of the magnetic field cause the electrons in the wire to flow in a specific direction. When the magnetic field moves in the opposite direction, electrons in the wire flow in the opposite direction.

    (a) True
    (b) False

3.4 AC Generator

4. A simple ac generator consists of a loop of wire rotating across the lines of force between the opposite poles of a magnet.

    (a) True
    (b) False

5. In ac generators that produce large quantities of electricity, the conductor coils are stationary and the magnetic field revolves within the coils.

    (a) True
    (b) False

6. Output voltage of a generator depends upon the _____.

    (a) number of turns of wire
    (b) strength of the magnetic field
    (c) speed at which the coil rotates
    (d) all of these

3.5 Waveform

7. A waveform image is used to display the level and direction of current, but not voltage.

    (a) True
    (b) False

8. The waveform for ac circuits displays the level and direction of the current and voltage for every instant of time for one full revolution of the rotor.

    (a) True
    (b) False

3.6 Sine Wave

9. The _____ wave is a waveform that is symmetrical with positive above and negative below the zero reference level.

    (a) nonsinusoidal
    (b) nonsymmetrical
    (c) sine
    (d) any of these

Unit 3 | Practice Questions

3.7 Frequency

10. The number of complete waveforms in one second is called the frequency. Frequency is expressed as _____ or cycles per second.

    (a) degrees
    (b) a sine wave
    (c) phase
    (d) Hertz

3.8 Phase

11. The term "Phase" is used to indicate the time or degree relationship between two waveforms, such as voltage-to-current or voltage-to-voltage.

    (a) True
    (b) False

12. In a purely resistive ac circuit, the current and voltage are _____. This means that they both reach their zero and peak values at the same time.

    (a) in-phase
    (b) out-of-phase
    (c) a or b
    (d) none of these

3.9 Degrees

13. Phase differences are expressed in degrees; one full waveform is equal to _____.

    (a) 90°
    (b) 120°
    (c) 180°
    (d) 360°

14. A three-phase generator has each of its windings out-of-phase with each other by _____.

    (a) 90°
    (b) 120°
    (c) 180°
    (d) 360°

15. Phase differences are expressed in _____.

    (a) sine waves
    (b) phases
    (c) Hertz
    (d) degrees

3.10 Lead or Lag

16. When describing the relationship between voltage and current, the reference waveform is always _____.

    (a) current
    (b) resistance
    (c) voltage
    (d) none of these

17. If the voltage waveform finishes before the current waveform, the voltage is said to _____ the current waveform.

    (a) lead
    (b) lag
    (c) be in-phase with
    (d) none of these

18. When the current waveform finishes before the voltage waveform, the voltage _____ the current waveform.

    (a) leads
    (b) lags
    (c) is in-phase with
    (d) none of these

3.11 Values of Alternating Current

19. _____ is the value of the voltage or current at a moment of time.

    (a) "Peak"
    (b) "Root-mean-square"
    (c) "Effective"
    (d) "Instantaneous"

20. The peak value is equal to the effective value _____.

    (a) times 0.707
    (b) times 1.41
    (c) divided by 1.41
    (d) times 0.58

21. _____ is the maximum value that ac current or voltage reaches, both for positive and negative polarity.

    (a) "Peak"
    (b) "Root-mean-square"
    (c) "Instantaneous"
    (d) none of these

Practice Questions Unit 3

22. The effective value is equal to the peak value _____.

    (a) times 0.707
    (b) times 1.41
    (c) divided by 2
    (d) times 0.58

23. Effective ac voltage or effective ac current is the equivalent value of dc voltage or dc current that will produce the same amount of heat in a resistor.

    (a) True
    (b) False

24. _____ describes the steps necessary to determine the effective voltage or current value.

    (a) "Peak"
    (b) "Root-mean-square"
    (c) "Instantaneous"
    (d) none of these

PART B—CAPACITANCE

Introduction

25. _____ is a property of an electrical circuit that enables it to store electrical energy by means of an electric field and to release this energy at a later time.

    (a) "Capacitance"
    (b) "Induction"
    (c) "Self-induction"
    (d) none of these

26. A half-wave rectifier can be used to convert ac voltage into dc voltage allowing it to continuously charge a capacitor during both alternations of the ac sine wave.

    (a) True
    (b) False

3.12 Charged Capacitor

27. When a capacitor has a potential difference between the plates, it is said to be _____. One plate has an excess of free electrons, and the other plate has a lack of them.

    (a) induced
    (b) charged
    (c) discharged
    (d) shorted

3.13 Electrical Field

28. If a capacitor is overcharged, the electrons from the negative plate could be pulled through the insulation to the positive plate. The capacitor is said to have _____.

    (a) charged
    (b) discharged
    (c) induced
    (d) shorted

3.14 Discharging a Capacitor

29. To discharge a capacitor, all that is required is a(n) _____ path between the terminals of the capacitor. The free electrons on the negative plate then flow through the external circuit to the positive plate.

    (a) conductive
    (b) insulating
    (c) open
    (d) all of these

30. Even when power is removed from the circuit, capacitors can store large amounts of energy for a long period of time, and can discharge and arc if inadvertently shorted or grounded out.

    (a) True
    (b) False

3.15 Determining Capacitance

31. Factors that determine the capacitance of a capacitor are the _____.

    (a) surface area of the plates
    (b) distance between the plates
    (c) dielectric between the plates
    (d) all of these

3.16 Uses of Capacitors

32. Capacitors are used to start single-phase ac motors and to prevent arcing across the contacts of electric switches.

    (a) True
    (b) False

Unit 3 | Practice Questions

3.17 Phase Relationship

33. A capacitor is a device that resists changes in current. Because a capacitor introduces reactance to the circuit, it shifts the current waveform to _____.

    (a) lead the applied voltage by 90°
    (b) lag the applied voltage by 90°
    (c) lead the applied voltage by 180°
    (d) lag the applied voltage by 180°

34. The opposition offered to the flow of ac current by a capacitor is called capacitive reactance, which is expressed in ohms and abbreviated _____.

    (a) X_C
    (b) X_L
    (c) Z
    (d) none of these

PART C—INDUCTION

Introduction

35. When electrons move, the magnetic fields of the individual electrons combine to produce an overall magnetic field. The greater the current flow, the greater the overall magnetic field around the conductor.

    (a) True
    (b) False

36. The movement of electrons caused by an external magnetic field is called _____ current, and the associated potential that is established is called _____ voltage.

    (a) circuit
    (b) applied
    (c) induced
    (d) none of these

37. In order to induce voltage, relative motion is required between a conductor and a _____ field. This is the basis of the generator and transformer.

    (a) voltage
    (b) current
    (c) magnetic
    (d) none of these

3.18 Self-Induction

38. As the ac current through a conductor increases, an expanding and collapsing electromagnetic field through the conductor induces a voltage within the conductor. This is known as _____ voltage.

    (a) applied
    (b) circuit
    (c) self-induced
    (d) none of these

3.19 Induced Voltage and Applied Current

39. The induced voltage in a conductor carrying alternating current opposes the change in current flowing through the conductor. The induced voltage that opposes the current flow is called _____.

    (a) CEMF
    (b) counter-electromotive force
    (c) back-EMF
    (d) all of these

40. The waveform of the CEMF is _____ out-of-phase with the applied voltage.

    (a) 90°
    (b) 120°
    (c) 180°
    (d) 360°

41. When alternating current increases, the polarity of the CEMF within the conductor tries to prevent the current from increasing.

    (a) True
    (b) False

42. When alternating current decreases, the polarity of the induced voltage within the conductor tries to prevent the current from decreasing.

    (a) True
    (b) False

3.20 Conductor AC Resistance

43. In dc circuits, the only property that affects current and voltage flow is _____.

    (a) impedance
    (b) reactance
    (c) resistance
    (d) none of these

44. Conductor resistance is directly proportional to the conductor's length and cross-sectional area.

    (a) True
    (b) False

45. For ac circuits, the _____ of a conductor oppose(s) the current flow in the circuit.

    (a) eddy currents
    (b) skin effect
    (c) resistance
    (d) all of these

46. Eddy currents are small independent currents induced within the conductor because of direct current.

    (a) True
    (b) False

47. The expanding and collapsing magnetic field within the conductor induces a voltage in the conductors (CEMF) that repels the flowing electrons toward the surface of the conductor. This is called _____.

    (a) eddy currents
    (b) induced voltage
    (c) impedance
    (d) skin effect

3.21 Conductor Shape

48. The magnitude of self-induced voltage within a winding is directly proportional to the current flow, the winding, and the frequency at which magnetic fields cut through the winding.

    (a) True
    (b) False

3.22 Magnetic Cores

49. Because an iron core provides an easy path for magnetic flux, windings with soft iron cores produce a greater self-inductance than windings with air cores.

    (a) True
    (b) False

3.23 Self-Induced and Applied Voltage

50. Self-induced voltage is 180° out-of-phase with the _____. When the applied voltage is at its maximum in one direction, the induced voltage is at its maximum in the opposite direction.

    (a) applied current
    (b) applied voltage
    (c) induced voltage
    (d) induced current

3.24 Inductive Reactance

51. Self-induced voltage opposes the change in current flowing in the conductor. This is called inductive reactance and is abbreviated _____.

    (a) X_L
    (b) X_C
    (c) Z
    (d) none of these

3.25 Phase Relationship

52. In a purely inductive circuit, the CEMF waveform is _____ out-of-phase with the applied voltage waveform.

    (a) 90°
    (b) 120°
    (c) 180°
    (d) 360°

53. In an inductive circuit, the CEMF waveform is 90° out-of-phase with the current waveform and 180° out-of-phase with the applied voltage waveform, making the voltage waveform lead the current waveform by _____.

    (a) 90°
    (b) 120°
    (c) 180°
    (d) 360°

3.26 Uses of Induction

54. A common practical use of induction is for _____.

    (a) motors
    (b) transformers
    (c) generators
    (d) all of these

| Unit 3 | **Practice Questions** |

PART D—POWER FACTOR

Introduction

55. Energy is stored in the electromagnetic field of an inductor and the electric field of a capacitor.

    (a) True
    (b) False

3.27 Apparent Power (Volt-Amperes)

56. If you measure voltage and current in an inductive or capacitive circuit and then multiply them together, you obtain the circuit's _____.

    (a) true power
    (b) power factor
    (c) apparent power
    (d) none of these

3.28 True Power (Watts)

57. To determine the true power consumed by a dc circuit, multiply the volts by the amperes.

    (a) True
    (b) False

58. True power of an ac circuit equals volts times amperes.

    (a) True
    (b) False

3.29 Power Factor

59. AC inductive or capacitive reactive loads cause the voltage and current to be in-phase with each other.

    (a) True
    (b) False

60. Power factor is a measurement of how far the current is out-of-phase with the voltage.

    (a) True
    (b) False

3.30 Unity Power Factor

61. When an ac circuit supplies power to a purely _____ load, the circuit voltage and current are in-phase with each other. This condition is called "Unity Power Factor."

    (a) capacitive
    (b) inductive
    (c) resistive
    (d) any of these

3.32 Cost of True Power

62. What does it cost per year (at 9 cents per kWh) for ten 150W recessed luminaires to operate if they are turned on for six hours a day?

    (a) $150
    (b) $300
    (c) $500
    (d) $800

3.33 Effects of Power Factor

63. Because apparent power (VA) is greater than true power (W), more loads can be placed on a circuit, so fewer circuits and panels, and smaller transformers might be required.

    (a) True
    (b) False

64. When sizing circuits or equipment, always size the circuit components and transformers according to the apparent power (VA), not the true power (W).

    (a) True
    (b) False

65. What is the true power of a 10A circuit operating at 120V with unity power factor?

    (a) 1,200 VA
    (b) 2,400 VA
    (c) 1,200W
    (d) 2,400W

66. What size transformer is required for a 100A, 240V, single-phase noncontinuous load with unity power factor?

    (a) 15 kVA
    (b) 25 kVA
    (c) 37.50 kVA
    (d) 50 kVA

67. What size transformer is required for a 100A, 240V, single-phase noncontinuous load that has a power factor of 85 percent?

    (a) 15 kVA
    (b) 25 kVA
    (c) 37.50 kVA
    (d) 50 kVA

68. How many 20A, 120V circuits are required for forty-two, 300W incandescent luminaires (noncontinuous load)?

    (a) 3 circuits
    (b) 4 circuits
    (c) 5 circuits
    (d) 6 circuits

69. How many 20A, 120V circuits are required for forty-two, 300W HID luminaires (assume this is a noncontinuous inductive load) that have a power factor of 85 percent?

    (a) 5 circuits
    (b) 6 circuits
    (c) 7 circuits
    (d) 8 circuits

PART E—EFFICIENCY

Introduction

70. In the transfer of electrical energy, there are power losses in the conductors, the power supply, and the load itself.

    (a) True
    (b) False

71. Efficiency describes how much input energy is used for its intended purpose.

    (a) True
    (b) False

3.34 Efficiency Formulas

72. If the output power is 1,320W and the input power is 1,800W, what is the efficiency of the equipment?

    (a) 62%
    (b) 73%
    (c) 80%
    (d) 100%

73. If the input power of a 1 hp dc motor is 1,128W and the output power is 746W, what is the efficiency of the motor?

    (a) 66%
    (b) 74%
    (c) 87%
    (d) 100%

74. If the output power is 1,600W and the equipment is 88 percent efficient, what are the input amperes at 120V?

    (a) 10A
    (b) 15A
    (c) 20A
    (d) 25A

75. If a transformer is 97 percent efficient, for every 1 kW input, there will be _____ output.

    (a) 970W
    (b) 1,000W
    (c) 1,030W
    (d) 1,200W

Unit 3 — Challenge Questions

UNIT 3—UNDERSTANDING ALTERNATING CURRENT CHALLENGE QUESTIONS

(•Indicates that 75% or fewer of those who took this exam answered the question correctly.)

PART A—UNDERSTANDING ALTERNATING CURRENT

3.2 Why Alternating Current Is Used

1. Alternating current is primarily used because it can be transmitted inexpensively due to the ease of transforming to high-transmission voltage and then transforming this voltage back to low-distribution voltage.

   (a) True
   (b) False

2. One of the advantages of a higher-voltage system as compared to a lower-voltage system (for the same wattage loads) is _____.

   (a) reduced voltage drop
   (b) reduced power use
   (c) large currents
   (d) lower electrical pressure

3. The advantage of ac over dc is that ac provides for _____.

   (a) better speed control
   (b) ease of voltage variation
   (c) lower resistance at high currents
   (d) none of these

3.5 Waveform

4. A waveform represents _____.

   (a) the magnitude and direction of current or voltage
   (b) how current or voltage can vary with time
   (c) how output voltage can vary with the generator armature
   (d) all of these

3.7 Frequency

5. The frequency of an ac waveform is the number of times the current or voltage goes through 360 degrees in _____.

   (a) $1/10$ second
   (b) 1 second
   (c) 5 seconds
   (d) 60 seconds

6. How much time does it take for 60 Hz ac to travel through 180 degrees?

   (a) $1/120$ second
   (b) $1/90$ second
   (c) $1/80$ second
   (d) $1/40$ second

3.11 Values of Alternating Current

7. •The heating effects of 10A of ac as compared to 10A of dc is _____.

   (a) the same
   (b) less
   (c) greater
   (d) none of these

8. If the peak value of an ac system is 50A, the RMS value will be approximately _____.

   (a) 25A
   (b) 30A
   (c) 35A
   (d) 40A

9. •The maximum value of 120V dc is equal to the peak value of an equivalent 120V ac.

   (a) True
   (b) False

Challenge Questions Unit 3

10. •120V is the reading shown on a voltmeter. This is an indication of the _____ value of the voltage source.

    (a) average
    (b) peak
    (c) effective
    (d) instantaneous

PART B—CAPACITANCE

3.15 Determining Capacitance

11. The insulating material between the surface plates of a capacitor is called the _____.

    (a) inhibitor
    (b) electrolyte
    (c) dielectric
    (d) regulator

12. If three capacitors are to be connected together, which connection will provide the highest capacitance?

    (a) Connect all three in series.
    (b) Connect all three in parallel.
    (c) Connect the capacitors in a combination of series and parallel.
    (d) none of these

3.16 Uses of Capacitors

13. •In a circuit that has only capacitive reactance (X_C), the voltage and current are said to be out-of-phase to each other because the voltage _____.

    (a) leads the current by 90°
    (b) lags the current by 90°
    (c) leads the current
    (d) none of these

14. Resonance occurs when _____.

    (a) only resistance occurs in the system
    (b) the power factor is equal to zero
    (c) $X_L = X_C$
    (d) R = Z

3.17 Phase Relationship

15. Capacitive reactance is measured in _____.

    (a) ohms
    (b) volts
    (c) watts
    (d) Henrys

PART C—INDUCTION

3.19 Induced Voltage and Applied Current

16. _____ Law states that a change in current produces a counter-electromotive force whose direction is such that it opposes the change in current.

    (a) Kirchoff's Second
    (b) Kirchoff's First
    (c) Lenz's
    (d) Hertz's

17. Inductive reactance is abbreviated as _____.

    (a) I^2R
    (b) LX
    (c) X_L
    (d) Z

18. Inductive reactance changes proportionately with frequency.

    (a) True
    (b) False

19. Inductive reactance is measured in _____.

    (a) farads
    (b) watts
    (c) ohms
    (d) coulombs

20. •If the frequency is constant, the inductive reactance of a circuit will _____.

    (a) remain constant regardless of the current and voltage changes
    (b) vary directly with the voltage
    (c) vary directly with the current
    (d) not affect the impedance

Unit 3 — Challenge Questions

3.21 Conductor Shape

21. The total opposition to current flow in an ac circuit is expressed in ohms and is called _____.

    (a) impedance
    (b) conductance
    (c) reluctance
    (d) none of these

22. Impedance is present in _____ type circuit(s).

    (a) resistance
    (b) direct-current
    (c) alternating-current
    (d) none of these

23. Conductor resistance to ac flow is _____ the resistance to dc.

    (a) higher than
    (b) lower than
    (c) the same as
    (d) none of these

PART D—POWER FACTOR

3.28 True Power (Watts)

24. If you multiply the voltage by the current in an inductive or capacitive circuit, the answer will be the _____ of the circuit.

    (a) watts
    (b) true power
    (c) apparent power
    (d) all of these

25. The apparent power of a 19.20A, 120V load is _____.

    (a) 0.233 kVA
    (b) 1.23 kVA
    (c) 2.30 kVA
    (d) 230 kVA

3.29 Power Factor

26. •A wattmeter is connected in _____ in the circuit.

    (a) series
    (b) parallel
    (c) series-parallel
    (d) none of these

27. •True power is always voltage times current for _____.

    (a) all ac circuits
    (b) dc circuits
    (c) ac circuits at unity power factor
    (d) b and c

28. Power consumed (watts) in either a single-phase ac or dc system is always equal to _____.

    (a) $I \times R^2$
    (b) $E \times R$
    (c) $I^2 \times R$
    (d) $E/(I \times R)$

29. •The power consumed by a 208V, 76A, three-phase circuit that has a power factor of 89 percent is _____.

    (a) 24,367W
    (b) 27,379W
    (c) 35,808W
    (d) 212,456W

30. •The true power of a 2.10 kVA, single-phase load with a power factor of 91 percent is _____.

    (a) 0.191 kW
    (b) 1.75 kW
    (c) 1.91 kW
    (d) 2.10 kW

3.30 Unity Power Factor

31. Power factor in an ac circuit is unity (100 percent), if the circuit contains only _____.

    (a) induction motors
    (b) transformers
    (c) reactance coils
    (d) resistive loads

32. •Three 8 kW electric-discharge lighting bank circuits, which have a 92 percent power factor, are connected to a 230V, three-phase source. The current flow of these lights is _____.

    (a) 37A
    (b) 50A
    (c) 65A
    (d) 75A

Challenge Questions — Unit 3

3.31 Power Factor Formulas

33. Assuming 100 percent efficiency, what is the power factor for each ballast rated 0.65A at 120V for a fixture containing two 34W lamps?

    (a) 67%
    (b) 77%
    (c) 87%
    (d) 97%

34. What is the apparent power in VA of a fluorescent ballast that has a power factor of 90 percent when connected to two 34W lamps?

    (a) 76 VA
    (b) 86 VA
    (c) 96 VA
    (d) 106 VA

35. What is the true power of a 12A load rated 120V, with a power factor of 80 percent?

    (a) 1,050W
    (b) 1,150W
    (c) 1,250W
    (d) 1,350W

3.32 Cost of True Power

36. The cost of electrical power is based on the true power consumed during a month, multiplied by the cost per kWh (1,000W for a period of one hour).

    (a) True
    (b) False

37. What is the cost of power consumed per month (at $0.10 per kWh) for a 10 ohm ground rod having a ground-fault voltage of 120V?

    (a) $75
    (b) $85
    (c) $95
    (d) $105

3.33 Effects of Power Factor

38. Apparent power (VA) is used for sizing circuits and equipment. Because the VA of the load is greater than the watts of the load, fewer loads can be supplied by each branch circuit.

    (a) True
    (b) False

39. What size transformer is required for fifty, 100W luminaires that have a power factor of 90 percent (noncontinuous load)?

    (a) 2 kVA
    (b) 4 kVA
    (c) 6 kVA
    (d) 8 kVA

40. How many 20A, 120V circuits are required for fifty, 100W luminaires that have a power factor of 90 percent (noncontinuous load)?

    (a) 3 circuits
    (b) 5 circuits
    (c) 7 circuits
    (d) 9 circuits

PART E—EFFICIENCY

3.34 Efficiency Formulas

41. •Motor efficiency can be determined by which of the following formulas?

    (a) hp x 746W
    (b) hp x 746W/VA Input
    (c) hp x 746W/W Input
    (d) Input W/(hp x 746W)

42. The efficiency ratio of a 4,000 VA transformer winding with a secondary VA of 3,600 VA is _____ percent.

    (a) 70
    (b) 80
    (c) 90
    (d) 110

UNIT 4 — Motors and Transformers

INTRODUCTION TO UNIT 4—MOTORS AND TRANSFORMERS

Electric motors are used to convert electrical energy into mechanical energy that can be used to do work and accomplish a productive goal. Motors are an essential component of our industrialized society, so it's important for electricians to understand the basic operational characteristics of motors and how to properly install and connect them.

There are many different types of specialized motors. This unit discusses the most commonly used direct-current (dc) motors and alternating-current (ac) motors. Some of the important things covered in this unit include the differences between full-load current rating, full-load ampere rating, and locked-rotor current.

Generators are used to convert mechanical energy into electrical energy. DC generators, single-phase ac generators, and three-phase ac generators are introduced in this unit. Generators can be found everywhere from very large units at utility generating plants, to emergency generators at hospitals and manufacturing plants, to small residential standby units.

The ability to distribute alternating-current (ac) at high voltages with low line losses and then transform that ac current to lower voltages for local distribution systems and utilization equipment is one of the most attractive features of using ac as opposed to dc for distribution systems. Transformers are also used within premises wiring systems to provide the necessary voltages for utilization equipment and applications such as control circuits.

Transformers operate using the principle of mutual induction, which accomplishes the transfer of energy from one system to a second system with no physical connection between the two. In this unit we'll explain some basic transformer principles and you'll be introduced to three-phase transformer configurations.

PART A—MOTOR BASICS

Introduction

Electric motors are among the most common loads on the electrical system of commercial and industrial occupancies. Think of the sheer number of things that must move in some way. Every pump, fan, compressor, conveyor, or other device that moves material or objects, requires either a motor or an engine powered by steam, air pressure, internal combustion, or some other force.

A motor is a device that converts electrical energy into motion. Computers contain many small motors (in the fans, DVDs, CD drives, hard drives, and other devices), as do automobiles (in which electrically-operated motors have steadily been replacing vacuum-operated and fluid-operated motors). Where there's motion, there's probably an electric motor.

Electricians typically work with various configurations of motors that are powered by 120V, 208V, 240V, or 480V supplies. A solid understanding of the fundamentals of motor operation, motor calculations, motor circuits, and motor controls is essential for success as an electrician. The goal of this unit is to provide you with that basic understanding and prepare you for learning more about this important subject. Motor calculations are covered by Unit 7 of this textbook.

Motors and Transformers — Unit 4

4.1 Motor Principles

A motor must have two opposing magnetic fields in order to rotate.

Stator. The stationary field winding of a motor is mounted on the stator, which is the part of the motor that doesn't turn. For some small permanent magnet motors, there is no field winding because the field is produced by permanent magnets.

Rotor. The rotating part of the motor is referred to as either the "rotor" or the "armature."

4.2 Dual-Voltage AC Motors

Dual-voltage ac motors are made with two field windings, each rated for the lower of two possible operating voltages. The field windings are connected in parallel for low-voltage operation and in series for high-voltage operation.

For example, a 230/460V dual-rated motor has its windings connected in series if supplied by a 460V source, or the windings are connected in parallel if supplied by a 230V source. Figure 4–1

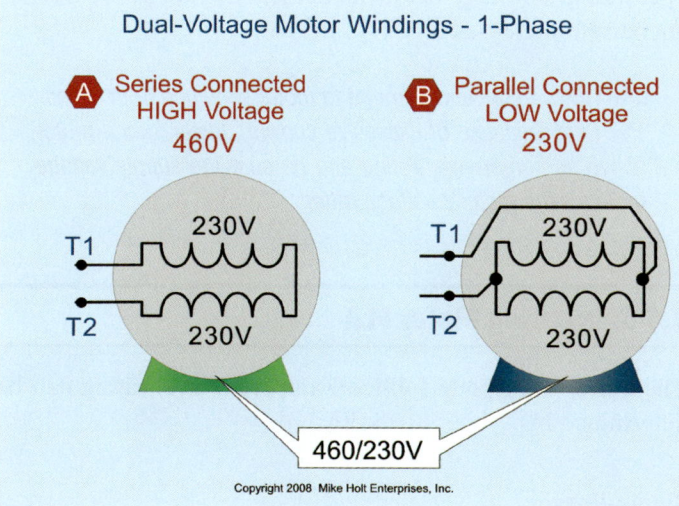

Figure 4–1

Author's Comment: The *NEC* describes voltages as either nominal or rated. "Nominal" voltage refers to the approximate expected utility voltage; for example, a 120/240V system. However, because of circuit voltage drop, the actual voltage at the load is less than 120V or 240V. Therefore, equipment is often rated at a value lower than the nominal system voltage, such as 115V or 230V.

4.3 Motor Horsepower Ratings

The output mechanical work of a motor is measured or rated in horsepower (hp).

Author's Comment: The conversion of mechanical work to electrical energy is measured in watts, where one horsepower = 746W.

CAUTION: *746W is the electrical output value for each horsepower. But this value is not the input electrical rating required to calculate the motor nameplate current rating.*

▶ **Horsepower Example**

Question: *What size motor, in horsepower, produces an output of 15 kW?* **Figure 4–2**

(a) 5 hp (b) 10 hp (c) 20 hp (d) 30 hp

Answer: *(c) 20 hp*

Horsepower = Output Watts/746
Horsepower = 15,000W/746W
Horsepower = 20 hp

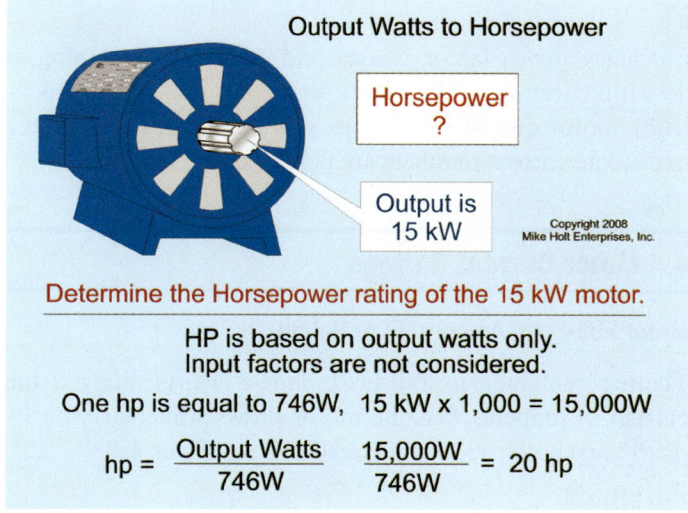

Figure 4–2

Unit 4 — Motors and Transformers

▶ **Output Watts Example**

Question: What is the output watts rating of a 10 hp, ac, 480V, three-phase motor with an efficiency rating of 75 percent and a power factor of 70 percent? Figure 4–3

(a) 5 kW (b) 6 kW (c) 7.50 kW (d) 8 kW

Answer: (c) 7.50 kW

Output Watts = hp x 746W
Output Watts = 10 hp x 746W
Output Watts = 7,460W

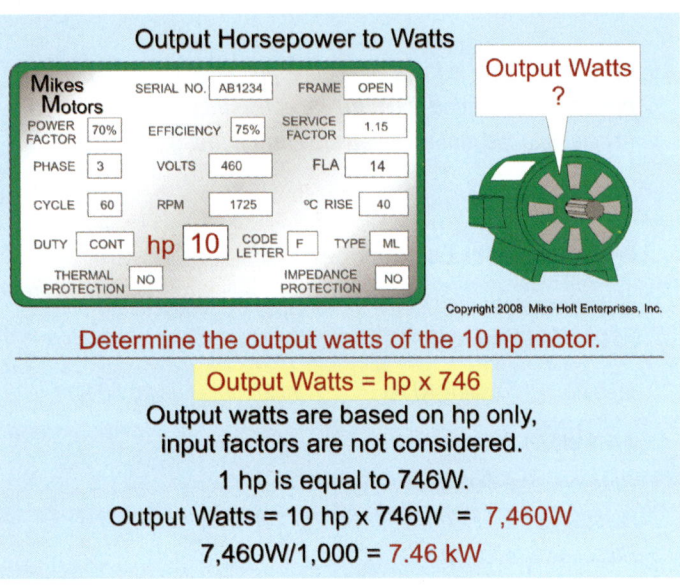

Figure 4–3

4.4 Motor Current Ratings

Motor Full-Load Ampere (FLA) Rating

The motor nameplate full-load ampere (FLA) rating is the current in amperes that the motor draws while carrying its rated horsepower load at its rated voltage. Figure 4–4

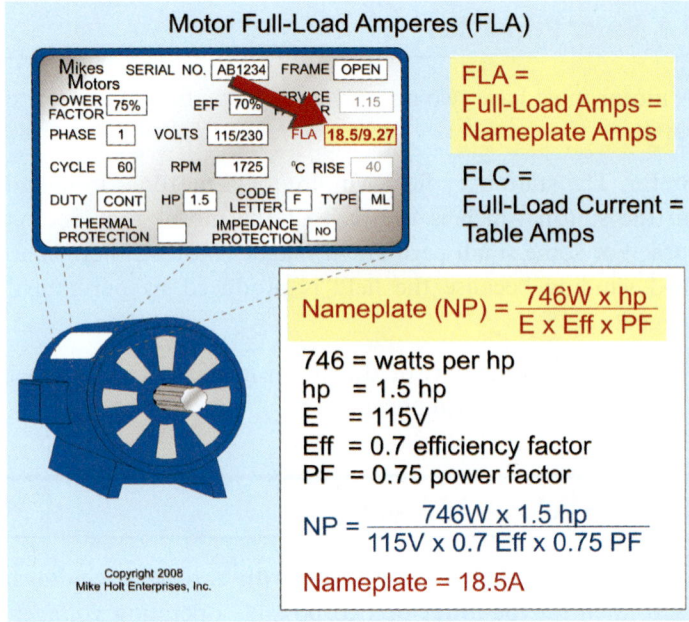

Figure 4–4

Actual Motor Current

The actual current drawn by a motor depends upon the load on the motor and the voltage at the motor terminals. If the load increases, the current also increases or if the motor operates at a voltage below its nameplate rating, the operating current increases.

> **CAUTION:** To prevent damage to motor windings from excessive heat (because of excessive current), never load a motor above its horsepower rating and be sure the supply voltage matches the motor's voltage rating.

4.5 Calculating Motor FLA

The motor nameplate full-load ampere (FLA) rating can be determined by:

Single-Phase

FLA = (Motor hp x 746W)/(E x Eff x PF)

Three-Phase

FLA = (Motor hp x 746W)/(E x 1.732 x Eff x PF)

Efficiency, power factor, phases, and voltage have nothing to do with determining the output watts of a motor! But calculating motor output watts helps you understand how motor nameplate current numbers are developed.

Motors and Transformers — Unit 4

▶ Single-Phase Motor FLA Example

Question: What is the motor nameplate FLA rating for a 7.50 hp motor, 230V, single-phase, having an efficiency rating of 93 percent and a power factor of 87 percent? **Figure 4–5**

(a) 16A   (b) 19A   (c) 24A   (d) 30A

Answer: (d) 30A

Step 1: Determine the motor output watts:

   Output Watts = hp x 746W
   Output Watts = 7.50 hp x 746W
   Output Watts = 5,595W

Step 2: Determine the motor input watts:

   Input = Output/Eff
   Input = 5,595W/0.93 Eff
   Input = 6,016W

Step 3: Determine the motor input VA:

   VA = Watts/PF
   VA = 6,016W/0.87 PF
   VA = 6,915 VA

Step 4: Determine the motor amperes:

   FLA = VA/E
   FLA = 6,915W/230V
   FLA = 30A

OR

   FLA = (hp x 746W)/(E x Eff x PF)
   FLA = (7.50 hp x 746W)/(230V x 0.93 Eff x 0.87 PF)
   FLA = 30A

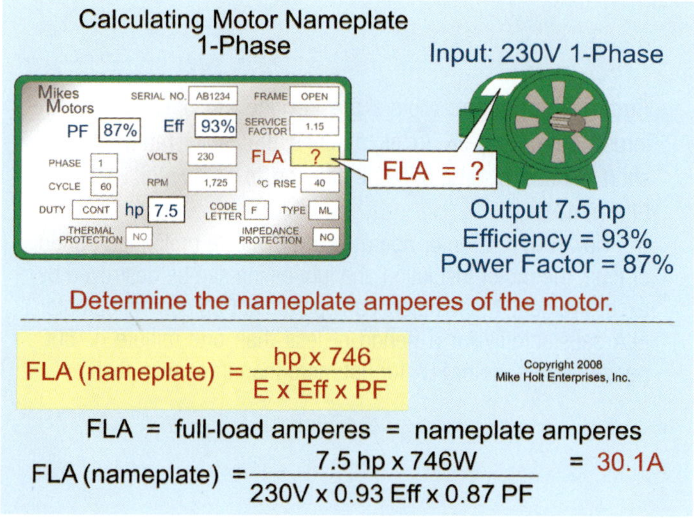

▶ Three-Phase Motor FLA Example

Question: What is the motor FLA rating for a 40 hp, 208V, three-phase motor having an efficiency rating of 80 percent and a power factor of 90 percent?

(a) 76A   (b) 84A   (c) 99A   (d) 115A

Answer: (d) 115A

Step 1: Determine the motor output watts:

   Output Watts = hp x 746W
   Output Watts = 40 hp x 746W
   Output Watts = 29,840W

Step 2: Determine the motor input watts:

   Input Watts = Output/Eff
   Input Watts = 29,840W/0.80 Eff
   Input Watts = 37,300W

Step 3: Determine the motor input VA:

   VA = Watts/PF
   VA = 37,300W/0.90 PF
   VA = 41,444 VA

Step 4: Determine the motor amperes:

   FLA = VA/(E x 1.732)
   FLA = 41,444VA/(208V x 1.732)
   FLA = 115A

OR

   FLA = (hp x 746W)/(E x 1.732 x Eff x PF)
   FLA = (40 hp x 746W)/
         (208V x 1.732 x 0.80 Eff x 0.90 PF)
   FLA = 115A

Figure 4–5

Calculating Motor Nameplate 1-Phase

Input: 230V 1-Phase

Mikes Motors — SERIAL NO. AB1234, FRAME OPEN, PF 87%, Eff 93%, SERVICE FACTOR 1.15, PHASE 1, VOLTS 230, FLA ?, CYCLE 60, RPM 1,725, °C RISE 40, DUTY CONT, hp 7.5, CODE LETTER F, TYPE ML, THERMAL PROTECTION NO, IMPEDANCE PROTECTION NO

FLA = ?
Output 7.5 hp
Efficiency = 93%
Power Factor = 87%

Determine the nameplate amperes of the motor.

FLA (nameplate) = (hp x 746) / (E x Eff x PF)

FLA = full-load amperes = nameplate amperes

FLA (nameplate) = (7.5 hp x 746W) / (230V x 0.93 Eff x 0.87 PF) = 30.1A

Copyright 2008 Mike Holt Enterprises, Inc.

4.6 Motor-Starting Current

When voltage is first applied to the field winding of an induction motor, only the conductor resistance opposes the flow of current through the motor winding. Because the conductor resistance is so low, the motor has a very large inrush current (a minimum of six times the full-load ampere rating). **Figure 4–6A**

4.7 Motor-Running Current

However, once the rotor begins turning, the rotor-bars (winding) are increasingly cut by the stationary magnetic field, resulting in an increasing counter-electromotive force. **Figure 4–6B**

Unit 4 — Motors and Transformers

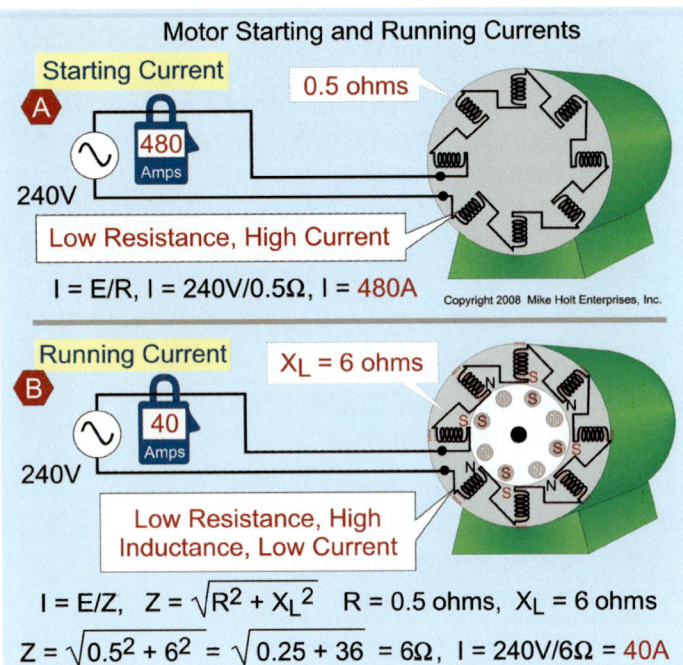

Figure 4–6

We learned previously that CEMF opposes the applied voltage, resulting in an increased opposition to current flow within the conductor. This is called inductive reactance. The increase in inductive reactance, because of self-induction, causes the impedance of the conductor winding to increase which results in reduced current flow.

4.8 Motor Locked-Rotor Current (LRC)

If the rotating part of the motor winding (armature) becomes jammed so that it can't rotate, no more CEMF is produced in the motor winding. This results in decreased conductor impedance to the point that it's effectively a short circuit. Result—the motor operates at locked-rotor current (LRC), which is about six times the running current value, and causes the motor winding to overheat to the point that it will be destroyed if the current is not quickly stopped.

Author's Comment: The *National Electrical Code* requires that most motors be provided with overcurrent protection to prevent damage to the motor winding because of locked-rotor current.

4.9 Motor Overload Protection

Motors must be protected against excessive winding heat. Motors must not be overloaded; they must operate near their nameplate voltage, and measures must be taken to prevent the motor from jamming (LRC). If a motor is overloaded or if it operates at a voltage below its rating, the operating current can increase to a value above the motor full-load amperes (FLA) rating. The excessive operating current may damage the motor winding from excessive heat.

Figure 4–7 shows a type of overload device called a melting-alloy overload. Other types of overload devices that provide motor overload protection include the dashpot type and bimetallic type, as well as solid-state overload relays.

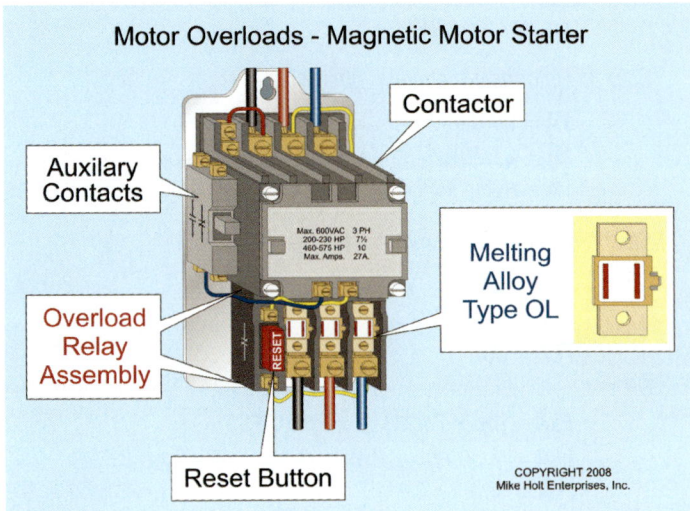

Figure 4–7

Author's Comment: Motors are designed to operate with an inrush current of six to eight times the motor-rated FLA for short periods of time without damage to the motor windings.
Figure 4–8
However, if a motor operates at LRC for a prolonged period of time, the motor insulation and lubrication can be destroyed by excessive heat. Most motors can operate at 600 percent of motor FLA satisfactorily for a period of less than one minute or 300 percent of the motor FLA for not more than three minutes.

Motors and Transformers   Unit 4

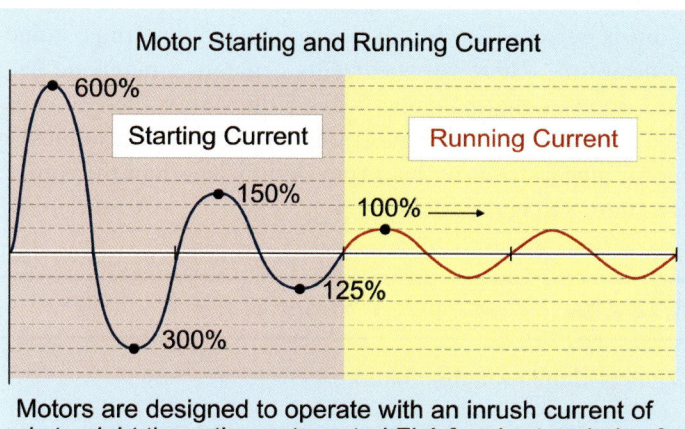

Figure 4–8

NEC Requirement

To protect against excessive heat from an overload, the *National Electrical Code* requires standard overload protection devices to be sized at 115 to 125 percent of the motor FLA rating [430.32(A)(1)]. **Figure 4–9**

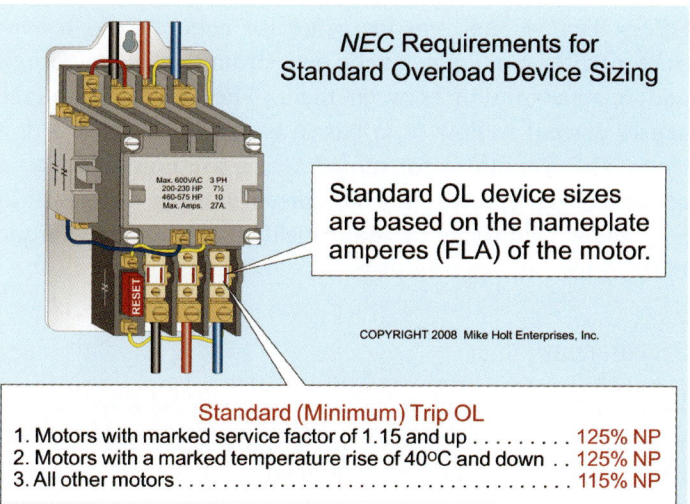

Figure 4–9

4.10 Direct-Current Motor Principles

DC motors use brushes (pieces of conductive carbon) with springs to hold them in contact with a commutator. A commutator is a ring around the rotor shaft that has segments with insulation between them.

Commutators are necessary for dc motors because the maximum torque on the rotor develops when the active rotor winding opposes the magnetic field of the stator. As the winding passes this point, less torque is developed. As the commutator turns, the carbon brushes come into contact with the next set of rotor windings, which are in the proper position to oppose the stator field and keep the rotor moving. As the rotor turns, the brush-commutator set always supplies power to the rotor windings that are in the correct position to develop maximum torque.

4.11 Direct-Current Motor Types

Shunt-Wound DC Motor

One of the great advantages of the shunt-wound dc motor (whose armature and field are in parallel) is its ability to maintain a constant speed under load. If the speed of a dc motor begins to increase, the armature cuts through the electromagnetic field at an increasing rate. This results in a greater armature CEMF, which acts to cut down on the increased armature current, resulting in the motor slowing back down.

Placing a load on a dc motor causes the motor to slow down, which reduces the rate at which the armature is cut by the field flux lines. As a result, the armature CEMF decreases, resulting in an increase in the applied armature voltage and current. The increase in current results in an increase in motor speed. This gives shunt-wound dc motors a built-in system for regulating their own speed. For these reasons, computer disk drives and recording equipment use dc motors exclusively.

Series-Wound DC Motor

Series-wound dc motors (those in which the field winding and the armature winding are connected in series) have poor speed regulation, and slow down considerably when a load is applied. When unloaded, they often run at a very high rpm and in some cases they tend to "run away" and increase speed to the point of damaging the motor. However, series dc motors have very good torque characteristics. One of the best examples of a series-wound dc motor is the starter motor on your automobile, which has very high torque to start the engine and is connected to a load so that it doesn't overspeed.

4.12 Reversing the Rotation of a DC Motor

To reverse the rotation of a dc motor, you must reverse either the stator field or the armature magnetic field. This is accomplished by reversing either the field or armature current. Because most dc motors have the field and armature windings fed from the same dc power supply, reversing the polarity of the power supply simultaneously changes the field and armature. This results in the motor continuing to run in the same direction.

The polarity of either the field or armature must be changed by disconnecting and interchanging the polarity of the leads of the field winding (while leaving the armature polarity unchanged), or interchanging the polarity of the armature (while leaving the field polarity unchanged). **Figure 4–10**

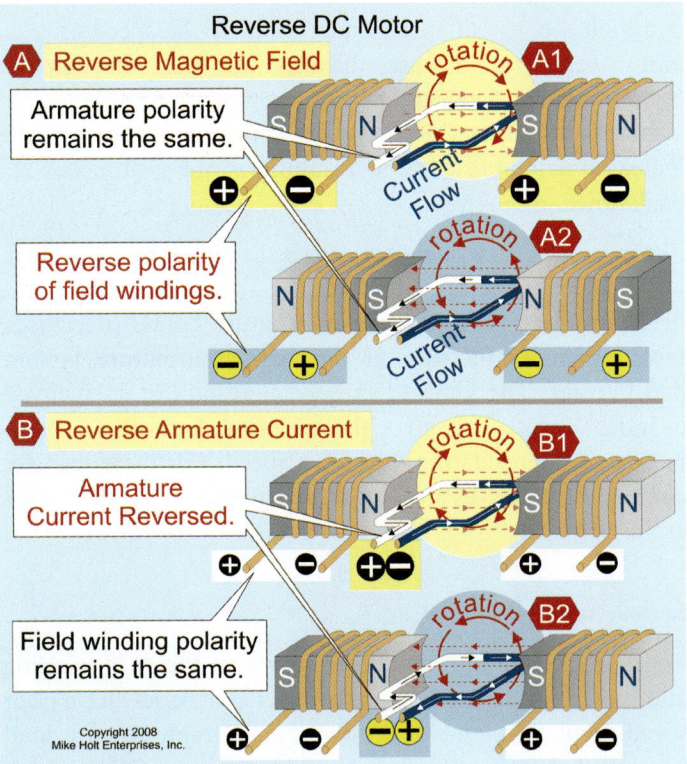

Figure 4–10

4.13 AC Induction Motor

In an ac induction motor, the stator produces a rotating magnetic field.

The rotor is a series of coils (windings) that are connected in a closed loop. The rotating magnetic field of the stator induces currents flowing in the rotor windings (thus, induction motor). These rotor currents generate a magnetic field that opposes the magnetic field of the stator, thereby causing the rotor to turn. This is why the rotor of an induction motor always turns with an rpm slightly less than the rpm of the stator's magnetic field.

4.14 Alternating-Current Motor Types

Squirrel-Cage Induction Motor

The most common type of ac motor is the induction motor, which has no physical connection between its rotating member (the rotor), and the stationary member surrounding it (the stator).

Three-phase ac squirrel-cage induction motors are used in almost all major industrial applications. They're called squirrel-cage motors because the rotor consists of bars that are either parallel to the shaft or at a slight angle and are connected together at the ends by shorting rings. These bars would resemble a hamster or squirrel cage if you were to remove the core material around them.

As the magnetic field from the stator of an induction motor rotates, the field induces voltage in the rotor bars as it cuts across them. Conditions for generator action exist—a conductor (rotor bars), a magnetic field (from the field winding), and relative motion between them. The generated voltage causes current to flow from bar to bar through the shorting rings. The conditions for motor action are now present—a conductor (each rotor bar) has current flowing through it in a magnetic field (from the stator field). This produces torque in the rotor causing it to turn.

Synchronous Motor

In a synchronous motor, the rotor is actually locked in step with the rotating stator field and is dragged along at the synchronous speed of the rotating magnetic field. Synchronous motors maintain their speed with a high degree of accuracy. Small synchronous motors are used for clock motors.

Large synchronous motors are often found in large industrial facilities driving loads such as compressors, crushers, and large pumps. Sometimes they're operated unloaded, meaning that nothing is attached to their shafts. In this application, the synchronous motor is being "overexcited," meaning that a large amount of dc current is being fed into the rotor through the slip rings. The synchronous motor acts as though it were a large capacitor when it's in this condition and can be used for power factor correction.

Wound-Rotor Motor

Wound-rotor induction motors are used only in special applications because of their complexity. Wound-rotor induction motors only operate on three-phase ac power. They're similar to an induction motor; however, the rotor windings are connected in a wye configuration and the points of the wye are brought out through slip rings to an external controller.

Resistors are usually inserted into the rotor winding circuit during start-up in order to reduce the inrush current. As the motor increases speed, the value of the resistance is changed to lower values. Ultimately, the rotor windings are shorted, thereby allowing the motor to achieve full speed.

The wound-rotor motor can also be used in applications that require some speed control. Today, however, it's more common to use a Variable Speed Drive coupled to an induction motor.

Universal Motor

Universal motors are fractional horsepower motors that operate equally well on ac and dc. They're used for vacuum cleaners, electric drills, mixers, and light household appliances. These motors have the inherent disadvantage associated with dc motors, which is the need for commutation. The problem with commutators is that as the motor operates, motor parts rub against each other and wear out.

4.15 Reversing the Rotation of an AC Motor

Some ac single-phase motors are constructed so that they can be reversed easily and some are not. In order to reverse a single-phase ac induction motor, it's necessary to change the relative polarity of the start winding in relation to the run winding. The motor nameplate will contain the wiring connection information for forward and reverse operation.

A three-phase motor can be reversed by swapping any two of the three powerline conductors. Industry practice is to reverse Line 1 and Line 3. This can be done by reconnecting the motor, or a special reversing starter can be used to switch these connections for a motor that will be run in both directions during normal operation. **Figure 4–11**

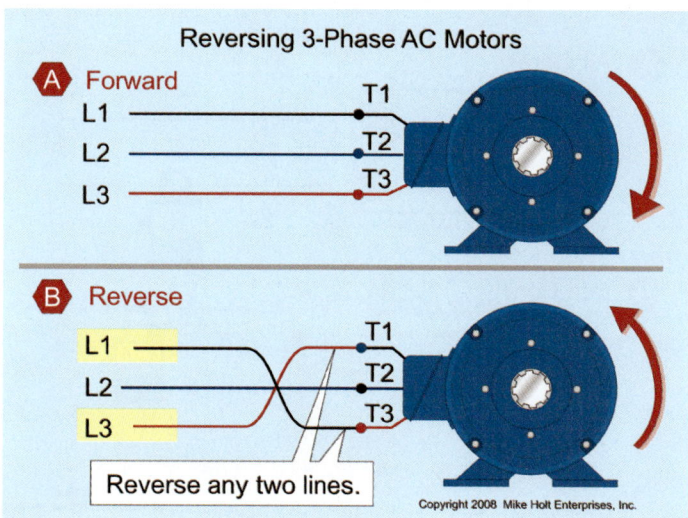

Figure 4–11

PART B—TRANSFORMERS

Introduction

A transformer is a device used to transfer electrical energy (power) from one system to another by induction with no physical connection between the two systems, except in the case of autotransformers. Other than isolation transformers, transformers are used to raise or lower voltages.

This unit will cover transformer basics. For more information on three-phase transformers, transformer calculations, and *Code* requirements for transformers, refer to Unit 12 in this textbook.

4.16 Transformer Basics

Primary versus Secondary

The transformer winding connected to the voltage source is called the primary. The transformer winding connected to the load is called the secondary. Transformers can be used to either raise or lower voltage.

Mutual Induction

The energy transfer of a transformer is accomplished because the electromagnetic lines of force from the primary winding induce a voltage in the secondary winding. This process is called mutual induction. **Figure 4–12**

Unit 4 — Motors and Transformers

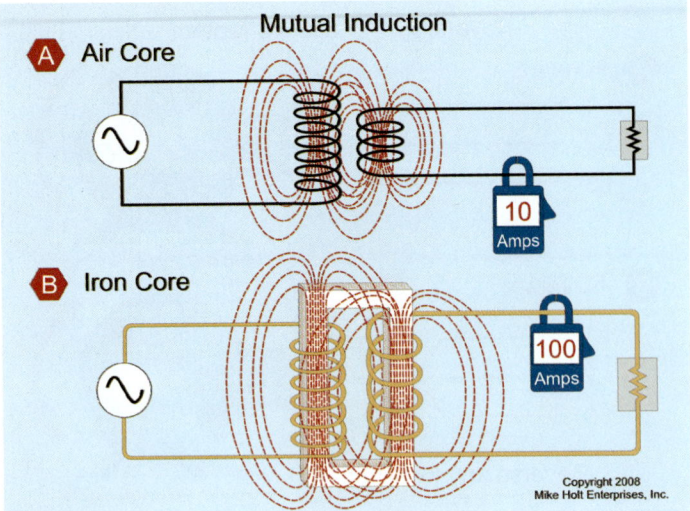

Figure 4–12

The voltage level that can be induced in the secondary winding, from the primary magnetic field, is a function of the number of secondary conductor loops (turns) that are cut by the primary electromagnetic field.

4.17 Secondary Induced Voltage

Voltage induced in the secondary winding of a transformer is equal to the sum of the voltages induced in each loop of the secondary winding. For example, if a transformer has three windings on the secondary and each secondary winding has 80V induced, the secondary voltage is 240V. **Figure 4–13**

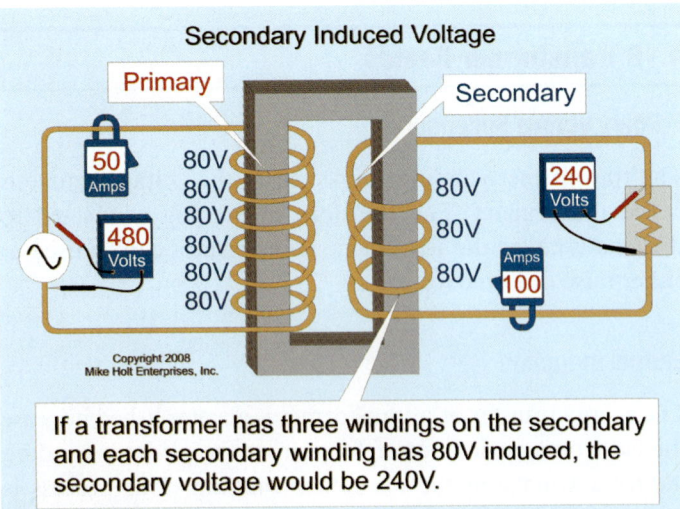

Figure 4–13

The induced voltage in each loop of the secondary winding depends on the number of secondary conductor loops (turns), as compared to the number of primary turns. **Figure 4–14**

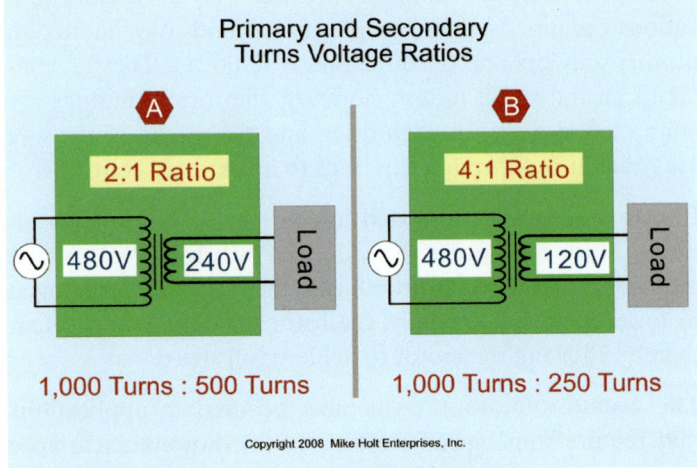

Figure 4–14

If both windings have the same number of turns, and if all of the magnetism set up by the primary passes through the secondary, the secondary will deliver the same voltage and power as the primary. This is how an isolation transformer works. **Figure 4–15**

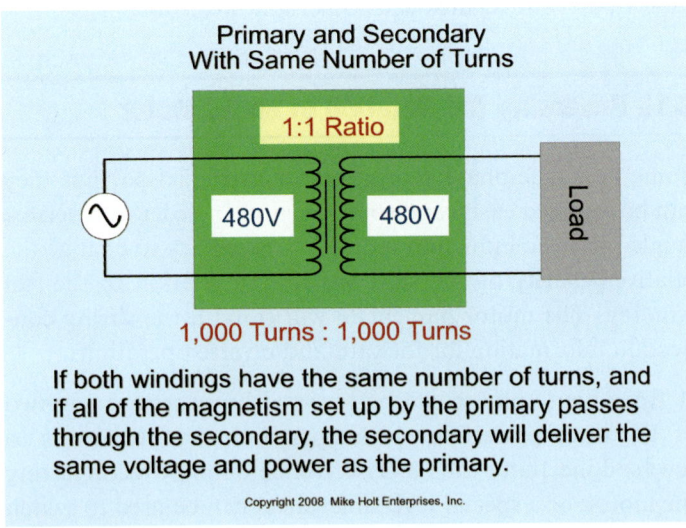

Figure 4–15

Motors and Transformers — Unit 4

4.18 Efficiency

Because of conductor resistance, flux leakage, eddy currents, and hysteresis losses, not all of the input power is transferred to the secondary winding of a transformer for useful purposes. Transformer losses are typically quite small with efficiency approaching 95 to 97 percent. Thus, for most practical purposes, transformer efficiency can be ignored. **Figure 4–16**

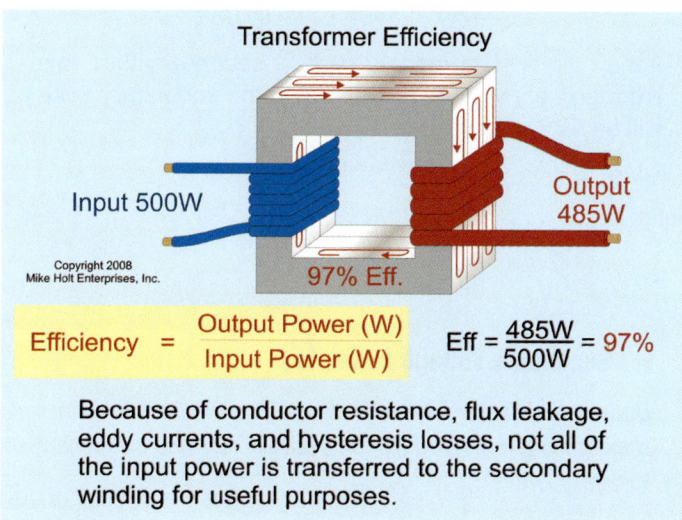

Figure 4–16

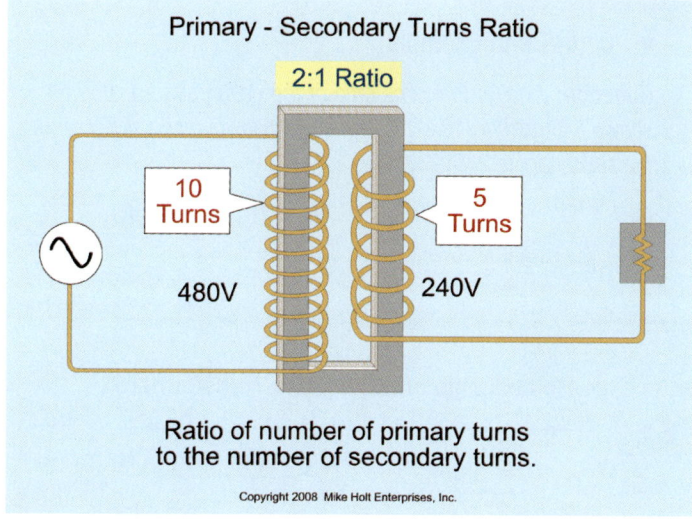

Figure 4–17

4.19 Transformer Turns Ratio

The relationship of the primary winding voltage to the secondary winding voltage is the same as the relationship between the number of turns of wire on the primary and the number of turns on the secondary. This relationship is called "turns ratio." **Figure 4–17**

▶ **Delta/Wye Example**

Question: What is the turns ratio of the delta/wye transformer shown in **Figure 4–18** if the primary phase voltage is 480V and the secondary is 120V?

(a) 1:4   (b) 1:2   (c) 2:1   (d) 4:1

Answer: (d) 4:1

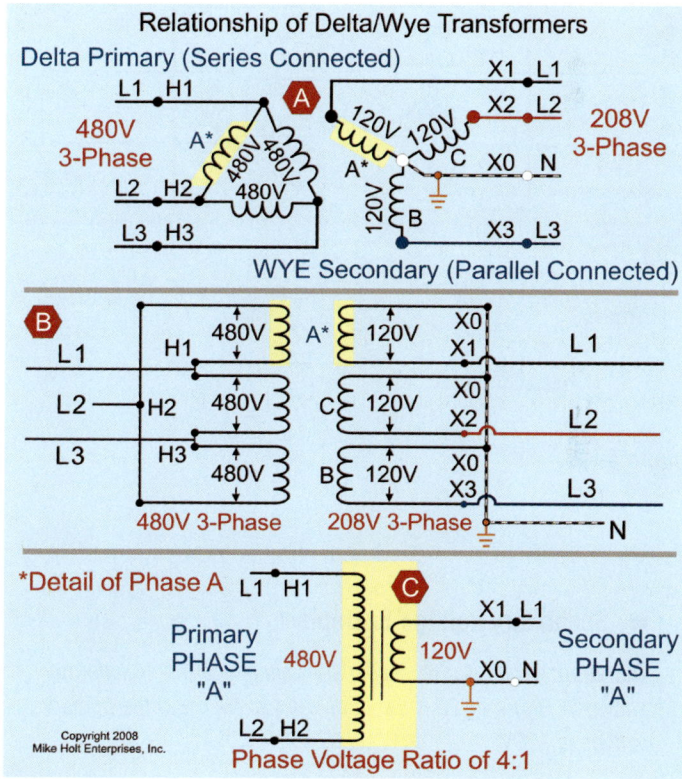

Figure 4–18

Unit 4 — Motors and Transformers

▶ **Delta/Delta Example**

Question: What is the turns ratio of the delta/delta transformer shown in **Figure 4–19** if the primary phase voltage is 480V and the secondary is 240V?

(a) 1:4 (b) 1:2 (c) 2:1 (d) 4:1

Answer: (c) 2:1

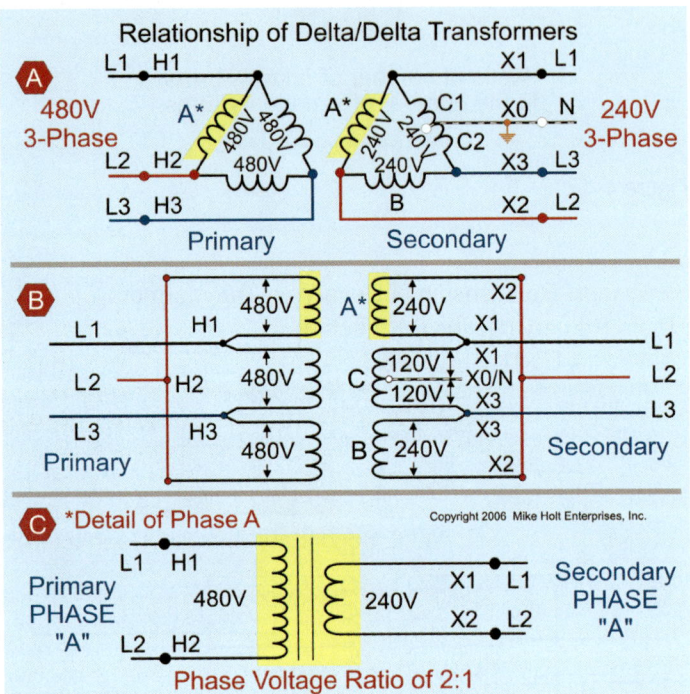

Figure 4–19

▶ **Secondary Voltage Example 1**

Question: What is the secondary voltage of the transformer shown in Figure 4–20 if the turns ratio is 10:1 and the primary voltage is 120V?

(a) 6V (b) 12V (c) 24V (d) 48V

Answer: (b) 12V

The turns ratio is 10:1. This means the voltage ratio is also 10:1. Since the primary voltage is 120V, the secondary voltage is 10 times smaller, 120V/10 = 12V.

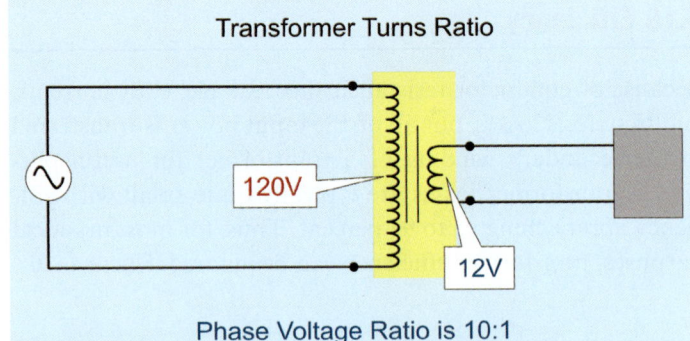

Figure 4–20

▶ **Secondary Voltage Example 2**

Question: What is the secondary voltage of the transformer shown in **Figure 4–21** if the turns ratio is 10:1 and the primary voltage is 240V?

(a) 6V (b) 24V (c) 48V (d) 120V

Answer: (b) 24V

Since the primary voltage is 240V, the secondary voltage is 10 times smaller, 240V/10 = 24V.

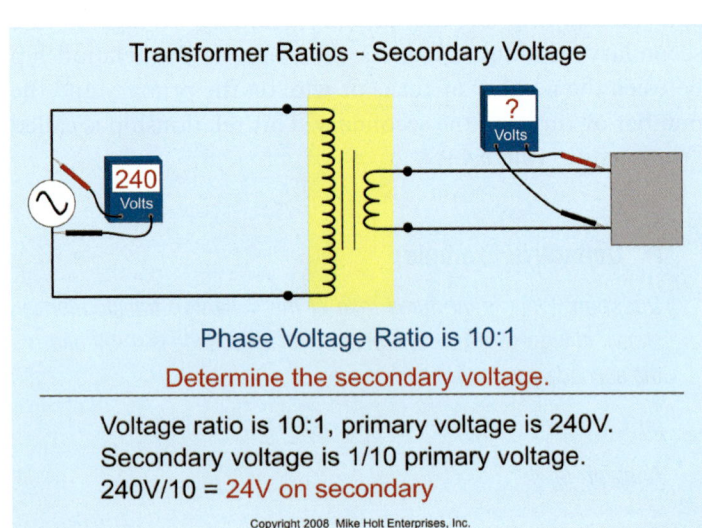

Figure 4–21

Mike Holt's Illustrated Guide to NEC Exam Preparation

Motors and Transformers — Unit 4

4.20 Autotransformers

Autotransformers use a single winding for both the primary and secondary.

Autotransformers are often used to step the voltage up from 208V to 240V, or down from 240V to 208V. **Figure 4–22**

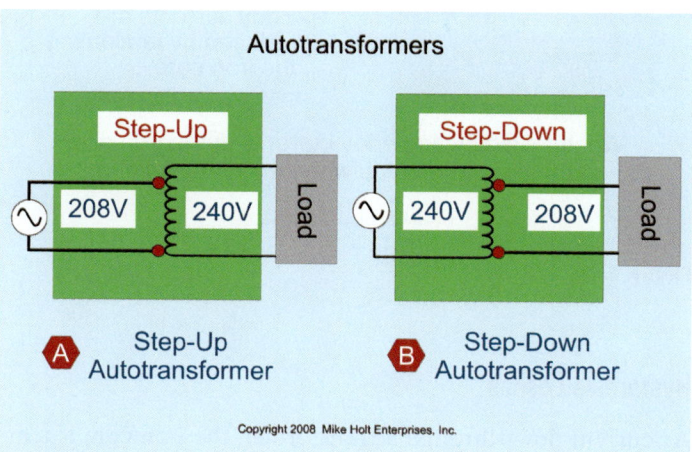

Figure 4–22

Step-Down Transformer

The secondary winding of a step-down transformer has fewer turns than the primary winding, resulting in a lower secondary voltage as compared to the primary.

Step-Up Transformer

The secondary winding of a step-up transformer has more turns than the primary winding, resulting in a higher secondary voltage as compared to the primary.

Lack of Isolation

The disadvantage of an autotransformer is the lack of electrical isolation between the primary and secondary conductors. But they're often used because they're inexpensive.

4.21 Power Losses

In an ideal transformer, all the primary power is transferred from the primary winding to the secondary winding. But real-world transformers have power losses because of conductor resistance, flux leakage, eddy currents, and hysteresis losses.

Conductor Resistance Loss

Transformer primary and secondary windings are generally made of many turns of copper. Conductor resistance is directly proportional to the length of a conductor and inversely proportional to the cross-sectional area of a conductor.

▶ **Primary Power Loss Example**

Question: What is the primary conductor power loss of a 75 kVA transformer if the primary current rating is 90A and the winding has a resistance of 0.08 ohms? **Figure 4–23A**

(a) 400W (b) 648W (c) 1,100W (d) 1,300W

Answer: (b) 648W

Power = $I^2 \times R$

$I = 90A$

$R = 0.08$ ohms

$P = 90A^2 \times 0.08$ ohms

$P = 648W$

Figure 4–23

Unit 4 | Motors and Transformers

> ▶ **Secondary Power Loss Example**
>
> **Question:** What is the secondary conductor power loss of a 75 kVA transformer if the secondary current rating is 208A and the winding has a resistance of 0.0313 ohms? **Figure 4–23B**
>
> (a) 1,354W   (b) 1,800W   (c) 2,100W   (d) 2,700W
>
> **Answer:** (a) 1,354W
>
> Power = $I^2 \times R$
>
> I = 208A
> R = 0.0313 ohms
> P = $208A^2 \times 0.0313$ ohms
> P = 1,354W

Flux Leakage Loss

The leakage of the electromagnetic flux lines between the primary and secondary windings of transformers also represents wasted energy. **Figure 4–24**

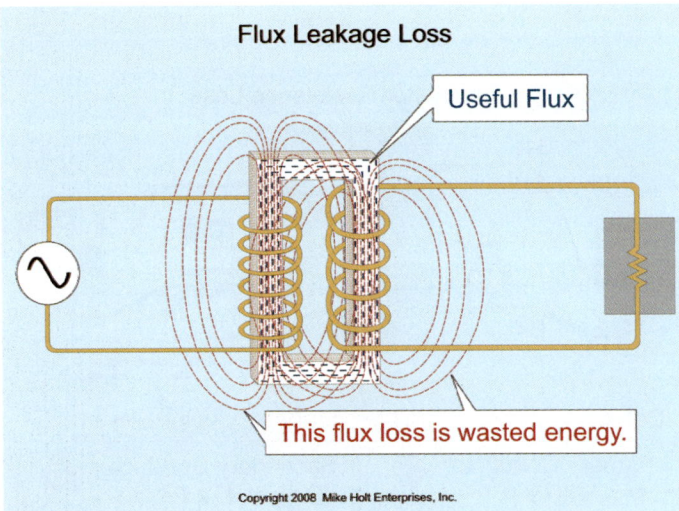

Figure 4–24

Eddy Currents

Iron is the only metal used for transformer cores because it is easy to magnetize. However, the expanding and collapsing electromagnetic field from alternating current induces a voltage in the iron core. These circulating eddy currents cause the core to heat up without any useful purpose. To reduce losses because of eddy currents, long laminated iron cores, separated by insulation (usually lacquer) are used. **Figure 4–25**

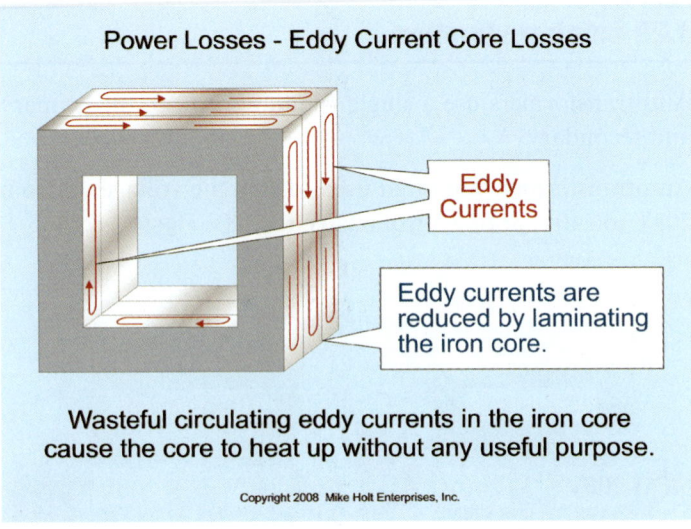

Figure 4–25

Hysteresis Losses

As current flows through a transformer, the iron core is temporarily magnetized by the electromagnetic field created by the alternating current. Each time the primary magnetic field expands and collapses, the core molecules realign themselves to the changing polarity of the electromagnetic field. The energy required to realign the core molecules to the changing electromagnetic field is called the hysteresis loss of the core. **Figure 4–26**

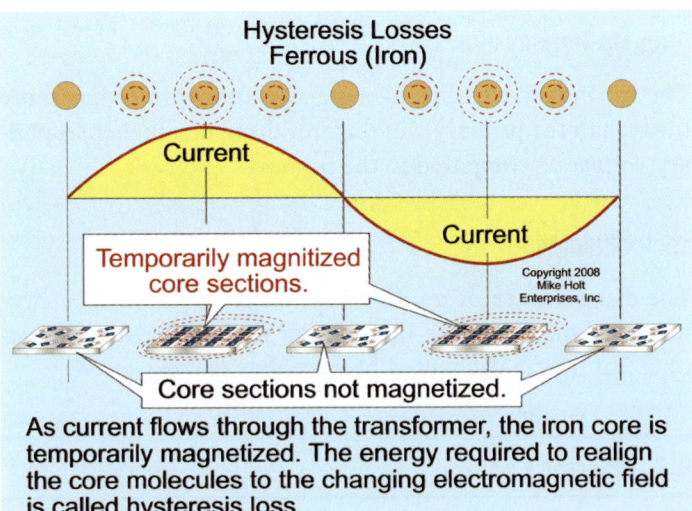

Figure 4–26

Hysteresis losses are directly proportional to the alternating-current frequency. The higher the frequency, the more times per second the molecules must realign. Hysteresis loss is one of the main reasons why solid iron-core transformers are not used in applications involving high frequencies and nonlinear loads.

4.22 Transformer kVA Rating

Transformers are rated in kilovolt-amperes (kVA), where 1 kilovolt-ampere = 1,000 volt-amperes (VA). **Figure 4–27** lists common standard transformer sizes used in commercial and industrial applications. However, there are also much larger transformers.

Standard Transformer Ratings - In kVA	
Single-Phase	Three-Phase
1.0	3.0
1.5	6.0
2.0	9.0
3.0	15.0
5.0	22.5
7.5	25.0
10.0	30.0
15.0	37.5
25.0	45.0
37.5	50.0
50.0	75.0
75.0	112.5
100.0	150.0
125.0	225.0

Copyright 2008 Mike Holt Enterprises, Inc.

Figure 4–27

4.23 Current Flow

The following steps explain the process of primary and secondary current flow in a transformer.

Step 1: When a load is connected to the secondary of a transformer, secondary voltage induced from the primary magnetic field causes current to flow through the secondary conductor winding.

Step 2: The secondary current flow in the secondary winding creates an electromagnetic field that opposes the primary electromagnetic field.

Step 3: The flux lines from the secondary magnetic field effectively reduce the strength of the primary flux lines, and as a result, less CEMF is generated in the primary winding conductors. With less CEMF to oppose the primary applied voltage, the primary current increases in direct proportion to the secondary current. **Figure 4–28**

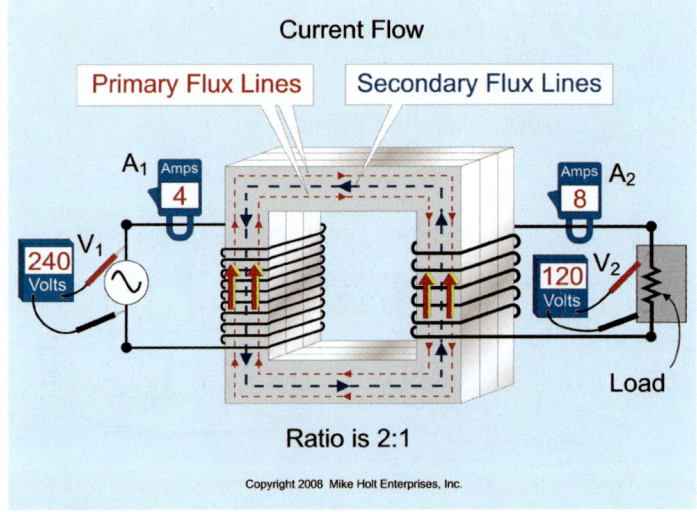

Figure 4–28

4.24 Current Rating

A transformer's primary and secondary line current can be determined by the formulas:

Single-Phase
I = VA/E

Three-Phase
I = VA/(E x 1.732)

Unit 4 — Motors and Transformers

▶ **Single-Phase Example**

Question: What is the maximum primary and secondary line current at full load for a 480/240V, 25 kVA, single-phase transformer? **Figure 4–29**

(a) 52/104A (b) 104/52A
(c) 104/208A (d) 208/104A

Answer: (a) 52/104A

$I_{primary} = VA/E$
$I_{primary} = 25{,}000\ VA/480V$
$I_{primary} = 52A$

$I_{secondary} = VA/E$
$I_{secondary} = 25{,}000\ VA/240V$
$I_{secondary} = 104A$

▶ **Three-Phase Example**

Question: What is the maximum primary and secondary line current at full load for a 480/208V, 37.50 kVA, three-phase transformer? **Figure 4–30**

(a) 45/104A (b) 104/40A
(c) 140/120A (d) 208/140A

Answer: (a) 45/104A

$I_{primary} = VA/(E \times 1.732)$
$I_{primary} = 37{,}500\ VA/(480V \times 1.732)$
$I_{primary} = 45A$

$I_{secondary} = VA/(E \times 1.732)$
$I_{secondary} = 37{,}500\ VA/(208V \times 1.732)$
$I_{secondary} = 104A$

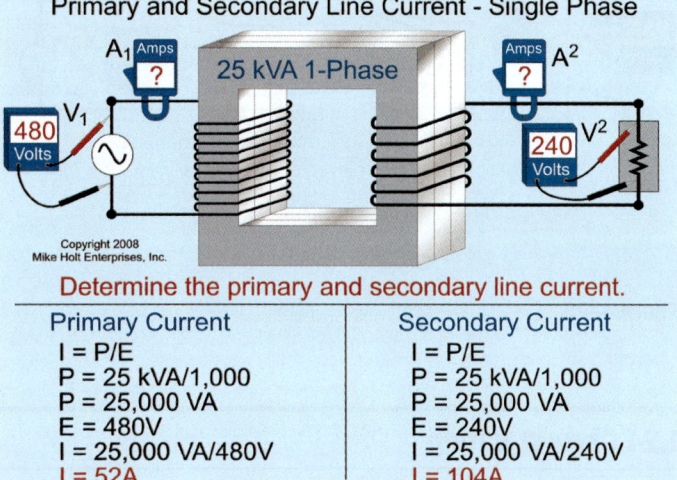

Figure 4–29

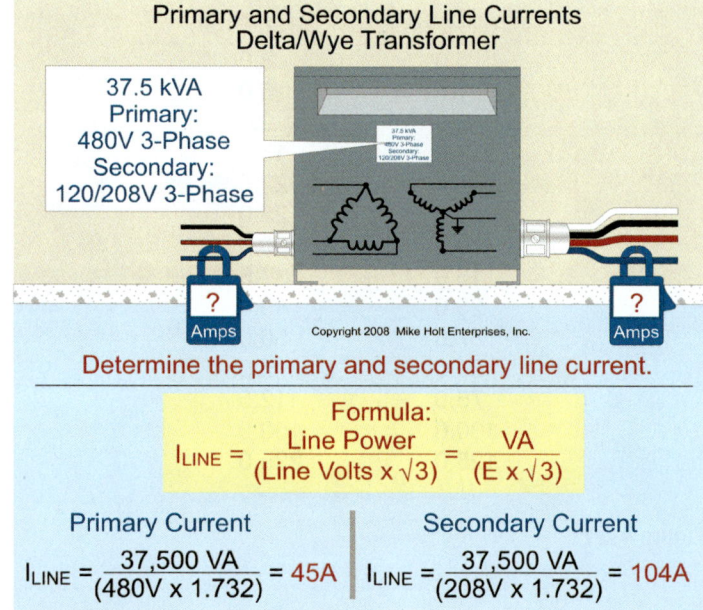

Figure 4–30

Unit 4 — Conclusion

CONCLUSION TO UNIT 4—MOTORS AND TRANSFORMERS

Article 430 of the *National Electrical Code* addresses motors, motor circuits, and motor controllers. It's one of the largest of all the *NEC* articles. Now that you have studied basic motor theory, motor calculations, motor circuits, and motor controls, you can see why this is so. Although you have learned much here, you will benefit from learning even more, especially if you're going to do commercial or industrial work.

Unit 7 of this textbook provides a study of motor calculations, and Unit 12 provides more on transformer calculations and the *Code* requirements for transformer installations and grounding. Be sure to continue your study with these important topics.

Unit 4 Practice Questions

UNIT 4—MOTORS AND TRANSFORMERS PRACTICE QUESTIONS

PART A—MOTOR BASICS

4.1 Motor Principles

1. A motor must have two opposing magnetic fields in order to rotate. The stationary field winding is mounted on the stator, and the rotating part is referred to as the armature.

    (a) True
    (b) False

4.2 Dual-Voltage AC Motors

2. Dual-voltage ac motors are made with two field windings. The field windings are connected in _____ for low-voltage operation and in _____ for high-voltage operation.

    (a) series, parallel
    (b) parallel, series
    (c) series, series
    (d) parallel, parallel

3. For a dual-voltage 230/460V motor, the field windings are connected in parallel for _____ operation and in series for _____ operation.

    (a) 230V, 460V
    (b) 460V, 230V
    (c) 230V, 230V
    (d) 460V, 460V

4.3 Motor Horsepower Ratings

4. Motors are used to convert electrical energy into mechanical work and the output mechanical work of a motor is rated in horsepower. 1 hp = _____.

    (a) 476W
    (b) 674W
    (c) 746W
    (d) 840W

5. What size motor, in horsepower, is required to produce approximately 30 kW of output?

    (a) 20 hp
    (b) 30 hp
    (c) 40 hp
    (d) 50 hp

6. What is the approximate output of a 15 hp motor?

    (a) 11 kW
    (b) 15 kW
    (c) 22 kW
    (d) 31 kW

7. What is the approximate output of a 5 hp, 480V, three-phase motor?

    (a) 3.75 kW
    (b) 4.75 kW
    (c) 6.75 kW
    (d) 7.75 kW

4.4 Motor Current Ratings

8. The nameplate motor FLA rating describes the motor current rating when it carries its rated horsepower load at its rated _____.

    (a) power
    (b) resistance
    (c) CEMF
    (d) voltage

9. The actual motor current depends upon the load on the motor and the operating voltage at the motor terminals.

    (a) True
    (b) False

10. The motor FLA rating is used when sizing motor conductors or circuit protection.

    (a) True
    (b) False

Practice Questions — Unit 4

4.5 Calculating Motor FLA

11. What is the nameplate FLA for a 5 hp, 230V, single-phase motor with 93 percent power factor and an 87 percent efficiency rating?

    (a) 10A
    (b) 20A
    (c) 28A
    (d) 35A

12. What is the nameplate FLA for a 20 hp, 208V, three-phase motor with 90 percent power factor and an 80 percent efficiency rating?

    (a) 51A
    (b) 58A
    (c) 65A
    (d) 80A

4.6 Motor-Starting Current

13. When a motor starts, the current drawn is at least _____ times the motor FLA; this is known as motor locked-rotor amperes (LRA).

    (a) 0.80
    (b) 1.25
    (c) 3
    (d) 6

4.7 Motor-Running Current

14. Once a motor begins turning, the rotor windings are increasingly cut by the stationary magnetic field, resulting in an increasing counter-electromotive force.

    (a) True
    (b) False

4.8 Motor Locked-Rotor Current (LRC)

15. If the rotating part of the motor winding is jammed so that it can't rotate, no CEMF is produced in the motor winding. As a result, the motor operates at _____ and the windings will be destroyed by excessive heat.

    (a) FLA
    (b) FLC
    (c) LRC
    (d) any of these

4.9 Motor Overload Protection

16. Motors must be protected against excessive winding heat by a properly sized overload protection device, based on the motor _____ current rating.

    (a) FLA
    (b) FLC
    (c) LRC
    (d) any of these

4.10 Direct-Current Motor Principles

17. DC motors use brushes with springs to hold them in contact with a(n) _____.

    (a) rotor
    (b) stator
    (c) armature
    (d) commutator

4.11 Direct-Current Motor Types

18. If the speed of a shunt-wound dc motor begins to increase, the armature cuts through the electromagnetic field at an increasing rate. This results in a greater armature CEMF, which acts to cut down on the increased armature current, resulting in the motor speeding up.

    (a) True
    (b) False

19. Placing a load on a shunt-wound dc motor causes the motor to slow down, which increases the rate at which the field flux lines are cut by the armature. As a result, the armature CEMF increases resulting in an increase in the applied armature voltage and current. The increase in current results in increased motor speed.

    (a) True
    (b) False

4.12 Reversing the Rotation of a DC Motor

20. To reverse the rotation of a dc motor, you must reverse the _____.

    (a) stator field
    (b) rotor field
    (c) a and b
    (d) a or b

4.13 AC Induction Motor

21. In an ac induction motor, the stator produces a rotating magnetic field that induces current in the rotor windings. The rotor current generates a magnetic field in opposition to the magnetic field of the stator, thereby causing the rotor to turn.

    (a) True
    (b) False

4.14 Alternating-Current Motor Types

22. In a(n) _____ motor, the rotor is actually locked in step with the rotating stator field and is dragged along at the speed of the rotating magnetic field.

    (a) wound-rotor
    (b) induction
    (c) synchronous
    (d) squirrel-cage

23. _____ motors are fractional horsepower motors that operate equally well on ac and dc and are used for vacuum cleaners, electric drills, mixers, and light household appliances.

    (a) AC
    (b) Universal
    (c) Wound-rotor
    (d) Synchronous

4.15 Reversing the Rotation of an AC Motor

24. Swapping _____ of the line conductors can reverse a three-phase ac motor's rotation.

    (a) one
    (b) two
    (c) three
    (d) four

PART B—TRANSFORMERS

4.16 Transformer Basics

25. A _____ is used to raise or lower voltage and it has the ability to transfer electrical energy from one system to another with no physical connection between the two systems.

    (a) capacitor
    (b) motor
    (c) relay
    (d) transformer

26. The transformer winding that is connected to the source is called the _____ winding and the transformer winding that is connected to the load is called the _____.

    (a) secondary, primary
    (b) primary, secondary
    (c) depends on the wiring
    (d) none of these

27. The energy transfer ability of a transformer is accomplished because the primary electromagnetic lines of force induce a voltage in the secondary winding.

    (a) True
    (b) False

4.17 Secondary Induced Voltage

28. Voltage induced in the secondary winding of a transformer is dependent on the number of secondary turns as compared to the number of primary turns.

    (a) True
    (b) False

4.18 Efficiency

29. Because of conductor resistance, flux leakage, eddy currents, and hysteresis losses, not all of the input power is transferred to the secondary winding for useful purposes.

    (a) True
    (b) False

30. For most practical purposes, transformer efficiency can be ignored.

    (a) True
    (b) False

4.19 Transformer Turns Ratio

31. The relationship of the number of turns of wire on the _____ as compared to the number of turns on the _____ is called "turns ratio."

    (a) primary, secondary
    (b) secondary, primary
    (c) primary, primary
    (d) secondary, secondary

Practice Questions — Unit 4

32. If the primary phase voltage is 480V and the secondary phase voltage is 240V, the turns ratio is _____.

    (a) 1:4
    (b) 1:2
    (c) 2:1
    (d) 4:1

4.20 Autotransformers

33. Autotransformers use separate windings for the primary and secondary.

    (a) True
    (b) False

4.21 Power Losses

34. Wasteful circulating _____ in the iron core cause(s) the core to heat up without any useful purpose.

    (a) conductor resistance
    (b) flux leakage
    (c) eddy currents
    (d) hysteresis losses

35. _____ can be reduced by dividing the core into many flat sections or laminations.

    (a) Conductor resistance
    (b) Flux leakage
    (c) Eddy currents
    (d) Hysteresis losses

36. As current flows through the transformer, the iron core is temporarily magnetized. The energy required to realign the core molecules to the changing electromagnetic field is called _____.

    (a) conductor resistance
    (b) flux leakage
    (c) eddy currents
    (d) hysteresis loss

37. The leakage of electromagnetic flux lines between the primary and secondary windings of a transformer represents wasted energy.

    (a) True
    (b) False

4.22 Transformer kVA Rating

38. Transformers are rated in _____.

    (a) Henrys
    (b) kW
    (c) W
    (d) kVA

4.23 Current Flow

39. The primary electromagnetic field induces a voltage in the secondary. As the secondary current flows, it produces an electromagnetic field that reduces the strength of the primary flux lines. This results in an increase in primary current.

    (a) True
    (b) False

40. Current flow in a secondary transformer winding creates an electromagnetic field that opposes the primary electromagnetic field resulting in less primary CEMF. The primary current automatically increases in direct proportion to the secondary current.

    (a) True
    (b) False

Unit 4 Challenge Questions

UNIT 4—MOTORS AND TRANSFORMERS CHALLENGE QUESTIONS

(•Indicates that 75% or fewer of those who took this exam answered the question correctly.)

PART A—MOTOR BASICS

4.3 Motor Horsepower Ratings

1. The input VA of a 5 hp (15.20A), 230V, three-phase motor is closest to _____.

   (a) 4,600 VA
   (b) 5,300 VA
   (c) 6,100 VA
   (d) 7,500 VA

2. The input VA of a 1 hp (16A), 115V, single-phase motor is _____.

   (a) 1,650 VA
   (b) 1,840 VA
   (c) 2,960 VA
   (d) 3,190 VA

4.11 Direct-Current Motor Types

3. A(n) _____ type of electric motor tends to "run away" if it isn't always connected to its load.

   (a) dc series
   (b) dc shunt
   (c) ac induction
   (d) ac synchronous

4. •A(n) _____ motor has a wide speed range.

   (a) ac
   (b) dc
   (c) synchronous
   (d) induction

4.12 Reversing the Rotation of a DC Motor

5. •If the two line (supply) leads of a dc series motor are reversed, the motor will _____.

   (a) not run
   (b) run backwards
   (c) run the same as before
   (d) become a generator

6. •To reverse a dc series motor, we may simply reverse the supply (power) leads.

   (a) True
   (b) False

4.14 Alternating-Current Motor Types

7. The _____ induction motor is used only in special applications and is always operated on three-phase ac power.

   (a) compound
   (b) synchronous
   (c) split-phase
   (d) wound-rotor

8. •The rotating part of a dc motor or generator is called the _____.

   (a) shaft
   (b) rotor
   (c) armature
   (d) b or c

PART B—TRANSFORMERS

4.19 Transformer Turns Ratio

9. •A transformer winding has a voltage turns ratio of 2:1. The current flowing through the secondary winding will be _____ the current flowing through the primary winding.

   (a) higher than
   (b) lower than
   (c) the same as
   (d) none of these

10. •The primary winding of a transformer has 100 turns and the secondary winding has 10 turns. What is the current flowing through the primary winding if the current flowing through the secondary winding is 5A?

    (a) 0.50A
    (b) 5A
    (c) 10A
    (d) 25A

11. •Which winding of a current transformer (CT) carries more current?

    Tip: An induction clamp-on ammeter operates on the principle of the meter acting as the secondary winding.

    (a) Primary
    (b) Secondary
    (c) Interwinding
    (d) Tertiary

12. If a transformer primary winding has 900 turns and the secondary winding has 90 turns, which winding of the transformer has a larger conductor in cross-sectional area?

    (a) Primary
    (b) Secondary
    (c) Interwinding
    (d) Autowinding

13. If the primary transformer winding operates at 480V and the secondary at 240V, which winding has a larger conductor in cross-sectional area?

    (a) Tertiary
    (b) Secondary
    (c) Primary
    (d) The windings are equal.

14. If the transformer winding voltage turns ratio is 5:1 and the current flowing through the secondary winding is 10A, the current flowing through the primary winding will be _____.

    (a) 1A
    (b) 2A
    (c) 10A
    (d) 25A

15. A transformer primary winding operates at 240V, the secondary winding operates at 12V, and the secondary load is two 100W lamps. What is the secondary current flow if the transformer winding is rated 80 percent efficient?

    (a) 1A
    (b) 17A
    (c) 28A
    (d) 200A

16. A transformer primary winding operates at 240V, the secondary at 120V, and the load is 1,500W. If the transformer is rated 92 percent efficient, what is the current flowing through the primary winding of this transformer?

    (a) 6.80A
    (b) 7.80A
    (c) 8.60A
    (d) 9.90A

17. •The approximate primary kVA of a transformer that supplies a 208V, 100A, three-phase load is _____. See **Figure 4-31**.

    (a) 21 kVA
    (b) 30 kVA
    (c) 42 kVA
    (d) 72.50 kVA

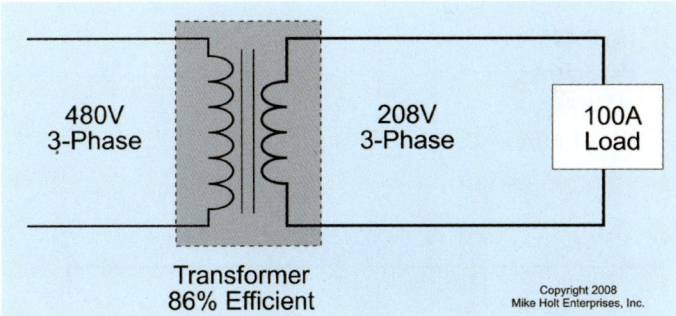

Figure 4–31

18. •The primary current of the 480/208V transformer in **Figure 4–31** is _____.

    (a) 12A
    (b) 25A
    (c) 50A
    (d) 90A

Unit 4 Challenge Questions

19. •The secondary voltage of the transformer in **Figure 4–32** is _____.

    (a) 6V
    (b) 12V
    (c) 24V
    (d) 30V

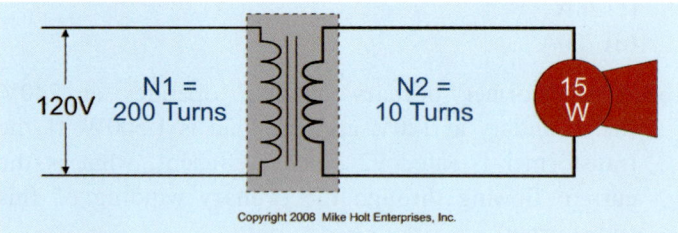

Figure 4–32

20. The primary load in watts for a transformer (rated 95 percent efficient) that supplies a 500W load is _____. See **Figure 4–33**.

    (a) 400W
    (b) 475W
    (c) 526W
    (d) 550W

21. The primary current of the transformer is _____. See **Figure 4–33**.

    (a) 0.416A
    (b) 3.56A
    (c) 4.38A
    (d) 41.60A

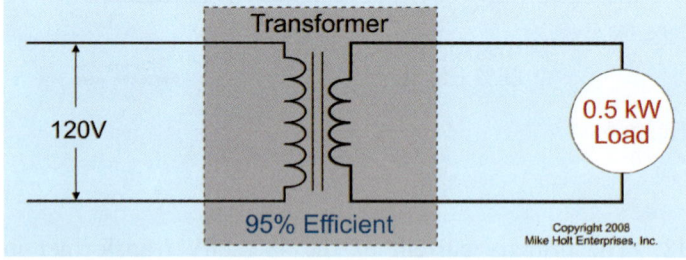

Figure 4–33

22. The primary winding of a transformer operates at 240V, the secondary at 12V, and the load consists of two 100W lamps. What is the load on the secondary winding if the transformer winding is rated 92 percent efficient?

    (a) 200 VA
    (b) The same as the primary VA
    (c) Cannot be calculated
    (d) none of these

23. •The secondary winding of a transformer operates at 24V and it supplies a load that has a rating of 5A. If the transformer is rated 90 percent efficient, what is the secondary VA load rating?

    (a) 120 VA
    (b) The same as the primary VA
    (c) Cannot be calculated
    (d) a and b

24. A primary transformer winding operates at 240V, the secondary at 12V, and the load is two 100W lamps. What is the primary VA load if the transformer is rated 92 percent efficient?

    (a) 0.217 VA
    (b) 185 VA
    (c) 217 VA
    (d) 275 VA

25. The input power for a 20 kW load will be approximately _____ if it operates at an efficiency level of 65 percent.

    (a) 11 kW
    (b) 20 kW
    (c) 31 kW
    (d) 33 kW

4.21 Power Losses

26. •When the current in the steel core of a transformer has risen to a point where high flux density has been reached and additional increases in current produce few additional flux lines, the metal core is said to be _____.

    (a) maximized
    (b) saturated
    (c) full
    (d) none of these

4.22 Transformer kVA Rating

27. What is the primary kVA rating for a transformer winding that is 100 percent efficient if a 5A load on the secondary operates at 12V?

    (a) 0.06 kVA
    (b) 6 kVA
    (c) 30 kVA
    (d) 600 kVA

CHAPTER 1 Notes

CHAPTER 2: NEC Calculations

- **Unit 5** — Raceway and Box Calculations
- **Unit 6** — Conductor Sizing and Protection Calculations
- **Unit 7** — Motor and Air-Conditioning Calculations
- **Unit 8** — Voltage-Drop Calculations
- **Unit 9** — Dwelling Unit Calculations

Chapter 2 Notes

UNIT 5 — Raceway and Box Calculations

INTRODUCTION TO UNIT 5—RACEWAY AND BOX CALCULATIONS

Anyone who has ever pulled wire into a conduit understands the reason for maximum limits on the wire fill for raceways. Trying to pull too many conductors into a raceway can damage the conductor insulation due to the friction and mechanical abuse. We have all heard a joke about tying a wire-pulling rope onto the hitch of the service truck and "locking in the hubs." At least, we hope this is a joke and not an accurate recounting of an installation.

Chapter 9, Table 1 provides the maximum limits the *Code* recognizes for wire fill in terms of percentage of the raceway's interior cross-sectional area. This unit explains those limits and provides instruction regarding the use of the associated tables in Chapter 9 to figure conductor fill. How to use the tables in Annex C when all of the conductors in the raceway are the same size (total cross-sectional area including insulation) is also covered.

Wireway abuse results from disregard to the limits on the radius of bends required to make transitions into and out of the wireway and the number of conductors and splices allowed. There are specific rules in the *NEC* to help plan wireway installations that are in compliance and much easier to work with.

The *NEC* provides a limit to the number of conductors allowed in outlet boxes, based on Table 314.16(A). This limit is often joked about as being the "maximum number of conductors that can be installed in the outlet box while using the persuasion of your hammer handle." This method doesn't follow the *NEC's* guidance set forth in 314.16(B). In this unit, you'll learn how to properly calculate the maximum number of conductors and "conductor equivalents" to be installed in an outlet box. Be sure to read this material carefully so you'll understand what the *Code* means by "conductor equivalents."

An explanation of the sizing requirements of 314.28(A)(1) and (2) for larger pull boxes, junction boxes, and conduit bodies which enclose conductors 4 AWG and larger, rounds out the information provided here in Unit 5.

PART A—RACEWAY FILL

Introduction

Raceways must be large enough to hold conductors without overheating, and to avoid damaging the insulation when conductors are pulled into the raceway. Chapter 9 and Annex C of the *NEC* are the primary references for determining allowable conductor fill in raceways. For the most common condition, where multiple conductors of the same size are installed together in a raceway, the maximum number of conductors permitted can be determined from the tables in Annex C. For situations where conductors of different sizes are mixed together in a raceway, Chapter 9 contains the information necessary to calculate the required raceway size.

Because different conductor types (THW, TW, THHN, etc.) have different thicknesses of insulation, the number and size of conductors permitted in a given raceway often depend on the conductor type used.

5.1 Understanding the *NEC*, Chapter 9 Tables

Table 1—Conductor Percent Fill

The maximum percentage of allowable conductor fill is listed in Chapter 9, Table 1. It's based on common conditions where the length of the conductor and number of raceway bends are within reasonable limits [Chapter 9, Table 1, FPN No. 1]. **Figure 5–1**

Unit 5 — Raceway and Box Calculations

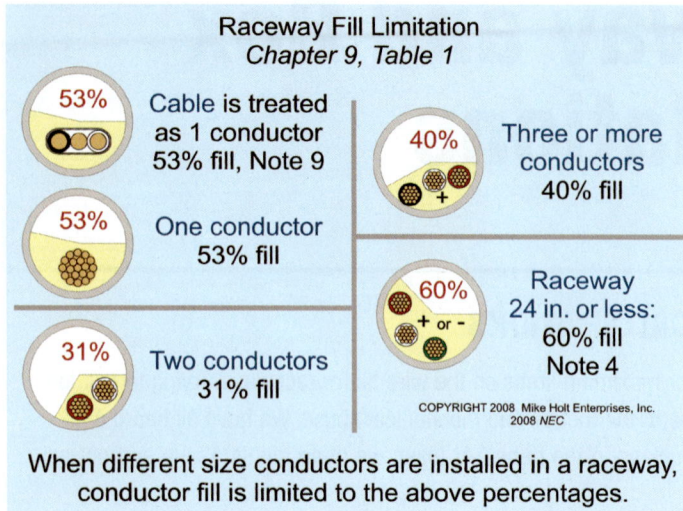

Figure 5–1

Table 1, Note 1—Conductors all the Same Size

When all conductors are the same size (total cross-sectional area including insulation), the number of conductors permitted in a raceway can be determined by simply looking at the tables located in Annex C—Raceway Fill Tables for Conductors and Fixture Wires of the Same Size.

Table 1 of Chapter 9, Maximum Percent Conductor Fill

Number of Conductors	Percent Fill Permitted
1 conductor	53% fill
2 conductors	31% fill
3 or more conductors	40% fill
Raceway 24 inches or less	60% fill Note 4

- Tables C.1 through C.12(a) are based on maximum percent fill as listed in Chapter 9, Table 1.
- Table C.1—Conductors and fixture wires in electrical metallic tubing (EMT)
- Table C.1(A)—Compact conductors in electrical metallic tubing (EMT)
- Table C.2—Conductors and fixture wires in electrical nonmetallic tubing (ENT)
- Table C.2(A)—Compact conductors in electrical nonmetallic tubing (ENT)
- Table C.3—Conductors and fixture wires in flexible metal conduit (FMC)
- Table C.3(A)—Compact conductors in flexible metal conduit (FMC)
- Table C.4—Conductors and fixture wires in intermediate metal conduit (IMC)
- Table C.4(A)—Compact conductors in intermediate metal conduit (IMC)
- Table C.5—Conductors and fixture wires in liquidtight flexible nonmetallic conduit (gray type) (LFNC-B)
- Table C.5(A)—Compact conductors in liquidtight flexible nonmetallic conduit (gray type) (LFNC-B)
- Table C.6—Conductors and fixture wires in liquidtight flexible nonmetallic conduit (orange type) (LFNC-A)
- Table C.6(A)—Compact conductors in liquidtight flexible nonmetallic conduit (orange type) (LFNC-A)

Author's Comment: The annex does not have a table for LFNC of the black type (LFNC-C).

- Table C.7—Conductors and fixture wires in liquidtight flexible metal conduit (LFMC)
- Table C.7(A)—Compact conductors in liquidtight flexible metal conduit (LFMC)
- Table C.8—Conductors and fixture wires in rigid metal conduit (RMC)
- Table C.8(A)—Compact conductors in rigid metal conduit (RMC)
- Table C.9—Conductors and fixture wires in rigid PVC conduit, Schedule 80
- Table C.9(A)—Compact conductors in rigid PVC conduit, Schedule 80
- Table C.10—Conductors and fixture wires in rigid PVC conduit, Schedule 40
- Table C.10(A)—Compact conductors in rigid PVC conduit, Schedule 40
- Table C.11—Conductors and fixture wires in Type A, rigid PVC conduit
- Table C.11(A)—Compact conductors in Type A, rigid PVC conduit
- Table C.12—Conductors and fixture wires in Type EB, PVC conduit
- Table C.12(A)—Compact conductors in Type EB, PVC conduit

Raceway and Box Calculations — Unit 5

▶ Annex C—Table C.1—EMT

Question: How many 14 RHH conductors (without cover) can be installed in trade size 1 EMT? **Figure 5–2**

(a) 13 (b) 16 (c) 19 (d) 25

Answer: (b) 16 conductors [Annex C, Table C.1]

Note 2 at the end of Annex C, Table C.1 indicates that an asterisk (*) with conductor insulation types RHH*, RHW*, and RHW-2* means that these types don't have an outer covering. Insulation types RHH, RHW, and RHW-2 (without the asterisk) do have an outer cover. This is a cover (which may be a fiberous material) that increases the dimensions of the conductor more than the thin nylon cover encountered with conductors such as THHN.

▶ Annex C—Table C.3—FMC

Question: If trade size 1¼ FMC has three THHN conductors (not compact), what is the largest conductor permitted to be installed? **Figure 5–3**

(a) 1 THHN (b) 1/0 THHN (c) 2/0 THHN (d) 3/0 THHN

Answer: (a) 1 THHN [Annex C, Table C.3]

It's common to see conductors with a dual insulation rating, such as THHN/THWN. This type of conductor can be used in a dry location at the THHN 90°C ampacity, or if used in a wet location, the THWN ampacity rating of the 75°C column of Table 310.16 for THWN insulation types must be adhered to.

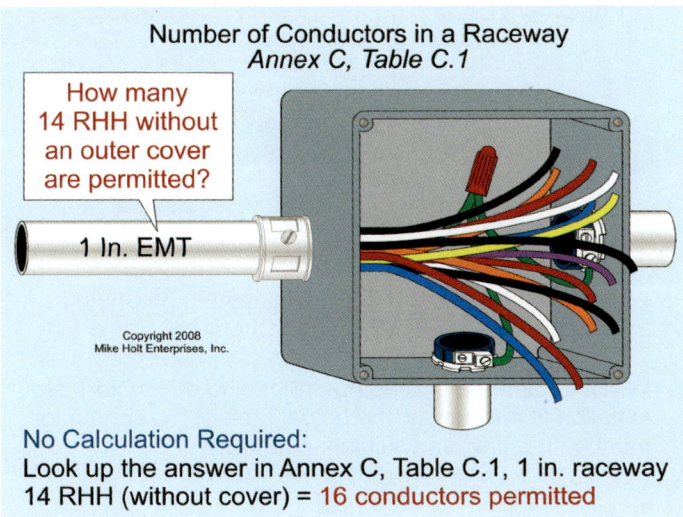

Figure 5–2

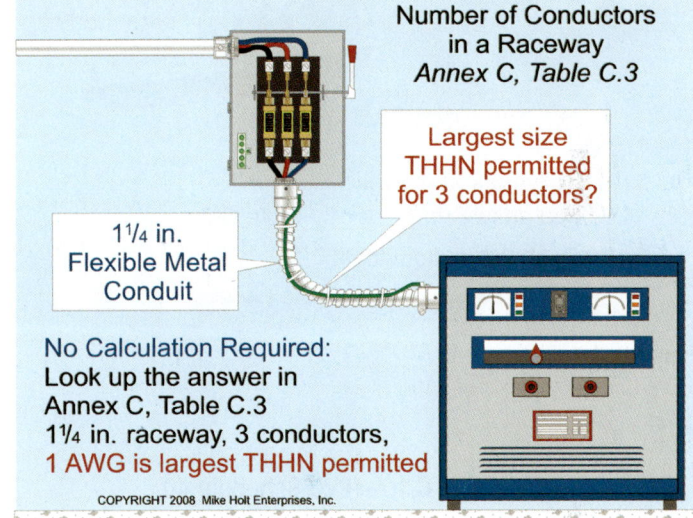

Figure 5–3

▶ Annex C—Table C.2A—Compact Conductors in ENT

Question: How many 6 XHHW compact conductors can be installed in trade size 1¼ ENT?

(a) 6 (b) 10 (c) 13 (d) 16

Answer: (b) 10 conductors [Annex C, Table C.2A]

Compact stranding is the result of a manufacturing process where the standard conductor is compressed to the extent that the voids between strand wires are virtually eliminated [Annex C, Table C.1(a) footnote]. Unless the question specifically states compact conductors, assume that the conductors are not the compact type.

▶ Annex C—Table C.4—IMC

Question: How many 4/0 RHH conductors with an outer cover can be installed in trade size 2 IMC?

Note: RHH insulation with an outer cover has no asterisk (*).

(a) 1 conductor (b) 2 conductors
(c) 3 conductors (d) 4 conductors

Answer: (c) 3 conductors [Annex C, Table C.4]

Unit 5 Raceway and Box Calculations

▶ **Annex C—Table C.7—Fixture Wire in LFMC**

Question: How many 18 TFFN conductors can be installed in trade size ¾ LFMC? **Figure 5–4**

(a) 14 conductors (b) 26 conductors
(c) 30 conductors (d) 39 conductors

Answer: (d) 39 conductors [Annex C, Table C.7]

Figure 5–4

▶ **Annex C—Table C.10—PVC Schedule 40**

Question: What is the smallest trade size PVC Schedule 40 raceway that can be used for the installation of a single 1/0 THHN as a grounding electrode conductor?

(a) Trade size ½ (b) Trade size ¾
(c) Trade size 1 (d) Trade size 1 ¼

Answer: (a) Trade size ½ [Annex C, Table C.10]

▶ **Annex C—Table C.10(a)—Compact Conductors in PVC Schedule 40**

Question: If a trade size 3 PVC Schedule 40 raceway has three XHHW compact conductors, what is the largest conductor permitted to be installed?

(a) 1/0 XHHW (b) 4/0 XHHW
(c) 500 kcmil XHHW (d) 750 kcmil XHHW

Answer: (d) 750 kcmil XHHW [Annex C, Table C.10(A)]

Table 1, Note 2—Used for Physical Protection

The percentages listed in Table 1 apply only to complete raceway systems and are not intended to apply to sections of raceways used to protect wiring from physical damage.

Table 1, Note 3—Equipment Grounding Conductors

When equipment grounding and bonding conductors are installed in a raceway, the actual area of the conductor must be used to calculate raceway fill, **Figure 5–5**. Chapter 9, Table 5 can be used to determine the cross-sectional area of insulated conductors and Chapter 9, Table 8 can be used to determine the cross-sectional area of bare conductors (see Chapter 9, Table 1, Note 8).

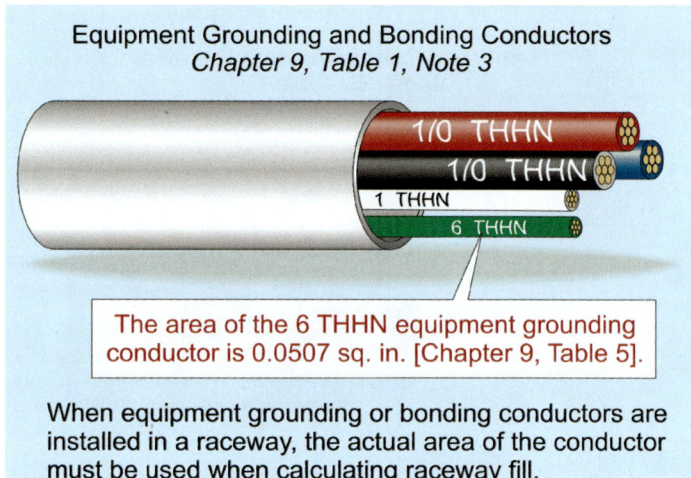

Figure 5–5

Table 1, Note 4—Raceways not Exceeding 24 Inches

When a raceway doesn't exceed 24 in. in length, it's permitted to be filled to 60 percent of its total cross-sectional area as identified in Table 4 of Chapter 9. **Figure 5–6**

Table 1, Note 5—Multiconductor Cables

For multiconductor cables, the actual cross-sectional area of the cable is to be used for raceway sizing.

Raceway and Box Calculations Unit 5

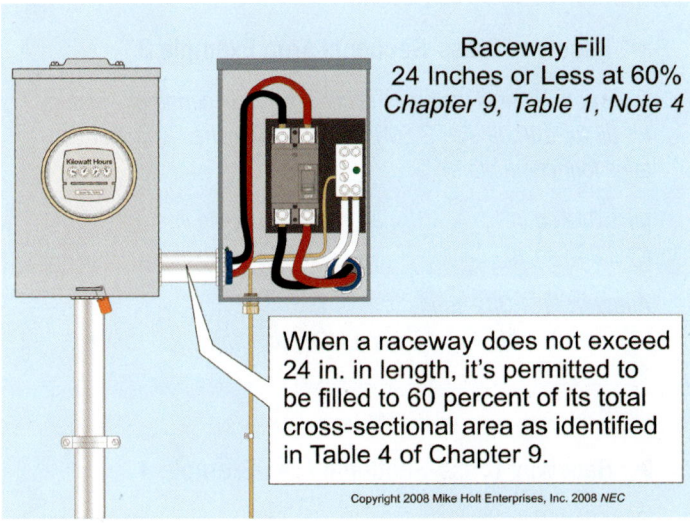

Figure 5–6

Table 1, Note 9

A multiconductor cable or flexible cord is treated as a single conductor for calculating percentage conduit fill area. For cables with elliptical cross sections, the cross-sectional area is based on using the major diameter of the ellipse as the circle diameter.

> ▶ **Example: NM cable in EMT**
>
> **Question:** What size EMT is needed for one 12-2 with ground NM cable that measures 7/16 on the major diameter?
>
> (a) Trade size ½ (b) Trade size ¾
> (c) Trade size 1 (d) Trade size 1¼
>
> **Answer:** (a) Trade size ½
>
> Area = 3.14 x r²
>
> D = 7/16 in.
> D = 0.4375 in.
> R = D/2
> R = 0.4375 in./2
> R = 0.22 in.
>
> Area = 3.14 x 0.22 in. x 0.22 in.
> Area = 0.1520 sq in.
>
> Using the one wire column (53%) of Chapter 9, Table 4 for EMT:
>
> Trade size ½ EMT provides 0.161 sq in., which is larger than 0.1520 sq in. and will meet the minimum requirement.

Table 1, Note 6—Different Size Conductors

When sizing raceways for conductors of different sizes, use Table 5 for the conductor's cross-sectional area, and Table 4 for the raceway cross-sectional area. This is also the method to use when doing calculations for raceways of 24 in. and shorter, as Annex C doesn't account for the 60 percent fill allowance provided by Table 4.

Table 1, Note 7—Rounding

When the calculated number of conductors (all of the same size including insulation) results in 0.80 or more, the next higher whole number of conductors can be used.

Table 1, Note 8—Bare Conductors

The dimensions for bare conductors are listed in Chapter 9, Table 8.

Table 4—Raceway Cross-Sectional Area

The sixth column of this table (Total Area 100%) gives the total cross-sectional area in square inches of the raceway. There are also 31% (2 wires), 40% (3 or more wires), 53% (1 wire), and 60% (nipple) cross-sectional area columns based on the number of conductors in accordance with Chapter 9, Table 1.

> ▶ **Raceway Cross-Sectional Area Example 1**
>
> **Question:** What is the cross-sectional area of permitted conductor fill for a trade size 1 EMT raceway 30 inches long that contains four conductors? **Figure 5–7**
>
> (a) 0.346 sq in. (b) 1.013 sq in.
> (c) 2.067 sq in. (d) 3.356 sq in.
>
> **Answer:** (a) 0.346 sq in.
> [Chapter 9, Table 1, Note 4 and Table 4, 40% column]

Mike Holt Enterprises, Inc. • www.MikeHolt.com • 1.888.NEC.CODE (1.888.632.2633)   123

Unit 5 Raceway and Box Calculations

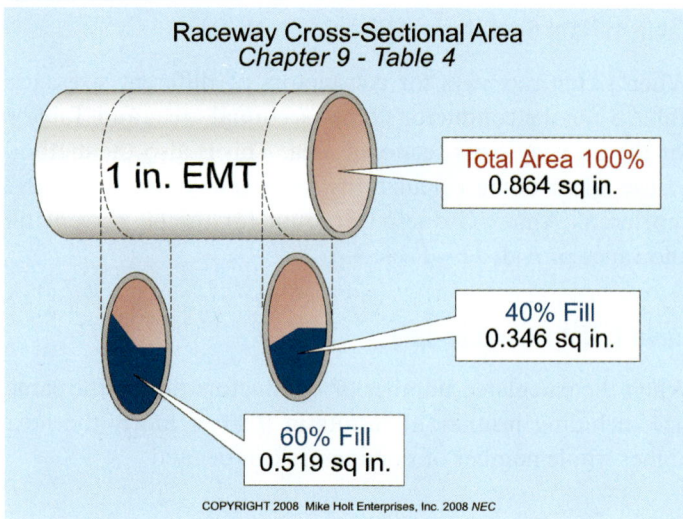

Figure 5–7

▶ **Raceway Cross-Sectional Area Example 2**

Question: What is the cross-sectional area of permitted conductor fill for a trade size 2 EMT raceway that is 20 inches long?
Figure 5–8

(a) 1.342 sq in. (b) 2.013 sq in.
(c) 2.067 sq in. (d) 3.356 sq in.

Answer: (b) 2.013 sq in.
[Chapter 9, Table 1, Note 4 and Table 4, 60% column]

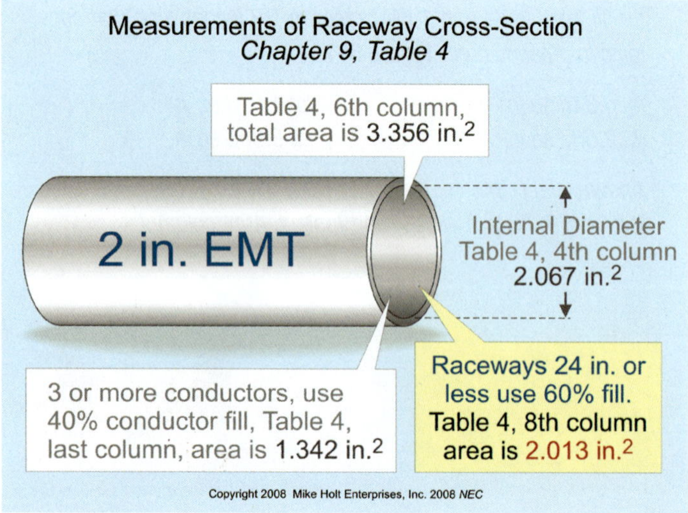

▶ **Raceway Cross-Sectional Area Example 3**

Question: What is the cross-sectional area of permitted conductor fill for a trade size 2 EMT raceway 30 inches long that contains four conductors?

(a) 1.342 sq in. (b) 2.013 sq in.
(c) 2.067 sq in. (d) 3.356 sq in.

Answer: (a) 1.342 sq in.
[Chapter 9, Table 1 and Table 4, 40% column]

▶ **Raceway Cross-Sectional Area Example 4**

Question: What is the minimum size EMT raceway required for three conductors with a wire fill of 0.25 sq in.?

(a) Trade size ½ (b) Trade size 1
(c) Trade size 1¼ (d) Trade size 1½

Answer: (b) Trade size 1

▶ **Raceway Cross-Sectional Area Example 5**

Question: What is the minimum size Schedule 80 PVC raceway required for three conductors with a wire fill of 0.35 sq in.?

(a) Trade size ½ (b) Trade size 1
(c) Trade size 1¼ (d) Trade size 1½

Answer: (c) Trade size 1¼

Table 5—Dimensions of Insulated Conductors and Fixture Wires

Chapter 9, Table 5 lists the cross-sectional area of insulated conductors and fixture wires (see **Table 5–1**).

Figure 5–8

Raceway and Box Calculations — Unit 5

Table 5–1: Commonly Used Conductor Cross-Sectional Area

Column 1	Column 2	Column 3	Column 4	Column 5	Column 6	Chapter 9 Table 8
	RHH/RHW With Cover	RHH/RHW Without Cover	TW or THW	THHN THWN	XHHW	BARE Stranded Conductors
	Chapter 9, Table 5					
Size AWG/kcmil	Approximate Cross-Sectional Area — Square Inches					
14	0.0293	0.0209	0.0139	0.0097	0.0139	0.004
12	0.0353	0.0260	0.0181	0.0133	0.0181	0.006
10	0.0437	0.0333	0.0243	0.0211	0.0243	0.011
8	0.0835	0.0556	0.0437	0.0366	0.0437	0.017
6	0.1041	0.0726	0.0726	0.0507	0.0590	0.027
4	0.1333	0.0973	0.0973	0.0824	0.0814	0.042
3	0.1521	0.1134	0.1134	0.0973	0.0962	0.053
2	0.1750	0.1333	0.1333	0.1158	0.1146	0.067
1	0.2660	0.1901	0.1901	0.1562	0.1534	0.087
0	0.3039	0.2223	0.2223	0.1855	0.1825	0.109
00	0.3505	0.2624	0.2624	0.2233	0.2190	0.137
000	0.4072	0.3117	0.3117	0.2679	0.2642	0.173
0000	0.4754	0.3718	0.3718	0.3237	0.3197	0.219

▶ **Table 5—THHN**

Question: What is the cross-sectional area for one 14 THHN conductor? **Figure 5–9**

(a) 0.0097 sq in. (b) 0.0172 sq in.
(c) 0.0206 sq in. (d) 0.0278 sq in.

Answer: (a) 0.0097 sq in.

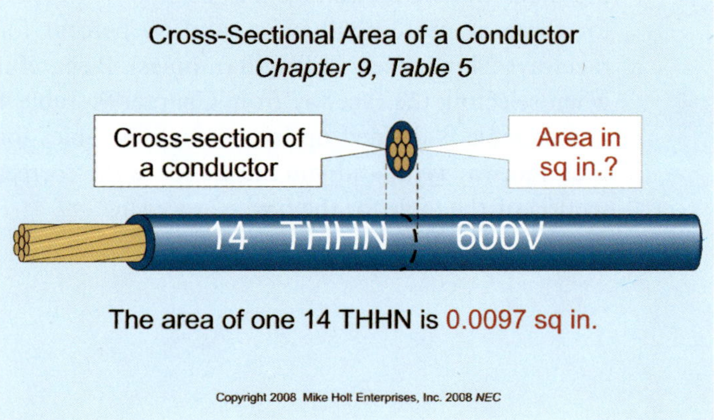

▶ **Table 5—RHW With an Outer Cover**

Question: What is the cross-sectional area for one 12 RHW conductor with an outer cover?

(a) 0.0172 sq in. (b) 0.0206 sq in.
(c) 0.0278 sq in. (d) 0.0353 sq in.

Answer: (d) 0.0353 sq in.

▶ **Table 5—RHH Without an Outer Cover**

Question: What is the cross-sectional area for one 10 RHH without an outer cover?

(a) 0.0117 sq in. (b) 0.0252 sq in.
(c) 0.0278 sq in. (d) 0.0333 sq in.

Answer: (d) 0.0333 sq in.

The note at the end of Table 5 states that conductor Types RHH, RHW, and RHW-2 without outer coverings are identified with an asterisk (*).

Cross-Sectional Area of a Conductor
Chapter 9, Table 5

Cross-section of a conductor — Area in sq in.?

14 THHN 600V

The area of one 14 THHN is 0.0097 sq in.

Figure 5–9

Unit 5 — Raceway and Box Calculations

Table 5A—Dimensions of Compact Insulated Conductors

Chapter 9, Table 5A, lists the cross-sectional areas for compact copper and aluminum building wires. These conductors use specially shaped strands so that the overall size of the conductor is more compact. The outer covering is labeled as a compact conductor.

▶ **Table 5—THHN Compact Conductors**

Question: What is the cross-sectional area for one 1 THHN compact conductor?

(a) 0.0117 sq in.   (b) 0.1352 sq in.
(c) 0.2733 sq in.   (d) 0.5216 sq in.

Answer: (b) 0.1352 sq in.

▶ **Table 5—Compact Conductors**

Question: What is the cross-sectional area for one 1/0 XHHW compact conductor?

(a) 0.0117 sq in.   (b) 0.1352 sq in.
(c) 0.2733 sq in.   (d) 0.5216 sq in.

Answer: (c) 0.2733 sq in.

Table 8—Conductor Properties

Chapter 9, Table 8 contains conductor properties such as cross-sectional area in circular mils, number of strands per conductor, cross-sectional area in square inches for bare conductors, and dc resistance at 75°C for both copper and aluminum conductors.

▶ **Bare Conductor—Cross-Sectional Area**

Question: What is the cross-sectional area for one 10 AWG bare conductor with seven strands? **Figure 5–10**

(a) 0.008 sq in.   (b) 0.011 sq in.
(c) 0.038 sq in.   (d) a or b

Answer: (b) 0.011 sq in. [Chapter 9, Table 8]

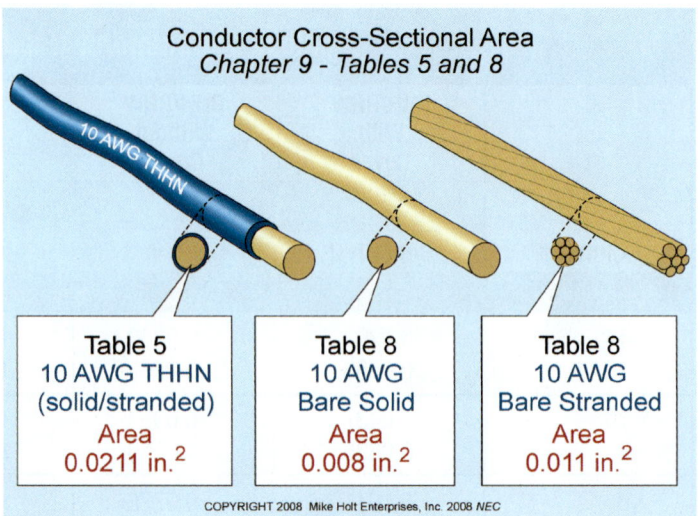

Figure 5–10

5.2 Raceway Calculations

Annex C—Tables 1 through 12 can't be used to determine raceway sizing when conductors of different sizes are installed in the same raceway. When this situation is encountered, the following steps are used to determine the raceway size and nipple size:

Step 1: Determine the cross-sectional area (in square inches) for each conductor from Chapter 9, Table 5 for insulated conductors and from Chapter 9, Table 8 for bare conductors.

Step 2: Determine the total cross-sectional area for all conductors.

Step 3: Size the raceway according to the percent fill as listed in Chapter 9, Table 1. Chapter 9, Table 4 includes the various types of raceways with columns representing the allowable percentage fills; such as 40 percent for three or more conductors, and 60 percent for raceways 24 in. or less in length (nipples). Be careful when selecting the raceway from Chapter 9, Table 4 as this table is divided up into numerous tables for each raceway type, and you must choose the correct section of the table for the type of raceway.

Raceway and Box Calculations — Unit 5

▶ Raceway Size

Question: What is the minimum size Schedule 40 PVC raceway required for three 500 kcmil THHN conductors, one 250 kcmil THHN conductor, and one 3 THHN conductor. **Figure 5–11**

(a) Trade size 2
(b) Trade size 2½
(c) Trade size 3
(d) Trade size 3½

Answer: (c) Trade size 3

Step 1: Determine the cross-sectional area of the conductors [Chapter 9, Table 5].

500 THHN 0.7073 sq in. x 3 wires = 2.1219 sq in.
250 THHN 0.3970 sq in. x 1 wire = 0.3970 sq in.
3 THHN 0.0973 sq in. x 1 wire = 0.0973 sq in.

Step 2: Total cross-sectional area of all conductors = 2.6162 sq in.

Step 3: Size the conduit at 40 percent fill [Chapter 9, Table 1] using Chapter 9, Table 4 (be sure to select the table for PVC Schedule 40).

Trade size 3 Schedule 40 PVC has an allowable cross-sectional area of 2.907 sq in. for over two conductors in the 40 percent column.

▶ Nipple Size

Question: What size RMC nipple is required for three 3/0 THHN conductors, one 1 THHN conductor, and one 6 THHN conductor? **Figure 5–12**

(a) Trade size ½
(b) Trade size 1
(c) Trade size 1½
(d) Trade size 2

Answer: (c) Trade size 1½

Step 1: Determine the cross-sectional area of the conductors [Chapter 9, Table 5].

3/0 THHN 0.2679 sq in. x 3 wires = 0.8037 sq in.
1 THHN 0.1562 sq in. x 1 wire = 0.1562 sq in.
6 THHN 0.0507 sq in. x 1 wire = 0.0507 sq in.

Step 2: Total cross-sectional area of the conductors = 1.0106 sq in.

Step 3: Size the conduit at 60 percent fill [Chapter 9, Table 1, Note 4] using Chapter 9, Table 4.

Trade size 1¼ nipple = 0.0916 sq in., too small
Trade size 1½ nipple = 1.243 sq in., just right
Trade size 2 nipple = 2.045 sq in., larger than required

Figure 5–11

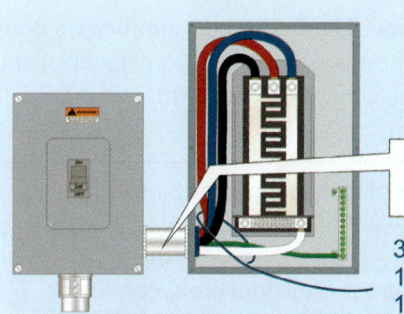

Figure 5–12

5.3 Wireways

Metal wireways are commonly used where access to the conductors within the raceway is required to make terminations, splices, or taps to several devices at a single location. Their high cost precludes their use for other than short distances, except in some commercial or industrial occupancies where the wiring is frequently revised.

Author's Comment: Both metal wireways and nonmetallic wireways are often called "troughs" or "gutters" in the field.

Definition—Metal Wireway [376.2]

A sheet metal raceway with hinged or removable covers for housing and protecting electric conductors and cable, and in which conductors are placed after the wireway has been installed.

Conductors—Maximum Size [376.21]

The maximum size conductor permitted in a wireway must not be larger than that for which the wireway is designed.

Number of Conductors and Ampacity [376.22]

(a) Number of Conductors. The maximum number of conductors permitted in a wireway is limited to 20 percent of the cross-sectional area of the wireway. Figure 5–13

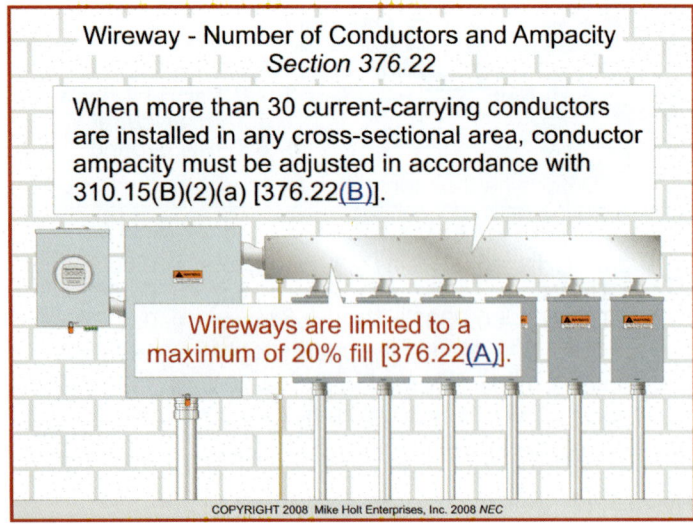

Figure 5–13

Author's Comment: Splices and taps must not fill more than 75 percent of the wiring space at any cross section [376.56].

(b) Conductor Ampacity Adjustment Factors. When more than 30 current-carrying conductors are installed in any cross-sectional area of the wireway, the conductor ampacity, as listed in Table 310.16, must be adjusted in accordance with Table 310.15(B)(2)(a).

Signaling and motor control conductors between a motor and its starter used only for starting duty aren't considered current carrying for conductor ampacity adjustment.

Wireway Sizing [376.23]

(a) Sizing for Conductor Bending Radius. Where conductors are bent within a metal wireway, the wireway must be sized to meet the bending radius requirements contained in Table 312.6(A), based on one wire per terminal. Figure 5–14

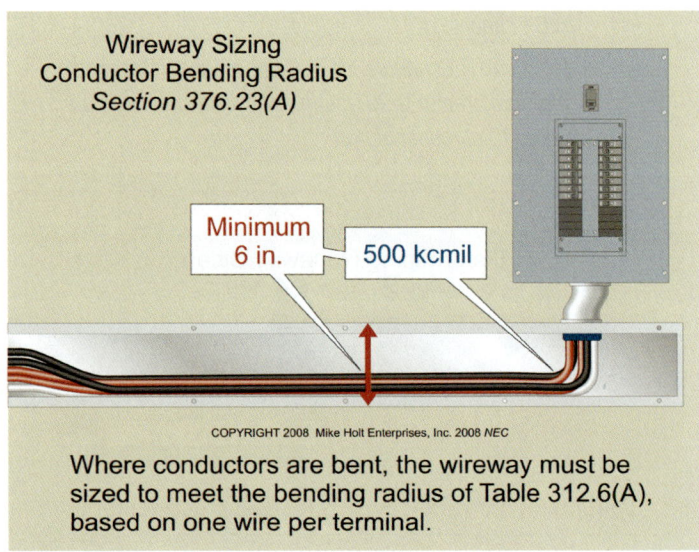

Where conductors are bent, the wireway must be sized to meet the bending radius of Table 312.6(A), based on one wire per terminal.

Figure 5–14

(b) Wireway Used as Pull Box. Where insulated conductors 4 AWG or larger are pulled through a metal wireway, the distance between raceway and cable entries enclosing the same conductor must not be less than required by 314.28(A)(1) and 314.28(A)(2). Figure 5–15

Author's Comments:

- Straight Pulls. The minimum distance from where the conductors enter to the opposite wall must not be less than eight times the trade size of the largest raceway [314.28(A)(1)].

Raceway and Box Calculations — Unit 5

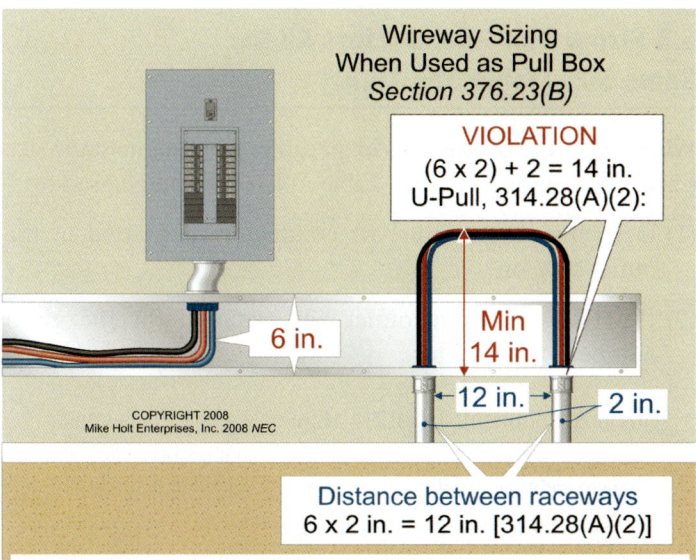

Where conductors 4 AWG or larger are pulled through a wireway, the distance between conductor entries must not be less than required by 314.28(A)(1) and 314.28(A)(2).

Figure 5–15

- Angle Pulls. The distance from the raceway entry to the opposite wall must not be less than six times the trade diameter of the largest raceway, plus the sum of the trade sizes of the remaining raceways in the same row on the same wall [314.28(A)(2)].

- U Pulls. When a conductor enters and leaves from the same wall, the distance from where the raceways enter to the opposite wall must not be less than six times the trade size of the largest raceway, plus the sum of the trade sizes of the remaining raceways on the same wall and row [314.28(A)(2)].

- The distance between raceways enclosing the same conductor must not be less than six times the trade size of the largest raceway [314.28(A)(2)].

▶ **Wireway Cross-Sectional Area**

Question: What is the cross-sectional area of a 6 in. x 6 in. wireway?

(a) 6 sq in.   (b) 16 sq in.   (c) 36 sq in.   (d) 66 sq in.

Answer: (c) 36 sq in.

The cross-sectional area is found by multiplying height by depth: 6 in. x 6 in. = 36 sq in.

▶ **Wireway Allowable Conductor Fill Area**

Question: What is the maximum allowable conductor fill in square inches for a 6 in. x 6 in. wireway?

(a) 6 sq in.   (b) 16 sq in.   (c) 36 sq in.   (d) 66 sq in.

Answer: (c) 36 sq in.

36 sq in. x 0.20 = 7.20 sq in. [376.22(A)]

▶ **Wireway Conductor Fill**

Question: What is the maximum number of 500 kcmil THHN conductors that can be installed in a 6 in. x 6 in. wireway?

(a) 4   (b) 6   (c) 10   (d) 20

Answer: (c) 10

36 sq in. x 0.20 = 7.20 sq in. [376.22(A)]

500 kcmil THHN = 0.7073 sq in. [Chapter 9, Table 5]

Maximum Allowable Area/Area per Conductor = Number of Conductors

7.20 sq in./0.7073 sq in. = 10.17 conductors

10 conductors can be installed.

Note: Conductor ampacity adjustment for bundling isn't required because there are fewer than 30 conductors [376.22(B)].

▶ **Wireway Conductor Fill**

Question: What size wireway is required for three 500 kcmil THHN, one 250 kcmil THHN, and four 4/0 THHN conductors?

(a) 4 in. x 4 in.   (b) 6 in. x 6 in.
(c) 8 in. x 8 in.   (d) 10 in. x 10 in.

Answer: (a) 4 in. x 4 in.

Find the conductor area [Chapter 9, Table 5]

500 kcmil THHN = 0.7073 sq in. x 3 = 2.319 sq in.
250 kcmil THHN = 0.3970 sq in.
4/0 THHN = 0.3237 x 4 = 0.1295
Total Conductor Area = 2.8455

The wireway must not be filled to over 20 percent of its cross-sectional area [376.22(A)]. Twenty percent is equal to one-fifth, so we can multiply the required conductor area by five to find the minimum square inch area required.

Conductor Area x 5 = Required Wireway Minimum Area
2.8455 x 5 = 14.2275 sq in.

A 4 in. x 4 in. wireway has a cross-sectional area of 16 sq in. and will be large enough.

Unit 5 — Raceway and Box Calculations

5.4 Tips for Raceway Calculations

Tip 1: Take your time.

Tip 2: Use a ruler or a straightedge when using tables, and highlight key words and important sections.

Tip 3: Watch out for different types of raceways and conductor insulations, particularly RHH/RHW with or without an outer cover.

PART B—OUTLET BOX FILL CALCULATIONS [314.16]

Introduction

Boxes must be of sufficient size to provide free space for all conductors. An outlet box is generally used for the attachment of devices and luminaires and has a specific amount of space (volume) for conductors, devices, and fittings. The volume taken up by conductors, devices, and fittings in a box must not exceed the box fill capacity. The volume of a box is the total volume of its assembled parts, including plaster rings, industrial raised covers, and extension rings. The total volume includes only those parts that are marked with their volumes in cubic inches [314.16(A)] or included in *NEC* Table 314.16(A). **Figure 5–16**

5.5 Sizing Box—Conductors All the Same Size [Table 314.16(A)]

When all of the conductors in an outlet box are the same size (insulation doesn't matter), Table 314.16(A) can be used to:

(1) Determine the number of conductors permitted in the outlet box, or

(2) Determine the size outlet box required for the given number of conductors.

Author's Comment: Table 314.16(A) only applies if the outlet box contains no switches, receptacles, luminaire studs, luminaire hickeys, manufactured cable clamps, or equipment grounding conductors. This is rarely the case.

▶ **Outlet Box Size**

Question: What is the minimum size square outlet box required for six 12 THHN conductors and three 12 THW conductors? **Figure 5–17**

(a) 4 x 1¼ in. square (b) 4 x 1½ in. square
(c) 4 x 2⅛ in. square (d) 4 x 2⅛ in. with extension

Answer: (b) 4 x 1½ in. square

Table 314.16(a) permits nine 12 AWG conductors; the insulation type isn't a factor when calculating box fill.

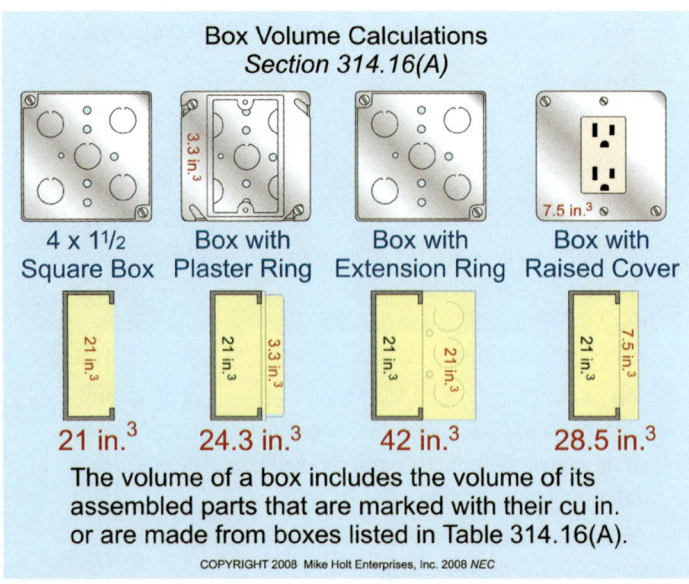

Figure 5–16

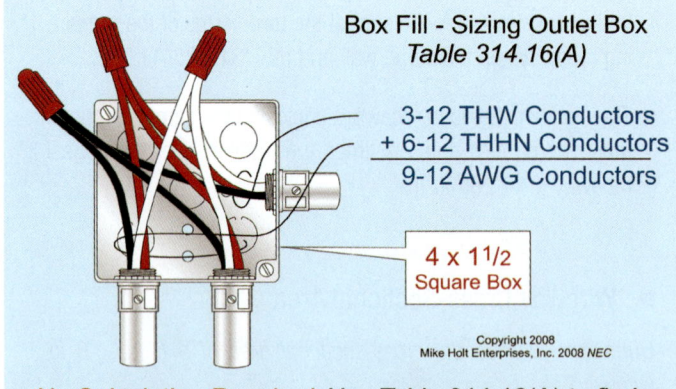

No Calculation Required: Use Table 314.16(A) to find a box that will permit at least 9-12 AWG conductors and match it to the multiple choice selections of the question being answered.

Figure 5–17

Raceway and Box Calculations — Unit 5

▶ Number of Conductors in an Outlet Box

Question: Using Table 314.16(A), how many 14 THHN conductors are permitted in a 4 x 1½ in. round box?

(a) 7 conductors
(b) 9 conductors
(c) 10 conductors
(d) 11 conductors

Answer: (a) 7 conductors

Table 314.16(b) Volume Allowance Required per Conductor

Conductor AWG	Volume cu in.
18	1.50
16	1.75
14	2.00
12	2.25
10	2.50
8	3.00
6	5.00

5.6 Conductor Equivalents

Box Fill Calculations [314.16(B)].

The calculated conductor volume determined by (1) through (5) and Table 314.16(B) are added together to determine the total volume of the conductors, devices, and fittings. Raceway and cable fittings, including locknuts and bushings, are not counted for box fill calculations. **Figure 5–18**

(1) Conductor Volume. Each unbroken conductor that runs through a box and each conductor that terminates in a box is counted as a single conductor volume in accordance with Table 314.16(B). Each loop or coil of unbroken conductor having a length of at least twice the minimum length required for free conductors in 300.14 must be counted as two conductor volumes. Conductors that originate and terminate within the box, such as pigtails, aren't counted at all. **Figure 5–19** and **Figure 5–20**

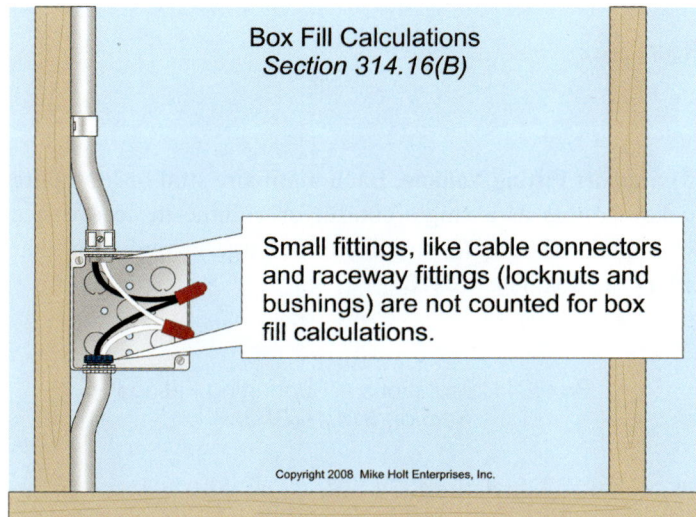

Figure 5–18

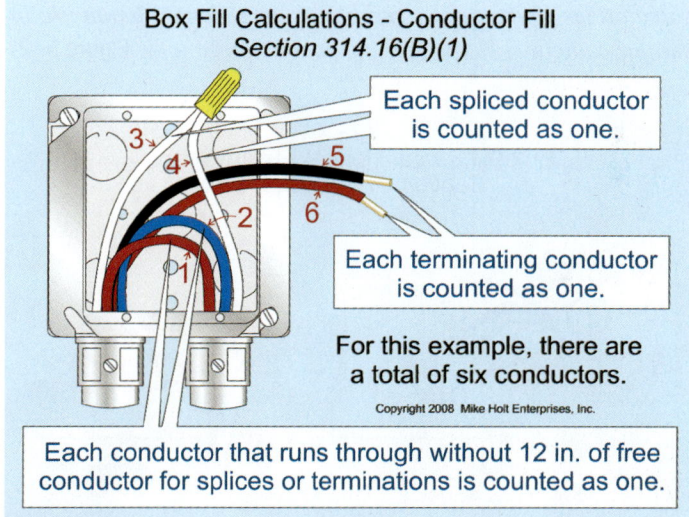

Figure 5–19

Unit 5 — Raceway and Box Calculations

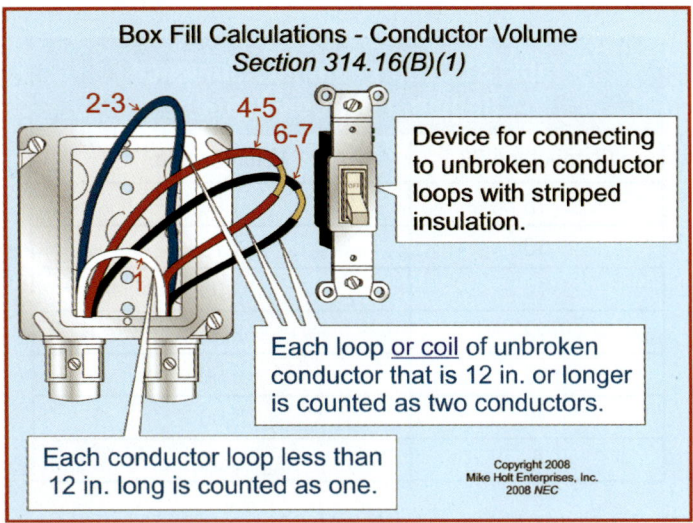

Figure 5–20

Author's Comment: According to 300.14, at least 6 in. of free conductor, measured from the point in the box where the conductors enter the enclosure, must be left at each outlet, junction, and switch point for splices or terminations of luminaires or devices.

Exception: Equipment grounding conductors, and up to four 16 AWG and smaller fixture wires, can be omitted from box fill calculations if they enter the box from a domed luminaire or similar canopy, such as a ceiling paddle fan canopy. **Figure 5–21**

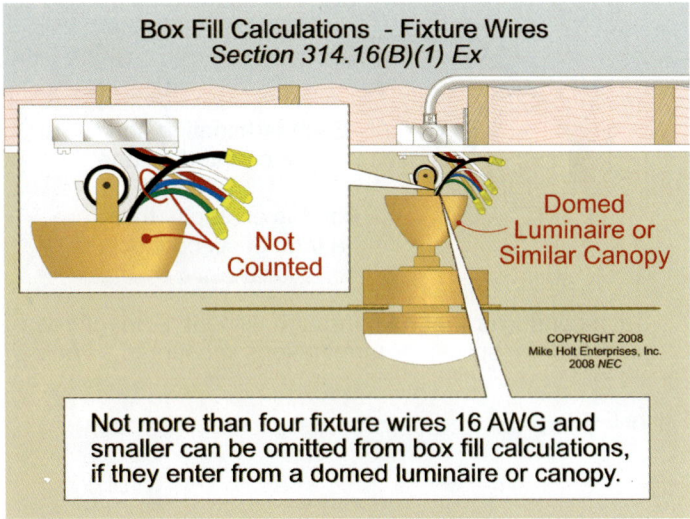

Figure 5–21

(2) Cable Clamp Volume. One or more internal cable clamps count as a single conductor volume in accordance with Table 314.16(B), based on the largest conductor that enters the box. Cable connectors that have their clamping mechanism outside the box aren't counted. **Figure 5–22**

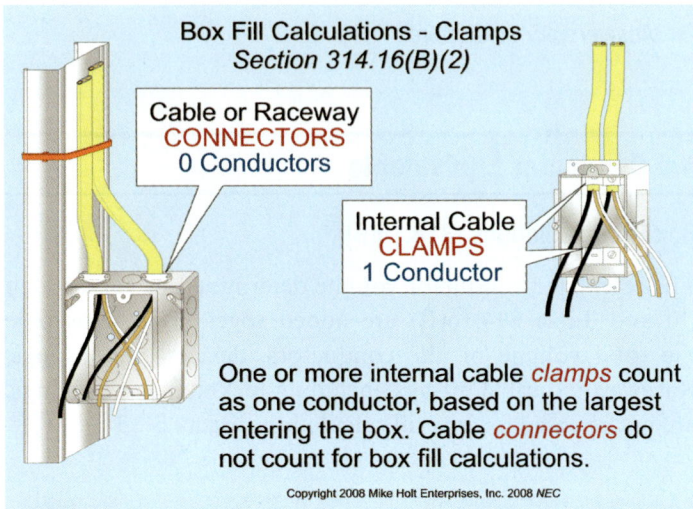

Figure 5–22

(3) Support Fitting Volume. Each luminaire stud or luminaire hickey counts as a single conductor volume in accordance with Table 314.16(B), based on the largest conductor that enters the box. **Figure 5–23**

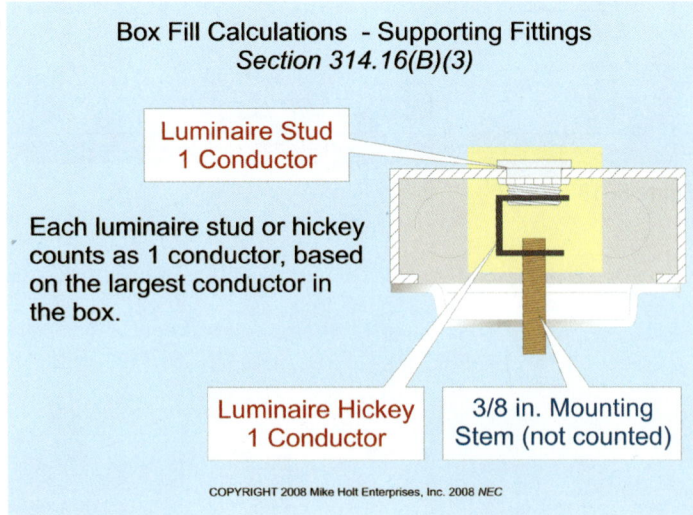

Figure 5–23

Raceway and Box Calculations — Unit 5

(4) Device Yoke Volume. Each single gang device yoke (regardless of the ampere rating of the device) counts as two conductor volumes based on the largest conductor that terminates on the device in accordance with Table 314.16(B).

Each multigang device yoke counts as two conductor volumes for each gang based on the largest conductor that terminates on the device in accordance with Table 314.16(B). **Figure 5–24**

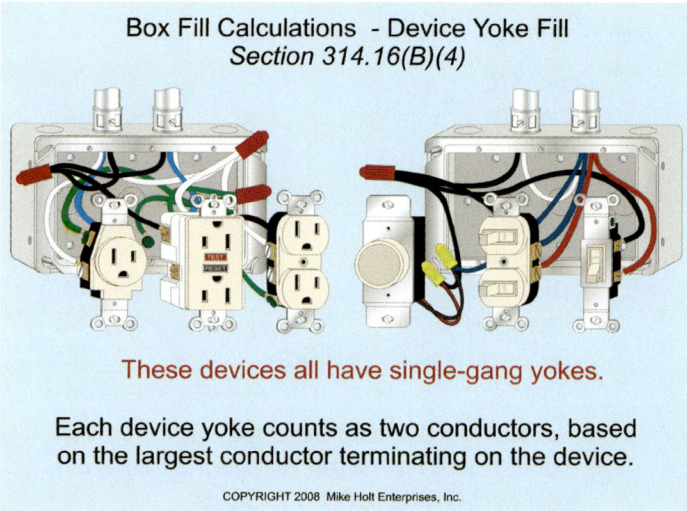

Figure 5–24

Author's Comment: A device that's too wide for mounting in a single gang box is counted based on the number of gangs required for the device. **Figure 5–25**

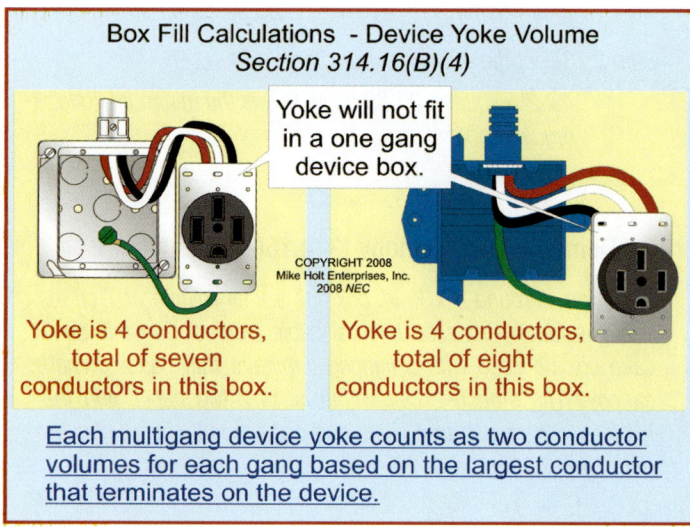

Figure 5–25

(5) Equipment Grounding Conductor Volume. All equipment grounding conductors in a box count as a single conductor volume in accordance with Table 314.16(B), based on the largest equipment grounding conductor that enters the box. Insulated equipment grounding conductors for receptacles having insulated grounding terminals (isolated ground receptacles) [250.146(D)] count as a single conductor volume in accordance with Table 314.16(B). **Figure 5–26**

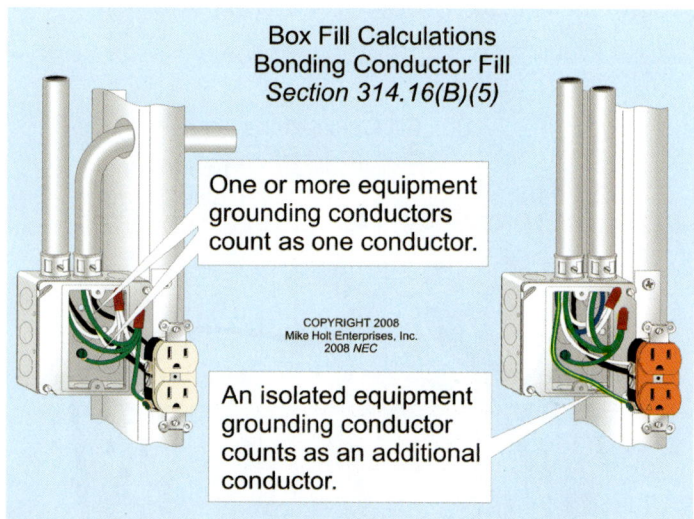

Figure 5–26

What's Not Counted

Wire connectors, cable connectors, raceway fittings, and conductors that originate and terminate within the outlet box (such as equipment bonding jumpers and pigtails) are not counted for box fill calculations [314.16(A)].

▶ **Number of Conductors**

Question: What is the total number of conductors used for the box fill calculations in **Figure 5–27**?

(a) 5 conductors (b) 7 conductors
(c) 9 conductors (d) 11 conductors

Answer: (d) 11 conductors

Switch and conductors	5 – 14 AWG †
Receptacles and conductors	4 – 14 AWG ††
Equipment grounding conductor	1 – 14 AWG
Cable clamps	+ 1 – 14 AWG
Total	11 – 14 AWG

(continued in next column)

Mike Holt Enterprises, Inc. • www.MikeHolt.com • 1.888.NEC.CODE (1.888.632.2633) 133

Unit 5 Raceway and Box Calculations

†two conductors for the device and three conductors terminating
††two conductors for the device and two conductors terminating
Each 14 AWG counts as 2 cu in. [Table 314.16(B)].

11 conductors x 2 cu in. = 22 cu in.

If the cubic inch volume of the mud ring is not stamped on it, or given in the problem, we can't include it in the box volume. Without knowing the mud ring volume, a 4 in. square by 2⅛ in. deep box will be the minimum required for this example.

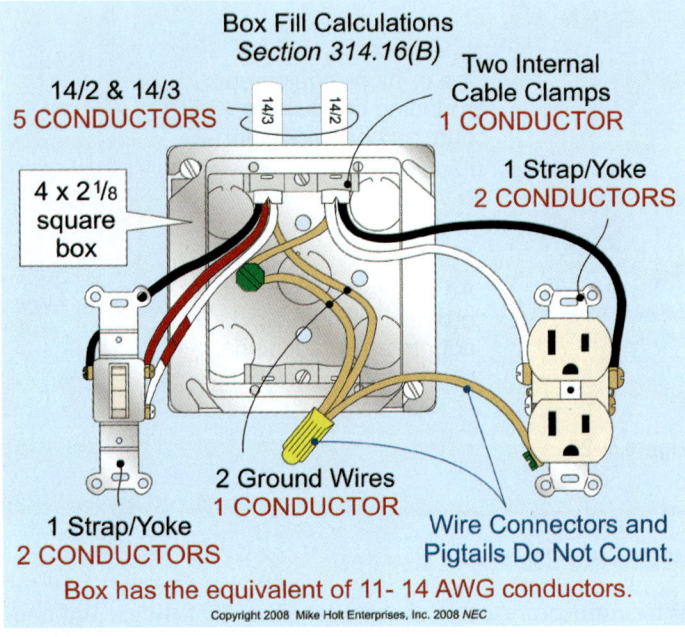

Figure 5–27

5.7 Outlet Box Sizing [314.16(B)]

To determine the size of the outlet box when the conductors are of different sizes (insulation is not a factor), follow these steps:

Step 1: Determine the number and size of conductor equivalents in the box.

Step 2: Determine the volume of the conductor equivalents from Table 314.16(B).

Step 3: Size the box by using Table 314.16(A).

▶ **Example: Calculating Different Size Conductors**

Question: What is the minimum size square outlet box required for one 14/3 Type NM cable that terminates on a 3-way switch, and one 12/2 Type NM cable that terminates on a receptacle? The box has internally installed cable clamps. **Figure 5–28**

(a) 4 x 1¼ square (b) 4 x 1½ square
(c) 4 x 2⅛ square (d) any of these

Answer: (c) 4 x 2⅛ square

Step 1: Determine the number of each size conductor.

14 AWG	
14/3 NM =	3 – 14 AWG
Switch	+ 2 – 14 AWG
Total	5 – 14 AWG

12 AWG	
12/2 NM	2 – 12 AWG
Cable clamp	1 – 12 AWG
Receptacle	2 – 12 AWG
Equipment grounding conductor	+ 1 – 12 AWG
Total	6 – 12 AWG

All equipment grounding conductors count as one conductor, based on the largest equipment grounding conductor entering the box [314.16(B)(5)].

Step 2: Determine the volume of the conductors [Table 314.16(B)].

14 AWG	2 cu in. each
2 cu in. x 5 conductors	10 cu in.
12 AWG	2.25 cu in. each
2.25 cu in. x 6 conductors	13.50 cu in.
Total Volume	10 cu in. + 13.50 cu in.
Total Volume	23.50 cu in.

Step 3: Select the outlet box from Table 314.16(A).

4 x 2⅛ square, 30.30 cu in. meets the minimum cu in. requirements

▶ **Domed Fixture Canopy [314.16(B)(1) Ex]**

Question: A round 4 x ½ in. box has a total volume of 7 cu in. and has factory-installed internal cable clamps. Can this pancake box be used with a lighting luminaire that has a domed canopy? The branch-circuit wiring is 14/2 NM cable, and the luminaire has two fixture wires and one ground wire all smaller than 14 AWG. **Figure 5–29**

(a) Yes (b) No

(continued on next page)

Raceway and Box Calculations — Unit 5

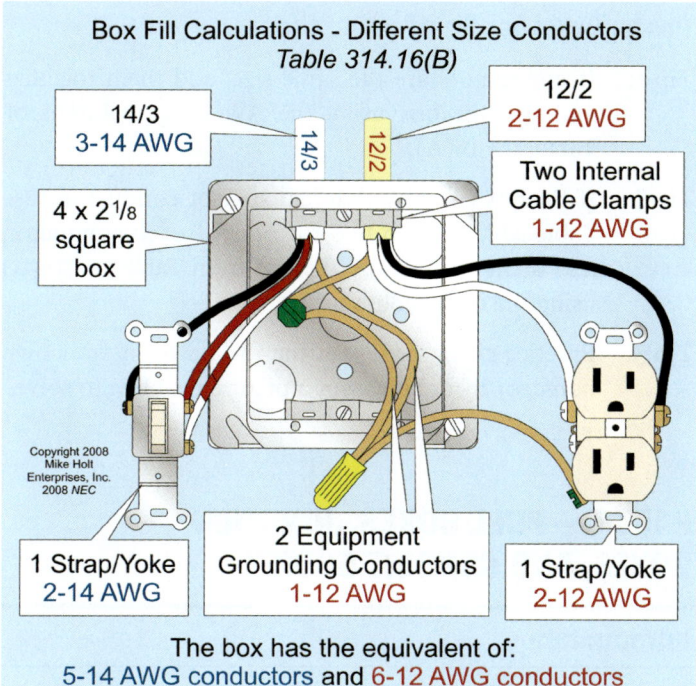

Figure 5–28

Answer: (b) No

The box is limited to 7 cu in., and the conductor equivalents total 8 cu in. [314.16(B)(1) Ex].

Step 1: Determine the number and size of conductors within the box.

Fixture wires, including the fixture equipment grounding conductor are not counted when the fixture has a domed canopy.

14/2 NM	2 – 14 AWG
Cable clamps	1 – 14 AWG
Ground wire	+ 1 – 14 AWG
Total	4 – 14 AWG conductors

Step 2: Determine the volume of the conductors [Table 314.16(B)].

14 AWG = 2 cu in.

Four 14 AWG conductors = 4 wires x 2 cu in. = 8 cu in.

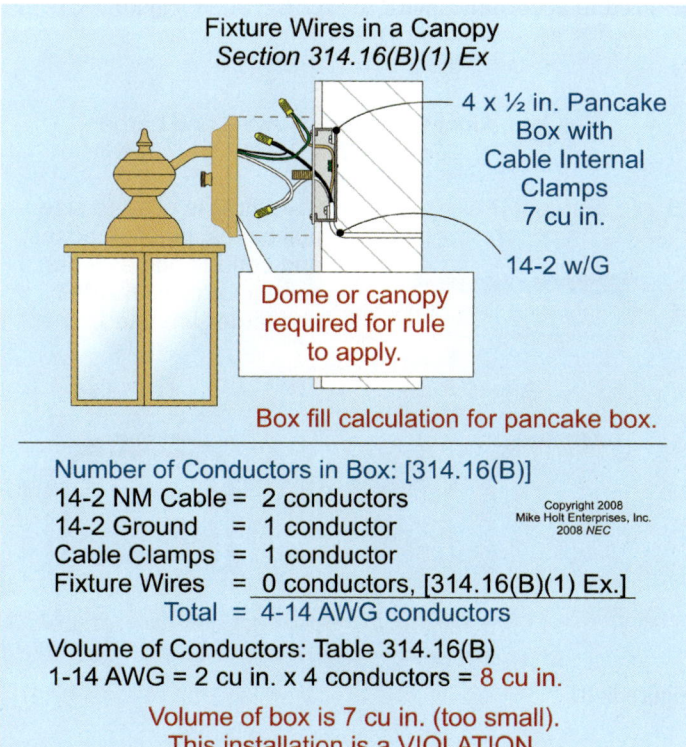

Figure 5–29

▶ **Conductors Added to an Existing Box**

Question: How many 14 AWG conductors can be pulled through a 4 x 2$\frac{1}{8}$ square box that has a plaster ring of 3.60 cu in.? The box already contains two receptacles, five 12 AWG conductors, and one 12 AWG equipment grounding conductor. **Figure 5–30**

(a) 4 conductors (b) 5 conductors
(c) 6 conductors (d) 7 conductors

Answer: (b) 5 conductors

Step 1: Determine the number and size of the existing conductors.

Two Receptacles (2 yokes x 2 conductors)	4 – 12 AWG conductors
Five 12 AWG	5 – 12 AWG conductors
One equipment grounding conductor	+ 1 – 12 AWG conductors
Total	10 – 12 AWG conductors

Step 2: Determine the volume of the existing conductors [Table 314.16(B)].

12 AWG conductor = 2.25 cu in.

10 wires x 2.25 cu in. = 22.50 cu in.

(continued in next column)

UNIT 5 — Raceway and Box Calculations

Step 3: Determine the space remaining for the additional 14 AWG conductors.

Remaining space = Total space less existing conductors

Total space = 30.30 cu in. (box) [Table 314.16(A)] + 3.60 cu in. (ring) = 33.90 cu in.

Remaining space = 33.90 cu in. − 22.50 cu in. (ten 12 AWG conductors)

Remaining space = 11.40 cu in.

Step 4: Determine the number of 14 AWG conductors permitted in the spare space.

Conductors added =
Remaining space/added conductors' volume

Conductors added = 11.40 cu in./2 cu in. = 5.70 [Table 314.16(B)]

Conductors added = 5
(Rounding up does not apply to box fill.)

Tips for Outlet Box Sizing

Tip 1: If conductors are the same size, add them together and size the box using the AWG size columns of Table 314.16(A).

Tip 2: If the box contains different sizes of conductors, use Table 314.16(B) to find the area of each conductor, add them up, and size the box from Table 314.16(A) using the cu in. column.

Tip 3: Practice sizing boxes on the jobsite or in your own home, or by drawing out a picture problem to solve.

PART C—PULL BOXES, JUNCTION BOXES, AND CONDUIT BODIES

Introduction

Pull boxes, junction boxes, and conduit bodies must be sized to permit conductors to be installed so that the conductor insulation is not damaged. For conductors 4 AWG and larger, pull boxes, junction boxes, and conduit bodies must be sized in accordance with the *NEC* 314.28. **Figure 5–31**

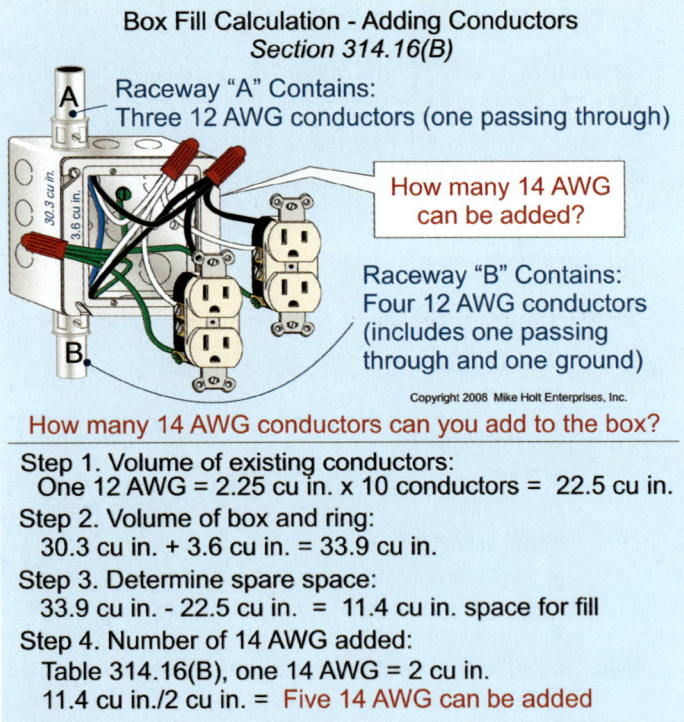

Figure 5–30

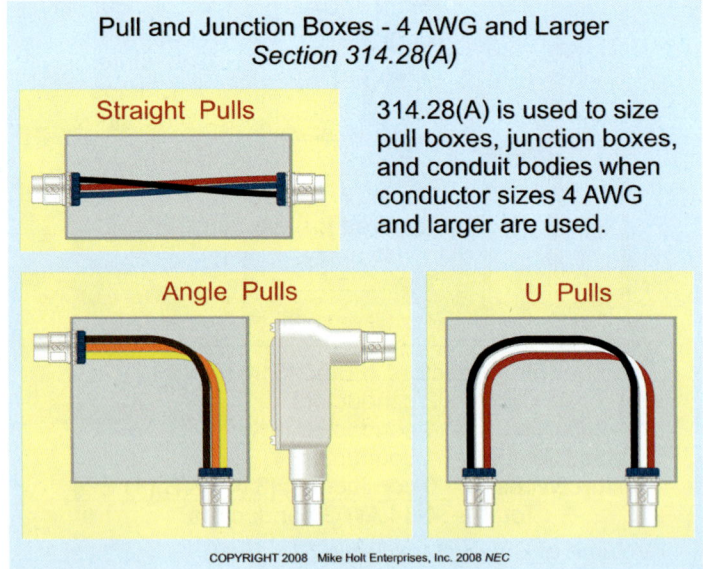

Figure 5–31

Raceway and Box Calculations — Unit 5

5.8 Pull/Junction Box Sizing Requirements

Boxes and Conduit Bodies for Conductors 4 AWG and Larger [314.28]

Boxes and conduit bodies containing conductors 4 AWG and larger that are required to be insulated must be sized so the conductor insulation will not be damaged.

(a) Minimum Size. For raceways containing conductors 4 AWG or larger, the minimum dimensions of boxes and conduit bodies must comply with the following:

(1) Straight Pulls. The minimum distance from where the conductors enter to the opposite wall must not be less than eight times the trade size of the largest raceway. Figure 5–32

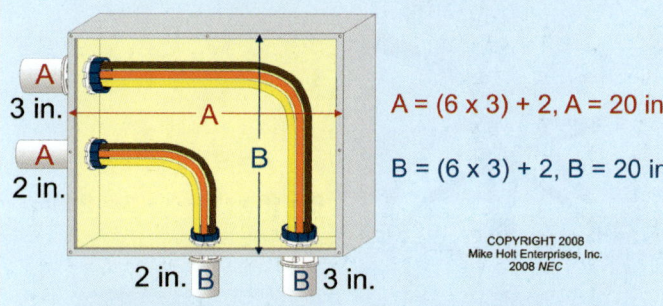

For angle pulls, the distance (measured from conductor wall entry to the opposite wall) must not be less than 6 times the largest raceway, plus the sum of the diameters of the remaining raceways on the same wall and row.

Figure 5–33

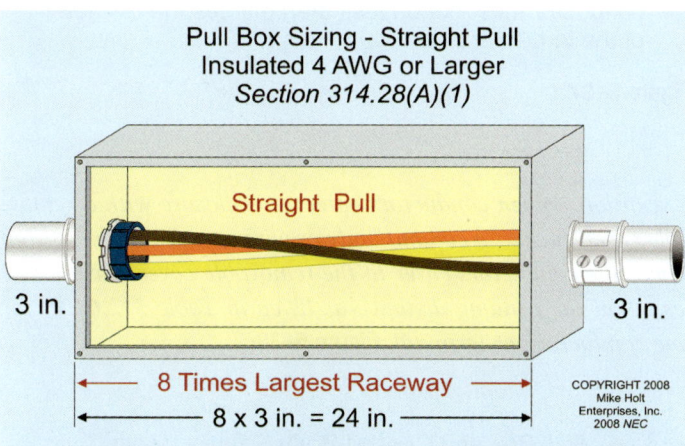

Figure 5–32

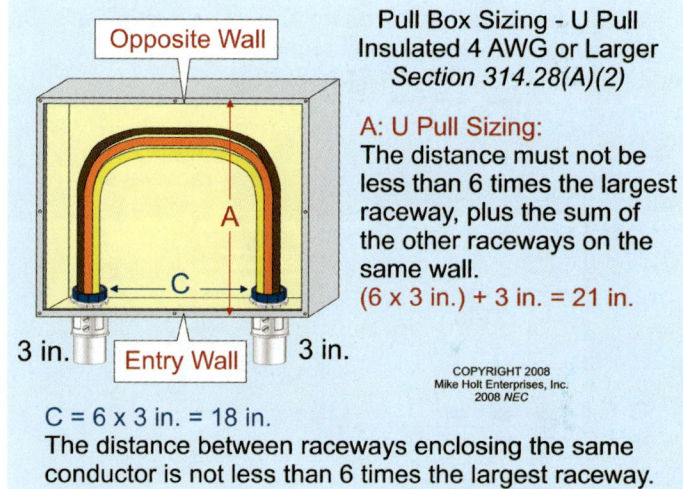

The distance between raceways enclosing the same conductor is not less than 6 times the largest raceway.

Figure 5–34

(2) Angle Pulls, U Pulls, or Splices.

- Angle Pulls. The distance from the raceway entry to the opposite wall must not be less than six times the trade size of the largest raceway, plus the sum of the trade sizes of the remaining raceways on the same wall and row. Figure 5–33

- U Pulls. When a conductor enters and leaves from the same wall, the distance from where the raceways enter to the opposite wall must not be less than six times the trade size of the largest raceway, plus the sum of the trade sizes of the remaining raceways on the same wall and row. Figure 5–34

- Splices. When conductors are spliced, the distance from where the raceways enter to the opposite wall must not be less than six times the trade size of the largest raceway, plus the sum of the trade sizes of the remaining raceways on the same wall and row. Figure 5–35

- Rows. Where there are multiple rows of raceway entries, each row is calculated individually and the row with the largest distance must be used. Figure 5–36

Mike Holt Enterprises, Inc. • www.MikeHolt.com • 1.888.NEC.CODE (1.888.632.2633)

Unit 5 Raceway and Box Calculations

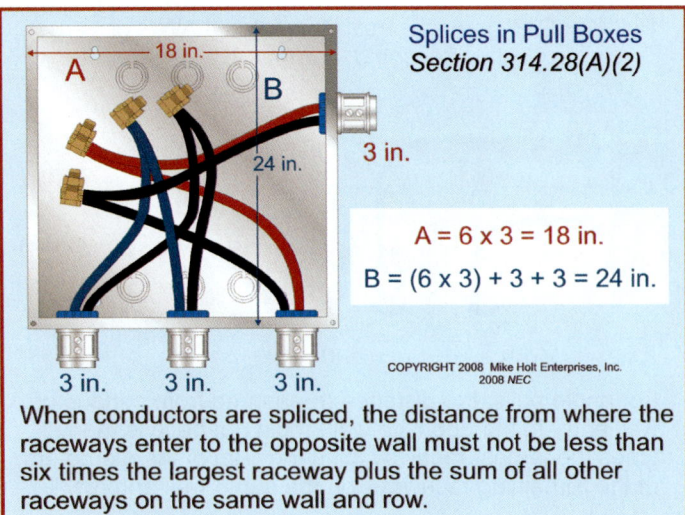

Figure 5–35

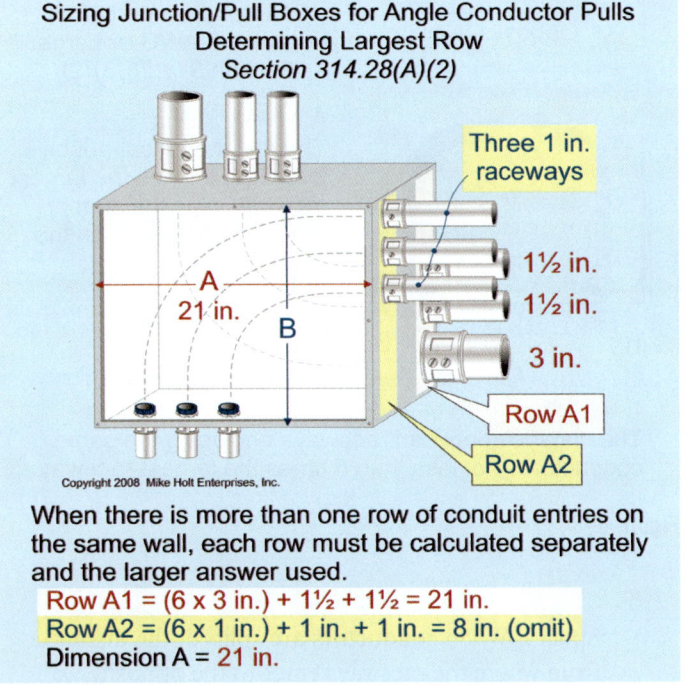

Figure 5–36

- Distance Between Raceways. The distance between raceways enclosing the same conductor must not be less than six times the trade size of the largest raceway, measured from the raceways' nearest edge-to-nearest edge. See **Figure 5–37**.

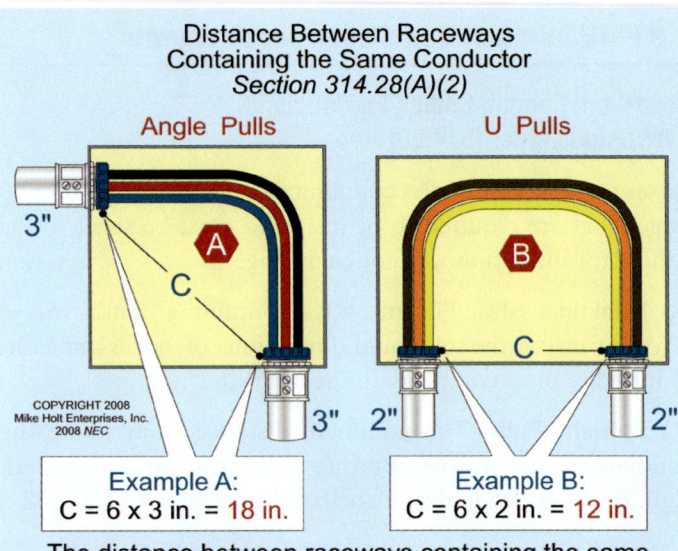

Figure 5–37

Exception: When conductors enter an enclosure with a removable cover, such as a conduit body or wireway, the distance from where the conductors enter to the removable cover must not be less than the bending distance as listed in Table 312.6(A) for one conductor per terminal. **Figure 5–38**

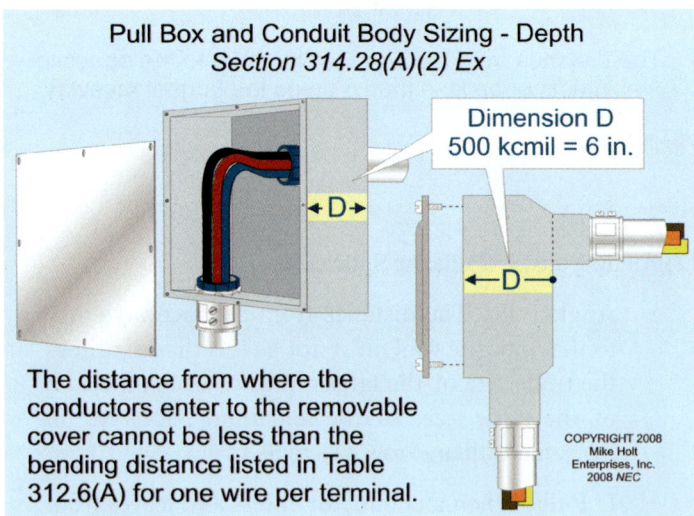

Figure 5–38

Raceway and Box Calculations — Unit 5

5.9 Pull/Junction Box Sizing Tips

When sizing pull and junction boxes, follow these suggestions:

Step 1: Always draw out the problem.

Step 2: Calculate the HORIZONTAL distance(s):
- Left to right straight calculation
- Right to left straight calculation
- Left to right angle or U pull calculation
- Right to left angle or U pull calculation

Step 3: Calculate the VERTICAL distance(s):
- Top to bottom straight calculation
- Bottom to top straight calculation
- Top to bottom angle or U pull calculation
- Bottom to top angle or U pull calculation

Step 4: Calculate the distance between raceways enclosing the same conductors.

5.10 Pull Box Examples

Pull Box Sizing

A junction box contains two trade size 3 raceways on the left side and one trade size 3 raceway on the right side. The conductors from one of the trade size 3 raceways on the left wall are pulled through the trade size 3 raceway on the right wall. The conductors from the other trade size 3 raceways on the left wall are pulled through a trade size 3 raceway at the bottom of the pull box.

▶ **Horizontal Dimension**

Question: What is the horizontal dimension of this box?

(a) 18 in.   (b) 21 in.   (c) 24 in.   (d) 30 in.

Answer: (c) 24 in. [314.28], **Figure 5–39**

Left to right straight pull	8 x 3 in. = 24 in.
Right to left straight pull	8 x 3 in. = 24 in.
Left to right angle pull	(6 x 3 in.) + 3 in. = 21 in.
Right to left angle pull	No calculation

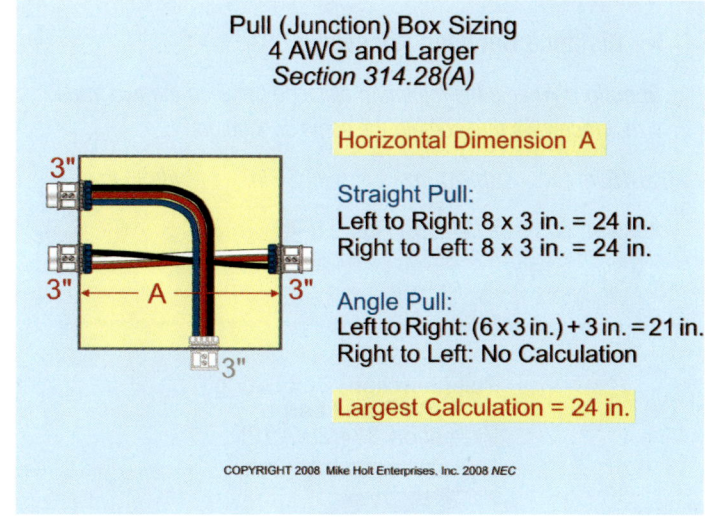

Pull (Junction) Box Sizing
4 AWG and Larger
Section 314.28(A)

Horizontal Dimension A

Straight Pull:
Left to Right: 8 x 3 in. = 24 in.
Right to Left: 8 x 3 in. = 24 in.

Angle Pull:
Left to Right: (6 x 3 in.) + 3 in. = 21 in.
Right to Left: No Calculation

Largest Calculation = 24 in.

Figure 5–39

▶ **Vertical Dimension**

Question: What is the vertical dimension of this box?

(a) 18 in.   (b) 21 in.   (c) 24 in.   (d) 30 in.

Answer: (a) 18 in. [314.28], **Figure 5–40**

Top to bottom straight	No calculation
Bottom to top straight	No calculation
Top to bottom angle	No calculation
Bottom to top angle	6 x 3 in. = 18 in.

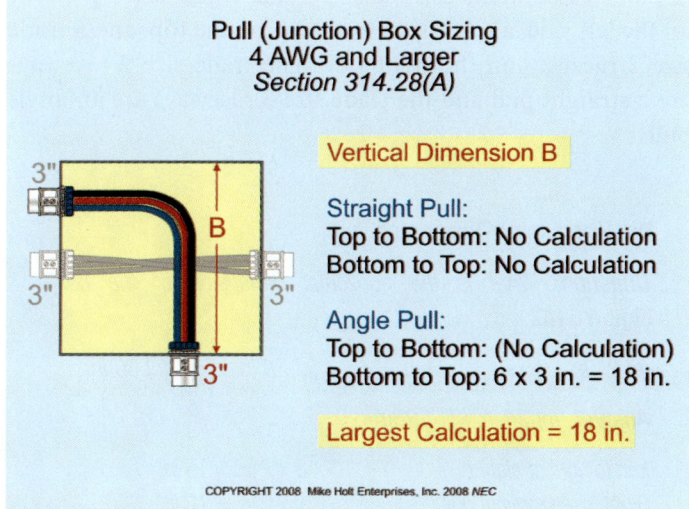

Pull (Junction) Box Sizing
4 AWG and Larger
Section 314.28(A)

Vertical Dimension B

Straight Pull:
Top to Bottom: No Calculation
Bottom to Top: No Calculation

Angle Pull:
Top to Bottom: (No Calculation)
Bottom to Top: 6 x 3 in. = 18 in.

Largest Calculation = 18 in.

Figure 5–40

Unit 5 Raceway and Box Calculations

▶ Distance Between Raceways

Question: What is the minimum distance between the two trade size 3 raceways that contain the same conductors?

(a) 18 in.   (b) 21 in.   (c) 24 in.   (d) 30 in.

Answer: (a) 18 in. [314.28] Figure 5–41

6 x 3 in. = 18 in.

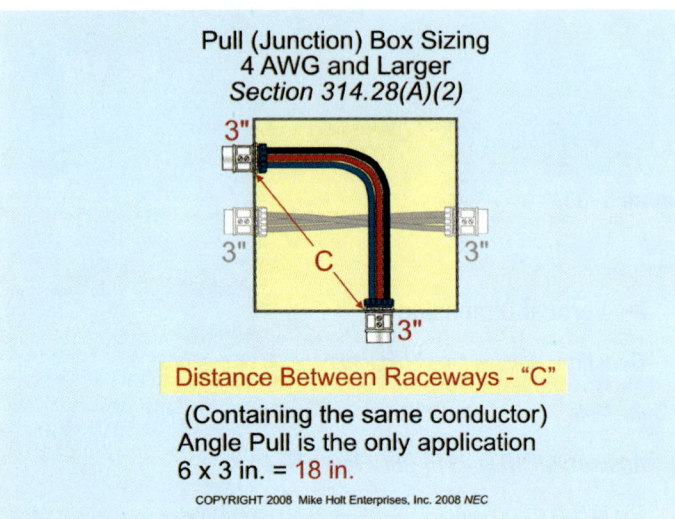

Figure 5–41

Pull Box Sizing

A pull box contains a trade size 2 and trade size 3 raceway on the left side, a trade size 3 raceway on the top, and a trade size 2 raceway on the right side. The trade size 2 raceways are a straight pull and the trade size 3 raceways are an angle pull.

▶ Horizontal Dimension

Question: What is the horizontal dimension of the box?
Figure 5–42

(a) 21 in.   (b) 24 in.   (c) 28 in.   (d) 30 in.

Answer: (b) 24 in. [314.28(A)(2)]

Left to right straight pull	8 x 2 in. = 16 in.
Right to left straight pull	8 x 2 in. = 16 in.
Left to right angle pull	(6 x 3 in.) + 2 in. = 20 in.
Right to left angle pull	No calculation

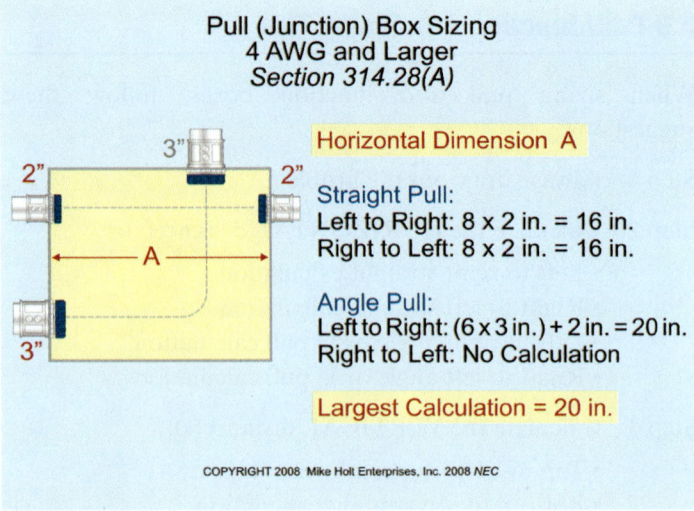

Figure 5–42

▶ Vertical Dimension

Question: What is the vertical dimension of the box?
Figure 5–43

(a) 14 in.   (b) 18 in.   (c) 21 in.   (d) 26 in.

Answer: (b) 18 in. [314.28(A)(2)]

Top to bottom straight	No calculation
Bottom to top straight	No calculation
Top to bottom angle	6 x 3 in. = 18 in.
Bottom to top angle	No calculation

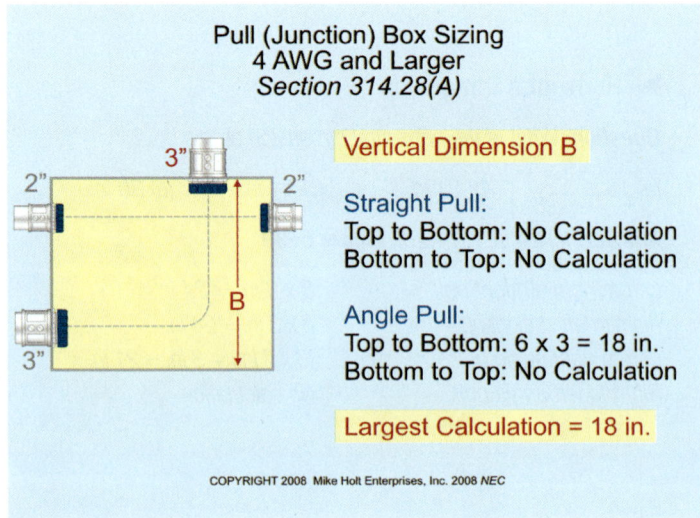

Figure 5–43

Raceway and Box Calculations — Unit 5

▶ **Distance Between Raceways**

Question: If the two trade size 3 raceways contain the same conductors, what is the minimum distance between these raceways? **Figure 5–44**

(a) 18 in.   (b) 21 in.   (c) 24 in.   (d) 30 in.

Answer: (a) 18 in. [314.28(A)(2)]
6 x 3 in. = 18 in.

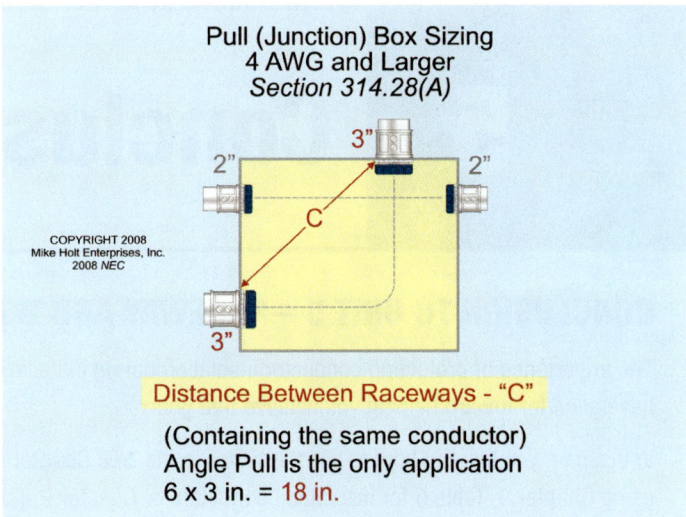

Figure 5–44

Summary

1. Slow down and take your time on these calculations. It's easy to make simple mistakes.

2. Pull box calculations can be tricky—remember which wall you're calculating.

3. Box calculations become more familiar with more practice. Use practical field examples to practice calculations.

4. Draw up some sample pull boxes and calculate them for practice.

Unit 5 Conclusion

CONCLUSION TO UNIT 5—RACEWAY AND BOX CALCULATIONS

The importance of protecting conductor insulation during installation is understood by every electrician. The principles covered in this unit have gone far toward helping you achieve that goal.

In this unit, you learned how to use the tables in the *NEC* Chapter 9 when calculating raceway fill. First, the area of the conductor is found, using Chapter 9, Table 5 for insulated conductors or Chapter 9, Table 8 for bare conductors. Then after totaling the cross-sectional area of all the conductors, the raceway trade size from Chapter 9, Table 4 is selected using the available cross-sectional area of the raceway. Take your time when working out these calculations to select the correct insulation type and the correct raceway for your problem.

Annex C provides a quicker method for sizing raceways than the Chapter 9 Tables when the conductors are all the same AWG size, and the same size in cross-sectional area (including the insulation). There are numerous tables in Annex C, based on the raceway type, that can be used to reduce the time spent in calculations, so once again be sure to take your time and be certain you're using the correct table.

Metal wireways must be sized using the requirements of Article 376. The maximum fill for wireways isn't just whatever you can squeeze into them, there are specific limits on conductor cross-sectional area, number of conductors, and wire bending radius. Following the rules of Article 376 will provide a much safer installation and takes the mystery out of sizing wireways.

Maintaining proper outlet box fill is important as you rough-in the installation. There are a number of details that were covered in determining the correct box size to be used that will be important on an exam. Be sure to remember what items are counted in box fill calculations, and review the information if necessary.

Junction box and pull box calculations only come into play when the conductors are 4 AWG and larger. Straight pulls, angle pulls, and U pulls were all covered in this unit, as well as an often forgotten requirement that applies to two raceways that contain the same conductors. Always draw out a problem involving junction or pull boxes so that you'll be able to visualize it and properly apply the calculations you've learned.

Unit 5 Practice Questions

UNIT 5—RACEWAY AND BOX CALCULATIONS PRACTICE QUESTIONS

(•Indicates that 75% or fewer of those who took this exam answered the question correctly.)

PART A—RACEWAY FILL

5.1 Understanding the NEC, Chapter 9 Tables

1. When all the conductors are the same size (total cross-sectional area including insulation), the number of conductors permitted in a raceway can be determined by simply looking at the Tables in the NEC _____.

   (a) Chapter 9
   (b) Annex B
   (c) Annex C
   (d) Annex D

2. When equipment grounding conductors are installed in a raceway, the actual area of the conductor must be used when calculating raceway fill.

   (a) True
   (b) False

3. When a raceway does not exceed 24 in. in length, the raceway is permitted to be filled to _____ percent of its cross-sectional area.

   (a) 31
   (b) 40
   (c) 53
   (d) 60

4. How many 16 TFFN conductors can be installed in trade size ¾ electrical metallic tubing?

   (a) 26
   (b) 29
   (c) 30
   (d) 40

5. How many 6 RHH conductors (without outer cover) can be installed in trade size 1¼ electrical nonmetallic tubing?

   (a) 7
   (b) 13
   (c) 16
   (d) 25

6. How many 1/0 XHHW conductors can be installed in trade size 2 flexible metal conduit?

   (a) 6
   (b) 7
   (c) 13
   (d) 16

7. How many 12 RHH conductors (with outer cover) can be installed in a trade size 1 IMC raceway?

   (a) 4
   (b) 5
   (c) 7
   (d) 11

8. Three THHN compact conductors are needed in a trade size 2 rigid metal conduit. What is the largest compact conductor that can be installed?

   (a) 4/0 AWG
   (b) 250 kcmil
   (c) 350 kcmil
   (d) 500 kcmil

9. The actual area of conductor fill is dependent on the raceway size and the number of conductors installed. If there are three or more conductors installed in a raceway, the total area of conductor fill is limited to _____ percent.

   (a) 31
   (b) 40
   (c) 53
   (d) 60

Unit 5 — Practice Questions

10. •What is the cross-sectional area in square inches for 10 THW?

    (a) 0.0172
    (b) 0.0243
    (c) 0.0252
    (d) 0.0278

11. What is the cross-sectional area in square inches for 14 RHW (without an outer cover)?

    (a) 0.0172
    (b) 0.0209
    (c) 0.0252
    (d) 0.0278

12. What is the cross-sectional area in square inches for 10 THHN?

    (a) 0.0117
    (b) 0.0172
    (c) 0.0211
    (d) 0.0252

13. What is the cross-sectional area in square inches for 12 RHH (with an outer cover)?

    (a) 0.0117
    (b) 0.0252
    (c) 0.0327
    (d) 0.0353

14. •What is the cross-sectional area in square inches for an 8 AWG bare solid conductor?

    (a) 0.013
    (b) 0.027
    (c) 0.038
    (d) 0.045

5.2 Raceway Calculations

15. The number of conductors permitted in a raceway is dependent on the _____.

    (a) area of the raceway
    (b) percent area fill as listed in Chapter 9, Table 1
    (c) area of the conductors as listed in Chapter 9, Tables 5 and 8
    (d) all of these

16. A 200A feeder installed in Schedule 80 PVC has three 3/0 THHN conductors, one 2 THHN conductor, and one 6 THHN conductor. What size raceway is required?

    (a) Trade size 2
    (b) Trade size 2½
    (c) Trade size 3
    (d) Trade size 3½

17. What size rigid metal nipple is required for three 4/0 THHN conductors, one 1/0 THHN conductor, and one 4 THHN conductor?

    (a) Trade size 1½
    (b) Trade size 2
    (c) Trade size 2½
    (d) Trade size 3

18. •An existing trade size ¾ rigid metal conduit nipple contains four 10 THHN conductors and one 10 AWG (bare stranded) ground wire. How many additional 10 THHN conductors can be installed?

    (a) 5
    (b) 7
    (c) 9
    (d) 11

5.3 Wireways

19. What is the cross-sectional area of a 4 in. x 4 in. wireway?

    (a) 6 sq in.
    (b) 16 sq in.
    (c) 36 sq in.
    (d) 66 sq in.

20. What is the maximum allowable sq in. of conductor fill for a 4 in. x 4 in. wireway?

    (a) 2.40 sq in.
    (b) 3.20 sq in.
    (c) 5.30 sq in.
    (d) 12 sq in.

21. What is the maximum number of 400 kcmil THHN conductors that can be installed in a 6 in. x 6 in. wireway?

    (a) 4
    (b) 6
    (c) 10
    (d) 12

Practice Questions — Unit 5

PART B—OUTLET BOX FILL CALCULATIONS [314.16]

5.5 Sizing Box—Conductors All the Same Size [Table 314.16(A)]

22. What size box is the minimum required for six 14 THHN conductors and three 14 THW conductors?

    (a) 4 x 1¼ square
    (b) 4 x 1½ round
    (c) 4 x 1¼ round
    (d) 4 x 2⅛ square

23. How many 10 AWG conductors are permitted in a 4 x 1½ square box?

    (a) 8 conductors
    (b) 9 conductors
    (c) 10 conductors
    (d) 11 conductors

5.6 Conductor Equivalents

24. Table 314.16(A) does not take into consideration the volume of _____.

    (a) switches and receptacles
    (b) luminaire studs and hickeys
    (c) internal cable clamps
    (d) all of these

25. When determining the number of conductors for box fill calculations, which of the following statements is(are) true?

    (a) A luminaire stud or hickey is considered as one conductor for each type, based on the largest conductor that enters the outlet box.
    (b) Internal factory cable clamps are considered as one conductor for one or more cable clamps, based on the largest conductor that enters the outlet box.
    (c) The single gang device yoke is considered as two conductors, based on the largest conductor that terminates on the strap (device mounting fitting).
    (d) all of these

26. •When determining the number of conductors for box fill calculations, which of the following statements is(are) true?

    (a) Each conductor that runs through the box without a splice or leaving a loop long enough to splice is considered as one conductor.
    (b) Each conductor that originates outside the box and terminates in the box is considered as one conductor.
    (c) Wirenuts, cable connectors, raceway fittings, and conductors that originate and terminate within the outlet box (equipment bonding jumpers and pigtails) are not counted for box fill calculations.
    (d) all of these

27. It is permitted to omit one equipment grounding conductor and not more than _____ that enter a box from a luminaire canopy.

    (a) five fixture wires
    (b) four 16 AWG fixture wires
    (c) four 18 AWG fixture wires
    (d) b and c

28. Can a round 4 x ½ in. box marked as 8 cu in. with manufactured cable clamps supplied with 14/2 NM be used with a luminaire that has two 18 TFN conductors and a canopy cover?

    (a) Yes
    (b) No

5.7 Outlet Box Sizing [314.16(B)]

29. •What size outlet box is required for one 12/2 NM cable that terminates on a switch, one 12/3 NM cable that terminates on a receptacle, and the box has manufactured cable clamps?

    (a) 4 x 1¼ square
    (b) 4 x 1½ square
    (c) 4 x 2⅛ square
    (d) 3 x 2 x 3½ device box

30. •How many 14 AWG conductors can be pulled through a 4 x 1½ square box with a plaster ring marked 3.60 cu in.? The box contains two duplex receptacles, five 14 AWG conductors, and two grounding conductors.

    (a) one
    (b) two
    (c) three
    (d) four

Unit 5 — Practice Questions

PART C—PULL BOXES, JUNCTION BOXES, AND CONDUIT BODIES

5.8 Pull/Junction Box Sizing Requirements

31. When conductors 4 AWG and larger are installed in boxes and conduit bodies, the enclosure must be sized according to which of the following requirements?

    (a) The minimum distance for straight pull calculations from where the conductors enter to the opposite wall must not be less than eight times the trade size of the largest raceway.
    (b) The distance for angle pull calculations from the raceway entry to the opposite wall must not be less than six times the trade size diameter of the largest raceway, plus the sum of the diameters of the remaining raceways on the same wall and row.
    (c) The distance between raceways enclosing the same conductor(s) must not be less than six times the trade size diameter of the largest raceway.
    (d) all of these

32. When conductors enter an enclosure opposite a removable cover, the distance from where the conductors enter to the removable cover must not be less than _____.

    (a) six times the largest raceway
    (b) eight times the largest raceway
    (c) a or b
    (d) none of these

The following information applies to the next three questions.

A junction box contains two trade size 2½ raceways on the left side and one trade size 2½ raceway on the right side. The conductors from one trade size 2½ raceway (left wall) are pulled through the raceway on the right wall. The other trade size 2½ raceway conductors (on the left wall) are pulled through a trade size 2½ raceway at the bottom of the pull box.

33. What is the minimum distance from the left wall to the right wall?

    (a) 18 in.
    (b) 20 in.
    (c) 21 in.
    (d) 24 in.

34. •What is the minimum distance from the bottom wall to the top wall?

    (a) 15 in.
    (b) 18 in.
    (c) 21 in.
    (d) 24 in.

35. What is the minimum distance between the raceways that contain the same conductors?

    (a) 15 in.
    (b) 18 in.
    (c) 21 in.
    (d) 24 in.

The following information applies to the next three questions.

A junction box contains two trade size 2 raceways on the left side, and two trade size 2 raceways on the top.

36. What is the minimum distance from the left wall to the right wall?

    (a) 14 in.
    (b) 21 in.
    (c) 24 in.
    (d) 28 in.

37. What is the minimum distance from the bottom wall to the top wall?

    (a) 14 in.
    (b) 18 in.
    (c) 21 in.
    (d) 24 in.

38. What is the minimum distance between the trade size 2 raceways that contain the same conductors?

    (a) 12 in.
    (b) 18 in.
    (c) 21 in.
    (d) 24 in.

Challenge Questions

UNIT 5—RACEWAY AND BOX CALCULATIONS CHALLENGE QUESTIONS

(•Indicates that 75% or fewer of those who took this exam answered the question correctly.)

PART A—RACEWAY FILL

5.2 Raceway Calculations

1. •A trade size 3 Schedule 40 PVC raceway contains seven 1 RHW conductors without outer cover. How many 2 THW conductors may be installed in this raceway with the existing conductors?

    (a) 11
    (b) 15
    (c) 20
    (d) 25

5.3 Wireways

2. What size wireway is required for three 400 kcmil THHN, one 250 kcmil THHN, four 4/0 THHN conductors, and three 8 THHN conductors?

    (a) 4 in. x 4 in.
    (b) 6 in. x 6 in.
    (c) 8 in. x 8 in.
    (d) 10 in. x 10 in.

PART B—OUTLET BOX FILL CALCULATIONS [314.16]

5.7 Outlet Box Sizing [314.16(B)]

3. •Determine the minimum cubic inches required for two 10 AWG conductors passing through a box, four 14 AWG conductors spliced in the box, two 12 AWG conductors terminating to a receptacle, and one 12 AWG equipment bonding jumper from the receptacle to the box.

    (a) 18.50 cu in.
    (b) 20 cu in.
    (c) 21.75 cu in.
    (d) 22 cu in.

4. •When determining the number of conductors in a box fill that has two 18 AWG fixture wires from a domed luminaire, one 14/3 nonmetallic-sheathed cable with ground, one duplex switch, and two internal cable clamps, the count will equal _____.

    (a) 6 conductors
    (b) 7 conductors
    (c) 8 conductors
    (d) 10 conductors

PART C—PULL BOXES, JUNCTION BOXES, AND CONDUIT BODIES

5.8 Pull/Junction Box Sizing Requirements

Figure 5–45 applies to the next four questions.

5. The minimum horizontal dimension for the junction box shown in the diagram in Figure 5–45 is _____.

    (a) 18 in.
    (b) 20 in.
    (c) 21 in.
    (d) 24 in.

Unit 5 Challenge Questions

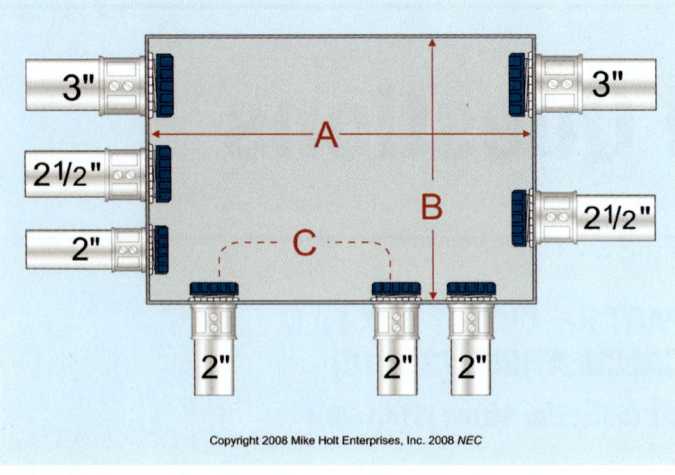

Figure 5–45

6. •The minimum vertical dimension for the junction box shown in **Figure 5–45** is _____.

   (a) 16 in.
   (b) 18 in.
   (c) 20 in.
   (d) 24 in.

7. The minimum distance between the two trade size 2 raceways that contain the same Conductor C, as illustrated in **Figure 5–45** is _____.

   (a) 12 in.
   (b) 18 in.
   (c) 24 in.
   (d) 30 in.

8. •If a trade size 3 raceway entry (250 kcmil) is in the wall opposite a removable cover, the distance from that wall to the cover must not be less than _____. **Figure 5–45**

   (a) 4 in.
   (b) 4½ in.
   (c) 5 in.
   (d) 6 in.

Unit 6: Conductor Sizing and Protection Calculations

INTRODUCTION TO UNIT 6—CONDUCTOR SIZING AND PROTECTION CALCULATIONS

Every conductor has a certain amount of resistance. Any time current flows through a conductor there is resistance, and heat is generated. The amount of heat produced is directly proportional to the resistance of the wire and the square of the current flow. In order to determine what size conductor is needed to carry a certain amount of current, the heat that will be generated must be taken into consideration. The number of conductors in the same raceway that are carrying current can affect this decision, as can the ambient or surrounding temperature.

This unit explains how to evaluate these factors and make the proper choice in sizing electrical conductors, and how to select the proper conductor based on the temperature rating of the equipment terminals. This can become a complicated choice, and you should study this unit carefully to learn how to make the right conductor sizing decisions.

The amount of current that a conductor is allowed to carry must be limited to a level that won't cause destructive overheating of the conductor or of the equipment or structure it may be in contact with. In order to make sure that the amount of current on a conductor is kept within safe limits, overcurrent protection is used to open the circuit. Generally this protection is provided by installing a fuse or circuit breaker, of a size based on the conductor's ampacity, at the point where the conductor receives its supply. The general rules for overcurrent protection are covered in this unit, along with exceptions to the general rules, such as those for tap conductors and special cases such as motor circuits. Motor circuits will be covered in more depth in Unit 7. Some specific overcurrent applications and tap rules will be covered to help you become familiar with the rules for sizing overcurrent devices.

PART A—GENERAL CONDUCTOR REQUIREMENTS

Introduction

All conductors used for general applications covered by the *NEC* are made from copper and aluminum. But beyond the conductor material, there are other characteristics that affect how conductors can be used and how their ampacity is calculated. These include the temperature rating of the insulation, the conditions under which the conductors are installed (i.e., in free air, in a raceway, direct buried), the ambient temperature, and the temperature rating of the terminal, device, or equipment to which the conductors are connected.

6.1 Conductor Insulation [Table 310.13]

Table 310.13 of the *NEC* provides information on conductor properties such as permitted use, maximum operating temperature, and other insulation details. **Figure 6–1**

The following abbreviations and explanations are helpful in understanding Tables 310.13 and 310.16.

- -2   Conductor is permitted to be used at a continuous 90°C operating temperature
- F   Fixture wire (solid or 7-strand) [Table 402.3]
- FF   Flexible fixture wire (19-strand) [Table 402.3]
- H   75°C insulation rating
- HH   90°C insulation rating
- N   Nylon outer cover
- T   Thermoplastic insulation
- W   Wet or damp location

Unit 6 — Conductor Sizing and Protection Calculations

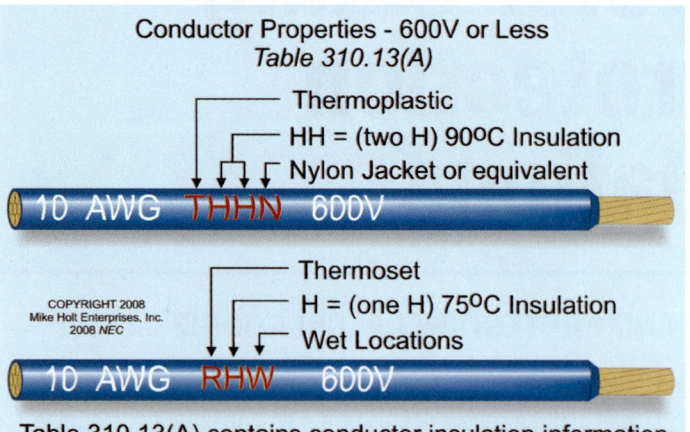

Figure 6–1

For more information about fixture wires, see Article 402, Table 402.3, and Table 402.5. For more information on flexible cords and flexible cables, see Article 400, Table 400.4, Table 400.5(A), and Table 400.5(B).

It is common to see conductors with a dual insulation rating, such as THHN/THWN. This type of conductor can be used in a dry location at the THHN 90°C ampacity, or if used in a wet location, the THWN ampacity rating of the 75°C column of Table 310.16 for THWN insulation types must be adhered to. When a -2 is added at the end of an insulation type, such as THWN-2, that means that the conductor can be used in a wet or dry location at the 90°C ampacity rating.

	Table 310.13 Conductor Information				
	Column 2	Column 3	Column 4	Column 5	Column 6
Type Letter	Insulation	Maximum Operating Temperature	Application	Sizes Available AWG or Kcmil	Outer Covering
THHN	Heat-resistant thermoplastic	90°C	Dry and damp locations	14 – 1,000	Nylon jacket or equivalent
THHW	Moisture- and heat-resistant thermoplastic	75°C 90°C	Wet locations Dry and damp locations	14 – 1,000	None
THW	Moisture- and heat-resistant thermoplastic	75°C 90°C	Dry, damp, and wet locations Within electrical discharge lighting equipment See Section 410.33	14 – 2,000	None
THWN	Moisture- and heat-resistant thermoplastic	75°C	Dry and wet locations	14 – 1,000	Nylon jacket or resistant equivalent
TW	Moisture-resistant thermoplastic	60°C	Dry and wet locations	14 – 2,000	None
XHHW	Moisture-resistant thermoset	90°C 75°C	Dry and damp locations Wet locations	14 – 2,000	None

Conductor Sizing and Protection Calculations — Unit 6

Table 310.16. Allowable Ampacities of Insulated Conductors
Based On Not More Than Three Current-Carrying Conductors and Ambient Temperature of 30°C (86°F)

Size AWG kcmil	Temperature Rating of Conductor, See Table 310.13						Size AWG kcmil
	60°C (140°F) TW UF	75°C (167°F) THHW THW THWN XHHW Wet Location	90°C (194°F) THHW XHHW Dry Location	60°C (140°F) TW UF	75°C (167°F) THHN THW THWN XHHW Wet Location	90°C (194°F) THHN THHW XHHN Dry Location	
	Copper			Aluminum/Copper-Clad Aluminum			
14*	20	20	25				12*
12*	25	25	30	20	20	25	10*
10*	30	35	40	25	30	35	8*
8	40	50	55	30	40	45	8
6	55	65	75	40	50	60	6
4	70	85	95	55	65	75	4
3	85	100	110	65	75	85	3
2	95	115	130	75	90	100	2
1	110	130	150	85	100	115	1
1/0	125	150	170	100	120	135	1/0
2/0	145	175	195	115	135	150	2/0
3/0	165	200	225	130	155	175	3/0
4/0	195	230	260	150	180	205	4/0
250	215	255	290	170	205	230	250
300	240	285	320	190	230	255	300
350	260	310	350	210	250	280	350
400	280	335	380	225	270	305	400
500	320	380	430	260	310	350	500

*See 240.4(D)

▶ **Table 310.13**

Question: Which of the following describe type TW insulation? **Figure 6–2**

(a) thermoplastic insulation
(b) suitable for dry or wet locations
(c) maximum operating temperature of 60°C
(d) all of these

Answer: (d) all of these

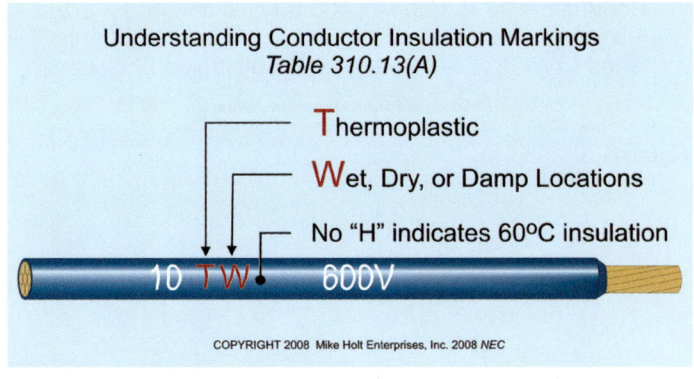

Understanding Conductor Insulation Markings
Table 310.13(A)

- **T**hermoplastic
- **W**et, Dry, or Damp Locations
- No "H" indicates 60°C insulation

Figure 6–2

UNIT 6 — Conductor Sizing and Protection Calculations

> ▶ **Table 402.3**
>
> *Question:* Which of the following describe type TFFN insulation?
>
> (a) stranded fixture wire
> (b) thermoplastic insulation with a nylon outer cover
> (c) suitable for dry and wet locations
> (d) both a and b
>
> *Answer:* (d) both a and b

6.2 Conductor Sizing [110.6]

Conductors are sized according to the American Wire Gage (AWG) from 40 AWG through 4/0 AWG. The smaller conductors are represented by the larger numbers, with the AWG size numbers decreasing as the conductor size increases. Above 1 AWG are sizes 0, 2/0, 3/0, and 4/0. Conductors larger than 4/0 AWG are identified according to their cross-sectional area in circular mils, such as 250,000 cmil, 300,000 cmil, 500,000 cmil, etc. The circular mil size is usually expressed in kcmil (1,000 circular mils), such as 250 kcmil, 300 kcmil, and 500 kcmil. **Figure 6–3**

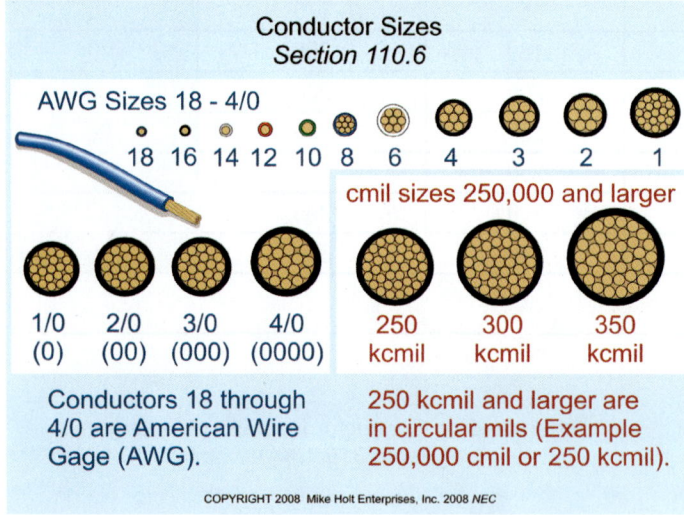

Figure 6–3

6.3 Smallest Conductor Size [310.5]

The smallest size conductors permitted by the *NEC* for branch circuits, feeders, or services are 14 AWG copper [Table 310.5]. However, some local codes require a minimum 12 AWG for commercial and industrial installations. Conductors smaller than 14 AWG are permitted for:

Class 1 remote-control circuits [725.43]
Fixture wire [402.6]
Motor control circuits [Table 430.72(B)]

6.4 Conductor Size—Terminal Temperature Rating [110.14(C)]

Conductors are to be sized to the lowest temperature rating of any terminal, device, or conductor of the circuit in accordance with the equipment terminal temperature rating as follows:

Circuits Rated 100A and Less [110.14(C)(1)(a)]

Equipment terminals rated 100A or less and pressure connector terminals for 14 AWG through 1 AWG conductors, must have the conductor sized to the 60°C temperature rating listed in Table 310.16, unless the terminals are marked for 75°C conductors.

> **Author's Comment:** Conductors are sized to prevent the overheating of terminals, in accordance with listing standards. For example, a 50A circuit with 60°C terminals requires the circuit conductors to be sized not smaller than 6 AWG, in accordance with the 60°C ampacity listed in Table 310.16. An 8 THHN insulated conductor has a 90°C ampacity of 50A in a dry location, but 8 AWG can't be used for this circuit because the conductor's operating temperature at full-load ampacity (50A) will be near 90°C, which is well in excess of the 60°C terminal rating.

> ▶ **Terminal Rated 60°C [110.14(C)(1)(a)(2)]**
>
> *Question:* What size THHN conductor is required for a 50A circuit where the terminals are not marked with a temperature rating? **Figure 6–4**
>
> (a) 10 AWG   (b) 8 AWG   (c) 6 AWG   (d) 4 AWG
>
> *Answer:* (c) 6 AWG

Conductor Sizing and Protection Calculations — Unit 6

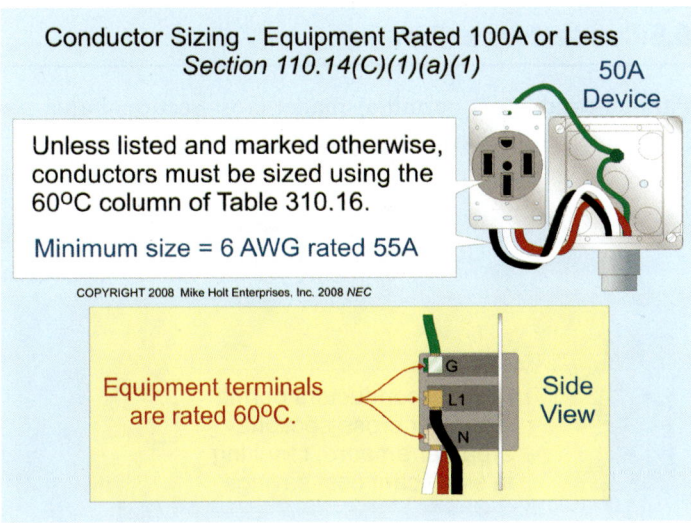

Figure 6–4

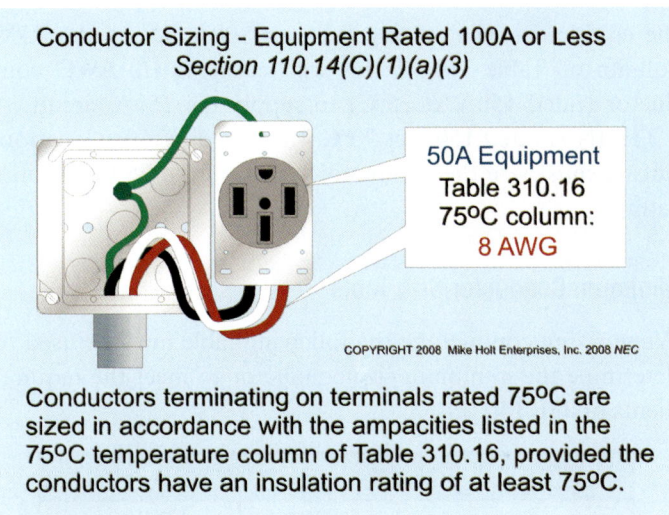

Figure 6–5

Conductors must be sized to the lowest temperature rating of either the equipment or the conductor. Equipment terminals for 100A and less that are not marked are to be sized using the 60°C column. THHN insulation can be used, but the conductor size must be selected based on the 60°C terminal rating of the equipment, not the 90°C rating of the insulation. Using the 60°C column of Table 310.16, this 50A circuit requires a 6 AWG conductor (rated 55A at 60°C).

▶ **Terminal Rated 75°C [110.14(C)(1)(a)(3)]**

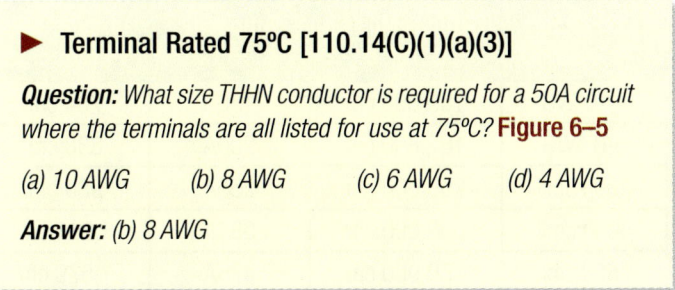

Conductors must be sized according to the lowest temperature rating of either the equipment or the conductor. THHN insulation can be used, but the conductor size must be selected based on the 75°C terminal rating of the equipment, not the 90°C rating of the insulation for dry locations. Using the 75°C column of Table 310.16, this installation will permit 8 AWG (rated 50A at 75°C) to supply the 50A circuit.

Circuits Over 100A [110.14(C)(1)(b)(1) and (2)]

Terminals for equipment rated over 100A and pressure connector terminals for conductors larger than 1 AWG must have the conductor sized according to the 75°C temperature rating listed in Table 310.16.

▶ **Over 100A [110.14(C)(1)(b)(2)]**

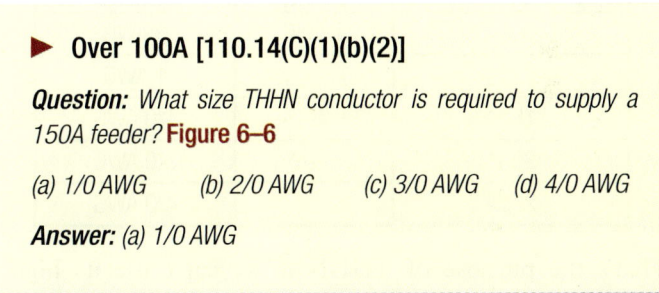

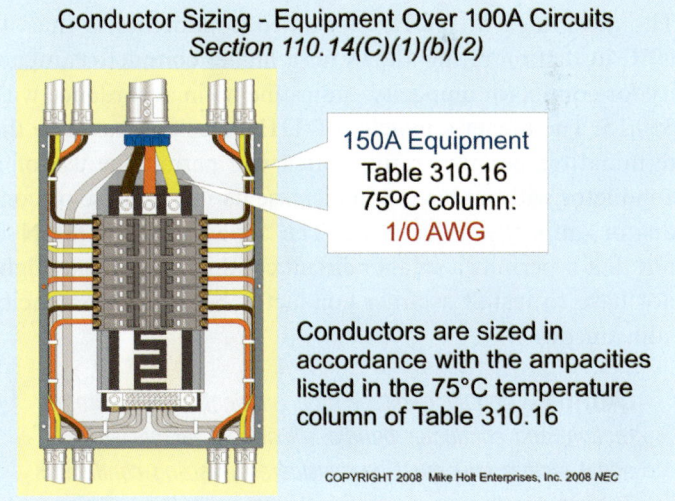

Figure 6–6

The conductors in this example must be sized to the lowest temperature rating of either the equipment or the conductor. THHN conductors can be used, but the conductor size must be selected according to the 75°C terminal rating of the equipment when sizing for over 100A. Using the 75°C column of Table 310.16, this will require a 1/0 AWG conductor (rated 150A at 75°C) to supply the 150A circuit. A 1 THHN is rated 150A at 90°C when used in a dry location, but we must size the conductor based on the 75°C terminal rating.

Minimum Conductor Size Table

When sizing conductors, the following table must be used to determine the minimum size conductor to meet the requirements of 110.14(C).

Circuit Ampacity	60°C	75°
50A	6 AWG	8 AWG
60A	4 AWG	6 AWG
70A	4 AWG	4 AWG
100A	1 AWG	3 AWG
125A	—	1 AWG
150A	—	1/0 AWG
200A	—	3/0 AWG
225A	—	4/0 AWG

What's the purpose of THHN if we can't use its higher ampacity at 90°C for conductor sizing?

The advantage of THHN insulated conductors is that its 90°C insulation rating allows for a higher conductor ampacity for conductor ampacity "adjustment" in accordance with 310.15. The higher ampacity of THHN, as compared to the terminal temperature rating, often will permit the use of a conductor without having to increase its size because of conductor ampacity adjustment. The advantage of THHN is not that it permits a smaller circuit conductor, but you might not have to install a larger conductor because of ampacity adjustments.

CAUTION: *Conductor voltage drop, ambient temperature correction, and conductor bundle adjustment factors are additional factors that must be considered in sizing conductors. These subjects are covered later in this textbook.*

Author's Comment: This is explained in detail in Part B of this unit.

6.5 Conductors in Parallel

Parallel conductors permit a smaller cross-sectional area per ampere, resulting in significant cost savings. **Figure 6–7**

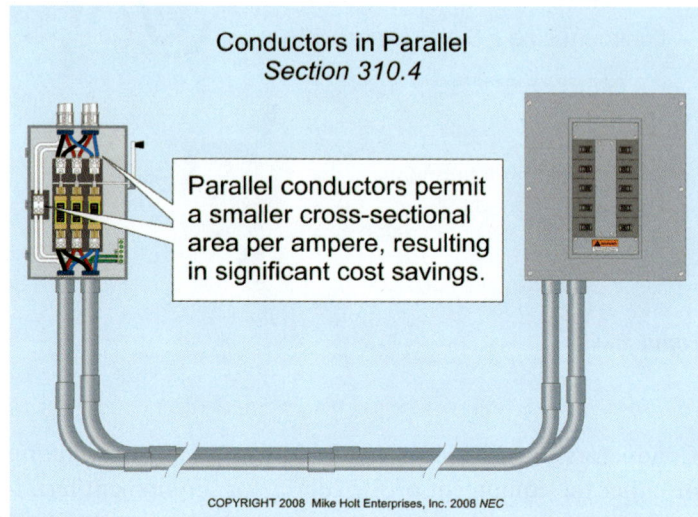

Figure 6–7

The following table demonstrates the circular mil area required per ampere for copper conductors.

Conductor Size	Circular Mils Ch 9, Tbl. 8	Ampacity 75°C	Circular Mils Per Ampere
1/0 AWG	105,600 cm	150A	704 cm
3/0 AWG	167,800 cm	200A	838 cm
250 kcmil	250,000 cm	255A	980 cm
500 kcmil	500,000 cm	380A	1,316 cm
750 kcmil	750,000 cm	475A	1,579 cm

As can be seen in the above table, 750 kcmil requires more than twice as much copper (1,579 cm) to carry one ampere of current as compared to 1/0 AWG (704 cm). With the high cost of copper, one needs to determine when it's cost effective to parallel an installation, however the *NEC* only permits the paralleling of conductors when the following requirements of Section 310.4 are met.

Conductor Sizing and Protection Calculations — Unit 6

6.6 *NEC* Requirements for Conductors in Parallel [310.4]

(A) General. Ungrounded and neutral conductors sized 1/0 AWG and larger can be connected in parallel. When circuit conductors are run in parallel, the current must be evenly distributed between the individual parallel conductors.

(B) Conductor Characteristics. All circuit conductors within a parallel set must:

(1) Be the same length.

(2) Be made of the same conductor material (copper/aluminum).

(3) Be the same size in circular mil area (minimum 1/0 AWG).

(4) Have the same insulation (like THHN).

(5) Terminate in the same method (set screw versus compression).

(C) Separate Raceways or Cables. Raceways or cables containing parallel conductors must have the same electrical characteristics and the same number of conductors. **Figure 6–8**

Paralleling is done in sets. Parallel sets of conductors aren't required to have the same physical characteristics as those of another set to achieve balance.

Author's Comment: For example, a 400A feeder with a neutral load of 240A can be paralleled as follows: **Figure 6–9**

- Phase A, Two—250 kcmil THHN aluminum, 100 ft
- Phase B, Two—3/0 THHN copper, 104 ft
- Phase C, Two—3/0 THHN copper, 102 ft
- Neutral, Two—1/0 THHN aluminum, 103 ft
- Equipment Grounding Conductor, Two—3 AWG copper, 101 ft

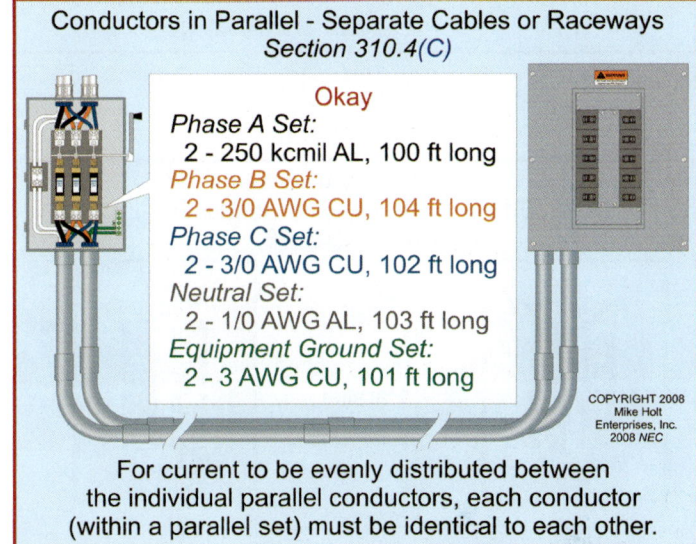

Figure 6–9

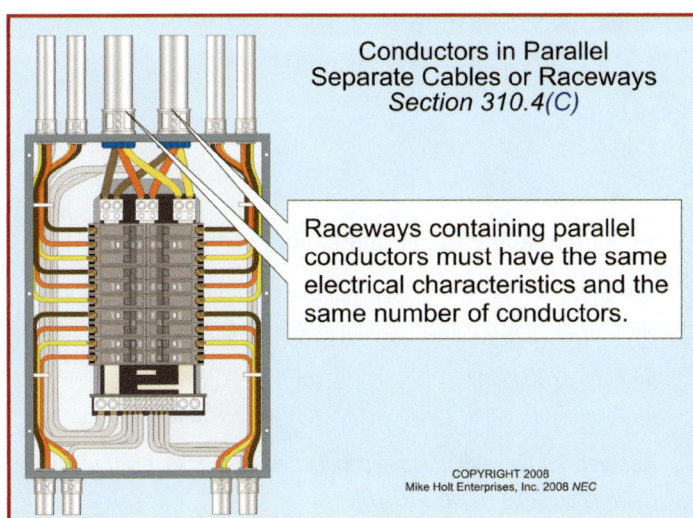

Figure 6–8

Author's Comment: If one set of parallel conductors is run in a metallic raceway and the other conductors are run in PVC conduit, the conductors in the metallic raceway will have an increased opposition to current flow (impedance) as compared to the conductors in the nonmetallic raceway. This results in an unbalanced distribution of current between the parallel conductors. Without getting into the details, this isn't good.

(D) Conductor Ampacity Adjustment. Each current-carrying conductor of a paralleled set of conductors must be counted as a current-carrying conductor for the purpose of conductor ampacity adjustment, in accordance with Table 310.15(B)(2)(a). **Figure 6–10**

(E) Equipment Grounding Conductors. The circuit equipment grounding conductors for circuits in parallel must be identical to each other in length, material, size, insulation, and termination. Where raceways requiring equipment grounding conductors are installed in parallel, they each must have an equipment grounding conductor sized in accordance with 250.122. The minimum 1/0 AWG rule for parallel conductors of 310.4 doesn't apply to equipment grounding conductors run in parallel [250.122(F)(1)]. **Figure 6–11**

Unit 6 — Conductor Sizing and Protection Calculations

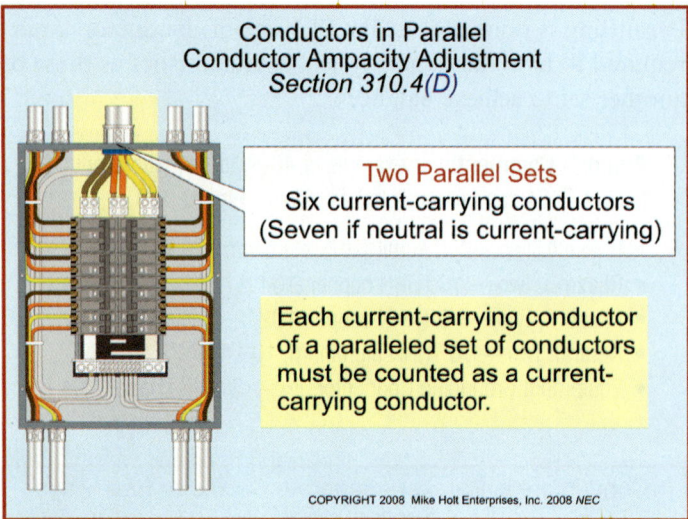

Figure 6–10

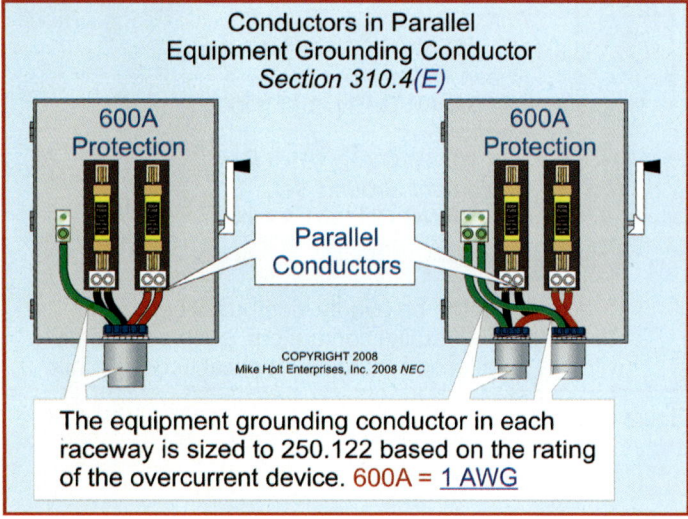

Figure 6–11

Sizing Parallel Conductors

The next two examples show the difference between single conductors and parallel conductors for a 600A service with a 550A calculated load.

▶ **Example 1: Single Conductors**

Question: *What size conductor is required for a 600A service?* **Figure 6–12**

(a) 750 kcmil (b) 1,000 kcmil
(c) 1,250 kcmil (d) 1,500 kcmil

Answer: *(d) 1,500 kcmil conductor, rated 625A at 75°C [Table 310.16].*

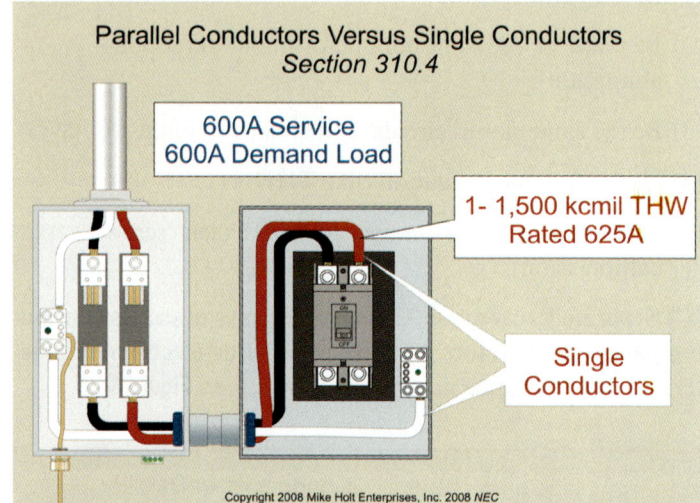

Figure 6–12

▶ **Example 2: Parallel Conductors**

Question: *What size parallel conductors are required for a 600A service, using two parallel conductors per phase?* **Figure 6–13**

(a) Two—250 kcmil (b) Two—300 kcmil
(c) Two—350 kcmil (d) Two—750 kcmil

Answer: *(c) Two—350 kcmil conductors per phase, each rated 310A is suitable [Table 310.16].*

If we parallel the conductors, we can use two 350 kcmil conductors (total 700 kcmil) instead of one 1,500 kcmil conductor per phase.

Conductor Sizing and Protection Calculations — Unit 6

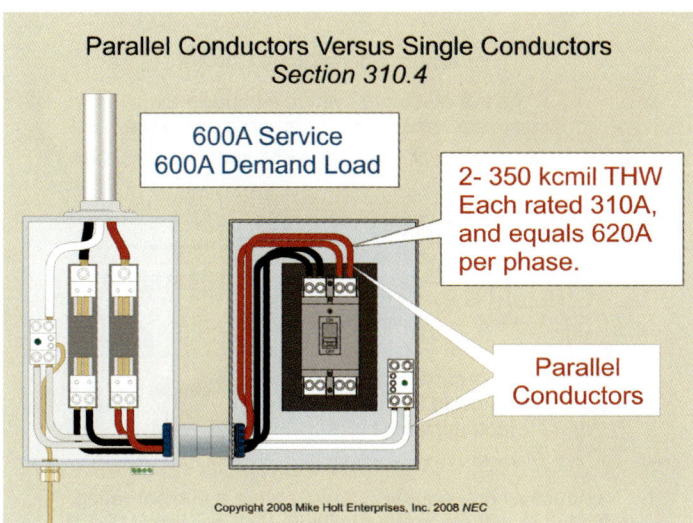

Figure 6–13

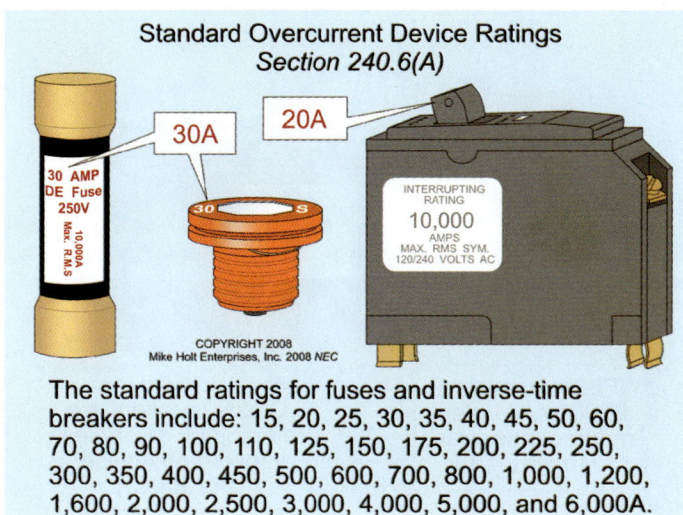

Figure 6–14

▶ **Example 3: Parallel Conductors**

Question: What size parallel conductors are required for a 600A service, using three parallels per phase?

(a) Three—3/0 AWG (b) Three—4/0 AWG
(c) Three—250 kcmil (d) Three—300 kcmil

Answer: (a) Three—3/0 AWG conductors per phase, each rated 200A are suitable [Table 310.16].

If we parallel the conductors, we can use three 3/0 AWG conductors per phase.

6.7 Overcurrent Protection [Article 240]

An overcurrent device protects the circuit by opening the device when the current reaches a value that will cause excessive or dangerous temperature rise (overheating) in conductors. This is the actual ampere rating of the overcurrent device, such as 15A, 20A, or 30A [240.6(A)]. **Figure 6–14**

The *NEC* lists standard sized overcurrent devices: 15, 20, 25, 30, 35, 40, 45, 50, 60, 70, 80, 90, 100, 110, 125, 150, 175, 200, 225, 250, 300, 350, 400, 450, 500, 600, 700, 800, 1,000, 1,200, and 1,600 [240.6(A)].

Interrupting Protection Rating [110.9]

Overcurrent devices such as circuit breakers and fuses are intended to interrupt the circuit, and they must have an interrupting rating sufficient for the short-circuit current available at the line terminals of the equipment [110.9]. **Figure 6–15**

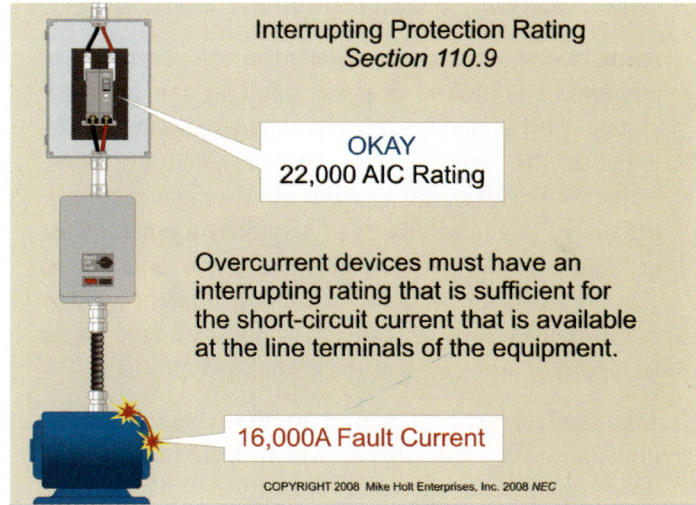

Figure 6–15

Author's Comments:

- See the definition of "Interrupting Rating" in Article 100.

- Unless marked otherwise, the ampere interrupting rating for circuit breakers is 5,000A [240.83(C)], and for fuses it's 10,000A [240.60(C)(3)]. **Figure 6–16**

Unit 6 — Conductor Sizing and Protection Calculations

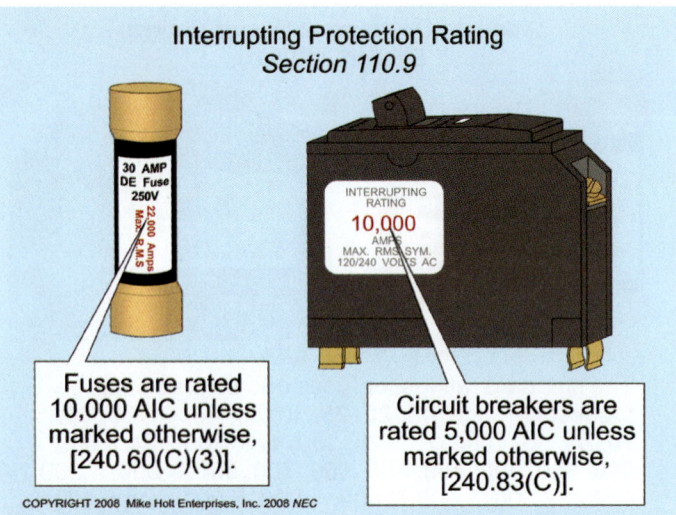

Figure 6–16

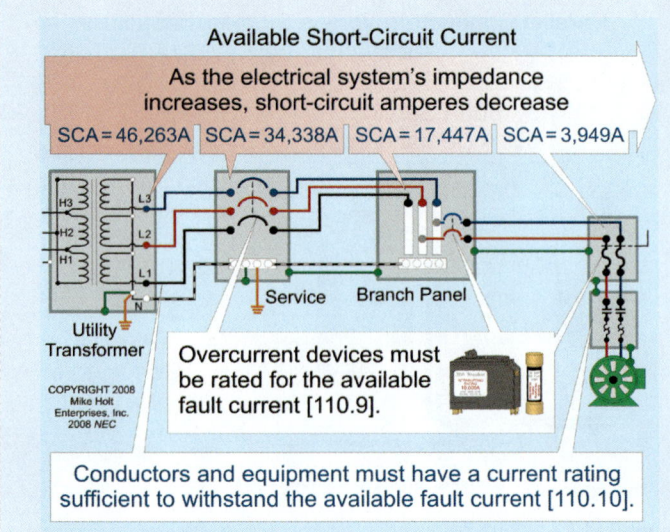

Figure 6–17

AVAILABLE SHORT-CIRCUIT CURRENT

Available short-circuit current is the current, in amperes, available at a given point in the electrical system. This available short-circuit current is first determined at the secondary terminals of the utility transformer. Thereafter, the available short-circuit current is calculated at the terminals of service equipment, then at branch-circuit panelboards and other equipment. The available short-circuit current is different at each point of the electrical system. It's highest at the utility transformer and lowest at the branch-circuit load.

The available short-circuit current depends on the impedance of the circuit, which increases moving downstream from the utility transformer. The greater the circuit impedance (utility transformer and the additive impedances of the circuit conductors), the lower the available short-circuit current. **Figure 6–17**

Factors that impact the available short-circuit current at the utility transformer include the system voltage, the transformer kVA rating, and its impedance (expressed in a percentage on the equipment nameplate). Properties that affect the impedance of the circuit include the conductor material (copper versus aluminum), conductor size, conductor length, and motor-operated equipment supplied by the circuit.

Author's Comment: Many people in the industry describe Amperes Interrupting Rating (AIR) as "Amperes Interrupting Capacity" (AIC).

DANGER: *Extremely high values of current flow (caused by short circuits or ground faults) produce tremendously destructive thermal and magnetic forces. If the circuit overcurrent device isn't rated to interrupt the current at the available fault values at its listed voltage rating, it can explode while attempting to open the circuit overcurrent device resulting from a short circuit or ground fault, which can cause serious injury or death, as well as property damage.* **Figure 6–18**

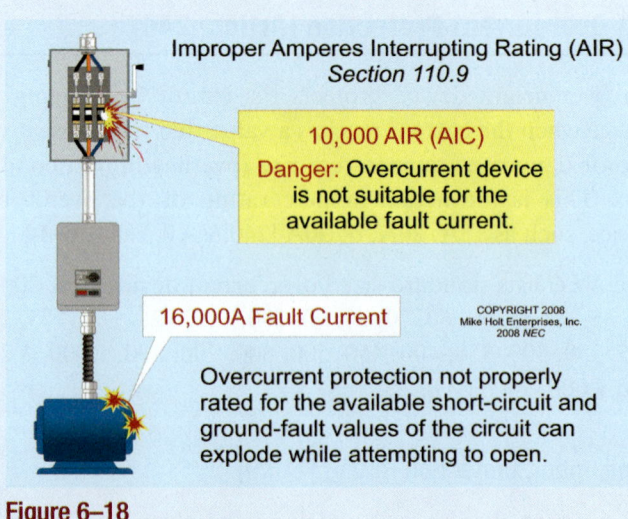

Figure 6–18

Conductor Sizing and Protection Calculations — Unit 6

Short-Circuit Calculation

The available short circuit current at the transformer is found with the formula:

Isca = VA/(E x 1.732 x Z%)

Isca = Short-Circuit Current Available
VA = Transformer VA Rating
E = Voltage
Z% = Transformer Percent Impedance
1.732 multiplier is included for three-phase transformers.

Author's Comment: This calculation is made assuming a condition called "infinite buss" which means that the impedance on the primary side of the transformer, including the impedance of the utility lines to that point, is ignored. This results in a higher short-circuit current than will be experienced in an actual situation, so it provides a "worst case scenario," and staying within these calculations will allow an error on the side of safety.

▶ **Example 1: Short-Circuit Calculation**

Question: What is the available short-circuit current for a 112.50 kVA, 208V, three-phase transformer with 3.20 percent impedance?

(a) 220A   (b) 400A   (c) 5,450A   (d) 9,759A

Answer: (d) 9,759A

Isca = VA/(E x 1.732 x Z%)
Isca = 112,500VA/(208V x 1.732 x 0.032)
Isca = 9,759A

▶ **Example 2: Short-Circuit Calculation**

Question: What is the available short-circuit current for a 30 kVA, 208V, three-phase transformer with 6.30 percent impedance?

(a) 275A   (b) 950A   (c) 1,321A   (d) 5,450A

Answer: (c) 1,321A

Isca = VA/(E x 1.732 x Z%)
Isca = 30,000VA/(208V x 1.732 x 0.063)
Isca = 1,321A

6.8 Overcurrent Protection of Conductors—General Requirements [240.4]

There are many different rules for protecting conductors and equipment. It's not simply 12 AWG wire and a 20A breaker. The general rule is that conductors must be protected at the point where they receive their supply in accordance with their ampacities, as listed in Table 310.16, **Figure 6–19**. Other methods of protection are permitted or required as listed in Subsections (A) through (G) of Section 240.4.

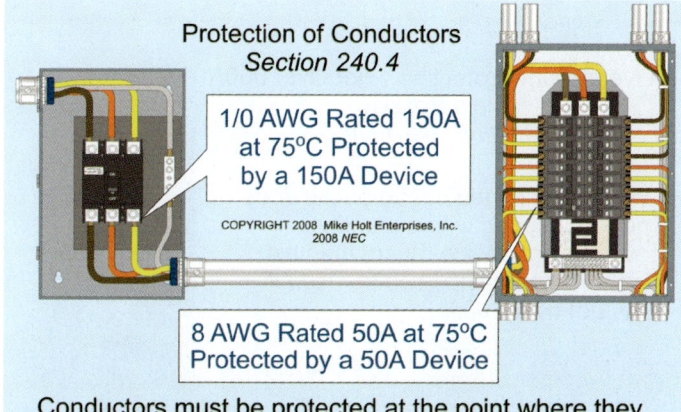

Conductors must be protected at the point where they receive their supply in accordance with their ampacities, as listed in Table 310.16, except as permitted or required by 240.4(A) through (G).

Figure 6–19

(A) Power Loss Hazard. Conductor overload protection is not required, but short-circuit protection is required where the interruption of the circuit will create a hazard; such as in a material-handling electromagnet circuit or fire pump circuit.

(B) Overcurrent Devices Rated 800A or Less. The next higher standard rating of overcurrent device (above the ampacity of the ungrounded conductors being protected) is permitted, provided all of the following conditions are met:

(1) The conductors don't supply multioutlet receptacle branch circuits.

(2) The ampacity of a conductor, after ampacity adjustment and/or correction, doesn't correspond with the standard rating of a fuse or circuit breaker in 240.6(A).

(3) The overcurrent device rating doesn't exceed 800A.

Unit 6 — Conductor Sizing and Protection Calculations

▶ **Overcurrent Protection Under 800A Example**

Question: What is the maximum size overcurrent device that can be used to protect 500 kcmil conductors, where each conductor has an ampacity of 380A at 75°C, in accordance with Table 310.16. **Figure 6–20**

(a) 300A   (b) 350A   (c) 400A   (d) 500A

Answer: (c) 400A

This "next size up" rule doesn't apply to feeder tap conductors [240.21(B)], or transformer secondary conductors [240.21(C)].

Three sets of 600 kcmil conductors per phase, where each conductor has an ampacity of 420A at 75°C, in accordance with Table 310.16.

420A x 3 = 1,260A

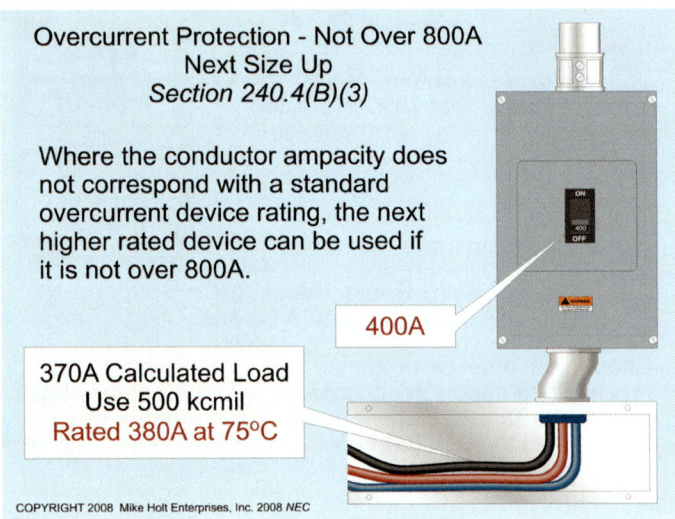

Figure 6–20

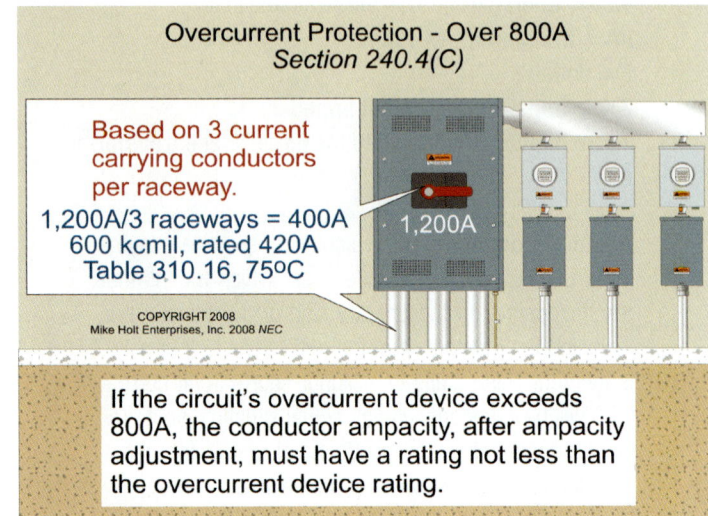

Figure 6–21

(C) Overcurrent Devices Rated Over 800A. If the circuit's overcurrent device exceeds 800A, the conductor ampacity, after ampacity adjustment and/or correction, must have a rating of not less than the rating of the overcurrent device.

▶ **Overcurrent Protection Over 800A Example**

Question: What is the minimum size of conductors allowed to be protected by A 1,200A overcurrent device using three sets of conductors per phase with terminals rated at 75°C? **Figure 6–21**

(a) 400 kcmil   (b) 500 kcmil
(c) 600 kcmil   (d) 750 kcmil

Answer: (c) 600 kcmil

(D) Small Conductors. Unless specifically permitted in 240.4(E) or (G), overcurrent protection must not exceed the following after ampacity adjustment and/or correction according to 310.15: **Figure 6–22**

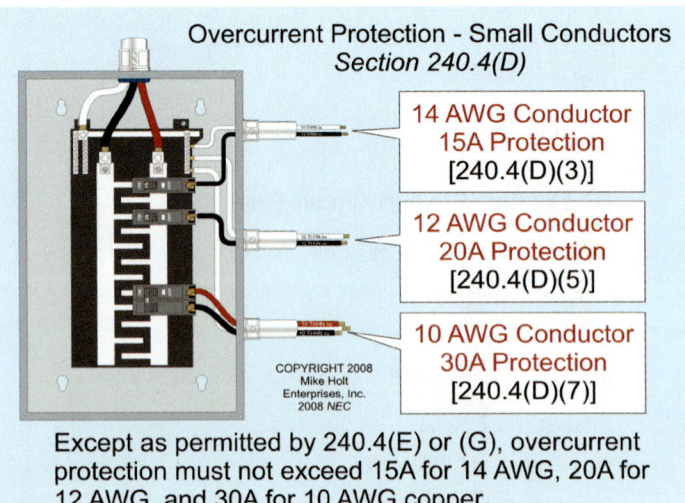

Except as permitted by 240.4(E) or (G), overcurrent protection must not exceed 15A for 14 AWG, 20A for 12 AWG, and 30A for 10 AWG copper.

Figure 6–22

Conductor Sizing and Protection Calculations | Unit 6

(1) 18 AWG Copper—7A

(2) 16 AWG Copper—10A

(3) 14 AWG Copper—15A

(4) 12 AWG Aluminum/Copper-Clad Aluminum—15A

(5) 12 AWG Copper—20A

(6) 10 AWG Aluminum/Copper-Clad Aluminum—25A

(7) 10 AWG Copper—30A

(E) Tap Conductors. Tap conductors must be protected against overcurrent as follows:

(1) Household Ranges and Cooking Appliances and Other Loads, 210.19(A)(3) and (4)

(2) Fixture Wire, 240.5(B)(2)

(3) Location in Circuit, 240.21

(4) Reduction in Ampacity Size of Busway, 368.17(B)

(5) Feeder or Branch Circuits (busway taps), 368.17(C)

(6) Single Motor Taps, 430.53(D)

(F) Transformer Secondary Conductors. The primary overcurrent device sized in accordance with 450.3(B) is considered suitable to protect the secondary conductors of a 2-wire (single voltage) system, provided the primary overcurrent device doesn't exceed the value determined by multiplying the secondary conductor ampacity by the secondary-to-primary transformer voltage ratio.

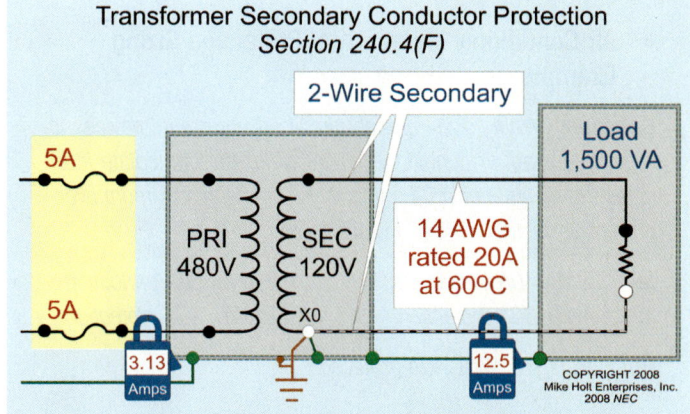

Figure 6–23

▶ **Transformer Secondary Conductor Sizing Example**

Question: What is the minimum secondary conductor size required for a 2-wire, 480V to 120V transformer rated 1.50 kVA? **Figure 6–23**

(a) 16 AWG   (b) 14 AWG   (c) 12 AWG   (d) 10 AWG

Answer: (b) 14 AWG

Primary Current = VA/E
VA = 1,500 VA
E = 480V

Primary Current = 1,500 VA/480V
Primary Current = 3.13A

Primary Protection [450.3(B)] = 3.13A x 1.67
Primary Protection = 5.22A or 5A Fuse

(continued in next column)

Secondary Current = 1,500 VA/120V
Secondary Current = 12.50A

Secondary Conductor = 14 AWG, rated 20A at 60C, Table 310.16

The 5A primary overcurrent device can be used to protect 14 AWG secondary conductors because it doesn't exceed the value determined by multiplying the secondary conductor ampacity by the secondary-to-primary transformer voltage ratio (5A = 20A x 120V/480V).

(G) Overcurrent Protection for Specific Applications. Overcurrent protection for specific equipment and conductors must comply with that referenced in Table 240.4(G).

Air-Conditioning and Refrigeration [Article 440]. Air-conditioning and refrigeration equipment, and their circuit conductors, must be protected against overcurrent in accordance with 440.22.

Author's Comment: Typically, the branch-circuit ampacity and protection size is marked on the equipment nameplate [440.4(A)].

Unit 6 — Conductor Sizing and Protection Calculations

▶ **Air Conditioner Overcurrent Protection Sizing Example**

Question: What size branch-circuit overcurrent device is required for an air conditioner (18A) when the nameplate indicates the minimum circuit ampacity is 23A, with maximum overcurrent protection of 40A? **Figure 6–24**

(a) 12 AWG, 40A protection (b) 12 AWG, 50A protection
(c) 12 AWG, 60A protection (d) 12 AWG, 70A protection

Answer: (a) 12 AWG, 40A protection

▶ **Motor Branch-Circuit Conductor, Short-Circuit, and Ground-Fault Protection Sizing Example**

Question: What size branch-circuit conductor and overcurrent device (circuit breaker) is required for a 7½ hp, 230V, three-phase motor? **Figure 6–25**

(a) 10 AWG, 30A breaker (b) 10 AWG, 60A breaker
(c) 10 AWG, 70A breaker (d) none of these

Answer: (b) 10 AWG, 60A breaker

Step 1: Determine the branch-circuit conductor size [Table 310.16, 430.22, and Table 430.250]:

22A x 1.25 = 28A, 10 AWG, rated 30A at 60°C

Step 2: Determine the branch-circuit protection size [240.6(A), 430.52(C)(1) Ex 1, Table 430.250].

Inverse Time Breaker: 22A x 2.50 = 55A
Next size up = 60A

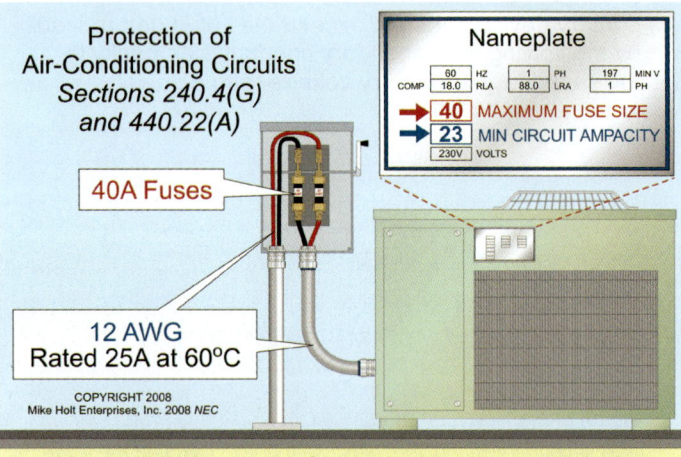

Figure 6–24

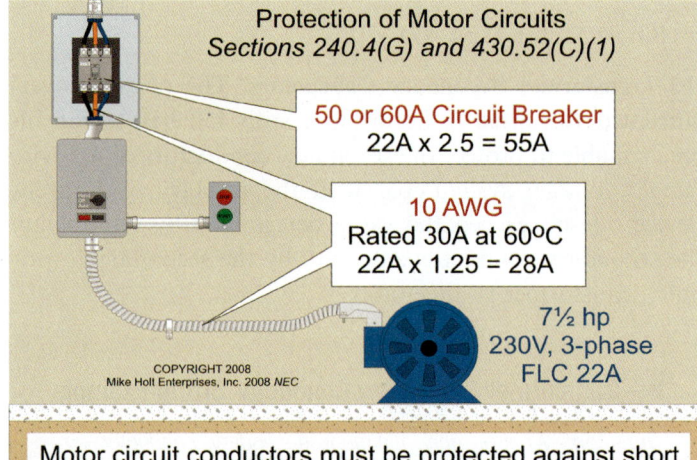

Figure 6–25

Author's Comment: Air-conditioning and refrigeration nameplate values are calculated by the manufacturer according to the following:

- Branch-Circuit Conductor Size [440.32]
  18A x 1.25 = 22.50A, 12 AWG rated 25A at 60°C

- Branch-Circuit Protection Size [440.22(A)]
  18A x 2.25 = 40.50A, 40A maximum fuse size (fuses in accordance with the manufacturer's instructions) [110.3(b) and 240.6(A)]

- Motors [Article 430]. Motor circuit conductors must be protected against short circuits and ground faults in accordance with 430.52 and 430.62 [430.51].

Motor Control [430].

Motor control circuit conductors must be sized and protected in accordance with 430.72.

Remote-Control, Signaling, and Power-Limited Circuits [725].

Remote-control, signaling, and power-limited circuit conductors must be protected against overcurrent in accordance with 725.43.

Conductor Sizing and Protection Calculations — Unit 6

6.9 Overcurrent Protection of Conductors—Specific Requirements

When sizing and protecting conductors for equipment, be sure to apply the specific *NEC* requirement.

- Air-Conditioning [440.22 and 440.32]
- Appliances [422.10 and 422.11]
- Branch Circuits [210.20]
- Electric Heating Equipment [424.3]
- Motors [Article 430]
  - Branch Circuits [430.22 and 430.52]
  - Feeders [430.24 and 430.62]
  - Remote Control [430.72]
- Panelboards [408.36(A)]
- Transformers [240.21 and 450.3]
- Feeder Conductors [215.2 and 215.3]
- Service Conductors [230.42 and 230.90(A)]
- Tap Conductors [240.21(B)]
- Transformer Secondary Conductors [240.21(B)]

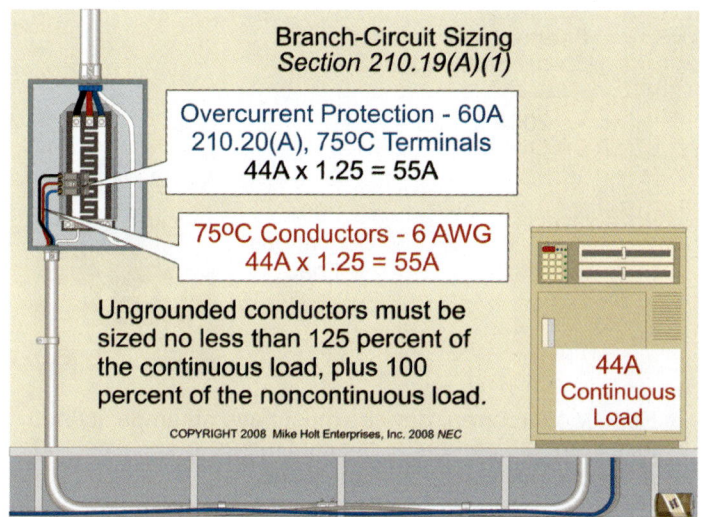

Figure 6–26

Branch-Circuit Overcurrent Protection

Conductors must be sized no less than 125 percent of the continuous loads, plus 100 percent of the noncontinuous loads, based on the terminal temperature rating ampacities as listed in Table 310.16, before any ampacity adjustment [110.14(C) and 210.19(A)(1)].

▶ **Branch-Circuit Overcurrent Protection Example**

Question: What size branch-circuit conductor and overcurrent protection is required for a 44A continuous load if the equipment terminals are rated 75°C as permitted by 110.14(C)(1)(a)? **Figure 6–26**

(a) 10 AWG and 30A circuit breaker
(b) 8 AWG and 50A circuit breaker
(c) 6 AWG and 60A circuit breaker
(d) 4 AWG and 80A circuit breaker

Answer: (c) 6 AWG and 60A circuit breaker

Since the load is 44A continuous, the conductor must be sized to have an ampacity of not less than 55A (44A x 1.25) [210.19(A)(1)]. According to Table 310.16, 75°C column, a 6 AWG conductor is suitable, because it has an ampere rating of 65A at 75°C before any conductor ampacity adjustment and/or correction. The overcurrent device must be sized at 60A (44 x 1.25) which protects the 6 AWG conductor within its ampacity rating [210.20(A)].

Feeder Overcurrent Protection

The minimum feeder-circuit conductor ampacity, before the application of any adjustment and/or correction factors, must be no less than 125 percent of the continuous load, plus 100 percent of the noncontinuous load, based on the terminal temperature rating ampacities as listed in Table 310.16 [110.14(C) and 215.2(A)(1)]. The feeder overcurrent devices must have an ampacity not less than 125 percent of the continuous loads, plus 100 percent of the noncontinuous loads [215.3].

▶ **Feeder Overcurrent Protection Example**

Question: What size feeder overcurrent device is required for a 200A continuous load? **Figure 6–27**

(a) 200A (b) 225A (c) 250A (d) 300A

Answer: (c) 250A

Overcurrent Protection Location in Circuit [240.21]

Except as permitted by (A) through (G), overcurrent devices must be placed at the point where the branch or feeder conductors receive their power. Tap and secondary conductors are not permitted to supply another conductor (tapping a tap is not permitted).

(A) Branch-Circuit Taps. Branch-circuit taps are permitted in accordance with 210.19.

UNIT 6 — Conductor Sizing and Protection Calculations

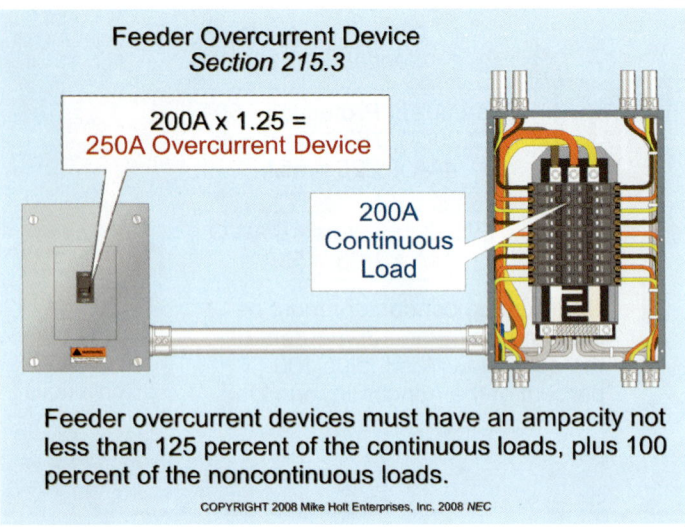

Figure 6–27

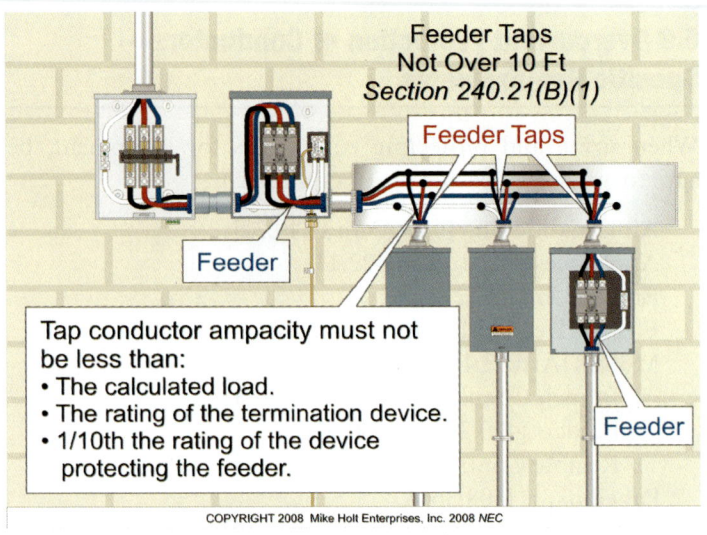

Figure 6–28

(B) Feeder Tap Conductors. Conductors can be tapped from a feeder as specified in 240.21(B)(1) through (B)(5).

> ▶ **Branch-Circuit Tap Conductors Example**
>
> **Question:** What size tap conductors will be required for a 150A circuit breaker if the calculated continuous load is 100A?
>
> (a) 3 AWG, rated 100A (b) 2 AWG, rated 115A
> (c) 1 AWG, rated 130A (d) 1/0 AWG, rated 150A
>
> **Answer:** (d) 1/0 AWG, rated 150A

(1) 10-Foot Feeder Tap. Feeder tap conductors up to 10 ft long are permitted without overcurrent protection at the tap location if installed as follows: **Figure 6–28**

(1) The ampacity of the tap conductor must not be less than:

   a. The calculated load in accordance with Article 220, and

   b. The rating of the device or overcurrent device supplied by the tap conductors.

(2) The tap conductors must not extend beyond the equipment they supply.

(3) The tap conductors must be installed in a raceway if they leave the enclosure.

(4) The tap conductors must have an ampacity not less than 10 percent of the ampacity of the overcurrent device that protects the feeder.

FPN: See 408.36 for the overcurrent protection requirements for panelboards.

For example, 1/0 AWG feeder taps rated 150A at 75°C, can be used to feed a 150A disconnect if all other tap rules are followed. **Figure 6–29**

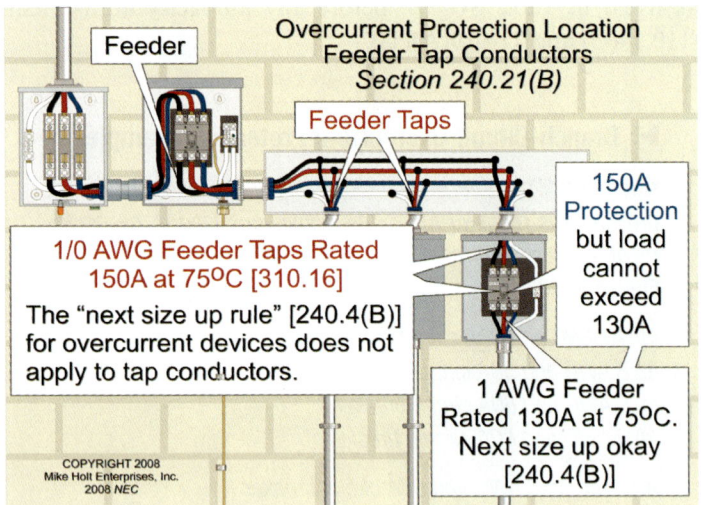

Figure 6–29

10-Foot Feeder Tap

A 400A breaker protects a set of 500 kcmil feeder conductors. There are three taps fed from the 500 kcmil feeders that supply disconnects with 200A, 150A, and 100A overcurrent devices.

Conductor Sizing and Protection Calculations   Unit 6

▶ 10-Foot Feeder Tap Example 1

Question: Using the 10-ft tap rule, what are the minimum size conductors required to feed the 200A tap?

(a) 3 AWG   (b) 1/0 AWG   (c) 3/0 AWG   (d) 250 kcmil

Answer: (c) 3/0 AWG

3/0 AWG is rated 200A at 75°C, and is greater than 10 percent of the ampacity of 500 kcmil, which is rated 380A at 75°C [Table 310.16].

▶ 10-Foot Feeder Tap Example 2

Question: Using the 10-ft tap rule, what are the minimum size conductors required to feed the 150A tap?

(a) 3 AWG   (b) 1/0 AWG   (c) 3/0 AWG   (d) 250 kcmil

Answer: (b) 1/0 AWG

1/0 AWG is rated 150A at 75°C, and is greater than 10 percent of the ampacity of 500 kcmil, which is rated 380A at 75°C [Table 310.16].

▶ 10-Foot Feeder Tap Example 3

Question: Using the 10-ft tap rule, what are the minimum size conductors required to feed the 100A tap?

(a) 3 AWG   (b) 1/0 AWG   (c) 3/0 AWG   (d) 250 kcmil

Answer: (a) 3 AWG

3 AWG is rated 100A at 75°C, and is greater than 10 percent of the ampacity of 500 kcmil, which is rated 380A at 75°C [Table 310.16].

A 400A breaker protects a set of 500 kcmil feeder conductors. There are three taps fed from the 500 kcmil feeders that supply disconnects with 200A, 150A, and 30A overcurrent devices.

▶ 10-Foot Feeder Tap Example 4

Question: Using the 10-ft tap rule, what are the minimum size conductors required to feed the 30A tap?

(a) 12 AWG   (b) 8 AWG   (c) 1/0 AWG   (d) 3/0 AWG

Answer: (b) 8 AWG

(continued in next column)

8 AWG is rated 50A at 75°C, and is greater than 10 percent of the ampacity of 500 kcmil, which is rated 380A at 75°C [Table 310.16]. Anything smaller than 8 AWG can't be used, as it will have an ampacity of less than 10 percent of 380A (38A) in the 75°C column.

(2) 25-Foot Feeder Tap. Feeder tap conductors up to 25 ft long are permitted without overcurrent protection at the tap locaton if installed as follows: **Figure 6–30**

(1) The ampacity of the tap conductors must not be less than one-third the ampacity of the overcurrent device that protects the feeder.

(2) The tap conductors terminate in a single circuit breaker, or set of fuses rated no more than the tap conductor ampacity in accordance with 310.15 [Table 310.16].

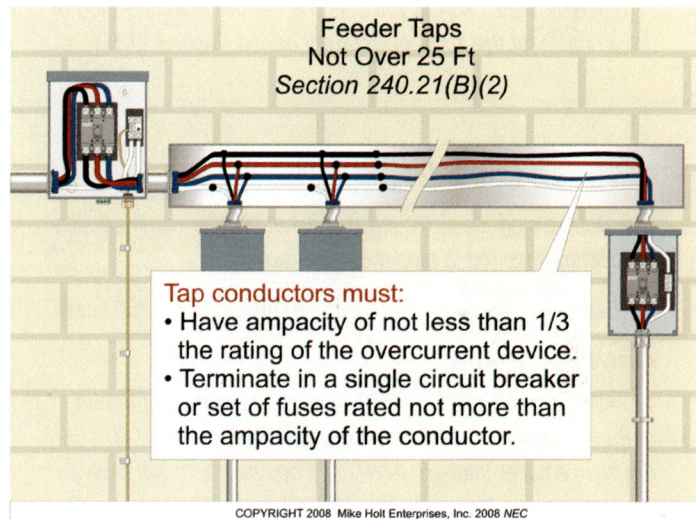

Figure 6–30

(3) The tap conductors must be protected from physical damage by being enclosed in a manner approved by the authority having jurisdiction, such as within a raceway.

25-Foot Feeder Tap

A 400A breaker protects a set of 500 kcmil feeder conductors. There are three taps fed from the 500 kcmil feeders that supply disconnects with 200A, 150A, and 100A overcurrent devices.

Unit 6   Conductor Sizing and Protection Calculations

▶ **25-Foot Feeder Tap Example 1**

Question: Using the 25-ft tap rule, what are the minimum size conductors required to feed the 200A tap?

(a) 3 AWG   (b) 1/0 AWG   (c) 3/0 AWG   (d) 250 kcmil

Answer: (c) 3/0 AWG

3/0 AWG is rated 200A at 75°C, and is greater than one-third of the ampacity of the 400A overcurrent device [Table 310.16].

▶ **25-Foot Feeder Tap Example 2**

Question: Using the 25-ft tap rule, what are the minimum size conductors required to feed the 150A tap?

(a) 3 AWG   (b) 1/0 AWG   (c) 3/0 AWG   (d) 250 kcmil

Answer: (b) 1/0 AWG

1/0 AWG is rated 150A at 75°C, and is greater than one-third of the ampacity of the 400A overcurrent device [Table 310.16].

▶ **25-Foot Feeder Tap Example 3**

Question: Using the 25-ft tap rule, what are the minimum size conductors required to feed the 100A tap?

(a) 3 AWG   (b) 1/0 AWG   (c) 3/0 AWG   (d) 250 kcmil

Answer: (b) 1/0 AWG

1/0 AWG is rated 150A at 75°C, and is greater than one-third of the ampacity of the 400A overcurrent device [Table 310.16]. Anything smaller than 1/0 AWG can't be used, as it will have an ampacity of less than one-third of 400A in the 75°C column.

Author's Comment: Taps may originate at the overcurrent device using multiple terminals per line. For instance, a 400A breaker can feed two 3/0 AWG feeder conductors using double-barrel terminal lugs, and each feeder may then feed two individual 200A panelboards as long as the appropriate tap rules for a 10-foot or 25-foot tap rule are followed.

(3) Taps Supplying a Transformer. Feeder tap conductors that supply a transformer must be installed as follows:

(1) The primary tap conductors must have an ampacity not less than one-third the ampacity of the overcurrent device.

(2) The secondary conductors must have an ampacity that, when multiplied by the ratio of the primary-to-secondary voltage, is at least one-third the rating of the overcurrent device that protects the feeder conductors.

(3) The total length of the primary and secondary conductors must not exceed 25 ft.

(4) Primary and secondary conductors must be protected from physical damage by being enclosed in a manner approved by the authority having jurisdiction, such as within a raceway.

(5) Secondary conductors terminate in a single circuit breaker, or set of fuses rated no more than the tap conductor ampacity in accordance with 310.15 [Table 310.16].

(5) Outside Feeder Tap of Unlimited Length Rule. Outside feeder tap conductors can be of unlimited length without overcurrent protection at the point they receive their supply if installed as follows: **Figure 6–31**

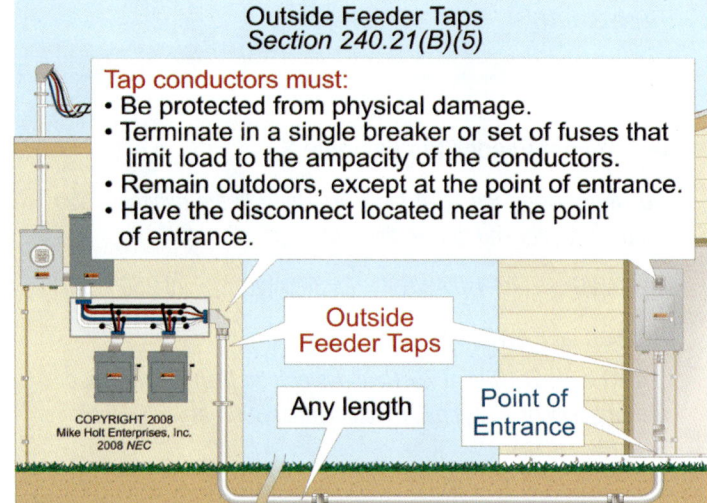

Figure 6–31

(1) The tap conductors must be suitably protected from physical damage in a raceway or manner approved by the authority having jurisdiction.

(2) The tap conductors terminate at a single circuit breaker or a single set of fuses that limits the load to the ampacity of the conductors.

Conductor Sizing and Protection Calculations — Unit 6

(3) The overcurrent device for the tap conductors must be an integral part of the disconnecting means or it must be located immediately adjacent to it.

(4) The disconnecting means must be located at a readily accessible location either outside the building or structure, or nearest the point of entry of the conductors.

(C) Transformer Secondary Conductors. A set of conductors supplying single or separate loads is permitted to be connected to a transformer secondary without overcurrent protection at the secondary, in accordance with (1) through (6):

(1) Protection by Primary Overcurrent Device. The primary overcurrent device sized in accordance with 450.3(B) is considered suitable to protect the secondary conductors of a 2-wire system, provided the primary protection device doesn't exceed the value determined by multiplying the secondary conductor ampacity by the secondary-to-primary transformer voltage ratio.

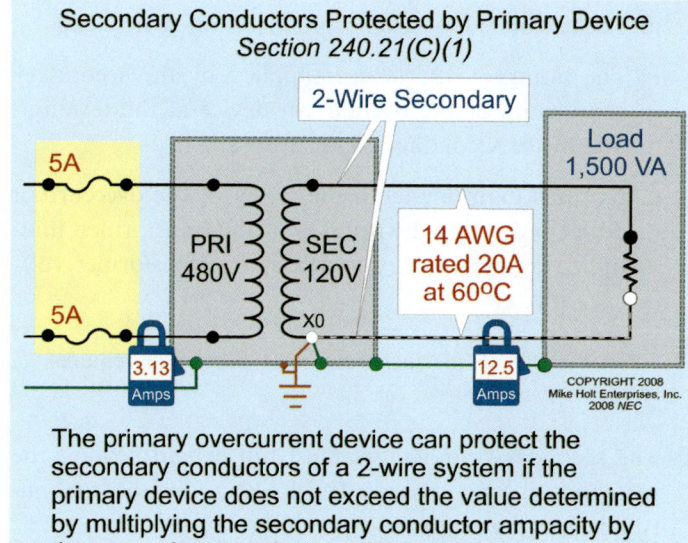

Figure 6–32

▶ **Transformer Secondary Conductor Sizing Example**

Question: What is the minimum size secondary conductor required for a 2-wire, 480V to 120V transformer rated 1.50 kVA? **Figure 6–32**

(a) 16 AWG   (b) 14 AWG   (c) 12 AWG   (d) 10 AWG

Answer: (b) 14 AWG

Primary Current = VA/E

VA = 1,500 VA
E = 480V

Primary Current = 1,500 VA/480V
Primary Current = 3.13A

Primary Protection [450.3(B)] = 3.13A x 1.67
Primary Protection [450.3(B)] = 5.22A or 5A Fuse

Secondary Current = 1,500 VA/120V
Secondary Current = 12.50A

Secondary Conductor = 14 AWG, rated 20A at 60°C, Table 310.16

The 5A primary protection device can be used to protect 14 AWG secondary conductors because it doesn't exceed the value determined by multiplying the secondary conductor ampacity by the secondary-to-primary transformer voltage ratio (5A = 20A x 120V/480V).

(2) 10-Foot Secondary Conductors. Secondary conductors can be run up to 10 ft without overcurrent protection if installed as follows:

(1) The ampacity of the secondary conductor must not be less than: **Figure-6-33**

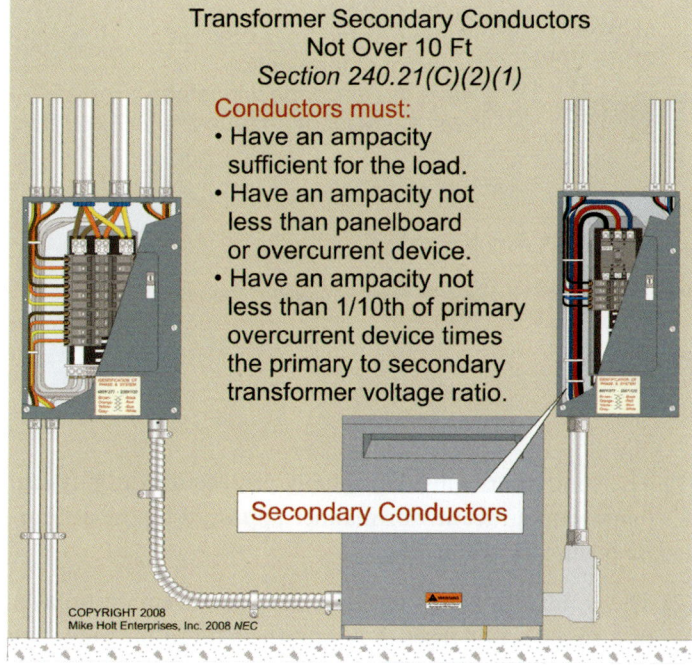

Figure 6–33

UNIT 6 — Conductor Sizing and Protection Calculations

a. The calculated load in accordance with Article 220,

b. The rating of the device supplied by the secondary conductors or the overcurrent device at the termination of the secondary conductors, and

c. Not less than one-tenth the rating of the overcurrent device protecting the primary of the transformer, multiplied by the primary-to-secondary transformer voltage ratio.

FPN: See 408.36 for the overcurrent protection requirements for panelboards.

(2) The secondary conductors must not extend beyond the switchboard, panelboard, disconnecting means, or control devices they supply.

(3) The secondary conductors must be enclosed in a raceway.

(4) Outside Secondary Conductors of Unlimited Length. Outside secondary conductors can be of unlimited length, without overcurrent protection at the point they receive their supply, if they are installed as follows: **Figure 6–34**

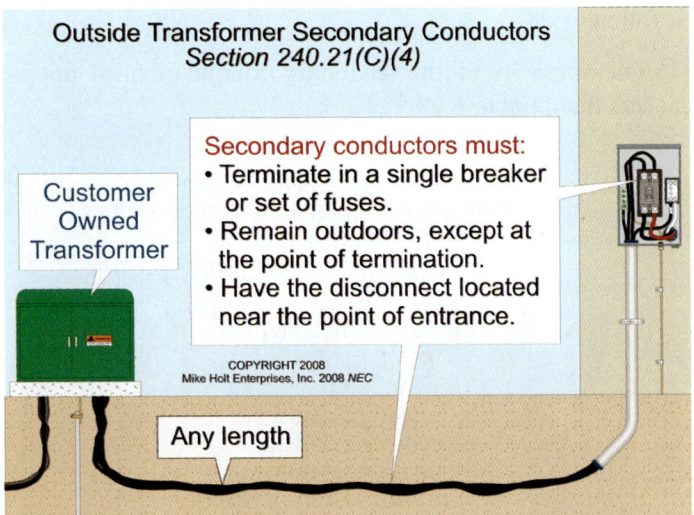

Figure 6–34

(1) The conductors must be suitably protected from physical damage in a raceway or manner approved by the authority having jurisdiction.

(2) The conductors must terminate at a single circuit breaker or a single set of fuses that limit the load to the ampacity of the conductors.

(3) The overcurrent device for the ungrounded conductors must be an integral part of a disconnecting means or it must be located immediately adjacent thereto.

(4) The disconnecting means must be located at a readily accessible location that complies with one of the following:

a. Outside of a building or structure.

b. Inside, nearest the point of entrance of the conductors.

c. Where installed in accordance with 230.6, nearest the point of entrance of the conductors.

▶ **Transformer Outdoor Secondary Conductor Sizing Example**

Question: When outdoor secondary conductors are used to feed a 100A circuit breaker which is backfeeding a 225A main lug only panelboard, what is the minimum size required for the secondary conductors?

(a) 3 AWG (b) 1 AWG (c) 4/0 AWG (d) 250 kcmil

Answer: (a) 3 AWG

(5) Secondary Conductors from a Feeder Tapped Transformer. Transformer secondary conductors must be installed in accordance with 240.21(B)(3).

(6) Secondary Conductors Not Over 25 ft Long. Where the length of secondary conductors does not exceed 25 ft and complies with the following: **Figure 6–35**

(1) The secondary conductors must have an ampacity not less than the value of the primary-to-secondary voltage ratio multiplied by one-third of the rating of the overcurrent device protecting the primary of the transformer.

(2) The conductors must terminate at a single circuit breaker or a single set of fuses that limit the load to the ampacity of the conductors.

(3) The conductors must be suitably protected from physical damage in a raceway or manner approved by the authority having jurisdiction.

Conductor Sizing and Protection Calculations — Unit 6

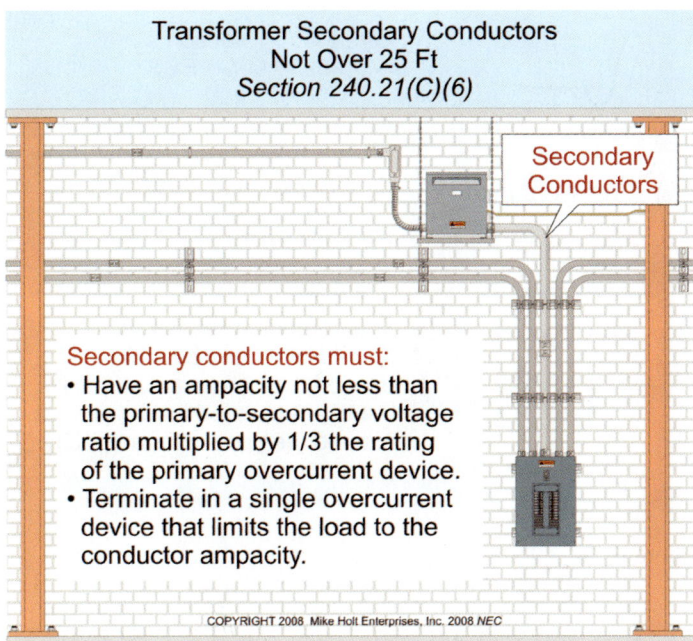

Figure 6–35

than 125 percent of the continuous loads, plus 100 percent of the noncontinuous loads, based on the terminal temperature rating ampacities as listed in Table 310.16, before any ampacity adjustment [110.14(C)].

Author's Comment: See 215.3 for the sizing requirements of feeder overcurrent devices for continuous and noncontinuous loads.

▶ **Service Conductor Sizing Example**

Question: What size service-entrance conductors are required for a 200A continuous load, if the terminals are rated 75ºC? Figure 6-36

(a) 2/0 AWG (b) 3/0 AWG (c) 4/0 AWG (d) 250 kcmil

Answer: (d) 250 kcmil

Since the load is 200A continuous, the service-entrance conductors must have an ampacity not less than 250A (200A x 1.25). According to the 75ºC column of Table 310.16, 250 kcmil conductors are suitable as they have an ampere rating of 255A at 75ºC.

▶ **Transformer Secondary Conductor Sizing Example**

Question: What size secondary conductors are the minimum required from a 45 kVA 120/208V three-phase transformer to feed a 175A main breaker panelboard less than 25 ft away, serving 125A of continuous load?

(a) 1/0 AWG (b) 2/0 AWG (c) 4/0 AWG (d) 250 kcmil

Answer: (b) 2/0 AWG

125A x 1.25 = 156A which is the minimum ampacity required for a continuous load [215.2(A)(1)]. At least a 175A conductor is required to feed the 175A circuit breaker in the panel [240.21(C)(6)]. 2/0 AWG is rated 175A at 75°C [Table 310.16].

(D) Service Conductors. Service conductors must be protected against overload by the service disconnect overcurrent device in accordance with 230.91.

Size and Rating [230.42]

(A) Load Calculations. Service-entrance conductors must have sufficient ampacity for the loads to be served in accordance with Parts III, IV, or V of Article 220.

(1) Continuous and Noncontinuous Loads. The ampacity of the service-entrance conductors, before the application of any adjustment or correction factors, must be sized no smaller

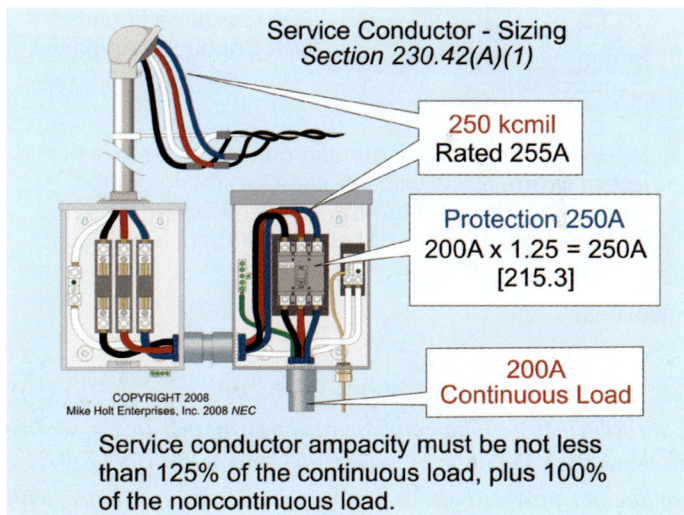

Figure 6–36

Author's Comment: The *NEC* doesn't require service conductors to be provided with short-circuit or ground-fault protection, but the feeder overcurrent device provides overload protection for the service conductors.

| Unit 6 | Conductor Sizing and Protection Calculations |

Overload Protection Required [230.90]

Each ungrounded service conductor must have overload protection at the point where the service conductors terminate [240.21(D)].

(A) Overcurrent Device Rating. The rating of the overcurrent device must not be more than the ampacity of the conductors.

Exception No. 2: Where the ampacity of the ungrounded conductors doesn't correspond with the standard rating of overcurrent devices as listed in 240.6(A), the next higher overcurrent device can be used, if it doesn't exceed 800A [240.4(B)].

> **Author's Comment:** Two sets of parallel 500 kcmil THHN conductors (each rated 380A at 75°C) can be protected by an 800A overcurrent device. **Figure 6–37**

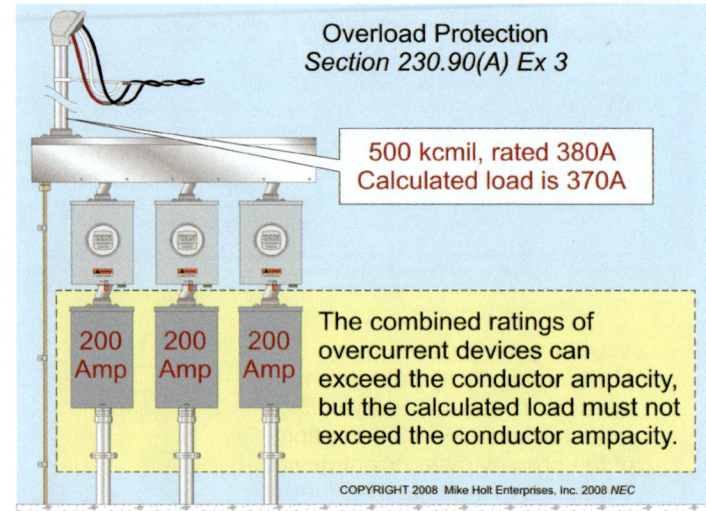

Figure 6–38

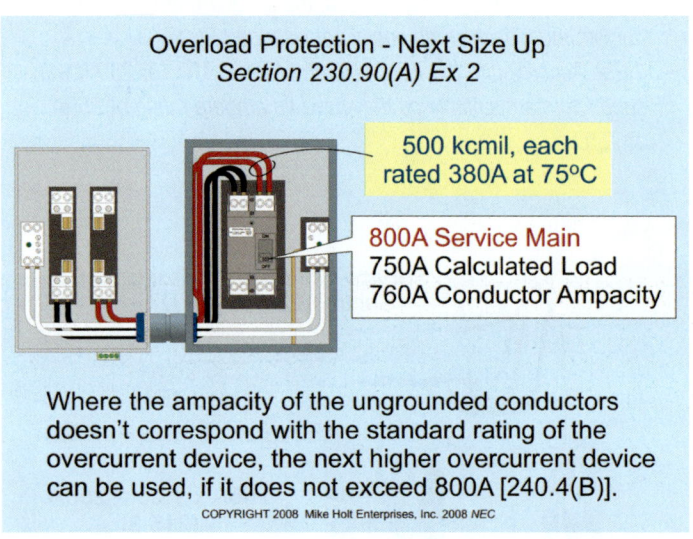

Figure 6–37

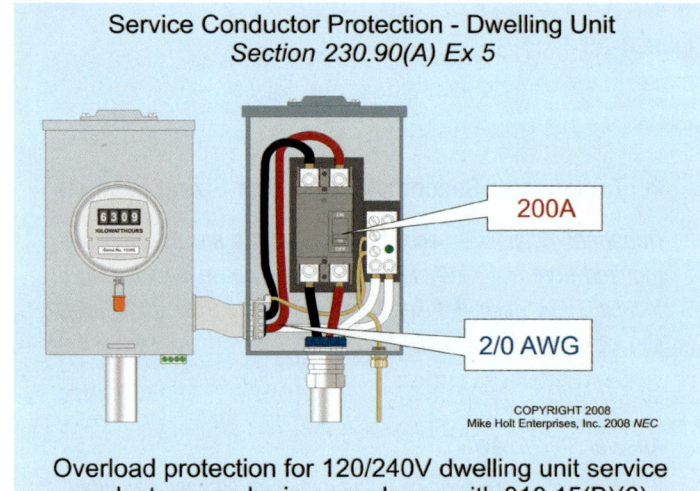

Figure 6–39

Exception No. 3: The combined ratings of two to six service disconnecting means can exceed the ampacity of the service conductors provided the calculated load, in accordance with Article 220, doesn't exceed the ampacity of the service conductors. **Figure 6–38**

Exception No. 5: Overload protection for 3-wire, single-phase, 120/240V dwelling unit service conductors can be in accordance with 310.15(B)(6). **Figure 6–39**

Overcurrent Protection of Panelboards [408.36]

Each panelboard must be provided with overcurrent protection located within, or at any point on the supply side of, the panelboard. The overcurrent device must have a rating not greater than that of the panelboard, and it can be located within or on the supply side of the panelboard. **Figure 6–40**

Exception No. 1: Individual overcurrent protection is not required for panelboards used as service equipment in accordance with 230.71.

Conductor Sizing and Protection Calculations — Unit 6

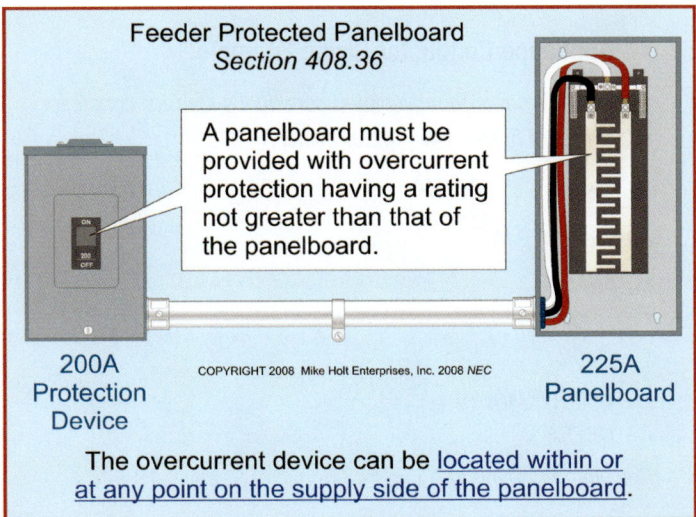

Figure 6–40

(B) Panelboards Supplied Through a Transformer. When a panelboard is supplied from a transformer, as permitted in 240.21(C), the overcurrent protection for the panelboard must be on the secondary side of the transformer. The required overcurrent protection can be in a separate enclosure ahead of the panelboard, or it can be in the panelboard. Figure 6-41

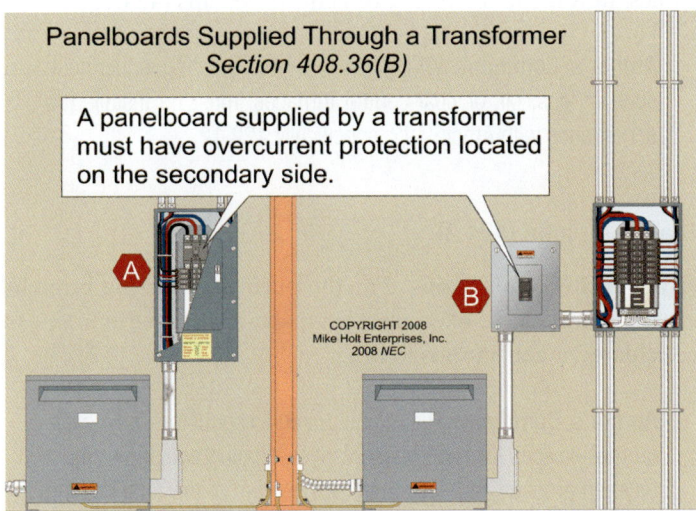

Figure 6–41

(D) Back-Fed Devices. Plug-in circuit breakers that are back-fed from field-installed conductors must be secured in place by an additional fastener that requires other than a pull to release the breaker from the panelboard. Figure 6–42

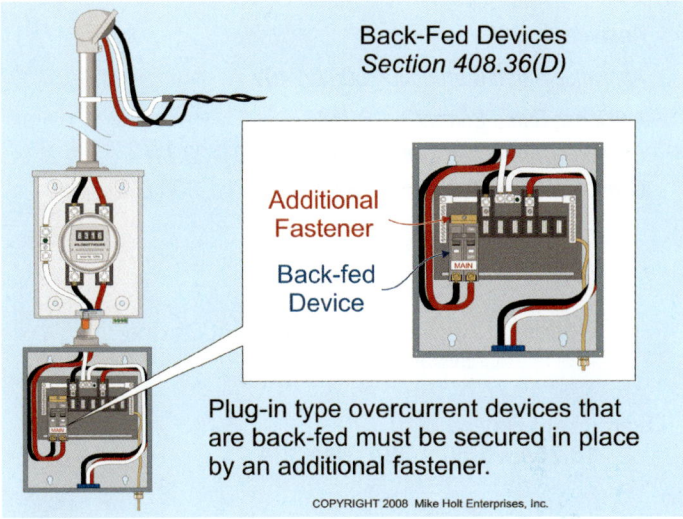

Figure 6–42

Overcurrent Protection [422.11]

(A) Branch-Circuits. Branch-circuit conductors must have overcurrent protection in accordance with 240.4, and the overcurrent device rating must not exceed the rating marked on the appliance.

(E) Nonmotor Appliances. The appliance overcurrent device must:

(1) Not exceed the rating marked on the appliance,

(2) Not exceed 20A if the overcurrent device rating isn't marked, and the appliance is rated 13.30A or less, or

(3) Not exceed 150 percent of the appliance rated current if the overcurrent device rating isn't marked, and the appliance is rated over 13.30A. Where 150 percent of the appliance rating doesn't correspond to a standard overcurrent device ampere rating listed in 240.6(A), the next higher standard rating is permitted.

Unit 6 — Conductor Sizing and Protection Calculations

▶ **Appliance Overcurrent Protection Example**

Question: What is the maximum size overcurrent protection for a 4,500W, 240V water heater? Figure 6–43

(a) 20A   (b) 30A   (c) 40A   (d) 50A

Answer: (b) 30A

Appliance Rated Current = 4,500W/240V
Appliance Rated Current = 18.75A
Conductor/Protection Size = 18.75A x 1.50 [422.11]
Conductor/Protection Size = 28A, next size up, 30A [240.6(A)]

▶ **Appliance Conductor Sizing Example**

Question: What is the calculated continuous load for conductor sizing for a 4,500W, 240V water heater?

(a) 15A   (b) 24A   (c) 30A   (d) 35A

Answer: (b) 24A

I = P/E
P = 4,500W
E = 240V
I = 4,500W/240V
I = 18.75A
Calculated Continuous Load for Conductor Sizing = 18.57A x 1.25
Calculated Continuous Load for Conductor Sizing = 23.43A

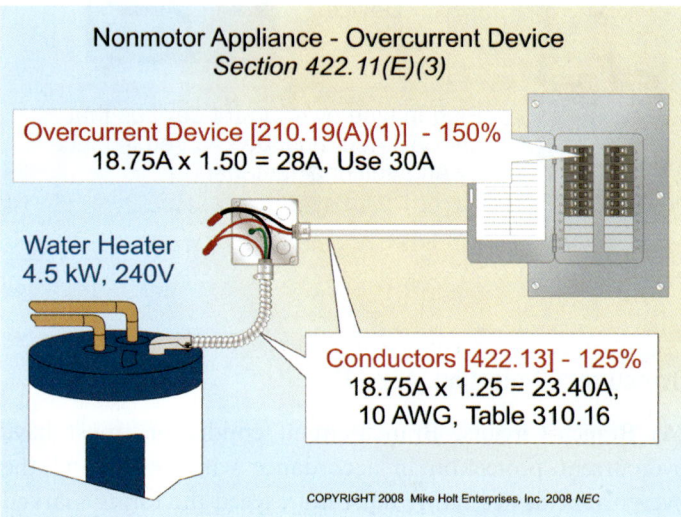

Figure 6–43

Author's Comment: In the previous example, the overcurrent device was sized at 30A, so even though a 12 AWG conductor provides the necessary ampacity, it can't be used on a 30A overcurrent device for appliance circuits, so 10 AWG conductors would be required [240.4(D)]

Article 424 contains the installation requirements for fixed electrical equipment used for space heating, such as heating cables, unit heaters, boilers, or central systems [424.1]

Author's Comment: Wiring for fossil-fuel heating equipment, such as gas, oil, or coal central furnaces, must be installed in accordance with Article 422, specifically 422.12.

Storage Water Heaters [422.13]

An electric water heater having a capacity of 120 gallons or less is considered a continuous load for the purpose of sizing branch circuits.

Author's Comment: Branch-circuit conductors and overcurrent devices must have an ampacity of at least 125 percent of the ampere rating of a continuous load [210.19(A)(1) and 210.20(A).]

Branch Circuits [424.3]

(B) Branch-Circuit Sizing. For the purpose of sizing branch-circuit conductors, fixed electric space-heating equipment is considered a continuous load.

Author's Comment: The branch-circuit conductors and overcurrent devices for fixed electric space-heating equipment must have an ampacity not less than 125 percent of the total heating load [210.19(A)(1) and 210.20(A)].

Conductor Sizing and Protection Calculations — Unit 6

▶ **Electric Space Heating Branch Circuit Sizing Example**

Question: What size conductor and overcurrent device (with 75°C terminals) are required for a 10 kW, 240V fixed electric space heater that has a 3A blower motor? **Figure 6–44**

(a) 10 AWG, 30A
(b) 8 AWG, 40A
(c) 6 AWG, 60A
(d) 4 AWG, 80A

Answer: (c) 6 AWG, 60A

Step 1: Determine the total load:

$I = VA/E$
$I = 10,000\ VA/240V$
$I = 41.67A + 3A = 44.67A$, round to 45A [220.5(B)]

Step 2: Size the conductors at 125 percent of the total current load [110.14(C), 210.19(A)(1), and Table 310.16]:

Conductor = 45A x 1.25
Conductor = 56A, 6 AWG, rated 65A at 75°C

Step 3: Size the overcurrent device at 125 percent of the total current load [210.20(A), 240.4(b) and 240.6(A)]:

Overcurrent Protection = 45A x 1.25
Overcurrent Protection = 56A, next size up is 60A

Table 450.3(B) Overcurrent Protection for Transformer

Primary Current Rating	Maximum Protection
9A or More	125%, see Note 1
Less Than 9A	167%
Less Than 2A	300%

Note 1. Where 125 percent of the primary current doesn't correspond to a standard rating of a fuse or nonadjustable circuit breaker, the next higher rating is permitted [240.6(A)].

▶ **Transformer Primary Overcurrent Protection Example 1**

Question: What is the primary overcurrent device rating and conductor size required for a 45 kVA, three-phase, 480V transformer that is fully loaded? Terminals are rated 75°C. **Figure 6–45**

(a) 8 AWG, 40A
(b) 6 AWG, 50A
(c) 6 AWG, 60A
(d) 4 AWG, 70A

Answer: (d) 4 AWG, 70A

Step 1: Determine the primary current:

$I = VA/(E\ x\ 1.732)$
$I = 45,000\ VA/(480V\ x\ 1.732)$
$I = 54A$

Step 2: Determine the primary overcurrent device rating [240.6(A)]:

54A x 1.25 = 68A, next size up 70A, Table 450.3(B), Note 1

Step 3: The primary conductor must be sized to carry 54A continuously (54A x 1.25 = 68A) [215.2(A)(1)] and be protected by a 70A overcurrent device [240.4(B)]. A 4 AWG conductor rated 85A at 75°C meets all of the requirements [110.14(C)(1) and 310.16].

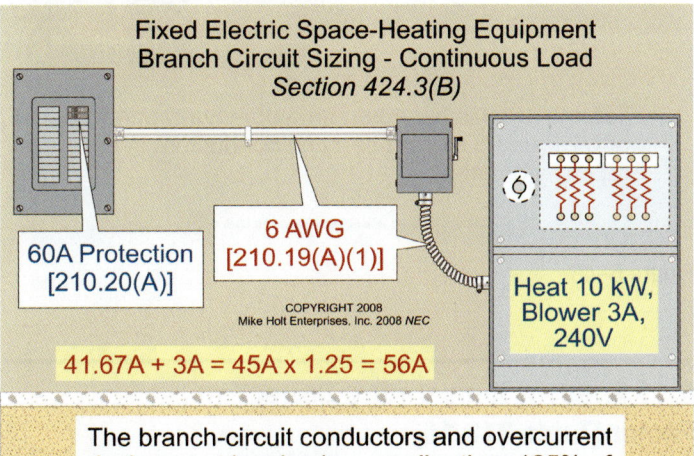

Figure 6–44

Secondary protection is not required for transformers with more than 9A of secondary current if the primary is protected at 125 percent [Table 450.3(B)].

Overcurrent Protection for Transformers Not Over 600V. The primary winding of a transformer must be protected against overcurrent in accordance with the percentages listed in Table 450.3(B) and all applicable notes.

Unit 6 | Conductor Sizing and Protection Calculations

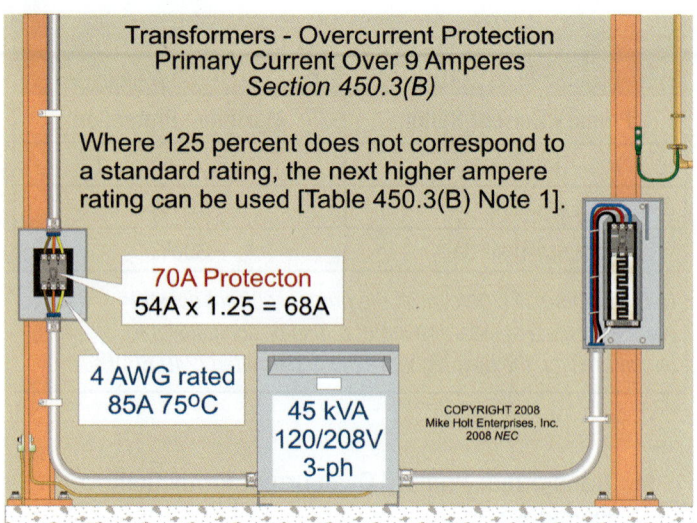

Figure 6–45

can withstand over a prolonged time period without experiencing serious degradation. The main factors to consider for conductor operating temperature are ambient temperature, heat generated internally from current flow through the conductor, the rate at which heat can dissipate, and adjacent load-carrying conductors [310.10 FPN No. 1].

6.10 Conductor Ampacity

The ampacity of a conductor is the current the conductors can carry continuously under specific conditions of use [Article 100 definition]. Conductor ampacities are listed in Table 310.16 under the condition of no more than three current-carrying conductors bundled together in an ambient temperature of 86°F. **Figure 6–46**

▶ **Transformer Primary Overcurrent Protection Example 2**

Question: If a 45 kVA, three-phase transformer has the 480V primary protected at 125 percent, what is the maximum size overcurrent device required on the primary? (Table 450.3(B), Note 1 says that where 125 percent is not a standard size, the next higher rating is permitted.)

(a) 30A   (b) 70A   (c) 90A   (d) 100A

Answer: (b) 70A

I=P/E
45,000 VA/480V x 1.732 = 54A
54A x 1.25 = 68A, round up to 70A [240.6(A)]

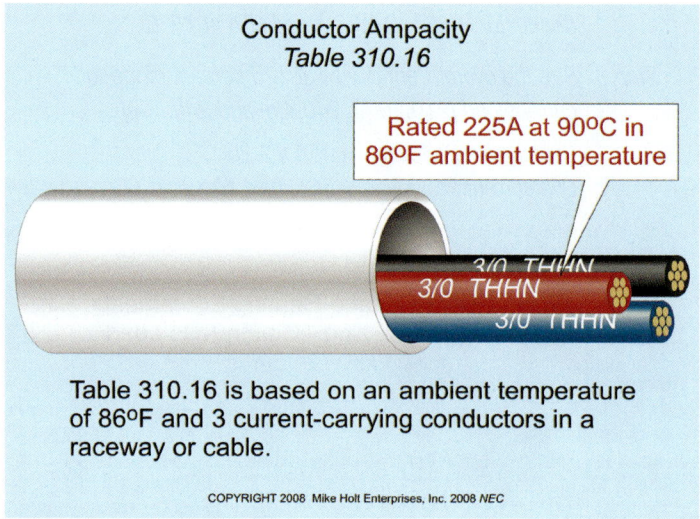

Figure 6–46

PART B—CONDUCTOR AMPACITY

Introduction

The insulation temperature rating of a conductor is limited to an operating temperature that prevents serious heat damage to the conductor's insulation. If the conductor carries excessive current, the I²R heating within the conductor can destroy the conductor insulation. To limit elevated conductor operating temperatures, the current flow (amperes) in the conductors must be limited.

The temperature rating of a conductor is the maximum temperature at any location along its length that the conductor

6.11 Ambient Temperature Correction Factor [Table 310.16]

The ampacity of a conductor as listed in Table 310.16 is based on the conductor operating at an ambient temperature of 86°F (30°C). When the ambient temperature is other than 86°F (30°C) for a prolonged period of time, the conductor ampacity listed in Table 310.16 must be corrected [310.15(B)].

Even though 90°C rated conductor ampacities can't be used for sizing circuit conductors, they offer the opportunity of having a higher conductor ampacity for conductor ampacity adjustment and correction purposes.

Mike Holt's Illustrated Guide to NEC Exam Preparation

Conductor Sizing and Protection Calculations — Unit 6

The temperature correction factors used to determine the new conductor ampacity are listed at the bottom of Table 310.16.

Table 310.16 Ambient Temperature Correction			
Ambient Temperature °F	Ambient Temperature °C	Correction Factor 75°C Conductors	Correction Factor 90°C Conductors
70–77°F	21–25°C	1.05	1.04
78–86°F	26–30°C	1.00	1.00
87–95°F	31–35°C	0.94	0.96
96–104°F	36–40°C	0.88	0.91
105–113°F	41–45°C	0.82	0.87
114–122°F	46–50°C	0.75	0.82
123–131°F	51–55°C	0.67	0.76
132–140°F	56–60°C	0.58	0.71
141–158°F	61–70°C	0.33	0.58
159–176°F	71–80°C		0.41

Author's Comment: When correcting conductor ampacity for elevated ambient temperature, the correction factor used for THHN conductors is based on the 90°C rating of the conductor in a dry location and the 75°C rating of the conductor in a wet location, based on the conductor ampacity listed in Table 310.16 [110.14(C)].

The following formula can be used to determine the conductor's new ampacity when the ambient temperature is not 86°F (30°C). **Figure 6–47**

◆ **Corrected Conductor Ampacity—Ambient Temperature Correction Formula**

Corrected Ampacity = Table 310.16 Ampacity x Ambient Temperature Correction Factor

Author's Comment: When different ampacities apply to a conductor length, the higher ampacity can be used for the entire circuit if the reduced ampacity length isn't in excess of 10 ft and its length doesn't exceed 10 percent of the length of the part of the circuit with the higher ampacity [310.15(A)(2) Ex].

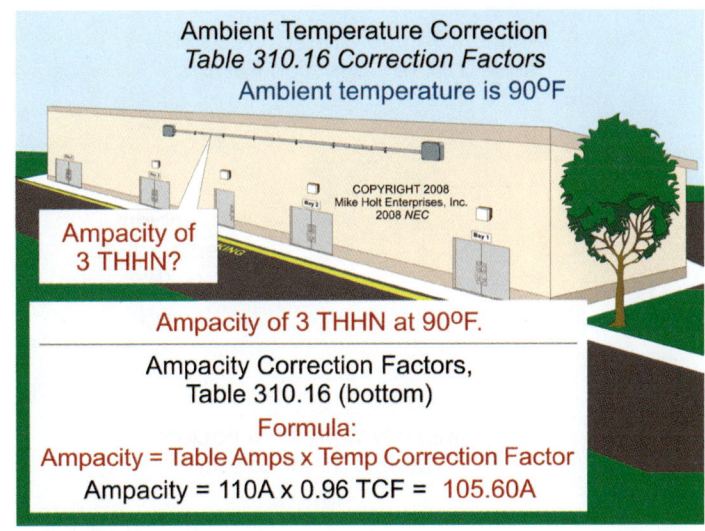

Figure 6–47

▶ **Ambient Temperature Below 86°F**

Question: What is the ampacity of 12 THHN when installed in a location that has an ambient temperature of 70°F? **Figure 6–48**

(a) 20A (b) 25A (c) 31A (d) 35A

• **Answer:** (c) 31A

Corrected Ampacity = Table 310.16 Ampacity x Ambient Temperature Correction Factor

Table 310.16 ampacity for 12 THHN installed in a dry location is 30A at 90°C.

Temperature Correction Factor for a 90°C conductor installed in an ambient temperature of 70°F is 1.04.

Corrected Ampacity = 30A x 1.04 = 31.20A

Note: Ampacity increases when the ambient temperature is less than 86°F (30°C).

Unit 6 — Conductor Sizing and Protection Calculations

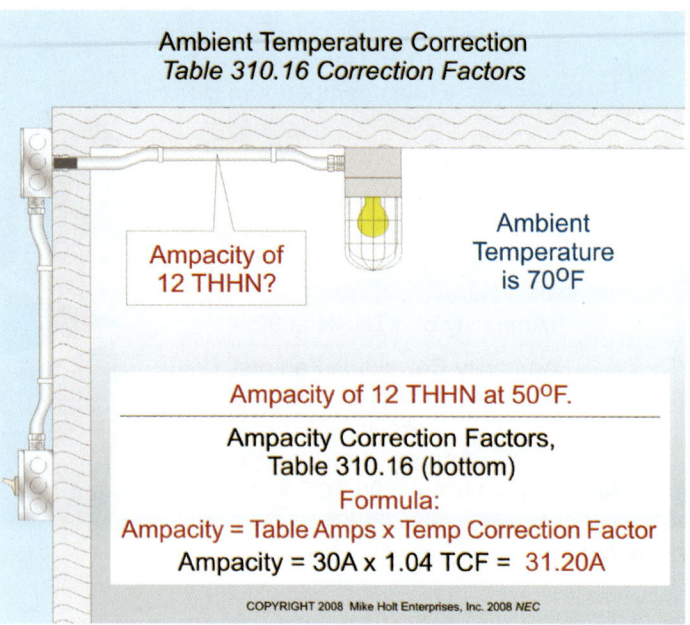

Figure 6–48

FPN: See ASHRAE Handbook—Fundamentals (www.ashrae.org) as a source for the average ambient temperatures in various locations.

Table 310.15(B)(2)(c) Ambient Temperature Adjustment for Conduits Exposed to Sunlight On or Above Rooftops		
Distance of Raceway Above Roof	C°	F°
0 to ½ in.	33	60
Above ½ in. to 3½ in.	22	40
Above 3½ in. to 12 in.	17	30
Above 12 in. to 36 in.	14	25

FPN to Table 310.15(B)(2)(c): The temperature adders in Table 310.15(B)(2)(c) are based on the results of averaging the ambient temperatures.

Author's Comment: This section requires the ambient temperature used for ampacity correction to be adjusted where conductors or cables are installed in conduit on or above a rooftop and the conduit is exposed to direct sunlight. The reasoning is that the air inside conduits in direct sunlight is significantly hotter than the surrounding air, and appropriate ampacity corrections must be made in order to comply with 310.10.

For example, a conduit with three 6 THHN conductors with direct sunlight exposure that is ¾ in. above the roof will require 40°F to be added to the correction factors at the bottom of Table 310.16. Assuming an ambient temperature of 90°F, the temperature to use for conductor correction will be 130°F (90°F + 40°F), and the 6 THHN conductor ampacity after correction will be 57A (75A x 0.75). **Figure 6–49**

When adjusting conductor ampacity, use the conductor ampacity as listed in Table 310.16 based on the conductors' insulation rating; in this case it's 75A at 90°F. Conductor ampacity adjustment is not based on the temperature terminal rating as per 110.14(C).

▶ **Ambient Temperature Above 86°F (30°C)**

Question: What is the ampacity of 6 THWN-2 when installed in an outdoor location that has an ambient temperature of 60°C? Note: The interior of a raceway installed in a wet location is considered to be a wet location [300.9]. THWN-2 is rated 90°C in wet or dry locations.

(a) 35A (b) 53A (c) 60A (d) 75A

Answer: (b) 53A

Corrected Ampacity = Table 310.16 Ampacity x Ambient Temperature Correction Factor

Table 310.16 ampacity for 6 THWN-2 is 75A at 90°C.

Temperature Correction Factor, 90°C conductor rating installed at 60°C is 0.71.

Corrected Ampacity = 75A x 0.71 = 53.25A

Note: Ampacity decreases when the ambient temperature is more than 86°F.

(c) Raceways Exposed to Sunlight on Rooftops. The ambient temperature adjustment contained in Table 310.15(B)(2)(c) is added to the outdoor ambient temperature for conductors or cables that are installed in raceways exposed to direct sunlight on or above rooftops when applying ampacity adjustment correction factors contained in Table 310.16.

Conductor Sizing and Protection Calculations — Unit 6

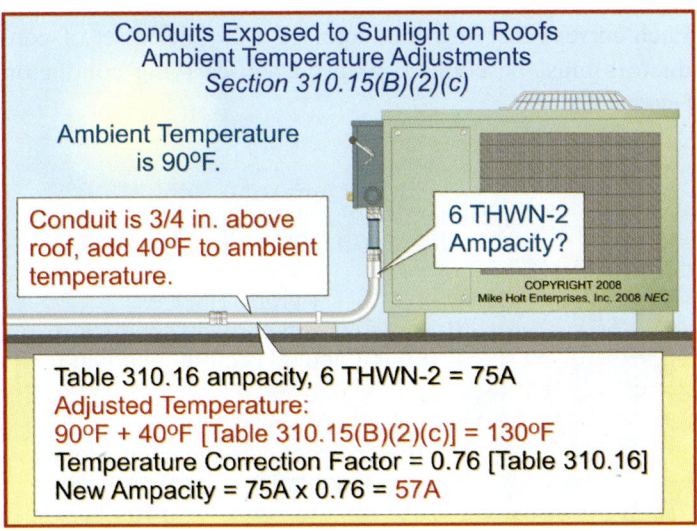

Figure 6–49

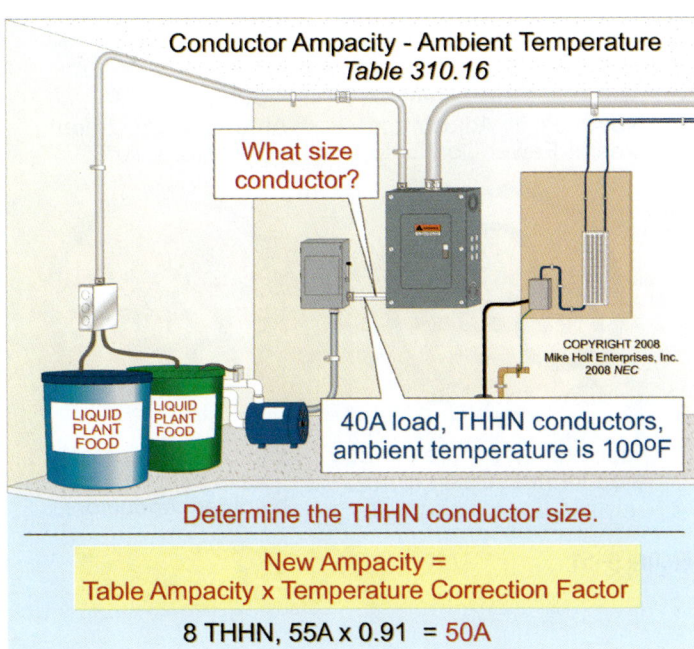

Figure 6–50

▶ Conductor Size

Question: What size THHN/THWN conductor is required to supply a 40A noncontinuous load in a dry location if the conductors pass through an ambient temperature of 100°F? Figure 6–50

(a) 10 AWG (b) 8 AWG (c) 6 AWG (d) 4 AWG

Answer: (b) 8 AWG

Corrected Ampacity = Table 310.16 Amperes x Ambient Temperature Correction Factor

For a dry location, use the 90°C column for THHN. For a wet location, use the 75°C column for THWN

Ambient Temperature Correction Factor for 100°F = 0.91 for THHN

Table 310.16 ampacity for 10 THHN is 40A at 90°C in a dry location

10 THHN = 40A x 0.91 = 36.40A

Table 310.16 ampacity for 8 THHN/THWN is 55A at 90°C in a dry location: use the THHN column

8 THHN = 55A x 0.91 = 50A

If this were in a wet location, would 8 THHN/THWN still be adequate?

Table 310.16 ampacity for 8 THHN/THWN is 50A at 75°C in a wet location: use the THWN column

Ambient Temperature Correction Factor for 100°F = 0.88 for THWN

8 THWN = 50A x 0.88 = 44A

Author's Comment: The conductor to the load must have an ampacity of 40A after applying the ambient temperature correction factor. In this example, the 8 THHN/THWN has sufficient ampacity after correction in either a wet or dry location. That is not always the case, so when using dual-rated insulation types, watch the wet or dry location situation carefully and use the column that corresponds to the location.

6.12 Conductor Bundling Ampacity Adjustment Factor [Table 310.15(B)(2)(a)]

When conductors are bundled together, the ability of the conductors to dissipate heat is reduced. The *NEC* requires that the ampacity of a conductor be reduced whenever four or more current-carrying conductors are bundled together, Figure 6–51. The higher insulation temperature rating of 90°C rated conductors provides a greater conductor ampacity for use in ampacity adjustment, even though conductors must be sized based on the column that corresponds to the temperature listing of the terminals [110.14 (C)(1)].

Conductor Bundle. Where the number of current-carrying conductors in a raceway or cable exceeds three, or where single conductors or multiconductor cables are installed without maintaining spacing for a continuous length longer than 24 in., the allowable ampacity of each conductor, as

Unit 6   **Conductor Sizing and Protection Calculations**

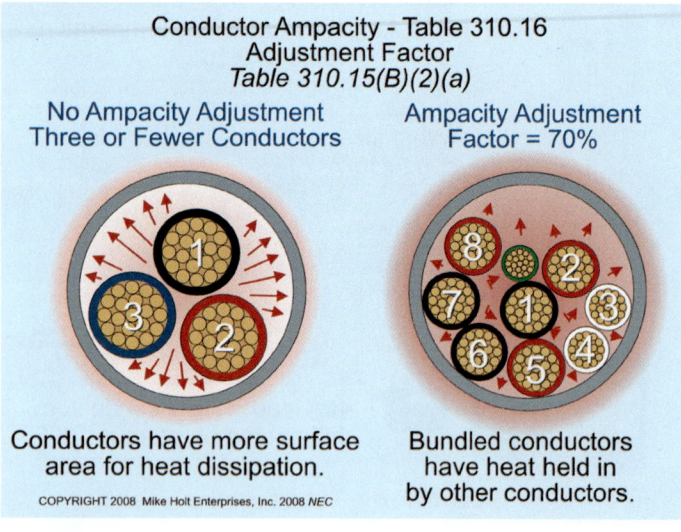

Figure 6–51

listed in Table 310.16, must be adjusted in accordance with the adjustment factors contained in Table 310.15(B)(2)(a). **Figure 6–52**

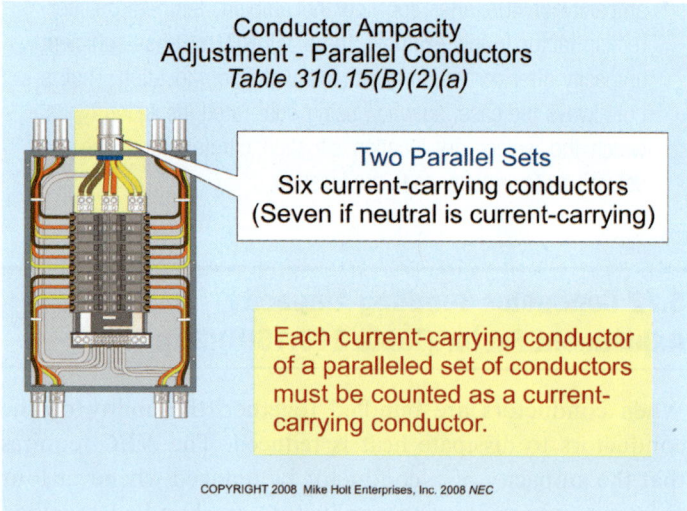

Figure 6–52

Each current-carrying conductor of a paralleled set of conductors must be counted as a current-carrying conductor. **Figure 6–53**

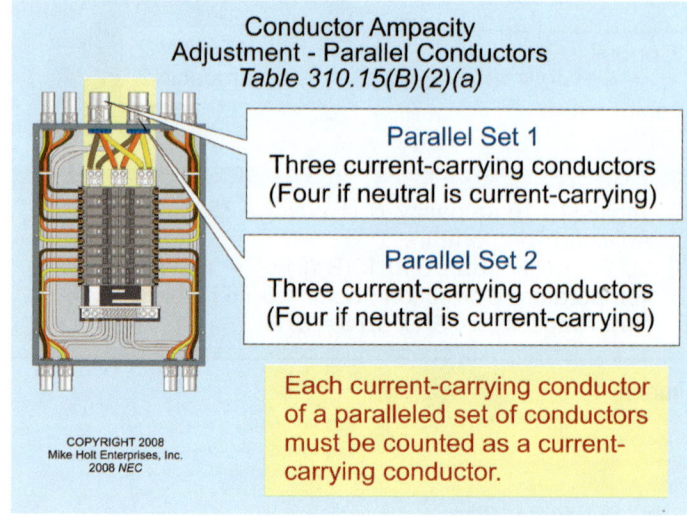

Figure 6–53

The ampacity adjustment factor used to determine the new ampacity is listed in Table 310.15(B)(2)(a). The following formula can be used to determine the new conductor ampacity when more than three current-carrying conductors are bundled together:

◆ **Conductor Bundling Ampacity Adjustment Formula**

Adjusted Ampacity = Table 310.16 Ampacity x Bundled Ampacity Adjustment Factor

Table 310.15(B)(2)(a)	
Number of Current–Carrying	Adjustment Factor
1–3 Conductors	1.00
4–6 Conductors	0.80
7–9 Conductors	0.70
10–20 Conductors	0.50

Author's Comments:

- The neutral conductor might be a current-carrying conductor, but only under the conditions specified in 310.15(B)(4). Equipment grounding conductors are never considered current carrying [310.15(B)(5)].

Conductor Sizing and Protection Calculations — Unit 6

- When correcting or adjusting conductor ampacity, the ampacity is based on the temperature insulation rating of the conductor as listed in Table 310.16, not the temperature rating of the terminal [110.14(C)].

- Where more than three current-carrying conductors are present and the ambient temperature isn't between 78°F and 86°F, the ampacity listed in Table 310.16 must be corrected and adjusted for both conditions.

▶ **Conductor Ampacity**

Question: What is the ampacity of four current-carrying 10 THHN conductors installed in a raceway or cable in a dry location? **Figure 6–54**

(a) 20A (b) 24A (c) 29A (d) 32A

Answer: (d) 32A

Adjusted Ampacity = Table 310.16 Ampacity x Bundled Ampacity Adjustment Factor

Table 310.16 ampacity for 10 THHN is 40A at 90°C.

Bundled adjustment factor for four current-carrying conductors is 0.80.

Adjusted Ampacity = 40A x 0.80 = 32A

▶ **Conductor Size**

Question: A raceway contains four current-carrying conductors. What size conductor is required to supply a 40A noncontinuous load in a dry location? **Figure 6–55**

(a) 12 THHN (b) 10 THHN (c) 8 THHN (d) 6 THHN

Answer: (c) 8 THHN

The conductor must have an ampacity of 40A after applying the bundled adjustment factor.

Adjusted Ampacity = Table 310.16 Ampacity x Bundled Ampacity Adjustment Factor

Table 310.16 ampacity for 10 THHN is 40A at 90°C

Bundled Adjustment Factor for 4 conductors = 0.80 [Table 310.15(B)(2)(a)]
10 THHN = 40A x 0.80 = 32A—too small

Table 310.16 ampacity for 8 THHN is 55A at 90°C
8 THHN = 55A x 0.80 = 44A—just right

Table 310.16 ampacity for 6 THHN is 75A at 90°C
6 THHN = 75A x 0.80 = 60A—larger than required by the NEC

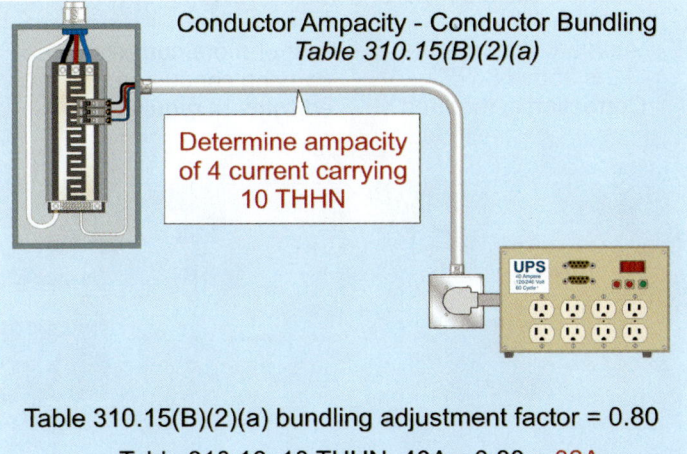

Figure 6–54

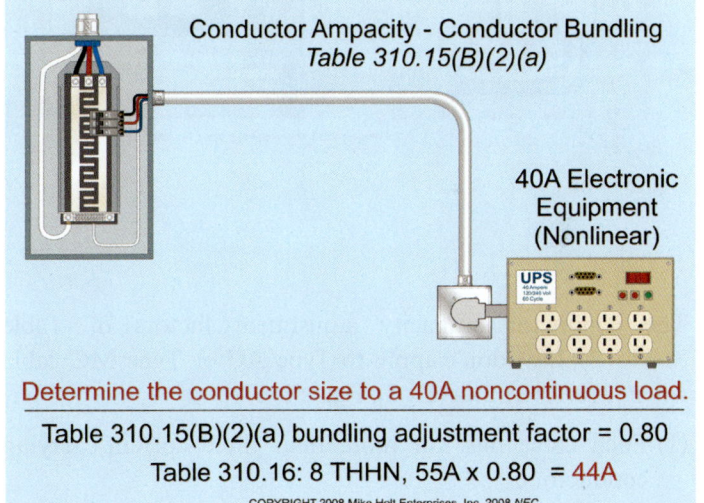

Figure 6–55

Unit 6 Conductor Sizing and Protection Calculations

Author's Comment: Not all conductors are considered current-carrying. For example, an equipment grounding conductor doesn't normally carry current and isn't counted as a current-carrying conductor [310.15(B)(5)]. Some neutral conductors are considered current-carrying and others aren't, depending on the conditions described in 310.15(B)(4). Of course, all ungrounded conductors are considered current-carrying. See Section 6.15 of this unit for specific information on how to determine the number of current-carrying conductors. Conductor bundling ampacity adjustment factors don't apply to conductors in a nipple less than 24 in. long. [310.15(B)(2)(a) Ex 3]. Figure 6–56

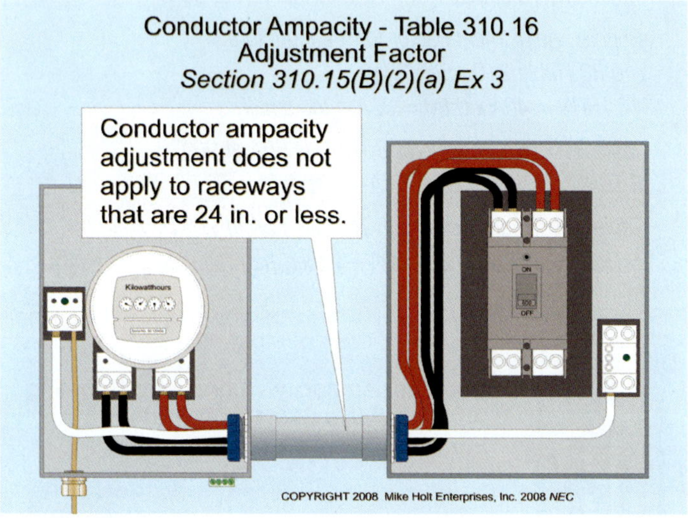

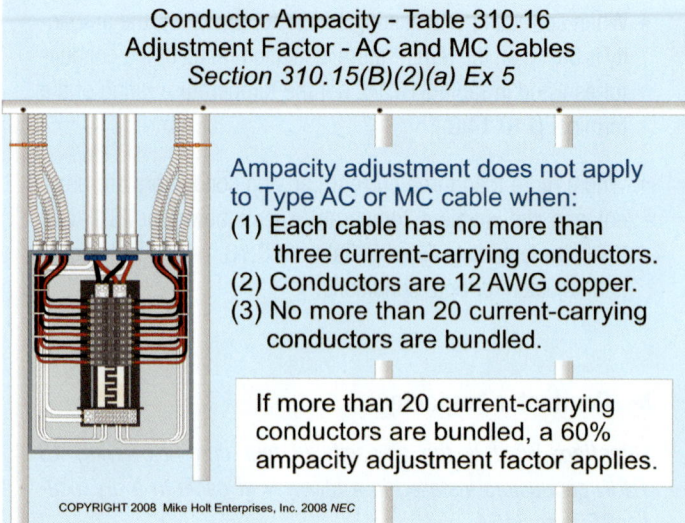

Figure 6–57

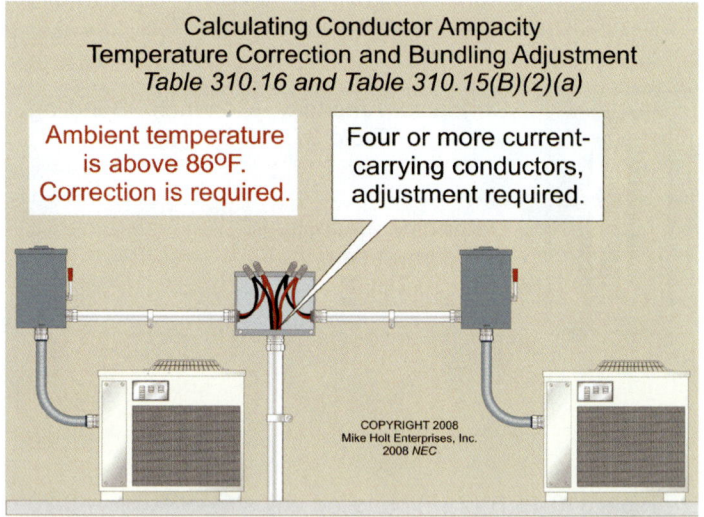

Figure 6–56

The conductor ampacity adjustment factors of Table 310.15(B)(2)(a) don't apply to Type AC or Type MC cable when [310.15(B)(2)(a) Ex 5]: **Figure 6–57**

(1) Each cable has not more than three current-carrying conductors,

(2) The conductors are 12 AWG copper, and

(3) No more than 20 current-carrying conductors (ten 2-wire cables or six 3-wire cables) are bundled.

Author's Comment: When 11 or more 2-wire cables or 7 or more 3-wire cables (more than 20 current-carrying conductors) are bundled or stacked for more than 24 in., an ampacity adjustment factor of 60 percent must be applied.

6.13 Ambient and Conductor Bundling Adjustment

If the ambient temperature is other than 86°F and there are more than three current-carrying conductors bundled together, then the ampacity listed in Table 310.16 must be adjusted and corrected for both conditions. **Figure 6–58**

Figure 6–58

The following formula can be used to determine the new conductor ampacity when both ambient temperature correction and bundled adjustment factors apply:

Adjusted/Corrected Ampacity = Table 310.16 Ampacity x Temperature Correction Factor x Bundled Adjustment Factor

Conductor Sizing and Protection Calculations — Unit 6

▶ **Conductor Ampacity Example**

Question: What is the ampacity of four current-carrying 8 THHN conductors installed in an ambient temperature of 90°F in a dry location? **Figure 6–59**

(a) 25A   (b) 40A   (c) 55A   (d) 60A

Answer: (b) 40A

Adjusted/Corrected Ampacity = Table 310.16 Ampacity x Temperature Factor x Bundled Adjustment Factor

Table 310.16 ampacity of 8 THHN is 55A at 90°C.

Temperature correction factor for 90°C conductor insulation at 90°F is 0.96 [Table 310.16].

Bundled adjustment factor for four conductors is 0.80 [Table 310.15(B)(2)(a)].

Adjusted/Corrected Ampacity = 55A x 0.961 x 0.80 = 42.24A

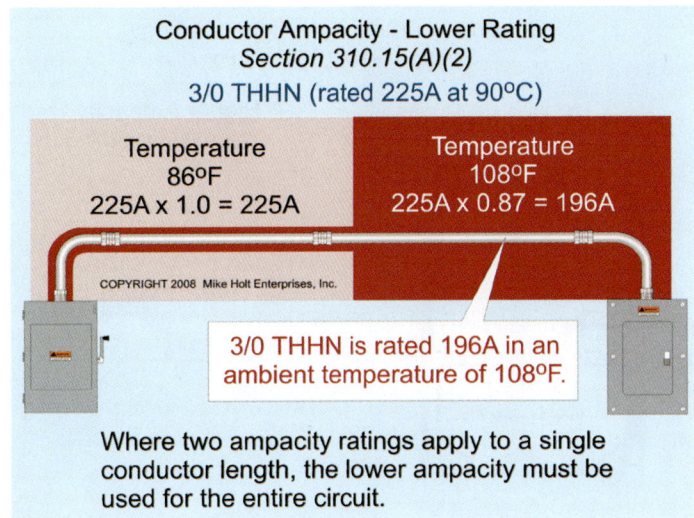

Where two ampacity ratings apply to a single conductor length, the lower ampacity must be used for the entire circuit.

Figure 6–60

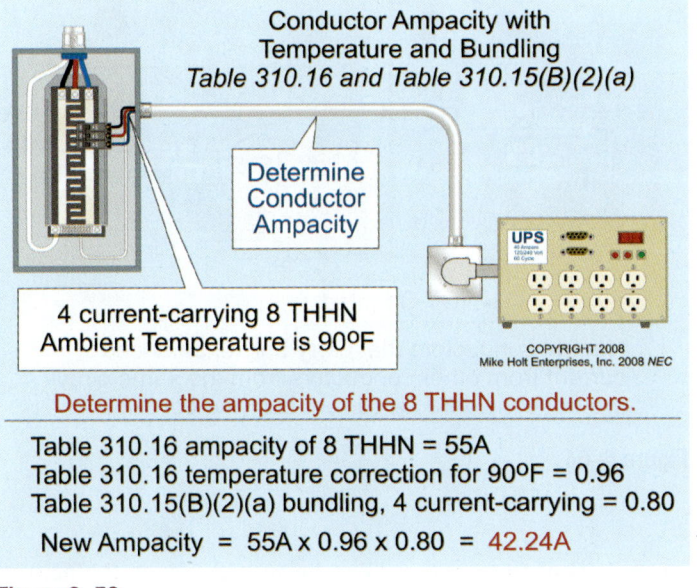

Figure 6–59

If a single length of conductor has two ampacities, the lower ampacity must be used for the entire circuit [310.15(B)]. Figure 6–60

When different ampacities apply to a conductor length, the higher ampacity can be used for the entire circuit if the reduced ampacity length is not in excess of 10 ft and its length does not exceed 10 percent of the length of the part of the circuit with the higher ampacity [310.15(b) Ex]. Figure 6-61 and Figure 6-62

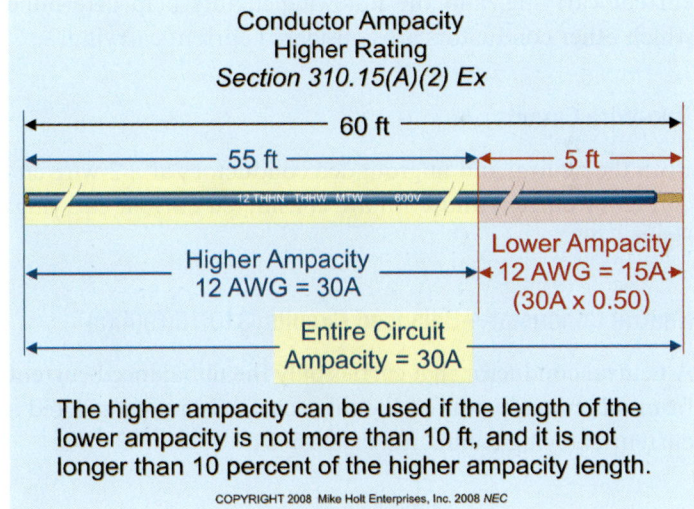

The higher ampacity can be used if the length of the lower ampacity is not more than 10 ft, and it is not longer than 10 percent of the higher ampacity length.

Figure 6–61

Mike Holt Enterprises, Inc. • www.MikeHolt.com • 1.888.NEC.CODE (1.888.632.2633)   181

Unit 6 — Conductor Sizing and Protection Calculations

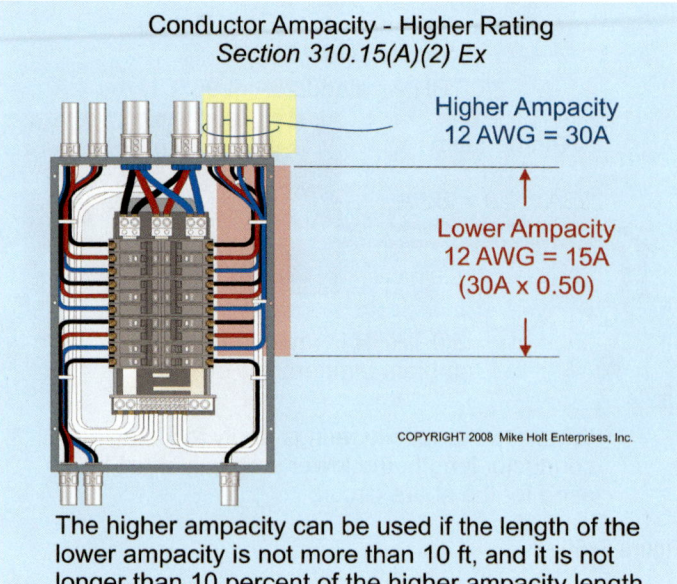

Figure 6–62

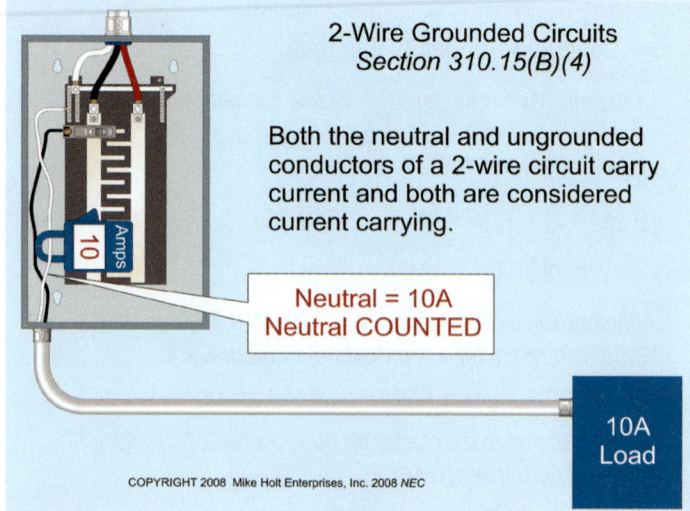

Figure 6–63

6.14 Current-Carrying Conductors

Table 310.15(B)(2)(a) adjustment factors only apply when there are more than three current-carrying conductors bundled together. Naturally, all phase conductors are considered current carrying, and the following factors help determine which other conductors are considered current carrying:

Two-Wire Circuits

Both the neutral and ungrounded conductors of a 2-wire circuit carry current and both are considered current carrying. **Figure 6–63**

Neutral Conductor—Balanced Circuits, 310.15(B)(4)(a)

A neutral conductor that carries only the unbalanced current from other conductors of the same circuit is not considered a current-carrying conductor. **Figure 6–64**

Neutral Conductor—Unbalanced 3-Wire Wye Circuit, 310.15(B)(4)(b)

The neutral conductor of a 3-wire circuit of a 4-wire, three-phase, wye-connected system carries approximately the same current as the line-to-neutral load currents of the other conductors and is considered a current-carrying conductor. **Figure 6–65** In such a situation, one of the line-to-neutral

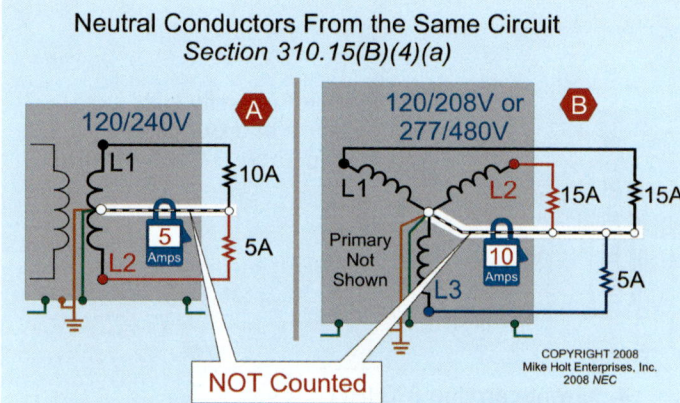

Figure 6–64

currents is not present and can be zeroed out of the neutral current formula, resulting in the following formula:

◆ **Unbalanced 3-Wire Neutral Current Formula**

$$I_{Neutral} = \sqrt{(I_{Line1}^2 + I_{Line2}^2) - (I_{Line1} \times I_{Line2})}$$

Conductor Sizing and Protection Calculations — Unit 6

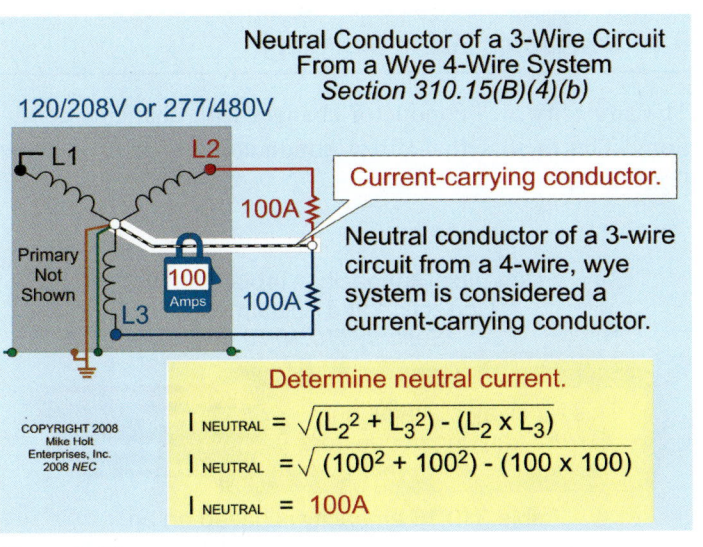

Figure 6–65

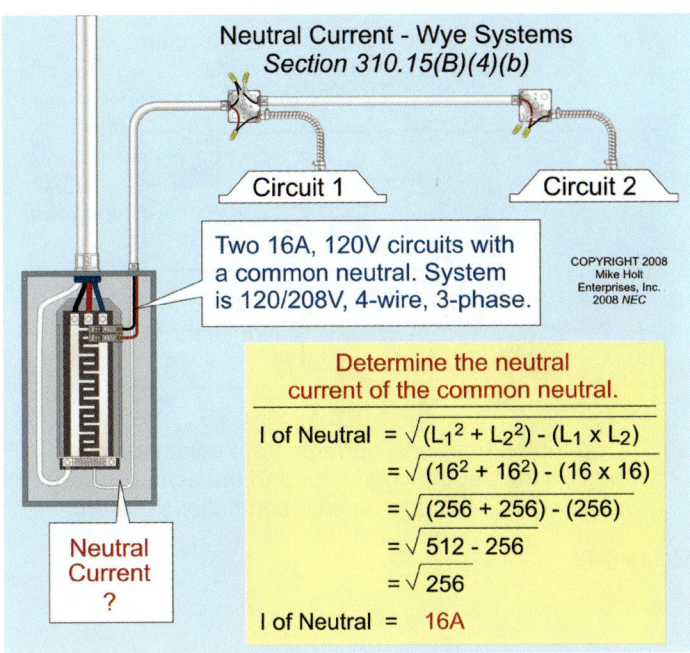

Figure 6–66

▶ **Neutral Conductor Current Example**

Question: What is the neutral current for two 16A, 120V circuits with a common neutral? The system is a 120/208V, three-phase, 4-wire, wye-connected system that supplies fluorescent lighting. **Figure 6–66**

(a) 8A (b) 16A (c) 32A (d) 40A

Answer: (b) 16A

$$I_{Neutral} = \sqrt{(I_{Line1}^2 + I_{Line2}^2) - (I_{Line1} \times I_{Line2})}$$
$$I_{Neutral} = \sqrt{(16^2 + 16^2) - (16 \times 16)}$$
$$I_{Neutral} = \sqrt{(512 - 256)}$$
$$I_{Neutral} = \sqrt{256}$$
$$I_{Neutral} = 16A$$

Neutral Conductor—Nonlinear Loads, 310.15(B)(4)(c)

The neutral conductor for a 4-wire, 3-phase wye circuit is considered a current-carrying conductor where the major portion of the load consists of nonlinear loads. **Figure 6–67**

> **CAUTION:** *Nonlinear loads produce harmonic currents that add on the neutral conductor, and the current on the neutral can be as much as twice the current on the ungrounded conductors.* **Figure 6–68**

Author's Comment: Nonlinear loads supplied by a 4-wire, three-phase, 120/208V or 277/480V wye-connected system can

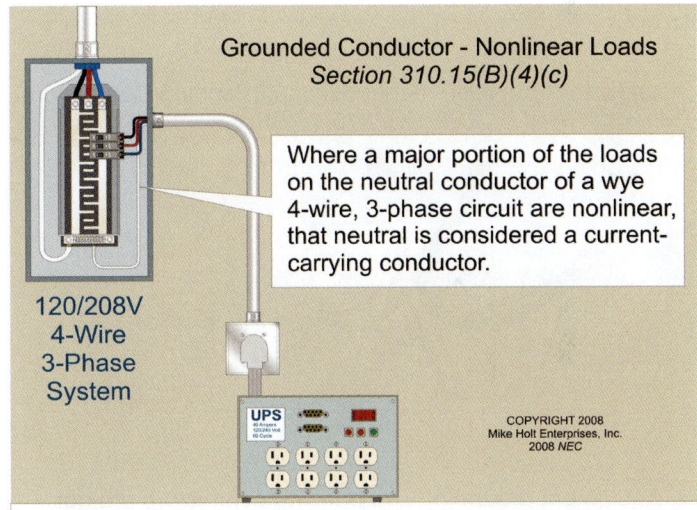

Figure 6–67

produce unwanted and potentially hazardous triplen harmonic currents (3rd, 9th, 15th, etc.) that can add on the neutral conductor. To prevent fire or equipment damage from excessive harmonic neutral current, the designer should consider increasing the size of the neutral conductor or installing a separate neutral for each phase. For more information, visit www.MikeHolt.com, click on the "Technical" link, then the "Power Quality" link. Also see 210.4(a) FPN, 220.61 FPN No. 2, and 450.3 FPN No. 2.

Unit 6 — Conductor Sizing and Protection Calculations

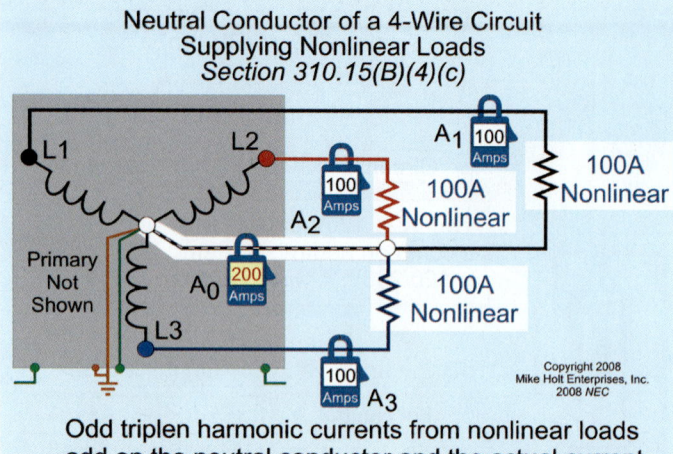

Figure 6–68

Grounding and Bonding Conductors, 310.15(B)(5)

Grounding and bonding conductors do not normally carry current and are not considered current carrying. **Figure 6–69**

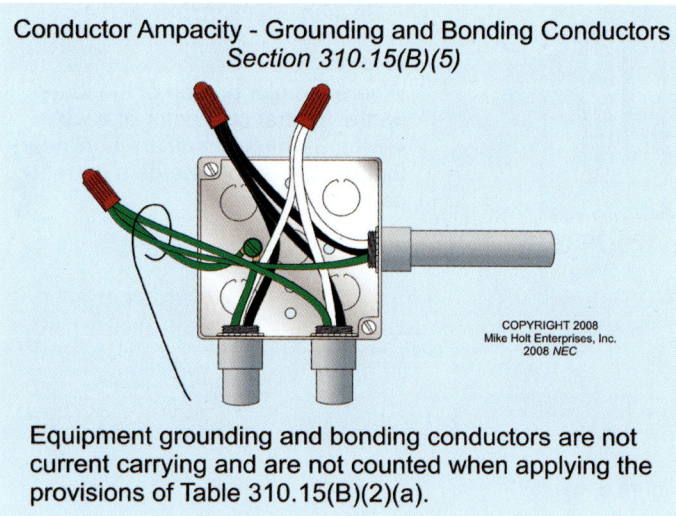

Figure 6–69

Author's Comment: Grounding and bonding conductors aren't counted when adjusting conductor ampacity for the effects of conductor bundling. [310.15(B)(5)] They do occupy space in the raceway and are included in raceway fill calculations (see Chapter 9, Table 1, Note 3).

6.15 Conductor Sizing Summary

The ampacity of a conductor changes with changing conditions. The factors that affect conductor ampacity include: **Figure 6–70**

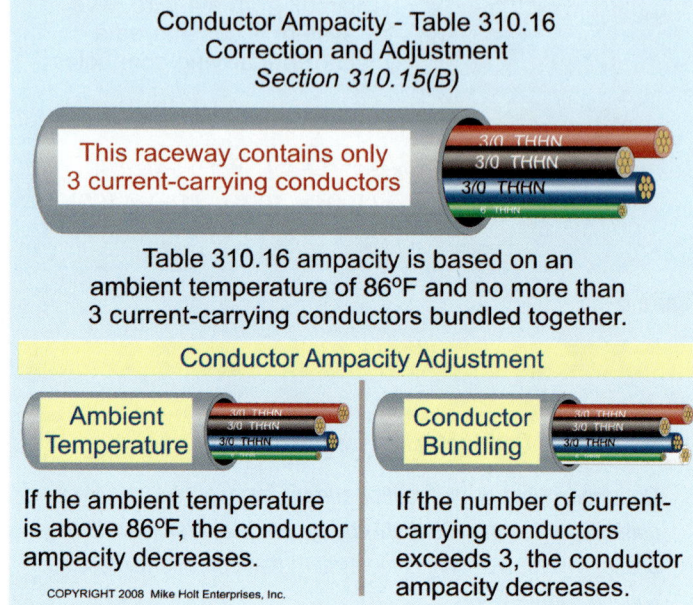

Figure 6–70

- The allowable ampacity as listed in Table 310.16.
- The ambient temperature correction factors, if the ambient temperature is not 86°F.
- Conductor ampacity adjustment factors apply if four or more current-carrying conductors are bundled together.

Terminal Ratings, 110.14(C)

Equipment rated 100A or less must have the conductor sized no smaller than the 60°C column of Table 310.16 unless the equipment terminals are rated for a higher temperature [110.14(C)(1)(a)]. Equipment rated over 100A must have the conductors sized no smaller than the 75°C column of Table 310.16 [110.14(C)(1)(b)]. The higher insulation temperature rating of conductors such as THHN allow the use of the 90°C column for ampacity correction and adjustment calculations.

Unit 6: Conclusion

CONCLUSION TO UNIT 6—CONDUCTOR SIZING AND PROTECTION CALCULATIONS

The sizing of conductors and overcurrent protection is much more complicated than it appears at first. This unit has provided an insight into the numerous factors that must be taken into account in properly sizing conductors and providing for their protection.

Temperature is a key factor in the resistance of a conductor, and consequently temperature affects a conductor's current-carrying capacity. High temperature on conductors can lead to the breakdown of conductor insulation and damage to equipment terminals. Therefore, when sizing conductors, the ampacity values given in the *NEC* tables must be corrected for ambient temperatures that differ from the ambient of 86° F, on which the *Code* tables are based. A bundle of current-carrying conductors will have increased operating temperatures, so the *NEC* also requires that an adjustment be made when bundling more than three current-carrying conductors together.

This unit also taught you that conductors must not be sized to a higher temperature column of the ampacity table than the rating that the equipment terminals allow. This means that even though you install a 90°C rated conductor such as THHN, a lower ampacity column must be used to match the terminal rating. This confusing requirement results in mistakes on exam problems, so review this part of the unit again if you're not sure how to apply Section 110.14(C) of the *Code*.

Overcurrent protection must be sized correctly to protect the circuit wiring and equipment from damage that can result from overheating. This unit covered common overcurrent sizing requirements as well as some of the exceptions to the general rules of Article 240. A number of specific examples of overcurrent protection were explained along with examples of how to size the overcurrent devices in order to help you understand that there are specific rules for many types of equipment or situations that don't follow the general overcurrent protection requirements.

Unit 6 Practice Questions

UNIT 6—CONDUCTOR SIZING AND PROTECTION CALCULATIONS PRACTICE QUESTIONS

(•Indicates that 75% or fewer of those who took this exam answered the question correctly.)

PART A—GENERAL CONDUCTOR REQUIREMENTS

6.1 Conductor Insulation [Table 310.13]

1. THHN can be described as _____.

    (a) thermoplastic insulation with a nylon outer cover
    (b) suitable for dry and wet locations
    (c) having a maximum operating temperature of 90°C
    (d) a and c

6.2 Conductor Sizing [110.6]

2. Conductor sizes are expressed in American Wire Gage (AWG) from 40 AWG through 4/0 AWG. Conductors larger than _____ are expressed in circular mils.

    (a) 1 AWG
    (b) 1/0 AWG
    (c) 3/0 AWG
    (d) 4/0 AWG

3. The smallest size conductor permitted for branch circuits, feeders, and services for residential, commercial, and industrial locations is _____.

    (a) 14 AWG copper
    (b) 12 AWG aluminum
    (c) 12 AWG copper
    (d) a and b

6.4 Conductor Size—Terminal Temperature Rating [110.14(C)]

4. Equipment terminals rated 100A or less (receptacles, switches, circuit breakers, fuses, etc.) and pressure connector terminals for 14 AWG through 1 AWG conductors must have the conductor sized according to the 60°C temperature rating, as listed in Table 310.16.

    (a) True
    (b) False

5. •What is the minimum size THHN conductor permitted to terminate on a 70A circuit breaker or fuse if the circuit breaker and equipment terminals are listed for 60°C conductor sizing?

    (a) 8 AWG
    (b) 6 AWG
    (c) 4 AWG
    (d) 2 AWG

6. •What is the minimum size THHN conductor that is permitted to terminate on a 50A circuit breaker or fuse if the circuit breaker and equipment terminals are listed for 75°C conductor sizing?

    (a) 10 AWG
    (b) 8 AWG
    (c) 6 AWG
    (d) 4 AWG

7. Terminals for equipment rated over 100A and pressure connector terminals for conductors larger than 1 AWG must have the conductor sized according to the 75°C temperature rating, as listed in Table 310.16.

    (a) True
    (b) False

8. What size THHN/THWN conductor is required for an air-conditioning unit if the nameplate requires a conductor ampacity of 34A? The terminals of all the equipment and circuit breakers are rated 75°C and this is installed in a wet location.

    (a) 14 AWG
    (b) 12 AWG
    (c) 10 AWG
    (d) 8 AWG

9. What is the minimum size THHN conductor required for a 150A circuit breaker or fuse for a nonmotor circuit? Be sure to comply with the requirements of 110.14(C)(1)(b).

    (a) 1/0 AWG
    (b) 2/0 AWG
    (c) 3/0 AWG
    (d) 4/0 AWG

10. In general, THHN (90°C) conductor ampacities cannot be used when sizing conductors. When more than three current-carrying conductors are bundled together, or if the ambient temperature is greater than 86°F, the allowable conductor ampacity must be decreased. THHN offers the opportunity of having a greater ampacity for conductor adjustment purposes, which permits the same conductor type to be used without having to increase the conductor size.

    (a) True
    (b) False

11. What size conductor is required to supply a 190A non-continuous load in a dry location? The terminals are rated 75°C.

    (a) 2/0 AWG
    (b) 3/0 AWG
    (c) 4/0 AWG
    (d) 300 kcmil

6.5 Conductors in Parallel

12. •Phase and grounded conductors sized _____ AWG and larger are permitted to be connected in parallel.

    (a) 1
    (b) 1/0
    (c) 2/0
    (d) 3/0

13. To ensure that currents are evenly distributed between parallel conductors, each conductor within a parallel set must be installed in the same type of raceway (metallic or nonmetallic) and must be the same length, material, circular mils, insulation type, and must terminate in the same method.

    (a) True
    (b) False

14. When equipment grounding conductors are installed in parallel, each raceway must have a full-size equipment grounding conductor sized according to the overcurrent device rating of that circuit.

    (a) True
    (b) False

15. All parallel equipment grounding conductors are required to be a minimum 1/0 AWG.

    (a) True
    (b) False

16. What size equipment grounding conductor is required in each raceway for an 800A, 500 kcmil feeder paralleled in two raceways?

    (a) 1 AWG
    (b) 2 AWG
    (c) 3 AWG
    (d) 1/0 AWG

17. If an 800A service has a calculated load of 750A, what size 75°C conductors are required if the conductors are paralleled in two raceways?

    (a) 4/0 AWG
    (b) 250 kcmil
    (c) 500 kcmil
    (d) 750 kcmil

18. •What size conductors are required for a 250A feeder paralleled in two raceways?

    (a) 3 AWG
    (b) 2 AWG
    (c) 1/0 AWG
    (d) 2/0 AWG

Unit 6 — Practice Questions

6.7 Overcurrent Protection [Article 240]

19. One of the purposes of conductor overcurrent protection is to protect the conductors against excessive or dangerous heat.

    (a) True
    (b) False

20. Overcurrent devices must be designed and rated to clear fault current and must have a short-circuit interrupting rating sufficient for the available fault levels. The minimum interruption rating for circuit breakers is _____ and _____ for fuses.

    (a) 5,000A, 5,000A
    (b) 5,000A, 10,000A
    (c) 10,000A, 5,000A
    (d) 10,000A, 10,000A

21. Which of the following is a standard size for circuit breakers and fuses?

    (a) 25A
    (b) 90A
    (c) 350A
    (d) any of these

22. Where a circuit supplies continuous loads or any combination of continuous and noncontinuous loads, the rating of the overcurrent device must not be less than the noncontinuous load plus _____ of the continuous load.

    (a) 80 percent
    (b) 100 percent
    (c) 125 percent
    (d) 150 percent

6.8 Overcurrent Protection of Conductors—General Requirements [240.4]

23. The maximum overcurrent device size for 14 AWG is 15A, 12 AWG is 20A, and 10 AWG is 30A. This is a general rule, but it does not apply to motors or air conditioners according to 240.4(G).

    (a) True
    (b) False

24. If the ampacity of a conductor does not correspond with the standard ampere rating of a fuse or circuit breaker, the next size up overcurrent device is permitted. This applies only if the conductors supply multioutlet receptacles for portable cord-and-plug-connected loads.

    (a) True
    (b) False

25. What size conductor is required for a 70A breaker that supplies a 70A noncontinuous load where the terminals are rated 75°C?

    (a) 8 AWG
    (b) 6 AWG
    (c) 4 AWG
    (d) any of these

6.9 Overcurrent Protection of Conductors—Specific Requirements

26. What size feeder overcurrent device is required for a 100A continuous load?

    (a) 100A
    (b) 125A
    (c) 150A
    (d) 200A

27. Tap conductors not over 25 ft shall be permitted, providing the _____.

    (a) ampacity of the tap conductors is not less than one-third the rating of the overcurrent device protecting the feeder conductors being tapped
    (b) tap conductors terminate in a single circuit breaker or set of fuses that limit the load to the ampacity of the tap conductors
    (c) tap conductors are suitably protected from physical damage
    (d) all of these

28. Outside secondary conductors can be of unlimited length without overcurrent protection at the point they receive their supply if:

    (a) The conductors are suitably protected from physical damage.
    (b) The conductors terminate at a single circuit breaker or a single set of fuses that limits the load to the ampacity of the conductors.
    (c) a and b
    (d) none of these

Practice Questions — Unit 6

PART B—CONDUCTOR AMPACITY

6.10 Conductor Ampacity

29. The temperature rating of a conductor is the maximum operating temperature the conductor insulation can withstand (without serious damage) over a prolonged period of time. The _____ provide guidance for adjusting conductor ampacities for different conditions.

    (a) conductor allowable ampacities
    (b) ambient temperature correction factors
    (c) over three current-carrying conductors adjustment factors
    (d) all of these

6.11 Ambient Temperature Correction Factor [Table 310.16]

30. •The ampacities listed in Table 310.16 apply only when the ambient temperature is 40°C and there are no more than two current-carrying conductors bundled together. If the ambient temperature is not 40°C, or there are more than two current-carrying conductors in a raceway, the allowable ampacities must be adjusted to reflect the ampacity under the condition of use.

    (a) True
    (b) False

31. •What is the ampacity of an 8 THHN conductor when installed in a walk-in cooler (dry location) if the ambient temperature is 50°F?

    (a) 40A
    (b) 50A
    (c) 55A
    (d) 57A

32. What size THHN conductor is required to feed a 16A noncontinuous load when the conductors are in an ambient temperature of 100°F in a dry location? The circuit is protected with a 20A overcurrent device.

    (a) 14 AWG
    (b) 12 AWG
    (c) 10 AWG
    (d) 8 AWG

6.12 Conductor Bundling Ampacity Adjustment Factor [Table 310.15(B)(2)(a)]

33. When four or more current-carrying conductors are bundled together for more than _____, the conductor allowable ampacity must be reduced according to the factors listed in Table 310.15(B)(2)(a).

    (a) 12 in.
    (b) 24 in.
    (c) 36 in.
    (d) 48 in.

34. •What is the ampacity of four 1/0 THHN current-carrying conductors in a raceway in a dry location?

    (a) 111A
    (b) 136A
    (c) 153A
    (d) 171A

35. A raceway contains eight current-carrying conductors. What size conductor is required to feed a 21A noncontinuous lighting load in a dry location? The overcurrent device is rated 30A.

    (a) 14 THHN
    (b) 12 THHN
    (c) 10 THHN
    (d) any of these

6.13 Ambient and Conductor Bundling Adjustment

36. What is the ampacity of eight current-carrying 10 THHN conductors installed in an ambient temperature of 100°F in a dry location?

    (a) 21A
    (b) 25A
    (c) 32A
    (d) 40A

6.14 Current-Carrying Conductors

37. •The grounded conductor of a balanced 3-wire delta circuit, or 4-wire, three-phase wye circuit, is considered a current-carrying conductor for the purpose of applying the adjustment factors of Table 310.15(B)(2)(a).

    (a) True
    (b) False

Unit 6 — Practice Questions

38. The grounded conductor of a balanced 4-wire, three-phase wye circuit, that is at least 50 percent loaded with nonlinear loads (electric-discharge lighting, electronic ballasts, dimmers, controls, computers, laboratory test equipment, medical test equipment, recording studio equipment, etc.) is not considered a current-carrying conductor for the purpose of applying bundle adjustment factors.

    (a) True
    (b) False

39. •The grounded conductor of a 3-wire wye circuit from a 4-wire, three-phase, wye system is not considered a current-carrying conductor for the purpose of applying bundle adjustment factors.

    (a) True
    (b) False

6.15 Conductor Sizing Summary

40. •The ampacity of a conductor can be different along the length of the conductor. The higher calculated ampacity can be used if the length of the lower ampacity is no more than 10 ft, or no more than 10 percent of the length of the higher ampacity circuit conductors, whichever is less.

    (a) True
    (b) False

41. Terminals are presumed to be rated 60°C for equipment 100A or less and 75°C for equipment terminals rated over 100A unless the terminal temperature rating is specified. Regardless of the conductor ampacity, conductors must be sized no smaller than required by the terminal temperature rating.

    (a) True
    (b) False

Unit 6 Challenge Questions

UNIT 6—CONDUCTOR SIZING AND PROTECTION CALCULATIONS CHALLENGE QUESTIONS

(•Indicates that 75% or fewer of those who took this exam answered the question correctly.)

PART A—GENERAL CONDUCTOR REQUIREMENTS

6.7 Overcurrent Protection [Article 240]

1. •A continuous load of 27A requires the circuit overcurrent device to be sized at _____.

   (a) 20A
   (b) 30A
   (c) 35A
   (d) 40A

2. •What size overcurrent device is required for a 45A continuous load? The circuit is in a raceway with 14 current-carrying conductors.

   (a) 45A
   (b) 50A
   (c) 60A
   (d) 70A

3. •A 65A continuous load requires a _____ overcurrent device.

   (a) 60A
   (b) 70A
   (c) 75A
   (d) 90A

4. A department store feeder (continuous load) supplies a lighting load of 103A. The minimum size overcurrent device permitted for this feeder is _____.

   (a) 110A
   (b) 125A
   (c) 150A
   (d) 175A

PART B—CONDUCTOR AMPACITY

6.11 Ambient Temperature Correction Factor [Table 310.16]

5. A 2 TW conductor is installed in a location where the ambient temperature is expected to be 102°F. The temperature correction factor for conductor ampacity in this location is _____.

   (a) 0.71
   (b) 0.82
   (c) 0.88
   (d) 0.96

6. •If the ambient temperature is 71°C, the minimum insulation temperature rating that a conductor must have and still have the capacity to carry current is _____.

   (a) 60°C
   (b) 90°C
   (c) 105°C
   (d) 120°C

6.12 Conductor Bundling Ampacity Adjustment Factor [Table 310.15(B)(2)(a)]

7. The ampacity of six current-carrying 4/0 XHHW aluminum conductors installed in a ground floor slab (wet location) is _____.

   (a) 135A
   (b) 144A
   (c) 185A
   (d) 210A

Unit 6 — Challenge Questions

6.13 Ambient and Conductor Bundling Adjustment

8. The ampacity of 15 current-carrying 10 RHW aluminum conductors in a raceway in an ambient temperature of 75°F is _____.

   (a) 12A
   (b) 16A
   (c) 22A
   (d) 30A

9. •A(n) _____ THHN conductor is required for a 19.70A noncontinuous load if the ambient temperature is 75°F and there are nine current-carrying conductors in the raceway in a dry location.

   (a) 14
   (b) 12
   (c) 10
   (d) 8

10. •The ampacity of nine current-carrying 10 THW conductors installed in a 20 in. long raceway is _____.

    (a) 25A
    (b) 30A
    (c) 35A
    (d) 40A

11. •The ampacity of 10 current-carrying 6 THHW conductors installed in an 18 in. long conduit in a dry location having an ambient temperature of 39°C is _____.

    (a) 40A
    (b) 68A
    (c) 75A
    (d) 90A

6.14 Current-Carrying Conductors

12. A raceway contains the following: one 4-wire, multiwire branch circuit that supplies a balanced incandescent 120V lighting load; one 4-wire, multiwire branch circuit that supplies a balanced 120V fluorescent lighting load; two conductors that supply a receptacle; and one equipment grounding conductor. The system is 120/208V, three-phase. Taking these factors into consideration, how many of these conductors are considered current-carrying?

    (a) 7 conductors
    (b) 8 conductors
    (c) 9 conductors
    (d) 11 conductors

13. •There is a total of nine 10 THW conductors in a raceway. The system voltage is 120/208V, three-phase. One conductor is an equipment grounding conductor, four conductors supply a 4-wire, multiwire branch circuit for balanced electric-discharge luminaires, and the remaining conductors supply a 4-wire, multiwire branch circuit for balanced incandescent luminaires. Taking all of these factors into consideration, how many of these conductors are considered current-carrying?

    (a) 6 conductors
    (b) 7 conductors
    (c) 9 conductors
    (d) 10 conductors

UNIT 7 — Motor and Air-Conditioning Calculations

INTRODUCTION TO UNIT 7—MOTOR AND AIR-CONDITIONING CALCULATIONS

Motor circuits have special requirements that affect how the overcurrent protection is sized and installed. Motors typically draw about six times as much current at start-up as they draw during normal operation. Article 430 provides guidance on how to properly protect the motor from overcurrent and still avoid nuisance tripping of the fuse or circuit breaker protecting the motor. Similar rules are included in Article 440 for air conditioners.

The *Code* definition of "Overcurrent" is made up of three factors: short circuits, ground faults, and overloads. For motors, the function of overcurrent protection is divided into two parts. The short-circuit and ground-fault protection of a motor is usually provided by a fuse or circuit breaker which is sized large enough to let the motor start, but too large to provide overload protection. Overload protection is provided to protect the motor and wiring at a value close to the actual running current of the motor, but with sufficient time delay to allow the motor to start. This protection is often provided by the "heaters" which are overload sensing devices in a magnetic starter.

Article 430 spells out the minimum sizing of conductors for motor branch circuits and feeders as well. When conductors and short-circuit ground-fault protection are sized based on Article 430, the fuse or circuit breaker may appear to be much larger than it should be for the conductors selected. Make sure everything is sized correctly based on Article 430, then don't be concerned. The overcurrent protection rules of Article 240 don't apply to motors or air conditioners, so often an installation may not "look right," even though it complies with Article 430 requirements for motors and Article 440 for air conditioners.

Careful study of this unit will help you understand the sometimes confusing requirements of Articles 430 and 440.

PART A—MOTOR CALCULATIONS

7.1 Scope of Article 430

Scope [430.1]. Article 430 covers motors, motor branch-circuit and feeder conductors and their protection, motor overload protection, motor control circuits, motor controllers, and motor control centers. This article is divided into many parts, the most important being: **Figure 7–1**

- General—Part I
- Circuit Conductors—Part II
- Overload Protection—Part III
- Branch-Circuit Short-Circuit and Ground-Fault Protection—Part IV
- Feeder-Circuit Short-Circuit and Ground-Fault Protection—Part V
- Control Circuits—Part VI
- Controllers—Part VII
- Motor Control Centers—Part VIII
- Disconnecting Means—Part IX

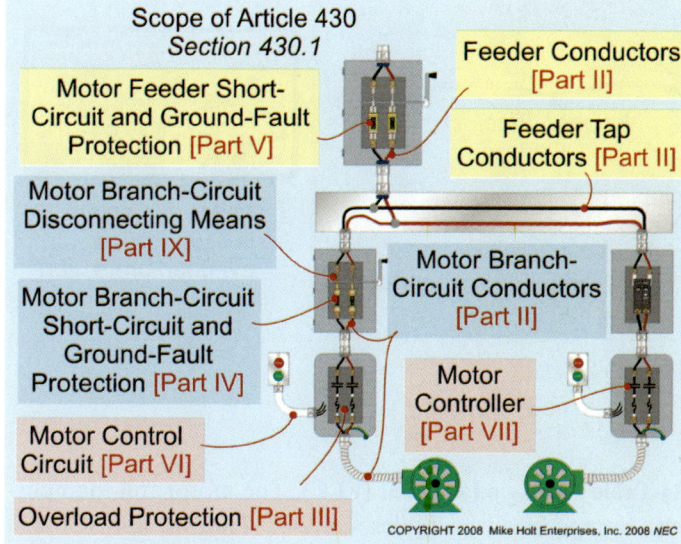

Figure 7–1

Unit 7 — Motor and Air-Conditioning Calculations

FPN No. 1: Article 440 contains the installation requirements for electrically driven air-conditioning and refrigeration equipment [440.1]. Also see 110.26(F) for dedicated space requirements for motor control centers.

7.2 FLC Versus Motor Nameplate

Motors and their associated equipment must be protected against overcurrent (overload, short circuit, or ground fault) [Article 100]. Due to the special characteristics of induction motors, overcurrent protection is generally accomplished by having the overload protection separated from the short-circuit and ground-fault device.

Part III of Article 430 contains the requirements for motor overload protection, and Part IV of Article 430 contains the requirements for motor short-circuit and ground-fault protection.

Table FLC Versus Motor Nameplate Current Rating [430.6]

(A) General Requirements. Figure 7–2

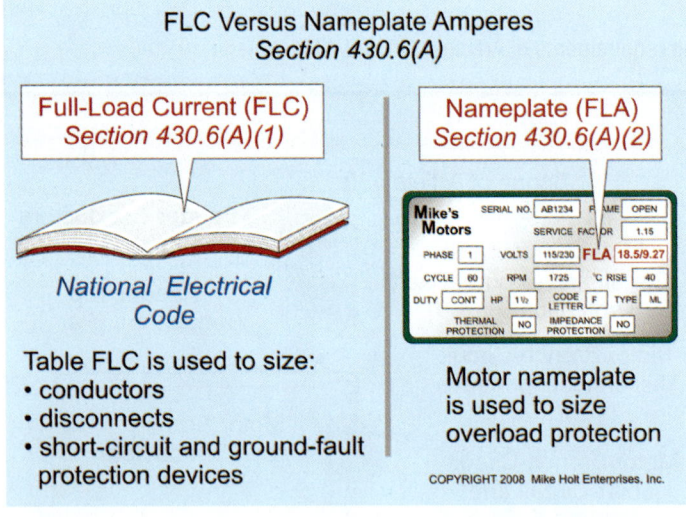

Figure 7–2

(1) Table Full-Load Current (FLC). The motor full-load current ratings listed in Tables 430.247, 430.248, and 430.250 are used to determine the conductor ampacity [430.22], the circuit short-circuit and ground-fault overcurrent device size [430.52 and 430.62], and the ampere rating of disconnecting switches [430.110].

Author's Comment: The actual current rating on the motor nameplate full-load amperes (FLA) [430.6(A)(2)] isn't permitted to be used to determine the conductor ampacity, the branch-circuit short-circuit and ground-fault overcurrent device size, nor the ampere rating of disconnecting switches.

Motors built to operate at less than 1,200 RPM or that have high torques may have higher full-load currents, and multi-speed motors have full-load current varying with speed, in which case the nameplate current ratings must be used.

Exception No. 3: For a listed motor-operated appliance, the motor full-load current marked on the nameplate of the appliance must be used instead of the horsepower rating on the appliance nameplate to determine the ampacity or rating of the disconnecting means, the branch-circuit conductors, the controller, and the branch-circuit short-circuit and ground-fault protection.

(2) Motor Nameplate Current Rating (FLA). Overload devices must be sized based on the motor nameplate current rating in accordance with 430.31.

Author's Comment: The motor nameplate full-load ampere rating is identified as full-load amperes (FLA). The FLA rating is the current in amperes the motor draws while producing its rated horsepower load at its rated voltage, based on its rated efficiency and power factor. **Figure 7–3**

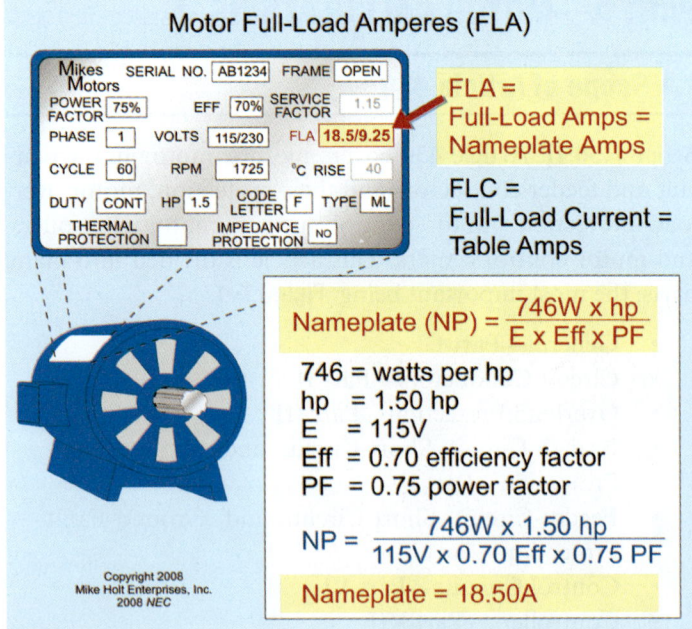

Figure 7–3

Motor and Air-Conditioning Calculations — Unit 7

The actual current drawn by the motor depends upon the load on the motor and on the actual operating voltage at the motor terminals. That is, if the load increases, the current also increases, or if the motor operates at a voltage below its nameplate rating, the operating current will increase.

Author's Comment: Other factors that may affect a motor's performance and output include varying duty and environment factors such as installation at higher elevations. Check with motor manufacturers for more information on these topics.

CAUTION: *To prevent damage to motor windings from excessive heat (caused by excessive current), never load a motor above its horsepower rating, and be sure the voltage source matches the motor's voltage rating.*

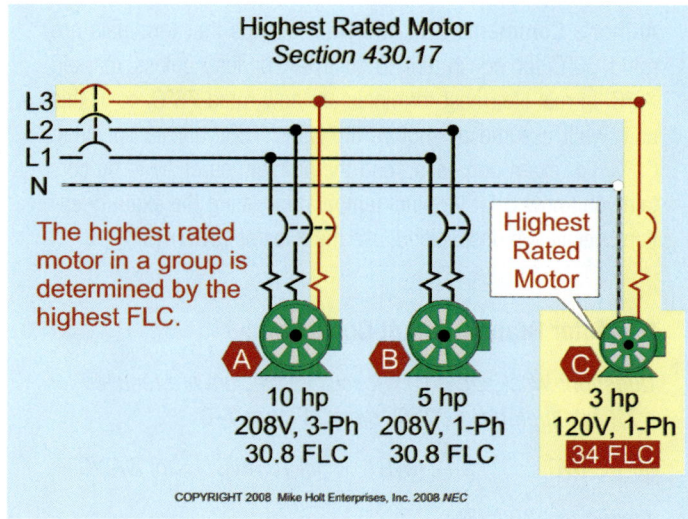

Figure 7–4

7.3 Highest Rated Motor [430.17]

When selecting the feeder conductors or the feeder short-circuit ground-fault protection device, the highest rated motor must be the highest rated motor FLC (not the highest rated horsepower) [430.17].

▶ **Highest Rated Motor**

Question: Which is the highest rated motor of the following? Figure 7–4

(a) 10 hp, 208V, three-phase (b) 3 hp, 208V, single-phase
(c) 3 hp, 120V, single-phase (d) any of these

Answer: (c) 3 hp, 120V, single-phase

10 hp, 208V, three-phase = 30.80A [Table 430.250]
3 hp, 208V, single-phase = 18.70A [Table 430.248]
3 hp, 120V, single-phase = 34A [Table 430.248]

7.4 Branch-Circuit Conductor Size

Branch-circuit conductors to a single motor must have an ampacity of not less than 125 percent of the motor's full-load current (FLC) as listed in Tables 430.247 through 430.250 [430.6(A)(1)]. **Figure 7–5** When selecting motor current from one of these tables, note that the last sentence above each table allows us to use the ampacity columns for a range of

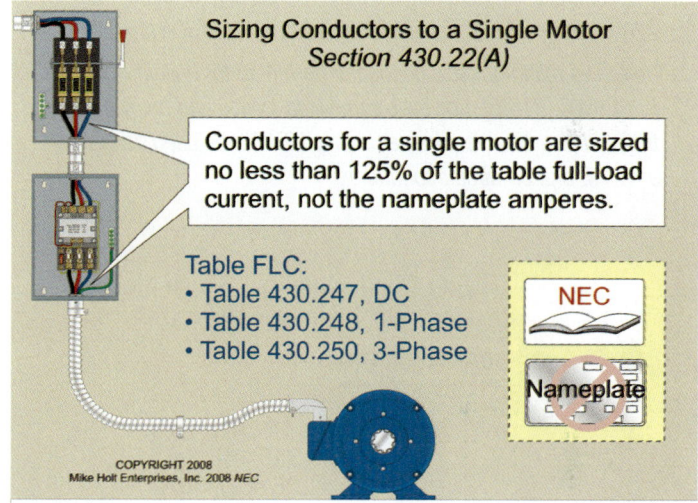

Figure 7–5

system voltages without any adjustment. The actual conductor size must be selected from Table 310.16 according to the terminal temperature rating (60°C or 75°C) of the equipment [110.14(C)].

THHN/THWN is a common conductor insulation type which can be used in a dry location at the THHN 90°C ampacity, or in a wet location using the 75°C ampacity for the THWN insulation type. Regardless of the conductor insulation type, the conductor must be sized based on the rules of 110.14(C).

Unit 7 | Motor and Air-Conditioning Calculations

Author's Comment: 110.14(C)(1)(a) tells us that terminals are rated 60ºC for equipment rated 100A or less unless marked 75ºC. In real life, most terminals are now rated 75ºC, so in this unit, we'll assume all motors are rated 75ºC unless specified 60ºC. For exam purposes, read the problem carefully to be certain you know what terminal temperature rating the exam question specifies. If unspecified, use the rules of 110.14(C).

▶ Motor Branch-Circuit Conductors

Question: What size branch-circuit conductors are required for a 7½ hp, three-phase, 230V motor? **Figure 7–6**

(a) 14 AWG   (b) 12 AWG   (c) 10 AWG   (d) 8 AWG

Answer: (c) 10 AWG

Motor FLC – Table 430.248:
7 ½ hp, 230V, three-phase FLC = 22A

The conductor is sized no less than 125 percent of motor FLC:
22A x 1.25 = 27.50A, Table 310.16, 10 AWG rated 30A at 75ºC

Note: The minimum size conductor permitted for building wiring is 14 AWG [310.5]; however, some local codes and many industrial facilities have requirements that 12 AWG be used as the smallest branch-circuit conductor.

7.5 Feeder Conductor Size [430.24]

Conductors that supply several motors must have an ampacity of not less than:

(1) 125 percent of the highest rated motor FLC [430.17], plus

(2) The sum of the FLCs of the other motors (on the same line). The FLC is found using the *NEC* Tables [430.6(A)(1)].

Author's Comment: The highest rated motor is based on the motor with the highest full-load current [430.17]. The "other motors in the group" value (on the same line) is determined by balancing the motors' FLCs on the feeder being sized, then selecting the line that has the highest rated motor on it (refer to Section 7.3 and **Figure 7–4**).

▶ Feeder Conductor Size

Question: What size feeder conductor is required for two 7½ hp, three-phase, 230V motors? The terminals are rated for 75ºC. **Figure 7– 7**

(a) 40A   (b) 50A   (c) 60A   (d) 76A

Answer: (b) 50A

(22A x 1.25) + 22A = 49.50A

Note: An 8 AWG conductor is rated 50A at 75ºC [Table 310.16].

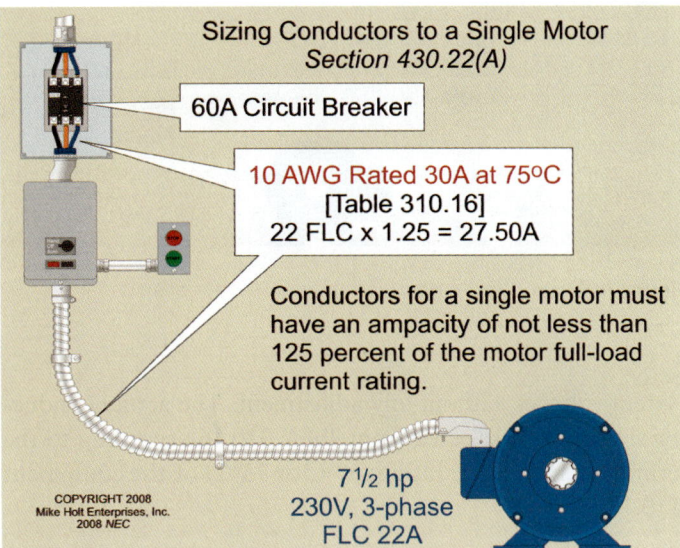

Figure 7–6

Author's Comment: The motor full-load current (FLC) is from the *Code* Tables [430.6(A)(1)] which isn't the same thing as the motor full-load amperes (FLA), which is the motor's nameplate rating [430.6(A)(2)].

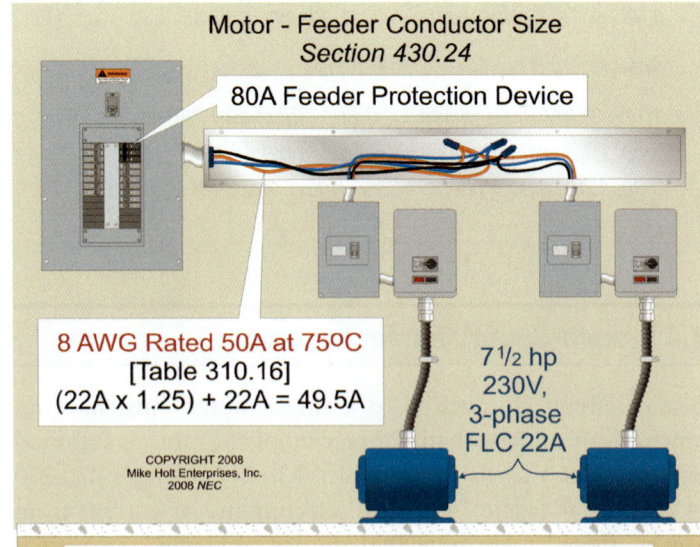

Figure 7–7

Motor and Air-Conditioning Calculations — Unit 7

7.6 Overload Protection [430.6(A)(2) and 430.32(A)]

Overload Protection

Overload is the condition where current exceeds the equipment ampere rating, which can result in equipment damage due to dangerous overheating [Article 100]. Overload protection devices, sometimes called "heaters" are intended to protect the motor, the motor control equipment, and the branch-circuit conductors from excessive heating due to motor overload [430.31].

Overload protection is not intended to protect against short circuits or ground-fault currents. **Figure 7–8**

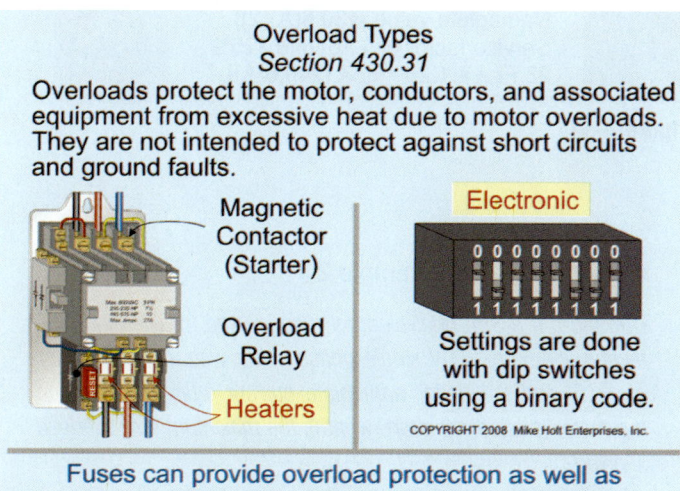

Figure 7–8

If overload protection is accomplished using fuses, a fuse must be installed for each ungrounded conductor [430.36 and 430.55].

Short-Circuit and Ground-Fault Protection

Branch-circuit short-circuit and ground-fault protection devices are intended to protect the motor, the motor control apparatus, and the conductors against short circuits or ground faults, but they are not intended to protect against an overload [430.51]. **Figure 7–9**

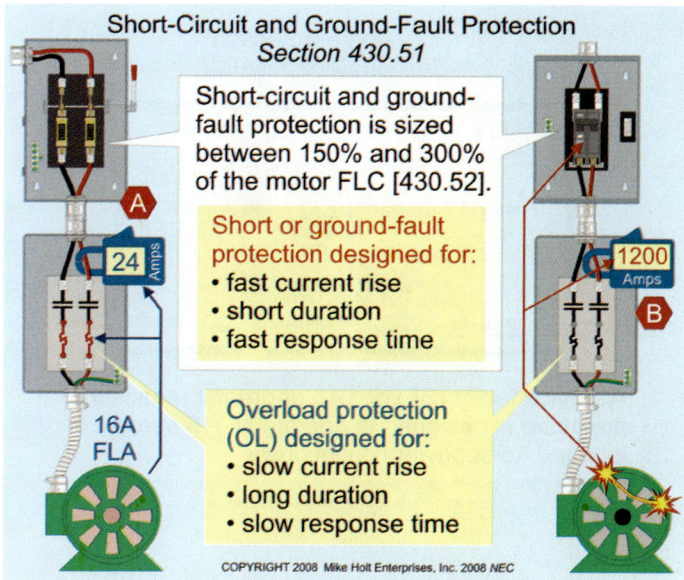

Figure 7–9

Author's Comment: The short-circuit and ground-fault protection device required for motor circuits isn't the same thing as the ground-fault circuit-interrupter (GFCI) protection required for personnel [210.8], temporary wiring for receptacles [590.6], or ground-fault protection installed on feeders [215.9 and 240.13] or services [230.95].

In addition to short-circuit and ground-fault protection, motors must be protected against overload. Generally, the motor overload device is part of the motor starter; however, a separate overload device like a dual-element fuse can be used [430.55]. Motors rated more than 1 hp without integral thermal protection, and motors 1 hp or less (automatically started) [430.32(B)], must have an overload device sized in response to the motor nameplate current rating [430.6(A)(2)]. The overload device must be sized no larger than required by 430.32.

Service Factor

Motors with a nameplate service factor (SF) rating of 1.15 or more must have their overload protection device sized no more than 125 percent of the motor nameplate current rating [430.6(A)(2)]. A service factor of 1.15 means that the motor is designed to operate at 115 percent of its rated horsepower continuously. **Figure 7–10**

Unit 7 — Motor and Air-Conditioning Calculations

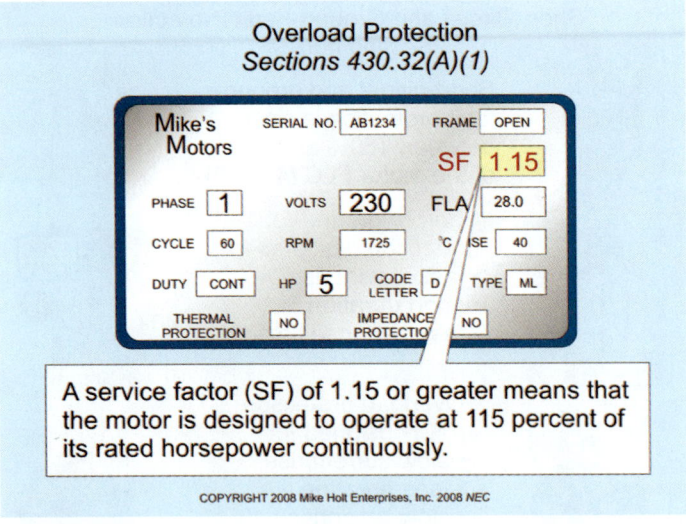

Figure 7–10

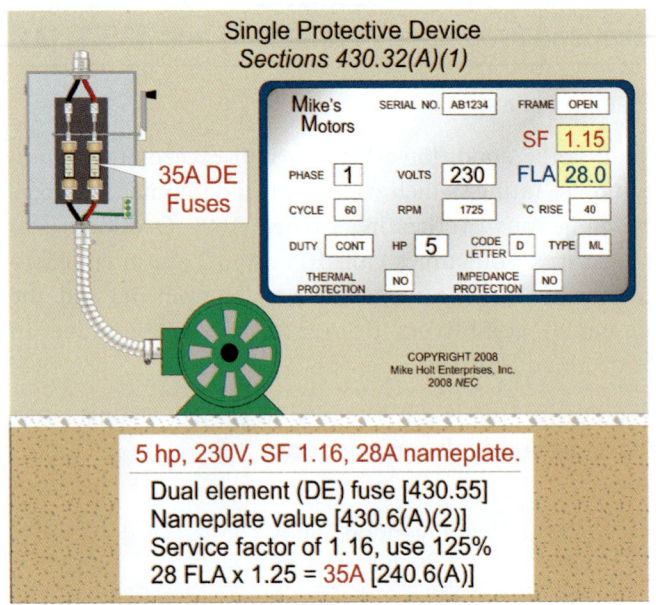

Figure 7–11

Author's Comment: Operation at a higher elevation may require a derating of the motor's service factor. For instance, a motor with a 1.15 service factor at an elevation of 3,300 ft or below is derated to a service factor of 1.0 at 9,000 ft unless the nameplate states that it was designed for a specific elevation.

▶ **Service Factor Example 1**

Question: If a dual-element fuse is used for overload protection, what size fuse is required for a 5 hp, 230V, single-phase motor, with a service factor of 1.16, if the motor nameplate current rating is 28A? **Figure 7–11**

(a) 25A (b) 30A (c) 35A (d) 40A

Answer: (c) 35A

Overload protection is sized to the motor nameplate current rating [430.6(A), 430.32(A)(1), and 430.55].

28A x 1.25 = 35A [240.6(A)]

In a case where the motor overload isn't of sufficient size to allow the motor to start, the overload size can be increased, but not to exceed 140 percent for motors with a marked service factor of 1.15 or greater.

▶ **Service Factor Example 2**

Question: If a 35A dual-element fuse is used for overload protection of a 5 hp, 230V, single-phase motor, with a service factor of 1.16, and the motor nameplate current rating is 28A, and the motor is unable to start, what is the maximum size overload allowed?

(a) 25A (b) 30A (c) 35A (d) 40A

Answer: (c) 35A

28A x 1.40 = 39.20A, use a 35A fuse maximum [240.6(A)].

Temperature Rise

Motors with a nameplate temperature rise rating not over 40°C must have the overload protection device sized no more than 125 percent of the motor nameplate current rating. A temperature rise of 40°C means that the motor is designed to operate so that it will not heat up more than 40°C above its rating. **Figure 7–12**

Mike Holt's Illustrated Guide to NEC Exam Preparation

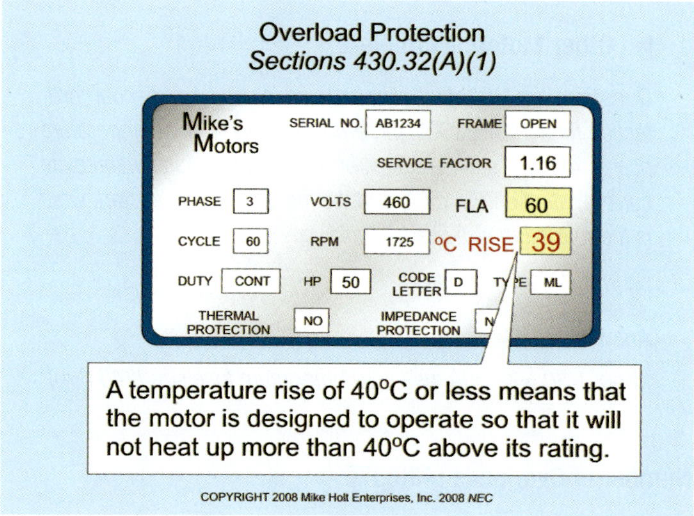

Motor and Air-Conditioning Calculations — Unit 7

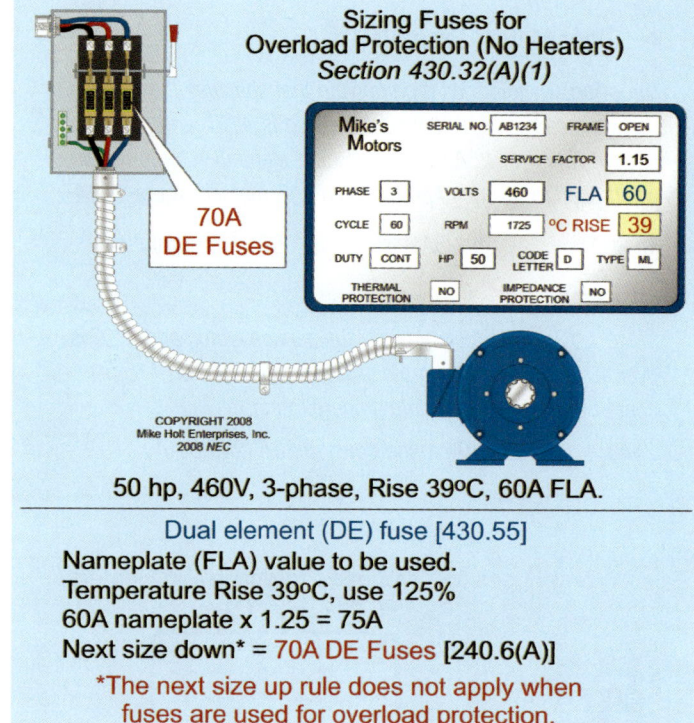

Figure 7–12

Figure 7–13

▶ **Temperature Rise Example 1**

Question: If a dual-element fuse is used for the overload protection, what size fuse is required for a 50 hp, 460V, three-phase motor, with a temperature rise of 39°C, and a motor nameplate current rating of 60A (FLA)? **Figure 7–13**

(a) 40A (b) 50A (c) 60A (d) 70A

Answer: (d) 70A

Overloads are sized according to the motor nameplate current rating, not the motor FLC rating from the Code book tables.

60A x 1.25 = 75A, use a 70A fuse [240.6(a) and 430.32(A)(1)]

When using a fuse for overload protection, the values of 430.32(A)(1) are not to be exceeded, so we must round down to the next smaller fuse. Standard fuse sizes are listed in 240.6(A).

In a case where the motor overload is not of sufficient size to allow the motor to start, the overload size can be increased, but not to exceed 140 percent for motors with a marked temperature rise of 40°C or less.

▶ **Temperature Rise Example 2**

Question: If a 70A dual-element fuse is used for overload protection for a 50 hp, 460V, three-phase motor, with a temperature rise of 39°C, and a motor nameplate current rating of 60A (FLA) and the motor is unable to start, what is the maximum size overload allowed?

(a) 50A (b) 60A (c) 70A (d) 80A

Answer: (d) 80A

60A x 1.40 = 84A, use an 80A fuse maximum [240.6(A)].

All Other Motors

Motors that do not have a service factor rating of 1.15 and up, or a temperature rise rating of 40°C and less, must have the overload protection device sized at not more than 115 percent of the motor nameplate ampere rating.

Unit 7 — Motor and Air-Conditioning Calculations

▶ **Other Motors Example 1**

Question: *A motor has a nameplate that specifies the following: service factor is 1.12, temperature rise is 41°C, and nameplate current rating of 25A. What size dual-element fuse is required when used for overload protection of this motor?* **Figure 7–14**

(a) 20A (b) 25A (c) 30A (d) 40A

Answer: *(b) 25A*

Since 1.12 is less than 1.15 for the service factor, and 41°C is over 40°C, the overload protection is sized based on 115 percent of the motor nameplate ampere rating.

25A x 1.15 = 28.75A, round down to 25A [240.6(A)]

▶ **Other Motors Example 2**

Question: *If a 25A dual-element fuse is used for overload protection for a 10 hp, 230V, three-phase motor, with a temperature rise of 41°C, a service factor of 1.12, and a motor nameplate current rating of 25A (FLA) and the motor is unable to start, what is the maximum size overload allowed?*

(a) 20A (b) 25A (c) 30A (d) 40A

Answer: *(c) 30A*

25A x 1.30 = 32.50A, use a 30A maximum fuse size [240.6(A)]

Number of Overloads [430.37]

An overload protection device must be installed in each ungrounded conductor according to the requirements of Table 430.37. If fuses are used for overload protection, a fuse must be installed in each ungrounded conductor [430.36].

7.7 Branch-Circuit Short-Circuit and Ground-Fault Protection [430.51]

General [430.51]

A branch-circuit short-circuit and ground-fault protective device protects the motor, the motor control apparatus, and the conductors against short circuits or ground faults, but not against overload. **Figure 7–15**

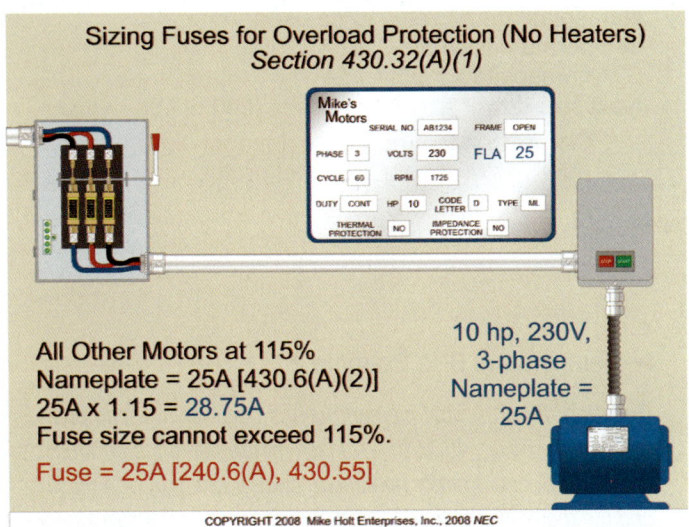

Figure 7–14

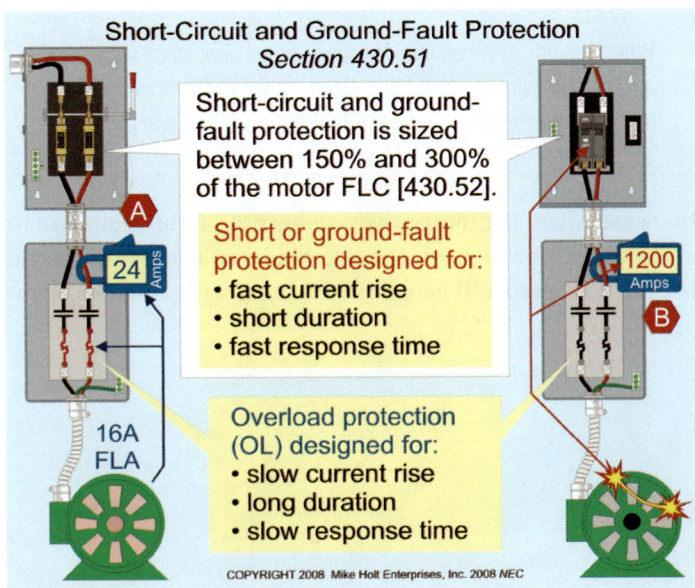

Figure 7–15

In a case where the motor overload is not of sufficient size to allow the motor to start, the overload size can be increased, but not to exceed 130 percent for motors with a service factor of less than 1.15 and a marked temperature rise of over 40°C.

Motor and Air-Conditioning Calculations — Unit 7

Author's Comment: Overload protection must comply with the requirements contained in 430.32.

Motor-Starting Current. When voltage is first applied to the field winding of an induction motor, only the conductor resistance opposes the flow of current through the motor winding. Because the conductor resistance is so low, the motor will have a very large inrush current. **Figure 7–16**

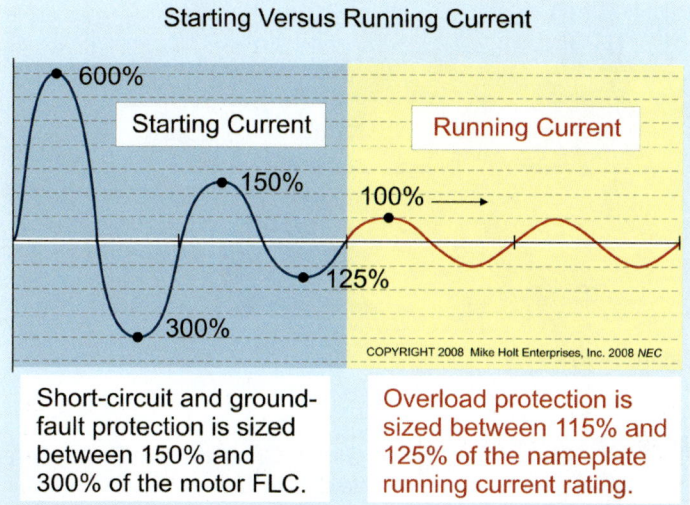

Figure 7–16

Motor-Running Current. Once the rotor begins turning, there is an increase in counter-electromotive force which reduces the starting current to running current.

Motor Locked-Rotor Current (LRC). If the rotating part of the motor winding (armature) becomes jammed so it can't rotate, no counter-electromotive force (CEMF) will be produced in the motor winding. This results in a decrease in conductor impedance to the point that it's effectively a short circuit. Result—the motor operates at locked-rotor current (LRC), often six times the full-load ampere rating, depending on the motor's code letter rating [430.7(B)], and this will cause the motor winding to overheat and be destroyed if the current isn't quickly reduced or removed.

Author's Comment: The *National Electrical Code* requires that most motors be provided with overcurrent protection to prevent damage to the motor winding because of locked-rotor current.

Branch-Circuit Short-Circuit and Ground-Fault Protection [430.52]

(A) General. The motor branch-circuit short-circuit and ground-fault protective device must comply with 430.52(b) and 430.52(C).

(B) All Motors. A motor branch-circuit short-circuit and ground-fault protective device must be capable of carrying the motor's starting current.

(C) Rating or Setting.

(1) Table 430.52. Each motor branch circuit must be protected against short circuit and ground faults by an overcurrent device sized no greater than the following percentages listed in Table 430.52.

Branch Circuit Protection Table 430.52			
Circuit Motor Type	Nontime Delay	Dual-Element Fuse	Inverse Time Breaker
Wound Rotor	150%	150%	150%
Direct Current	150%	150%	150%
All Other Motors	300%	175%	250%

▶ **Branch-Circuit Short-Circuit and Ground-Fault Protection Example 1**

Question: What size conductor and inverse time circuit breaker are required for a 2 hp, 230V, single-phase motor? **Figure 7–17**

(a) 14 AWG, 30A breaker
(b) 14 AWG, 35A breaker
(c) 14 AWG, 40A breaker
(d) 14 AWG, 45A breaker

Answer: (a) 14 AWG, 30A breaker

Step 1: Determine the branch-circuit conductor [Table 310.16, 430.22(A), and Table 430.248]:

12A x 1.25 = 15A, 14 AWG, rated 20A at 75°C [Table 310.16]

Step 2: Determine the branch-circuit protection [240.6(A), 430.52(C)(1), and Table 430.248]:

12A x 2.50 = 30A

Unit 7 — Motor and Air-Conditioning Calculations

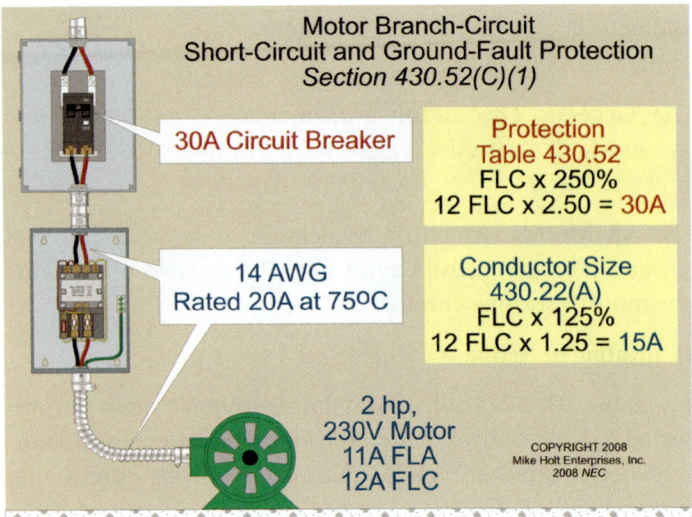

Figure 7–17

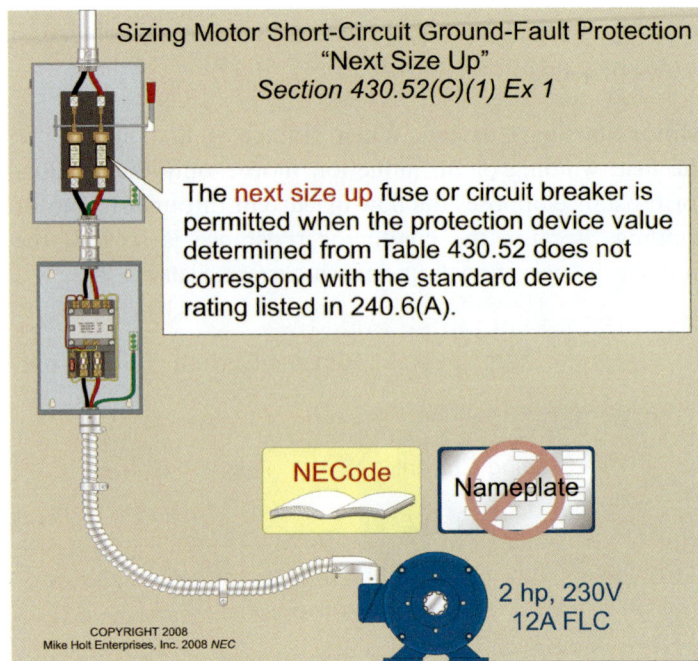

Figure 7–18

Author's Comment: I know it bothers many in the electrical industry to see a 14 AWG conductor protected by a 30A circuit breaker, but branch-circuit conductors are protected against overloads by the overload device, which is sized between 115 and 125 percent of the motor nameplate current rating [430.32]. See 240.4(G) for details.

Exception No. 1: Where the motor short-circuit and ground-fault protective device values derived from Table 430.52 don't correspond with the standard overcurrent device ratings listed in 240.6(A), the next higher overcurrent device rating can be used. **Figure 7–18**

▶ **Branch-Circuit Short-Circuit and Ground-Fault Protection Example 2**

Question: What size nontime-delay fuses are required for a 2 hp, 230V, single-phase motor? **Figure 7–19**

(a) 30A fuse (b) 35A fuse
(c) 40A fuse (d) 45A fuse

Answer: (c) 40A fuse

Determine the branch-circuit protection [240.6(A), 430.52(C)(1), and Table 430.248]:

12A x 3 = 36A, use a 40A nontime-delay fuse

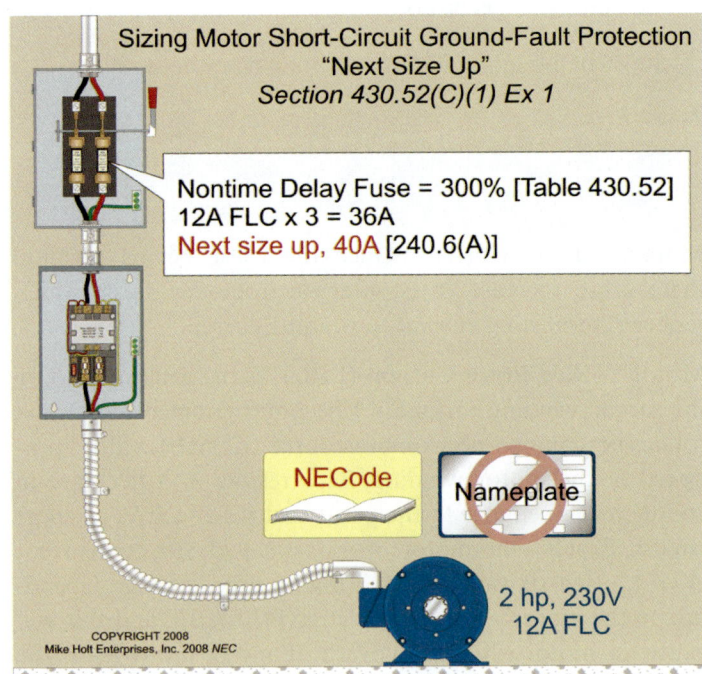

Figure 7–19

Motor and Air-Conditioning Calculations — Unit 7

▶ **Branch-Circuit Short-Circuit and Ground-Fault Protection Example 3**

Question: What size time-delay fuses are required for a 2 hp, 230V, single-phase motor?

(a) 25A fuse (b) 30A fuse (c) 35A fuse (d) 40A fuse

Answer: (a) 25A fuse

Determine the branch-circuit protection [240.6(A), 430.52(C)(1), and Table 430.248]:

12A x 1.75 = 21A, use a 25A time-delay fuse

▶ **Branch-Circuit Short-Circuit and Ground-Fault Protection Example 4**

Question: What size conductor and inverse time circuit breaker are required for a 7½ hp, 230V, three-phase motor? Figure 7–20

(a) 10 AWG, 50A breaker (b) 10 AWG, 60A breaker
(c) 8 AWG, 70A breaker   (d) 8 AWG, 80A breaker

Answer: (b) 10 AWG, 60A breaker

Step 1: Determine the branch-circuit conductor [Table 310.16, 430.22(A), and Table 430.250]:

22A x 1.25 = 27.50A, 10 AWG, rated 30A at 75°C [Table 310.16]

(continued in next column)

Step 2: Determine the branch-circuit protection [240.6(A), 430.52(C)(1) Ex 1, and Table 430.250]:

22A x 2.50 = 55A, next size up = 60A

7.8 Branch-Circuit Summary

Branch-Circuit Conductors [430.22(A)] Figure 7–21

Branch-circuit conductors to a single motor must have an ampacity of not less than 125 percent of the motor FLC as listed in Tables 430.247 through 430.250 [430.6(A)].

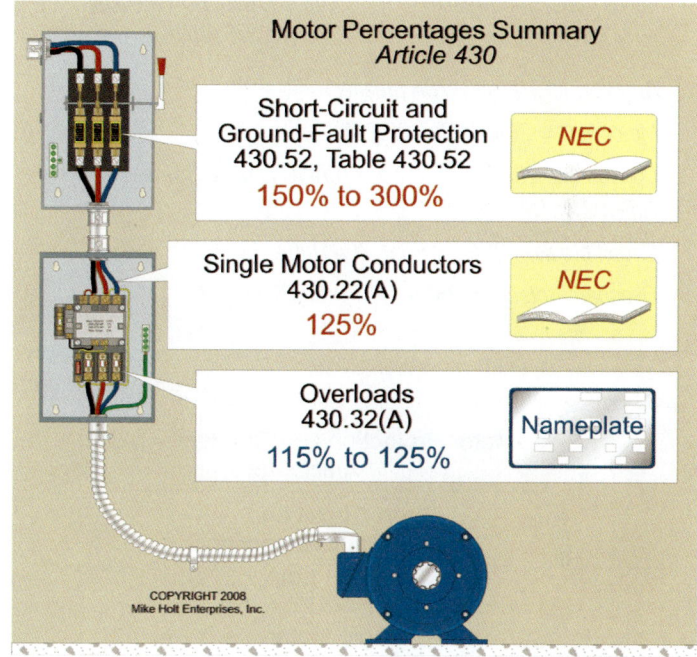

Figure 7–21

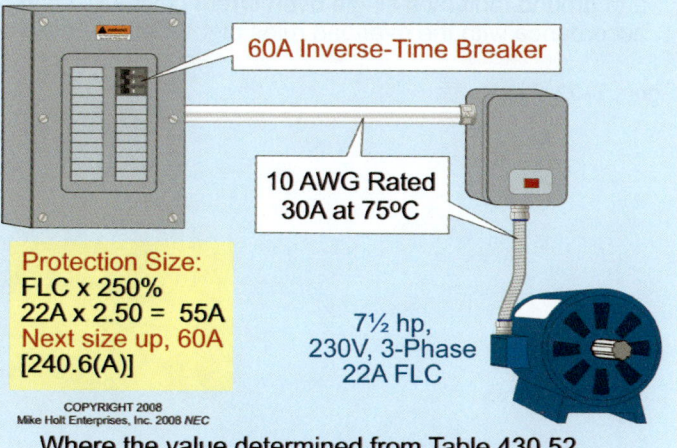

Figure 7–20

Branch-Circuit Short-Circuit Protection [430.52(C)(1)]

The branch-circuit short-circuit and ground-fault protection device protects the motor, the motor control apparatus, and the conductors against overcurrent due to short circuits or ground faults, but not against overloads [430.51]. The branch-circuit short-circuit and ground-fault protection devices are sized by considering the type of motor and the type of protection device and applying the percent of motor FLC listed in Table 430.52. When the protection device values determined from Table 430.52 do not correspond

Unit 7 — Motor and Air-Conditioning Calculations

with the standard rating of overcurrent devices as listed in 240.6(A), the next higher overcurrent device can be installed [430.52(C)(1) Ex 1].

Overload Protection [430.32(A)]

Overload protection is sized based on the motor nameplate rating.

▶ **Branch Circuit Summary Example 1**

Question: If an inverse time circuit breaker is used for short-circuit and ground-fault protection, what size circuit breaker and conductor is required for a 5 hp, 230V, single-phase motor having a nameplate current rating of 26A? **Figure 7–22**

(a) 10 AWG, 50A breaker (b) 10 AWG, 60A breaker
(c) 8 AWG, 70A breaker   (d) 8 AWG, 80A breaker

Answer: (c) 10 AWG, 70A breaker

Motor FLC = 28A [Table 430.248].

Conductors: 28A x 1.25 = 35A. 10AWG is rated 35A at 75°C [Table 310.16].

Circuit breaker: 28A x 2.50 = 70A [Table 430.52]

Overload Protection: 26A x 1.15 = 29.90 [430.32]

Single Overcurrent Device [430.55]

A motor can be protected against overload, short circuit, and ground faults by a single overcurrent device sized to the overload requirements contained in 430.32.

▶ **Branch Circuit Summary Example 2**

Question: What size dual-element fuse is permitted to protect a 5 hp, 230V, single-phase motor with a service factor of 1.15 and a nameplate current rating of 28A from overloads as well as short-circuits and ground-faults? **Figure 7–23**

(a) 20A   (b) 25A   (c) 30A   (d) 35A

Answer: (d) 35A

Overload Protection [430.32(A)(1) and 430.55]
28A x 1.25 = 35A

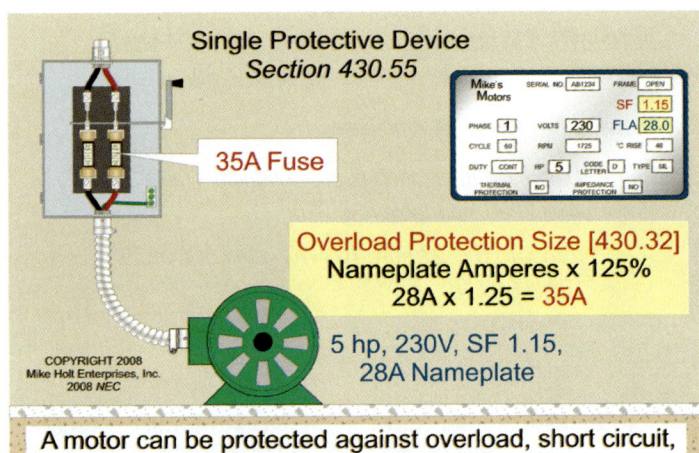

A motor can be protected against overload, short circuit, and ground fault by a single overcurrent device sized in accordance with the overload rules of 430.32.

Figure 7–23

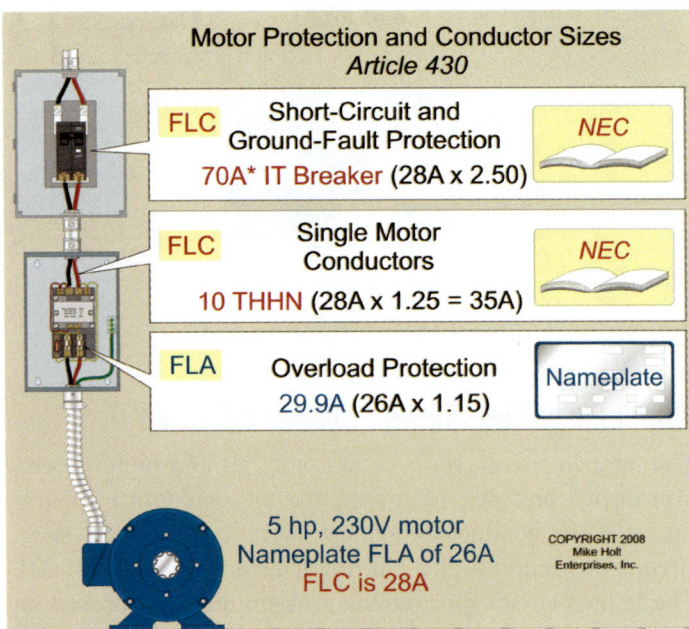

Figure 7–22

Motor and Air-Conditioning Calculations — Unit 7

▶ **Branch Circuit Summary Example 3**

Question: What size dual-element fuse is permitted to protect a ½ hp, 230V, single-phase motor with a service factor of 1.20 from overloads as well as short-circuits and ground-faults? Figure 7–24

(a) 3A (b) 6A (c) 10A (d) 15A

Answer: (b) 6A

Overload Protection [430.32(A)(1) and 430.55] (when the nameplate FLA is not given, we can use the FLC from Tables)

FLC = 4.90A [Table 430.248]
4.90A x 1.25 = 6.125A. Use a 6A fuse [240.6(A)]

The conductor will be sized at: 4.90A x 1.25 = 6.125A, 14AWG is rated 20A at 75°C [Table 310.16].

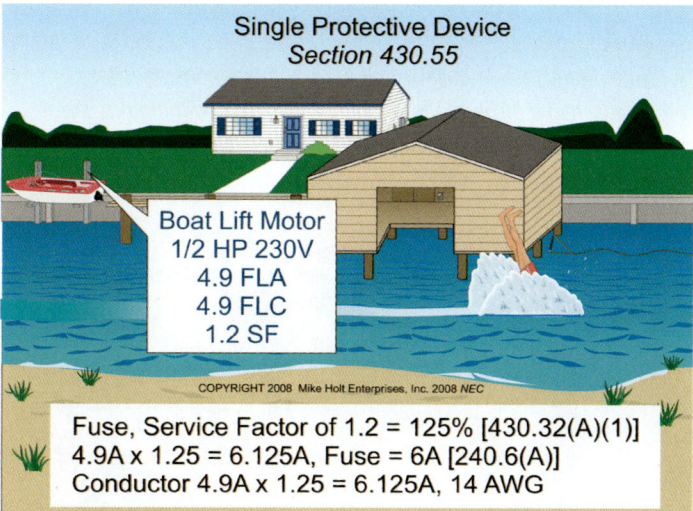

Figure 7–24

▶ **Motor Feeder Protection Example**

Question: What size feeder protection (inverse time breakers with 75°C terminals) and conductors are required for the following two motors? Figure 7–25

Motor 1—20 hp, 460V, three-phase: FLC = 27A
 [Table 430.250]
Motor 2—10 hp, 460V, three-phase: FLC = 14A
 [Table 430.250]

(a) 8 AWG, 70A breaker (b) 8 AWG, 80A breaker
(c) 8 AWG, 90A breaker (d) 10 AWG, 90A breaker

Answer: (b) 8 AWG, 80A breaker

Step 1: Determine the feeder conductor size [430.24]:

 (27A x 1.25) + 14A = 48A
 8 AWG rated 50A at 75°C [110.14(C) and Table 310.16]

Step 2: Feeder protection [430.62(A)] is not greater than the largest branch-circuit ground-fault and short-circuit overcurrent device plus the other motor FLC.

Step 3: Determine the largest branch-circuit ground-fault and short-circuit overcurrent device [430.52(C)(1) Ex]:

 20 hp Motor = 27A x 2.50 = 68, next size up = 70A
 10 hp Motor = 14A x 2.50 = 35A

Step 4: Determine the size feeder protection:

 Not more than 70A + 14A, = 84A, next size down = 80A [240.6(A)]

The "next size up protection" rule for branch circuits [430.52(C)(1) Ex 1] doesn't apply to the motor feeder overcurrent device rating.

7.9 Feeder Protection [430.62]

(A) Motors Only. Feeder conductors must be protected against short circuits and ground faults by a protective device sized not more than the largest rating of the branch-circuit short-circuit and ground-fault protective device for any motor, plus the sum of the full-load currents of the other motors in the group.

Unit 7 | Motor and Air-Conditioning Calculations

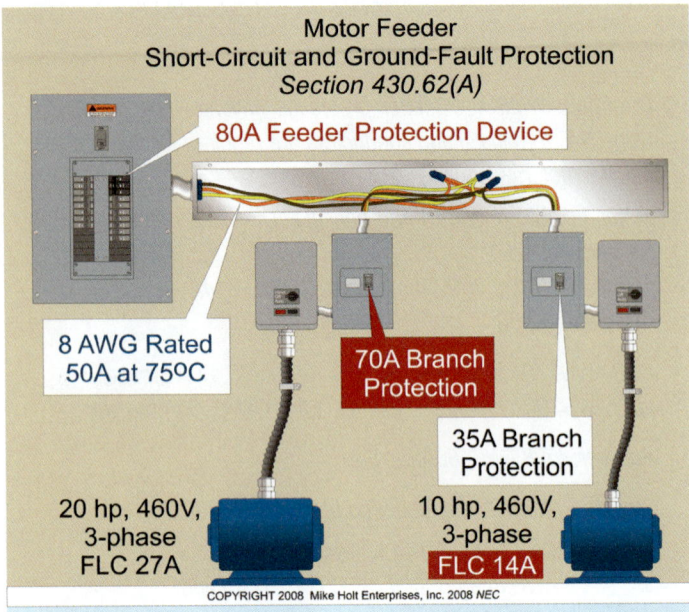

Figure 7–25

7.10 Motor VA Calculations

The measure of mechanical output of a motor is horsepower. Horsepower can be converted to watts by multiplying the horsepower rating by 746 watts per horsepower. Since this is a measure of a motor's output, don't confuse it with the input power required by a motor.

▶ **Single-Phase Motor VA Output Example**

Question: What is the output VA for a dual voltage, 1 hp motor, rated 115/230V?

(a) 746W (b) 1,000W (c) 1,400W (d) 1,840W

Answer: (a) 746W

746W x 1 hp = 746W Output power

The input VA of a motor is determined by multiplying the motor volts by the motor amperes. To determine the motor VA rating, the following formulas can be used:

◆ **Motor Single-Phase VA Formula**

Motor Single-Phase VA =
Motor Volt Rating x Motor Ampere Rating

◆ **Motor Three-Phase VA Formula**

Motor Three-Phase VA =
Motor Volt Rating x Motor Ampere Rating x 1.732

▶ **Motor VA—Single-Phase Example**

Question: What is the input VA for a dual voltage, 1 hp motor rated 115/230V? **Figure 7–26**

(a) 920 VA at 230V (b) 1,840 VA at 115V
(c) 1,840 VA at 230V (d) b and c

Answer: (d) b and c

Motor VA = Volts x FLC
Table 430.248, 115V FLC = 16A, 230V FLC = 8A

VA at 230V = 230V x 8A
VA at 230V = 1,840 VA

VA at 115V = 115V x 16A
VA at 115V = 1,840 VA

Note: Many people believe that a 230V motor consumes less power than a 115V motor (I thought that), but both motors consume the same amount of power.

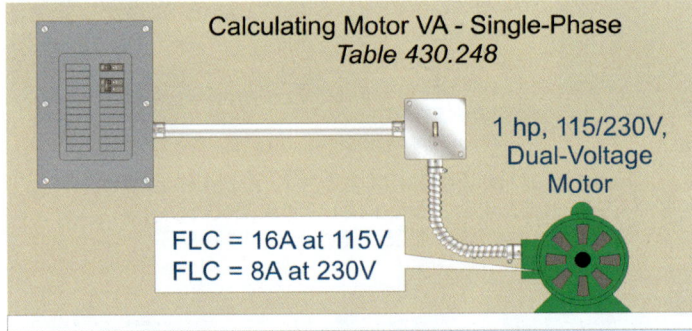

Figure 7–26

206 | Mike Holt's Illustrated Guide to NEC Exam Preparation

Motor and Air-Conditioning Calculations — Unit 7

▶ **Motor VA—Three-Phase Example**

Question: What is the input VA for a 5 hp, 230V, three-phase motor? **Figure 7–27**

(a) 3,730 VA   (b) 6,055 VA   (c) 6,440 VA   (d) 8,050 VA

Answer: (b) 6,055 VA [Table 430.250]

Motor Three-Phase VA = Volts x FLC x 1.732
Table 430.250, FLC = 15.20A

Motor VA = 230V x 15.20A x 1.732
Motor VA = 6,055 VA

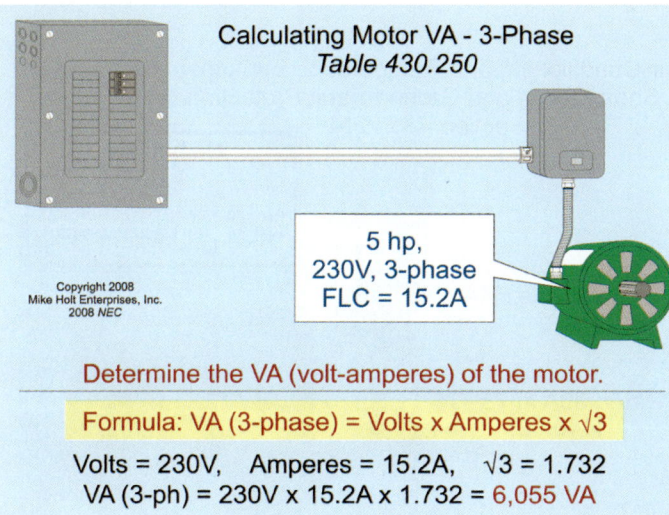

Figure 7–27

PART B—AIR-CONDITIONING CALCULATIONS

7.11 Scope of Article 440

Scope [440.1]

Article 440 applies to electrically driven air-conditioning and refrigeration equipment.

Definitions [440.2]

Hermetic Refrigerant Motor-Compressor. A compressor and motor enclosed in the same housing, operating in the refrigerant.

Rated-Load Current. The current resulting when the motor-compressor operates at rated load and rated voltage.

7.12 Other Articles

Other Articles [440.3]

(B) Equipment with No Hermetic Motor-Compressors. Air-conditioning and refrigeration equipment that do not have hermetic refrigerant motor-compressors, such as furnaces with evaporator coils, must comply with Article 422 for appliances, Article 424 for electric space-heating, and Article 430 for motors.

(C) Household Refrigerant Motor-Compressor Appliances. Household refrigerators and freezers, drinking water coolers, and beverage dispensing machines are listed as appliances, and their installation must also comply with Article 422 for appliances.

Ampacity and Rating [440.6]

(A) Hermetic Refrigerant Motor-Compressor. For a hermetic refrigerant motor-compressor, the rated-load current marked on the nameplate of the equipment is to be used in determining the rating of the disconnecting means, the branch-circuit conductors, the controller, and the branch-circuit short-circuit and ground-fault protection.

Exception No. 1: The branch-circuit selection current must be used instead of the rated-load current if provided on the equipment nameplate.

7.13 Short-Circuit and Ground-Fault Protection

General [440.21]

The branch-circuit conductors, control apparatus, and circuits supplying hermetic refrigerant motor-compressors must be protected against short circuits and ground faults in accordance with 440.22.

> **Author's Comment:** If the equipment nameplate specifies "Maximum Fuse Size," then a one-time or dual-element fuse must be used.

Short-Circuit and Ground-Fault Protection Device Size [440.22]

(A) Single Motor-Compressors. The short-circuit and ground-fault protection device must not be more than 175 percent of the motor-compressor current rating.

Unit 7 — Motor and Air-Conditioning Calculations

▶ **Air-Conditioner Protection Example 1**

Question: What size conductor and protection is required for a 24A motor-compressor connected to a 240V circuit? **Figure 7–28**

(a) 10 AWG, 40A
(b) 10 AWG, 60A
(c) a or b
(d) 10 AWG, 90A

Answer: (a) 10 AWG, 40A

Step 1: Determine the branch-circuit conductor [Table 310.16 and 440.32]:

24A x 1.25 = 30A, 10 AWG is rated 35A at 75°C [Table 310.16]

Step 2: Determine the branch-circuit protection [240.6(A) and 440.22(A)]:

24A x 1.75 = 42A, next size down = 40A

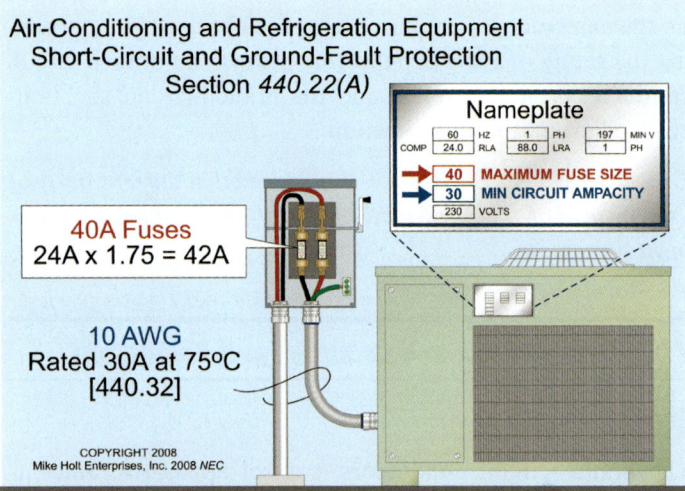

Short-circuit and ground-fault protection device must not be greater than 175% of the equipment load current rating.

Figure 7–28

▶ **Air-Conditioner Protection Example 2**

Question: What size protection is required for a 24A motor-compressor connected to a 240V circuit if the fuse sized at 175 percent isn't sufficient to allow the air-conditioner compressor to start? **Figure 7–29**

(a) 30A (b) 40A (c) 50A (d) 90A

Answer: (c) 50A

Determine the branch-circuit protection [240.6(A) and 440.22(A)]:

24A x 2.25 = 54A, next size down 50A

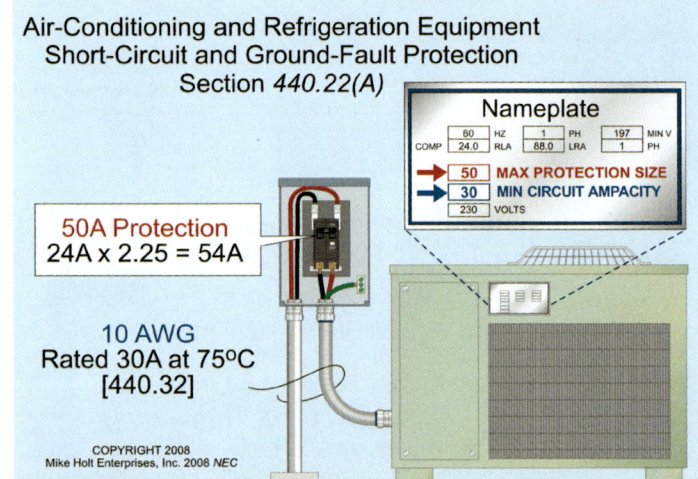

If 175% can't carry the staring current, the overcurrent device can be sized at a maximum of 225%.

Figure 7–29

If the overcurrent device sized at 175 percent isn't capable of carrying the starting current of the motor-compressor, the next size larger overcurrent device can be used, but in no case can it exceed 225 percent of the motor-compressor current rating.

(B) Several Motor-Compressors. Where the equipment incorporates more than one hermetic refrigerant motor-compressor, or a hermetic refrigerant motor-compressor and other motors or other loads, the equipment short-circuit and ground-fault protection must be sized as follows:

(1) Motor-Compressor Largest Load. The rating of the branch-circuit short-circuit and ground-fault protective device must not be more than the largest motor-compressor short-circuit ground-fault overcurrent device, plus the sum of the rated-load currents of the other compressors.

> **Author's Comment:** The branch-circuit conductors are sized at 125 percent of the larger motor-compressor current, plus the sum of the rated-load currents of the other compressors [440.33].

Motor and Air-Conditioning Calculations — Unit 7

7.14 Conductor Sizing for Single Motor-Compressor

Conductor Size for Single Motor-Compressors [440.32]

Branch-circuit conductors to a single motor-compressor must have an ampacity not less than 125 percent of the motor-compressor rated-load current or the branch-circuit selection current, whichever is greater.

Author's Comment: Branch-circuit conductors for a single motor-compressor must have short-circuit and ground-fault protection sized between 175 percent and 225 percent of the rated-load current [440.22(A)].

▶ **Air-Conditioner Conductor Sizing Example**

Question: What size conductor and overcurrent device is required for an 18A motor compressor? **Figure 7–30**

(a) 12 AWG, 30A
(b) 10 AWG, 50A
(c) a or b
(d) 10 AWG, 60A

Answer: (a) 12 AWG, 30A

Step 1: Determine the branch-circuit conductor [Table 310.16 and 440.32]:

18A x 1.25 = 22.50A, 12 AWG, rated 25A at 75°C [Table 310.16]

Step 2: Determine the branch-circuit protection [240.6(a) and 440.22(A)]:

18A x 1.75 = 31.50A, next size down = 30A

If the 30A overcurrent device isn't capable of carrying the starting current, then the overcurrent device can be sized up to 225 percent of the equipment load current rating. 18A x 2.25 = 40.50A, next size down 40A

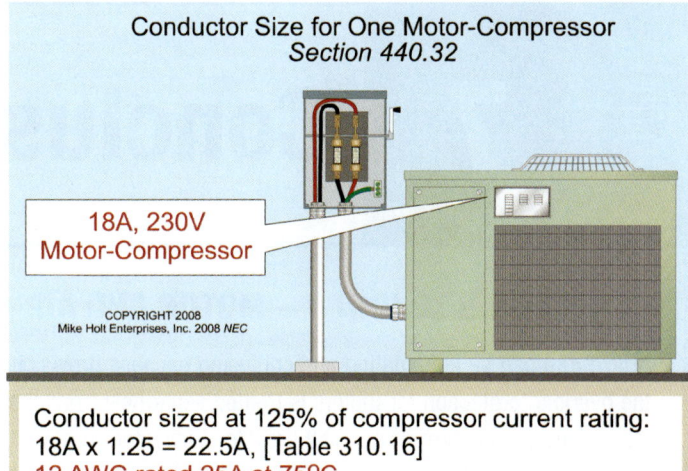

Figure 7–30

Author's Comment: A 30A or 40A overcurrent device is permitted to protect a 12 AWG conductor for an air-conditioning circuit. See 240.4(G) for details.

Conductor Size for Several Motor-Compressors [440.33]

Conductors that supply several motor-compressors must have an ampacity not less than 125 percent of the highest motor-compressor current of the group, plus the sum of the rated load or branch-circuit selection current ratings of the other compressors.

Author's Comment: These conductors must be protected against short circuits and ground faults in accordance with 440.22(B)(1).

UNIT 7 Conclusion

CONCLUSION TO UNIT 7—MOTOR AND AIR-CONDITIONING CALCULATIONS

Article 430 can be intimidating and confusing because provisions for motor circuits differ from general *NEC* rules. This unit explains that the overload protection for motors is treated separately from short-circuit and ground-fault protection. Motor circuits must accommodate increased starting current and still provide satisfactory running protection. For this reason, the overcurrent protection rules you learned in Article 240 don't apply to motor circuits. Article 430 contains the rules that are used to size protection as well as conductors for motor branch circuits and feeders.

A common mistake made in motor circuits is the use of the motor nameplate current for calculations that should be made using the full-load current from the Article 430 tables. The nameplate full-load amperes is only used for overload protection; Tables 430.248, 430.249, and 430.250 are used for sizing conductors, short-circuit and ground-fault protection, and disconnect switches. Review the material in this unit regarding 430.6(A) if this isn't clear to you.

Be sure you understand the differences between motor branch circuits and feeders, and how to apply the requirements of Article 430 to size the conductors and short-circuit and ground-fault protection for each.

The rules for sizing conductors and protection for air conditioners are similar to those for motor circuits. Apply the rules from Article 440 when sizing air-conditioner circuits, and be sure to comply with the instructions on the nameplate. Usually sizing the wiring and protection for an air conditioner is as simple as reading the information from the nameplate and trusting that the manufacturer followed the requirements of Article 440 in calculating that information for you.

Practice Questions

UNIT 7—MOTOR AND AIR-CONDITIONING CALCULATIONS PRACTICE QUESTIONS

(•Indicates that 75% or fewer of those who took this exam answered the question correctly.)

PART A—MOTOR CALCULATIONS

7.2 FLC Versus Motor Nameplate

1. Motors and their associated equipment must be protected against overcurrent (overload, short circuit, or ground fault), but because of the special characteristics of induction motors, overcurrent protection is generally accomplished by having the overload protection separate from the short-circuit and ground-fault protection.

    (a) True
    (b) False

2. Which part of Article 430 contains the requirements for motor short-circuit and overload protection?

    (a) Motor and Branch-Circuit Overload Protection—Part III
    (b) Motor Branch-Circuit Short-Circuit and Ground-Fault Protection—Part IV
    (c) a and b
    (d) none of these

7.3 Highest Rated Motor [430.17]

3. When selecting the feeder conductors and short-circuit ground-fault protection device, the highest rated motor must be the highest rated _____.

    (a) horsepower
    (b) full-load current
    (c) nameplate current
    (d) any of these

4. •Which is the highest rated FLC of the following?

    (a) 25 hp, 460V, synchronous, three-phase
    (b) 20 hp, 460V, three-phase
    (c) 15 hp, 460V, three-phase
    (d) 3 hp, 120V, single-phase

7.4 Branch-Circuit Conductor Size

5. What AWG size conductor is required for a 5 hp, 230V, single-phase motor? The terminals are rated 75°C.

    (a) 14 AWG
    (b) 12 AWG
    (c) 10 AWG
    (d) 8 AWG

7.5 Feeder Conductor Size [430.24]

6. Feeder conductors that supply several motors must have an ampacity of not less than _____.

    (a) 125 percent of the highest rated motor's FLC
    (b) the sum of the full-load currents of the other motors on the same phase
    (c) a or b
    (d) a and b

7.6 Overload Protection [430.6(A)(2) and 430.32(A)]

7. •Overload is the condition where the current is greater than the equipment ampacity rating, resulting in equipment damage due to dangerous overheating [Article 100]. Overload protection devices, sometimes called heaters, are intended to protect the _____ from dangerous overheating.

    (a) motor
    (b) motor control equipment
    (c) branch-circuit conductors
    (d) all of these

Unit 7 Practice Questions

8. •The *NEC* requires motor overload protection devices to be sized according to the motor full-load current ratings listed in Tables 430.247, 430.248, or 430.250.

    (a) True
    (b) False

9. If the motor overload protection relay sized according to 430.32 is not capable of carrying the motor starting and running current, a larger overload can be used if sized according to the requirements of 430.32(C).

    (a) True
    (b) False

10. Motors with a nameplate service factor (SF) rating of 1.15 or more must have the overload protection device sized at no more than _____ percent of the motor nameplate current rating.

    (a) 100
    (b) 115
    (c) 125
    (d) 135

11. Motors with a nameplate temperature rise rating not over 40°C must have the overload protection device sized at no more than _____ percent of the motor nameplate current rating.

    (a) 100
    (b) 115
    (c) 125
    (d) 135

12. Motors that have a service factor rating of 1.10 must have the overload protection device sized at not more than _____ percent of the motor nameplate ampere rating.

    (a) 100
    (b) 115
    (c) 125
    (d) 135

13. If a dual-element fuse is used for overload protection, what size fuse is required for a 5 hp, 208V, three-phase motor with a service factor of 1.16 and a motor nameplate current rating of 16A (FLA)?

    (a) 20A
    (b) 25A
    (c) 30A
    (d) 35A

14. If a dual-element fuse is used for overload protection, what size fuse is required for a 30 hp, 460V, three-phase synchronous motor, with a temperature rise of 39°C?

    (a) 20A
    (c) 25A
    (b) 30A
    (d) 40A

7.7 Branch-Circuit Short-Circuit and Ground-Fault Protection [430.51]

15. The branch-circuit short-circuit and ground-fault protection device is intended to protect the motor, the motor control apparatus, and the conductors against overcurrent due to _____.

    (a) short circuits
    (b) ground faults
    (c) overloads
    (d) a and b

16. In addition to overload protection, each motor and its accessories require short-circuit and ground-fault protection according to 430.52. When sizing the branch-circuit protection device, which of the following factors must be considered?

    (a) The motor type, such as induction, synchronous, wound-rotor, single-phase ac, three-phase ac, etc.
    (b) Motor full-load current from the *Code* tables must be used.
    (c) The type of protection device to be used; nontime-delay fuse, dual-element fuse, or inverse time breaker.
    (d) all of these

17. The *NEC* requires motor branch-circuit short-circuit and ground-fault protection to be sized not greater than the percentages listed in Table 430.52. When the short-circuit ground-fault protection device value determined from Table 430.52 does not correspond with the standard rating of overcurrent devices as listed in 240.6(A), the next _____ device size is permitted to be used.

    (a) smaller
    (b) larger
    (c) a or b
    (d) none of these

18. To determine the percentage of the motor FLC from Table 430.52 that is to be used to size the motor branch-circuit short-circuit and ground-fault protection device, which of the following steps should be used?

    (a) Locate the motor type on Table 430.52, such as dc, wound-rotor, single-phase, or polyphase ac motors.
    (b) Select the percentage from Table 430.52 according to the type of protection device, such as nontime-delay fuse, dual-element fuse or inverse time circuit breaker.
    (c) After making the necessary calculations, round up to the next standard size overcurrent device.
    (d) all of these

19. Where the next size up time-delay (dual-element) fuse is not sufficient for the starting current of the motor, the rating of the fuse is permitted to be increased, but must in no case exceed _____ percent of the motor full-load current.

    (a) 125
    (b) 150
    (c) 175
    (d) 225

20. Conductors are sized at _____ percent of the motor full-load current [430.22(A)], overloads from _____ percent, and the motor short-circuit, ground-fault protection device (inverse time circuit breaker) is sized up to _____ percent of the motor full-load current.

    (a) 100, 100, 100
    (b) 100, 125, 150
    (c) 125, 125, 125
    (d) 125, 115, 250

21. Which of the following statements are true for a 10 hp, 208V, three-phase motor with a nameplate current of 29A?

    (a) The branch-circuit conductors can be 8 THHN.
    (b) Overload protection is 33A.
    (c) Short-circuit and ground-fault protection can be an 80A circuit breaker.
    (d) all of these

7.9 Feeder Protection [430.61]

22. Motor feeder conductors (sized according to 430.24) must have a feeder protection device to protect against short circuits and ground faults (not overloads), sized not greater than the _____.

    (a) largest branch-circuit short-circuit ground-fault protection device [430.52] of any motor of the group
    (b) sum of the full-load currents of the other motors on the same phase
    (c) a or b
    (d) a and b

7.10 Motor VA Calculations

23. What is the VA input of a dual voltage, 5 hp, three-phase motor rated 460/230V?

    (a) 3,027 VA at 460V
    (b) 6,055 VA at 230V
    (c) 6,055 VA at 460V
    (d) b and c

24. What is the input VA of a 3 hp, 208V, single-phase motor?

    (a) 3,890 VA
    (b) 6,440 VA
    (c) 6,720 VA
    (d) 9,422 VA

PART B—AIR-CONDITIONING CALCULATIONS

7.13 Short-Circuit and Ground-Fault Protection

25. What size protection is required for a 24A motor-compressor connected to a 240V circuit?

    (a) 40A
    (b) 60A
    (c) 70A
    (d) 90A

26. What size protection is required for an 18A motor-compressor connected to a 240V circuit?

    (a) 30A
    (b) 40A
    (c) 50A
    (d) 60A

7.14 Conductor Sizing for Single Motor-Compressor

27. What size conductor is required for a 24A motor-compressor connected to a 240V circuit?

    (a) 14 AWG
    (b) 12 AWG
    (c) 10 AWG
    (d) 8 AWG

28. What size conductor is required for an 18A motor-compressor connected to a 240V circuit?

    (a) 14 AWG
    (b) 12 AWG
    (c) 10 AWG
    (d) 8 AWG

Unit 7 Challenge Questions

UNIT 7—MOTOR AND AIR-CONDITIONING CALCULATIONS CHALLENGE QUESTIONS

(•Indicates that 75% or fewer of those who took this exam answered the question correctly.)

PART A—MOTOR CALCULATIONS

7.4 Branch-Circuit Conductor Size

1. •The branch-circuit conductors of a 5 hp, 230V motor with a nameplate rating of 25A must have an ampacity of not less than _____.

   Note: The motor is used for intermittent duty and, due to the nature of the apparatus it drives, it can't run for more than five minutes at any one time.

   (a) 21A
   (b) 23A
   (c) 33A
   (d) 37A

7.5 Feeder Conductor Size [430.24]

2. •The motor feeder conductor size for three 15 hp, 208V, three-phase motors; three 3 hp, 208V, single-phase motors; and three 1 hp, 120V, single-phase motors will be _____.

   (a) 2/0 AWG
   (b) 3/0 AWG
   (c) 4/0 AWG
   (d) 250 kcmil

7.6 Overload Protection [430.6(A)(2) and 430.32(A)]

3. The standard overload protection device for a 2 hp, 115V motor that has a full-load current rating of 24A and a nameplate rating of 21.50A must not exceed _____.

   (a) 20.60A
   (b) 24.70A
   (c) 29.90A
   (d) 33.80A

7.7 Branch-Circuit Short-Circuit and Ground-Fault Protection [430.51]

4. A 2 hp, 120V motor requires a _____ branch-circuit short-circuit protection device.

   Note: Use an inverse time breaker.

   (a) 20A
   (b) 30A
   (c) 40A
   (d) 60A

5. •The branch-circuit short-circuit and ground-fault protection device for a 10 hp, 230V, single-phase motor must not exceed _____.

   Note: Use an inverse time breaker.

   (a) 50A
   (b) 75A
   (c) 80A
   (d) 125A

6. The branch-circuit short-circuit and ground-fault protection (circuit breaker) for a 125 hp, 240V, dc motor is _____.

   (a) 400A
   (b) 600A
   (c) 700A
   (d) 800A

Unit 7 — Challenge Questions

7.9 Feeder Protection [430.61]

7. •There are three motors; one 5 hp, 230V, single-phase motor with a service factor of 1.20, and two 1½ hp, 120V, single-phase motors. Using an inverse time breaker, the 3-wire feeder conductor protection device after balancing all three motors will be _____.

   (a) 60A
   (b) 70A
   (c) 80A
   (d) 90A

8. •If an inverse time breaker is used for the feeder short-circuit protection, what size protection is required for the following 460V, three-phase motors?

   (1) Motor 1 = 40 hp, 52 FLC
   (2) Motor 2 = 20 hp, 27 FLC
   (3) Motor 3 = 10 hp, 14 FLC
   (4) Motor 4 = 5 hp, 7.60 FLC

   (a) 125A
   (b) 175A
   (c) 200A
   (d) 225A

9. •The feeder protection for one 25 hp, 208V, three-phase motor, and three 3 hp, 120V, single-phase motors will be _____ after balancing.

   Note: Use inverse time breakers.

   (a) 200A
   (b) 225A
   (c) 250A
   (d) 300A

10. •Which of the following statements about a 30 hp, 460V, three-phase synchronous motor and a 10 hp, 460V, three-phase motor are true?

    (a) The 30 hp motor is supplied by 8 AWG conductors with an 80A breaker.
    (b) The 10 hp motor is supplied by 14 AWG conductors with a 35A breaker.
    (c) The feeder conductors must be 6 AWG conductors with a 90A breaker.
    (d) all of these

UNIT 8: Voltage-Drop Calculations

INTRODUCTION TO UNIT 8—VOLTAGE-DROP CALCULATIONS

When electrical current flows through a conductor, there's a certain amount of voltage drop in the conductor due to its inherent resistance. This is another example of Ohm's Law at work. The results of voltage drop, however, aren't useful work. One of the effects of high voltage drop is a large amount of heat given off by the conductors. This is wasted power that must be paid for on the utility bill. This wasted power heats up any other conductors in the same raceway or in close proximity. This results in inefficiency at the least, and overheated circuit conductors that can result in damage as a worst-case scenario.

Excessive voltage drop can also result in delivering a voltage to the circuit load that's less than the rated voltage of the equipment. In some cases this may not be a serious problem, but for circuits such as sensitive electronic equipment, it may be critical. The manufacturer's nameplate and installation information should be consulted to determine the acceptable voltage limitations of any particular piece of equipment.

The amount of voltage drop encountered in a particular circuit is directly proportional to the amount of current flow in the circuit and directly proportional to the resistance of the circuit, as predicted by Ohm's Law. The resistance of the circuit is determined by the material of the conductor (copper or aluminum), the cross-sectional area of the conductor, and the length of the conductor.

In this unit, we'll show you how to find all the information necessary to make an accurate voltage-drop calculation for either single-phase or three-phase circuits. This will be helpful in designing practical circuits as well as in preparing for an exam.

PART A—CONDUCTOR RESISTANCE CALCULATIONS

Introduction

Ohm's Law expresses the relationship between current, voltage, and resistance in an electric circuit. Most *NEC* applications deal primarily with current (conductor ampacity, equipment ratings) and voltage (nominal system voltage, equipment ratings, transformer primary and secondary voltages). For most everyday purposes, resistance is ignored in the *Code*. However, all conductor materials have opposition to current flow, and understanding voltage-drop calculations requires knowledge of this phenomenon and how to deal with it. Opposition to current flow is called resistance in dc circuits and impedance in ac circuits. Both dc and ac circuits will be covered in this unit.

8.1 Conductor Resistance

Metals intended to carry electric current are called conductors or wires, and by their nature, they oppose the flow of electrons. Electrical conductors for applications covered by the *NEC* can be solid or stranded, and are made from copper or aluminum. All materials oppose the flow of electrons. This opposition to electron movement is known as resistance. All conductors offer some resistance to electron movement

The conductor's opposition to the flow of current depends on the conductor material (copper or aluminum), its cross-sectional area (wire size), its length, and its operating temperature.

Unit 8 Voltage-Drop Calculations

Material

Aluminum is often used when weight or costs are important considerations, but copper is the most common type of metal used for electrical conductors. Gold doesn't tarnish so it is used mostly for plating terminals (electroplating) for some electronic equipment.

Cross-Sectional Area

The cross-sectional area of a conductor is the conductor's surface area expressed in circular mils (cmils). **Figure 8–1**

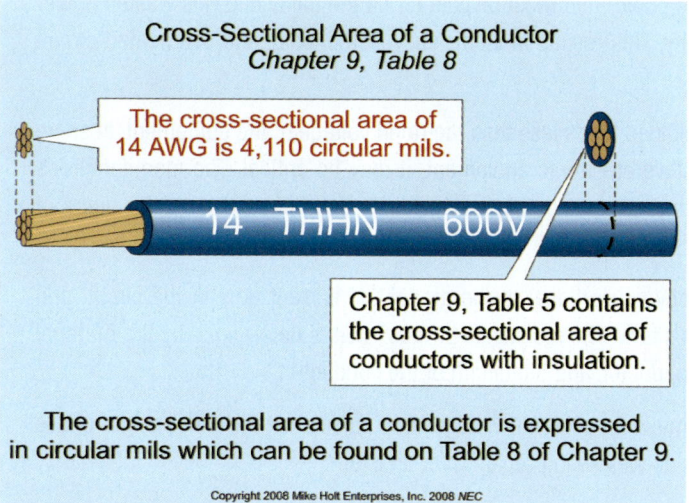

Figure 8–1

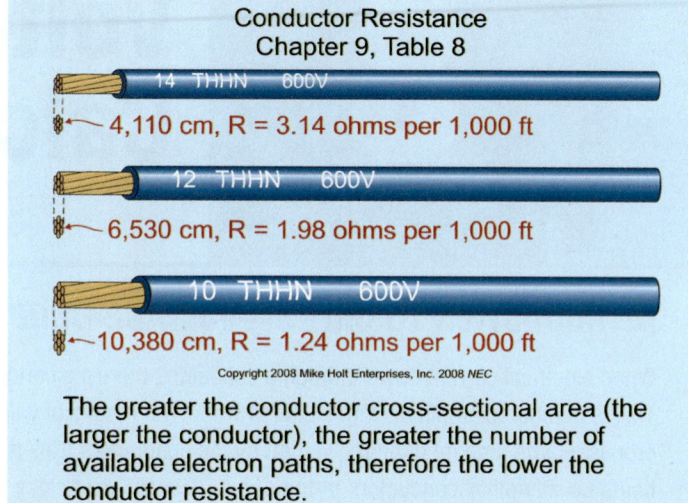

Figure 8–2

The greater the conductor cross-sectional area (the larger the conductor), the greater the number of available electron paths and the lower the conductor resistance. **Figure 8–2**

Conductors are sized according to American Wire Gage (AWG), which ranges from 40 AWG to 4/0 AWG. Conductors larger than 4/0 are identified in thousand circular mils (kcmil), such as 250 kcmil, 350 kcmil, and 500 kcmil. **Figure 8–3**

Conductor resistance varies inversely with the conductor's diameter; the smaller the wire size, the greater the resistance, and the larger the wire size, the lower the resistance. **Figure 8–4**

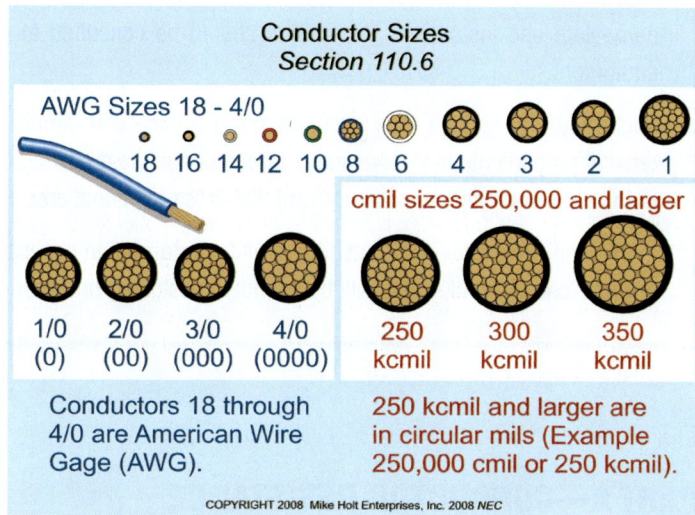

Figure 8–3

Conductor Length

The resistance of a conductor is directly proportional to its length. **Table 8–1** provides examples of conductor resistance and circular mils area for conductors 1,000 ft long. Longer or shorter lengths will naturally have different conductor resistances.

Mike Holt's Illustrated Guide to NEC Exam Preparation

Voltage-Drop Calculations — Unit 8

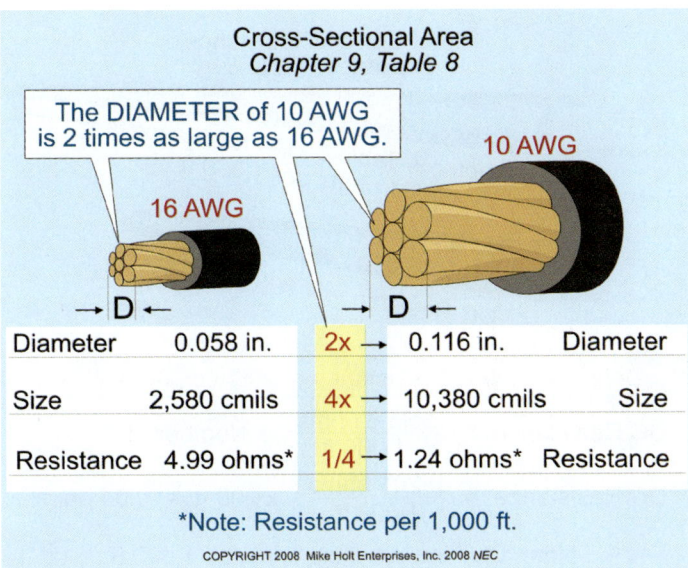

Figure 8–4

Table 8–1. Conductor Properties, NEC Chapter 9, Table 8

Conductor Size American Wire Gage	Conductor Resistance Per 1,000 Feet at 75°C	Conductor Diameter Inches	Conductor Area Circular Mils
14 AWG	3.140 ohms (stranded)	0.073	4,110
12 AWG	1.980 ohms (stranded)	0.092	6,530
10 AWG	1.240 ohms (stranded)	0.116	10,380
8 AWG	0.778 ohms (stranded)	0.146	16,510
6 AWG	0.491 ohms (stranded)	0.184	26,240

Temperature

The resistance of a conductor changes with temperature. The amount of change per degree is called the temperature coefficient. Positive temperature coefficient indicates that as the temperature rises, the conductor resistance also rises; this reduces conductor ampacity. Copper and aluminum conductors have a positive temperature coefficient.

The conductor resistances listed in the *NEC* Chapter 9, Tables 8 and 9, are based on an operating temperature of 75°C. A three-degree rise in temperature results in a 1 percent increase in conductor resistance for both copper and aluminum conductors. The formula to determine the change in conductor resistance with changing temperature is listed at the bottom of Table 8 in Note 2. For example, the resistance of copper at 90°C is about 5 percent more than at 75°C.

Temperature Adjustment, Table 8, Note 2:

R for CU = Table R x [1 + (0.00323 x (Temp°C − 75°))]
R for AL = Table R x [1 + (0.00333 x (Temp°C − 75°))]

8.2 Conductor Resistance—Direct-Current Circuits [Chapter 9, Table 8]

The *NEC* lists the resistance and area in circular mils for both dc and ac circuit conductors. Direct-current circuit conductor resistances are listed in Chapter 9, Table 8, and alternating-current circuit conductor resistances are listed in Chapter 9, Table 9. The tables include both solid and stranded conductors. Stranded conductors will be used in this textbook unless specified otherwise.

The dc conductor resistances listed in Chapter 9, Table 8 apply to conductors 1,000 ft long. The following formula can be used to determine the conductor resistance for conductor lengths other than 1,000 ft:

◆ **DC Conductor Resistance Formula**

DC Conductor Resistance =
(Conductor Resistance Ohms/1,000 ft) x Conductor Length

▶ **Conductor Resistance Copper**

Question: What is the dc resistance of 200 ft of 12 AWG copper?

(a) 0.21 ohms (b) 0.29 ohms
(c) 0.396 ohms (d) 0.72 ohms

Answer: (c) 0.396 ohms

The dc resistance of 12 AWG copper 1,000 ft long is 1.98 ohms [Chapter 9, Table 8].

The dc resistance of 200 ft is:
(1.98 ohms/1,000 ft) x 200 ft = 0.396 ohms

Unit 8   Voltage-Drop Calculations

▶ **Conductor Resistance Copper**

Question: What is the dc resistance of 200 ft of 6 AWG copper? Figure 8–5

(a) 0.21 ohms (b) 0.29 ohms
(c) 0.49 ohms (d) 0.98 ohms

Answer: (d) 0.98 ohms

The dc resistance of 6 AWG copper 1,000 ft long is 0.491 ohms [Chapter 9, Table 8].

The dc resistance of 420 ft is:
(0.491 ohms/1,000 ft) x 200 ft = 0.0982 ohms

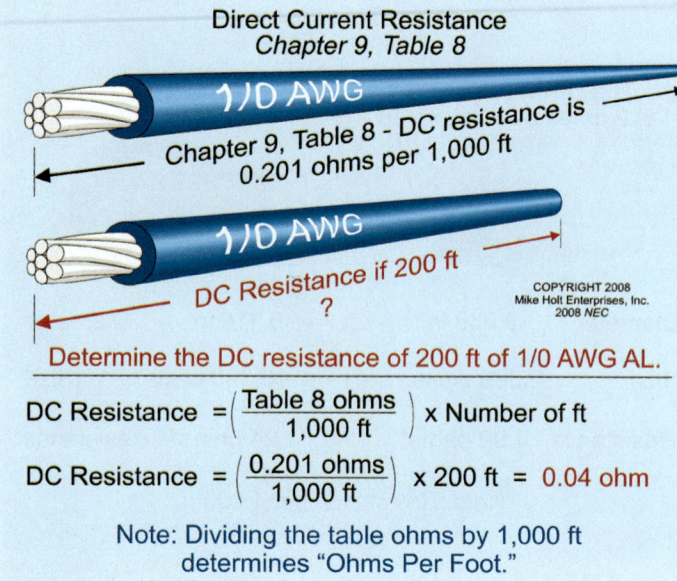

Figure 8–6

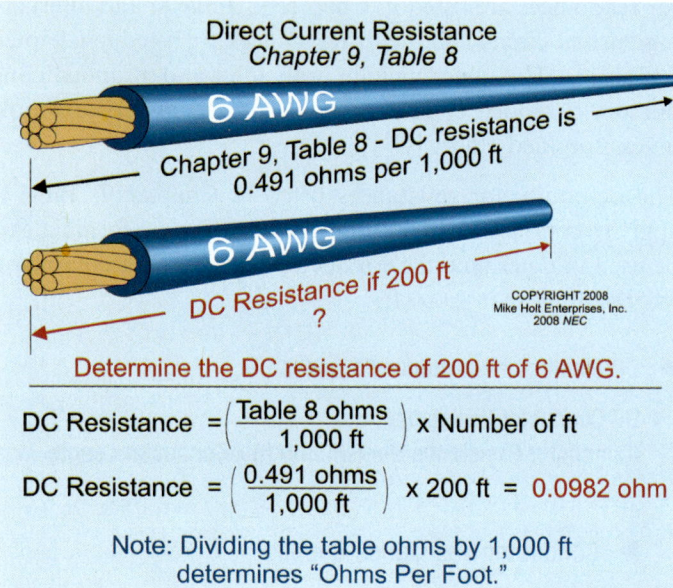

Figure 8–5

▶ **Conductor Resistance Aluminum**

Question: What is the resistance of 200 ft of 1/0 AWG aluminum? Figure 8–6

(a) 0.04 ohms (b) 0.29 ohms
(c) 0.60 ohms (d) 0.72 ohms

Answer: (a) 0.04 ohms

The resistance of 1/0 AWG aluminum 1,000 ft long is 0.201 ohms [Chapter 9, Table 8].

The resistance of 200 ft is:
(0.201 ohms/1,000 ft) x 200 ft = 0.04 ohms

8.3 Conductor Resistance—Alternating-Current Circuits

In dc circuits, the only property that opposes the flow of electrons is resistance. In ac circuits, the expanding and collapsing magnetic field within the conductor induces an electromotive force that opposes the flow of the alternating current. This opposition to the flow of ac is called inductive reactance.

In addition, ac flowing through a conductor generates small, erratic, independent currents called eddy currents. Figure 8–7 Eddy currents are strongest in the center of the conductors and repel the flowing electrons toward the conductor surface. This is known as skin effect. Figure 8–8

Because of skin effect, the effective cross-sectional area of an ac conductor is reduced, which results in an increased opposition to current flow. The total opposition to the movement of electrons in an ac circuit (resistance plus inductive reactance plus skin effect) is called impedance.

8.4 Alternating-Current Resistance

An alternating-current conductor's opposition to current flow (resistance and reactance) is listed in Chapter 9, Table 9 of the *NEC*. The total opposition to current flow in an ac circuit is called impedance and depends on the conductor mate-

Voltage-Drop Calculations — Unit 8

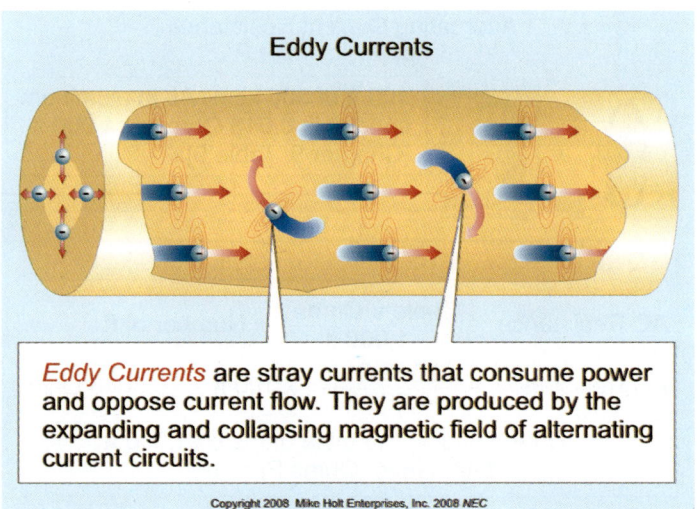

Figure 8–7

▶ **Alternating-Current Ohms-to-Neutral Resistance Per 1,000 Ft**

Question: What is the ac ohms-to-neutral resistance of a 250 kcmil conductor that is 1,000 ft long in each of the following situations?

Copper conductor in nonmetallic raceway	0.052 ohms
Copper conductor in aluminum raceway	0.057 ohms
Copper conductor in steel raceway	0.054 ohms
Aluminum conductor in nonmetallic raceway	0.085 ohms
Aluminum conductor in aluminum raceway	0.090 ohms
Aluminum conductor in steel raceway	0.086 ohms

Alternating-Current Conductor Resistance Formula

The following formula can be used to determine conductor resistance:

◆ **Alternating-Current Resistance Formula**

Alternating-Current Resistance = (Conductor Ohms-to-Neutral Resistance/ 1,000 ft) x Conductor Length

▶ **Ohms-to-Neutral Resistance Example 1**

Question: What is the ac ohms-to-neutral resistance of 100 ft of 2/0 AWG copper installed in a steel raceway? **Figure 8–9**

(a) 0.010 ohms   (b) 0.042 ohms
(c) 0.069 ohms   (d) 0.072 ohms

Answer: (a) 0.010 ohms

The ohms-to-neutral resistance of 2/0 AWG copper is 0.10 ohms per 1,000 ft [Chapter 9, Table 9].

The ohms-to-neutral resistance of 100 ft of 2/0 AWG is: (0.10 ohms/1,000 ft) x 100 ft = 0.010 ohms

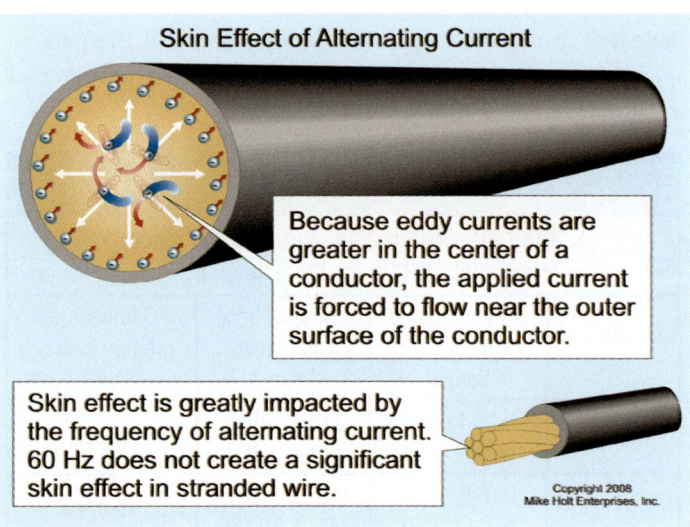

Figure 8–8

rial (copper or aluminum) and on the magnetic property of the raceway or cable in which the conductors are installed.

The header row of Chapter 9, Table 9 shows, "Ohms to Neutral per Kilometer" and directly under that it reads, "Ohms to Neutral per 1,000 Feet." Each entry on Table 9 follows this format. For example, the ac resistance (impedance) for 2 AWG in PVC is 0.62 ohms per kilometer and 0.19 ohms per 1,000 ft. In this textbook, we will use the "Ohms to Neutral per 1,000 Feet" for all calculations.

Unit 8 — Voltage-Drop Calculations

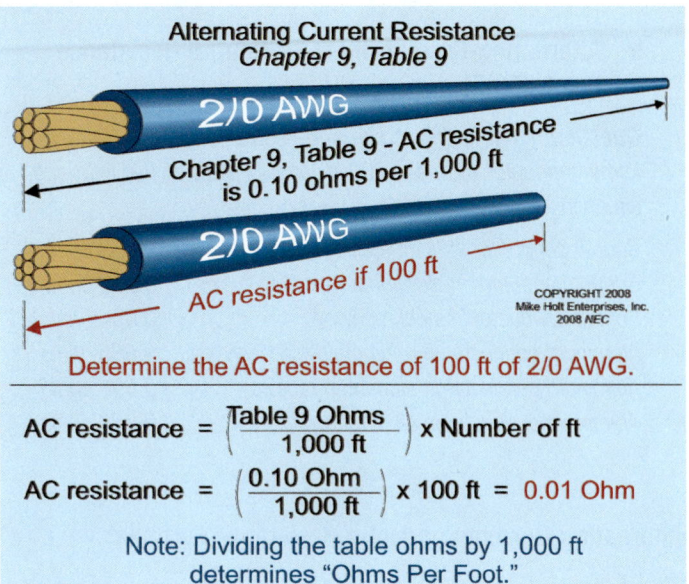

Figure 8–9

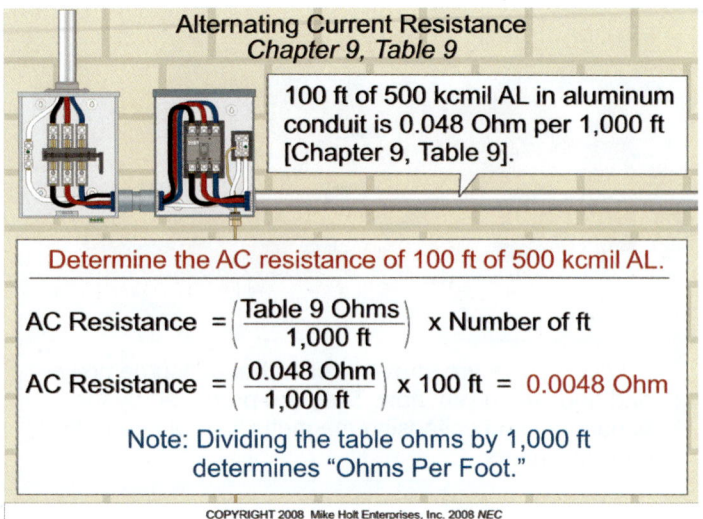

Figure 8–10

▶ **Ohms-to-Neutral Resistance Example 2**

Question: What is the ac ohms-to-neutral resistance of 100 ft of 500 kcmil aluminum conductors installed in an aluminum conduit? **Figure 8–10**

(a) 0.0021 ohms
(b) 0.0029 ohms
(c) 0.0048 ohms
(d) 0.0081 ohms

Answer: (c) 0.0048 ohms

The ohms-to-neutral resistance of 500 kcmil installed in an aluminum conduit is 0.048 ohms per 1,000 ft.

The ohms-to-neutral resistance of 100 ft of 500 kcmil in an aluminum conduit is: (0.048 ohms/1,000 ft) x 100 ft = 0.0048 ohms

8.5 AC Resistance as Compared to DC Resistance

The opposition to current flow (impedance) is greater for ac circuits compared to the resistance of dc circuits because of inductive reactance, eddy currents, and skin effect in addition to the resistance of the conductor. **Table 8–2** provides some comparisons between ac resistance and dc resistance. Chapter 9, Table 8 provides dc resistance values for conductors, and Chapter 9, Table 9 provides ac resistance values for conductors. These values are presented in a graphical comparison in **Figure 8–11**. Note that for conductors smaller than 1/0 AWG, the two values are essentially equal.

Table 8–2 shows the larger conductor sizes, and the percentage by which the resistance of an ac circuit increases over dc resistance as the conductor size increases.

Table 8–2			
COPPER – Alternating-Current Impedance versus Direct-Current Resistance at 75°C			
Conductor Size	AC Impedance Chapter 9, Table 9	DC Resistance Chapter 9, Table 8	AC impedance greater than dc resistance by %
250,000	0.054 ohms per 1,000 ft	0.0515 ohms per 1,000 ft	4.85%
500,000	0.029 ohms per 1,000 ft	0.0258 ohms per 1,000 ft	12.40%
1,000,000	0.018 ohms per 1,000 ft	0.0129 ohms per 1,000 ft	39.50%
ALUMINUM – Alternating-Current Impedance versus Direct-Current Resistance at 75°C			
Conductor Size	AC Impedance Chapter 9, Table 9	DC Resistance Chapter 9, Table 8	AC impedance greater than dc resistance by %
250,000	0.086 ohms per 1,000 ft	0.0847 ohms per 1,000 ft	1.50%
500,000	0.045 ohms per 1,000 ft	0.0424 ohms per 1,000 ft	6.13%
1,000,000	0.025 ohms per 1,000 ft	0.0212 ohms per 1,000 ft	17.92%

Voltage-Drop Calculations — Unit 8

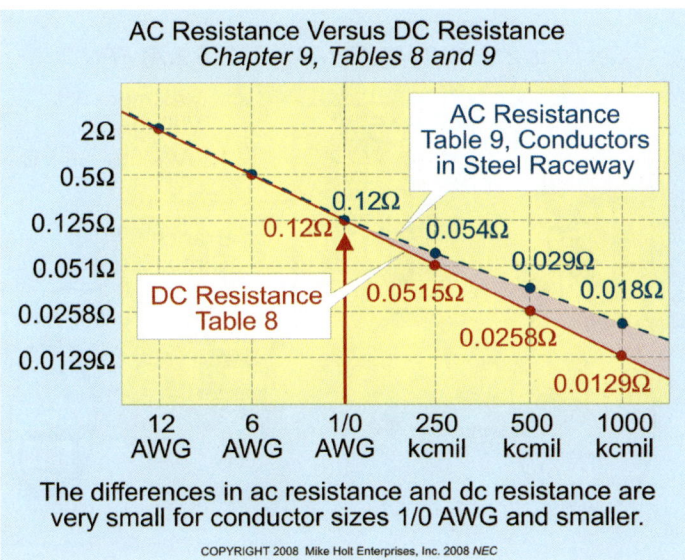

Figure 8–11

PART B—VOLTAGE-DROP CONSIDERATIONS

Introduction

Contrary to many beliefs, the *NEC* generally doesn't require conductors to be increased in size to accommodate voltage drop. However, it does recommend that we consider the effects of voltage drop when sizing conductors. **Figure 8–12**

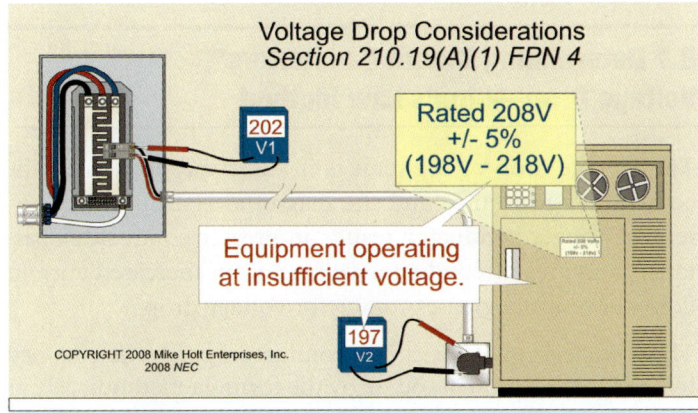

Figure 8–12

8.6 *NEC* Voltage-Drop Recommendations

Although the *NEC* generally doesn't include rules for voltage drop, it does contain several recommendations in 210.19(A)(1) FPN No. 4, 215.2(A)(3) FPN No. 2, and 310.15(A)(1) FPN No. 1. Be aware that Fine Print Notes in the *NEC* are recommendations, not requirements [90.5(C)]. There may be other standards besides the *NEC* that specify voltage-drop requirements on a specific project. The *Code* recommends that the maximum combined voltage drop for both the feeder and branch circuit should not exceed 5 percent, and the maximum on the feeder or branch circuit should not exceed 3 percent. **Figure 8–13**

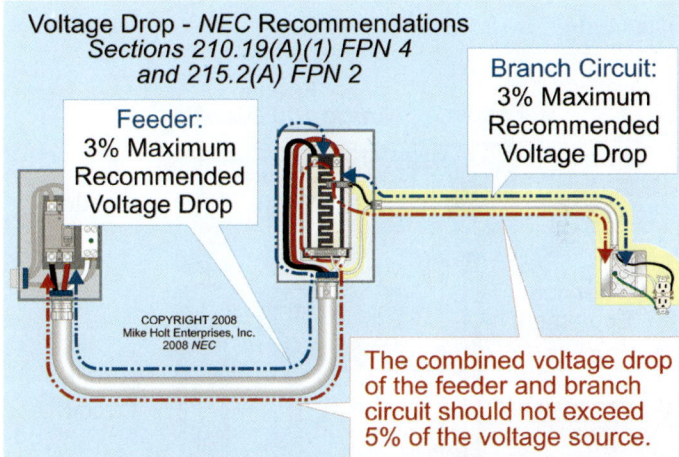

Figure 8–13

The Article 100 definitions for a branch circuit and a feeder are:

Branch Circuit [Article 210]. The conductors between the final overcurrent device and the receptacle outlets, lighting outlets, or other outlets as defined in this article. **Figure 8–14**

Feeder [Article 215]. The conductors between the service equipment, a separately derived system, or other power supply and the final branch-circuit overcurrent device. **Figure 8–15**

Unit 8 — Voltage-Drop Calculations

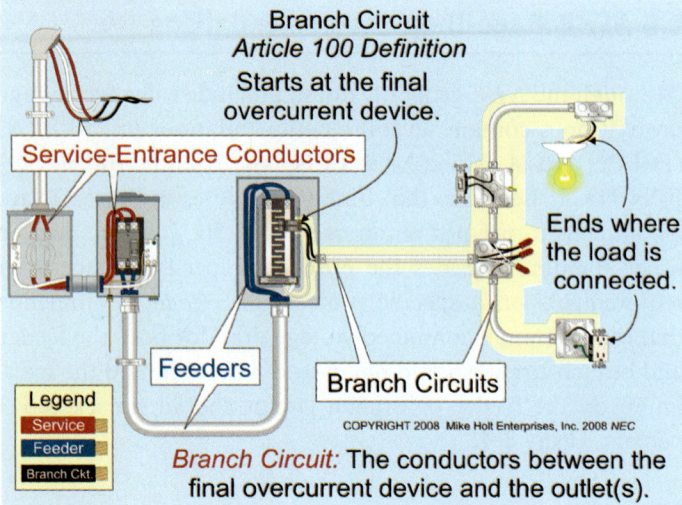

Figure 8–14

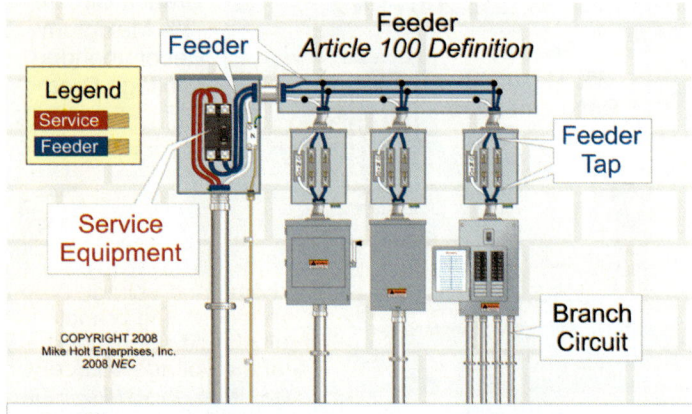

Figure 8–15

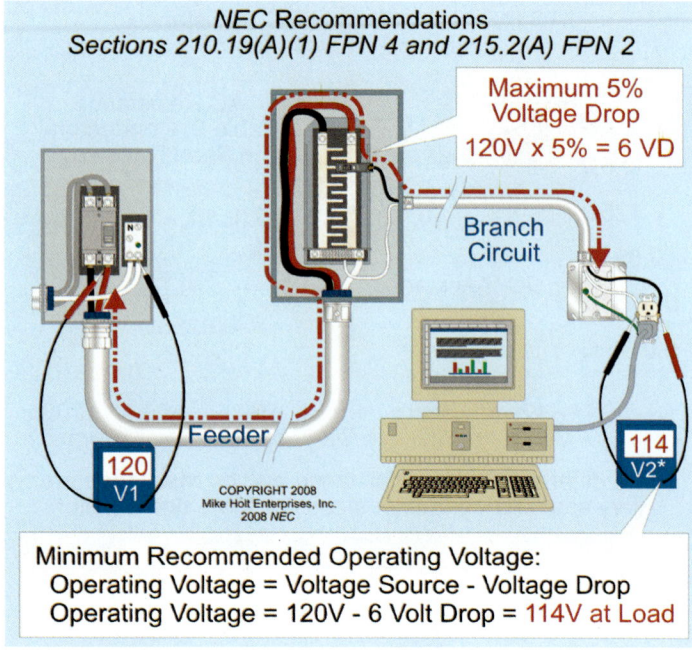

Figure 8–16

▶ **NEC Voltage-Drop Recommendation**

Question: What is the minimum NEC recommended operating volts for a 115V rated load that is connected to a 120V source? Figure 8–16

(a) 114V (b) 115V (c) 116V (d) 120V

Answer: (a) 114V

The maximum conductor voltage drop recommended for both the feeder and branch circuit is 5 percent of the voltage source (120V). The total conductor voltage drop (feeder and branch circuit) should not exceed 120V x 0.05 = 6V. The operating voltage at the load is calculated by subtracting the conductor's voltage drop from the voltage source: 120V – 6V = 114V.

Undervoltage harms the performance of electrical equipment. For inductive loads such as motors and heaters it can cause inefficiency, overheating, and shortened life spans. Solid-state equipment such as TVs and computers are also affected by voltage drop. When conductor resistance causes circuit voltage to drop below an acceptable value, the conductor size should be increased.

8.7 Determining Circuit Conductors' Voltage Drop—Ohm's Law Method

The voltage drop of a circuit is in direct proportion to the conductor's resistance and the magnitude (size) of the current. A longer conductor results in greater conductor resistance, and greater conductor voltage drop. Increased current flow will result in greater conductor voltage drop.

The voltage drop of the circuit conductors can be determined by the Ohm's Law method or by the formula method:

◆ **Ohm's Law Method—Single-Phase**

$E_{vd} = I \times R$

E_{vd} = Conductor voltage drop expressed in volts.

I = The load in amperes at 100 percent, not at 125 percent, for motors or continuous loads.

Voltage-Drop Calculations — Unit 8

R* = Conductor Resistance, Chapter 9, Table 8 for dc resistance or Chapter 9, Table 9 for ac resistance.

*For conductors 1/0 AWG and smaller, the difference in resistance between dc and ac circuits is so little that it can be ignored. In addition, you can ignore the small difference in resistance between stranded and solid conductors.

Author's Comment: Conductor resistance is based on the total length of both circuit conductors. So if a 2-wire branch circuit is 80 ft long, you must calculate voltage drop for a total conductor length of 160 ft.

▶ Voltage Drop 120V

Question: What is the voltage drop of two 12 AWG conductors that supply a 16A, 120V, single-phase load located 100 ft from the power supply? **Figure 8–17**

(a) 3.20V (b) 6.40V (c) 9.60V (d) 12.80V

Answer: (b) 6.40V

$E_{vd} = I \times R$
$I = 16A$
$R = 2$ ohms per 1,000 ft [Chapter 9, Table 9]
$R = (2$ ohms per 1,000 ft/1,000 ft$) \times 200$ ft
$R = 0.40$ ohms

$E_{vd} = I \times R$
$E_{vd} = 16A \times 0.40$ ohms
$E_{vd} = 6.40V$

▶ Voltage Drop 240V

Question: A 240V, 24A, single-phase load is located 100 ft from the panelboard and is wired with 10 AWG. What is the voltage drop of the circuit conductors? **Figure 8–18**

(a) 3.64V (b) 4.53V (c) 5.76V (d) 9.22V

Answer: (c) 5.76V

$E_{vd} = I \times R$
$I = 24A$
$R = 1.20$ ohms per 1,000 ft [Chapter 9, Table 9]
$R = (1.20$ ohms per 1,000 ft/1,000 ft$) \times 200$ ft
$R = 0.24$ ohms

$E_{vd} = I \times R$
$E_{vd} = 24A \times 0.24$ ohms
$E_{vd} = 5.76V$

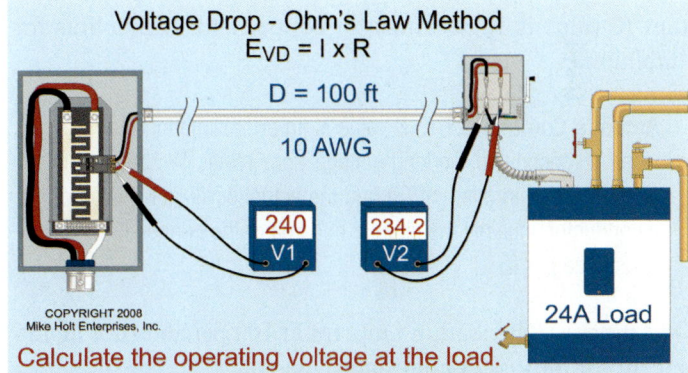

Figure 8–18

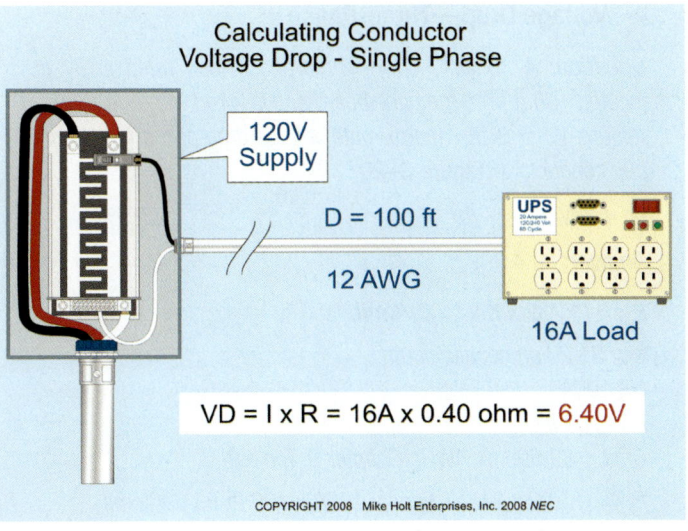

Figure 8–17

Unit 8 — Voltage-Drop Calculations

8.8 Determining Circuit Conductors' Voltage Drop—Formula Method

Voltage Drop Using the Formula Method

In addition to the Ohm's Law method, the following formula can be used to determine the conductor voltage drop:

- Single-Phase Circuit Voltage Drop

  $E_{vd} = (2 \times K \times I \times D)/Cmil$

- Three-Phase Circuit Voltage Drop

  $E_{vd} = (1.732 \times K \times I \times D)/Cmil$

E_{vd} = Volts Dropped: The voltage drop of the circuit expressed in volts.

K = Direct-Current Constant. This constant K represents the dc resistance for a 1,000 circular mils conductor that is 1,000 ft long, at an operating temperature of 75°C. The constant K value is 12.90 ohms for copper and 21.20 ohms for aluminum.

Author's Comment: This K value is an approximation that provides a reasonable working voltage drop result. We'll use this approximation in this unit. To find the actual or "True K" value of a conductor, multiply the circular mil area of the conductor by its resistance per ft.

I = Amperes: The load in amperes at 100 percent (not at 125 percent for motors or continuous loads).

D = Distance: The distance the load is from the power supply. Do not use length of wire for this formula.

Cmil = Circular Mils: The circular mils of the circuit conductor as listed in the *NEC* Chapter 9, Table 8.

▶ K Value Example 1

Question: What is the actual K value for a 400 kcmil copper conductor in a PVC raceway?

(a) 11.0 (b) 12.5 (c) 13.20 (d) 14.50

Answer: (c) 13.20

400 kcmil copper has a resistance of 0.033 ohms per kft [Chapter 9, Table 9].

400,000 cmils x (0.033 ohms/1,000 ft) = 13.20.

▶ K Value Example 2

Question: What is the actual K value for a 6 AWG copper conductor in a steel raceway?

(a) 11.0 (b) 12.86 (c) 13.20 (d) 14.50

Answer: (b) 12.86

6 AWG copper has a resistance of 0.49 ohms per kft [Chapter 9, Table 9].

6 AWG has a circular mil area of 26,240 cmil [Chapter 9, Table 8]

26,240 cmils x (0.49 ohms/1,000 ft) = 12.86.

▶ Voltage Drop—Single-Phase

Question: A 24A, 240V load is located 100 ft from a panelboard and is wired with 10 AWG. What is the approximate voltage drop of the branch-circuit conductors? **Figure 8–19**

(a) 4.25V (b) 5.97V (c) 3% (d) 5%

Answer: (b) 5.97V

$E_{vd} = (2 \times K \times I \times D)/Cmil$
K = 12.90, copper
I = 24A
D = 100 ft
Cmil = 10,380 cmil, (10 AWG) [Chapter 9, Table 8]

$E_{vd} = (2 \times K \times I \times D)/Cmil$
E_{vd} = (2 x 12.90 ohms x 24A x 100 ft)/10,380 cmil
E_{vd} = 5.97V

▶ Voltage Drop—Three-Phase

Question: A 36 kVA, 100A, three-phase load rated 208V is located 100 ft from the panelboard and is wired with 1 AWG aluminum. What is the approximate voltage drop of the feeder circuit conductors? **Figure 8–20**

(a) 4.39V (b) 7V (c) 3% (d) 5%

Answer: (a) 4.39V

$E_{vd} = (1.732 \times K \times I \times D)/Cmil$
K = 21.20 ohms, aluminum
I = 100A
D = 100 ft
Cmil = 83,690, (1 AWG) [Chapter 9, Table 8]

E_{vd} = (1.732 x 21.20 ohms x 100A x 100 ft)/83,690 cmil
E_{vd} = 4.39V

Voltage-Drop Calculations — Unit 8

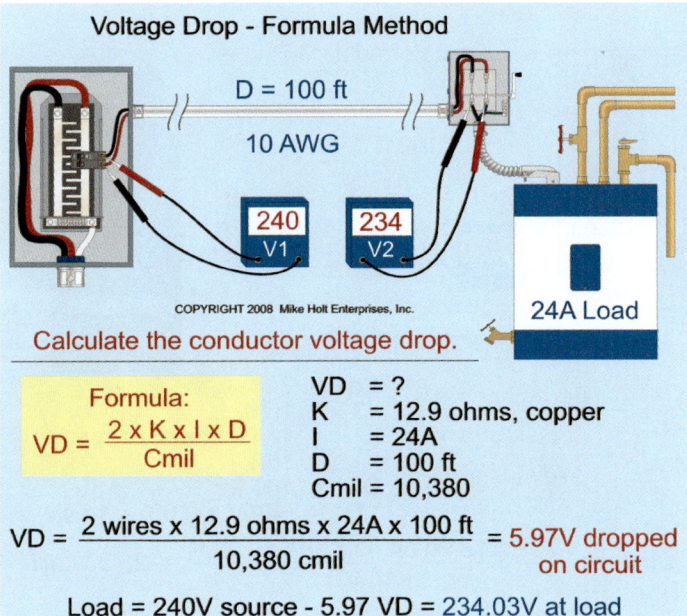

Figure 8–19

8.9 Sizing Conductors to Prevent Excessive Voltage Drop

The size of a conductor (actually its resistance) affects voltage drop. If we want to decrease the voltage drop of a circuit, we can increase the cross-sectional area of the conductor (reduce its resistance). When sizing conductors to prevent excessive voltage drop, use the following formulas:

◆ Single-Phase Circuit Voltage Drop

Cmil = (2 x K x I x D)/E_{vd}

◆ Three-Phase Circuit Voltage Drop

Cmil = (1.732 x K x I x D)/ E_{vd}

▶ Size Conductor—Single-Phase

Question: A 5 hp motor is located 100 ft from a 120/240V, single-phase panelboard. What size conductor should be used to limit the voltage drop to no more than 3 percent if the motor nameplate indicates 26A at 230V? The terminals are rated for 75°C. Figure 8–21

(a) 10 AWG   (b) 8 AWG   (c) 6 AWG   (d) 4 AWG

Answer: (a) 10 AWG

Cmil = (2 x K x I x D)/ E_{vd}

K = 12.90 ohms, copper
I = 26A at 230V
D = 100 ft

E_{vd} = 240V x 0.03
E_{vd} = 7.20V

Cmil = (2 x 12.90 ohms x 26A x 100 ft)/7.20V
Cmil = 9,317 cmil, (10 AWG) [Chapter 9, Table 8]

Note: Section 430.22(A) requires that the motor conductors be sized not less than 125 percent of the motor FLCs as listed in Table 430.248. The motor FLC is 28A so the conductor must be sized at: 28A x 1.25 = 35A.

The 10 AWG conductor required for voltage drop is rated for 35A at 75°C according to 110.14(C) and Table 310.16.

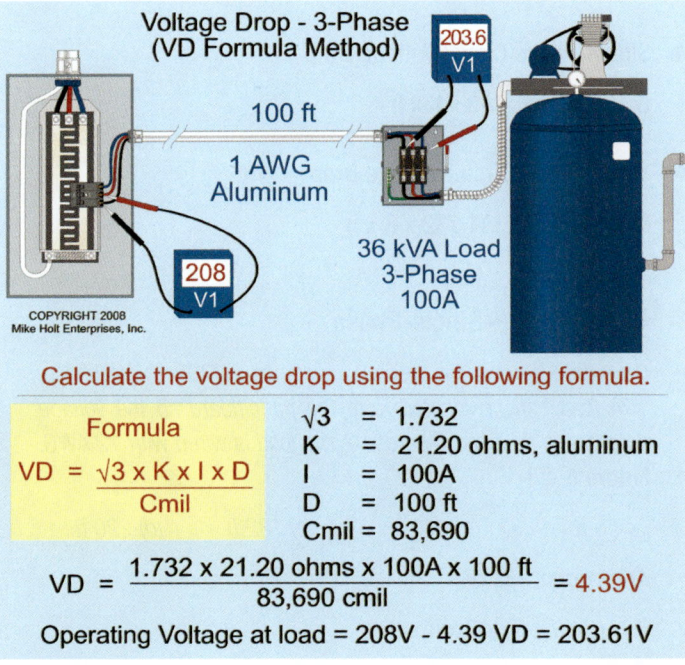

Figure 8–20

Unit 8 — Voltage-Drop Calculations

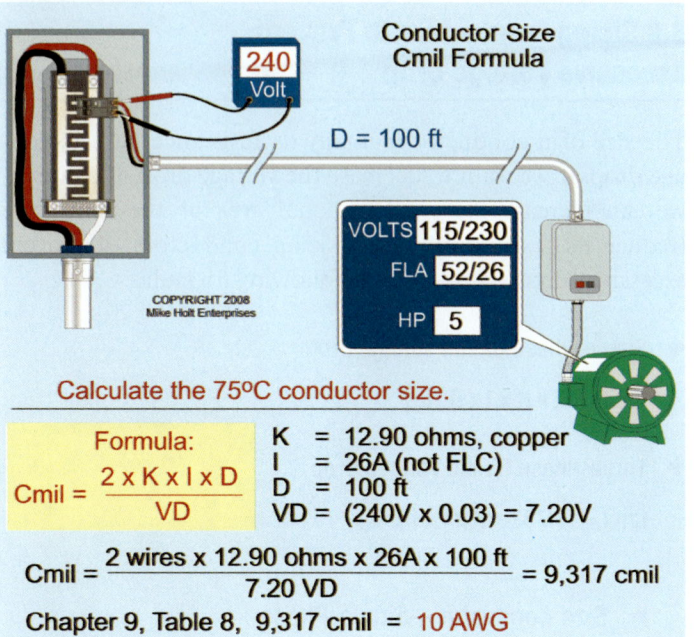

Figure 8–21

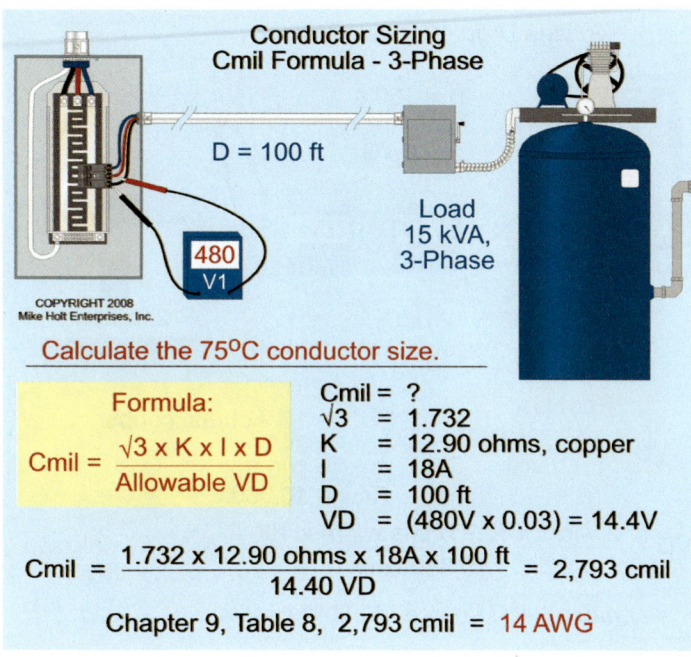

Figure 8–22

▶ **Size Conductor—Three-Phase**

Question: A 15 kVA, 18A three-phase load rated 460V is located 100 ft from the panelboard. What size conductor is required to prevent the voltage drop from exceeding 3 percent? Figure 8–22

(a) 14 AWG (b) 10 AWG (c) 8 AWG (d) 6 AWG

Answer: (a) 14 AWG

$Cmil = (1.732 \times K \times I \times D)/E_{vd}$

$K = 12.90$ ohms, copper
$I = 18A$
$D = 100$ ft
$E_{vd} = 480V \times 0.03 = 14.40V$

$Cmil = (1.732 \times 12.90 \text{ ohms} \times 18A \times 100 \text{ ft})/14.40V$
$Cmil = 2,793$ cmil, (14 AWG) [Chapter 9, Table 8]

8.10 Limiting Conductor Length to Minimize Voltage Drop

Limiting the length of conductors can also reduce voltage drop. The following formulas can be used to help determine maximum conductor length to limit the voltage drop to *NEC* recommendations:

◆ Single-Phase Circuit Voltage Drop

$D = (Cmil \times E_{vd})/(2 \times K \times I)$

◆ Three-Phase Circuit Voltage Drop

$D = (Cmil \times E_{vd})/(1.732 \times K \times I)$

▶ **Distance—Single-Phase**

Question: What is the maximum distance a 240V, single-phase, 10A load can be located from the panelboard so the voltage drop doesn't exceed 3 percent? The load is wired with 12 AWG. Figure 8–23

(a) 55 ft (b) 110 ft (c) 182 ft (d) 220 ft

Answer: (c) 182 ft

$D = (Cmil \times E_{vd})/(2 \times K \times I)$

$Cmil = 6,530$ (12 AWG) [Chapter 9, Table 8]
$E_{vd} = 240V \times 0.03$
$E_{vd} = 7.20V$
$K = 12.90$ ohms, copper
$I = 10A$
$D = (6,530 \text{ cmil} \times 7.20V)/(2 \times 12.90 \text{ ohms} \times 10A)$
$D = 182.20$ ft

Voltage-Drop Calculations   Unit 8

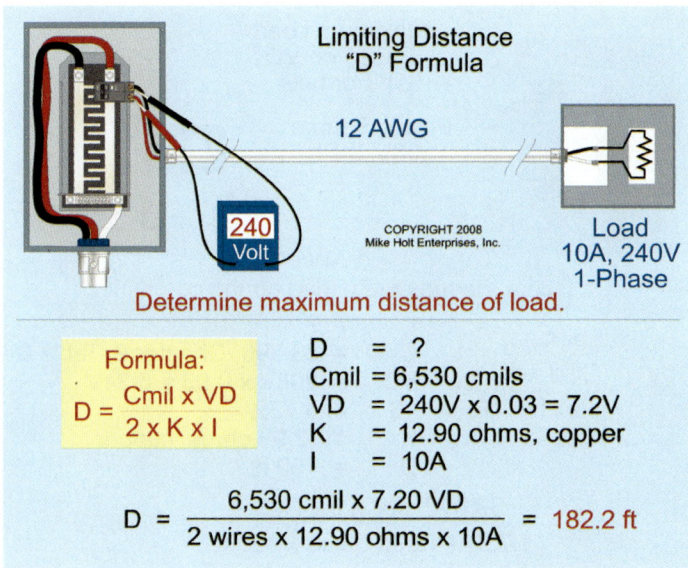

Figure 8–23

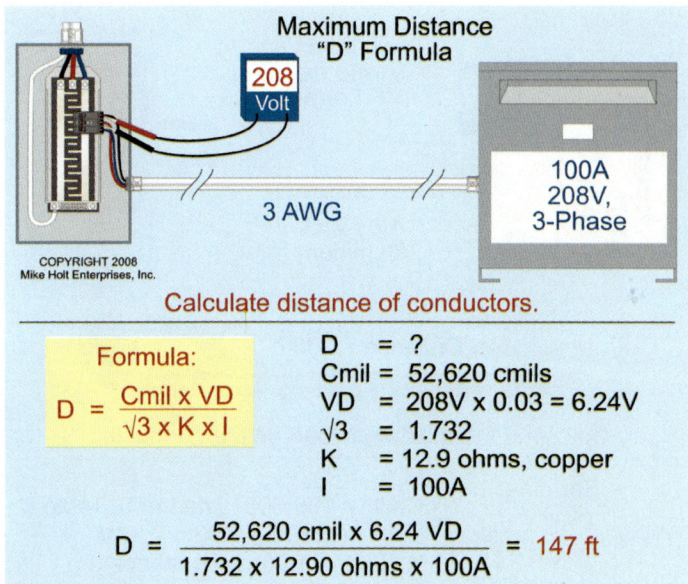

Figure 8–24

▶ Distance—Three-Phase

Question: What is the maximum distance a 100A, 208V, three-phase load, wired with 3 AWG conductors, can be located from the panelboard so the voltage drop doesn't exceed 3 percent? **Figure 8–24**

(a) 75 ft (b) 125 ft (c) 147 ft (d) 325 ft

Answer: (c) 147 ft

$D = (Cmil \times E_{vd})/(1.732 \times K \times I)$

Cmil = 52,620, (3 AWG) [Chapter 9, Table 8]

$E_{vd} = 208V \times 0.03$

$E_{vd} = 6.24V$

K = 12.90 ohms, copper

I = 100A

D = (52,620 cmil x 6.24V)/(1.732 x 12.90 ohms x 100A)

D = 147 ft

8.11 Limiting Current to Limit Voltage Drop

Sometimes the only method of limiting the circuit voltage drop is to limit the load on the conductors. The following formulas can be used to determine the maximum load:

◆ Single-Phase Circuit Voltage Drop

$I = (Cmil \times E_{vd})/(2 \times K \times D)$

◆ Three-Phase Circuit Voltage Drop

$I = (Cmil \times E_{vd})/(1.732 \times K \times D)$

▶ Maximum Load—Single-Phase

Question: What is the maximum recommended load that should be placed on 1/0 AWG aluminum conductors in a nonmetallic raceway to a panelboard located 200 ft from a 240V single-phase power source so the NEC recommendation for voltage drop isn't exceeded? **Figure 8–25**

(a) 51A (b) 89A (c) 94A (d) 115A

Answer: (b) 89A

$I = (Cmil \times E_{vd})/(2 \times K \times D)$

Cmil = 105,600 (1/0 AWG) [Chapter 9, Table 8]

$E_{vd} = 240V \times 0.03$

$E_{vd} = 7.20V$

K = 21.20 ohms, aluminum

D = 200 ft

I = (105,600 cmil x 7.20V)/(2 x 21.20 ohms x 200 ft)

I = 89.66A

Note: The maximum load permitted on 1/0 AWG aluminum at 75°C terminals is 120A [Table 310.16]. This circuit will safely carry the 89.66A without excessive voltage drop, but can't be loaded to the rated value of 120A.

Unit 8 — Voltage-Drop Calculations

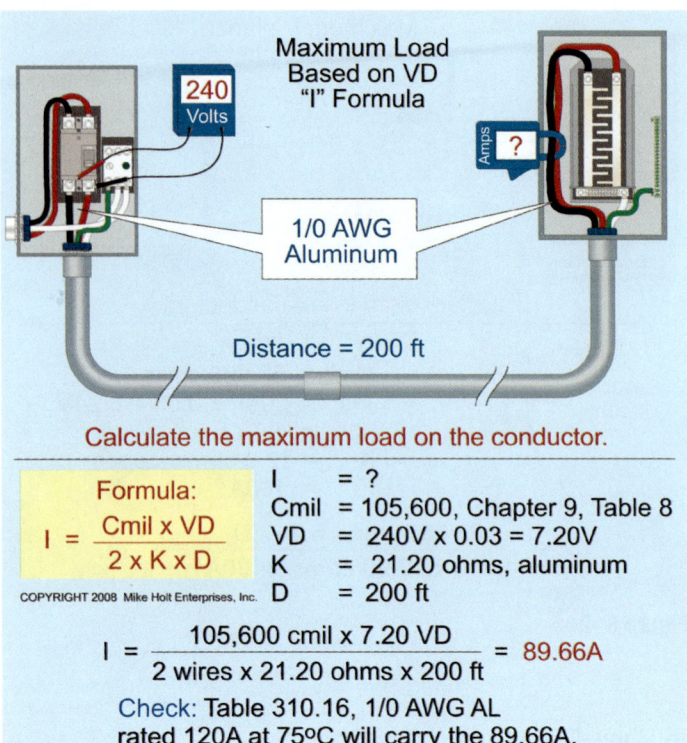

Figure 8–25

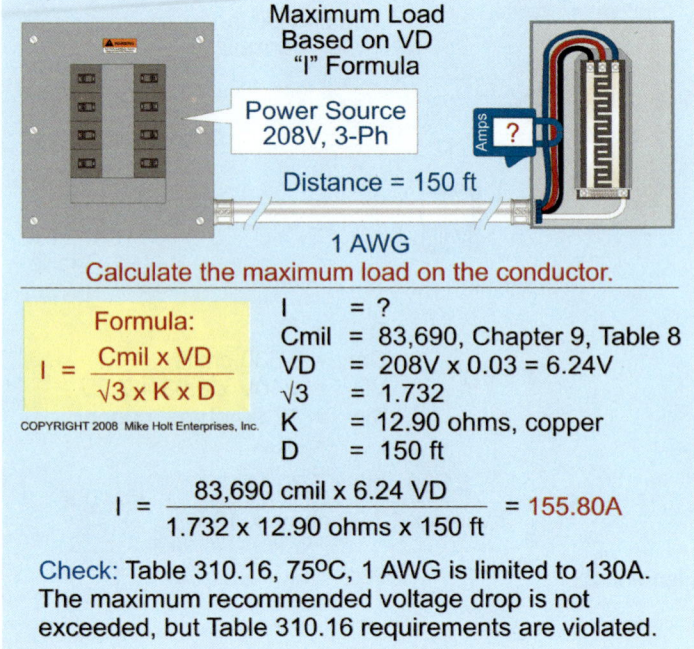

Figure 8–26

▶ **Maximum Load—Three-Phase**

Question: What is the maximum recommended load that should be placed on 1 AWG copper conductors in an aluminum raceway to a panelboard located 150 ft from a 208V, three-phase power source so the NEC recommendation for voltage drop isn't exceeded? **Figure 8–26**

(a) 155A (b) 190A (c) 210A (d) 240A

Answer: (a) 155A

$I = (Cmil \times E_{vd})/(1.732 \times K \times D)$
Cmil = 83,690 (1 AWG) [Chapter 9, Table 8]
E_{vd} = 208V x 0.03
E_{vd} = 6.24V
K = 12.90 ohms, copper
D = 150 ft
I = (83,690 cmils x 6.24V)/(1.732 x 12.90 x 150 ft)
I = 155.80A

Note: The maximum load of 180A limits the voltage drop to no more than 3 percent. When working this type of problem, don't lose sight of other Code requirements. Table 310.16 must also be consulted for the maximum ampacity permitted on 1 AWG at 75ºC, which is 130A, so 130A is the working limit for this circuit [110.14(C) and Table 310.16].

8.12 Fire Pump Motor Circuits

Power Wiring [695.6]

(C) Conductor Size.

(1) Fire Pump Motors and Other Equipment. Conductors supplying fire pump motors and accessory equipment must be sized no less than 125 percent of the sum of the motor full-load current as listed in Tables 430.248 or 430.250, plus 100 percent of the ampere rating of the fire pump's accessory equipment.

(2) Fire Pump Motors Only. Conductors supplying a single fire pump motor must be sized to the requirements of 430.22.

Author's Comment: This means that the branch-circuit conductors to a single fire pump motor must have an ampere rating of at least 125 percent of the fire pump motor full-load current (FLC), as listed in Table 430.248 or 430.250.

Voltage-Drop Calculations — Unit 8

▶ Fire Pump Conductor Sizing Example

Question: What size conductor is required for a 25 hp, 208V, three-phase fire pump motor?

(a) 4 AWG   (b) 3 AWG   (c) 2 AWG   (d) 1 AWG

Answer: (b) 3 AWG

Conductors are sized at 125 percent of the motor's FLC as listed in Table 430.250.

FLC of 25 hp = 74.80A [Table 430.250]

Conductor = 74.80A x 1.25

Conductor = 93.50A

3 AWG is rated 100A at 75°C

Note: The fire pump motor circuit overcurrent device size must be set to carry indefinitely the sum of the locked-rotor current of the fire pump motor. According to Table 430.251(B), the locked-rotor current of a 25 hp, 208V, three-phase motor is 404A.

CAUTION: The branch-circuit conductors for a fire pump motor must be sized to accommodate the voltage-drop requirements of 695.7.

Voltage Drop [695.7]

Controller. The voltage at the line terminals of the controller, when the motor starts (locked-rotor current), must not drop more than 15 percent below the controller's rated voltage. In addition, the voltage at the motor terminals must not drop more than 5 percent below the voltage rating of the motor when the motor operates at 115 percent of the fire pump full-load current rating.

The locked-rotor current (LRC) of a motor can range from 5 times the motor FLC to as high as 14 times the motor FLC. Typically, motors fall in the range of 5.50 and 7.50 times the FLC. Table 430.251(B) lists three-phase motor locked-rotor currents. For the purposes of this unit, LRC is considered to be 6 times the motor FLC.

▶ Fire Pump Example—Three-Phase

Question: A 25 hp, 208V, three-phase fire pump motor is located 175 ft from the service. The fire pump motor controller is located 150 ft from the service. To accommodate conductor voltage drop, what size conductor must be installed to the fire pump motor controller? **Figure 8–27**

(a) 4 AWG   (b) 3 AWG   (c) 2 AWG   (d) 1 AWG

(continued in next column)

Answer: (b) 3 AWG

VD = Motor Locked-Rotor Current x Conductor Impedance

VD = Cmil = (1.732 x Z x LRC x D)/VD

Cmil = Conductor cross-sectional area [Chapter 9, Table 8]

Z = Conductor impedance, 12.90 ohms, copper

Locked-Rotor Current = FLC x 6

FLC = 74.80A [Table 430.250]

LRC = 74.80 x 6

LRC = 449

Locked-Rotor Current = 449 LRC

D = 150 ft

VD (15% of Source Voltage) = 208V x 0.15

VD (15% of Source Voltage) = 31.20V

Cmil = (1.732 x 12.90 ohms x 449A x 150 ft)/31.20V

Cmil = 48,230 cmil

3 AWG is 52,620 cmil, so is large enough [Chapter 9, Table 8]

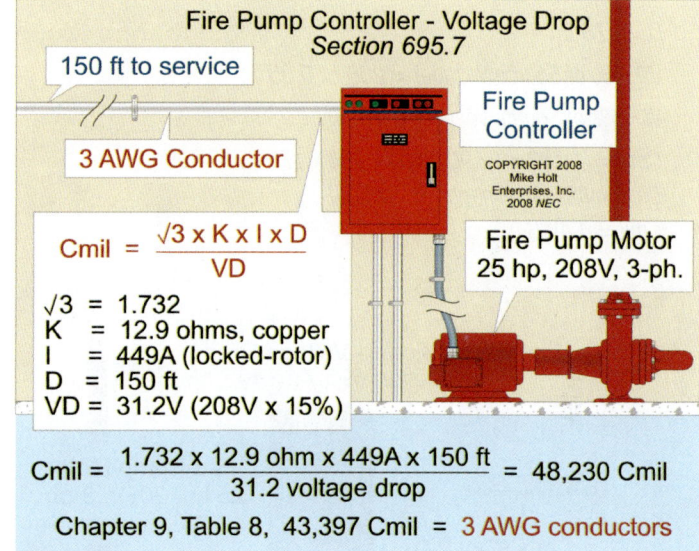

Figure 8–27

Author's Comment: Where ungrounded conductors are increased in size, equipment grounding conductors (where installed) must be proportionally increased in size according to the circular mil area of the ungrounded conductors [250.122(B)].

Unit 8 — Voltage-Drop Calculations

▶ **Example—Three-Phase**

Question: A 25 hp, 208V, three-phase fire pump motor is located 175 ft from the service. The fire pump motor controller is located 150 ft from the service. To accommodate conductor voltage drop, what size conductors must be installed to the fire pump motor if it's located 25 ft from the controller?
Figure 8–28

(a) 4 AWG (b) 3 AWG (c) 2 AWG (d) 1 AWG

Answer: (a) 4 AWG

The operating voltage at the terminals of the motor must not drop more than 5 percent below the voltage rating of the motor while the motor is operating at 115 percent of the full-load current rating of the motor.

Cmil = (1.732 x K x I x D)/VD
Cmil = Conductor size [Chapter 9, Table 8]
K = 12.90 ohms, copper
I (115% of Motor FLC) = 74.80A x 1.15 [Table 430.250]
I (115% of Motor FLC) = 86A
D = 175 ft
VD (5% of Source Voltage) = 208V x 0.05
VD (5% of Source Voltage) = 10.40V
Cmil = (1.732 x 12.90 ohms x 86A x 175 ft)/10.40V
Cmil = 32,332 [Chapter 9, Table 8]
Cmil = 4 AWG

CAUTION: *For voltage drop, the 4 AWG conductor is okay from the controller to the motor, but 695.6(C)(2) requires the branch-circuit conductors to be sized not smaller than 3 AWG.*
Figure 8–29 *and* Figure 8–30

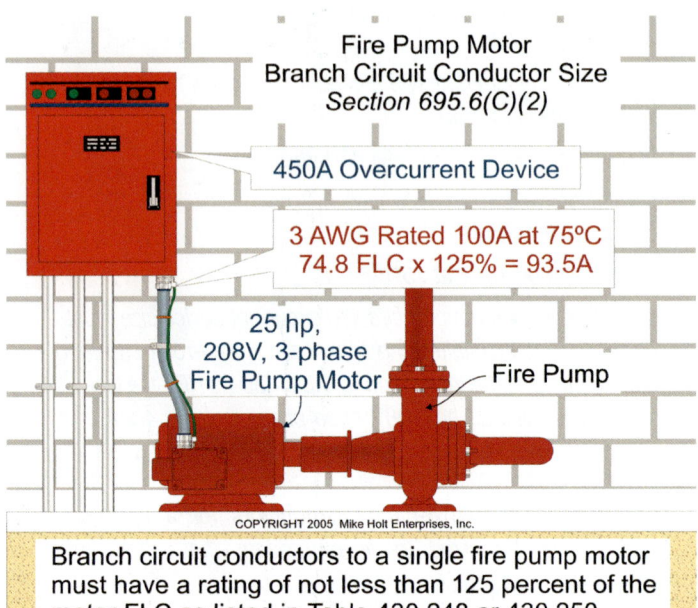

Branch circuit conductors to a single fire pump motor must have a rating of not less than 125 percent of the motor FLC as listed in Table 430.248 or 430.250.

Figure 8–29

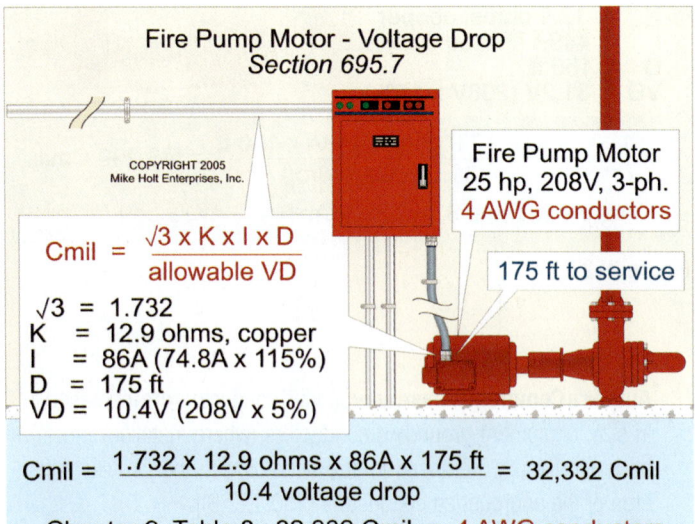

Figure 8–28

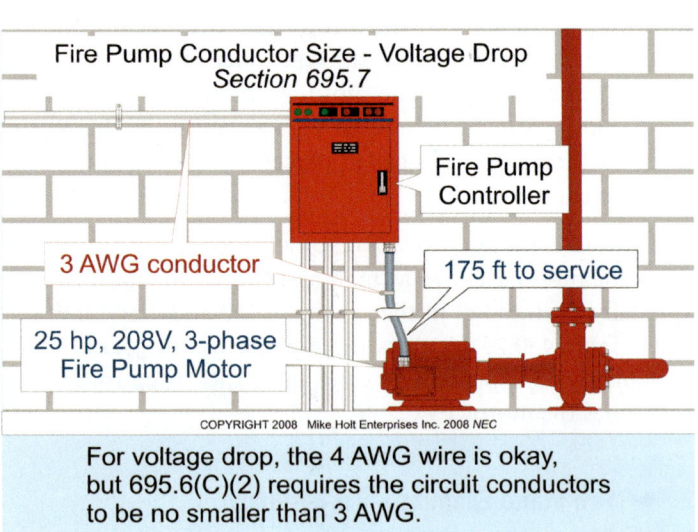

For voltage drop, the 4 AWG wire is okay, but 695.6(C)(2) requires the circuit conductors to be no smaller than 3 AWG.

Figure 8–30

232 Mike Holt's Illustrated Guide to NEC Exam Preparation

Voltage-Drop Calculations — Unit 8

8.13 Sizing Conductors with Loads at Different Distances

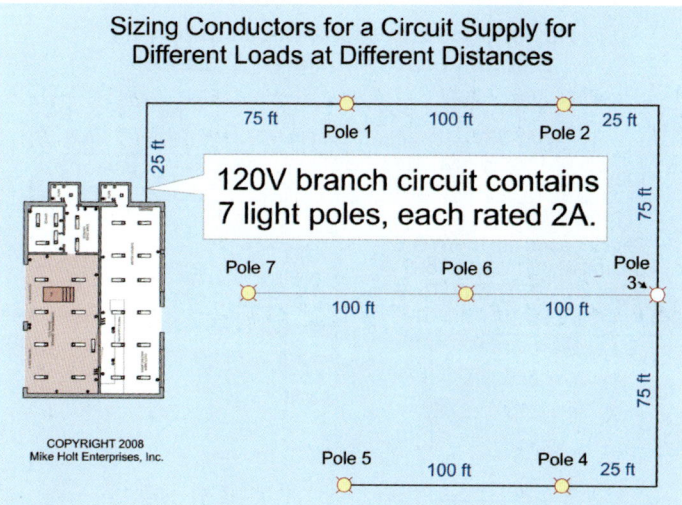

Figure 8–31

▶ **Advanced Example**

Question: Size the conductors to limit the voltage drop for seven light poles. Each luminaire is rated 120V, 2A and the poles are spaced as shown in **Figure 8–31**.

Step 1: Determine minimum circular mil area of conductor needed to supply each luminaire.

$Cmil = (2 \times K \times I \times D)/E_{vd}$

$K = 12.90$ ohms, copper

$I = 2A$

$D = 100$ ft

$E_{vd} = (120V \times 3\%) = 3.60V$

$Cmil = (2 \times 12.90 \times 2A \times 100 ft)/3.60V$

$Cmil = 1,433$ cmil

The first 100 ft run will need to be sized to carry the load of all 7 poles, so the wire must be sized at least 7 x 1,433 cmil

$Cmil = 7 \times 1,433$ cmil

$Cmil = 10,033$ cmil

10 AWG is 10,380 cmil [Chapter 9, Table 8]

Step 2: Determine the voltage drop of each segment of the circuit.

$E_{vd1} = I \times R$

Resistance of 10 AWG = 1.24 ohms per 1,000 ft [Chapter 9, Table 8]

Find the resistance of 100 ft of 10 AWG:

$R = (1.24$ ohms$/1,000) \times 100$ ft

$R = 0.124$ ohms

R for two wires = 0.124 x 2

$R = 0.248$ **Figure 8–32**

The voltage drop for the first segment will be:

$E_{vd1} = I \times R$

$E_{vd1} = 14A \times 0.248$ (two wires)

$E_{vd1} = 3.47V$

For each following segment, reduce the amperage to match the number of light poles remaining on that branch. For the second segment, the load will be less by 2A since we are past the first luminaire, so the voltage drop will be: **Figure 8–33**

$E_{vd}2 = 12A \times 0.248$

(continued in next column)

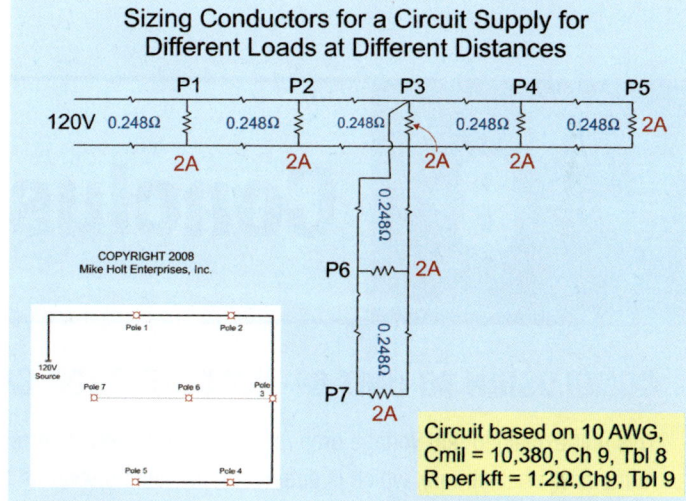

Figure 8–32

$E_{vd2} = 2.98V$

$E_{vd3} = 10A \times 0.248$

$E_{vd3} = 2.48V$

$E_{vd4} = 4A \times 0.248$

$E_{vd4} = 0.90V$

$E_{vd5} = 2A \times 0.248$

$E_{vd5} = 0.50V$

$E_{vd6} = 4A \times 0.248$

$E_{vd6} = 0.90V$

$E_{vd7} = 2A \times 0.248$

$E_{vd7} = 0.50V$

Unit 8 | Voltage-Drop Calculations

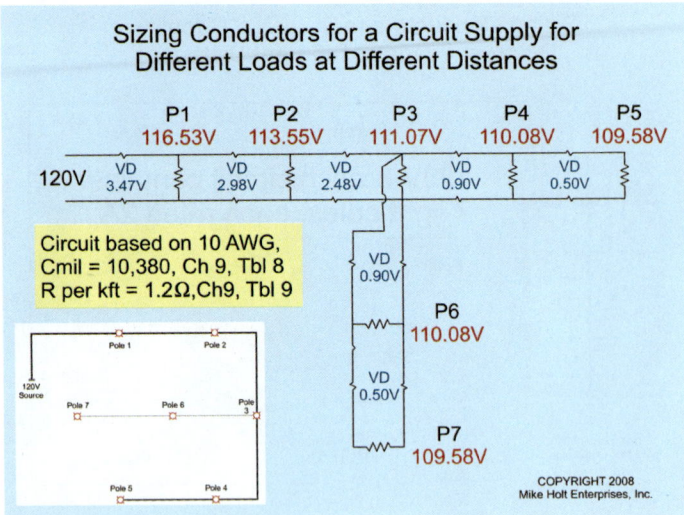

Figure 8–33

UNIT 8 Conclusion

CONCLUSION TO UNIT 8—VOLTAGE-DROP CALCULATIONS

This unit explained that voltage drop results anytime there's current flowing through a resistor. Every conductor has resistance (opposition to the flow of electrons), which is determined by factors such as the conductor material, cross-sectional area, length, and operating temperature. Excessive voltage drop results in power losses and possibly under-operating voltage for electrical equipment.

The *Code* provides specific voltage-drop requirements in some special cases, such as for fire pumps, but doesn't generally require conductors to be sized to limit conductor voltage drop. Manufacturers' instructions may require a minimum operating voltage, which means that voltage drop must be considered. In this unit, you learned how to correctly size conductors in order to stay within the *NEC* recommendations for voltage drop.

Transpositions of voltage-drop formulas were reviewed in this unit so that you'll be able to select the proper formula to solve for various unknowns. You now have the tools needed to properly install circuits for proper equipment operation.

If you're not yet fully comfortable with voltage-drop calculations, go back and review this unit carefully and see how each voltage-drop formula relates to Ohm's Law. It all can be traced back to the basics.

Unit 8 Practice Questions

UNIT 8—VOLTAGE-DROP CALCULATIONS PRACTICE QUESTIONS

(•Indicates that 75% or fewer of those who took this exam answered the question correctly.)

PART A—CONDUCTOR RESISTANCE CALCULATIONS

8.1 Conductor Resistance

1. The larger the cross-sectional area of a conductor, the _____ the number of paths for electrons, which lowers the resistance of the conductor.

   (a) greater
   (b) fewer

2. Conductor resistance is determined by the _____.

   (a) material type
   (b) cross-sectional area
   (c) conductor length
   (d) all of these

3. _____ is the best conductor, better than gold, but the high cost limits its use to special applications, such as fuse elements and some switch contacts.

   (a) Silver
   (b) Copper
   (c) Aluminum
   (d) none of these

4. _____ conductors are often used when weight or cost are important considerations.

   (a) Silver
   (b) Copper
   (c) Aluminum
   (d) none of these

5. •Conductor cross-sectional area is expressed in AWG or _____.

   (a) square inches
   (b) mils
   (c) circular mils
   (d) none of these

6. The resistance of a conductor is directly proportional to its length.

   (a) True
   (b) False

7. The resistance of a conductor changes with temperature. Temperature coefficient describes the effect that temperature has on the resistance of a conductor. Conductors with a _____ temperature coefficient have an increase in resistance with an increase in temperature.

   (a) positive
   (b) negative
   (c) neutral
   (d) none of these

8.2 Conductor Resistance—Direct-Current Circuits [Chapter 9, Table 8]

8. The *National Electrical Code* lists the resistance and area in circular mils for both dc and ac conductors. DC conductor resistances are listed in Chapter 9, Table _____ and ac conductor resistances are listed in Chapter 9, Table _____.

   (a) 1, 5
   (b) 3, 4
   (c) 9, 8
   (d) 8, 9

Unit 8 — Practice Questions

9. What is the dc resistance of a 6 AWG copper conductor 400 ft long?

   (a) 0.20 ohms
   (b) 0.30 ohms
   (c) 0.40 ohms
   (d) 0.50 ohms

10. What is the dc resistance of a 1 AWG copper conductor 200 ft long?

    (a) 0.0308 ohms
    (b) 0.0311 ohms
    (c) 0.0423 ohms
    (d) 0.0564 ohms

11. What is the dc resistance of a 3 AWG aluminum conductor 1,000 ft long?

    (a) 0.231 ohms
    (b) 0.313 ohms
    (c) 0.403 ohms
    (d) 0.422 ohms

12. What is the dc resistance of a 1/0 AWG aluminum conductor 800 ft long?

    (a) 0.08 ohms
    (b) 0.16 ohms
    (c) 0.23 ohms
    (d) 0.56 ohms

13. What is the dc resistance of a 1 AWG copper conductor 100 ft long?

    (a) 0.0154 ohms
    (b) 0.162 ohms
    (c) 0.233 ohms
    (d) 0.561 ohms

14. What is the dc resistance of a 3 AWG aluminum conductor 500 ft long?

    (a) 0.0231 ohms
    (b) 0.1614 ohms
    (c) 0.233 ohms
    (d) 0.2015 ohms

15. What is the dc resistance in ohms for three 300 kcmil conductors in a parallel run that is 1,000 ft long?

    (a) 0.014 ohms
    (b) 0.026 ohms
    (c) 0.047 ohms
    (d) 0.052 ohms

8.3 Conductor Resistance—Alternating-Current Circuits

16. The intensity of a magnetic field depends on the intensity of ac. The greater the current flow, the greater the overall magnetic field.

    (a) True
    (b) False

17. In ac circuits, the expanding and collapsing magnetic field within the conductor induces an electromotive force that opposes the flow of ac. This opposition to the flow of ac is called inductive reactance.

    (a) True
    (b) False

18. _____ currents are small independent currents that are produced as a result of the expanding and collapsing magnetic field. They flow erratically within the conductor opposing current flow and consuming power.

    (a) Lenz
    (b) Ohm's
    (c) Eddy
    (d) Kirchoff's

19. The expanding and collapsing magnetic field induces a counter voltage within the conductors, which repels the flowing electrons toward the conductor surface. This is known as _____ effect.

    (a) inductive
    (b) skin
    (c) surface
    (d) watt

8.4 Alternating-Current Resistance

20. The opposition to current flow is greater for ac circuits than for dc circuits because of _____.

    (a) eddy currents
    (b) skin effect
    (c) inductive reactance
    (d) all of these

21. The ac conductor resistances listed in Chapter 9, Table 9 of the *NEC* are different for copper and aluminum and for nonmagnetic and magnetic raceways.

    (a) True
    (b) False

22. What is the ac ohms-to-neutral resistance for a 2/0 AWG copper conductor 300 ft long?

    (a) 0.03 ohms
    (b) 0.04 ohms
    (c) 0.05 ohms
    (d) 0.06 ohms

23. What is the ac ohms-to-neutral resistance for a 500 kcmil copper conductor installed in an aluminum raceway 1,000 ft long?

    (a) 0.027 ohms
    (b) 0.029 ohms
    (c) 0.030 ohms
    (d) 0.032 ohms

24. What is the ac ohms-to-neutral resistance for a 3 AWG copper conductor 100 ft long?

    (a) 0.012 ohms
    (b) 0.025 ohms
    (c) 0.33 ohms
    (d) 0.43 ohms

25. What is the ac ohms-to-neutral resistance of a 500 kcmil copper conductor installed in PVC conduit 400 ft long?

    (a) 0.0108 ohms
    (b) 0.0204 ohms
    (c) 0.0333 ohms
    (d) 0.0431 ohms

26. A 2-wire circuit supplies a 36A load that is located 100 ft from the panelboard. The load is wired with 1 AWG aluminum in PVC conduit. What is the total ac ohms-to-neutral resistance of the circuit conductors?

    (a) 0.05 ohms
    (b) 0.25 ohms
    (c) 0.50 ohms
    (d) 0.62 ohms

27. What size copper conductors in a steel raceway can be used to replace 1/0 AWG aluminum that supplies a 110A load? Hint: find a copper conductor with the same resistance per foot as the existing aluminum conductor.

    Note: We do not want to increase the circuit voltage drop.

    (a) 3 AWG
    (b) 2 AWG
    (c) 1 AWG
    (d) 1/0 AWG

28. What is the circuit voltage if the conductor voltage drop is 3.30V?

    Note: Assume 3 percent voltage drop.

    (a) 110V
    (b) 115V
    (c) 120V
    (d) 240V

29. •What is the ac ohms-to-neutral resistance for three 1/0 AWG aluminum conductors run in parallel for 1,000 ft?

    (a) 0.067 ohms
    (b) 0.114 ohms
    (c) 0.231 ohms
    (d) 0.413 ohms

PART B—VOLTAGE-DROP CONSIDERATIONS

8.6 *NEC* Voltage-Drop Recommendations

30. Because of high demand for electricity, utilities sometimes are forced to reduce their output voltage. In addition, transformers, services, feeders, and branch-circuit conductors oppose the flow of current. The opposition to current flow results in voltage drop. All circuits have voltage drop, simply because all conductors have resistance.

    (a) True
    (b) False

31. When sizing conductors for feeders and branch circuits, the *NEC* _____ that we take voltage drop into consideration.

    (a) assumes
    (b) suggests
    (c) requires
    (d) demands

32. •_____ equipment such as motors and electromagnetic ballasts can overheat at reduced voltage. This results in reduced equipment operating life and inconvenience to the customer.

    (a) Inductive
    (b) Electronic
    (c) Resistive
    (d) all of these

Unit 8 — Practice Questions

33. •_____ equipment such as computers, laser printers, copy machines, etc., can suddenly power down because of reduced voltage, resulting in data losses.

    (a) Inductive
    (b) Electronic
    (c) Resistive
    (d) all of these

34. When a conductor resistance causes the voltage to be dropped below an acceptable point, the conductor size should be increased.

    (a) True
    (b) False

35. How can conductor voltage drop be reduced?

    (a) Reduce the conductor resistance.
    (b) Increase the conductor size.
    (c) Decrease the conductor length.
    (d) all of these

36. If the branch-circuit supply voltage is 208V, the maximum recommended voltage drop of the circuit should not be more than _____.

    (a) 3.60V
    (b) 6.24V
    (c) 6.90V
    (d) 7.20V

37. If the feeder supply voltage is 240V, the maximum recommended voltage drop of the feeder should not be more than _____.

    (a) 3.60V
    (b) 6.24V
    (c) 6.90V
    (d) 7.20V

8.7 Determining Circuit Conductors' Voltage Drop—Ohm's Law Method

38. What is the voltage drop of two 12 AWG conductors supplying a 12A continuous load?

    Note: The continuous load is located 100 ft from the power supply.

    (a) 3.20V
    (b) 4V
    (c) 4.76V
    (d) 12.80V

8.8 Determining Circuit Conductors' Voltage Drop—Formula Method

39. A 240V, 24A, single-phase load is located 160 ft from the panelboard. The load is wired with 10 AWG. What is the approximate voltage drop of the branch-circuit conductors?

    (a) 3.20V
    (b) 4.25V
    (c) 5.90V
    (d) 9.50V

40. A 100A, three-phase load is located 100 ft from the panelboard and is wired with 1 AWG aluminum. What is the approximate voltage drop of the circuit conductors?

    (a) 3V
    (b) 3.50V
    (c) 4.40V
    (d) 5V

8.9 Sizing Conductors to Prevent Excessive Voltage Drop

41. A single-phase, 5 hp motor is located 110 ft from a panelboard in a dry location. The nameplate indicates that the voltage is 115V and the FLA is 52/26A. What size conductor is required?

    Note: Apply the NEC recommended voltage-drop limit for this branch circuit.

    (a) 10 AWG
    (b) 8 AWG
    (c) 6 AWG
    (d) 3 AWG

42. •A single-phase, 5 hp motor is located 110 ft from a panelboard. The nameplate indicates that the voltage is 115/230 and the FLA is 26A. What size conductor is required?

    Note: Apply the NEC recommended voltage-drop limits.

    (a) 10 AWG
    (b) 8 AWG
    (c) 6 AWG
    (d) 4 AWG

Practice Questions — Unit 8

43. A 480V, 18A, three-phase load is located 300 ft from the panelboard. What size copper conductors are required to prevent the voltage drop from exceeding 3 percent?

    (a) 10 AWG
    (b) 8 AWG
    (c) 6 AWG
    (d) 4 AWG

8.10 Limiting Conductor Length to Minimize Voltage Drop

44. What is the approximate distance that a 240V, 31A, single-phase load can be located from the panelboard so the voltage drop does not exceed 3 percent? The load is wired with 8 AWG copper.

    (a) 55 ft
    (b) 110 ft
    (c) 145 ft
    (d) 220 ft

45. What is the approximate distance a 480V, 45A, three-phase transformer, wired with 6 AWG copper can be located from the panelboard so the voltage drop does not exceed 3 percent?

    (a) 250 ft
    (b) 300 ft
    (c) 375 ft
    (d) 400 ft

8.11 Limiting Current to Limit Voltage Drop

46. An existing installation consists of 1/0 AWG copper conductors in a nonmetallic raceway to a panelboard located 200 ft from a 240V, single-phase power source. What is the maximum load that can be placed on the panelboard so that the *NEC* recommendations for voltage drop are not exceeded?

    (a) 71A
    (b) 94A
    (c) 109A
    (d) 147A

47. An existing installation contains 2 AWG aluminum conductors in an aluminum raceway to a panelboard located 350 ft from a 480V, three-phase power source. What is the maximum load the conductors can carry without exceeding the *NEC* recommendation for voltage drop?

    (a) 49A
    (b) 64A
    (c) 74A
    (d) 85A

8.12 Fire Pump Motor Circuits

48. Feeder conductors supplying fire-pump motors and accessory equipment must be sized no less than ____ percent of the sum of the motor full-load currents as listed in Article 430, plus 100 percent of the ampere rating of the fire-pump accessory equipment.

    (a) 100
    (b) 125
    (c) 250
    (d) 600

49. The voltage at the line terminals of a fire-pump motor controller must not drop more than ____ percent below the controller's normal rated voltage under motor-starting conditions.

    (a) 5
    (b) 10
    (c) 15
    (d) 20

50. When a fire-pump motor is operating at 115 percent of its full-load current rating, the supply voltage at the motor terminals must not drop more than ____ percent below the voltage rating of the motor.

    (a) 5
    (b) 10
    (c) 15
    (d) 20

Unit 8 Challenge Questions

UNIT 8—VOLTAGE-DROP CALCULATIONS CHALLENGE QUESTIONS

PART A—CONDUCTOR RESISTANCE CALCULATIONS

8.1 Conductor Resistance

1. The resistance of a conductor is affected by temperature change. This is called the _____.

   (a) temperature correction factor
   (b) temperature coefficient
   (c) ambient temperature factor
   (d) none of these

8.3 Conductor Resistance—Alternating-Current Circuits

2. The total opposition to current flow in an ac circuit is expressed in ohms and is called _____.

   (a) impedance
   (b) conductance
   (c) reluctance
   (d) resistance

8.4 Alternating-Current Resistance

3. A 240V, 40A, single-phase load is located 150 ft from an existing junction box. The junction box is located 50 ft from the panelboard and is wired with 4 AWG aluminum wire. The total resistance of the two 4 AWG conductors from the panelboard to the junction box is approximately _____.

   (a) 0.03 ohms
   (b) 0.04 ohms
   (c) 0.05 ohms
   (d) 0.08 ohms

4. A load is located 100 ft from a 230V power supply and is wired with 4 AWG aluminum conductors. What size copper conductor can be used to replace the aluminum conductors and not increase the conductor voltage drop?

   (a) 8 AWG
   (b) 6 AWG
   (c) 2 AWG
   (d) 1/0 AWG

PART B—VOLTAGE-DROP CALCULATIONS

8.7 Determining Circuit Conductors' Voltage Drop—Ohm's Law Method

5. A 40A, 240V rated single-phase load is 150 ft from a junction box and the junction box is located 50 ft from a panelboard (for a total of 200 ft). If the voltage at the panelboard is 240V, what is the minimum voltage recommended by the *NEC* at the 40A load?

   (a) 117.70V
   (b) 228.20V
   (c) 232.80V
   (d) 236.20V

8.8 Determining Circuit Conductors' Voltage Drop—Formula Method

6. What is the voltage drop of two 4 AWG aluminum conductors that supply a 5 hp, 115V, single-phase motor that has a nameplate rating of 55A? The motor is located 95 ft from the power supply.

   (a) 3.25V
   (b) 5.31V
   (c) 6.24V
   (d) 7.26V

Challenge Questions — Unit 8

8.9 Sizing Conductors to Prevent Excessive Voltage Drop

7. A 240V, 40A, single-phase load is located 150 ft from an existing junction box. The junction box is located 50 ft from the panelboard. When the 40A load is on, the voltage at the junction box is calculated to be 236V. The NEC recommends that the voltage drop for this branch circuit not exceed 3 percent of the 240V source (7.20V). What size copper conductor can be installed from the junction box to the load and still meet the NEC recommendations for branch circuits?

    (a) 6 AWG
    (b) 3 AWG
    (c) 1 AWG
    (d) 1/0 AWG

8. An existing junction box is located 65 ft from the panelboard and contains 4 AWG aluminum conductors. What size copper conductor can be used to extend this circuit 85 ft and supply a 50A, 208V load?

    Note: Apply the NEC recommended voltage-drop limits.

    (a) 10 AWG
    (b) 8 AWG
    (c) 6 AWG
    (d) 4 AWG

8.10 Limiting Conductor Length to Minimize Voltage Drop

9. How far can a 230V, 50A, three-phase load be located from a panelboard if it is fed with 3 AWG conductors and still meet the NEC recommendations for voltage drop for branch circuits?

    (a) 275 ft
    (b) 300 ft
    (c) 325 ft
    (d) 350 ft

8.11 Limiting Current to Limit Voltage Drop

10. Two 8 AWG copper conductors supply a 120V load that is located 225 ft from the panelboard. What is the maximum load in amperes that can be applied to these conductors without exceeding the NEC recommendation on conductor voltage drop?

    (a) 0A
    (b) 5A
    (c) 10A
    (d) 15A

UNIT 8 Notes

UNIT 9 — Dwelling Unit Calculations

INTRODUCTION TO UNIT 9—DWELLING UNIT CALCULATIONS

Unit 9 focuses on sizing the service for a one-family dwelling unit. The *NEC* defines a dwelling unit as a single unit, providing complete and independent living facilities for one or more persons, including permanent provisions for living, sleeping, cooking, and sanitation. Figure 9–1

The *Code* provides two different calculation methods for this: standard and optional. Both methods are explained in this unit and you should be familiar with both for exam purposes. A good grasp of the material in this unit is important as a basis for the following units which will address multifamily dwellings and commercial calculations, so be sure to study the examples and complete the questions at the end to solidify your understanding.

Sizing a service isn't simply a matter of adding up all of the loads or all of the circuit breaker sizes. The *NEC* recognizes that not all loads will be used simultaneously under normal living conditions, so there are a number of "demand factors" that come into play in sizing an electrical service. It's important to be familiar with all of the requirements of Article 220 so the proper demand factors will be applied to a given installation. There are numerous tables and requirements to follow, but take them one at a time and you'll be able to complete these calculations.

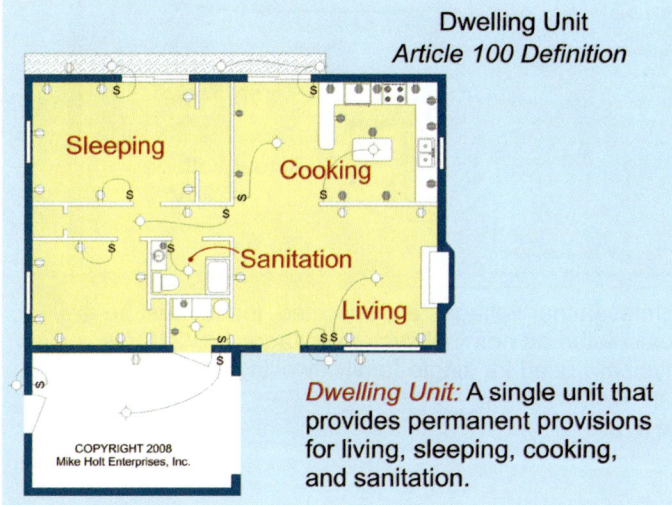

Figure 9–1

PART A—GENERAL REQUIREMENTS

Introduction

The *NEC* contains detailed requirements for calculating service and feeder loads in many types of occupancies (buildings). Although most of these are located in Article 220, other rules affecting service and feeder loads are scattered throughout the *Code*. When performing load calculations, dwelling units are a special class of occupancy. In the first place, the *Code* permits two different general approaches for load calculations: standard and optional. When using either of these general methods, the rules for dwelling unit calculations also contain a number of demand factors that can't be used in other types of buildings.

9.1 General Requirements

Article 220 provides the requirements for residential branch circuits, feeders, and service calculations [220.1]. Other important articles include: Branch Circuits—210, Feeders—215, Services—230, Overcurrent Protection—240, Wiring Methods—300, Conductors—310, Appliances—422, Electric Space-Heating Equipment—424, Motors—430, and Air-Conditioning Equipment—440.

Unit 9 — Dwelling Unit Calculations

9.2 Voltages [220.5(A)]

Unless other voltages are specified, branch-circuit, feeder, and service loads must be calculated using the nominal system voltage such as 120, 120/240, 120/208, 240, 347, 277/480, 480, 347/600, or 600. For single-family dwelling unit calculations, the nominal voltage is typically 120/240V. **Figure 9–2**

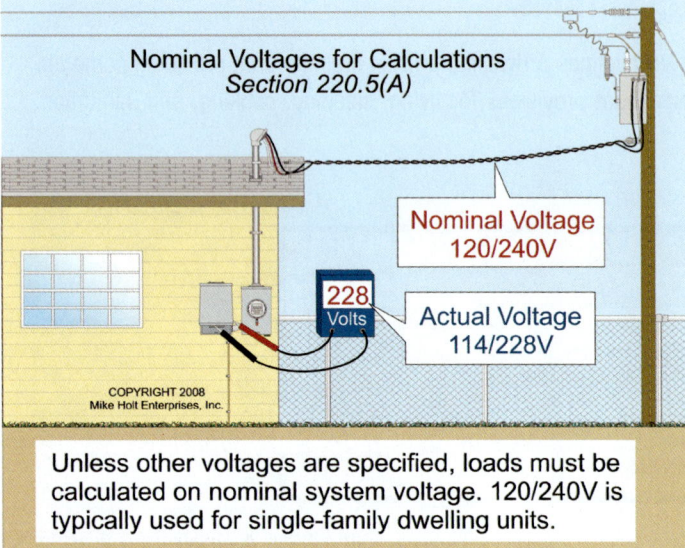

Figure 9–2

Author's Comments:

- A nominal value is assigned to a circuit for the purpose of conveniently designating its voltage class. The actual voltage at which a circuit operates can vary from the nominal within a range that permits satisfactory operation of equipment [100].

- Motor VA is based on motor table voltage and current values, such as 115V, 230V, or 460V—not 120V, 240V, or 480V [430.248 and 430.250].

9.3 Fraction of an Ampere [220.5(B)]

Where calculations result in a fraction of less than 0.50A, the fractions can be dropped.

Author's Comment: When do you round—after each calculation, or at the final calculation? The *NEC* isn't specific on this issue, so I guess it all depends on the answer you want to see!

Rounding

The *NEC* defines a continuous load as a load where the maximum current is expected to continue for 3 hours or more [Article 100]. When sizing branch-circuit conductors and overcurrent devices for a continuous load, the continuous load is first taken times 125 percent [210.19(A)(1) and 210.20(A)]. **Figure 9-3**

According to 424.3(B), fixed electric heating is considered to be a continuous load, and therefore the branch-circuit conductors and overcurrent device for electric space-heating equipment must be sized no less than 125 percent of the total load.

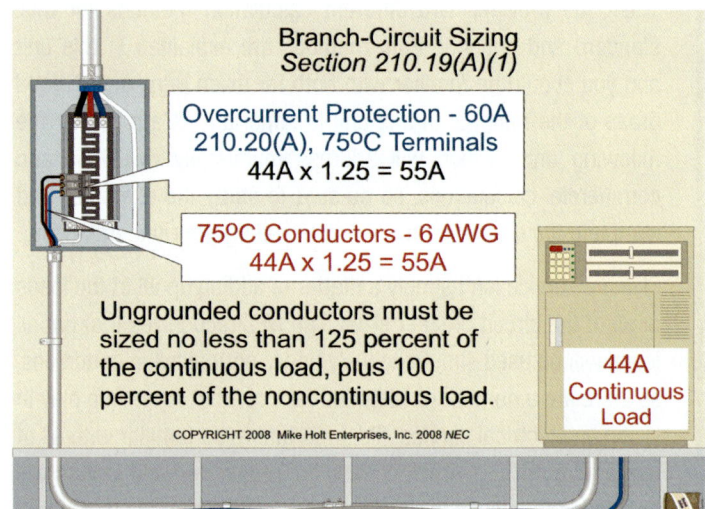

Figure 9–3

▶ **Electric Space Heating Conductor Sizing Example**

Question: What size conductor is required to supply a 9 kVA (37.50A), 240V, single-phase fixed space heater with a 3A blower motor if equipment terminals are rated 75°C? **Figure 9–4**

(a) 10 AWG   (b) 8 AWG   (c) 6 AWG   (d) 4 AWG

Answer: (c) 6 AWG

Step 1: Determine the total load.

$I = VA/E$

$I = 9{,}000\ VA/240V$

$I = 37.50A$

(continued in next column)

Dwelling Unit Calculations — Unit 9

Step 2: The conductor is sized at 125 percent of the load [210.19(A)(1)].

Conductor Size = (37.50A + 3A) x 1.25

Conductor Size = 50.63A, round up to 51A

If we rounded down, then 8 AWG rated 50A at 75°C could be used, but since we have to round up, 6 AWG rated 65A at 75°C is required.

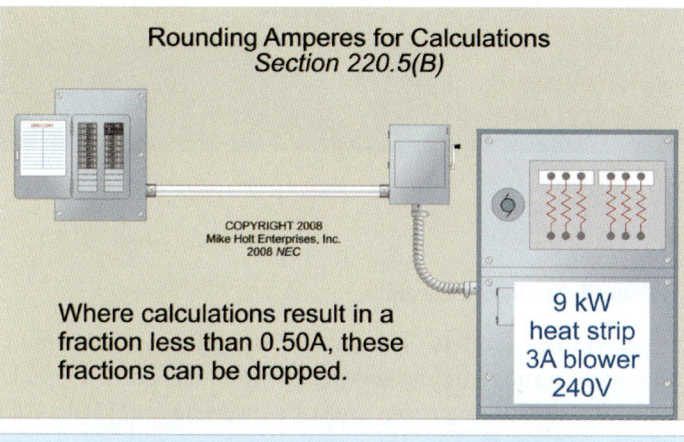

Figure 9–4

Author's Comment: Receptacle outlets in a dwelling unit must be installed so that no point measured horizontally along the floor line in any wall space is more than 6 ft from a receptacle outlet [210.52(A)(1)]. Switching one receptacle of a duplex receptacle can meet the lighting requirements of 210.70(A)(1) and the receptacle placement requirements of this section. **Figure 9-5**

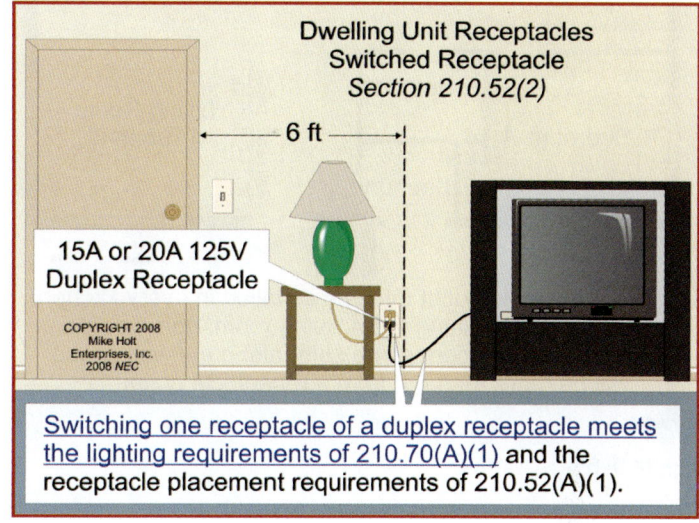

Figure 9–5

9.4 Small-Appliance Circuits [210.11(C)(1)]

The rules for receptacle outlet placement in a dwelling unit are included in 210.52.

Dwelling Unit Receptacle Outlet Requirements [210.52]

Receptacles rated 15A and 20A, 125V must be installed as required in (A) through (H), and the receptacles required by this section are in addition to any receptacle that is:

Author's Comment: Circuits are rated 120V and receptacles are rated 125V.

(1) Part of a luminaire or appliance,

(2) Controlled by a wall switch to meet the illumination requirements of 210.70(A)(1) Ex 1,

(3) Located within cabinets or cupboards, or

(4) Located more than 5½ ft above the floor.

(A) General Requirements—Dwelling Unit. A receptacle outlet must be installed in every kitchen, family room, dining room, living room, sunroom, parlor, library, den, bedroom, recreation room, and similar room or area in accordance with (1), (2), and (3): **Figure 9-6**

(1) Receptacle Placement. A receptacle outlet must be installed so that no point along the wall space is more than 6 ft, measured horizontally along the floor line, from a receptacle outlet.

Author's Comment: The purpose of this rule is to ensure that a general-purpose receptacle is conveniently located to reduce the chance that an extension cord will be used.

(2) Definition of Wall Space. Figure 9-7

(1) Any space 2 ft or more in width, unbroken along the floor line by doorways, fireplaces, and similar openings.

Unit 9 — Dwelling Unit Calculations

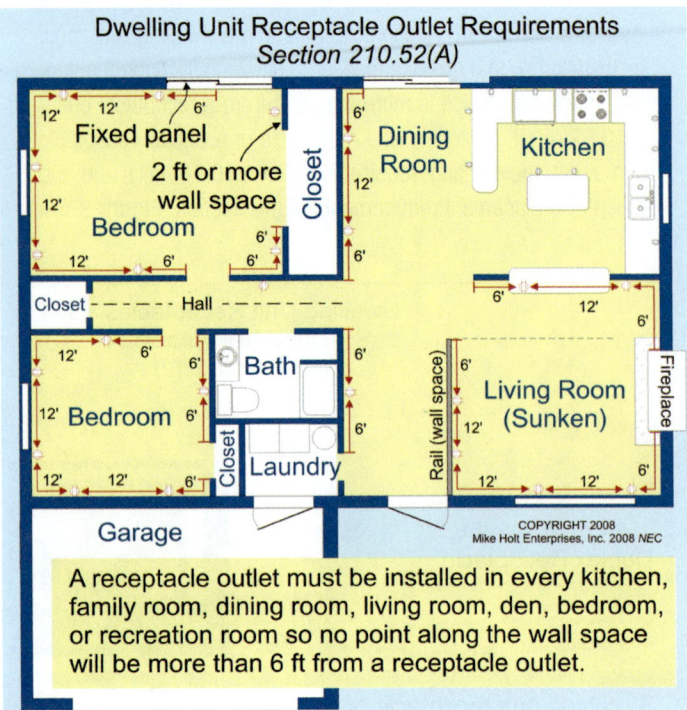

Figure 9–6

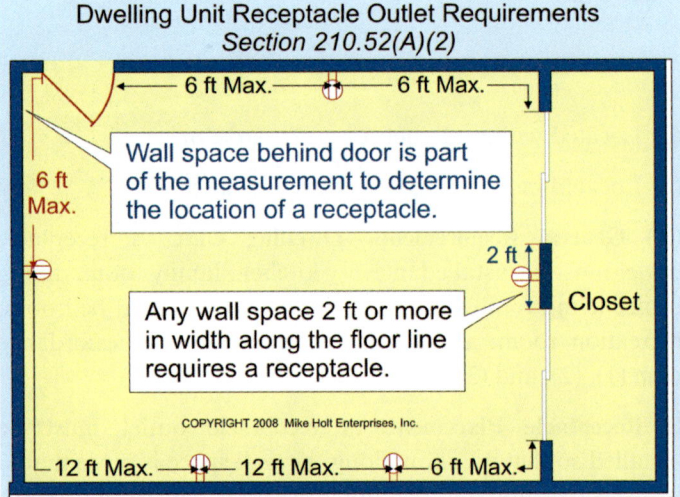

Figure 9–7

(2) The space occupied by fixed panels in exterior walls.

(3) The space occupied by fixed room dividers, such as free-standing bar-type counters or guard rails.

(3) Floor Receptacle Outlets. Floor receptacle outlets are not counted as the required receptacle wall outlet if they are located more than 18 in. from the wall. **Figure 9-8**

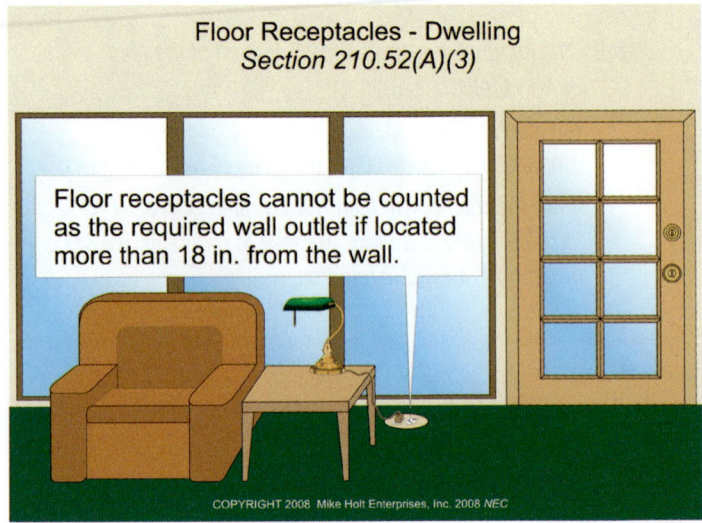

Figure 9–8

(B) Small-Appliance Circuits.

(1) Receptacle Outlets. The two or more 20A, 120V small-appliance branch circuits serving the kitchen, pantry, breakfast room, and dining room area of a dwelling unit [210.11(C)(1)] must serve all wall, floor, and countertop receptacle outlets [210.52(C)], and the receptacle outlet for refrigeration equipment. **Figure 9-9**

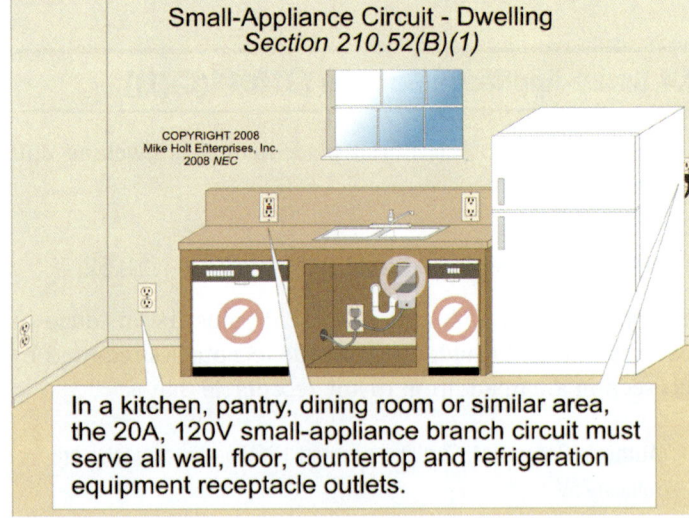

Figure 9–9

Exception No. 2: The receptacle outlet for refrigeration equipment can be supplied from an individual branch circuit rated 15A or greater. **Figure 9-10**

246   Mike Holt's Illustrated Guide to NEC Exam Preparation

Dwelling Unit Calculations — Unit 9

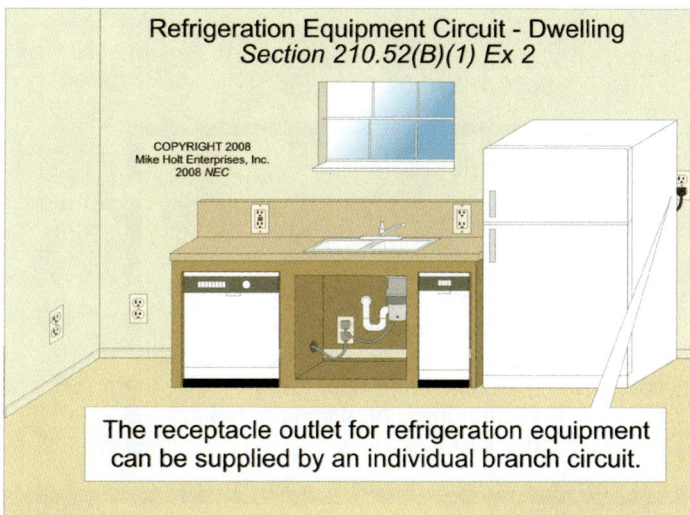

Figure 9–10

(2) Not Supply Other Outlets. The 20A, 120V small-appliance circuits required by 210.11(C)(1) must not supply outlets for luminaires or appliances.

Exception No. 1: The 20A, 120V small-appliance branch circuit can be used to supply a receptacle for an electric clock.

Exception No. 2: A receptacle can be connected to the small-appliance branch circuit to supply a gas-fired range, oven, or counter-mounted cooking unit. **Figure 9-11**

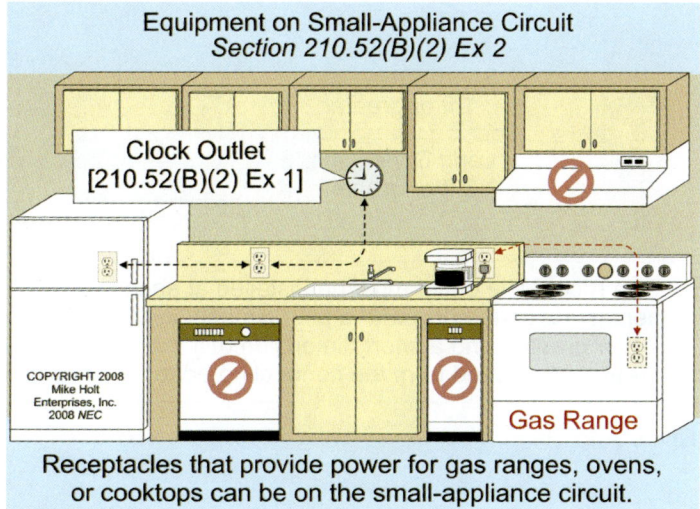

Figure 9–11

Author's Comment: A range hood or above the range microwave listed as a range hood must be supplied by an individual branch circuit if connected by cord and receptacle [422.16(B)(4)(5)].

(3) Kitchen Countertop Receptacles. Kitchen countertop receptacles, as required by 210.52(C), must be supplied by not less than two 20A, 120V small-appliance branch circuits [210.11(C)(1)]. Either or both of these circuits can supply receptacle outlets in the same kitchen, pantry, breakfast room, or dining room of the dwelling unit [210.11(C)(1) and 210.52(B)(1)].

(C) Countertop Receptacles. In kitchens, pantries, breakfast rooms, dining rooms and similar areas of dwelling units, receptacle outlets for countertop spaces must be installed according to (1) through (5) below.

Where a range, counter-mounted cooking unit, or sink is installed in an island or peninsular countertop, and the width of the counter behind the range, counter-mounted cooking unit, or sink is less than 12 in., the countertop space is considered to be two separate countertop spaces. **Figure 9-12**

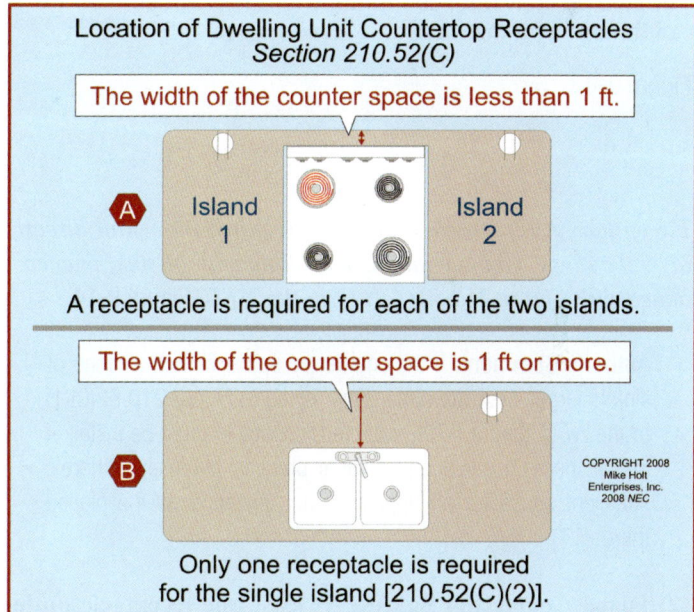

Figure 9–12

Author's Comment: GFCI protection is required for all 15A and 20A, 125V receptacles that supply kitchen countertop surfaces [210.8(A)(6)].

Unit 9 Dwelling Unit Calculations

(1) Wall Countertop Spaces. A receptacle outlet must be installed for each kitchen and dining area countertop wall space 1 ft or wider, and receptacles must be placed so that no point along the countertop wall space is more than 2 ft, measured horizontally, from a receptacle outlet. **Figure 9-13**

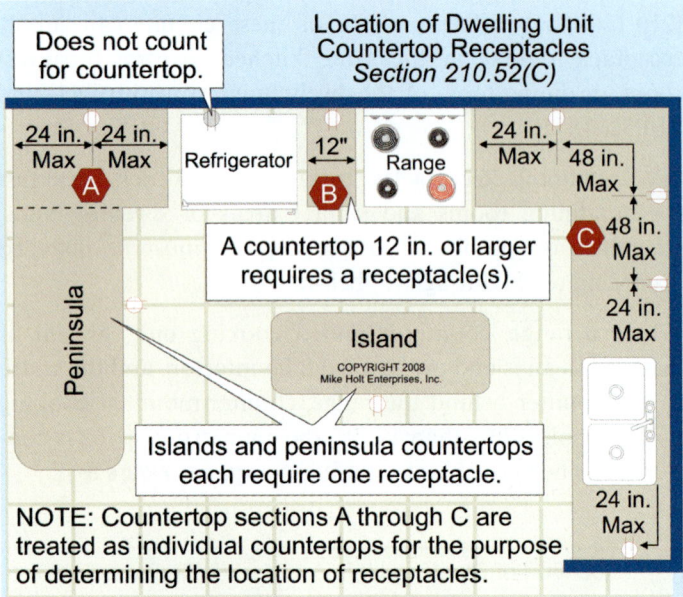

Figure 9–13

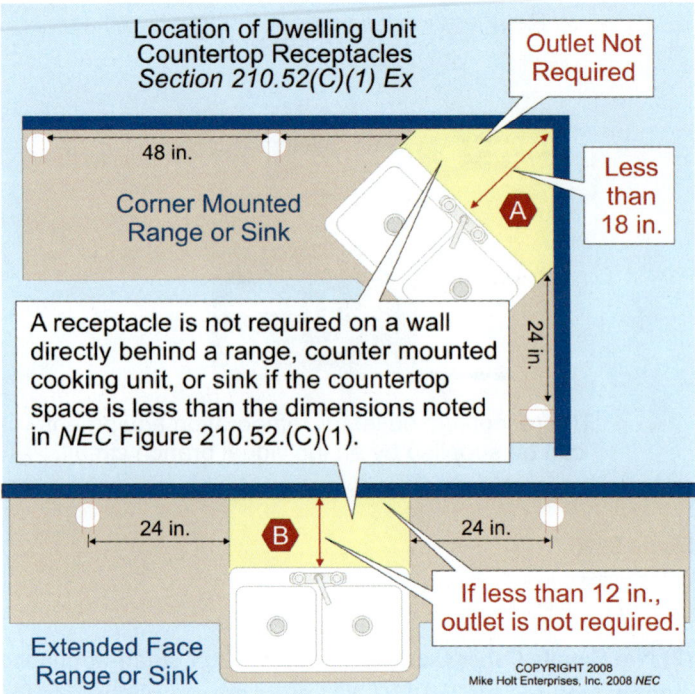

Figure 9–14

Exception: A receptacle outlet isn't required on a wall directly behind a range, counter-mounted cooking unit, or sink, in accordance with Figure 210.52(C)(1) in the NEC. **Figure 9-14**

Author's Comment: If the countertop space behind a range or sink is larger than the dimensions noted in Figure 210.52(C)(1) of the *NEC*, then a GFCI-protected receptacle must be installed in that space. This is because, for all practical purposes, if there's sufficient space for an appliance, an appliance will be placed there.

(2) Island Countertop Spaces. At least one receptacle outlet must be installed at each island countertop space with a long dimension of 2 ft or more, and a short dimension of 1 ft or more.

(3) Peninsular Countertop Spaces. At least one receptacle outlet must be installed at each peninsular countertop with a long dimension of 2 ft or more, and a short dimension of 1 ft or more, measured from the connecting edge. **Figure 9-15**

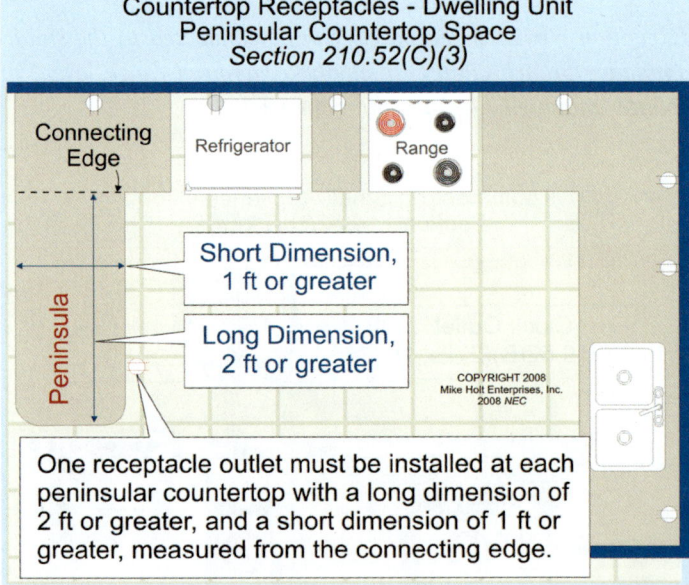

Figure 9–15

Author's Comment: The *Code* doesn't require more than one receptacle outlet in an island or peninsular countertop space, regardless of the length of the countertop, unless the countertop is broken as described in 210.52(C)(4).

Dwelling Unit Calculations — Unit 9

(4) Separate Countertop Spaces. When breaks occur in countertop spaces for rangetops, refrigerators, or sinks, each countertop space is considered as a separate countertop for determining receptacle placement. **Figure 9-16**

Exception: The receptacle outlet for the countertop space can be installed below the countertop only when wall space or a backsplash is not available, such as in an island or peninsular counter. Under these conditions, the required receptacle(s) must be located no more than 1 ft below the countertop surface and no more than 6 in. from the countertop edge, measured horizontally. **Figure 9-18**

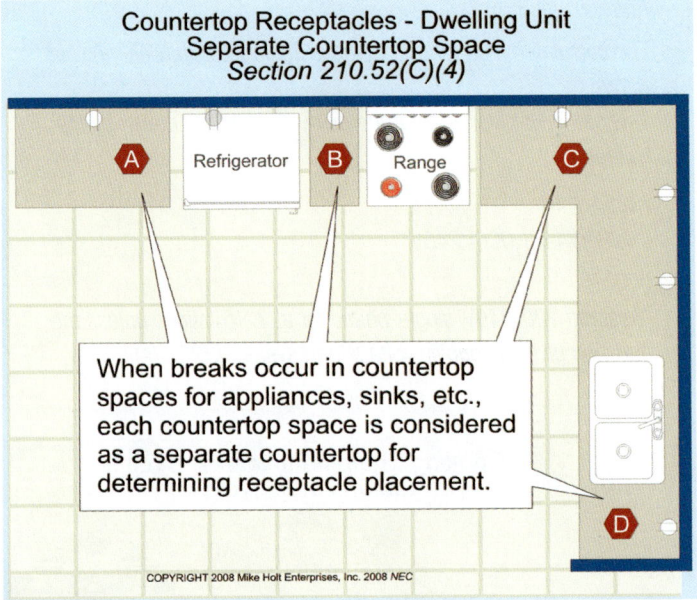

Figure 9–16

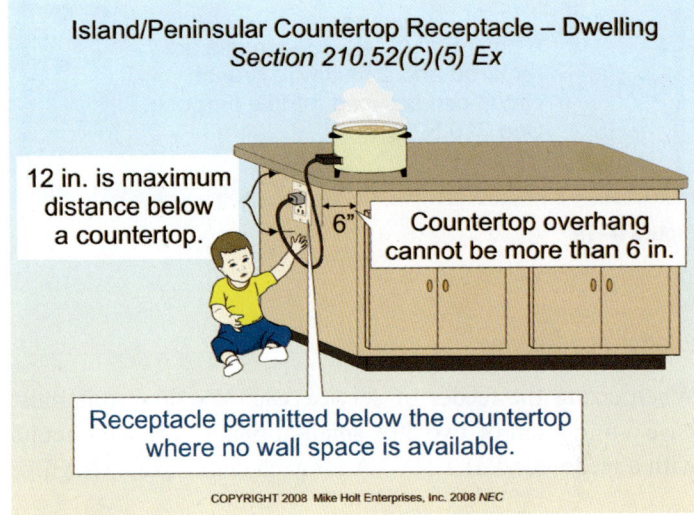

Figure 9–18

(5) Receptacle Location. Receptacle outlets required by 210.52(C)(1) for the countertop space must be located above, but not more than 20 in. above, the countertop surface. **Figure 9-17**

Receptacle outlets rendered not readily accessible by appliances fastened in place, located in an appliance garage, behind sinks, or rangetops [210.52(C)(1) Ex], or supplying appliances that occupy dedicated space do not count as the required countertop receptacles.

> **Author's Comment:** An "appliance garage" is an enclosed area on the countertop where an appliance can be stored and hidden from view when not in use. If a receptacle is installed inside an appliance garage, it doesn't count as a required countertop receptacle outlet.

A minimum of two 20A small-appliance branch circuits are required for receptacle outlets in the kitchen, dining room, breakfast room, pantry, or similar dining areas [220.11(C)(1)]. In general, receptacles from other areas or lighting outlets cannot be connected to these 20A small-appliance branch circuits [210.52(B)(2) Ex]. **Figure 9–19**

Figure 9–17

Mike Holt Enterprises, Inc. • www.MikeHolt.com • 1.888.NEC.CODE (1.888.632.2633)

Unit 9 Dwelling Unit Calculations

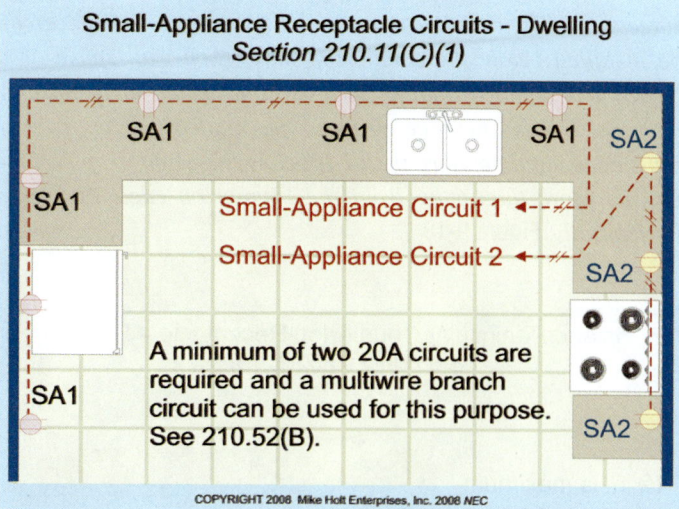

Figure 9–19

▶ Less Than 12 kVA, Table 220.55, Column C Example

Question: What is the branch-circuit calculated load (in amperes) for one 9 kVA range? **Figure 9–20**

(a) 21A   (b) 27A   (c) 33A   (d) 38A

Answer: (c) 33A

The calculated load for one range in Table 220.55, Column C is 8 kVA.

This can be converted to amperes by dividing the power by the voltage:

$I = VA/E$

$I = (8\ kVA \times 1,000)/240V^*$

$I = 33A$

*Assume 120/240V, single-phase for all calculations unless the question gives a specific voltage and system [220.5(A)].

Feeder and Service

When sizing the feeder or service, each dwelling unit must have a minimum of two 20A small-appliance branch circuits with a feeder load of 1,500 VA for each circuit [220.52(A)].

Other Related *Code* Sections

- 15A or 20A receptacles can be used on 20A circuits [210.21(B)(3)].

- Receptacle outlets required for kitchen countertops [210.52(C)].

- Areas supplied by the small-appliance circuits [210.52(B)(1)].

9.5 Cooking Equipment— Branch Circuit [Table 220.55, Note 4]

The branch-circuit calculated load for household ranges and cooking equipment can be determined using Table 220.55, Note 4. The branch-circuit calculated load for one range may be calculated according to the demand factors listed in Table 220.55, but a minimum 40A circuit is required for ranges rated 8.75 kVA and larger [210.19(A)(3)].

Author's Comment: Article 100 defines "Demand Factor" as the ratio of the maximum demand of a system, or part of a system, to the total connected load of that part of the system.

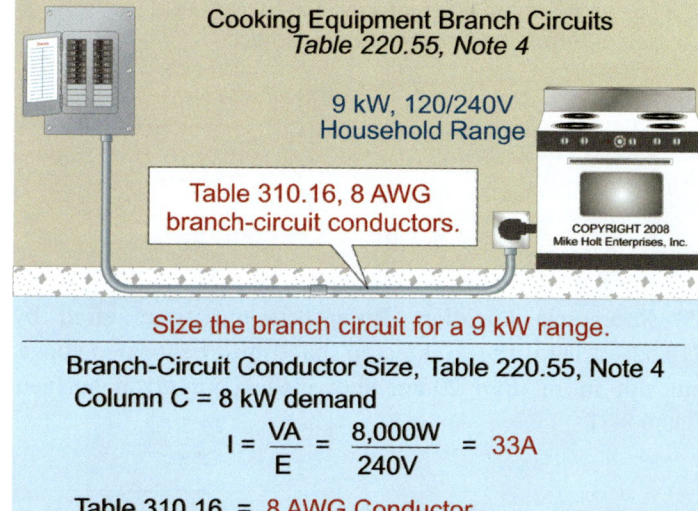

Figure 9–20

Author's Comment: A minimum 40A circuit is required for ranges rated 8.75 kVA and larger [210.19(A)(3)]. Also, notice the footnote to Table 310.16 that refers to 240.4(D). The rule in 240.4(D) puts a limit of 30A protection on a 10 AWG conductor, except in specific situations listed in 240.4(E) and 240.4(G).

Dwelling Unit Calculations — Unit 9

▶ **More than 12 kVA, Table 220.55, Note 1 Example**

Question: What is the branch-circuit load for one 14 kVA range? **Figure 9–21**

(a) 33A (b) 37A (c) 50A (d) 58A

Answer: (b) 37A

Step 1: Since the range exceeds 12 kVA, follow Note 1 of Table 220.55. The first step is to determine the calculated load as listed in Column C of Table 220.55 for one unit = 8 kVA.

Step 2: We must increase the Column C value (8 kVA) by 5% for each kVA that the range exceeds 12 kVA. In this case, 14 kVA exceeds 12 kVA by 2 kVA.

2 x 5% = 10% increase of the Column C value, resulting in 110%, or a 1.10 multiplier.

Calculated Load = Column C Value x Multiplier
Calculated Load = 8 kVA x 1.10
Calculated Load = 8,800 VA

Step 3: Convert the calculated load to amperes.

I = VA/E
I = 6,000 VA/240V
I = 36.70A

Author's Comment: A minimum 40A circuit is required for ranges rated 8.75 kVA and larger [210.19(A)(3)].

▶ **One Counter-Mounted Cooking Unit or One Wall-Mounted Oven Example**

Question: What is the branch-circuit load for one 6 kVA wall-mounted oven? **Figure 9–22**

(a) 15A (b) 20A (c) 25A (d) 30A

Answer: (c) 25A

P = 6 kVA nameplate value
E = 240V (assumed)

I = VA/E
I = 6,000 VA/240V
I = 25A

The branch-circuit load of one wall-mounted oven or one counter-mounted cooking unit must be the nameplate rating of the oven or cooking unit [Table 220.55, Note 4].

Author's Comment: 12 AWG is rated 25A in the 60°C column. However, the overcurrent device must be sized at a minimum of 25A. For this reason, 10 AWG is the minimum size allowed due to the limitations of 240.4(D).

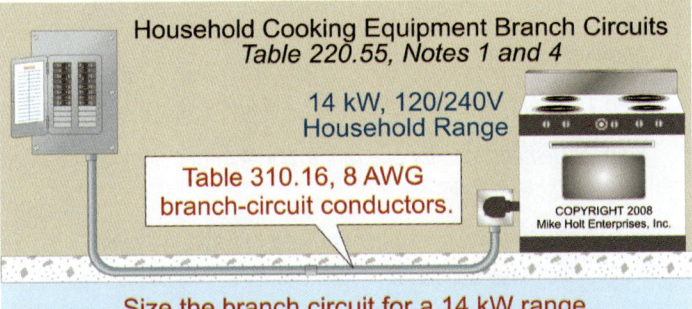

Figure 9–21

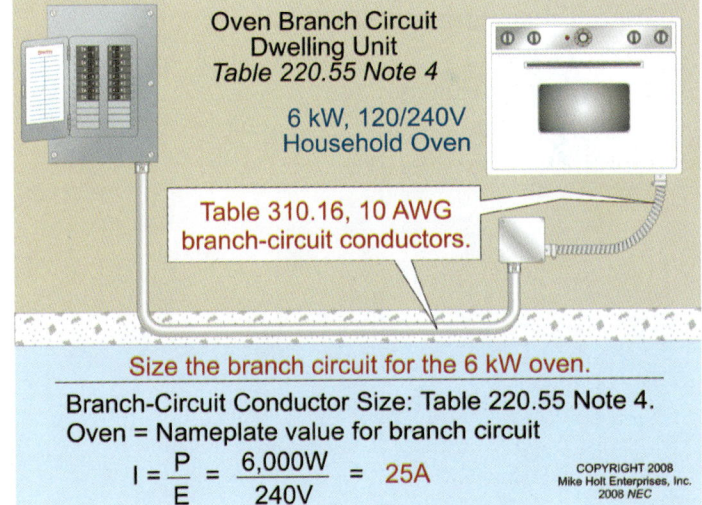

Figure 9–22

Unit 9 — Dwelling Unit Calculations

One Counter-Mounted Cooking Unit and Up to Two Wall-Mounted Ovens [Table 220.55, Note 4]

To calculate the load for one counter-mounted cooking unit (cooktop) and up to two wall-mounted ovens, complete the following steps:

Step 1: Add the nameplate ratings of the cooking appliances and treat this total as one range.

Step 2: Determine the kVA load for one unit from Table 220.55, Column C.

Step 3: If the total nameplate rating exceeds 12 kVA, increase Column C (8 kVA) 5% for each kVA, or major fraction (0.50 kVA), that the combined rating exceeds 12 kVA.

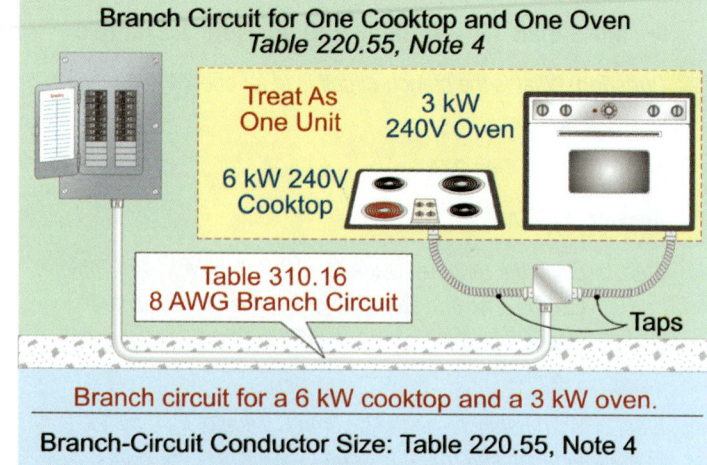

Figure 9–23

▶ **Cooktop and One Wall-Mounted Oven Example**

Question: What is the branch-circuit load for one 6 kVA counter-mounted cooking unit and one 3 kVA wall-mounted oven? **Figure 9–23**

(a) 25A   (b) 33A   (c) 38A   (d) 42A

Answer: (b) 33A

Step 1: Total connected load: 6 kVA + 3 kVA = 9 kVA

Step 2: kVA load for one range, Column C of Table 220.55: 8 kVA

Step 3: Convert the calculated load to amperes:

  I = VA/E
  I = 8,000 VA/240V
  I = 33.30A

▶ **Cooktop and Two Wall-Mounted Ovens Example**

Question: What is the branch-circuit load for one 6 kVA counter-mounted cooking unit (cooktop) and two 4 kVA wall-mounted ovens? **Figure 9–24**

(a) 22A   (b) 27A   (c) 33A   (d) 37A

Answer: (d) 37A

Step 1: Total connected load: 6 kVA + 4 kVA + 4 kVA = 14 kVA

Step 2: Demand load for one range, 8 kVA from Column C of Table 220.55

Step 3: Since the total (14 kVA) exceeds 12 kVA, we must increase Column C (8 kVA) 5% for each kVA that the total exceeds 12 kVA.

14 kVA exceeds 12 kVA by 2 kVA

2 x 5% = 10% increase of the Column C value, resulting in 110%, or a 1.10 multiplier.

Calculated Load = Column C Value x Multiplier
Calculated Load = 8 kVA x 1.10
Calculated Load = 8,800 VA

Step 4: Convert the calculated load to amperes:

  I = VA/E
  I = 8,800 VA/240V
  I = 36.70A

Dwelling Unit Calculations — Unit 9

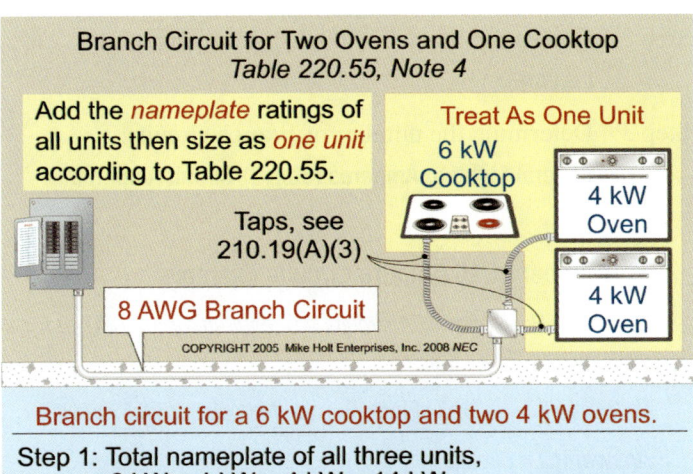

Branch circuit for a 6 kW cooktop and two 4 kW ovens.

Step 1: Total nameplate of all three units,
  6 kW + 4 kW + 4 kW = 14 kW
Step 2: Treat as one 14 kW range. Column C for one unit is 8 kW. Increase Column C answer (8 kW) by 5% for every kW over 12 kW; 14 kW - 12 kW = 2 kW = 10%, 8 kW × 1.10 = 8.8 kW
  I = VA/E = 8,800W/240V = 37A

Figure 9–24

9.6 Laundry Circuit [210.11(C)(2)]

One 20A branch circuit for the laundry receptacle outlet (or outlets) is required and this laundry circuit cannot serve any other outlet such as the laundry room lights [210.52(F) and 210.11(C)(2)]. The *NEC* does not require a separate circuit for the washing machine, but does require a separate circuit for the laundry room receptacle or receptacles. **Figure 9–25**

Figure 9–25

Feeder and Service Calculations

Each dwelling unit must have a feeder/service calculated load consisting of 1,500 VA for the 20A laundry receptacle circuit [220.52(B)].

Other Related *Code* Rules

- A laundry area receptacle outlet must be within 6 ft of a washing machine [210.50(C)].
- A laundry area receptacle outlet is required [210.52(F)].
- When a laundry area receptacle outlet is within 6 ft of the outside edge of a laundry sink, it must be GFCI protected [210.8(A)(7)].

9.7 Lighting and Receptacles

General Lighting Calculated Load [220.12]

The *NEC* requires a minimum 3 VA per sq ft for the general lighting and general-use receptacles for the purpose of determining branch circuits and feeder/service calculations. The dimensions for determining the area must be computed from the outside dimensions of the building and must not include open porches, garages, or spaces not adaptable for future use. **Figure 9–26**

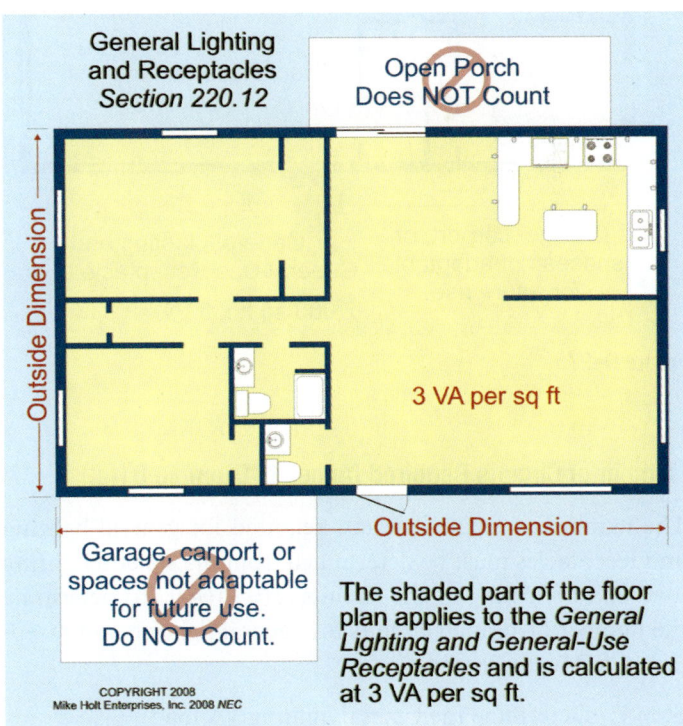

Figure 9–26

Unit 9   Dwelling Unit Calculations

Author's Comment: The 3 VA per sq ft rule for general lighting includes all 15A and 20A general-use receptacles, but it doesn't include the small-appliance or laundry circuit receptacles. See 220.14(J) for details.

▶ **General Lighting Load Example [Table 220.12]**

Question: What is the general lighting and receptacle load for a 2,000 sq ft dwelling unit that has 34 convenience receptacles and 12 luminaires rated 100W each? **Figure 9–27**

(a) 2,100 VA   (b) 4,200 VA   (c) 6,000 VA   (d) 8,400 VA

Answer: (c) 6,000 VA
2,000 sq ft x 3 VA = 6,000 VA

Note: No additional load is required for general-use receptacles and lighting outlets. See 220.14(J).

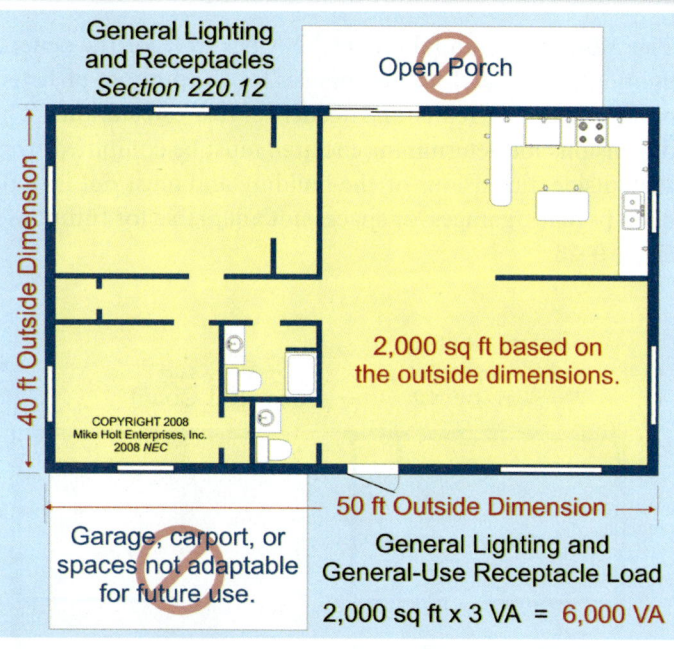

Figure 9–27

Number of Circuits Required [Annex D, Example D1(a)]

The number of branch circuits required for general lighting and receptacles must be determined from the general lighting load and the rating of the circuits [210.11(A)]. To determine the number of branch circuits for general lighting and receptacles, follow these steps:

Step 1: Determine the general lighting VA load:

     Living Area Square Footage x 3 VA

Step 2: Determine the general lighting ampere load:

     I = VA/E

Step 3: Determine the number of branch circuits:

     General Lighting Amperes (Step 2)/Circuit Amperes

▶ **Number of 15 Ampere Circuits Example**

Question: How many 15A circuits are required for a 2,000 sq ft dwelling unit? **Figure 9–28**

(a) 2 circuits   (b) 3 circuits   (c) 4 circuits   (d) 5 circuits

Answer: (c) 4 circuits

Step 1: General Lighting VA = 2,000 sq ft x 3 VA
         General Lighting VA = 6,000 VA

Step 2: General Lighting Amperes:

     I = VA/E
     I = 6,000 VA/120V*
     I = 50A

     *Use 120V, single-phase unless specified otherwise.

Step 3: Determine the number of circuits:

     Number of Circuits =
     General Lighting Amperes/Circuit Amperes
     Number of Circuits = 50A/15A
     Number of Circuits = 3.30 or 4 circuits. Any fraction of a circuit must be rounded up.

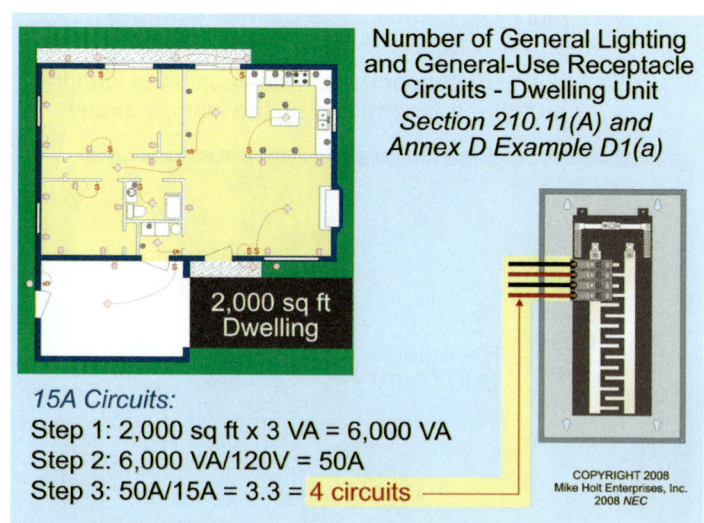

15A Circuits:
Step 1: 2,000 sq ft x 3 VA = 6,000 VA
Step 2: 6,000 VA/120V = 50A
Step 3: 50A/15A = 3.3 = **4 circuits**

Figure 9–28

Dwelling Unit Calculations — Unit 9

▶ Number of 20 Ampere Circuits Example

Question: How many 20A circuits are required for a 2,000 sq ft dwelling unit?

(a) 2 circuits (b) 3 circuits (c) 4 circuits (d) 5 circuits

Answer: (b) 3 circuits

Step 1: General Lighting VA = 2,000 sq ft x 3 VA = 6,000 VA

Step 2: General Lighting Amperes:

$I = VA/E$
$I = 6,000\ VA/120V$
$I = 50A$

Step 3: Determine the number of circuits:

Number of Circuits =
General Lighting Amperes/Circuit Amperes
Number of Circuits = 50A/20A
Number of Circuits = 2.50 or 3 circuits. Any fraction of a circuit must be rounded up.

PART B—STANDARD METHOD—FEEDER/SERVICE LOAD CALCULATIONS

Introduction

Article 220 of the *NEC* permits two distinctly different methods of calculating residential loads: the standard method in Part III and the optional method in Part IV. The two methods are different and typically give different results. You must use one method or the other; rules from both can't be mixed together. On an exam, you'll usually be told which method to use for a particular question, but if the question doesn't specify a method, use the standard calculation.

9.8 Dwelling Unit Feeder/Service Load Calculations (Article 220, Part III)

Follow these steps to determine the feeder or service size for a dwelling unit using the standard method contained in Article 220, Part III:

Step 1: General Lighting and Receptacles, Small-Appliance, and Laundry Circuits [Table 220.42]

The *NEC* recognizes that the general lighting and receptacle, small-appliance, and laundry circuits won't all be on, or loaded, at the same time and permits a demand factor to be applied to the total connected load [220.52]. To determine the feeder calculated load for these loads, use the following steps:

Step a: Total Connected Load. Determine the total connected load for: (1) general lighting and receptacles (3 VA per sq ft), (2) two small-appliance circuits each at 1,500 VA, and (3) one laundry circuit at 1,500 VA.

Step b: Demand Factor. Apply the Table 220.42 demand factors to the total connected load.

Step c: First 3,000 VA at 100 percent demand. Remaining VA at 35 percent demand.

Step 2: Air-Conditioning versus Heat

Because the air-conditioning and heating loads aren't on at the same time, the smaller of the two loads can be omitted [220.60]. The air-conditioning load is calculated at 100 percent [220.50] and the fixed electric heating load is calculated at 100 percent [220.51].

Author's Comment: Some believe that the largest motor in the air conditioners should be taken at 125 percent. Either way, it typically doesn't make a big difference in the end result.

Step 3: Appliances [220.53]

A 75 percent demand factor is permitted to be applied when four or more appliances are fastened in place, such as a waste disposal, dishwasher, trash compactor, water heater, etc., and are on the same feeder. This doesn't apply to space-heating equipment [220.51], clothes dryers [220.54], cooking appliances [220.55], or air-conditioning equipment.

Author's Comment: This demand factor is applied to the nameplate rating of the appliances.

Step 4: Clothes Dryer [220.54]

The feeder or service load for electric clothes dryers located in a dwelling unit must not be less than 5,000W or the nameplate rating if greater than 5,000W. A feeder or service dryer load isn't required if the dwelling unit doesn't contain an electric dryer!

Unit 9 Dwelling Unit Calculations

Step 5: Cooking Equipment [220.55]

Household cooking appliances rated over 1¾ kVA can have the feeder and service calculated according to the demand factors of Table 220.55, including Notes 1, 2, and 3.

Step 6: Feeder and Service Conductor Size

Step a: For individual dwelling units of one-family, two-family, and multifamily dwellings, Table 310.15(B)(6) can be used to size 3-wire, single-phase, 120/240V service or feeder conductors that supply all loads that are part of, or associated with, the dwelling unit. **Figure 9-29**

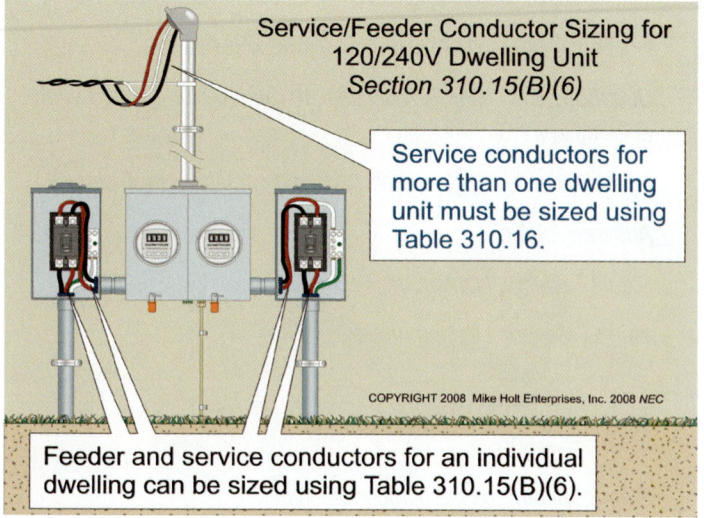

Figure 9–30

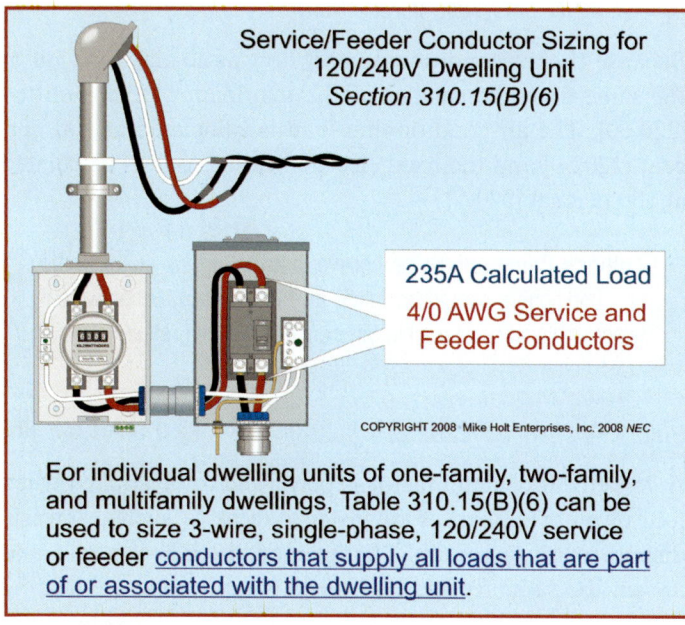

Figure 9–29

Step b: Over 400 Amperes. Size the ungrounded and neutral conductors according to Table 310.16.

Author's Comment: Table 310.15(B)(6) can't be used for service conductors for two-family or multifamily buildings. **Figure 9-30**

Feeder conductors are not required to have an ampacity rating more than the service conductors [215.2(A)(3)].

Table 310.15(B)(6) Conductor Sizes for 120/240V, 3-Wire, Single-Phase Dwelling Services and Feeders		
Amperes	Copper	Aluminum
100	4 AWG	2 AWG
110	3 AWG	1 AWG
125	2 AWG	1/0 AWG
150	1 AWG	2/0 AWG
175	1/0 AWG	3/0 AWG
200	2/0 AWG	4/0 AWG
225	3/0 AWG	250 kcmil
250	4/0 AWG	300 kcmil
300	250 kcmil	350 kcmil
350	350 kcmil	500 kcmil
400	400 kcmil	600 kcmil

WARNING: Table 310.15(B)(6) doesn't apply to 3-wire feeder/service conductors connected to three-phase, 120/208V systems, because the neutral conductor in these systems always carries neutral current, even when the load on the phases is balanced [310.15(B)(4)(b)]. For more information on this topic, see 220.61(C)(1). **Figure 9-31**

Dwelling Unit Calculations — Unit 9

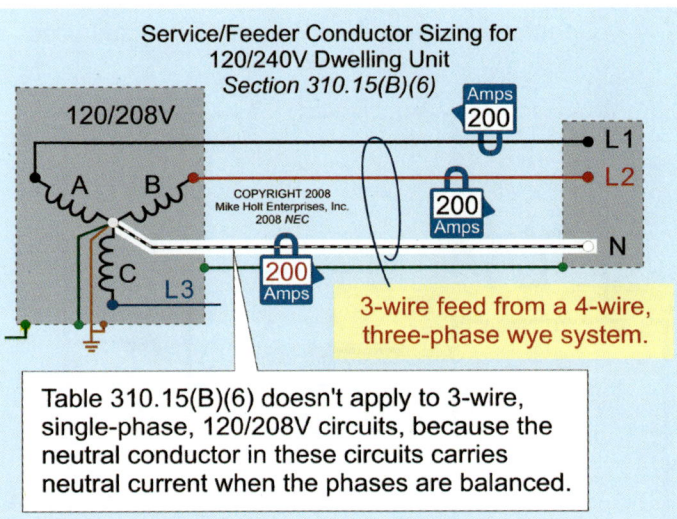

Figure 9–31

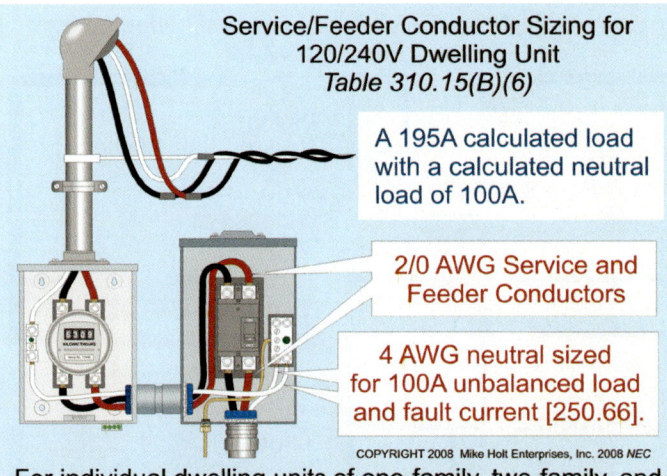

Figure 9–32

Neutral Conductor Sizing. Table 310.15(B)(6) can be used to size the neutral conductor of a 3-wire, single-phase, 120/240V service or feeder that carries all loads associated with the dwelling unit, based on the calculated load in accordance with 220.61.

CAUTION: *Because the service neutral conductor is required to serve as the effective ground-fault current path, it must be sized so it can safely carry the maximum fault current likely to be imposed on it [110.10 and 250.4(A)(5)]. This is accomplished by sizing the neutral conductor in accordance with Table 250.66, based on the area of the largest ungrounded service conductor [250.24(C)(1)].*

▶ **Dwelling Service Conductor Sizing Example**

Question: *What size service conductors are required if the calculated load for a dwelling unit equals 195A, and the maximum unbalanced neutral load is 100A?* **Figure 9-32**

(a) 1/0 AWG and 6 AWG   (b) 2/0 AWG and 4 AWG
(c) 3/0 AWG and 2 AWG   (d) 4/0 AWG and 1 AWG

Answer: *(b) 2/0 AWG and 4 AWG*

Service Conductor: 2/0 AWG rated 200A [Table 310.15(B)(6)]

Neutral Conductor: 4 AWG is rated 100A in accordance with Table 310.15(B)(6). In addition, 250.24(C) requires the neutral conductor to be sized no smaller than 4 AWG, based on 2/0 AWG service conductors in accordance with Table 250.66.

9.9 Dwelling Unit Example

Step 1: General Lighting and Receptacle, Small Appliance, and Laundry Demand [Table 220.42]

The demand factors in Table 220.42 apply to the 3 VA per sq ft for general lighting and general-use receptacles [Table 220.12], small-appliance circuits [220.52(A)], and the laundry circuit [220.52(B)].

▶ **General Lighting and Receptacle Example 1**

Question: *What is the general lighting and receptacle, small appliance, and laundry feeder/service calculated load for a 2,700 sq ft dwelling unit?* **Figure 9–33**

(a) 2,700 VA   (b) 6,360 VA   (c) 8,100 VA   (d) 12,600 VA

Answer: *(b) 6,360 VA*

General Lighting/Receptacles (2,700 sq ft x 3 VA) [220.12]	8,100 VA
Small-Appliance Circuits (1,500 VA x 2) [220.52(A)]	3,000 VA
Laundry Circuit (1,500 VA x 1) [220.52(B)]	+ 1,500 VA
Total Connected Load	12,600 VA
First 3,000 VA at 100% [Table 220.42]	−3,000 VA
x 1.00 =	3,000 VA
Remainder at 35%	9,600 VA
x 0.35 =	+ 3,360 VA
Calculated Load	6,360 VA

Mike Holt Enterprises, Inc. • www.MikeHolt.com • 1.888.NEC.CODE (1.888.632.2633)

Unit 9 — Dwelling Unit Calculations

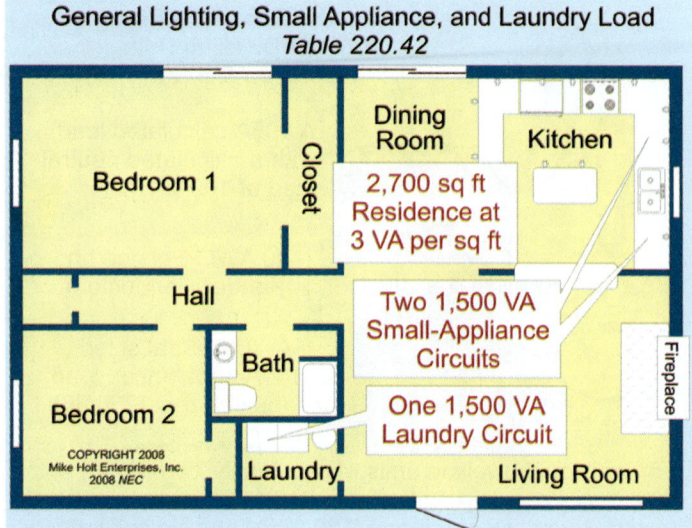

Figure 9–33

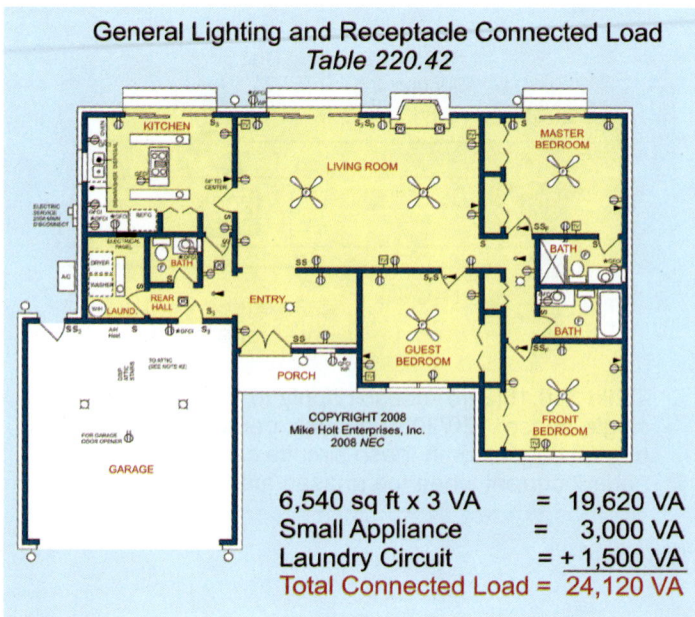

Figure 9–34

▶ **General Lighting and Receptacle Example 2**

Question: What is the general lighting and receptacle, small appliance, and laundry feeder/service calculated load for a 6,540 sq ft dwelling unit?

(a) 2,700 VA   (b) 8,100 VA   (c) 10,392 VA   (d) 12,600 VA

Answer: (c) 10,392 VA

General Lighting/Receptacles (6,540 sq ft x 3 VA) [220.12]	19,620 VA		
Small-Appliance Circuits (1,500 VA x 2) [220.52(A)]	3,000 VA		
Laundry Circuit (1,500 VA x 1) [220.52(B)]	+ 1,500 VA		
Total Connected Load **Figure 9-34**	24,120 VA		
First 3,000 VA at 100% [Table 220.42]	– 3,000 VA	x 1.00 =	3,000 VA
Remainder at 35%	21,120 VA	x 0.35 =	+ 7,392 VA
Calculated Load			10,392 VA

Dwelling Unit Calculations — Unit 9

Step 2: Air-Conditioning versus Heat [220.60]

Compare the A/C load at 100 percent [220.50] against the heating load at 100 percent [220.51] and omit the smaller of the two loads [220.60].

$VA = V \times A$
VA = 230V x 28A
VA = 6,440 VA
VA = 6,440 VA [220.60] (omit)
Heat [220.51]: 3,000 VA x 3 units = 9,000 VA

▶ **Air-Conditioning Calculated Load Example**

Question: What is the calculated VA load for a 240V single-phase air conditioner that contains a 16A compressor and a 1A fan? **Figure 9-35**

(a) 4 kVA (b) 5 kVA (c) 6 kVA (d) 7 kVA

Answer: (a) 4 kVA

A/C VA Load = Volts x Amperes
A/C VA Load = 240V x (16A + 1.00A)
A/C VA Load = 240V x 17A
A/C VA Load = 4,080 VA
4,080 VA/1,000 = 4.08 kVA

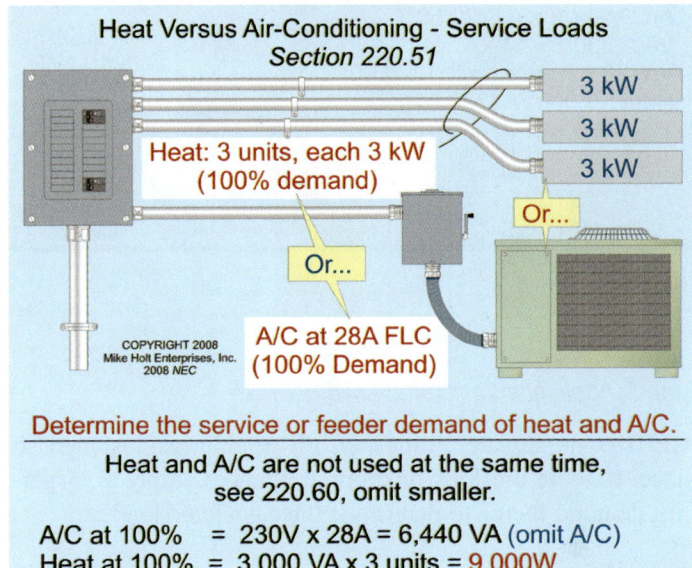

Figure 9–36

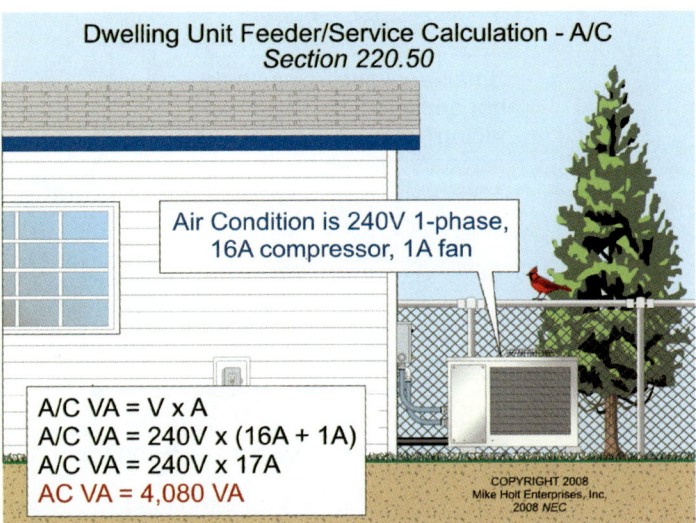

Figure 9–35

▶ **Air-Conditioning versus Heat Example 1**

Question: What is the feeder/service calculated load for a 28A, 230V A/C unit, versus three 3 kVA baseboard heaters? **Figure 9–36**

(a) 3,000 VA (b) 6,400W (c) 8,050 VA (d) 9,000W

Answer: (d) 9,000W

(continued in next column)

▶ **Air-Conditioning versus Heat Example 2**

Question: What is the feeder/service calculated load for a 14A, 230V A/C unit containing 10 kVA of electric heat? **Figure 9-37**

(a) 3,910 VA (b) 5,000 VA (c) 10,000 VA (d) 10,350 VA

Answer: (c) 10,000 VA

A/C Unit:
$VA = V \times A$
VA = 230V x 14A
VA = 3,220 VA
VA = 3,220 VA [220.60] (omit)

Heat [220.51]:
10,000 VA

Unit 9 | Dwelling Unit Calculations

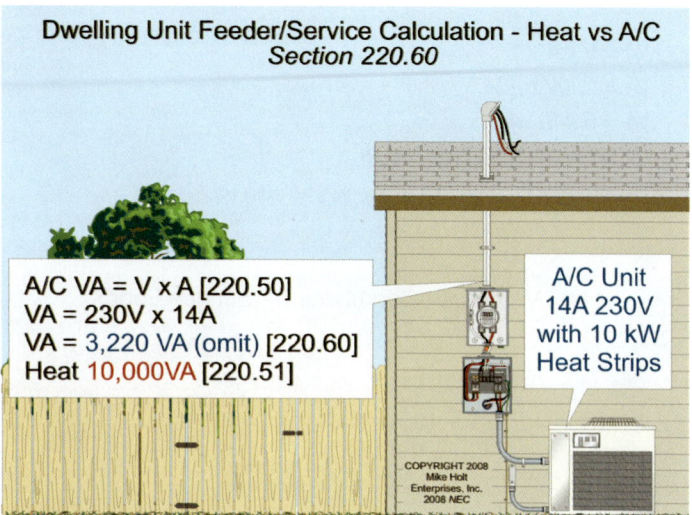

Figure 9–37

Step 3: Appliance Calculated Load [220.53]

Add the nameplate ratings of all appliances fastened in place. If there are four or more appliances, apply a 75 percent demand factor to determine the calculated load.

Author's Comment: This demand factor doesn't apply to motors [220.50], space-heating equipment [220.51], clothes dryers [220.54], electric ranges [220.55], or air-conditioning equipment.

▶ Appliance Example 1

Question: What is the feeder/service calculated load for a dishwasher (1,250 VA), waste disposal (940 VA), and a water heater (4,500 VA)?

(a) 5,018 VA (b) 6,272 VA (c) 6,690 VA (d) 8,363 VA

Answer: (c) 6,690 VA

Dishwasher	1,250 VA
Waste Disposal	940 VA
Water Heater	+ 4,500 VA
Calculated Load	6,690 VA

▶ Appliance Example 2

Question: What is the feeder/service calculated load for a dishwasher (1,250 VA), trash compactor (1,100 VA), waste disposal (940 VA), and water heater (4,500 VA)? **Figure 9–38**

(a) 5,843 VA (b) 7,303 VA (c) 7,790 VA (d) 9,738 VA

Answer: (a) 5,843 VA

Dishwasher	1,250 VA
Trash Compactor	1,100 VA
Waste Disposal	940 VA
Water Heater	+ 4,500 VA
Connected Load	7,790 VA

Calculated Load =
Connected Load x Demand Factor [220.53]
Calculated Load = 7,790 VA x 0.75
Calculated Load = 5,843 VA

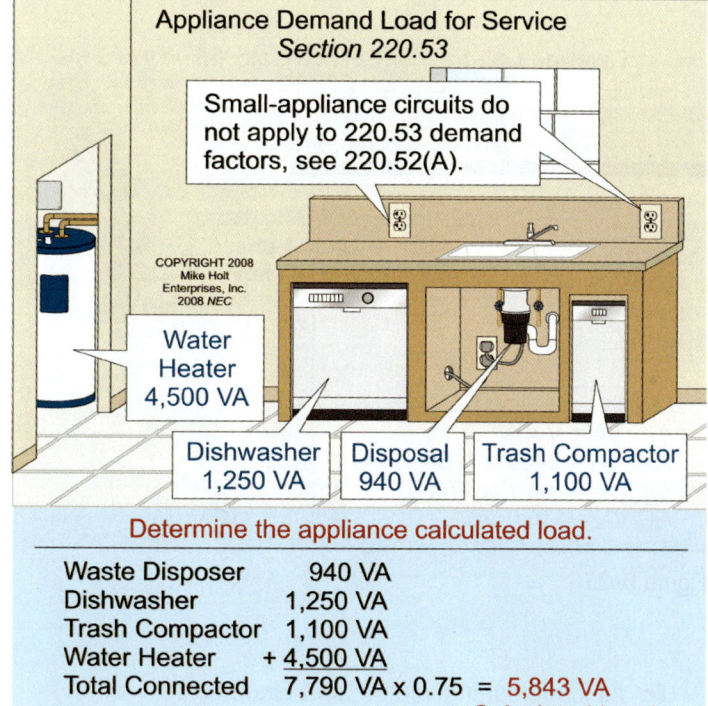

Figure 9–38

260   Mike Holt's Illustrated Guide to NEC Exam Preparation

Dwelling Unit Calculations — Unit 9

Step 4: Dryer Calculated Load [220.54]

The minimum calculated load for a household electric dryer is 5,000 VA or use the nameplate rating if higher.

▶ **Dryer Example 1**

Question: What is the feeder/service calculated load for a 4 kVA dryer? **Figure 9–39**

(a) 3,000 VA   (b) 4,000 VA   (c) 5,000 VA   (d) 5,500 VA

Answer: (c) 5,000 VA

The dryer load must not be less than 5,000 VA.

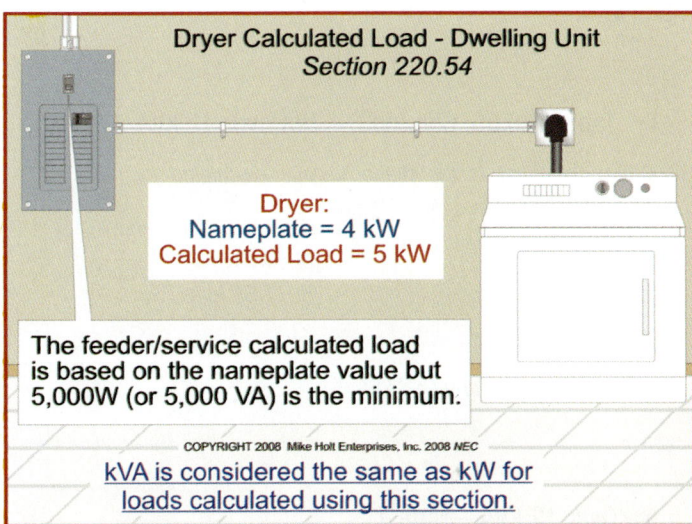

Figure 9–39

▶ **Dryer Example 2**

Question: What is the feeder/service calculated load for a 5.50 kVA dryer?

(a) 3,000 VA   (b) 4,000 VA   (c) 5,000 VA   (d) 5,500 VA

Answer: (d) 5,500 VA

The dryer load must not be less than the nameplate rating if greater than 5,000 VA.

Step 5: Cooking Equipment Calculated Load [220.55]

When using Table 220.55, note the following:

- Column A applies to cooking equipment rated over 1¾ kVA but less than 3½ kVA.
- Column B applies to cooking equipment rated 3½ to 8¾ kVA.
- Column C applies to cooking equipment rated 8¾ kVA to 12 kVA.

These columns are used with Notes 1, 2, and 3 for determining feeder/service calculated loads.

▶ **Note 3—Over 1¾ and Less Than 3½ kVA— Table 220.55, Column A Figure 9-40**

Question: What is the feeder/service calculated load for two 3 kVA cooking appliances in a dwelling unit?

(a) 3 kVA   (b) 3.90 kVA   (c) 4.50 kVA   (d) 4.80 kVA

Answer: (c) 4.50 kVA

Feeder/Service Connected Load = 3 kVA x 2 units
Feeder/Service Calculated Load = 6 kVA x 0.75
Feeder/Service Calculated Load = 4.50 kVA

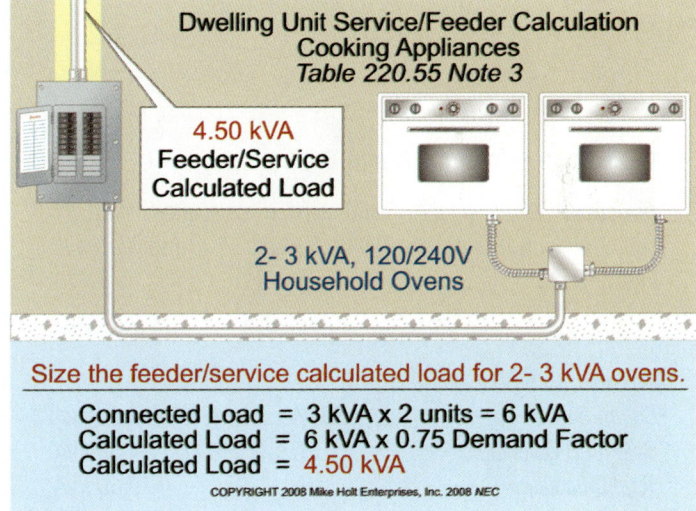

Figure 9–40

Unit 9 — Dwelling Unit Calculations

▶ **Note 3—Not over 8¾—Column B**

Question: What is the feeder/service calculated load for one 6 kVA cooking appliance in a dwelling unit? **Figure 9-41**

(a) 3.90 kVA (b) 4.50 kVA (c) 4.80 kVA (d) 6 kVA

Answer: (c) 4.80 kVA

Feeder/Service Calculated Load = 6 kVA x 0.80
Feeder/Service Calculated Load = 4.80 kVA

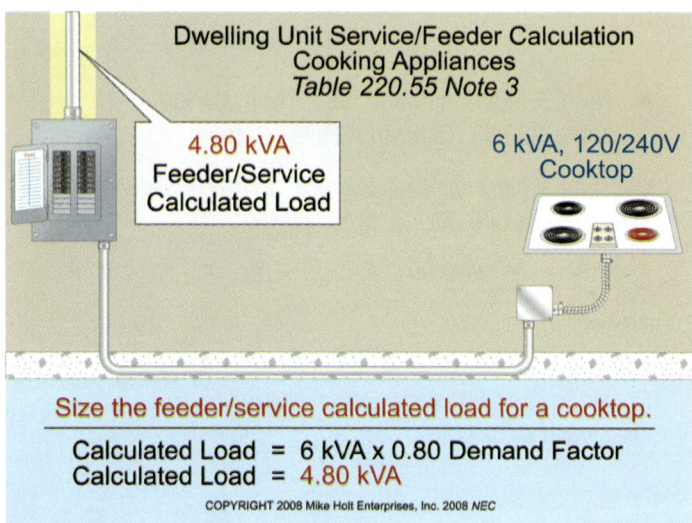

Figure 9–41

▶ **Note 3—Columns A and B**

Question: What is the feeder/service calculated load for two 3 kVA ovens and one 6 kVA cooktop in a dwelling unit? **Figure 9-42**

(a) 4.50 kVA (b) 4.80 kVA (c) 6 kVA (d) 9.30 kVA

Answer: (d) 9.30 kVA

Column A Calculated: (3 kVA x 2) x 0.75   4.50 kVA
Column B Calculated: 6 kVA x 0.80 + 4.80 kVA
Total Calculated   9.30 kVA

Figure 9–42

▶ **Table 220.55, Column C—Not over 12 kVA**

Question: What is the feeder/service calculated load for an 11.50 kVA range in a dwelling unit?

(a) 6 kVA (b) 8 kVA
(c) 9.20 kVA (d) 11.50 kVA

Answer: (b) 8 kVA

Table 220.55, Column C: 8 kVA

▶ **Over 12 kVA—Note 1**

Question: What is the feeder/service calculated load for a 13.60 kVA range in a dwelling unit?

(a) 6 kVA (b) 8 kVA (c) 8.80 kVA (d) 9.20 kVA

Answer: (c) 8.80 kVA

Step a: 13.60 kVA exceeds 12 kVA by 1 kVA and one major fraction of a kVA (total of 2 kVA). The Column C value (8 kVA) must be increased 5% for each kVA or major fraction of a kVA (0.50 kVA or larger) over 12 kVA [Table 220.55, Note 1].

2 x 5% = 10% increase of the Column C value, resulting in 110%, or a 1.10 multiplier.

Step b: **Calculated Load = Column C Value x Multiplier**
Calculated Load = 8 kVA x 1.10
Calculated Load = 8,800 VA

Dwelling Unit Calculations — Unit 9

Step 6: Service Conductor Size [Table 310.15(B)(6)]

Dwelling Unit Feeder/Service Conductors. For individual dwelling units of one-family, two-family, and multifamily dwellings, Table 310.15(B)(6) can be used to size 3-wire, single-phase, 120/240V service or feeder conductors (including neutral conductors) that serve as the main power feeder. Feeder conductors are not required to have an ampacity rating greater than the service conductors [215.2(A)(3)].

Author's Comment: Table 310.15(B)(6) permits a smaller feeder/service conductor size than listed in Table 310.16 for the same calculated load, but this reduced size conductor is only permitted for a 120/240V, 3-wire supplied dwelling unit up to 400A.

Neutral Conductor Sizing. Table 310.15(B)(6) can be used to size the neutral conductor of a 3-wire, single-phase, 120/240V service or feeder that serves as the main power feeder, based on the feeder calculated load in accordance with 220.61.

Author's Comment: Because the neutral service conductor is required to serve as the effective ground-fault current path, it must be sized so that it can safely carry the maximum fault current likely to be imposed on it [110.10 and 250.4(A)(5)]. This is accomplished by sizing the neutral conductor in accordance with Table 250.66, based on the total area of the largest ungrounded conductor [250.24(C)(1)].

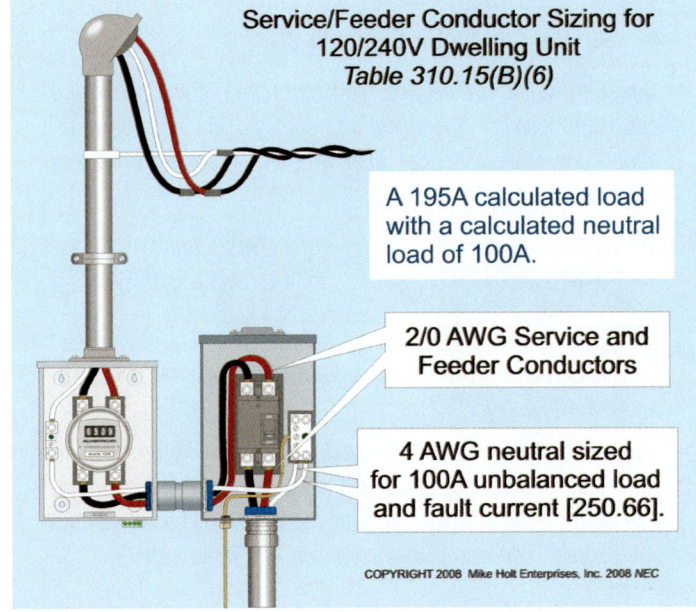

Figure 9–43

▶ **Service Conductor Example 1**

Question: What size service conductors (120/240V, single-phase) are required when the calculated load for a dwelling unit equals 195A and the neutral load is 100A? **Figure 9–43**

(a) 1/0 AWG and 6 AWG
(b) 2/0 AWG and 4 AWG
(c) 3/0 AWG and 2 AWG
(d) 4/0 AWG and 1 AWG

Answer: (b) 2/0 AWG and 4 AWG

Service Conductor: 2/0 AWG rated 200A [Table 310.15(B)(6)]

Neutral Conductor: 4 AWG is rated 100A in accordance with Table 310.15(B)(6). In addition, 250.24(C) requires the neutral conductor to be sized no smaller than 4 AWG, based on 2/0 AWG service conductors in accordance with Table 250.66.

Table 310.15(B)(6) doesn't apply to 3-wire, single-phase, 120/208V systems, because the neutral conductor in these systems carries neutral current even when the load on the phases is balanced [310.15(B)(4)(b)]. For more information on this topic, see the NEC 220.61(C)(1). **Figure 9–44**

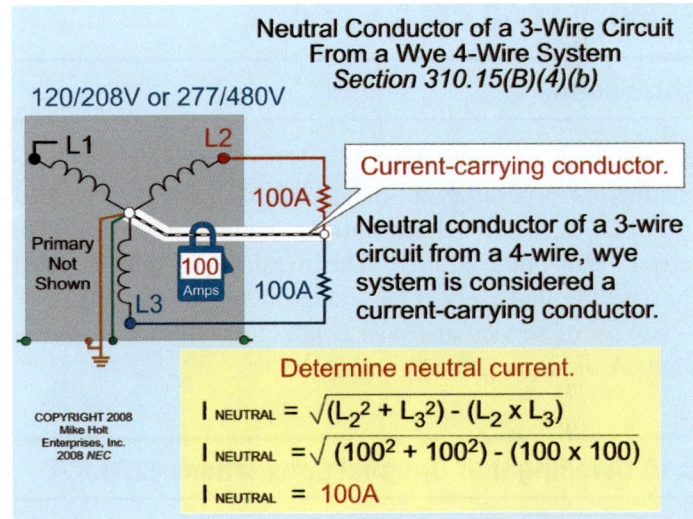

Figure 9–44

Author's Comment: 120/208V, single-phase feeders from a 120/208V, three-phase service are sometimes used to supply single-family dwelling units in a multifamily building.

Unit 9 Dwelling Unit Calculations

> ▶ **Service Conductor Example 2**
>
> **Question:** What size service conductors are required when the calculated load for a dwelling unit equals 195A and the maximum unbalanced neutral load is 65A, if supplied from a 120/208V system?
>
> (a) 1/0 AWG and 6 AWG
> (b) 2/0 AWG and 4 AWG
> (c) 3/0 AWG and 4 AWG
> (d) 4/0 AWG and 1 AWG
>
> **Answer:** (c) 3/0 AWG and 4 AWG
>
> Service Conductor: 3/0 AWG rated 200A at 75°C [110.14(C) and Table 310.16]
>
> Neutral Conductor: 6 AWG is rated 65A at 75°C in accordance with Table 310.16, but 250.24(C) requires the grounded neutral conductor to be sized no smaller than 4 AWG, based on 3/0 AWG service conductors in accordance with Table 250.66.

PART C—OPTIONAL METHOD—FEEDER/SERVICE LOAD CALCULATIONS

Introduction

Instead of sizing the dwelling unit feeder and/or service phase conductors according to the standard method described in Article 220, Part III, an optional method in Article 220, Part IV can often be used. This method can only be used for dwelling units served by a single 120/240V or 120/208V 3-wire set of service or feeder conductors with an ampacity of 100A or larger [220.82].

9.10 Dwelling Unit Optional Calculations [220.82]

The following steps can be used to determine the feeder or service size for a dwelling unit using the optional method contained in Article 220, Part IV:

Step 1: General Loads [220.82(B)]

The calculated load must not be less than 100 percent for the first 10 kVA, plus 40 percent of the remainder of the following loads:

(1) General Lighting and Receptacles: 3 VA per sq ft

(2) Small-Appliance and Laundry Branch Circuits: 1,500 VA for each 20A small-appliance and laundry branch circuit specified in 220.52.

(3) Appliances: The nameplate VA rating of all appliances and motors that are fastened in place (permanently connected) or located on a specific circuit.

Note: *Be sure to calculate the range and dryer at their nameplate ratings!*

Step 2: Heating and Air-Conditioning Load [220.82(C)]

Include the larger of (1) through (6):

(1) Air-Conditioning Equipment: 100 percent

(2) Heat-Pump Compressor without Supplemental Heating: 100 percent

(3) Heat-Pump Compressor and Supplemental Heating: 100 percent of the nameplate rating of the heat-pump compressor and 65 percent of the supplemental electric heating for central electric space-heating systems.

If the heat-pump compressor is prevented from operating at the same time as the supplementary heat, it can be omitted in the calculation.

(4) Space-Heating Units (three or fewer units): 65 percent

(5) Space-Heating Units (four or more units): 40 percent

(6) Thermal Storage Heating: 100 percent

> **Author's Comment:** One form of thermal storage heating involves heating bricks or water at night when the electric rates are lower. Then during the day, the building uses the thermally stored heat.

Step 3: Feeder/Service Conductors [310.15(B)(6)]

400 Amperes and Less. For individual dwelling units of one-family, two-family, and multifamily dwellings, Table 310.15(B)(6) can be used to size 3-wire, single-phase, 120/240V service or feeder conductors (including neutral conductors) that serve as the main power feeder. Feeder conductors are not required to have an ampacity rating greater than the service conductors [215.2(A)(3)]. The neutral conductor must be sized to carry the unbalanced load according to Table 310.15(B)(6).

Over 400 Amperes. Size ungrounded conductors and the neutral conductor according to Table 310.16.

Dwelling Unit Calculations — Unit 9

9.11 Optional Calculation Example

▶ **Optional Load Calculation Example 1**

Question: What size service conductor is required for a 1,500 sq ft dwelling unit containing the following loads? Figure 9-45

Cooktop	6,000 VA
Disposal	900 VA
Dishwasher	1,200 VA
Dryer	4,000 VA
Ovens (two each)	3,000 VA
Water Heater	4,500 VA
A/C	17A, 230V
Electric Heating (one control unit)	10 kVA

(a) 6 AWG   (b) 4 AWG   (c) 3 AWG   (d) 2 AWG

Answer: (c) 3 AWG

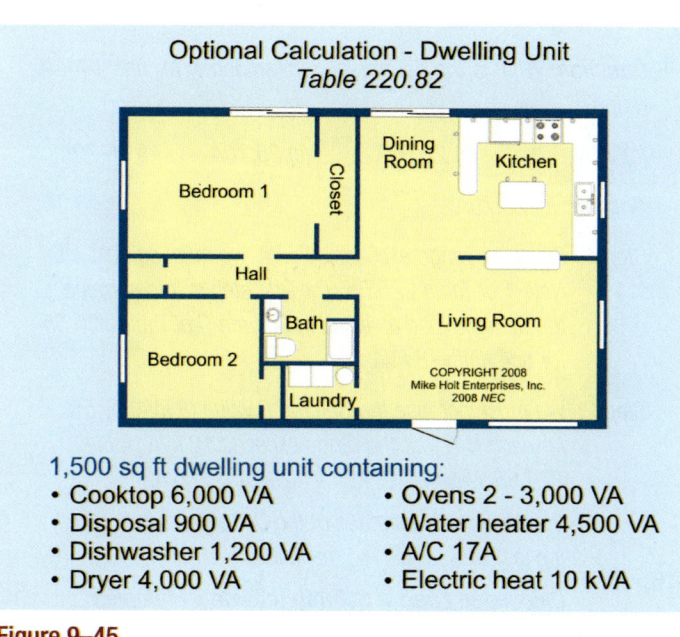

Figure 9–45

Step 1. General Loads [220.82(B)]

General Lighting (1,500 sq ft x 3 VA)		4,500 VA
Small-Appliance Circuits (1,500 VA x 2 circuits)		3,000 VA
Laundry Circuit		1,500 VA
Appliances (nameplate)		
Cooktop		6,000 VA
Disposal		900 VA
Dishwasher		1,200 VA
Dryer		4,000 VA
Ovens (each 3 kW)		6,000 VA
Water Heater		4,500 VA
Connected Load		31,600 VA
First 10 kW at 100%	− 10,000 VA x 1.00 =	10,000 VA
Remainder at 40%	21,600 VA x 0.40 =	+ 8,640 VA
Calculated General Load		18,640 VA

Step 2: Air-Conditioning versus Heat [220.82(C)]

Air-conditioning at 100 percent [220.82(C)(1)] versus electric space heating at 65 percent [220.82(C)(4)]

Air Conditioner [Table 430.248]:

A/C VA = V x A

A/C VA = 230V x 17A
A/C VA = 3,910 VA, (omit)
Electric Space Heat: 10,000 VA x 0.65 = 6,500 VA

Step 3: Feeder/Service Conductors [310.15(B)(6)]

Calculated General Load (Step 1)	18,640 VA
Heat Calculated Load (Step 2)	+ 6,500 VA
Total Calculated Load	25,140 VA

I = VA/E

I = 25,140 VA/240V
I = 105A

The feeder/service ungrounded conductor is sized to 110A, 3 AWG [310.15(B)(6)].

Unit 9 Dwelling Unit Calculations

PART D—OTHER TOPICS OF INTEREST

9.12 Neutral Calculations [220.61]

The feeder/service neutral load is the maximum unbalanced calculated load between the neutral conductor and any ungrounded conductor as determined by Article 220, Part III.

Because phase-to-phase loads aren't connected to the neutral conductor, they're not considered when sizing the neutral conductor.

▶ **Neutral Not Over 200A**

Question: What size 120/240V, single-phase ungrounded and neutral conductors are required for a one-family dwelling unit with a calculated load of 375A, of which 175A consist of 240V loads? **Figure 9–46**

(a) Two–350 kcmil and one–350 kcmil
(b) Two–400 kcmil and one–2/0 AWG
(c) Two–400 kcmil and one–350 kcmil
(d) Two–500 kcmil and one–500 kcmil

Answer: (b) Two—400 kcmil and one—2/0 AWG

The ungrounded conductor is sized at 375A (400 kcmil) according to Table 310.15(B)(6).

Neutral Load = Total Load Less Phase-to-Phase Loads
Neutral Load = 375A - 175A
Neutral Load = 200A

The neutral conductor must be sized to carry the 200A of neutral current.

Neutral Conductor: A 2/0 AWG conductor is rated 200A [Table 310.15(B)(6)], but 250.24(C) requires the neutral conductor to be sized no smaller than 1/0 AWG, based on the 400 kcmil service conductors in accordance with Table 250.66.

Cooking Appliance Feeder/Service Neutral Load [220.61(B)(1)]

The feeder/service cooking appliance neutral load for household cooking appliances, such as electric ranges, wall-mounted ovens, or counter-mounted cooking units is calculated at 70 percent of the calculated load as determined by 220.55.

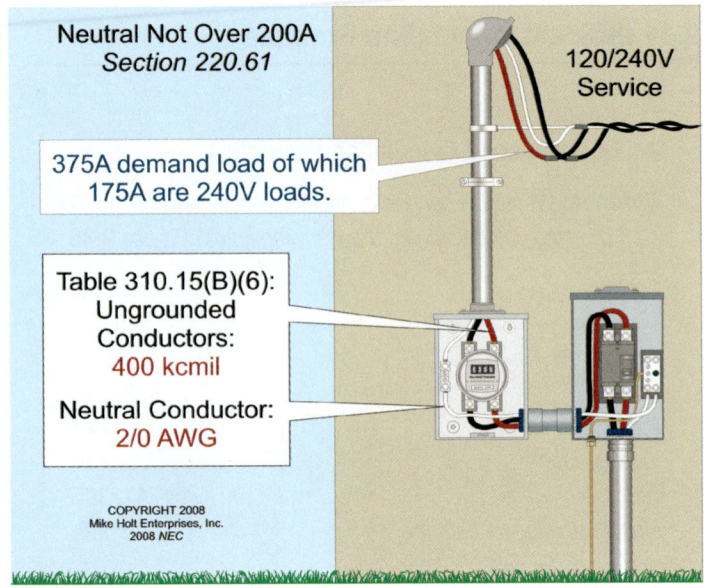

Figure 9–46

▶ **Range Neutral**

Question: What is the neutral load (in amperes) for one 14 kVA range?

(a) 16A   (b) 22.90A   (c) 25.70A   (d) 36.70A

Answer: (c) 25.70A

Step 1: Since the range exceeds 12 kVA, we must comply with Note 1 of Table 220.55. The first step is to determine the demand load as listed in Column C of Table 220.55 for one unit = 8 kVA.

Step 2: We must increase the Column C value (8 kVA) by 5% for each kVA that the range exceeds 12 kVA [Table 220.55, Note 1], 14 kVA – 12 kVA = 2 kVA.

2 x 5% = 10% increase of the Column C value, resulting in 110%, or a 1.10 multiplier.

Calculated Load = Column C Value x Multiplier
Calculated Load = 8 kVA x 1.10
Calculated Load = 8.80 kVA

I = VA/E
I = 8,800VA/240V
I = 36.70A

Neutral load at 70%:

Feeder/Service Neutral Calculated Load = 36.70A x 0.70 [220.61(B)(1)]
Feeder/Service Neutral Calculated Load = 25.70A

Dwelling Unit Calculations — Unit 9

Dryer Feeder/Service Neutral Load [220.61(B)(1)]

The feeder and service dryer neutral load for electric clothes dryers is calculated at 70 percent of the calculated load as determined by 220.54.

▶ **Dryer Over 5 kVA**

Question: What is the feeder/service neutral load for one 5.50 kVA dryer?

(a) 3.85 kVA (b) 4.60 kVA (c) 5 kVA (d) 5.50 kVA

Answer: (a) 3.85 kVA

Feeder/Service Neutral Calculated Load = 5.50 kVA x 0.70 [220.61(B)(1)]

Feeder/Service Neutral Calculated Load = 3.85 kVA

See the NEC Annex D Examples D1(a) and D1(b) for more neutral calculations.

▶ **Neutral Calculation Example**

Question: What is the minimum size feeder/service neutral required for a 2,700 sq ft single dwelling with:

- 17A, 230V A/C unit, 9 kVA 230V electric heater
- Dishwasher (1,250 VA), Waste disposal (940VA), Water Heater (4,500 VA)
- Clothes Dryer, (5,000VA) Electric Range (9,500 VA)
- The ungrounded conductors are 2/0 AWG copper.

For the neutral calculation, disregard the 240V loads such as electric heating and air-conditioning and the 240V water heater.

Find the general lighting/receptacle load using the Table 220.42 demand factors:

General Lighting/Receptacles (2,700 sq ft x 3 VA) [220.12]	8,100 VA		
Small-Appliance Circuits (1,500 VA x 2) [220.52(A)]	3,000 VA		
Laundry Circuit (1,500 VA x 1) [220.52(B)]	+ 1,500 VA		
Total Connected Load	12,600 VA		
First 3,000 VA at 100% [Table 220.42]	−3,000 VA	x 1.00	3,000 VA
Remainder at 35%	9,600 VA	x 0.35	+ 3,360 VA
Calculated Load			6,360 VA
Dishwasher			1,250 VA
Waste disposal			940 VA
Dryer calculated load from 220.54 times 70% [220.61(B)(1)]:	5,000 VA	x 0.70	3,500 VA
Range calculated load from Table 220.55 times 70% [220.61(B)(1)]:	8,000 VA	x 0.70	5,600 VA
			17,650 VA

Neutral Calculated Load: 17,650VA/240V = 73.54A

4 AWG is rated 85A at 75°C, which is also the minimum required neutral from Table 250.66, based on 2/0 AWG ungrounded conductors [250.24(C)(1)].

Unit 9 | Dwelling Unit Calculations

9.13 Grounding and Bonding of Service Equipment

Service Equipment Grounding and Bonding [250.24]

(A) Grounded System. Service equipment supplied from a grounded system must have the neutral conductor terminate in accordance with (1) through (5).

(1) Grounding Location. A grounding electrode conductor must connect the service neutral conductor to the grounding electrode at any accessible location, from the load end of the service drop or service lateral, up to and including the service disconnecting means. **Figure 9–47**

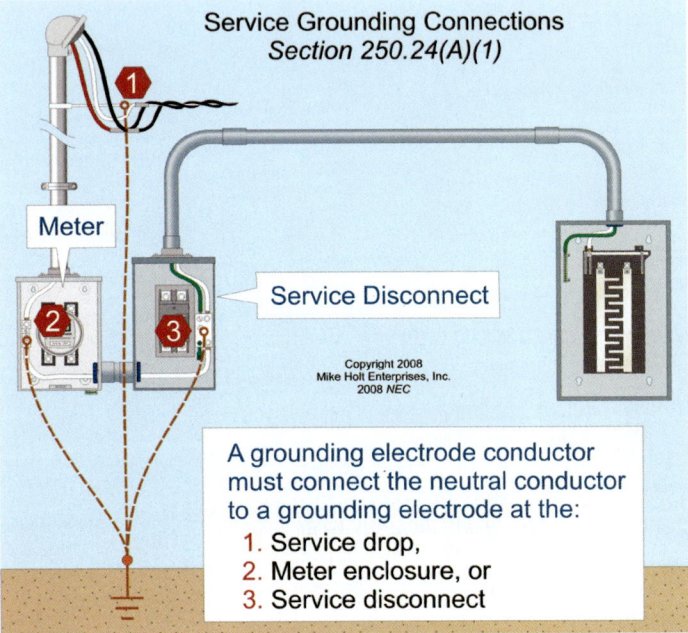

Figure 9–47

Author's Comment: Some inspectors require the service neutral conductor to be grounded (connected to the earth) from the meter socket enclosure, while other inspectors insist that the service neutral conductor be grounded (connected to the earth) only from the service disconnect.

(4) Grounding Termination. When the service neutral conductor is connected to the service disconnecting means [250.24(B)] by a wire or busbar [250.28], the grounding electrode conductor is permitted to terminate to either the neutral terminal or the equipment grounding terminal within the service disconnect.

(5) Neutral-to-Case Connection. A neutral-to-case connection is not permitted on the load side of service equipment, except as permitted by 250.142(B). **Figure 9–48**

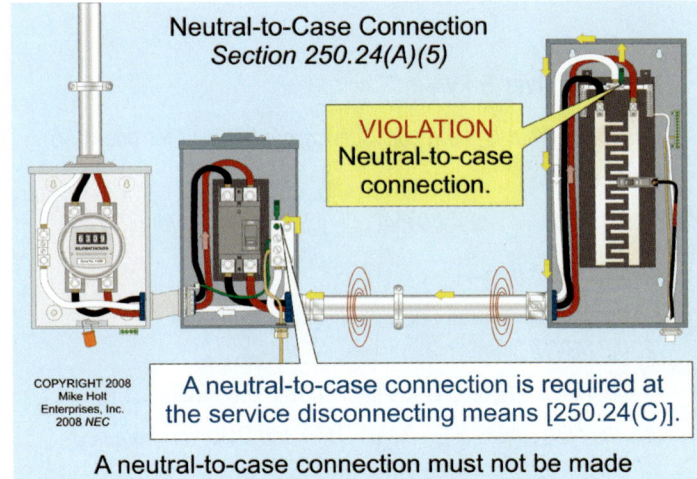

Figure 9–48

Author's Comment: If a neutral-to-case connection is made on the load side of service equipment, dangerous objectionable neutral current will flow on conductive metal parts of electrical equipment [250.6(A)]. Objectionable neutral current on metal parts of electrical equipment can cause electric shock and even death from ventricular fibrillation, as well as a fire. **Figures 9–49 and 9–50**

(B) Bonding. A main bonding jumper [250.28] must be installed for the purpose of connecting the neutral conductor to the metal parts of the service disconnecting means. **Figure 9–51**

(C) Neutral Conductor. A service neutral conductor from the electric utility must terminate to each service disconnecting means via a main bonding jumper [250.24(B)] that is installed between the service neutral conductor and each service disconnecting means enclosure. **Figures 9–52 and 9–53**

Author's Comment: The service neutral conductor provides the effective ground-fault current path to the power supply to ensure that dangerous voltage from a ground fault will be quickly removed by opening the overcurrent device [250.4(A)(3) and 250.4(A)(5)]. **Figure 9-54** An effective ground-fault current path is an intentionally constructed low-impedance conductive path designed to carry fault current from the point of a ground fault on a wiring system to the electrical supply source [Article 100]. **Figure 9-55**

Dwelling Unit Calculations — Unit 9

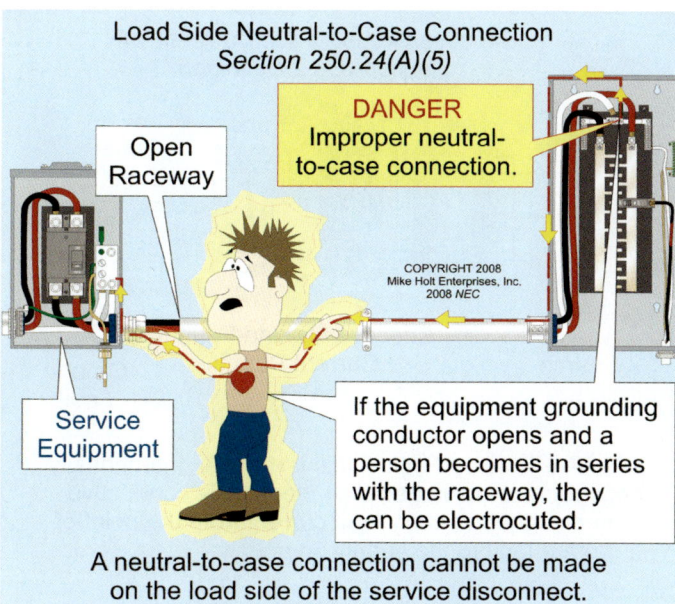

Figure 9–49

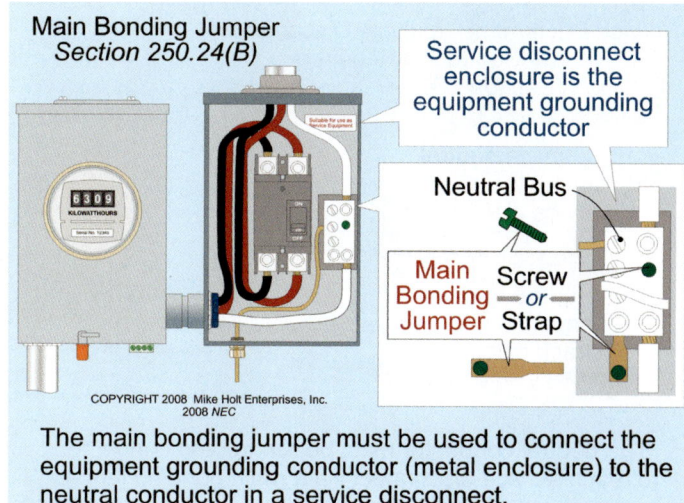

Figure 9–51

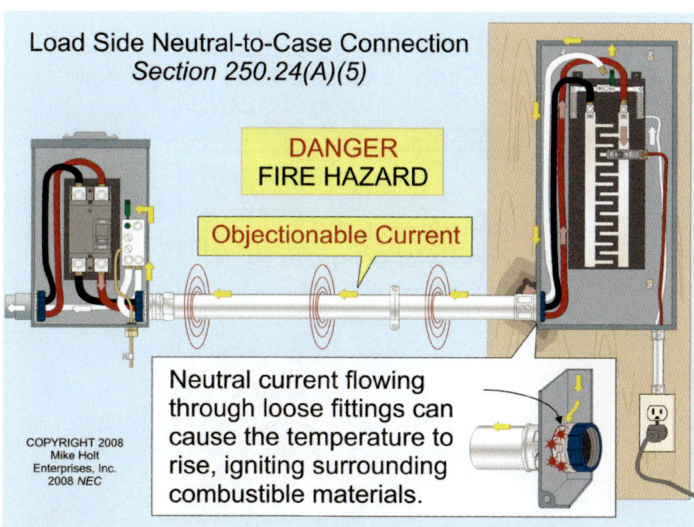

Figure 9–50

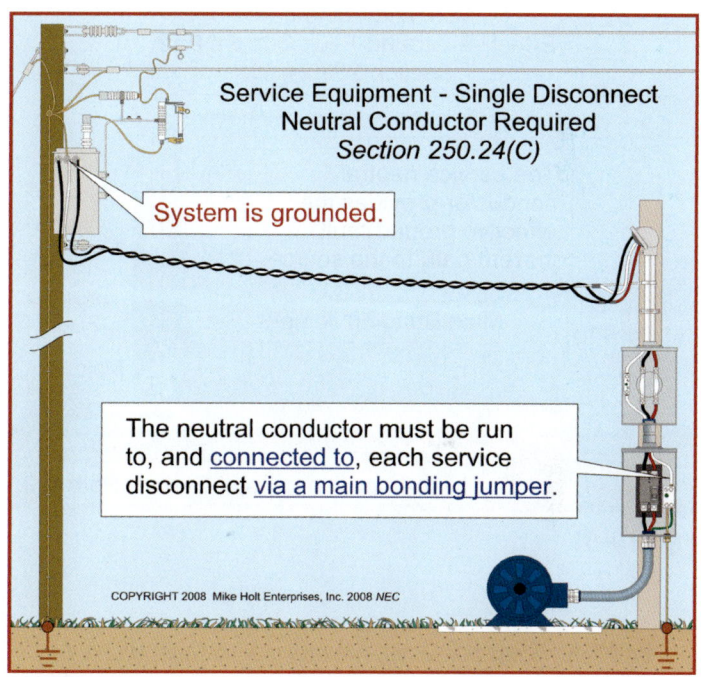

Figure 9–52

Unit 9 Dwelling Unit Calculations

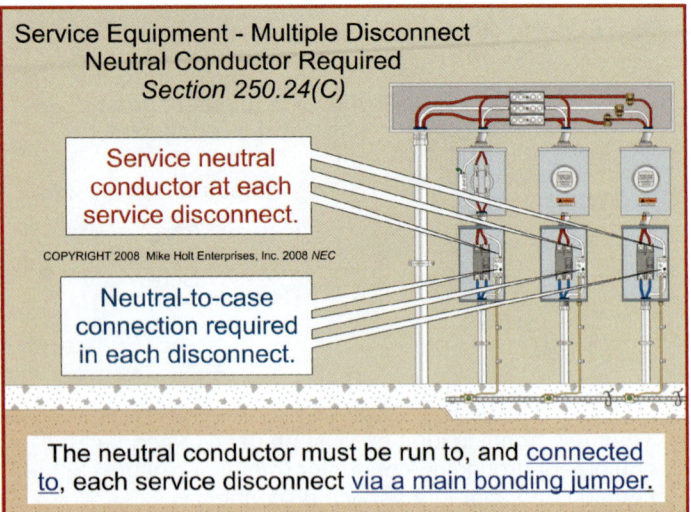

Figure 9–53

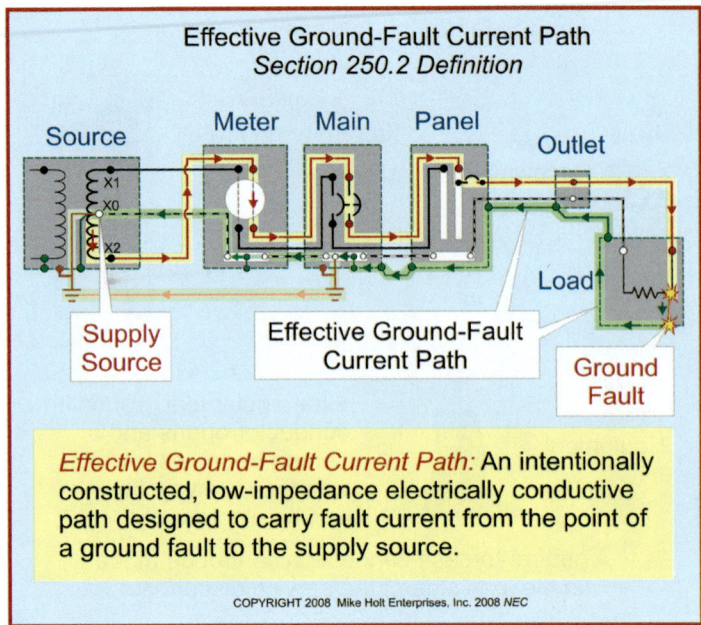

Figure 9–55

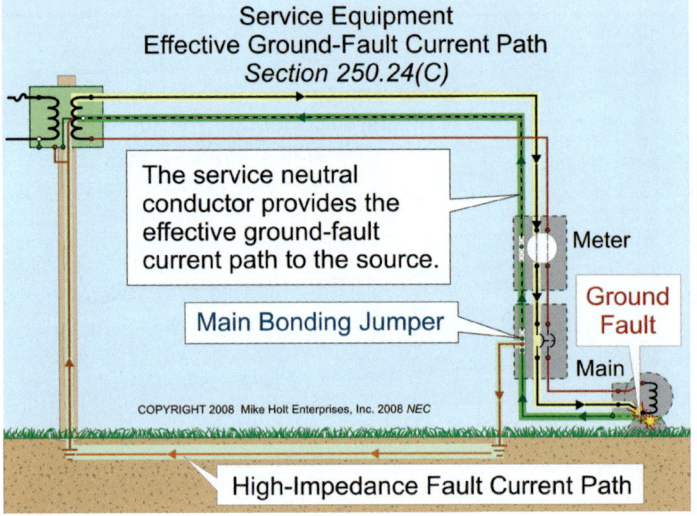

Figure 9–54

DANGER: Dangerous voltage from a ground fault will not be removed from metal parts, metal piping, and structural steel if the service disconnecting means enclosure is not connected to the service neutral conductor. This is because the contact resistance of a grounding electrode to the earth is so great that insufficient fault current returns to the power supply if the earth is the only fault current return path to open the circuit overcurrent device. Figure 9–56

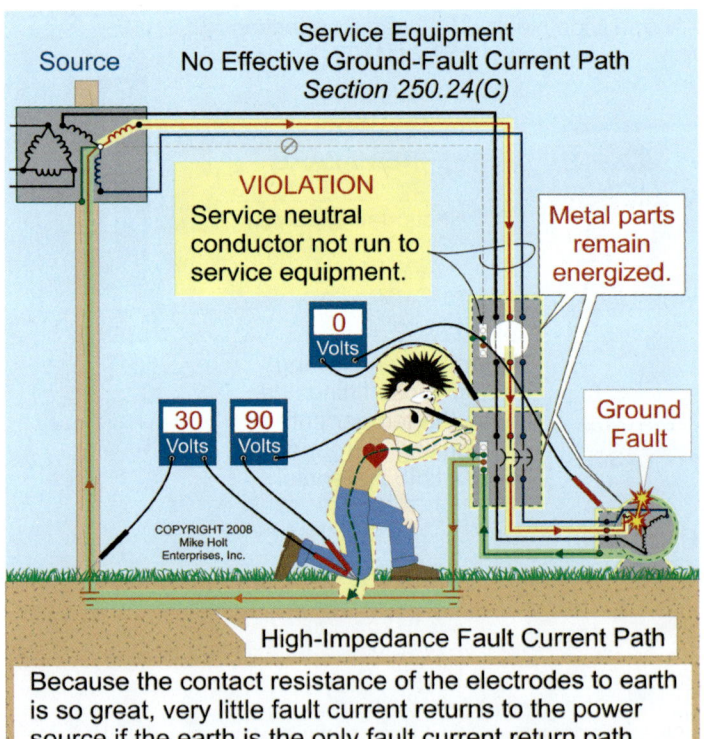

Figure 9–56

270   Mike Holt's Illustrated Guide to NEC Exam Preparation

Dwelling Unit Calculations — Unit 9

Author's Comment: For example, if the neutral conductor is opened, dangerous voltage will be present on metal parts under normal conditions, providing the potential for electric shock. If the earth's ground resistance is 25 ohms and the load's resistance is 25 ohms, the voltage drop across each of these resistors will be half of the voltage source. Since the neutral is connected to the service disconnect, all metal parts will be elevated to 60V above the earth's potential for a 120/240V system. **Figure 9–57**

To determine the actual voltage on the metal parts from an open service neutral conductor, you need to do some complex math calculations. Visit www.MikeHolt.com and go to the "Free Stuff" link to download a spreadsheet for this purpose.

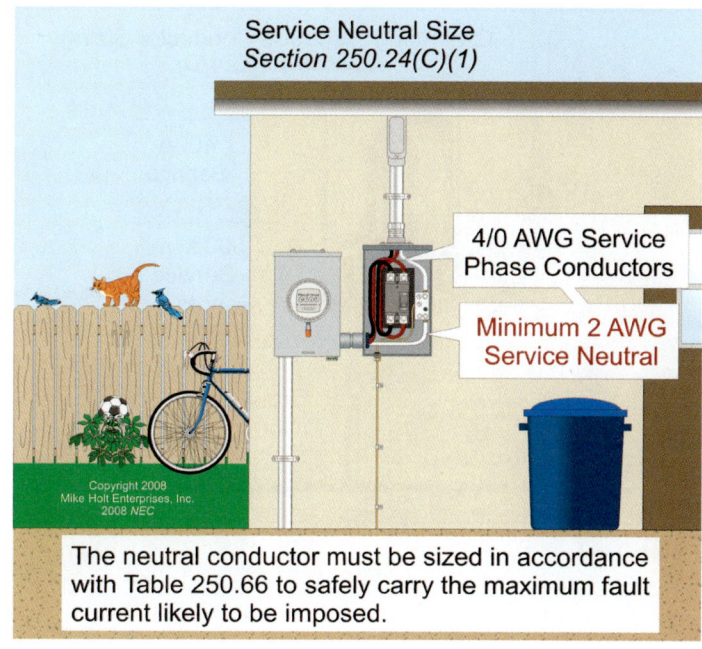

Figure 9–58

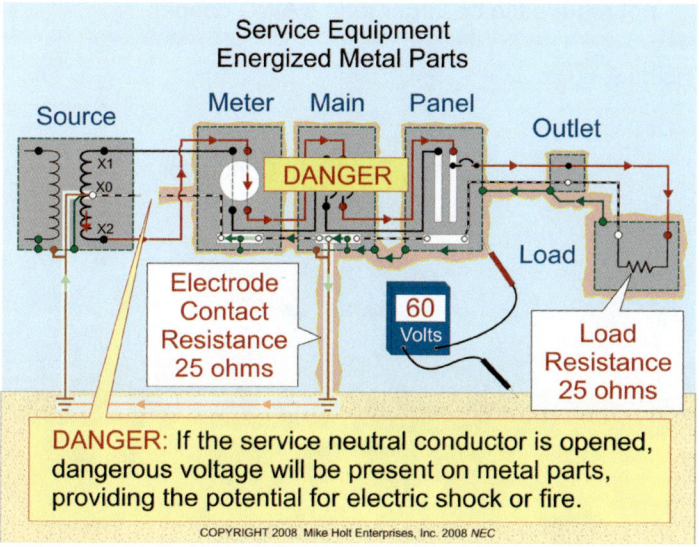

Figure 9–57

(1) Neutral Conductor Minimum Size. Because the service neutral conductor serves as the effective ground-fault current path to the source for ground faults, it must be sized so it can safely carry the maximum fault current likely to be imposed on it [110.10 and 250.4(A)(5)]. This is accomplished by sizing the neutral conductor in accordance with Table 250.66, based on the cross-sectional area of the ungrounded service conductor. **Figure 9-58**

Author's Comment: In addition, the neutral conductors must have the capacity to carry the maximum unbalanced neutral current in accordance with 220.61.

(2) Parallel Neutral Conductor. Where service conductors are paralleled, a neutral conductor must be installed in each of the parallel raceways and it must be sized in accordance with Table 250.66, based on the area of the largest service conductor in the raceway. In no case can the neutral conductor in each parallel set be sized smaller than 1/0 AWG [310.4(A)].

Author's Comment: In addition, the neutral conductors must have the capacity to carry the maximum unbalanced neutral current in accordance with 220.61.

(D) Grounding Electrode Conductor. A grounding electrode conductor, sized in accordance with 250.66 based on the area of the ungrounded service conductor, must connect the metal parts of service equipment enclosures to a grounding electrode in accordance with Part III of Article 250.

▶ **Grounding Electrode Conductor Sizing Example**

Question: What is the minimum size grounding electrode conductor for a 400A service where the ungrounded service conductors are sized at 500 kcmil? **Figure 9-59**

(a) 3 AWG (b) 2 AWG (c) 1 AWG (d) 1/0 AWG

Answer: (d) 1/0 AWG [Table 250.66]

Unit 9 Dwelling Unit Calculations

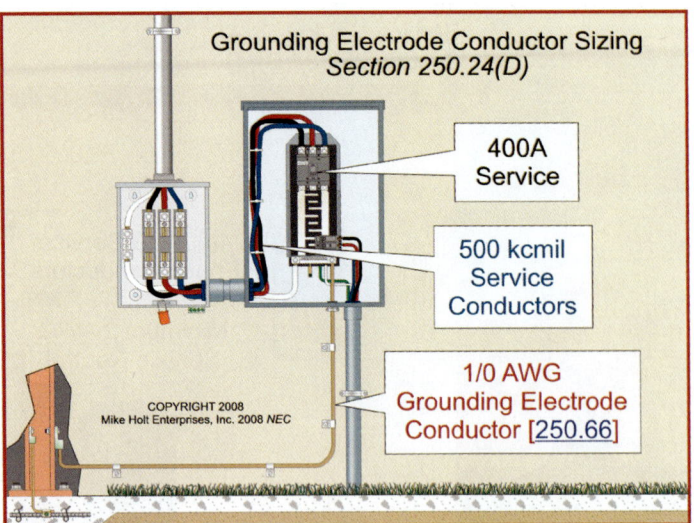

Figure 9–59

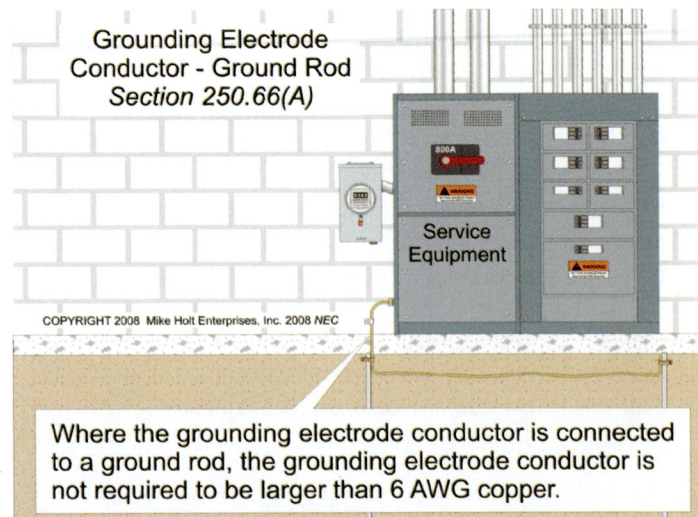

Figure 9–60

Author's Comment: Where the grounding electrode conductor is connected to a ground rod, that portion of the conductor that is the sole connection to the ground rod isn't required to be larger than 6 AWG copper [250.66(A)], **Figure 9–60**. Where the grounding electrode conductor is connected to a concrete-encased electrode, that portion of the conductor that is the sole connection to the concrete-encased electrode isn't required to be larger than 4 AWG copper [250.66(B)]. **Figure 9–61**

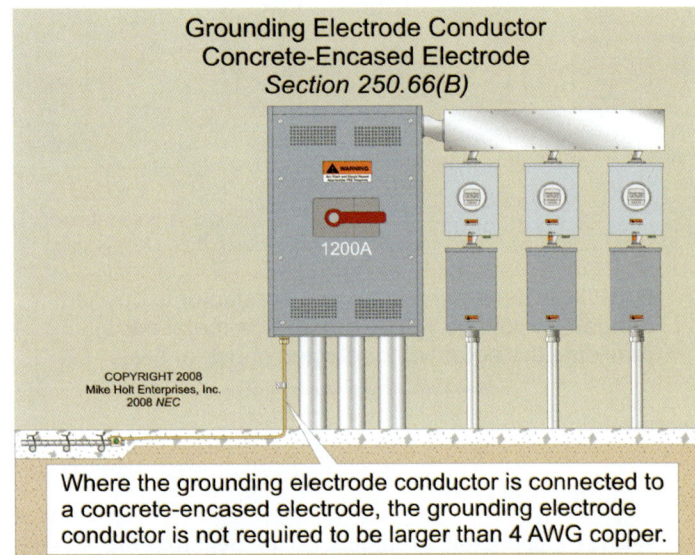

Figure 9–61

Unit 9 — Conclusion

CONCLUSION TO UNIT 9—DWELLING UNIT CALCULATIONS

The residential branch circuit, feeder, and service calculations in Unit 9 required you to understand and apply many different demand factors. Since not all loads are used simultaneously in a home, the service to a home is allowed to be sized not on the total possible load, but on a percentage of that, based on what we call demand factors.

The *Code* doesn't explain how demand factors were derived, and it's not essential that you understand this in order to apply them correctly. Be sure to work on some practice calculations so you understand how to apply the various demand factors to a dwelling unit calculation. It's important to have a good working knowledge of this before going on to the following unit on multifamily dwellings, which will build on the concepts developed in this unit.

The standard calculation and the optional calculation methods were both discussed in this Unit 9. These are two distinctly different calculation methods, so be careful not to mix them together. Which method is better to use? On an exam, you'll likely be told which method to use on a specific question, but if the question doesn't specify a method, use the standard calculation. The optional method is usually faster and easier to apply, so it has a natural advantage for daily use on the job.

Unit 9 Practice Questions

UNIT 9—DWELLING UNIT CALCULATIONS PRACTICE QUESTIONS

(•Indicates that 75% or fewer of those who took this exam answered the question correctly.))

PART A—GENERAL REQUIREMENTS

9.2 Voltages [220.5(A)]

1. Unless other voltages are specified, feeder and service loads must be computed at a nominal system voltage of _____.

   (a) 120V
   (b) 120/240V
   (c) 120/208V
   (d) any of these

9.3 Fraction of an Ampere [220.5(B)]

2. Where a calculation results in a fraction of an ampere that is less than _____, that fraction is permitted to be dropped.

   (a) 0.40
   (b) 0.45
   (c) 0.50
   (d) 0.55

9.4 Small-Appliance Circuits [210.11(C)(1)]

3. A minimum of _____ 20A small-appliance branch circuit(s) is(are) required for receptacle outlets in the kitchen, dining room, breakfast room, pantry, and similar areas.

   (a) one
   (b) two
   (c) three
   (d) four

4. When sizing the feeder or service, each dwelling unit must have a minimum feeder load of _____ for the two small-appliance branch circuits.

   (a) 1,000 VA
   (b) 2,000 VA
   (c) 3,000 VA
   (d) 4,000 VA

9.5 Cooking Equipment—Branch Circuit [Table 220.55, Note 4]

5. What is the branch-circuit calculated load (in amperes) for one 11 kW range?

   (a) 21A
   (b) 27A
   (c) 33A
   (d) 38A

6. What is the branch-circuit calculated load (in amperes) for one 13.50 kW range?

   (a) 33A
   (b) 37A
   (c) 50A
   (d) 58A

One Counter-Mounted Cooking Unit(s) and Up to Two Wall-Mounted Ovens [Table 220.55, Note 4]

7. •What is the branch-circuit load (in amperes) for one 6 kW wall-mounted oven rated 240V, single-phase?

   (a) 15A
   (b) 25A
   (c) 30A
   (d) 40A

8. •What is the branch-circuit load (in amperes) for one 4.80 kW counter-mounted cooking unit rated 240V, single-phase?

    (a) 15A
    (b) 20A
    (c) 25A
    (d) 30A

9. What is the branch-circuit load (in amperes) for one 6 kW counter-mounted cooking unit and one 3 kW wall-mounted oven rated 240V, single-phase?

    (a) 27A
    (b) 30A
    (c) 33A
    (d) 36A

10. What is the branch-circuit load (in amperes) for one 6 kW counter-mounted cooking unit and two 4 kW wall-mounted ovens rated 240V, single-phase?

    (a) 21A
    (b) 27A
    (c) 31A
    (d) 37A

9.6 Laundry Circuit [210.11(C)(2)]

11. The *NEC* does not require a separate circuit for the washing machine, but does require a separate circuit for the laundry room receptacle or receptacles, of which one can be for the washing machine.

    (a) True
    (b) False

12. Each dwelling unit must have a feeder load consisting of _____ for the 20A laundry circuit.

    (a) 1,500 VA
    (b) 2,000 VA
    (c) 3,000 VA
    (d) 4,500 VA

9.7 Lighting and Receptacles

13. The *NEC* requires a minimum of 3 VA for each square foot of living space for the required general lighting and general-use receptacles for dwelling units. The dimensions for determining the area are computed from the outside of the building and do not include _____.

    (a) open porches
    (b) garages
    (c) spaces not adaptable for future use
    (d) all of these

14. The 3 VA for each square foot of living space for general lighting includes receptacles rated 20A, but not the small-appliance or laundry circuit receptacles.

    (a) True
    (b) False

15. What is the general lighting load and general-use receptacle load for a 2,100 sq ft home that has 74 receptacle outlets and nine recessed luminaires rated at 100W each?

    (a) 2,100 VA
    (b) 4,200 VA
    (c) 6,300 VA
    (d) 8,400 VA

Number of Circuits Required [Annex D, Example D1(a)]

16. The number of branch circuits required for general lighting and receptacles must be determined from the general lighting load and the rating of the circuits.

    (a) True
    (b) False

17. How many 15A general lighting circuits are required for a 2,340 sq ft home?

    (a) 2 circuits
    (b) 3 circuits
    (c) 4 circuits
    (d) 5 circuits

18. How many 20A general lighting circuits are required for a 2,100 sq ft home?

    (a) 2 circuits
    (b) 3 circuits
    (c) 4 circuits
    (d) 5 circuits

Unit 9 Practice Questions

19. How many 20A circuits are required for the general lighting and receptacles for a 1,500 sq ft dwelling unit?

    (a) 2 circuits
    (b) 3 circuits
    (c) 4 circuits
    (d) 5 circuits

PART B—STANDARD METHOD—FEEDER/SERVICE LOAD CALCULATIONS

9.8 Dwelling Unit Feeder/Service Load Calculations [Article 220, Part III]

20. The *NEC* recognizes that the general lighting and receptacles, and small-appliance and laundry circuits will not all be on or loaded at the same time and permits a(n) _____ to be applied to the total of these loads.

    (a) demand factor
    (b) adjustment factor
    (c) correction factor
    (d) correction

21. Because the air-conditioning and heating loads are not on at the same time they are called noncoincidental, and it is permissible to omit the smaller of the two loads.

    (a) True
    (b) False

22. A load demand factor of 75 percent is permitted for _____ or more appliances fastened in place such as waste disposals, dishwashers, trash compactors, water heaters, etc.

    (a) one
    (b) two
    (c) three
    (d) four

23. The feeder or service calculated load for electric clothes dryers located in dwelling units must not be less than _____.

    (a) 5,000W
    (b) the nameplate rating
    (c) a or b
    (d) the greater of a or b

24. •A feeder or service dryer load is required even if the dwelling unit does not contain an electric dryer.

    (a) True
    (b) False

25. Household cooking appliances rated 1kW can have the feeder and service loads calculated according to the demand factors of Table 220.55.

    (a) True
    (b) False

26. Dwelling unit feeder and service ungrounded (hot) conductors are sized according to Table 310.15(B)(6) for _____.

    (a) 3-wire, single-phase, 120/240V systems up to 400A
    (b) 3-wire, single-phase, 120/208V systems up to 400A
    (c) 4-wire, single-phase, 120/240V systems up to 400A
    (d) none of these

27. What is the general lighting load for a 2,700 sq ft dwelling unit?

    (a) 2,700 VA
    (b) 6,840 VA
    (c) 8,100 VA
    (d) 12,600 VA

28. What is the total connected load for general lighting and receptacles, small-appliance, and laundry circuits for a 6,540 sq ft dwelling unit?

    (a) 2,700 VA
    (b) 8,100 VA
    (c) 12,600 VA
    (d) 24,120 VA

29. What is the feeder or service calculated load for one air conditioner (28A, 230V) and three baseboard heaters (3 kW each)?

    (a) 3,000 VA
    (b) 6,400 VA
    (c) 8,050 VA
    (d) 9,000 VA

30. What is the feeder or service calculated load for an air conditioner (28A, 230V) and electric space heater (10 kW)?

    (a) 3,910 VA
    (b) 6,440 VA
    (c) 10,000 VA
    (d) 10,350 VA

Practice Questions — Unit 9

31. What is the feeder or service calculated load for a waste disposal (940 VA), dishwasher (1,250 VA), and a water heater (4,500 VA)?

    (a) 5,018 VA
    (b) 6,272 VA
    (c) 6,690 VA
    (d) 8,363 VA

32. •What is the feeder or service calculated load for a waste disposal (940 VA), dishwasher (1,250 VA), a trash compactor (1,100 VA), and a water heater (4,500 VA)?

    (a) 5,843 VA
    (b) 7,303 VA
    (c) 7,790 VA
    (d) 9,738 VA

33. What is the feeder or service calculated load for a 4 kW dryer?

    (a) 3 kW
    (b) 4 kW
    (c) 5 kW
    (d) 6 kW

34. What is the feeder or service calculated load for a 5.50 kW dryer?

    (a) 3 kW
    (b) 4 kW
    (c) 5 kW
    (d) 5.50 kW

35. What is the feeder or service calculated load for two 3 kW cooking appliances?

    (a) 3.90 kW
    (b) 4.20 kW
    (c) 4.50 kW
    (d) 5.00 kW

36. What is the feeder or service calculated load for one 6 kW cooking appliance?

    (a) 3.90 kW
    (b) 4.50 kW
    (c) 4.80 kW
    (d) 6 kW

37. What is the feeder or service calculated load for one 6 kW and two 3 kW cooking appliances?

    (a) 4.50 kW
    (b) 4.80 kW
    (c) 6 kW
    (d) 9.30 kW

38. What is the feeder or service calculated load for an 11.50 kW range?

    (a) 7 kW
    (b) 8 kW
    (c) 9 kW
    (d) 10 kW

39. What is the feeder or service calculated load for a 13.60 kW range?

    (a) 6 kW
    (b) 8 kW
    (c) 8.80 kW
    (d) 9.20 kW

40. What AWG size 120/240V, single-phase feeder/service copper conductors are required for a dwelling unit that has a 190A service calculated load?

    (a) 1/0 AWG
    (b) 2/0 AWG
    (c) 3/0 AWG
    (d) 4/0 AWG

41. •What is the net calculated load for the general lighting, small-appliance, and laundry circuits for a 1,500 sq ft dwelling unit?

    (a) 3,500 VA
    (b) 4,200 VA
    (c) 5,100 VA
    (d) 5,700 VA

42. What is the feeder/service calculated load for an air conditioner (18A, 230V) and electric space heating (4 kW)?

    (a) 2,000 VA
    (b) 3,910 VA
    (c) 4,000 VA
    (d) 4,140 VA

Unit 9 — Practice Questions

43. What is the feeder/service calculated load for a dwelling unit that has one dishwasher (1.50 kW) and one water heater (4 kW)?

    (a) 4 kVA
    (b) 4.13 kVA
    (c) 5.50 kVA
    (d) 1,500 kVA

44. What is the calculated load for a 4.50 kW dryer?

    (a) 4.50 kW
    (b) 5 kW
    (c) 5.50 kW
    (d) 8 kW

45. What is the calculated load for a 14 kW range?

    (a) 8 kW
    (b) 8.40 kW
    (c) 8.80 kW
    (d) 14 kW

46. If the total calculated load is 30 kVA for a 120/240V dwelling unit, what is the feeder/service copper conductor size?

    (a) 4 AWG
    (b) 2 AWG
    (c) 1/0 AWG
    (d) 3/0 AWG

47. If a service raceway contains 2 AWG service conductors, what size copper bonding jumper is required for the service raceway?

    (a) 8 AWG
    (b) 6 AWG
    (c) 4 AWG
    (d) 3 AWG

48. If a service contains 2 AWG conductors, what is the minimum size grounding electrode conductor required?

    (a) 8 AWG
    (b) 6 AWG
    (c) 4 AWG
    (d) 3 AWG

PART C—OPTIONAL METHOD—FEEDER/SERVICE LOAD CALCULATIONS

9.10 Dwelling Unit Optional Calculations [220.82]

49. When sizing the feeder or service conductor according to the optional method we must:

    (a) Determine the total connected load of the general lighting and receptacles, small-appliance and laundry branch circuits, and the nameplate VA rating of all appliances and motors.
    (b) Determine the calculated load by applying the following demand factors to the total connected load: First 10 kVA at 100 percent, remainder at 40 percent.
    (c) Determine the air-conditioning versus heat calculated load.
    (d) all of these.

50. A 1,500 sq ft dwelling unit contains the following loads:

    Waste Disposal (1 kVA)
    Dishwasher (1.50 kVA)
    Water Heater (5 kW)
    Dryer (5.50 kW)
    Cooktop (6 kW)
    Two Ovens (each 3 kW)
    A/C (3 hp, 230V)
    Two separately controlled 5 kW space-heating units

    Use the optional method of calculation to determine what size aluminum conductors are required for the 120/240V, single-phase service.

    (a) 1 AWG AL
    (b) 1/0 AWG AL
    (c) 2/0 AWG AL
    (d) 3/0 AWG AL

Practice Questions — Unit 9

51. A dwelling unit has 2,330 sq ft of living space. In addition, there is 900 sq ft of porch, and 400 sq ft of carport (not a garage) with the following loads:

    Waste Disposal (1 kVA)

    Dishwasher (1.50 kW)

    Trash Compactor (1.50 kVA)

    Water Heater (6 kW)

    Range (14 kW)

    Dryer (4.50 kW)

    A/C Unit (5 hp 230V)

    Baseboard Heat (Four at 2.50 kW each)

    Use the optional method to determine what size aluminum conductors are required for the 120/240V, single-phase service.

    (a) 1 AWG AL
    (b) 1/0 AWG AL
    (c) 2/0 AWG AL
    (d) 3/0 AWG AL

PART D—OTHER TOPICS OF INTEREST

9.12 Neutral Calculations [220.61]

52. The feeder and service neutral load is the maximum unbalanced calculated load between the neutral conductor and any one ungrounded (hot) conductor as determined by Article 220. Since 240V loads cannot be connected to the neutral conductor, 240V loads are not considered for sizing the feeder grounded conductor.

    (a) True
    (b) False

53. •What size single-phase, 3-wire feeder is required for a 475A calculated load of which 275A consists of 240V loads? Size the conductors based on 75°C terminals in accordance with 110.14(C)(1)(b).

    (a) Two 500 kcmil and one 350 kcmil
    (b) Two 750 kcmil and one 3/0 AWG
    (c) Two 750 kcmil and one 350 kcmil
    (d) Two 750 kcmil and one 500 kcmil

54. The feeder or service neutral load for household cooking appliances, such as electric ranges, wall-mounted ovens, or counter-mounted cooking units, is permitted to be calculated at _____ percent of the load as determined by 220.55.

    (a) 50
    (b) 60
    (c) 70
    (d) 80

55. What is the dwelling unit service neutral load for one 9 kW range?

    (a) 3.50 kW
    (b) 5.60 kW
    (c) 6.50 kW
    (d) 12 kW

56. The service neutral load for household electric clothes dryers is permitted to be calculated at _____ percent of the load as determined by 220.54.

    (a) 50
    (b) 60
    (c) 70
    (d) 80

57. What is the feeder neutral load for one 6 kW household dryer?

    (a) 4.20 kW
    (b) 4.70 kW
    (c) 5.40 kW
    (d) 6.60 kW

Unit 9 Challenge Questions

UNIT 9—DWELLING UNIT CALCULATIONS CHALLENGE QUESTIONS

(•Indicates that 75% or fewer of those who took this exam answered the question correctly.)

PART A—GENERAL REQUIREMENTS

9.5 Cooking Equipment—Branch Circuit [Table 220.55, Note 4]

1. The branch-circuit calculated load for one 18 kW range is _____.

   (a) 8 kW
   (b) 10.40 kW
   (c) 12 kW
   (d) 18 kW

2. A dwelling unit kitchen has the following appliances: one 9 kW cooktop and one wall-mounted oven rated 5.30 kW. The branch-circuit calculated load for these appliances is _____.

   (a) 8 kW
   (b) 8.80 kW
   (c) 12 kW
   (d) 14.30 kW

PART B—STANDARD METHOD—FEEDER/SERVICE LOAD CALCULATIONS

9.8 Dwelling Unit Feeder/Service Load Calculations [Article 220, Part III]

3. An apartment building contains 20 units, each 840 sq ft. What is the general lighting and general-use receptacle feeder calculated load for each dwelling unit?

   Note: Laundry facilities are provided on the premises for all tenants and no laundry circuit is required in each unit [210.52(F) Ex 1].

   (a) 3,520 VA
   (b) 3,882 VA
   (c) 4,220 VA
   (d) 6,300 VA

4. What is the calculated load for the general lighting, receptacles, small-appliance circuits, and laundry circuits for a 1,800 sq ft dwelling unit?

   (a) 4,300 VA
   (b) 4,900 VA
   (c) 5,415 VA
   (d) 5,700 VA

5. How many 15A, 120V branch circuits are required for general-use receptacles and lighting in a 1,800 sq ft dwelling unit?

   (a) 1 circuit
   (b) 2 circuits
   (c) 3 circuits
   (d) 4 circuits

6. How many 20A, 120V branch circuits are required for general-use receptacles and lighting in a 2,800 sq ft dwelling unit?

   (a) 2 circuits
   (b) 4 circuits
   (c) 6 circuits
   (d) 7 circuits

Challenge Questions — Unit 9

Appliances [220.53]

7. A dwelling unit contains one of each of the following: dishwasher (1.20 kVA), trash compactor (1.50 kVA), and water heater (4 kVA). The appliance calculated load added to the service is _____.

   (a) 5.90 kVA
   (b) 6.70 kVA
   (c) 7.70 kVA
   (d) 8.80 kVA

8. •A dwelling unit contains one of each of the following: dishwasher (½ hp) and water heater (4 kVA). What is the appliance calculated load for the dwelling unit?

   (a) 4 kVA
   (b) 5.10 kVA
   (c) 9 kVA
   (d) 11 kVA

Clothes Dryer [220.54]

9. A dwelling unit contains a 5.50 kW electric clothes dryer. What is the feeder calculated load for the dryer?

   (a) 3.38 kW
   (b) 4.50 kW
   (c) 5 kW
   (d) 5.50 kW

Cooking Equipment [220.55]

10. The feeder load for twelve 1.75 kW cooktops is _____.

    (a) 10.50 kW
    (b) 12.50 kW
    (c) 13.50 kW
    (d) 21 kW

11. The minimum neutral for a branch circuit to an 8 kW household range is _____. (See 210.19(A)(3) Ex 2)

    (a) 10 AWG
    (b) 12 AWG
    (c) 8 AWG
    (d) 6 AWG

12. What is the maximum dwelling unit feeder or service calculated load for fifteen 8 kW cooking units?

    (a) 38.40 kW
    (b) 42 kW
    (c) 48 kW
    (d) 52 kW

13. What is the feeder calculated load for five 10 kW, five 14 kW, and five 16 kW household ranges?

    (a) 21 kW
    (b) 25 kW
    (c) 28 kW
    (d) 33 kW

14. A dwelling unit has one 6 kW cooktop and one 6 kW oven. What is the minimum feeder calculated load for the cooking appliances?

    (a) 6.50 kW
    (b) 7.10 kW
    (c) 7.80 kW
    (d) 8.20 kW

15. The maximum feeder calculated load for an 8 kW range is _____.

    (a) 6.30 kW
    (b) 8 kW
    (c) 8.50 kW
    (d) 12 kW

16. The minimum feeder calculated load for five 5 kW ranges, two 4 kW ovens, and four 7 kW cooking units is _____.

    (a) 16.50 kW
    (b) 17.10 kW
    (c) 19.50 kW
    (d) 28.20 kW

17. The minimum feeder calculated load for two 3 kW wall-mounted ovens and one 6 kW cooktop is _____.

    (a) 8.40 kW
    (b) 9.30 kW
    (c) 9.60 kW
    (d) 11 kW

18. •If each unit of a duplex apartment requires a 100A main, the resulting 200A service will require _____ AWG conductors.

    (a) 1 AWG
    (b) 1/0 AWG
    (c) 3/0 AWG
    (d) 250 kcmil

Unit 9 — Challenge Questions

19. After all demand factors have been taken into consideration, the calculated load for a service is a 24,221 VA, 120/240V, single-phase system. The minimum service size is _____.

    (a) 70A
    (b) 80A
    (c) 110A
    (d) 125A

PART C—OPTIONAL METHOD—FEEDER/SERVICE LOAD CALCULATIONS

9.10 Dwelling Unit Optional Calculations [220.82]

20. After all demand factors have been taken into consideration, the calculated load for a dwelling unit is 21,560 VA. The minimum service size for this residence will be _____ if the optional method of service calculations is used for a 120/240V, single-phase system. Be sure to read 220.82 carefully!

    (a) 80A
    (b) 100A
    (c) 125A
    (d) 150A

21. Using the optional calculations method, determine the calculated load for a 6 kW electric space-heating unit and a 4 kVA air-conditioning unit.

    (a) 3,900 VA
    (b) 4,000 VA
    (c) 5,000 VA
    (d) 9,000 VA

22. The total connected load of a dwelling unit is 25 kVA, not including heat or air-conditioning. If the heat is separately controlled in five rooms (10 kW) and the air-conditioning is 6 kVA, then the total service calculated load is _____ if the optional method is used.

    (a) 5 kVA
    (b) 22 kVA
    (c) 23 kVA
    (d) 24 kVA

23. Using the optional method, determine the service for the following loads: 1,200 sq ft first floor, plus 600 sq ft on the second floor, a 200 sq ft open porch, dishwasher (½ hp), water heater (4 kW), clothes dryer (4 kW), oven (6 kW), cooktop (6 kW), A/C (5 hp, 230V), electric space heating (6 kW), and pool pump (¾ hp, 115V). The service source voltage is 120/240V, single-phase.

    (a) 70A
    (b) 80A
    (c) 110A
    (d) 125A

24. An 1,800 sq ft residence contains the following: a 4 kW water heater, five separately controlled electric space-heating units totaling 10 kW, one 1.50 kW dishwasher, one 6 kW range, one 4.50 kW dryer, two 3 kW ovens, and one 6 kW air conditioner. The 120/240V, single-phase service for the loads will be _____. The optional method of service calculation is to be used.

    (a) 100A
    (b) 110A
    (c) 125A
    (d) 150A

25. A dwelling unit has 1,200 sq ft on the first floor, 600 sq ft upstairs (unfinished but adaptable for future use), a 200 sq ft open porch, and the following loads: pool pump (¾ hp), range (13.90 kW), dishwasher (1.20 kW), water heater (4 kW), dryer (4 kW), A/C (5 hp), and electric space heating (6 kW). Using the optional calculation method, what size feeder/service conductor is required for this 120/240V, single-phase service?

    (a) 100A service with 4 AWG
    (b) 110A service with 3 AWG
    (c) 125A service with 2 AWG
    (d) 150A service with 1 AWG

CHAPTER 3: ADVANCED *NEC* CALCULATIONS

Unit 10   Multifamily Dwelling Calculations

Unit 11   Commercial Calculations

Unit 12   Transformer Calculations

Chapter 3 Notes

UNIT 10 — Multifamily Dwelling Calculations

INTRODUCTION TO UNIT 10—MULTIFAMILY DWELLING CALCULATIONS

The *NEC* defines "Multifamily Dwellings" as buildings that contain three or more dwelling units. There are many similarities in the electrical requirements for multifamily and single-family dwellings, but there are differences also. The wiring method used for a single-family dwelling unit is often Type NM cable (unless local codes don't allow Type NM). Depending on several factors, such as the number of units, number of stories, the building construction type, and local building codes, a multifamily dwelling may use metal raceways or metallic-sheathed cable. The branch-circuit requirements of Article 210 are almost the same for each of these types of buildings, so much of the branch-circuit design is similar.

When sizing the service or feeder conductors for a single-family dwelling in Unit 9, the use of Table 310.15(B)(6) is allowed. When sizing conductors for the service or feeder to a multifamily dwelling, use Table 310.16 instead. However, Table 310.15(B)(6) may still be used for the feeders to an individual dwelling unit within the building.

To size the electrical feeders to individual dwelling units, the requirements of Article 220 that were discussed in Unit 9 may be followed, but the main service that feeds a multifamily dwelling is based on the sum of all of the individual dwelling unit loads with the appropriate demand factors applied.

Demand factors are allowed for multifamily dwellings because diversity in usage is expected, so the maximum connected load is not likely to be in use simultaneously. Many of the principles explained in Unit 9 will still be used for multifamily dwellings here in Unit 10.

The *NEC* defines a dwelling unit as a single unit, providing complete and independent living facilities for one or more persons, including permanent provisions for living, sleeping, cooking, and sanitation. **Figure 10–1**

A multifamily dwelling is a building that contains three or more dwelling units. **Figure 10–2** A two-family dwelling is not considered a multifamily dwelling [Article 100 Definitions].

Author's Comment: Hotels and motels without cooking areas are not considered dwelling units. **Figure 10–3**

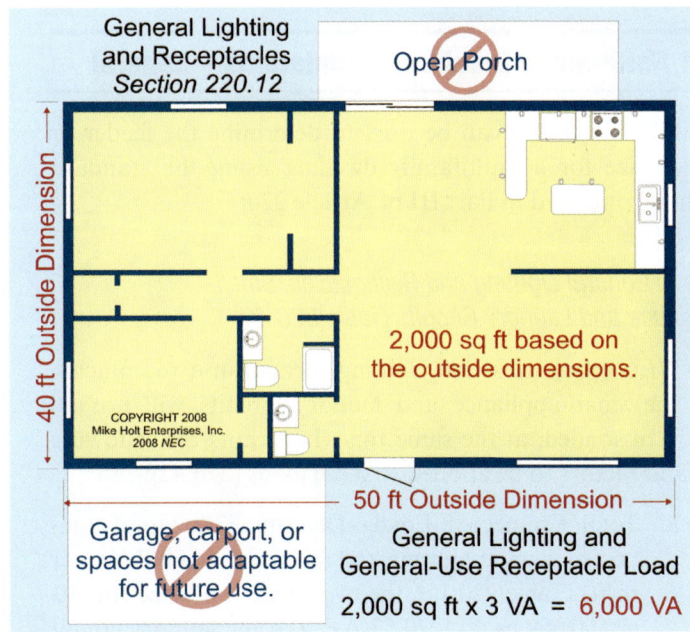

Figure 10–1

Mike Holt Enterprises, Inc. • www.MikeHolt.com • 1.888.NEC.CODE (1.888.632.2633)

Unit 10 — Multifamily Dwelling Calculations

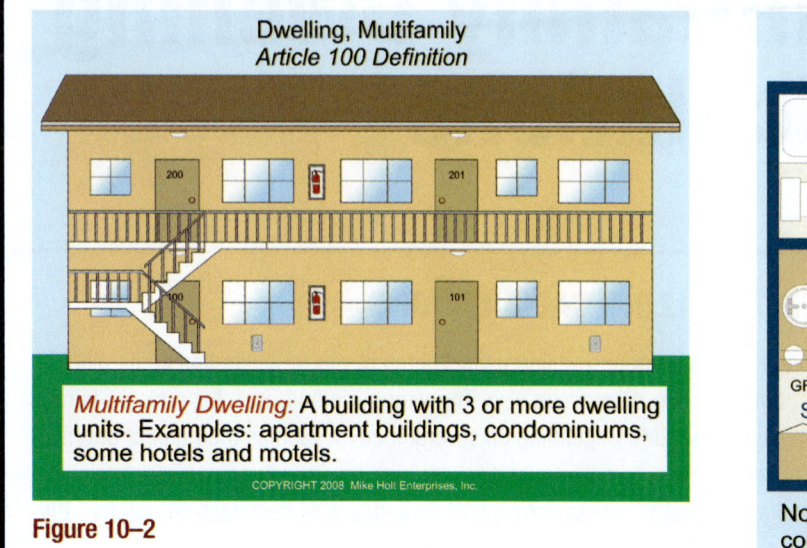

Figure 10–2

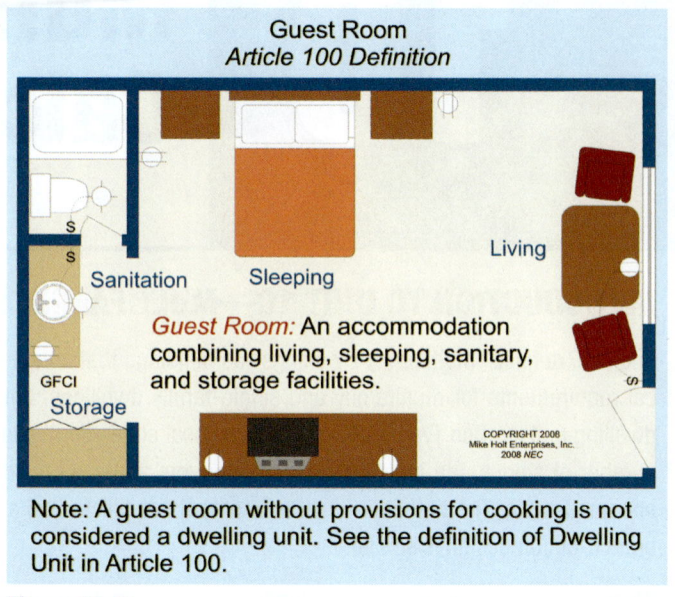

Figure 10–3

10.1 Multifamily Dwelling Calculations—General

The following steps can be used to determine the feeder or service size for a multifamily dwelling using the standard method contained in Part III of Article 220:

Step 1: General Lighting and Receptacles, Small-Appliance and Laundry Circuits [Table 220.42]

The *NEC* recognizes that lighting, general-use receptacles, and the small-appliance and laundry circuits will not all be on, or loaded, at the same time. It permits the following demand factors to be applied to these loads [220.42]:

Step a: Total Connected Load—Determine the total connected general lighting and receptacle load (3 VA per sq ft) [Table 220.12], the two small-appliance circuits (1,500 VA each) [220.52(A)], and the laundry circuit load (1,500 VA) [220.52(B)] of all dwelling units. The laundry load (1,500 VA) can be omitted if laundry facilities are provided on the premises and are available to all building occupants [210.52(F) Ex 1].

Step b: Demand Factor—Apply the Table 220.42 demand factors to the total connected load (Step a). The first 3,000 VA is calculated at 100 percent demand, the next 117,000 VA (120,000 - 3,000) is calculated at 35 percent demand, and the remainder at 25 percent according to Table 220.42.

Step 2: Air-Conditioning versus Heat [220.51 and 220.60]

When the air-conditioning and heating loads are not on at the same time (simultaneously), the smaller of the two loads can be omitted [220.60].

• *Air-Conditioning.* The air-conditioning load must be calculated at 100 percent [220.50].

• *Heat.* Electric space-heating loads must be calculated at 100 percent [220.51].

Step 3: Appliances [220.53]

A demand factor of 75 percent is permitted for four or more appliances fastened in place, such as a dishwasher, kitchen waste disposal, trash compactor, water heater, etc. This does not apply to space-heating equipment [220.51], clothes dryers [220.54], cooking appliances [220.55], or air-conditioning equipment [220.50].

Step 4: Clothes Dryers [220.54]

The feeder or service calculated load for household electric clothes dryers located in dwelling units must not be less than 5,000W (or 5,000 VA), or the nameplate rating (whichever is greater) and may be adjusted according to the demand factors listed in Table 220.54.

Multifamily Dwelling Calculations — Unit 10

Author's Comment: A dryer load isn't required if the dwelling unit does not contain an electric dryer. Dryers in common laundry rooms must not have their loads calculated according to this method.

Step 5: Cooking Equipment [220.55]

Household cooking appliances rated over 1¾ kVA can have their feeder and service loads calculated according to the demand factors of Table 220.55 and Notes.

Step 6: Feeder and Service Conductor Size

The feeder/service conductors are sized according to Table 310.16. Conductors are presumed to be copper unless otherwise stated, and systems are presumed to be single-phase unless otherwise stated.

Author's Comments:

- Table 310.15(B)(6) can be used to size the 120/240V feeder conductors to the individual dwelling units of a multifamily dwelling, but Table 310.16 must be used to size service conductors that supply two or more dwelling units.

- Section 110.14(C)(1)(a) states that terminals are rated 60°C for equipment rated 100A or less unless marked 75°C. In real life, most terminals are now rated 75°C, so in this unit, we assume all terminals are rated 75°C unless 60°C is specified. Insulated conductors will be presumed to be 90°C rated. For exam purposes, read the problem carefully to be certain you know what terminal rating the exam question specifies. If unspecified, use the rules of 110.14(C).

PART A—STANDARD METHOD—FEEDER/SERVICE LOAD CALCULATIONS

Introduction

In Unit 9 you learned how to calculate feeder and/or service loads for single dwelling units, using two different methods: the standard and the optional. In a multifamily building, as with single dwelling units, the *NEC* permits demand factors to be applied to account for the fact that not all connected loads in a multifamily dwelling will operate simultaneously. This part of the unit will provide the standard method of calculation for a multifamily dwelling.

10.2 Multifamily Dwelling Unit Calculation Examples—Standard Method

The following steps can be used to determine feeder and service sizes for a multifamily dwelling using the standard method contained in Article 220, Part III:

Unit 10 — Multifamily Dwelling Calculations

Step 1: General Lighting, Small-Appliance, and Laundry Demand [Table 220.42]

▶ **General Lighting Load Example 1**

Question: What is the general lighting and receptacle (including small-appliance circuits) calculated load for an apartment building that contains 20 units? Each apartment is 840 sq ft.

Note: Laundry facilities are provided on the premises for all tenants [210.52(F) Ex 1]. **Figure 10–4**

(a) 5,200 VA   (b) 25,500 VA   (c) 40,590 VA   (d) 110,400 VA

Answer: (c) 40,590 VA

General Lighting (840 sq ft x 3 VA)	2,520 VA		
Small-Appliance Circuits 2 x 1,500 VA	3,000 VA		
Laundry Circuit	+ 0 VA		
Connected Load for one unit	5,520 VA		
Connected Load [Table 220.42] (5,520 VA x 20 units)	110,400 VA		
First 3,000 VA at 100 percent	– 3,000 VA	x 1.00 =	3,000 VA
Next 117,000 VA at 35 percent	107,400 VA	x 0.35 =	+ 37,590 VA
Total Calculated Load			40,590 VA

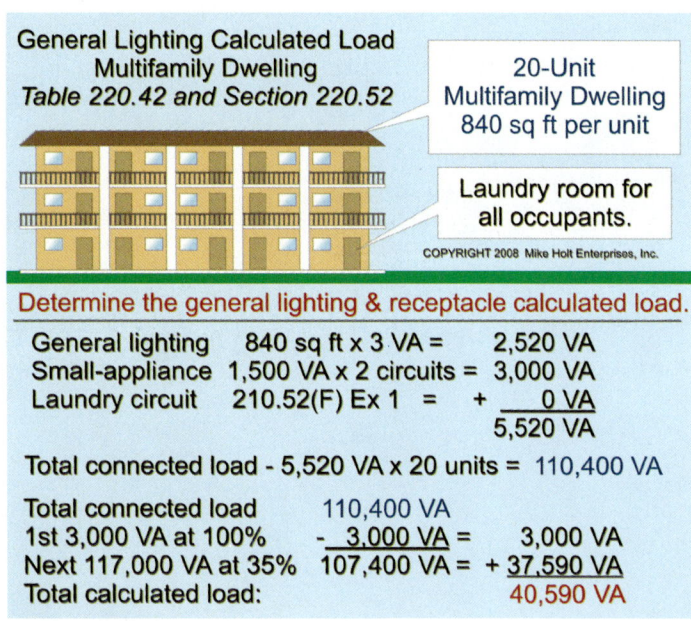

Figure 10–4

Multifamily Dwelling Calculations — Unit 10

▶ **General Lighting Load Example 2**

Question: *What is the general lighting and receptacle calculated load for a 20-unit apartment building? Each unit is 990 sq ft.* **Figures 10-5** and **10-6**

(a) 51,300 VA (b) 74,700 VA (c) 105,600 VA (d) 149,400 VA

Answer: *(a) 51,300 VA*

General Lighting (990 sq ft x 3 VA)	2,970 VA		
Small-Appliance Circuits	3,000 VA		
Laundry Circuit	+ 1,500 VA		
Connected Load for one unit	7,470 VA		
Connected Load [Table 220.42] (7,470 VA x 20 units)	149,400 VA		
First 3,000 VA at 100%	− 3,000 VA	x 1.00 =	3,000 VA
149,400 VA			
Next 117,000 VA at 35%	−117,000 VA	x 0.35 =	40,950 VA
Remainder VA at 25%	29,400 VA	x 0.25 =	+ 7,350 VA
Total Calculated Load			51,300 VA

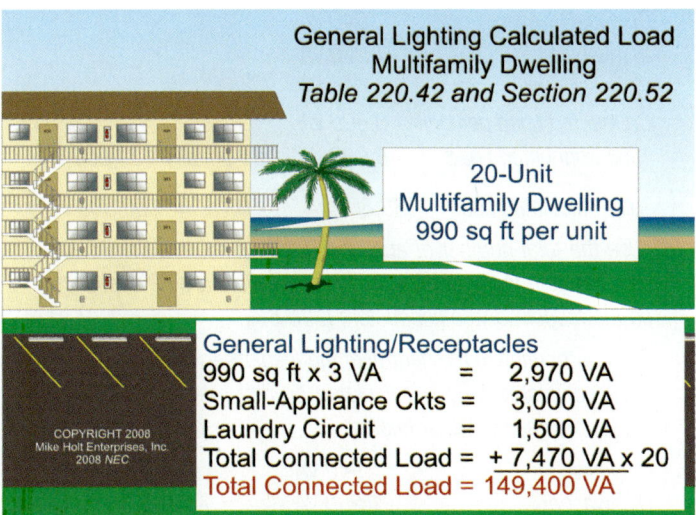

Figure 10–5

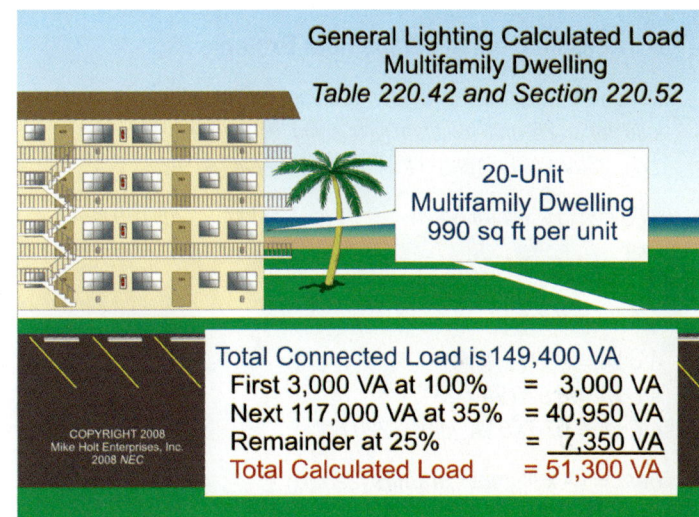

Figure 10–6

Unit 10 — Multifamily Dwelling Calculations

Step 2: Air-Conditioning versus Heat [220.51]

▶ Air-Conditioning versus Heat Example 1

Question: What is the calculated load for air-conditioning versus heat for a 40-unit multifamily building that has an A/C unit (14A, 230V) and two baseboard heaters (3 kVA) in each unit?

(a) 50 kVA   (b) 60 kVA   (c) 160 kVA   (d) 250 kVA

Answer: (d) 250 kVA

A/C VA = V x A
A/C VA = 230V x 14A
A/C VA = 3,220 VA x 40
A/C VA = 128,800 VA, (omit) [220.60]

Heat = VA x Number of Units [220.51]
Heat = 3 kVA x 2 Units
Heat = 6 kVA x 40 Units
Heat = 240,000 VA

▶ Air-Conditioning versus Heat Example 2

Question: What is the air-conditioning versus heat calculated load for a 25-unit multifamily building where each unit has a 17A, 230V A/C unit and 5 kVA of electric heat?

(a) 5 kVA   (b) 6 kVA   (c) 125 kVA   (d) 160 kVA

Answer: (c) 125 kVA

A/C VA = V x A
A/C VA = 230V x 17A
A/C VA = 3,910 VA x 25
A/C VA = 97,750 VA, (omit) [220.60]

Heat = VA x Number of Units [220.51]
Heat = 5 kVA x 25 Units
Heat = 125 kVA

▶ Air-Conditioning versus Heat Example 3

Question: What is the air-conditioning versus heat calculated load for a 25-unit multifamily building where each unit has a heat pump where the compressor load is 17A, 230V and there is a total of 18 kVA of electric supplemental heat in each heat pump air-handler?

(a) 255 kVA   (b) 388 kVA   (c) 548 kVA   (d) 727 kVA

Answer: (c) 548 kVA

(continued in next column)

Calculated Load = (Compressor Load + Supplemental Heat Load) x Number of Units

Compressor VA = V x A
Compressor VA = 230V x 17A
Compressor VA = 3,910
Calculated Load = (3,910 VA + 18,000 VA) x 25
Calculated Load = 547,750 VA

Step 3. Appliance Calculated Load [220.53]

▶ Appliance Load Example 1

Question: What is the appliance calculated load for the feeder/service to a 20-unit multifamily building that contains a 940 VA waste disposal, a 1,250 VA dishwasher, and a 4,500 VA water heater in each unit? **Figure 10–7**

(a) 5 kVA   (b) 7 kVA   (c) 100 kVA   (d) 134 kVA

Answer: (c) 100 kVA

Disposal	940 VA
Dishwasher	1,250 VA
Water Heater	+ 4,500 VA
Connected Load per Unit	6,690 VA
Total Connected Load	6,690 VA x 20 units = 133,800 VA
Total Calculated Load	133,800 VA x 0.75* = 100,350 VA

*Use the total number of appliances to determine if the 75 percent demand factor applies. In this case, there are 60 appliances on the feeder/service conductors [220.53].

The contribution to the neutral of the feeder/service for the appliance connected load is calculated by first removing the 240V loads, such as the water heater:

Disposal	940 VA
Dishwasher	1,250 VA
	2,190 VA x 20 units = 43,800 VA
Total Neutral Calculated Load	43,800 VA x 0.75 = 32,850 VA

▶ Appliance Load Example 2

Question: What is the appliance calculated load for the feeder/service to a 35-unit multifamily building that contains a 900 VA waste disposal, a 1,200 VA dishwasher, and a 5,000 VA water heater in each unit? **Figure 10–8**

(a) 71 kVA   (b) 107 kVA   (c) 142 kVA   (d) 186 kVA

(continued in next column)

Mike Holt's Illustrated Guide to NEC Exam Preparation

Multifamily Dwelling Calculations — Unit 10

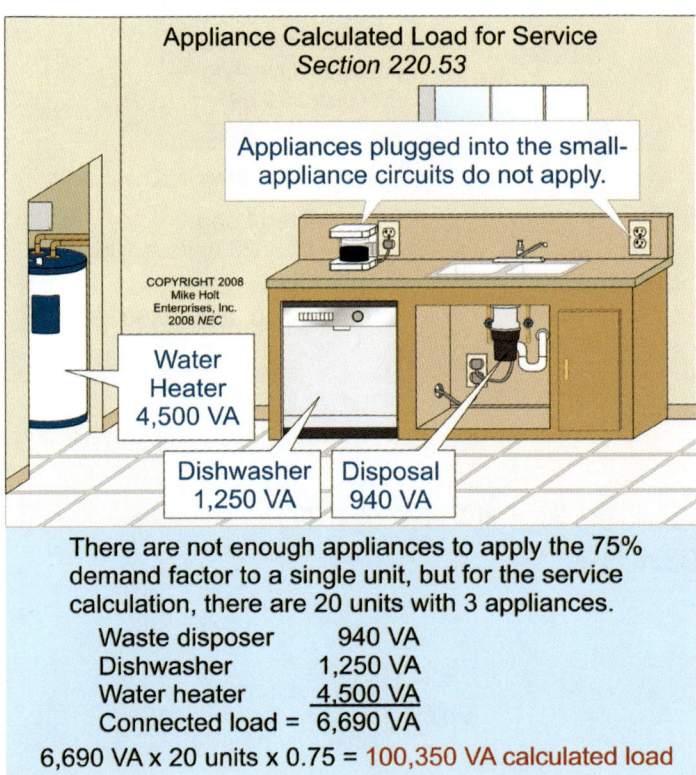

Figure 10–7

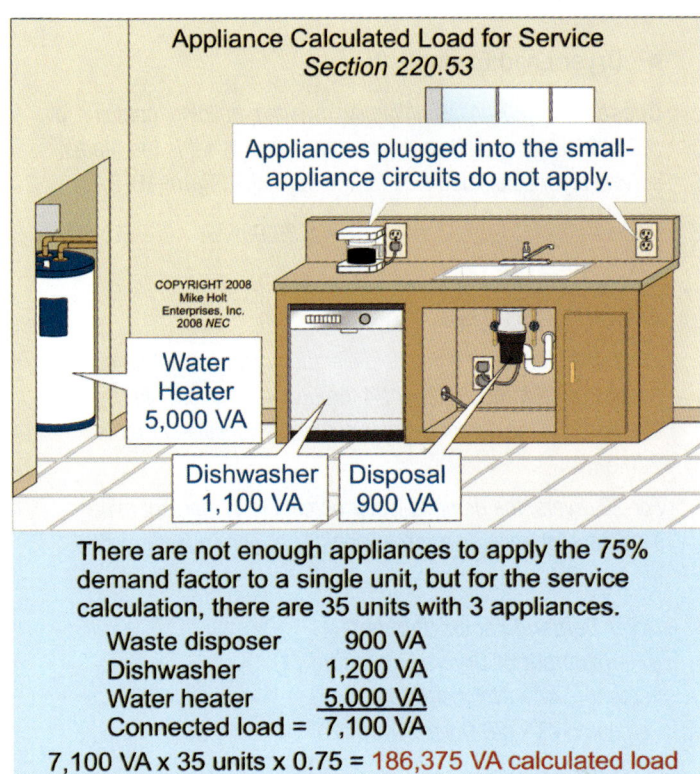

Figure 10–8

Answer: (d) 186 kVA

Disposal	900 VA
Dishwasher	1,200 VA
Water Heater	+ 5,000 VA
Connected Load per unit	7,100 VA
Total Connected Load	7,100 VA x 35 units = 248,500 VA
Calculated Load	248,500 VA x 0.75* = 186,375 VA

*Use the total number of appliances to determine if the 75 percent demand factor applies. In this case, there are 105 appliances on the feeder/service conductors [220.53].

The contribution to the neutral of the feeder/service for the appliance connected load is calculated by first removing the 240V loads, such as the water heater:

Disposal	900 VA
Dishwasher	1,200 VA
	2,100 VA x 35 units = 73,500 VA
Total Neutral	
Calculated Load	73,500 VA x 0.75 = 5,125 VA

Step 4: Household Dryer Calculated Load [220.54]

▶ **Dryer Load Example 1**

Question: A 10-unit multifamily dwelling building contains a 4.50 kVA electric clothes dryer in each unit. What is the feeder/service dryer calculated load for the building?

(a) 5 kVA   (b) 25 kVA   (c) 45 kVA   (d) 60 kVA

Answer: (b) 25 kVA

5 kVA is the minimum load for calculation [220.54]

Connected Load = 5 kVA x 10 Units
Connected Load = 50 kVA

Calculated Load = Connected Load x Demand Factor
[Table 220.54]
Calculated Load = 50 kVA x 0.50
Calculated Load = 25 kVA

The contribution to the neutral of the feeder/service for the clothes dryer load is calculated by first multiplying the dryer calculated load times 70 percent [220.61(B)(1)]:

25 kVA x 0.70 = 17.50 kVA

Unit 10 Multifamily Dwelling Calculations

▶ **Dryer Load Example 2**

Question: A 20-unit multifamily dwelling building contains a 5.25 kVA electric clothes dryer in each unit. What is the feeder/service dryer calculated load for the building? **Figure 10-9**

(a) 5 kVA   (b) 27 kVA   (c) 39.90 kVA   (d) 60 kVA

Answer: (c) 39.90 kVA

When the number of household electric dryers falls in the categories of 12 through 22, or 24 through 42, the formula used to determine the demand factor percent is specified in Table 220.54.

For 20 dryers, the demand factor percent is 47 percent minus 1 percent for each dryer exceeding 11, as shown by the following formula:

Dryer Demand Factor (Percent) =
47 – (Number of Dryers Exceeding 11)

Percent = 47 - (Number of Dryers - 11)
Percent = 47 - (20 Dryers - 11)
Percent = 47 - 9
Percent = 38%

Calculated Load = Connected Load x Demand Factor Percent
Connected Load = 5.25 kVA x 20 Units

Calculated Load = 5.25 kVA x 20 Units x 0.38
Calculated Load = 39.90 kVA

The contribution to the neutral of the feeder/service for the clothes dryer load is calculated by multiplying the dryer calculated load times 70 percent [220.61(B)(1)]:

39.90 kVA x 0.70 = 27.93 kVA

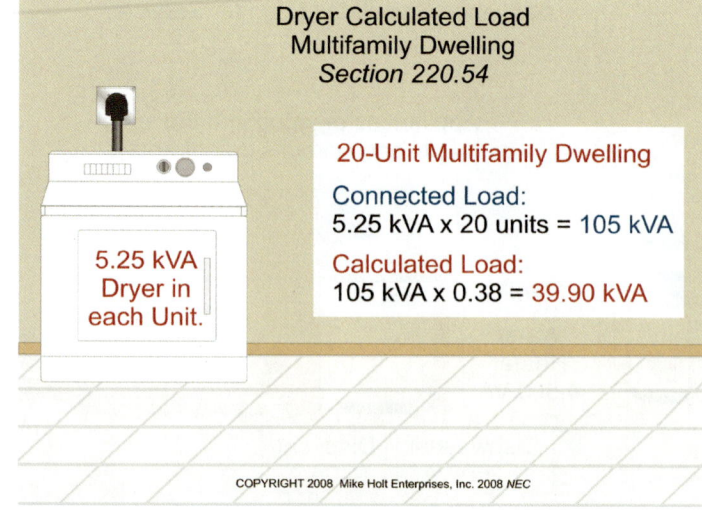

Figure 10–9

Step 5: Household Cooking Equipment Calculated Load [220.55]

▶ **Table 220.55 Column C—not Over 12 kVA Example**
   Figure 10–10

Question: What is the feeder/service calculated load for twenty 9 kVA ranges?

(a) 9 kVA   (b) 20 kVA   (c) 35 kVA   (d) 45 kVA

Answer: (b) 35 kVA

Over 8¾,and less than 12 kVA, look up the kW Demand Load in Table 220.55, Column C.

Figure 10–10

Multifamily Dwelling Calculations — Unit 10

▶ Table 220.55, Note 1—Over 12 kVA with Equal Ratings Example

Question: What is the feeder/service calculated load for twenty 15.60 kVA ranges?

(a) 28 kVA (b) 30 kVA (c) 38 kVA (d) 42 kVA

Answer: (d) 42 kVA

Step a: Column C Demand Load for 20 units: 35 kVA.

Step b: The 15.60 kVA range exceeds 12 kVA by 3.60 kVA. Increase the Column C Demand Load by 5% for each kVA or major fraction of kVA in excess of 12 kVA.

Step c: 3.60 kVA is equal to 3 kVA plus a major fraction of a kVA, so increase the Column C kVA value by 4 x 5% = 20% increase, resulting in 120%, or a 1.20 multiplier

Calculated Load = Column C Value x Multiplier
Calculated Load = 35 kVA x 1.20
Calculated Load = 42 kVA

The contribution to the neutral of the feeder/service for the range load is calculated by multiplying the range calculated load times 70 percent [220.61(B)(1)]:

42 kVA x 0.70 = 29.40 kVA

▶ Table 220.55, Note 2—Unequal Ratings Over 12 kVA Example

Question: What is the feeder/service calculated load for ten ranges rated 9 kVA and ten ranges rated 14 kVA?

(a) 22 kVA (b) 36 kVA (c) 42 kVA (d) 78 kVA

Answer: (b) 36 kVA

Step a: Determine the total connected load:
  9 kVA (min. 12 kVA) 10 ranges x 12 kVA = 120 kVA
  14 kVA   10 ranges x 14 kVA = 140 kVA
  Total Connected Load   260 kVA

Step b: Determine the average of range ratings:
  260 kVA/20 units = 13 kVA average rating.

Step c: Demand load from Table 220.55 Column C:
  20 ranges = 35 kVA

Step d: The average of the ranges (13 kVA) exceeds 12 kVA by 1 kVA. Increase the Column C Demand load (21 kVA) by 5%: resulting in 105%, or a 1.05 multiplier

(continued in next column)

Calculated Load = Column C Value x Multiplier
Calculated Load = 35 kVA x 1.05
Calculated Load = 36.75 kVA

The contribution to the neutral of the feeder/service for the range load is calculated by multiplying the range calculated load times 70 percent [220.61(B)(1)]:

36.75 kVA x 0.70 = 25.725 kVA

▶ Table 220.55, Note 3 Example

Question: What is the feeder/service calculated load for ten 3 kVA ovens?

(a) 10 kVA (b) 15 kVA (c) 20 kVA (d) 30 kVA

Answer: (b) 15 kVA (closest answer)

Connected Load = 3 kVA x 10 units = 30 kVA
Calculated Load = 30 kVA x 0.49 [Table 220.55, Column A]
Calculated Load = 14.70 kVA

Step 6: Service Conductor Size [Table 310.16]

▶ Service Conductor Size Example 1

Question: What size service copper conductors are required for a 120/240V, single-phase multifamily building that has a calculated load of 93 kVA? **Figure 10–11**

(a) 300 kcmil (b) 350 kcmil (c) 500 kcmil (d) 600 kcmil

Answer: (d) 600 kcmil

I = VA/E
I = 93,000 VA/240V
I = 388A

600 kcmil copper, rated 420A at 75°C [Table 310.16]

According to Annex C, Table C.10 a trade size 3 Schedule 40 PVC raceway will accommodate three 600 kcmil THWN conductors.

If the service conductors are paralleled in two sets:
- 388A/2 raceways = 194A
- 3/0 AWG is rated 200A at 75°C [Table 310.16], so two 3/0 AWG can be paralleled for this service. Trade size 1½ Schedule 40 PVC raceway will accommodate three 3/0 THWN conductors [Annex C, Table C.10].

Unit 10 Multifamily Dwelling Calculations

Figure 10–11

▶ Service Conductor Size Example 2

Question: What size aluminum service conductors are required for a 120/240V, single-phase multifamily building that has a calculated load of 93 kVA?

(a) 300 kcmil   (b) 350 kcmil   (c) 600 kcmil   (d) 800 kcmil

Answer: (d) 800 kcmil

I = VA/E
I = 93,000 VA/240V
I = 388A

800 kcmil aluminum, rated 395A at 75ºC [Table 310.16]

According to Annex C Table C.10, a trade size 3½ Schedule 40 PVC raceway will accommodate three 800 kcmil THWN conductors.

If the service conductors are paralleled in two sets:
- 388A/2 raceways = 194A
- 250 kcmil aluminum is rated 205A at 75ºC [Table 310.16], so two 250 kcmil aluminum conductors can be paralleled for this service. A trade size 2 Schedule 40 PVC raceway will accommodate three 250 kcmil THWN conductors [Annex C, Table C.10].

If the service conductors are paralleled in three sets:
- 388A/3 raceways = 129.33A

(continued in next column)

- 2/0 AWG aluminum is rated 135A at 75ºC [Table 310.16], so three 2/0 AWG aluminum conductors can be paralleled for this service. Trade size 1½ Schedule 40 PVC raceway will accommodate three 2/0 THWN conductors [Annex C, Table C.10].

Author's Comment: Table 310.15(B)(6) only applies to feeder/service conductors for individual 120/240V dwelling units with 400A or less for single-family, two-family, or multifamily buildings. In this building, Table 310.15(B)(6) can be used for the feeder to each apartment, but Table 310.16 must be used for sizing the feeder/service conductors that supply the entire building.

▶ Service Conductor Size Example 3

Question: What size copper service conductors are required for a multifamily building that has a calculated load of 270 kVA for a 120/208V, three-phase system? Service conductors are run in parallel in two raceways. **Figure 10–12**

(a) Two–300 kcmil per phase   (b) Two–350 kcmil per phase
(c) Two–500 kcmil per phase   (d) Two–600 kcmil per phase

Answer: (c) Two–500 kcmil per phase

I = VA/(1.732 x 208V)
I = 270,000 VA/(1.732 x 208V)
I = 270,000 VA/360V
I = 750A

Amperes per Parallel Set = Amperes/Number of Parallel Sets
750A/2 conductors in parallel = 375A per conductor.

A 500 kcmil conductor has an ampacity of 380A at 75ºC [Table 310.16]. Table 310.15(B)(6) doesn't apply to services or feeders for multifamily buildings.

Two sets of 500 kcmil conductors (380A x 2) can be protected by an 800A overcurrent device. Their combined ampacity is 760A, but we're allowed to round up to the next standard size overcurrent device [240.4(B) and 240.6(A)].

If the service conductors are paralleled in three sets:
- 750A/3 raceways = 250A
- 250 kcmil copper is rated 255A at 75ºC [Table 310.16], so three 250 kcmil conductors can be paralleled for this service. A trade size 2 Schedule 40 PVC raceway will accommodate three 250 kcmil THWN conductors [Annex C, Table C.10].

Author's Comment: Throughout this book, we use copper conductors unless otherwise specified [110.5].

Multifamily Dwelling Calculations — Unit 10

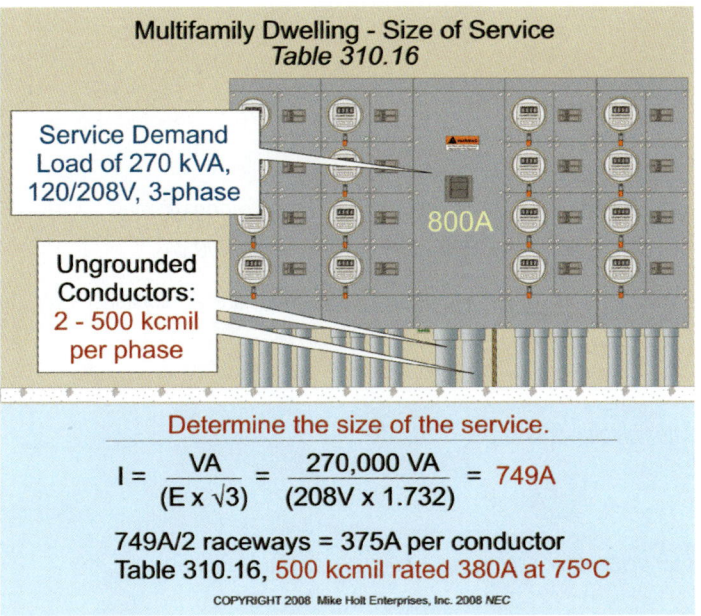

Figure 10–12

10.3 Multifamily Dwelling Calculations—Standard Method Example

The following example demonstrates the steps used to determine feeder and service sizes for a multifamily dwelling using the standard method contained in Article 220, Part III:

> ▶ **Standard Method Calculated Load Example**
>
> **Question:** What is the calculated load for a 25-unit apartment building. The system voltage is 120/240V, single-phase? Each apartment is 1,000 sq ft, and contains the following:
>
> | | |
> |---|---:|
> | Waste Disposal | 1.50 kVA |
> | Dishwasher | 1.20 kVA |
> | Water Heater | 4.50 kVA |
> | Dryer | 4.50 kVA |
> | Range | 15.50 kVA |
> | Air-conditioning (230V) | 28A |
> | Electric Space Heating | 7.50 kVA |
>
> (a) 200 kVA (b) 325 kVA (c) 475 kVA (D) 600 kVA
>
> **Answer:** (c) 475 kVA

Step 1: General Lighting, Small-Appliance, and Laundry Demand [Table 220.42]

General Lighting (1,000 sq ft x 3 VA) [220.12]	3,000 VA		
Small-Appliance Circuits [220.52(A)]	3,000 VA		
Laundry Circuit [220.52(B)]	+ 1,500 VA		
	7,500 VA		
Total Connected Load (7,500 VA x 25 units)	187,500 VA		
First 3,000 VA at 100 percent [Table 220.42]	− 3,000 VA	x 1.00 =	3,000 VA
	184,500 VA		
Next 117,000 VA at 35 percent	− 117,000 VA	x 0.35 =	40,950 VA
Remainder at 25 percent	67,500 VA	x 0.25 =	+ 16,875 VA
Total Calculated Load			60,825 VA

Step 2: Air-Conditioning versus Heat [220.51]

A/C VA Load = 28A x 230V x 25 Units
A/C VA Load = 161,000 VA, (omit) [220.60]

Heat Connected Load = VA x Number of Units [220.51]
Heat Connected Load = 7,500 VA x 25 Units
Heat Connected Load = 187,500 VA

(continued in next column)

Unit 10 — Multifamily Dwelling Calculations

Step 3: Appliance Calculated Load [220.53]

Waste Disposal	1,500 VA		
Dishwasher	1,200 VA		
Water Heater	+ 4,500 VA		
Connected Load	7,200 VA	x 25 units =	180,000 VA
Calculated Load	180,000 VA	x 0.75* =	135,000 VA

*Use the total number of appliances to determine if the 75 percent demand factor applies. In this case, there are 75 appliances on the feeder/service conductors [220.53]

Step 4: Dryer Calculated Load [220.54]

Dryers must be calculated at a minimum of 5,000 VA or the nameplate, whichever is larger [220.54].

Total Connected Load = 5 kVA x 25 units
Total Connected Load = 125,000 VA

Dryer Demand Factor for 25 dryers (Percent) = 35 percent minus 0.50 percent for each dryer exceeding 23 [Table 220.54].

Percent = 35 - 0.50 x (Number of Dryers Exceeding 23)
Percent = 35 - 0.50 x (25 Dryers - 23)
Percent = 35 - (0.50 x 2)
Percent = 35 - 1
Percent = 34%

Calculated Load = Connected Load x Demand Factor Percent
Calculated Load = 125,000 VA x 0.34
Dryer Calculated Load = 42,500 VA

Step 5: Cooking Equipment Calculated Load [220.55]

Step a: Column C Demand Load for 25 units = 40 kVA. [Table 220.54]

Step b: 15.50 kVA − 12 kVA = 3.50 kVA.
3.50 kVA is equal to 3 kVA plus a major fraction of a kVA, so increase to 4

Step c: Increase the Column C Demand Load (40 kVA) by 4 x 5% = 20%: resulting in 120%, or a 1.20 multiplier

Step d: Range calculated load = Column C Value x Multiplier
Range calculated load = 40 kVA x 1.20
Range calculated load = 48 kVA

Step 6: Service Conductor Size

Step a: Total general lighting, small-appliance, and laundry calculated load	60,825 VA
Step b: Total heat calculated load [220.51] (7,500 VA x 25 units)	187,500 VA
Step c: Total appliance calculated load	135,000 VA
Step d: Total dryer calculated load	42,500 VA
Step e: Range calculated load	+ 48,000 VA
Total Calculated Load	473,825 VA

Service Conductor Amperes = VA/E
Service Conductor Amperes = 473,825 VA/240V
Service Conductor Amperes = 1,974A

(continued in next column)

Multifamily Dwelling Calculations — Unit 10

PART B—OPTIONAL METHOD—FEEDER/SERVICE LOAD CALCULATIONS

Introduction

In a multifamily building consisting of three or more dwelling units, either the standard method or the optional method can be used to calculate the size of the main service. Part IV of Article 220 provides the optional method which permits demand factors to be applied in recognition of the diversity of usage of all the loads in all the separate units of a multifamily dwelling. In order to apply the optional method for multifamily dwellings, each dwelling unit must be equipped with electric cooking equipment, electric heating and/or air-conditioning, and supplied by no more than one feeder.

10.4 Multifamily Dwelling Unit Calculations [220.84]—Optional Method

The feeder/service calculated load for a building that has three or more dwelling units equipped with electric cooking equipment and either electric space heating or air-conditioning can be calculated in accordance with the demand factors of Table 220.84, based on the number of dwelling units. The feeder/service neutral calculated load must be determined in accordance with 220.61.

House loads for common areas are calculated in accordance with Article 220, Part III and then added to the Table 220.84 calculated load.

Author's Comment: House loads are those not directly associated with the individual dwelling units of a multifamily dwelling. Some examples of house loads might be landscape and parking lot lighting, hall and stairway lighting, common laundry facilities, common pool and recreation areas, etc.

The following steps can be used to determine feeder and service sizes for a multifamily dwelling using the optional method contained in Article 220, Part IV:

Step 1: Total Connected Load [220.84(C)]

The following connected loads from all the dwelling units are added together, and then the Table 220.84 demand factor is applied to determine the calculated load:

- 3 VA per sq ft for general lighting and general-use receptacles.
- 1,500 VA for each small-appliance circuit (minimum of 2 circuits) [220.52(A)].
- 1,500 VA for each laundry circuit [220.52(B)].
- The nameplate rating of all appliances.
- The nameplate rating of all motors.
- The larger of the air-conditioning load or the space-heating load.

Author's Comment: A laundry circuit isn't required for an individual dwelling unit if common laundry facilities are provided.

Step 2: Calculated Load

The calculated load is determined by applying the demand factor from Table 220.84 to the total connected load (Step 1). The calculated load (kVA) can be converted to amperes by:

◆ Single-Phase Formula: Three-Phase Formula:
  $I = VA/E$ $I = VA/(1.732 \times E)$

Step 3: Feeder and Service Conductor Size

The ungrounded conductors are sized according to Table 310.16, based on the calculated load.

10.5 Multifamily—Optional Method Example 1 [220.84]

A 120/240V, single-phase system supplies a 12-unit multifamily building. Each 1,500 sq ft unit contains: **Figure 10-13**

Dishwasher	1.50 kVA
Water Heater	4 kVA
Washing Machine	1.20 kVA
Dryer	4.50 kVA
Range	14.40 kVA
A/C (230V)	17A
Electric Space Heating	5 kVA

▶ **Parallel Service Conductor Sizing Example 1**

Question: What size conductor is required if the service is rated 120/240V, single-phase and the conductors are installed in parallel in two separate raceways?

(a) 350 kcmil   (b) 400 kcmil   (c) 500 kcmil   (d) 600 kcmil

Answer: (d) 600 kcmil

(continued in next column)

Unit 10   Multifamily Dwelling Calculations

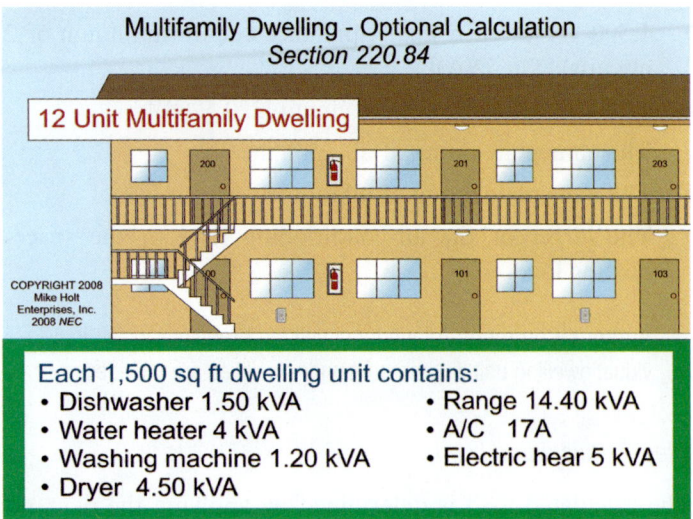

Figure 10–13

Step 1: Total Connected Load

    Step a: Determine the General Lighting Load:

General Lighting (1,500 sq ft x 3 VA)	4,500 VA
Small-Appliance Circuits (2 circuits x 1,500 VA)	3,000 VA
Laundry Circuit	+ 1,500 VA
	9,000 VA x 12 units = 108,000 VA

    Step b: Determine the Appliance Calculated Load:

Dishwasher	1,500 VA
Water Heater	4,000 VA
Dryer (nameplate)	4,500 VA
Range (nameplate)	+ 14,400 VA
	24,400 VA x 12 units = 292,800 VA

    Step c: Compare the Air-Conditioning versus Heat Load:

        A/C VA = 230V x 17A, 3,910 VA (omit)

        Heat   5,000 VA x 12 units = 60,000 VA

Step 2: Total Connected Loads

General Lighting, Receptacles	108,000 VA
Appliances Connected Load	292,800 VA
Heat	+ 60,000 VA
Total Connected Load	460,800 VA

Total Calculated Load = Total Connected Load x Demand Factor [Table 220.84]
Total Calculated Load = 460,800 VA x 0.41 [Table 220.84]
Total Calculated Load = 188,928 VA

Step 3: Service Conductor Size **Figure 10-14**

    I = VA/E

    I = 188,928 VA/240V

    I = 787A

(continued in next column)

Multifamily Dwelling Calculations Unit 10

Conductor size if paralleled in two raceways [240.4(B)]:

787A/2 raceways = 393A per conductor

Feeder/Service Conductors: Parallel 600 kcmil rated 420A at 75°C [Table 310.16].

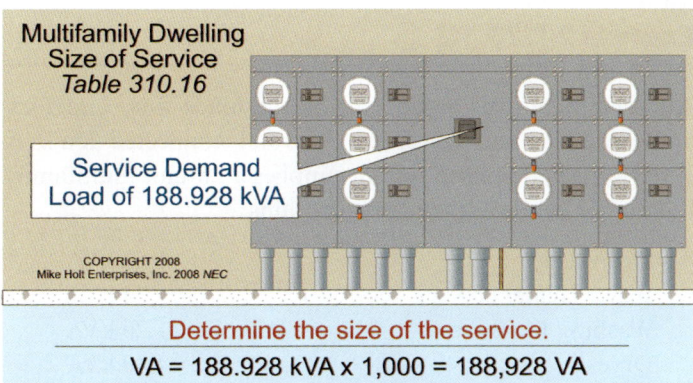

Figure 10–14

▶ **Grounding Electrode Conductor Sizing Example 1**

Question: What size grounding electrode conductor is required if the service ungrounded conductors are 600 kcmil with two conductors in parallel in two separate raceways?

(a) 4 AWG   (b) 1 AWG   (c) 3/0 AWG   (d) 4/0 AWG

Answer: (c) 3/0 AWG

600 kcmil x 2 = 1,200 kcmil equivalent area of the ungrounded conductors [Table 250.66, Note 1].

Over 1,100 kcmil for ungrounded conductors requires a grounding electrode conductor of 3/0 AWG [Table 250.66].

▶ **Parallel Service Conductor Sizing Example 2**

Question: What size conductor is required if the service with a calculated load of 787A is rated 120/240V, single-phase and the conductors are installed in parallel in three separate raceways?

(a) 300 kcmil   (b) 400 kcmil   (c) 500 kcmil   (d) 600 kcmil

Answer: (a) 300 kcmil

787A/3 raceways = 262.33A

300 kcmil copper is rated 285A at 75°C [Table 310.16], so three 300 kcmil conductors can be paralleled for this service.

▶ **Grounding Electrode Conductor Sizing Example 2**

Question: What size grounding electrode conductor is required if the service ungrounded conductors are 300 kcmil with three conductors in parallel in three separate raceways?

(a) 4 AWG   (b) 2/0 AWG   (c) 3/0 AWG   (d) 4/0 AWG

Answer: (b) 2/0 AWG

300 kcmil x 3 = 900 kcmil equivalent area of the ungrounded conductors [Table 250.66, note 1]

900 kcmil for ungrounded conductors requires a grounding electrode conductor of 2/0 AWG [Table 250.66]

▶ **Parallel Service Conductor Sizing Example 3**

Question: What size conductor is required if the service with a calculated load of 787A is rated 120/240V, single-phase and the conductors are installed in parallel in four separate raceways?

(a) 1/0 AWG   (b) 3/0 AWG   (c) 500 kcmil   (d) 600 kcmil

Answer: (b) 3/0 AWG

787A/4 raceways = 196.75A

3/0 AWG copper is rated 200A at 75°C [Table 310.16], so four 3/0 AWG conductors can be paralleled for this service.

Unit 10 Multifamily Dwelling Calculations

▶ **Grounding Electrode Conductor Sizing Example 3**

Question: What size grounding electrode conductor is required if the service ungrounded conductors are 3/0 AWG with four conductors in parallel in four separate raceways?

(a) 4 AWG (b) 2/0 AWG (c) 3/0 AWG (d) 4/0 AWG

Answer: (b) 2/0 AWG

3/0 AWG has an area of 167,800 cmil [Chapter 9, Table 8]

167,800 cmil x 4 = 671,000 cmil equivalent area of the ungrounded conductors [Table 250.66, Note 1].

671,000 cmil for ungrounded conductors requires a grounding electrode conductor of 2/0 AWG [Table 250.66].

Author's Comment: When installing feeders, an equipment grounding conductor is required in every raceway. Section 250.118 lists allowable equipment grounding conductors. To size this conductor using a wire-type equipment grounding conductor, go to Table 250.122 and select the equipment grounding conductor based on the overcurrent device protecting the conductors in the raceway [250.122(F)]. For instance, the equipment grounding conductor in each raceway of an 800A feeder which is paralleled using two 600 kcmil conductors per phase, will require a 1/0 AWG equipment grounding conductor in each raceway.

10.6 Multifamily—Optional Method Example 2 [220.84]

Another method is to add all of the loads first, and then apply the demand factor to the total connected load. A 120/240V, single-phase system supplies a 20-unit multifamily building. Each 1,500 sq ft unit contains:

Dishwasher	1.50 kVA
Water Heater	4 kVA
Washing Machine	1.20 kVA
Dryer	4.50 kVA
Range	14.40 kVA
A/C (230V)	17A
Electric Space Heating	5 kVA

▶ **Parallel Service Conductor Sizing Example 4**

Question: Using the optional method of calculation, what size conductor is required if the service is rated 120/240V, single-phase and the conductors are installed in parallel in three separate raceways?

(a) 350 kcmil (b) 400 kcmil (c) 500 kcmil (d) 600 kcmil

Answer: (d) 600 kcmil

Step 1: Total Connected Load

Step a: Determine the General Lighting Load:

General Lighting (1,500 sq ft x 3 VA)	4,500 VA
Small-Appliance Circuits (1,500 VA each x 2)	3,000 VA
Laundry Circuit	+ 1,500 VA
Connected Load per Unit	9,000 VA

Total Connected Load = Load per Unit x Number of Units
Total Connected Load = 9,000 VA x 20 units
Total Connected Load = 180,000 VA

Step b: Determine the Appliance Calculated Load:

Dishwasher	1,500 VA
Water Heater	4,000 VA
Dryer (nameplate)	4,500 VA
Range (nameplate)	+ 14,400 VA
	24,400 VA

(continued in next column)

Mike Holt's Illustrated Guide to NEC Exam Preparation

Total Appliance Connected Load = Load Per Unit x Number of Units
Total Appliance Connected Load = 24,400 VA x 20 units
Total Appliance Connected Load = 488,000 VA

Step c: Compare the Air-Conditioning versus Heat Load:

A/C VA = 240V x 17A
A/C VA Load = 4,080, (omit) [220.60]

Heat Load = VA x Number of Units
Heat = 5,000 VA x 12 Units
Heat = 60,000 VA

Step 2: Total Connected Loads

General Lighting, Receptacle	180,000 VA
Appliances Connected Load	488,800 VA
Heat	+ 60,000 VA
Total Connected Load	728,800 VA

Total Calculated Load = Total Connected Load x Demand Factor [Table 220.84]
Total Calculated Load = 728,800 VA x 0.38 [Table 220.84]
Total Calculated Load = 276,944 VA

Step 3: Service Conductor Size

I = VA/E
I = 276,944 VA/240V
I = 1,154A

Conductor size if paralleled in three raceways [240.4(B)]:

1,154A/3 raceways = 385A per conductor

Feeder/Service Conductors: Parallel 600 kcmil rated 420A at 75°C [Table 310.16].

When do you use the standard method versus the optional method? For the purpose of exam preparation, always use the standard load calculation unless the question specifies the optional method. In the field, you'll probably want to use the optional method because it's faster and easier to calculate.

Unit 10 — Conclusion

CONCLUSION TO UNIT 10—MULTIFAMILY DWELLING CALCULATIONS

As you saw in Unit 10, the sizing of branch circuits, feeders, and service conductors for multifamily dwellings is similar to the process for single-family dwellings. The feeders to individual dwelling units are sized in the same manner whether that dwelling unit is a single-family dwelling or an individual unit of an apartment building. The *Code* allows the use of Table 310.15(B)(6) for sizing the feeders or service conductors to an individual dwelling unit; however, the conductors that feed the service to a multifamily dwelling must be sized using Table 310.16.

Some additional demand factors for multifamily dwellings are allowed on the presumption that there will be diversity of usage between the various units. Dryers are allowed the demand factors of Table 220.54, which don't even begin to provide less than 100 percent demand until there are five units or more. Perhaps one of the most confusing single tables in the *NEC* is Table 220.55 for household ranges. This table is confusing because the first two columns are percentage multipliers, while the third column is a final kVA value. The notes to this table further complicate matters. Be sure you've studied this part of Unit 10 well and are able to complete the calculations using the range table as well as sizing the service for a multifamily dwelling unit.

Unit 10 Practice Questions

UNIT 10—MULTIFAMILY DWELLING CALCULATIONS PRACTICE QUESTIONS

(•Indicates that 75% or fewer of those who took this exam answered the question correctly.)

PART A—STANDARD METHOD—FEEDER/SERVICE LOAD CALCULATIONS

10.2 Multifamily Dwelling Unit Calculation Examples—Standard Method

1. •Each unit of a 20-unit apartment building is 840 sq ft. What is the general lighting and general-use receptacle feeder/service calculated load for the building?

   Note: Laundry facilities are provided on the premises for all tenants.

   (a) 5,200 VA
   (b) 40,590 VA
   (c) 60,250 VA
   (d) 110,400 VA

2. •Each unit of a 20-unit apartment building is 990 sq ft. What is the general lighting and general-use receptacle feeder/service calculated load for the building?

   (a) 51,300 VA
   (b) 74,700 VA
   (c) 105,600 VA
   (d) 149,400 VA

3. A 40-unit multifamily building has an air conditioner (17A, 230V) and two 3 kW baseboard heaters in each unit. What is the feeder/service calculated load for the air-conditioning and heat?

   (a) 50 kVA
   (b) 60 kVA
   (c) 160 kVA
   (d) 240 kVA

4. Each unit of a 25-unit multifamily building has air-conditioning (17A, 230V) and electric heat (5 kW). What is the air-conditioning and heat feeder/service calculated load?

   (a) 5 kVA
   (b) 6 kVA
   (c) 125 kVA
   (d) 160 kVA

5. In a 16-unit multifamily building, each unit contains a waste disposal (940 VA), a dishwasher (1,250 VA), and a water heater (4,500 VA). What is the feeder/service calculated load for these appliances?

   (a) 5 kVA
   (b) 80 kVA
   (c) 100 kVA
   (d) 134 kVA

6. Each unit in a 28-unit apartment building contains a waste disposal (900 VA), dishwasher (1,200 VA), and a water heater (5,000 VA). What is the feeder/service calculated load for the appliances?

   (a) 107 kVA
   (b) 142 kVA
   (c) 149 kVA
   (d) 186 kVA

7. A multifamily dwelling (40 units) contains a 4.50 kW electric clothes dryer in each unit. What is the feeder/service calculated load for all of the dryers?

   (a) 27 kW
   (b) 53 kW
   (c) 60 kW
   (d) 150 kW

Unit 10 — Practice Questions

8. What is the feeder/service calculated load for ten 5.25 kW dryers installed in dwelling units of a multifamily building?

   (a) 26 kW
   (b) 37 kW
   (c) 40 kW
   (d) 60 kW

9. What is the feeder/service calculated load for twelve 3.25 kW ovens?

   (a) 10 kW
   (b) 15 kW
   (c) 18 kW
   (d) 20 kW

10. What is the feeder/service calculated load for eight 7 kW counter-mounted cooktops?

    (a) 14.70 kW
    (b) 17 kW
    (c) 20 kW
    (d) 48 kW

11. What is the feeder/service calculated load for five 12.40 kW ranges?

    (a) 9 kW
    (b) 20 kW
    (c) 45 kW
    (d) 60 kW

12. •What is the feeder/service calculated load for three ranges rated 15.50 kW each?

    (a) 14 kW
    (b) 15 kW
    (c) 17 kW
    (d) 21 kW

13. What is the feeder/service calculated load for three ranges rated 11 kW and three ranges rated 14 kW?

    (a) 22 kW
    (b) 36 kW
    (c) 42 kW
    (d) 78 kW

14. •What size aluminum service conductors are required for a 120/240V, single-phase multifamily building that has a total calculated load of 90 kW?

    (a) 500 kcmil AL
    (b) 600 kcmil AL
    (c) 700 kcmil AL
    (d) 800 kcmil AL

15. •What size copper service conductors are required for a 120/240V, single-phase multifamily building that has a total calculated load of 90 kW?

    (a) 500 kcmil
    (b) 600 kcmil
    (c) 700 kcmil
    (d) 800 kcmil

16. What size copper service conductors are required for a multifamily building that has a total calculated load of 260 kW for a 120/208V wye, three-phase system? The conductors are paralleled in two raceways.

    (a) Two–300 kcmil
    (b) Two–350 kcmil
    (c) Two–500 kcmil
    (d) Two–700 kcmil

17. What size aluminum service conductors are required for a multifamily building that has a total calculated load of 260 kW for a 120/208V wye, three-phase system? The conductors are paralleled in two raceways.

    (a) Two–300 kcmil AL
    (b) Two–350 kcmil AL
    (c) Two–500 kcmil AL
    (d) Two–700 kcmil AL

The following information applies to the next five questions.
A multifamily building has 12 units. The system voltage is 120/240V, single-phase. Each 1,500 sq ft unit contains the following:

Dishwasher	1.50 kVA
Water Heater	4 kW
Dryer	4.50 kW
Washing Machine	1.20 kVA
Range	14.45 kW
A/C (230V)	3 hp
A/C Fan Motor (230V)	1/8 hp (334 VA)
Heat	5 kW

18. •What is the building's feeder/service calculated load for the general lighting and general-use receptacles, and the small-appliance and laundry circuits in kVA?

    (a) 40 kVA
    (b) 90 kVA
    (c) 105 kVA
    (d) 108 kVA

19. What is the building's feeder/service calculated load for the air conditioner (18A, 230V) versus electric heat (5 kW)?

    (a) 30 kVA
    (b) 52 kVA
    (c) 60 kVA
    (d) 105 kVA

20. What is the building's feeder/service calculated load for the appliances in kVA?

    (a) 40 kVA
    (b) 43 kVA
    (c) 50 kVA
    (d) 55 kVA

21. What is the building's feeder/service calculated load for the 4.50 kW dryers?

    (a) 27.60 kW
    (b) 45.70 kW
    (c) 55.30 kW
    (d) 60.40 kW

22. What is the building's feeder/service calculated load for the 14.45 kW range?

    (a) 27 kW
    (b) 30 kW
    (c) 35 kW
    (d) 168 kW

23. If the total calculated load of a multifamily dwelling unit is 206 kVA, what is the minimum feeder/service conductor size? The system voltage is 120/240, single-phase.

    (a) 600A
    (b) 800A
    (c) 1,000A
    (d) 1,200A

24. •If a service consists of three sets of parallel 400 kcmil conductors routed in three separate raceways, what size equipment bonding jumper is required for each service raceway?

    (a) 1 AWG
    (b) 1/0 AWG
    (c) 2/0 AWG
    (d) 3/0 AWG

25. •If three sets of 400 kcmil copper service conductors are paralleled in three raceways, what is the minimum size grounding electrode conductor required?

    (a) 1/0 AWG
    (b) 2/0 AWG
    (c) 3/0 AWG
    (d) 4/0 AWG

PART B—OPTIONAL METHOD—FEEDER/SERVICE LOAD CALCULATIONS

10.4 Multifamily Dwelling Unit Calculations [220.84]—Optional Method

26. •When determining the service (using the optional calculation method) for a multifamily dwelling, the total connected load must have the demand factors of Table 220.84 applied. When determining the total connected load, _____ must be used.

    (a) 125 percent of the air-conditioning load plus 100 percent of the space-heating load
    (b) the larger of the air-conditioning load or the space-heating load
    (c) the sum of the air-conditioning load and the space-heating load
    (d) 125 percent of the sum of the air-conditioning load and the space-heating load

27. A multifamily building has 60 units. Each unit is 1,500 sq ft. Using the optional dwelling unit calculations, what is the calculated load for the building's general lighting and general-use receptacles, and small-appliance and laundry circuits?

    (a) 108 kVA
    (b) 130 kVA
    (c) 145 kVA
    (d) 190 kVA

Unit 10 Practice Questions

28. A 60-unit multifamily dwelling has an air conditioner (3 hp with ⅛ hp blower) and electric heat (5 kW) in each unit. Using the optional calculation method, what is the air-conditioning versus heat calculated load?

    (a) 30 kVA
    (b) 50 kVA
    (c) 61 kVA
    (d) 72 kVA

29. Using the optional method for dwelling unit calculations, what is the fixed appliance portion of the 60-unit multifamily building's calculated load if each unit has a water heater (4 kW) and a dishwasher (1.50 kW)?

    (a) 30 kVA
    (b) 50 kVA
    (c) 60 kVA
    (d) 80 kVA

30. Each unit of a 60-unit apartment building has a 4 kW dryer. Using the optional calculation method, the calculated load added to the service for the dryers is _____.

    (a) 58 kW
    (b) 72 kW
    (c) 75 kW
    (d) 240 kW

31. Using the optional method for dwelling unit calculations, what is the dryer portion of a 60-unit multifamily building's calculated load if there is a 4.50 kW dryer in each unit?

    (a) 25 kW
    (b) 55 kW
    (c) 65 kW
    (d) 75 kW

32. Using the optional method for dwelling unit calculations, what is the electric range portion of a 60-unit multifamily building's calculated load if each apartment has a 14 kW range?

    (a) 50 kW
    (b) 100 kW
    (c) 150 kW
    (d) 200 kW

33. If the total calculated load for a multifamily dwelling is 270 kVA, what is the minimum feeder/service conductor size? The service is 120/208V, three-phase.

    (a) 600A
    (b) 800A
    (c) 1,000A
    (d) 1,200A

34. Each unit of a 20-unit multifamily dwelling has 900 sq ft of living space and contains one of each of the following: water heater (5 kW), range (14 kW), air-conditioning (5 hp), and heat (5 kW), The service for this apartment building is approximately _____ if the optional method of calculation is used. (120/240V)

    (a) 200 kVA
    (b) 250 kVA
    (c) 280 kVA
    (d) 320 kVA

35. •If two parallel service conductors per phase are installed in two raceways (500 kcmil in each raceway), what size equipment bonding jumper is required for each service raceway?

    (a) 2 AWG
    (b) 1 AWG
    (c) 1/0 AWG
    (d) 2/0 AWG

36. •If two parallel service conductors per phase are installed in two raceways (500 kcmil in each raceway), what is the minimum size copper grounding electrode conductor required?

    (a) 1/0 AWG
    (b) 2/0 AWG
    (c) 3/0 AWG
    (d) 4/0 AWG

Unit 10 Challenge Questions

UNIT 10—MULTIFAMILY DWELLING CALCULATIONS CHALLENGE QUESTIONS

(•Indicates that 75% or fewer of those who took this exam answered the question correctly.)

PART A—STANDARD METHOD—FEEDER/SERVICE LOAD CALCULATIONS

10.2 Multifamily Dwelling Unit Calculation Examples—Standard Method

General Lighting, Small Appliance, and Laundry Demand [Table 220.42]

1. Each dwelling unit of a 20-unit multifamily building has 900 sq ft of living space. What is the general lighting and general-use receptacle load, before demand factors are applied, for the multifamily apartment building?

    (a) 37 kVA
    (b) 45 kVA
    (c) 54 kVA
    (d) 60 kVA

2. A multifamily apartment building contains 20 units and each unit has 840 sq ft of living space. What is the general lighting service calculated load for the multifamily building if laundry facilities are provided on the premises for all tenants and no laundry circuit is installed in each unit?

    (a) 35 kVA
    (b) 41 kVA
    (c) 45 kVA
    (d) 63 kVA

Appliance Calculated Load [220.53]

3. A multifamily apartment building contains 20 units. Each unit contains a waste disposal (900 VA), a dishwasher (1,200 VA), and a water heater (5,000 VA). What is the feeder calculated load for the appliances in this building?

    (a) 60,000 VA
    (b) 106,500 VA
    (c) 117,100 VA
    (d) 137,000 VA

Household Dryer Calculated Load [220.54]

4. The nameplate rating for each household dryer in a 10-unit apartment building is 4 kW. This will add _____ to the service size.

    (a) 20 kW
    (b) 25 kW
    (c) 40 kW
    (d) 50 kW

Ranges [Table 220.55, Note 1]

5. •The calculated load for thirty 15.80 kW household ranges is _____.

    (a) 31 kW
    (b) 33 kW
    (c) 47 kW
    (d) 54 kW

Ranges [Table 220.55, Note 2]

6. What is the feeder calculated load for five 10 kW, five 14 kW, and five 16 kW household ranges?

    (a) 14 kW
    (b) 30 kW
    (c) 33 kW
    (d) 210 kW

Unit 10   Challenge Questions

7. What kW needs to be added to service loads for ten 12 kW, eight 14 kW, and two 9 kW household ranges?

    (a) 29.35 kW
    (b) 33 kW
    (c) 35 kW
    (d) 36.75 kW

Ranges [Table 220.55, Note 3]

8. What is the minimum service calculated load for five 5 kW cooktops, two 4 kW ovens, and four 7 kW ranges?

    (a) 8.80 kW
    (b) 15.50 kW
    (c) 18.20 kW
    (d) 19.50 kW

9. •What is the maximum dwelling unit feeder/service calculated load for fifteen 8 kW cooking units?

    (a) 30 kW
    (b) 38.40 kW
    (c) 60 kW
    (d) 120 kW

Neutral Calculation [220.61]

10. The service neutral load for household electric clothes dryers must be calculated at _____ percent of the load as determined by 220.54.

    (a) 50
    (b) 60
    (c) 70
    (d) 80

11. The feeder neutral demand for fifteen 9 kW cooking units is _____.

    (a) 21 kW
    (b) 30 kW
    (c) 38.40 kW
    (d) 120 kW

12. What is the feeder neutral load for ten 5 kW household dryers?

    (a) 17.50 kW
    (b) 21.50 kW
    (c) 27.50 kW
    (d) 32.50 kW

13. An 11-unit multifamily dwelling contains a 4 kW electric clothes dryer in each unit. What is the feeder/service neutral calculated load?

    (a) 18.10 kW
    (b) 22.30 kW
    (c) 29.30 kW
    (d) 32.90 kW

14. •What is the dwelling unit service neutral load for ten 9 kW ranges?

    (a) 12.40 kW
    (b) 13.50 kW
    (c) 15.50 kW
    (d) 17.50 kW

UNIT 11 Commercial Calculations

INTRODUCTION TO UNIT 11—COMMERCIAL CALCULATIONS

Commercial buildings have different use patterns than homes or multifamily dwellings. In a commercial building, certain loads (such as store lighting) are used for extended periods of time. Homes have a diversity of usage as discussed in Units 9 and 10, which allow us to apply demand factors to many of the loads used in dwelling units. There are some demand factors for commercial buildings, but they're different from those used in dwelling calculations. It's important to study this unit carefully so you understand the proper application of demand factors in commercial calculations.

Be careful when applying demand factors to commercial installations. For example, the demand factors for dryers in Table 220.54 are for dwelling units only and can't be used for laundromats or similar locations. The range table of 220.55 is allowed only for dwelling unit use and can't be applied to commercial kitchens. There are different rules, and in some cases different demand factors for commercial occupancies.

Commercial occupancies include many different types of businesses and uses. Some of the occupancies that are covered in this unit include banks, stores, restaurants, and office buildings. Some other locations with their own special requirements, such as marinas and mobile home parks, are also discussed.

Unit 11 includes some specific requirements for restaurant equipment, show-window lighting, sign lighting, multioutlet assemblies, and electric welders. As you study this unit, pay careful attention to when demand factors can be applied, and also to when a load must be considered a "continuous duty load" for proper conductor sizing.

PART A—GENERAL

Introduction

The *NEC* contains detailed requirements for calculating service and feeder loads in many types of occupancies (buildings). Although most of these requirements are located in Article 220, other rules affecting service and feeder loads are scattered throughout the *Code*. Units 9 and 10 covered the *NEC* rules for calculating loads in residential buildings. This unit covers commercial occupancies. It's an important and challenging subject, because there are many different types of commercial buildings and installations, and some have unique rules for calculating service and feeder loads.

11.1 General Requirements

Article 220 provides the requirements for branch circuits, feeders, and services. In addition to this article, other articles are applicable such as Branch Circuits—210, Feeders—215, Services—230, Overcurrent Protection—240, Wiring Methods—300, Conductors—310, Appliances—422, Electric Space-Heating Equipment—424, Motors—430, and Air-Conditioning Equipment—440.

11.2 Conductor Ampacity [Article 100]

The ampacity of a conductor is the rating, in amperes, that a conductor can carry continuously without exceeding its insulation temperature rating [Article 100]. Figure 11–1

Unit 11 | Commercial Calculations

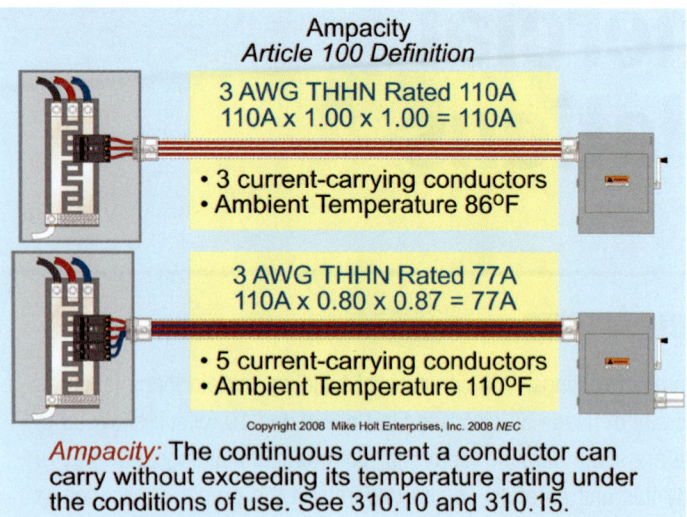

Figure 11–1

The allowable ampacities listed in Table 310.16 are affected by conductor insulation, ambient temperature, and conductor bundling [310.10 and 310.15(B)]. **Figure 11–2**

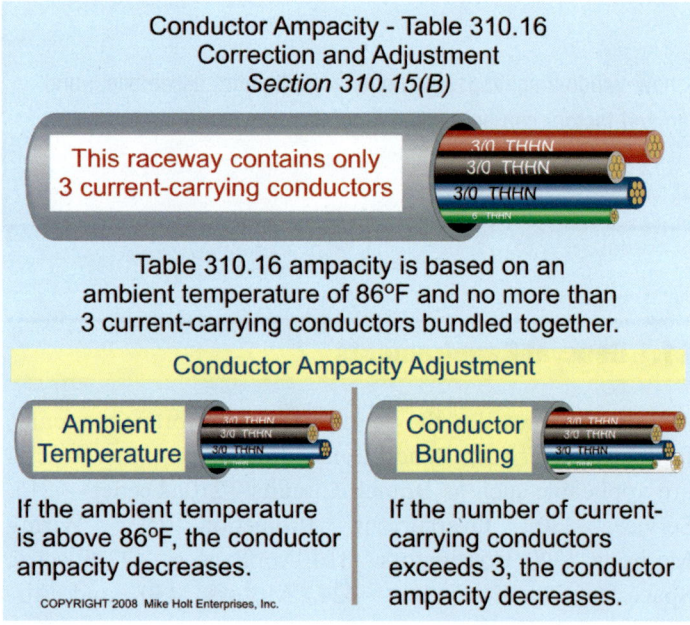

Figure 11–2

Author's Comment: Section 110.14(C)(1)(a) states that terminals are rated 60°C for equipment rated 100A or less unless marked 75°C. In real life, most terminals are now rated 75°C, so in this unit, we'll assume all terminals are rated 75°C unless specified 60°C. For exam purposes, and when answering questions at the end of this unit, read the problem carefully to be certain you know what terminal rating the question specifies. If unspecified, use the rules of 110.14(C).

Continuous Loads

Ungrounded conductors are sized at 125 percent of the continuous load before any adjustment factor and the overcurrent devices are also sized at 125 percent of the continuous load [210.19(A), 215.2(A)(1), and 230.42]. Neutral conductors that are't connected to an overcurrent device are permitted to be sized at 100 percent of the continuous and noncontinuous load [210.19(A)(1) Ex 2 and 215.2(A)(1) Ex 2].

Author's Comment: Article 100 defines a continuous load as a load where the maximum current is expected to continue for 3 hours or more. Some NEC sections tell you when certain loads are continuous, such as 422.13 for some types of water heaters, or 424.3(B) for fixed electric space-heating equipment. Most commercial lighting and electric signs are considered continuous loads.

Example: If a 60A continuous load with 75°C rated terminals is supplied by a feeder with four current-carrying conductors, it's necessary to adjust the conductor ampacity for four current-carrying conductors.

Step 1. Take the continuous load times 125 percent when sizing conductors and overcurrent protection.

60A x 1.25 = 75A

Step 2. Select the conductors using the column of Table 310.16 that corresponds with the temperature rating of the terminals [110.14(C)(1)].

4 AWG is rated 85A at 75°C [Table 310.16].

Step 3. Verify that the conductor is large enough for any necessary deratings, using the column that corresponds to the conductor's temperature rating. If you're using THHN conductors, use the 90°C column of Table 310.16 [110.14(C)].

Adjust the ampacity by 80 percent for four current-carrying conductors [Table 310.15(B)(2)(a)].

95A x 0.80 = 76A

This verifies that a 4 THHN conductor is large enough. It can be used on an 80A breaker because 240.4(B) allows rounding up to the next standard size listed in 240.6(A).

11.3 Conductor Overcurrent Protection [240.4]

The purpose of an overcurrent device is to protect conductors and equipment against excessive or dangerous temperatures [240.1 FPN]. There are many different rules for protecting conductors and equipment. It isn't simply 12 AWG wire and a 20A breaker. The general rule is that conductors must be protected at the point where they receive their supply in accordance with their ampacities, as listed in Table 310.16. Other methods of protection are permitted or required as listed in 240.4, such as:

(B) Overcurrent Devices Rated 800A or Less. The next higher standard rating of overcurrent device (above the ampacity of the ungrounded conductors being protected) is permitted, provided all of the following conditions are met:

(1) The conductors don't supply multioutlet receptacle branch circuits.

(2) The ampacity of a conductor, after ampacity adjustment and/or correction, doesn't correspond with the standard rating of a fuse or circuit breaker in 240.6(A).

(3) The overcurrent device rating doesn't exceed 800A.

Author's Comment: A 400A overcurrent device can protect 500 kcmil conductors, where each conductor has an ampacity of 380A at 75°C, in accordance with Table 310.16. **Figure 11-3**

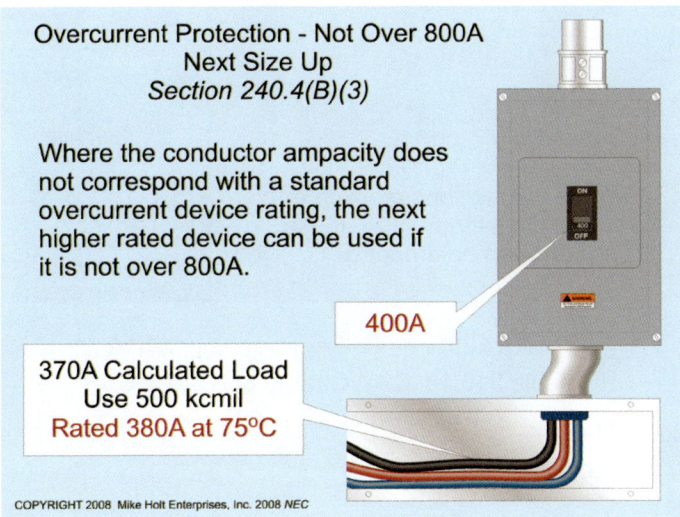

Figure 11–3

Author's Comment: This "next size up" rule doesn't apply to feeder tap conductors [240.21(B)], or transformer secondary conductors [240.21(C)].

(C) Overcurrent Devices Rated Over 800A. If the circuit's overcurrent device exceeds 800A, the conductor ampacity, after ampacity adjustment and/or correction, must have a rating of not less than the rating of the overcurrent device.

Author's Comment: A 1,200A overcurrent device can protect three sets of 600 kcmil conductors per phase, where each conductor has an ampacity of 420A at 75°C, in accordance with Table 310.16. **Figure 11-4**

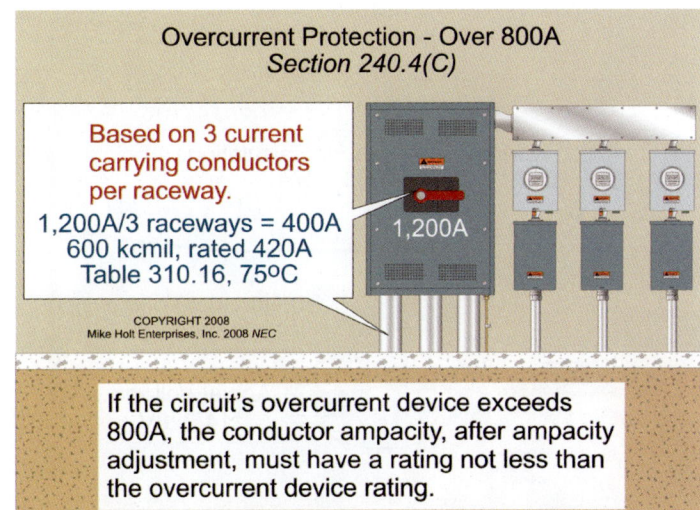

Figure 11–4

Standard Size Overcurrent Devices [240.6(A)]. The following is a list of some of the standard ampere ratings for fuses and inverse time circuit breakers: 15, 20, 25, 30, 35, 40, 45, 50, 60, 70, 80, 90, 100, 110, 125, 150, 175, 200, 225, 250, 300, 350, 400, 500, 600, 800, 1,000, 1,200, and 1,600. **Figure 11–5**

11.4 Voltages [220.5(A)]

Unless other voltages are specified, branch-circuit, feeder, and service loads must be computed at a nominal system voltage of 120V, 120/240V, 120/208V, 240V, 277/480V, or 480V. **Figure 11–6**

Unit 11 | Commercial Calculations

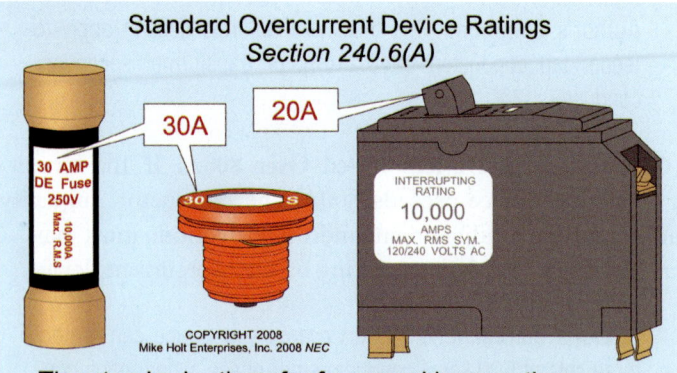

Figure 11–5

▶ Rounding Example

Question: According to 424.3(B), the branch-circuit conductors and overcurrent device for electric space-heating equipment must be sized no less than 125 percent of the total load. What size conductor is required to supply a 9 kVA (37.50A), 240V, single-phase fixed space heater with a 3A blower motor if the equipment terminals are rated 75°C? **Figure 11–7**

(a) 10 AWG   (b) 8 AWG   (c) 6 AWG   (d) 4 AWG

Answer: (c) 6 AWG

Step 1: Determine the total load

$I = VA/E$

$I = 9,000 \text{ VA}/240V$

$I = 37.50A$

Step 2: Conductor sized at 125 percent of the load.

Conductor Size = (37.50A + 3A) x 1.25

Conductor Size = 50.63A, round up to 51A

If we round down, then 8 AWG rated 50A at 75°C can be used, but since we have to round up, 6 AWG rated 65A at 75°C is required.

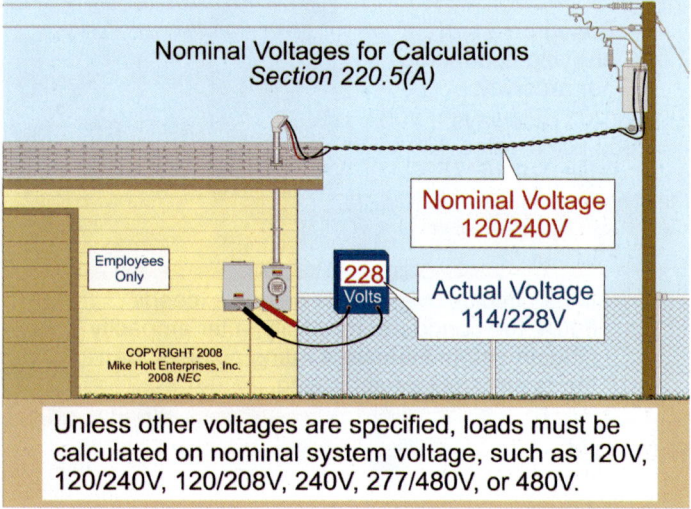

Figure 11–6

11.5 Rounding an Ampere [220.5(B)]

Where calculations result in a fraction of less than 0.50A, such fractions can be dropped.

> **Author's Comment:** When do you round—after each calculation, or at the final calculation? The *NEC* isn't specific on this issue, so I guess it all depends on the answer you want to see!

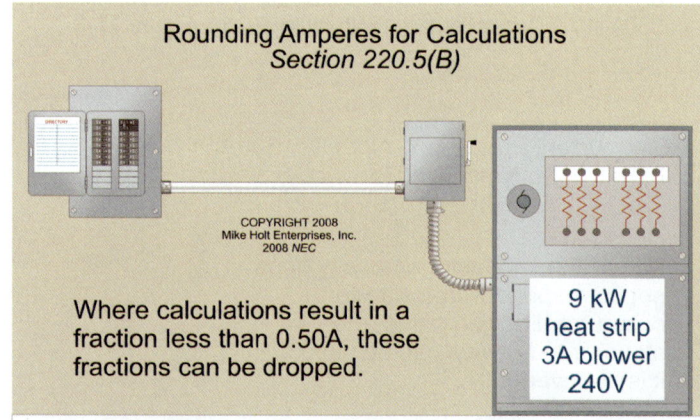

40.5A x 125% = 50.63A, Round up to 51A
Table 310.16, 6 AWG rated 65A at 75°C

Figure 11–7

Commercial Calculations — Unit 11

11.6 Lighting—Demand Factors
[Tables 220.12 and 220.42]

NEC Table 220.12 requires a minimum load per square foot for general lighting depending on the type of occupancy. For the guest rooms of hotels and motels, hospitals, and storage warehouses, the demand factors of Table 220.42 can be applied to the general lighting load.

Hotel or Motel Guest Rooms—General Lighting [Tables 220.12 and 220.42]

The general lighting unit load of 2 VA per sq ft [Table 220.12] for the guest rooms of hotels and motels is permitted to be reduced according to the demand factors listed in Table 220.42.

General Lighting Demand Factors

- First 20,000 VA at 50 percent demand factor
- Next 80,000 VA at 40 percent demand factor
- Remainder VA at 30 percent demand factor

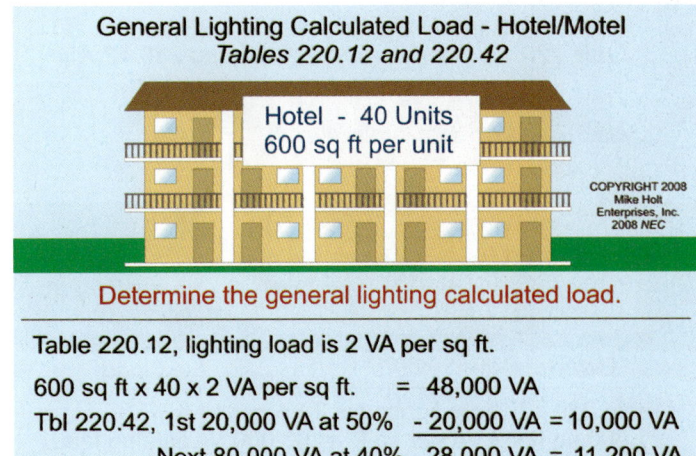

Figure 11–8

Author's Comment: The lighting loads listed in Table 220.12 are minimum requirements. If the actual lighting load is known and it's larger than the Table 220.12 value, then the actual load must be used.

▶ **Hotel General Lighting Calculated Load Example**

Question: What is the general lighting calculated load for a 40-room hotel? Each unit contains 600 sq ft of living area. Figure 11–8

(a) 20 kVA   (b) 21 kVA   (c) 24 kVA   (d) 48 kVA

Answer: (b) 21 kVA
[Tables 220.12 and 220.42]

```
40 units x 600 sq ft x 2 VA       48,000 VA
First 20,000 VA at 50%          − 20,000 VA  x 0.50   10,000 VA
Next 80,000 VA at 40%             28,000 VA  x 0.40  +11,200 VA
                                                      21,200 VA
```

▶ **Store Lighting Example**

Question: What is the general lighting load for a 21,000 sq ft store? Figure 11–9

(a) 40 kVA   (b) 63 kVA   (c) 79 kVA   (d) 81 kVA

Answer: (c) 79 kVA

Store is 3 VA per sq ft [Table 220.12]
General Lighting Load = 21,000 sq ft x 3 VA
General Lighting Load = 63,000 VA
General Lighting Calculated Load = 63,000 VA x 1.25
General Lighting Calculated Load = 78,750 VA

11.7 Lighting Without Demand Factors
[215.2(A)(1), 230.42(A)(1), and Table 220.12]

The feeder/service general lighting load for commercial occupancies other than guest rooms of motels and hotels, hospitals, and storage warehouses is assumed to be continuous and must be calculated at 125 percent of the general lighting load as listed in Table 220.12.

11.8 Lighting—Miscellaneous

The feeder/service calculated load for each linear foot of show-window lighting must be calculated at 200 VA per ft. Show-window lighting is assumed to be a continuous load. See Example D3 in Annex D of the *NEC* for an example calculation including show-window branch circuits.

Unit 11 Commercial Calculations

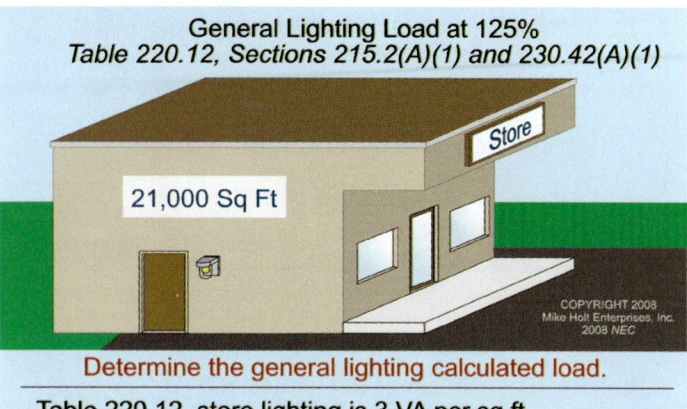

Determine the general lighting calculated load.

Table 220.12, store lighting is 3 VA per sq ft.
21,000 sq ft x 3 VA per sq ft = 63,000 VA lighting load
Store lighting is a continuous load at 125%
63,000 VA lighting load x 1.25 = 78,750 VA load

Figure 11–9

*In the example, 125% of the actual connected lighting load (8,500 VA x 1.25 = 10,650 VA) is less than 125% of the load from Table 220.12, so the minimum lighting load from Table 220.12 is used in the calculations. Had the actual lighting load been greater than the value calculated from Table 220.12, 125% of the actual connected lighting load would have been used. See NEC Annex D, Example D3 for the full calculation on this store building.

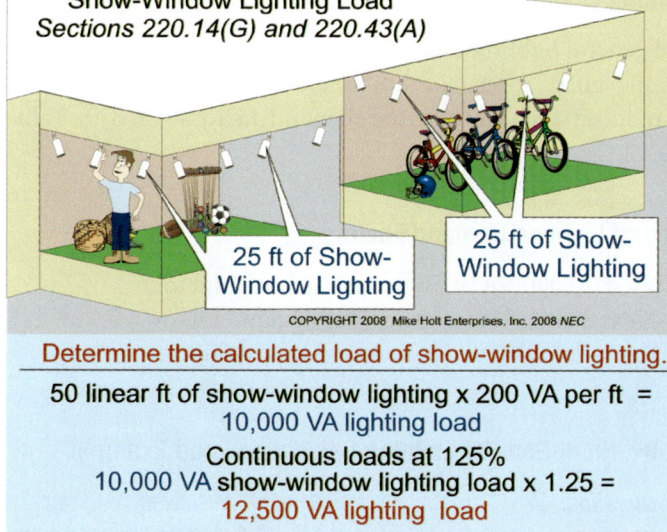

Determine the calculated load of show-window lighting.

50 linear ft of show-window lighting x 200 VA per ft = 10,000 VA lighting load

Continuous loads at 125%
10,000 VA show-window lighting load x 1.25 = 12,500 VA lighting load

Figure 11–10

▶ **Show-Window Load Example 1**

Question: What is the feeder/service calculated load in kVA for 50 ft of show-window lighting? **Figure 11–10**

(a) 6 kVA (b) 7.50 kVA (c) 9 kVA (d) 12.50 kVA

Answer: (d) 12.50 kVA

Lighting Load = 50 ft x 200 VA per ft
Lighting Load = 10,000 VA

Lighting Calculated Load = 10,000 VA x 1.25
Lighting Calculated Load = 12,500 VA

Lighting Calculated Load in kVA = 12,500 VA/1,000
Lighting Calculated Load in kVA = 12.50 kVA

▶ **Show-Window Load Example 2**

Example: A 3,000 sq ft store has 30 ft of show window. There are a total of 80 duplex receptacles. The service is 120/240V, single-phase. The actual connected lighting load is 8,500 VA [Annex D, Example D3].

General Lighting* = 3,000 sq ft at 3 VA per sq ft
General Lighting* = 9,000 VA

Window Lighting Load [220.14(G)] = 30 ft at 200 VA per ft
Window Lighting Load = 6,000 VA

Outside Sign Circuit [220.14(F)] = 1,200 VA

Lighting Total = 9,000 VA + 6,000 VA + 1,200 VA
Lighting Total = 16,200 VA

(continued in next column)

11.9 Multioutlet Receptacle Assembly [220.14(H)]

Each 5 ft or fraction thereof of multioutlet receptacle assembly must be considered to be 180 VA for feeder/service calculations where it is unlikely that appliances will be used simultaneously. When a multioutlet receptacle assembly is expected to have a number of appliances used simultaneously, each foot or fraction of a foot must be considered as 180 VA for feeder/service calculations. A multioutlet receptacle assembly is not generally considered to be a continuous load.

▶ **Multioutlet Receptacle Assembly Example**

Question: What is the calculated load for 10 work stations, each of which has 10 ft of multioutlet receptacle assembly not used simultaneously and 3 ft of multioutlet receptacle assembly simultaneously used? **Figure 11–11**

(continued in next column)

Commercial Calculations — Unit 11

(a) 5 kVA (b) 6 kVA (c) 7 kVA (d) 9 kVA

Answer: (d) 9 kVA

10 stations with 10 ft per station = 100 ft of multioutlet assembly not simultaneously used

10 stations with 3 ft per station = 30 ft of multioutlet assembly simultaneously used

100 ft/5 ft per section = 20 sections x 180 VA 3,600 VA
30 ft/1 ft per section = 30 sections x 180 VA + 5,400 VA
Calculated Load 9,000 VA

This example shows how to size the feeder to accommodate this load. To size the branch circuits using 20A branch circuits, first find the VA allowed per circuit:

- 120V x 20A = 2,400 VA for noncontinuous loads
- Then divide by 180 VA to find how many 180 VA sections can be served by a 20A circuit
- 2,400 VA/180 VA = 13 so a 20A branch circuit can serve two tables.

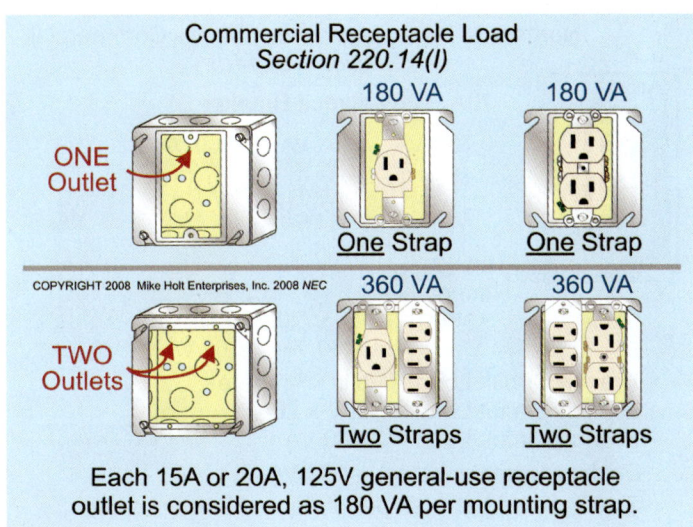

Figure 11–12

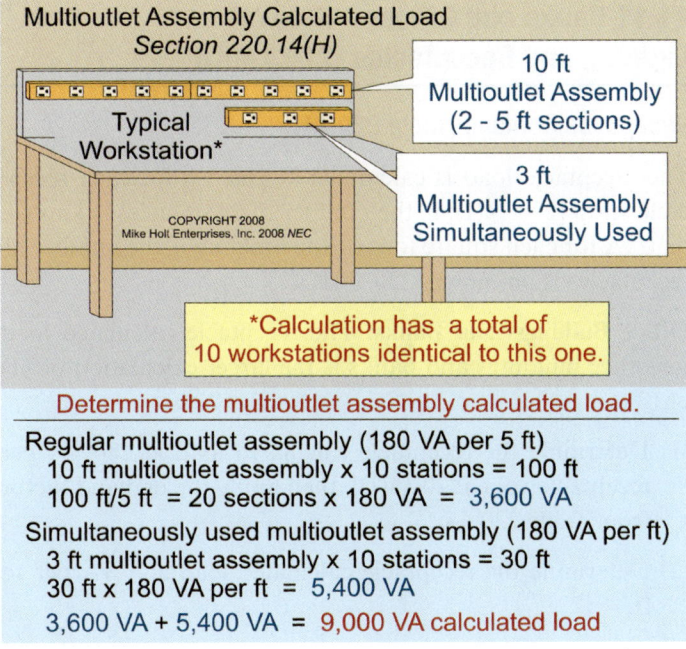

Figure 11–11

11.10 Receptacle VA Load

The load for a general-use receptacle outlet in a nondwelling occupancy is 180 VA per strap [220.14(I)]. Receptacles are not considered to be a continuous load. **Figure 11–12**

Number of Receptacles Permitted on a Circuit

The maximum number of receptacle outlets permitted on a commercial or industrial circuit depends on the circuit ampacity. The number of receptacles per circuit is calculated by dividing the VA rating of the circuit by 180 VA for each receptacle strap (also called a yoke). Based on the Article 100 definition, a duplex receptacle is two receptacles on the same yoke. For the purposes of this calculation, a single receptacle or a duplex receptacle each count as 180 VA.

▶ **Receptacles per 20A Circuit [220.14(I)] Example**

Question: How many 15A and 20A, 125V duplex receptacles are permitted on a 20A, 120V circuit in a nondwelling occupancy? **Figure 11–13**

(a) 10 (b) 13 (c) 15 (d) 20

Answer: (b) 13

Total circuit VA load for a 20A circuit = 120V x 20A
Total circuit VA load for a 20A circuit = 2,400 VA

Number of receptacle outlets (straps) per circuit = 2,400 VA/180 VA
Number of receptacle outlets (straps) per circuit = 13 receptacles

Unit 11 — Commercial Calculations

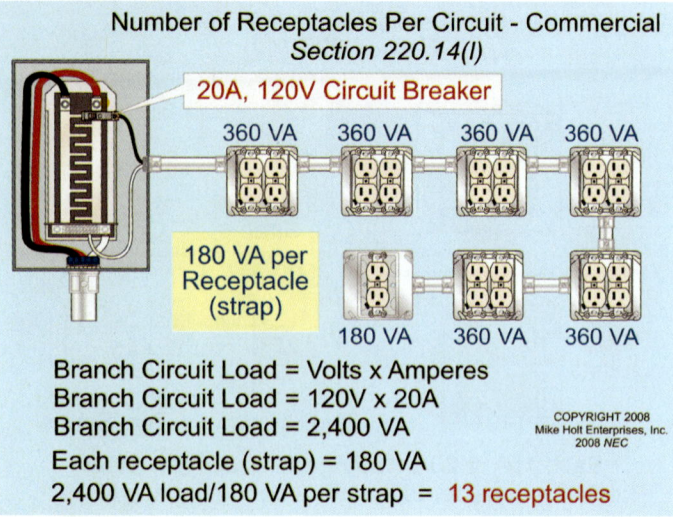

Figure 11–13

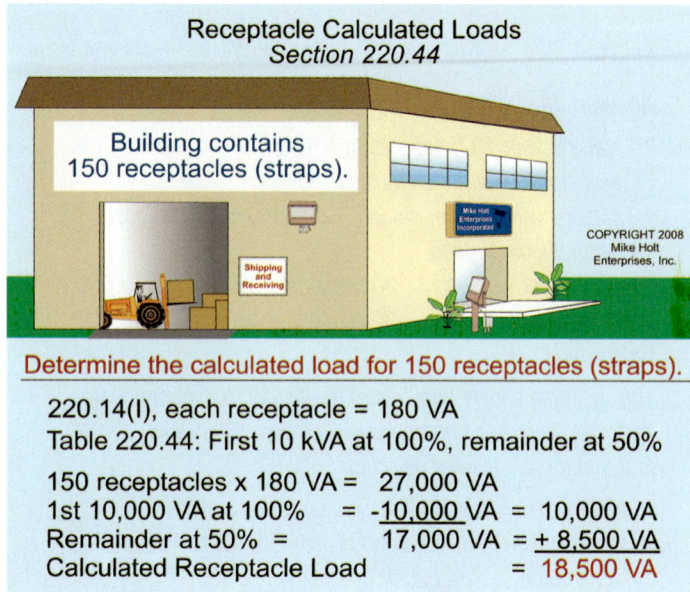

Figure 11–14

Receptacle Feeder/Service Calculated Load [220.44]

In other than dwelling units, receptacle loads computed at not less than 180 VA per outlet (strap) in accordance with 220.14(I), and fixed multioutlet assemblies computed in accordance with 220.14(H), can be added to the lighting loads and made subject to the demand factors given in Table 220.42. Or they are permitted to be made subject to the demand factors given in Table 220.44. Receptacles are generally not considered to be continuous loads.

Table 220.44 Receptacle Demand Factors

- First 10 kVA at 100 percent demand factor
- Remainder kVA at 50 percent demand factor

▶ **Receptacle Feeder/Service Calculated Load Using Table 220.44 Example**

Question: *What is the feeder/service calculated load for one hundred and fifty, 15A and 20A, 125V duplex receptacles in a nondwelling occupancy?* **Figure 11–14**

(a) 4 kVA   (b) 14 kVA   (c) 19 kVA   (d) 27 kVA

Answer: (c) 19 kVA

Total Receptacle Load
(150 receptacles x 180 VA)   27,000 VA
First 10 kVA at 100%   − 10,000 VA x 1.00   10,000 VA
Remainder at 50%   17,000 VA x 0.50 + 8,500 VA
Total Receptacle Calculated Load   18,500 VA

11.11 Banks and Offices—General Lighting and Receptacles [220.14(K)]

Receptacle Demand [Table 220.44]

The receptacle load is calculated at 180 VA for each receptacle strap [220.14(I)] if the number of receptacles is known, or 1 VA for each square foot of the building if the number of receptacles is unknown [220.14(K)(2)].

Office Buildings and Banks. The receptacle calculated load for office buildings and banks is the larger calculation of (1) or (2).

(1) Determine the receptacle calculated load at 180 VA per receptacle yoke [220.14(I)], then apply the demand factor from Table 220.44.

(2) Determine the receptacle calculated load at 1 VA per sq ft.

Author's Comment: Not knowing the exact number of receptacles that will eventually be installed in an office building or bank isn't unusual. The main structure is built first, then individual office space that's rented out to each tenant will often have a custom installation, or a new tenant will remodel. 1 VA per square foot allows a generic feeder/service demand for general receptacles.

Commercial Calculations — Unit 11

▶ **Bank/Office Building Receptacle Load Example 1**

Question: What is the receptacle calculated load for an 18,000 sq ft bank/office building containing one hundred sixty 15A and 20A, 125V receptacles (straps)? **Figure 11–15**

(a) 15,400 VA (b) 19,400 VA (c) 28,800 VA (d) 142 kVA

Answer: (b) 19,400 VA

[220.14(K)(1)]
160 receptacles (straps) x 180 VA [220.14(I)] = 28,800 VA

Total Receptacle Load =   28,800 VA
First 10,000 at 100%   – 10,000 VA x 1.00   10,000 VA
Remainder at 50%   18,800 VA x 0.50 + 9,400 VA
  [Table 220.44]
Receptacle Calculated Load   19,400 VA

Compare this to the 1 VA method [220.14(K)(2)]
18,000 x 1 VA per sq ft = 18,000 VA (smaller answer, omit)

▶ **Bank/Office Building Receptacle Load Example 2**

Question: What is the receptacle calculated load for an 18,000 sq ft Bank/Office Building containing one hundred forty 15A and 20A, 125V duplex receptacles? **Figure 11–16**

(a) 15,000 VA (b) 18,000 VA (c) 23,000 VA (d) 31,000 VA

Answer: (b) 18,000 VA

[220.14(K)(1)]
140 receptacles (straps) x 180 VA = 25,200 VA [220.14(I)]

Total Receptacle Load   25,200 VA
First 10,000 VA at 100%   – 10,000 VA x 1.00   10,000 VA
Remainder at 50%   15,200 VA x 0.50 + 7,600 VA
  [Table 220.44]
Receptacle Calculated Load   17,600 VA
                                                  (omit)

Compare this to the 1 VA method [220.14(K)(2)]
18,000 sq ft x 1 VA per sq ft = 18,000 VA

Figure 11–15

Figure 11–16

11.12 Sign Circuit [220.14(F) and 600.5]

The *NEC* requires each commercial occupancy accessible to pedestrians to have at least one 20A branch circuit for a sign [600.5(A)]. The load for the required exterior sign or outline lighting must be a minimum of 1,200 VA [220.14(F)]. A sign outlet is considered to be a continuous load and the

Unit 11 — Commercial Calculations

feeder/service conductor must be sized at 125 percent of the continuous load [215.2(A)(1) and 230.42].

▶ **Sign Calculated Load Example**

Question: What is the feeder/service conductor calculated load for one electric sign? **Figure 11–17**

(a) 1,200 VA   (b) 1,500 VA   (c) 1,920 VA   (d) 2,400 VA

Answer: (b) 1,500 VA

Feeder/Service Calculated Load = 1,200 VA x 1.25
Feeder/Service Calculated Load = 1,500 VA

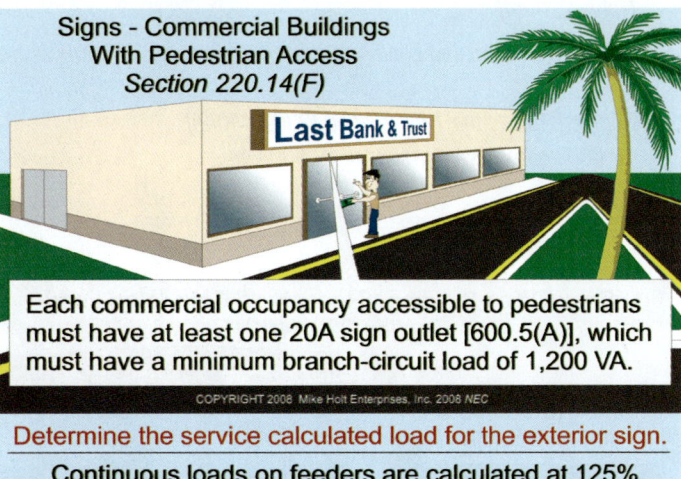

Figure 11–17

PART B—EXAMPLES

11.13 Bank/Office Building Example

Determine the feeder/service calculated load for an office building/bank with the loads listed in **Figure 11–18**.

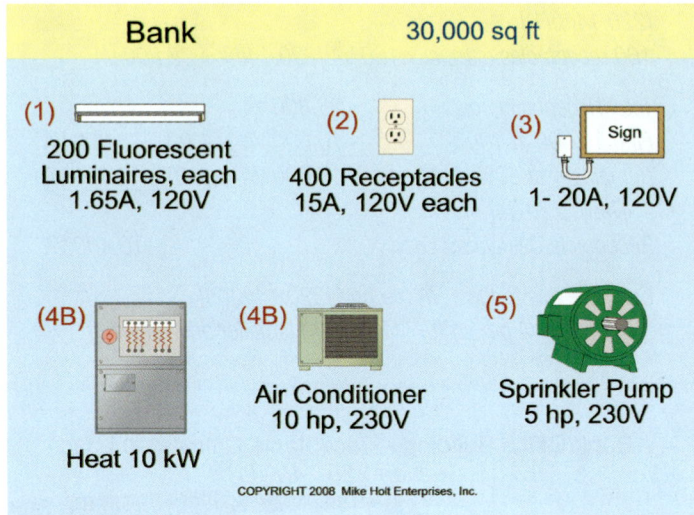

Figure 11–18

Determine the feeder/service calculated load:

Bold indicates neutral load. [215.2(A)(1) Ex 2].

Step 1a: Lighting [Table 220.12]
　　General Lighting = 30,000 sq ft x 3.50 VA
　　General Lighting = **105,000 VA**
　　General Lighting Calculated Load = 105,000 VA x 1.25　　　　　　　　　　131,250 VA

Step 1b: Actual Lighting
　　Actual Lighting Connected Load = 200 units x 120V x 1.65A
　　Actual Lighting Connected Load = 39,600 VA
　　Actual Lighting Calculated Load = 39,600 VA x 1.25　　　　　　　　　　49,500 VA (omit)

Commercial Calculations — Unit 11

Step 2a: The receptacle load is calculated by using the larger of the computed load from 220.44 or 1 VA per sq ft [220.14(K)].

    Computed Load from 220.44
    Receptacles [220.44] 400 receptacles (straps) x 180 VA   72,000 VA
    First 10 kVA at 100%   −10,000 VA   x 1.00   10,000 VA
    Remainder at 50%   62,000 VA   x 0.50   + 31,000 VA
                                                                                      41,000 VA

    Receptacle Calculated Load   41,000 VA
    Step 2b: 1 VA per sq. ft per 220.14(K)   30,000 sq ft   x 1 VA   30,000 VA (omit)

Step 3: Sign [220.14(F)]   1,200 VA   x 1.25   **1,500 VA**

Step 4: A/C versus Heat

Step 4a: Air-Conditioning 50A

    VA = (230V x 50A)
    VA = 11,500VA
    A/C Calculated Load   11,500 VA (omit)

    Note: Some exams use 240V (the system nominal voltage) instead of 230V which is the actual motor voltage of this problem, so be aware of this possibility.

Step 4b: Heat [220.51 and 220.60] 15,000 VA   15,000 VA

Step 5. Sprinkler Pump Motor 5 hp, single-phase, 230V FLC = 28A [Table 430.248]

    VA = V x A
    VA   230V x 28A   6,440 VA x 1.25   8,050 VA

    Note: Some exams use 240V (the system nominal voltage) instead of 230V which is the actual motor voltage of this problem, so be aware of this possibility.

Summary

	Feeder/Service	**Neutral**
Step 1. Lighting	131,250 VA	105,000 VA
Step 2. Receptacles	41,000 VA	41,000 VA
Step 3. Sign	1,500 VA	1,200 VA
Step 4. Heat	15,000 VA	0 VA
Step 5 Sprinkler Pump	8,050 VA	0 VA
	196,800 VA	147,200 VA *

*The neutral conductor is sized at 100 percent of the continuous load [215.2(A)(1) Ex. 1].

Feeder/Service Calculated Load:

I = VA/E
I = 196,800 VA/240V
I = 820A

Unit 11 — Commercial Calculations

Neutral Amperes: The neutral conductor is permitted to be reduced according to the requirements of 220.61(B)(2) for that portion of the unbalanced load that exceeds 200A. Since this system is a 120/240V, single-phase system, we're permitted to reduce the neutral by multiplying the portion that exceeds 200A by 70 percent.

Neutral Conductor:

$I = VA/E$
$I = 147{,}200\ VA/240V$
$I = 613A$

220.61(B)(2) allows reduction of the neutral load over 200A to 70%

First 200A at 100%	200A	x 1.00	200A
Remainder at 70%	413A	x 0.70	+ 289A
			489A

The service conductors and overcurrent device must be sized no less than 820A, and the neutral conductor must be sized to carry not less than 489A.

Summary Questions

▶ Overcurrent Protection Example

Question: What size service overcurrent device is required to supply an 820A calculated load?

(a) 800A   (b) 1,000A   (c) 1,200A   (d) 1,600A

Answer: (b) 1,000A [240.6(A)]

▶ Service Ungrounded Conductor Size Example

Question: What size service conductors are required in each raceway if the conductors are paralleled in four raceways to supply an 820A calculated load?

(a) 250 kcmil   (b) 300 kcmil   (c) 350 kcmil   (d) 400 kcmil

Answer: (a) 250 kcmil

When the overcurrent device is over 800A, the ampacity of the conductors must be equal to, or greater than, the rating of the overcurrent device. In this example, the overcurrent device is 1,000A, so the ampacity of the conductors must be 1,000A or more [240.4(C)].

1,000A/4 raceways = 250A per raceway, sized based on the 75°C terminal rating [110.14(C)(1)(b)]

250 kcmil rated 255A x 4 = 1,020A [Table 310.16]

▶ Parallel Service Conductors Sizing Example

Question: What size service conductors are required in each raceway if the conductors are paralleled in three raceways to supply an 820A calculated load?

(a) 250 kcmil   (b) 300 kcmil   (c) 350 kcmil   (d) 400 kcmil

Answer: (d) 400 kcmil

1,000A/3 raceways = 333A for each conductor, if fed by 3 parallel conductors

400 kcmil is rated 335A at 75°C [Table 310.16].

The neutral service conductor must have an ampacity of not less than 489A and it must not be smaller than 1/0 for paralleling [310.4], and meet the requirements of 250.24(C)(1).

If using three neutral conductors in parallel per phase:

- 489A/3 raceways = 163A
- 2/0 AWG is rated 175A at 75°C [Table 310.16]

▶ Service Neutral Conductor Size Example

Question: What size neutral service conductor is required in each of the four raceways to supply a 489A calculated neutral load if using four conductors in parallel per phase?

(a) 1/0 AWG   (b) 2/0 AWG   (c) 3/0 AWG   (d) 4/0 AWG

Answer: (b) 2/0 AWG

The neutral service conductor must have the capacity to carry 489A and be sized no smaller than 1/0 AWG to meet the paralleling requirements of 310.4. [220.61(B)(2)].

489A/4 raceways = 123A; 1 AWG rated 130A at 75°C [110.14(C)(1)(b), and Table 310.16], but the smallest size conductor permitted for paralleling is 1/0 AWG. The neutral must also be not smaller than called for by Table 250.66, based on the ungrounded conductors in the same raceway [250.24(C)(2)].

Commercial Calculations — Unit 11

▶ Grounding Electrode Conductor Size Example [250.66(B)]

Question: What size grounding electrode conductor is required for a 1,000A service fed by four sets of parallel 250 kcmil conductors per phase?

(a) 4 AWG (b) 1/0 AWG (c) 2/0 AWG (d) 3/0 AWG

Answer: (c) 2/0 AWG

The grounding electrode conductor is sized to the four sets of 250 kcmil conductors in accordance with Table 250.66 [250.24(C)(2)]. First find the equivalent size of all four conductors in each phase [250.24(C) and Table 250.66, Note 1]:

250 kcmil x 4 = 1,000 kcmil equivalent size

1,000 kcmil will require a 2/0 grounding electrode conductor [Table 250.66].

▶ Feeder Equipment Grounding Conductor Example

Question: If a 1,000A feeder is installed using four sets of parallel 250 kcmil conductors per phase in four raceways, what size equipment grounding conductor will be required in each raceway?

(a) 4 AWG (b) 1/0 AWG (c) 2/0 AWG (d) 3/0 AWG

Answer: (c) 2/0 AWG

The equipment grounding conductor for a feeder is based on the size of the overcurrent device protecting the feeder [250.122(F)].

For a feeder with 1,000A overcurrent protection, a 2/0 equipment grounding conductor is required [Table 250.122].

11.14 Marina [555.12]

The *NEC* permits a demand factor to apply to the receptacle outlets for boat slips at a marina. The demand factors of Table 555.12 are based on the number of receptacles on the feeder. The receptacles must also be balanced between the phases to determine the number of receptacles on any given line.

▶ Marina Receptacle Outlet Demand Example [555.12]

Question: What size 120/240V, single-phase service is required for a marina that has twenty 30A, 250V receptacles and twenty 20A, 125V receptacles? **Figure 11–19**

(a) 200A (b) 420A (c) 560A (d) 720A

Answer: (c) 560A

The calculated load is based on the number of receptacles. Try to divide the receptacles equally on each line in order to calculate the load per line.

	Line 1	Line 2
Twenty 30A, 250V	20 x 30A	20 x 30A
Ten 20A, 125V	+ 10 x 20A	+ 10 x 20A
Thirty receptacles	30—800A	30—800A

Table 555.12, 30 receptacles = 0.70 demand factor

Calculated Load Per Line =
Connected Load Per Line x Demand Factor
Calculated Load per Line = 800A x 0.70
Calculated Load per Line = 560A per line

▶ Overcurrent Device Sizing Example

Question: What size overcurrent device will be required for this service?

(a) 200A (b) 400A (c) 600A (d) 800A

Answer: (c) 600A

For overcurrent protection, use a 600A overcurrent device [240.6(A)]

Paralleling two conductors per phase: 560A/2 raceways = 280A

300 kcmil is rated 285A at 75°C: 285A x 2 = 570A

The minimum neutral size when paralleling conductors is 1/0 AWG [250.24(C)(2) and 310.4].

▶ Grounding Electrode Conductor Sizing Example

Question: What size grounding electrode conductor will be required for this service, using two sets of 300 kcmil conductors in parallel?

(a) 4 AWG (b) 1/0 AWG (c) 2/0 AWG (d) 3/0 AWG

Answer: (b) 1/0 AWG

300 kcmil x 2 = 600 kcmil [Table 250.66, Note 1]

600 kcmil requires a 1/0 grounding electrode conductor [250.66].

Unit 11   Commercial Calculations

Figure 11–19

Author's Comment: The largest size grounding electrode conductor required to a ground rod is 6 AWG and to a concrete encased electrode (ufer) is 4 AWG [250.66(A) and 250.66(B)].

11.15 Mobile/Manufactured Home Park [550.31]

The service calculated load for a mobile/manufactured home park is sized according to the demand factors of Table 550.31 to the larger of 16,000 VA for each mobile/manufactured home lot, or the calculated load for each mobile/manufactured home site according to 550.18. It is also permitted to size service and feeder conductors to an individual mobile home in accordance with 310.15(B)(6).

▶ **Mobile/Manufactured Home Park Example**

Question: What is the feeder/service calculated load for a mobile/manufactured home park that has facilities for 35 sites? The system is 120/240V, single-phase. **Figure 11–20**

(a) 400A   (b) 560A   (c) 800A   (d) 1,000A

(continued in next column)

Answer: (b) 560A

A mobile home park's electrical wiring system must be calculated at 120/240V based on the larger of the following [550.31]:

(1) 16,000 VA for each mobile home lot.

(2) The load calculated in accordance with 550.18 for the largest typical mobile home that each lot will accept.

The feeder or service load can be calculated in accordance with the demand factor percentages in Table 550.31

Since no calculated load in accordance with 550.18 is given, just use 16,000 VA per site [550.31(1)].

Calculated Load = VA x Sites x Demand Factor

Table 550.31, 35 sites = 0.24 demand factor.
Calculated Load = 16,000 VA x 35 sites x 0.24
Calculated Load = 134,400VA

I = VA/E

I = 134,000 VA/240V
I = 560A

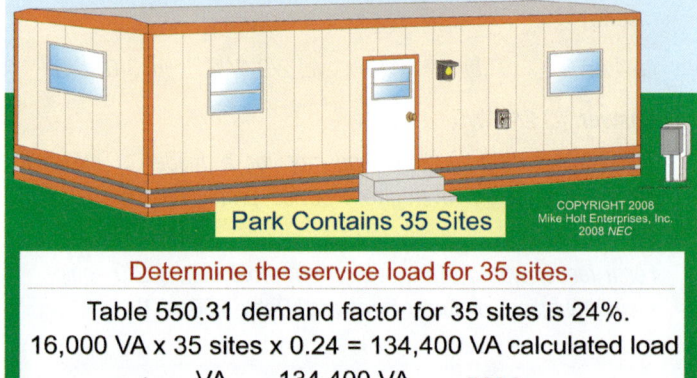

Figure 11–20

11.16 Recreational Vehicle Park [551.73]

Recreational vehicle parks are calculated according to the demand factors of Table 551.73. The total calculated load is based on:

- 2,400 VA for each site with 20A, 120V supply facilities,

Commercial Calculations — Unit 11

- 3,600 VA for each site with 20A and 30A, 120V supply facilities, and

- 9,600 VA for each site with 50A, 120/240V supply facilities.

▶ **Recreational Vehicle Park Example**

Question: What is the calculated load for a recreational vehicle park that has ten 20A supply facilities, fifteen 20A and 30A supply facilities, and twenty 30A and 50A supply facilities? The system is 120/240V, single-phase. **Figure 11–21**

(a) 420A   (b) 460A   (c) 520A   (d) 600A

Answer: (b) 460A

The service for the recreational vehicle park is sized according to the demand factors of Table 551.73.

Ten 20A sites (2,400 VA x 10 sites)	24,000 VA
Fifteen 20A and 30A sites (3,600 VA x 15 sites)	54,000 VA
Twenty 50A sites (9,600 VA x 20 sites)	+ 192,000 VA
Connected Load	270,000 VA

Calculated Load = 270,000 VA x 0.41 [Table 551.73]
Calculated Load = 110,700 VA

I = VA/E

I = 110,700 VA/240V
I = 461A

Paralleling two conductors per phase: 461A/2 raceways = 230A

250 kcmil is rated 255A at 75°C: 255A x 2 = 510A

The minimum neutral size when paralleling conductors is 1/0 AWG [250.24(C)(2) and 310.4].

Figure 11–21

PART C—OPTIONAL METHOD—FEEDER/SERVICE LOAD CALCULATIONS

Introduction

An optional method of calculating the service load is permitted for a new restaurant that has electric space heating or electric air-conditioning, or both.

11.17 New Restaurant—Optional Method [220.88]

The following steps can be used to determine the service size for a new restaurant using the optional method described in the *NEC* 220.88:

Step 1: Determine the total connected load. Add the nameplate rating of all loads at 100 percent and include both the air-conditioning and heat load [Table 220.88 Note].

Step 2: Apply the demand factors from Table 220.88 to the total connected load calculated in Step 1.

Unit 11 — Commercial Calculations

▶ All-Electric Restaurant Example

Question: What is the calculated load for an all-electric restaurant (120/208V, three-phase) that has a total connected load of 300 kVA?

(a) 420A (b) 472A (c) 520A (d) 600A

Answer: (b) 472A

Total Connected Load	300 kVA	
First 200 kVA at 80%	−200 kVA x 0.80	160 kVA
Next 201-325 kVA at 10%	100 kVA x 0.10	+ 10 kVA
Total Calculated Load		170 kVA

$I = VA/(E \times 1.732)$

$I = 170{,}000\ VA/(208V \times 1.732)$

$I = 170{,}000\ VA/360V$

$I = 472A$

Paralleling two conductors per phase: 472A/2 raceways = 236A

250 kcmil is rated 255A at 75°C: 255A x 2 = 510A

The minimum neutral size when paralleling conductors is 1/0 AWG [250.24(C)(2) and 310.4].

▶ Grounding Electrode Conductor Sizing Example

Question: What size grounding electrode conductor will be required for this service, using two sets of 250 kcmil conductors in parallel?

(a) 4 AWG (b) 1/0 AWG (c) 2/0 AWG (d) 3/0 AWG

Answer: (b) 1/0 AWG

250 kcmil x 2 = 500 kcmil [Table 250.66, Note 1]

500 kcmil requires a 1/0 AWG grounding electrode conductor [250.66].

Author's Comment: The largest size grounding electrode conductor required to a ground rod is 6 AWG and to a concrete encased electrode (uffer) is 4 AWG [250.66(A) and 250.66(B)].

▶ Not All-Electric Restaurant Example

Question: What is the calculated load for a not all-electric restaurant (120/208V, three-phase) that has a total connected load of 300 kVA?

(a) 420A (b) 460A (c) 520A (d) 695A

Answer: (d) 695A

Total Connected Load	300 kVA	
First 200 kVA at 100%	−200 kVA x 1.00	200 kVA
201-325 kVA at 50%	100 kVA x 0.50	+ 50 kVA
Total Calculated Load		250 kVA

$I = VA/(E \times 1.732)$

$I = 250{,}000\ VA/(208V \times 1.732)$

$I = 250{,}000\ VA/360V$

$I = 694A$

This will require an 800A overcurrent device [240.6(A)]

Paralleling two conductors per phase: 694A/2 raceways = 347A

500 kcmil is rated 380A at 75°C: 380A x 2 = 760A

The minimum neutral size when paralleling conductors is 1/0 AWG [250.24(C)(2) and 310.4]

▶ Grounding Electrode Conductor Sizing Example

Question: What size grounding electrode conductor will be required for this service, using two sets of 500 kcmil conductors in parallel?

(a) 4 AWG (b) 1/0 AWG (c) 2/0 AWG (d) 3/0 AWG

Answer: (c) 2/0 AWG

500 kcmil x 2 = 1,000 kcmil [Table 250.66, Note 1]

1,000 kcmil requires a 2/0 AWG grounding electrode conductor [250.66].

Author's Comment: The largest size grounding electrode conductor required to a ground rod is 6 AWG and to a concrete encased electrode (uffer) is 4 AWG [250.66(A) and 250.66(B)].

Commercial Calculations — Unit 11

PART D—WELDERS

Introduction

Electric welding equipment is a very specialized load, compared to occupancies such as schools, banks, stores, and dwellings. Electric welders operate either by creating an electric arc between two surfaces, or by heating a metal rod until it melts. Both techniques result in high momentary current draws. Article 630 of the *NEC* requires that information about current rating and duty cycle be marked on the nameplate of the equipment.

11.18 Arc Welders

Ampacity of Supply Conductors—Individual Welders

The supply conductors for arc welders must have an ampacity not less than the welder nameplate rating. If the nameplate rating isn't available, the supply conductors must have an ampacity not less than the rated primary current as adjusted by the multiplier in Table 630.11(A), based on the duty cycle of the welder.

Table 630.11(A) Duty Cycle Multiplication Factors for Arc Welders		
Duty Cycle	Nonmotor Generator	Motor Generator
100	1.00	1.00
90	0.95	0.96
80	0.89	0.91
70	0.84	0.86
60	0.78	0.81
50	0.71	0.75
40	0.63	0.69
30	0.55	0.62
20 or less	0.45	0.55

▶ **Individual Welder Example**

A nonmotor-generator arc welder has a primary current rating of 40A with a duty cycle of 50 percent. The branch-circuit conductor for the welder must not be sized less than: **Figure 11–22**

(a) 15A (b) 20A (c) 25A (d) 30A

Answer: (d) 30A

Calculated Load = Primary Rating x Multiplier [Table 630.11(A)]
Calculated Load = 40A x 0.71
Calculated Load = 28.40A

Conductor: 10 AWG is rated 35A at 75ºC [110.14(C) and Table 310.16].

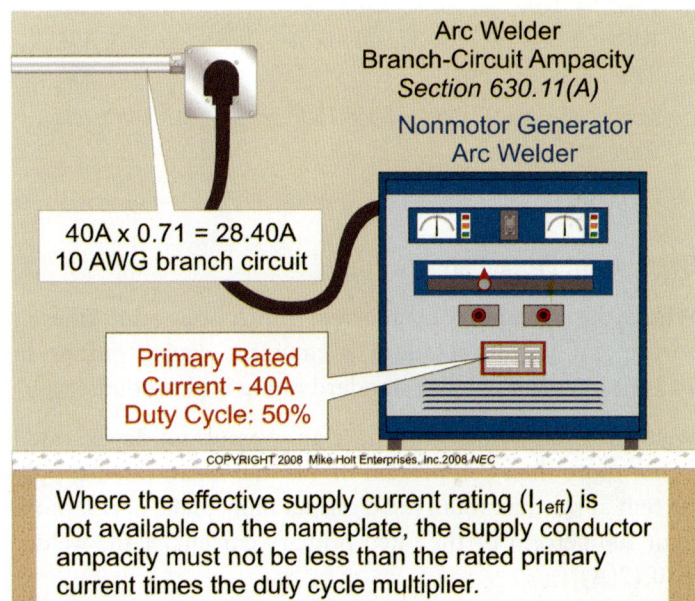

Figure 11–22

Ampacity of Supply Conductors—Group of Welders

Feeder conductors that supply a group of welders must have a minimum ampacity not less than the sum of the currents as determined in 630.11(A), based on 100 percent of the two largest welders, 85 percent for the third largest welder, 70 percent for the fourth largest welder, and 60 percent for all remaining welders [630.11(B)].

Author's Comment: This calculation method provides an ample margin of safety under high-production conditions.

> ▶ **Group of Welders Example**
>
> **Question:** What is the feeder conductor for five 50A nonmotor-generator welders with a duty cycle of 50 percent?
>
> (a) 4 AWG (b) 3 AWG (c) 1/0 AWG (d) 2/0 AWG
>
> **Answer:** (c) 1/0 AWG
>
> Calculated Load = Primary Rating x Multiplier [Table 630.11(A)] x Welder Percentage [630.11(B)]
>
> Welder 1: 50A x 0.71 = 35.50A x 100%   35.50A
> Welder 2: 50A x 0.71 = 35.50A x 100%   35.50A
> Welder 3: 50A x 0.71 = 35.50A x 85%   30.18A
> Welder 4: 50A x 0.71 = 35.50A x 70%   24.85A
> Welder 5: 50A x 0.71 = 35.50A x 60%   21.30A
> Total Calculated Load   147.33A
>
> Conductor size: 1/0 AWG is rated 150A at 75°C [110.14(C), and Table 310.16]

Overcurrent Protection

Where the calculated overcurrent protection value doesn't correspond to the standard overcurrent device ratings in 240.6(A), the next higher standard rating is permitted.

Welders. Each welder must have an overcurrent device rated at not more than 200 percent of the maximum rated supply current at the maximum rated output nameplate, or not more than 200 percent of the rated primary current of the welder [630.12(A)].

Conductors. The conductors must be protected by an overcurrent device rated at not more than 200 percent of the conductor rating [630.12(B)].

> ▶ **Overcurrent Protection Example**
>
> **Question:** What is the maximum overcurrent protection device rating for 10 THHN used for an arc welder branch circuit?
>
> (a) 30A (b) 40A (c) 50A (d) 70A
>
> **Answer:** (d) 70A
>
> 10 AWG is rated 35A at 75°C [110.14(C) and Table 310.16]
> 35A x 2 = 70A maximum overcurrent protection device size

Disconnecting Means

A disconnecting means is required for each arc welder that isn't equipped with an integral disconnect [630.13].

11.19 Resistance Welders

Ampacity of Supply Conductors—Individual Welders

The ampacity of the supply conductors for varied duty-cycle welders must not be less than 70 percent of the rated primary current for seam and automatically fed welders, and 50 percent of the rated primary current for manually operated welders [630.31(A)(1)].

The ampacity of the supply conductors for welders with a specific duty cycle and nonvarying current levels must have an ampacity not less than the rated primary current as adjusted by the multipliers in Table 630.31(A)(2), based on the duty cycle of the welder.

Table 630.31(A)(2) Duty Cycle Multiplication Factors for Resistance Welders	
Duty Cycle	Multiplier
50	0.71
40	0.63
30	0.55
25	0.50
20	0.45
15	0.39
10	0.32
7.50	0.27
5 or less	0.22

Commercial Calculations — Unit 11

> ### ▶ Individual Welders Example
>
> **Question:** What size branch-circuit conductors are required for a 50A resistance welder having a duty cycle of 50 percent?
>
> (a) 8 AWG (b) 6 AWG (c) 4 AWG (d) 2 AWG
>
> **Answer:** (a) 8 AWG
>
> **Calculated Load = Primary Rating x Multiplier**
> [Table 630.31(A)(2)]
> Calculated Load = 50A x 0.71
> Calculated Load = 35.50A
>
> 8 AWG is rated 50A at 75°C [110.14(C), and Table 310.16]

Ampacity of Supply Conductors—Group of Welders

Feeder conductors that supply a group of resistance welders must have an ampacity not less than the sum of the value determined using 630.31(A)(2) for the largest welder in the group, plus 60 percent of the values determined for all remaining welders [630.31(B)].

> ### ▶ Group of Welders Example
>
> **Question:** What is the minimum size feeder conductor for five 50A resistance welders with a duty cycle of 50 percent?
>
> (a) 4 AWG (b) 3 AWG (c) 2 AWG (d) 1 AWG
>
> **Answer:** (d) 1 AWG
>
> Conductor size: 1 AWG, rated 130A at 75°C [110.14(C) and Table 310.16].
>
> **Calculated Load = Primary Rating x Multiplier [Table 630.31(A)(2)] x Welder Percentage [630.31(B)]**
>
> | | |
> |---|---:|
> | Welder 1: 50A x 0.71 = 35.50A x 100% | 35.50A |
> | Welder 2: 50A x 0.71 = 35.50A x 60% | 21.30A |
> | Welder 3: 50A x 0.71 = 35.50A x 60% | 21.30A |
> | Welder 4: 50A x 0.71 = 35.50A x 60% | 21.30A |
> | Welder 5: 50A x 0.71 = 35.50A x 60% | 21.30A |
> | Total Calculated Load | 120.70A |
>
> 1 AWG is rated 130A at 75°C [110.14(C), and Table 310.16]

Overcurrent Protection

Where the calculated overcurrent protection value does not correspond with the standard overcurrent device ratings in 240.6(A), the next higher standard rating is permitted.

Welders. Each welder must have an overcurrent device set at not more than 300 percent of the rated primary current [630.32(A)].

Conductors. Branch-circuit conductors must be protected by an overcurrent device rated at not more than 300 percent of the conductor rating [630.32(B)].

Disconnecting Means

A switch or circuit breaker is required to disconnect each resistance welder and its control equipment from the supply circuit [630.33].

11.20 Light Industrial Calculation

Calculate the service size for a light industrial manufacturing building with the following loads:

- Lighting, 11,600 VA, comprised of electric-discharge luminaries operating at 277V
- Twenty-two 20A, 125V receptacle outlets on general-purpose branch circuits, supplied by a separately derived system
- Air compressor, 460V, three-phase, 7½ hp
- Grinder, 460V, three-phase, 1½ hp
- Five 50A nonmotor-generator arc welders with a duty cycle of 50 percent (nameplate: 50A, 480V)
- Three industrial process dryers, 480V, three-phase, 15 kVA each (assume continuous duty during shifts)

Unit 11   Commercial Calculations

Step 1: Determine the noncontinuous loads and motor loads. The noncontinuous and motor loads can be combined [430.24].

(1) Noncontinuous Loads

    (a) Receptacle Load [220.44] 22 receptacles at 180 VA   3,960 VA
    (b) Welder Load [630.11(A) and Table 630.11(A)]

        A multiplication factor of 0.71 is allowed for 50% duty cycle
        Each welder: 480V x 50A x 0.71 = 17,040 VA

        Welder demand factors for 5 welders: [630.11(B)]

First welder	100%
Second welder	100%
Third welder	85%
Fourth welder	70%
Fifth welder	60%

17,040 VA x 100%	17,040 VA
17,040 VA x 100%	17,040 VA
17,040 VA x 85%	14,484 VA
17,040 VA x 70%	11,928 VA
17,040 VA x 60%	+ 10,224 VA
	70,716 VA

    Welder Total Calculated Load =   + 70,716 VA
    Total Noncontinuous Loads (receptacles and welders) =   74,676 VA   74,676 VA

(2) Motor Loads [430.24 and Table 430.250]

    (a) Air Compressor: 7½ hp FLC = 11A [Table 430.250] 460V x 11A x 1.732   8,764 VA
    (b) Grinder: 1½ hp FLC = 3A [Table 430.250] 3A x 460V x 1.732 =   2,390 VA
    (c) Largest Motor, Additional 25 percent: 8,764 x 25% =   + 2,191 VA
    Total Motor Loads =   13,345 VA   + 13,345

Total Noncontinuous Loads and the Motor Loads [430.24] =   **88,021 VA**

Step 2: Determine the continuous loads:

(1) General Lighting   11,600 VA
(2) Three Industrial Process Dryers 15 kVA each   +45,000 VA

Total Continuous Loads   **56,600 VA**   **56,600 VA**

Step 3: Determine the overcurrent protection [215.3]:

The overcurrent device must be sized at 125 percent of the continuous loads, plus the noncontinuous loads:

(1) Noncontinuous Loads =   88,021 VA
(2) Continuous Loads (56,600 VA x 1.25) =   + 70,750 VA

Total VA   **158,771 VA**

Step 4: Convert to amperes to size the overcurrent device:

I = VA/(E x 1.732)
I = 158,771 VA/(480V x 1.732)
I = 158,771 VA/831V
I = 191A

Overcurrent Device = 200A [240.6]
Conductors: 3/0 AWG rated 225A at 75°C [110.14(C)(1)(b) and Table 310.16]

UNIT 11 Conclusion

CONCLUSION TO UNIT 11—COMMERCIAL CALCULATIONS

This unit showed that the calculations used to size feeders and service conductors for commercial installations don't include the same demand factors allowed for dwelling units. The nature of nondwelling occupancies don't create the same kinds of diversity found in homes and apartment buildings. Accordingly, Unit 11 provided a variety of examples to help you learn the *Code* requirements for commercial service sizing.

Article 220 isn't the only *Code* article where service calculation requirements are contained. We referred to Article 550 for mobile home parks, Article 551 for recreational vehicle parks, Article 555 for marinas, Article 630 for electric welders, Article 440 for hermetic air-conditioning equipment, and Article 430 for motors. When performing commercial service calculations, be careful to look at all of the components of the installation to determine if some other *NEC* article includes information pertinent to your installation.

Certain demand factors may be applied to commercial services. When dealing with individual branch circuits and feeders for continuous loads, remember to use a 125 percent multiplier for the ungrounded conductors. This carries over to the continuous duty portions of loads when determining the actual feeder and service conductors for commercial occupancies. The neutral conductor is allowed to be sized at 100 percent of the continuous and noncontinuous load for branch circuits and feeders when it's not connected to an overcurrent device [210.19(A)(1) Ex 2 and 215.2(A)(1) Ex 2]. Optional calculations are allowed for some commercial installations, so be certain to read problems carefully so you'll employ the correct method.

Unit 11 Practice Questions

UNIT 11—COMMERCIAL CALCULATIONS PRACTICE QUESTIONS

(•Indicates that 75% or fewer of those who took this exam answered the question correctly.)

PART A—GENERAL

11.2 Conductor Ampacity [Article 100]

1. The _____ of a conductor is the rating in amperes it can carry continuously without exceeding its insulation temperature rating. The allowable ampacities listed in Table 310.16 are affected by ambient temperature, current flow, conductor insulation, and conductor bundling [310.10].

   (a) load rating
   (b) ampacity
   (c) calculated load
   (d) continuous factor

11.3 Conductor Overcurrent Protection [240.4]

2. The purpose of _____ is to protect circuit conductors against excessive or dangerous temperatures. If the ampacity of a conductor does not correspond with the standard ampere rating of a fuse or circuit breaker, the next size up overcurrent device is permitted. This applies only if the conductors do not supply branch circuits for multiple receptacles and if the next size up overcurrent device does not exceed 800A.

   (a) short-circuit protection
   (b) ground-fault protection
   (c) overload protection
   (d) overcurrent protection

3. The following is a partial list of standard ampere ratings for overcurrent devices (fuses and inverse time circuit breakers): 15, 25, 35, 45, 80, 90, 110, 175, 250, and 350A.

   (a) True
   (b) False

11.5 Rounding an Ampere [220.5(B)]

4. Where a calculation results in a fraction of an ampere that is _____ or less, that fraction can be dropped.

   (a) 0.05
   (b) 0.49
   (c) 0.50
   (d) 0.51

11.6 Lighting—Demand Factors [Tables 220.12 and 220.42]

5. What is the general lighting and general-use receptacle feeder/service calculated load for a 24-room motel, where each unit is 685 sq ft?

   (a) 10 kVA
   (b) 15 kVA
   (c) 20 kVA
   (d) 25 kVA

6. What is the general lighting and general-use receptacle feeder/service calculated load for a 250,000 sq ft storage warehouse that has 200 receptacles?

   (a) 25 kVA
   (b) 56 kVA
   (c) 95 kVA
   (d) 105 kVA

11.7 Lighting Without Demand Factors [215.2(A)(1), 230.42(A)(1), and Table 220.12]

7. What is the feeder/service general lighting calculated load for a 3,200 sq ft dance club?

   (a) 3,200 VA
   (b) 6,400 VA
   (c) 8,000 VA
   (d) 12,000 VA

330   Mike Holt's Illustrated Guide to NEC Exam Preparation

Practice Questions — Unit 11

8. What is the feeder/service general lighting load for a 90,000 sq ft school?

    (a) 90 kVA
    (b) 120 kVA
    (c) 230 kVA
    (d) 338 kVA

11.8 Lighting—Miscellaneous

9. •What is the VA demand feeder/service load for 130 ft of show-window lighting?

    (a) 29 kVA
    (b) 33 kVA
    (c) 35 kVA
    (d) 38 kVA

11.9 Multioutlet Receptacle Assembly [220.14(H)]

10. What is the feeder calculated load for 50 ft of multioutlet assembly and 10 ft of multioutlet assembly simultaneously used?

    (a) 3,600 VA
    (b) 5,500 VA
    (c) 7,200 VA
    (d) 12,000 VA

11.10 Receptacle VA Load

11. How many receptacle outlets are permitted on a 20A, 120V circuit in a commercial occupancy?

    (a) 10
    (b) 13
    (c) 15
    (d) 20

12. What is the service calculated load for 110 receptacles (15A or 20A, 125V) in a commercial building?

    (a) 5 kVA
    (b) 10 kVA
    (c) 15 kVA
    (d) 20 kVA

11.11 Banks and Offices—General Lighting and Receptacles [220.14(K)]

13. What is the service calculated load for the general lighting and receptacles for a 30,000 sq ft bank?

    (a) 111 kVA
    (b) 123 kVA
    (c) 162 kVA
    (d) 175 kVA

14. •What is the general lighting and receptacle calculated load for a 10,000 sq ft office building with 75 receptacles?

    (a) 25 kVA
    (b) 35 kVA
    (c) 45 kVA
    (d) 56 kVA

11.12 Sign Circuit [220.14(F) and 600.5]

15. What is the feeder calculated load for sizing the overcurrent device for one electric sign?

    (a) 1,400 VA
    (b) 1,500 VA
    (c) 1,920 VA
    (d) 2,400 VA

PART B—EXAMPLES

11.14 Marina [555.12]

16. A marina has 24 slips with 20A, 250V receptacles, and 30 slips with 30A, 250V receptacles. What size service is required for the marina?

    (a) 400A
    (b) 550A
    (c) 900A
    (d) 1,000A

17. What size conductor is required for a service if the total calculated load for each phase is 570A and the service is paralleled in two raceways?

    (a) 4/0 AWG
    (b) 250 kcmil
    (c) 300 kcmil
    (d) 350 kcmil

Unit 11 | Practice Questions

11.15 Mobile/Manufactured Home Park [550.31]

18. What is the feeder and service calculated load for a mobile home park that has the facilities for 42 sites? The system is 120/240V, single-phase.

    (a) 452A
    (b) 512A
    (c) 644A
    (d) 732A

11.16 Recreational Vehicle Park [551.73]

19. A recreational vehicle park has 17 dedicated tent sites with only 20A, 240V receptacles; 35 sites with 20A and 30A, 240V receptacles; and 10 sites with 50A, 240V receptacles. The feeder/service calculated load is approximately _____.

    (a) 355A
    (b) 400A
    (c) 789A
    (d) 1,114A

PART C—OPTIONAL METHOD—FEEDER/SERVICE LOAD CALCULATIONS

11.17 New Restaurant—Optional Method [220.88]

20. A new restaurant has a total connected load of 400 kVA and is all electric. What is the calculated load for the service?

    (a) 210 kVA
    (b) 275 kVA
    (c) 300 kVA
    (d) 325 kVA

21. A new restaurant has a total connected load of 400 kVA and is not all electric. What is the calculated load for the service?

    (a) 275 kVA
    (b) 296 kVA
    (c) 300 kVA
    (d) 325 kVA

Unit 11 Challenge Questions

UNIT 11—COMMERCIAL CALCULATIONS CHALLENGE QUESTIONS

(•Indicates that 75% or fewer of those who took this exam answered the question correctly.)

PART A—GENERAL

11.6 Lighting—Demand Factors [Tables 220.12 and 220.42]

1. •Each unit of a 100-unit hotel is 12 x 15 ft. In addition, there is a 60 x 20 ft office and there are hallways of 120 sq ft. What is the calculated load for general lighting and receptacles?

   (a) 23 kVA
   (b) 25 kVA
   (c) 29 kVA
   (d) 37 kVA

11.10 Receptacle VA Load

2. If in the hall and other areas of a motel there are 100 receptacle outlets (not in the motel rooms), the calculated load added to the service for these receptacles is _____.

   (a) 0 VA
   (b) 10,000 VA
   (c) 14,000 VA
   (d) 20,000 VA

3. What is the receptacle calculated load for a 20,000 sq ft office building?

   (a) 10,000 VA
   (b) 20,000 VA
   (c) 30,000 VA
   (d) 40,000 VA

PART B—EXAMPLES

11.13 Bank/Office Building Example

4. What is the general lighting and general-use receptacle load for a 30,000 sq ft bank?

   (a) 123 kVA
   (b) 136 kVA
   (c) 161 kVA
   (d) 173 kVA

11.14 Marina [555.12]

5. A marina shore power facility has twenty 20A, 240V receptacles; seventeen 30A, 240V receptacles; and seven 50A, 240V receptacles. After applying demand factors, the service calculated load for the shore power boxes is _____.

   (a) 625A
   (b) 630A
   (c) 1,160A
   (d) 1,260A

11.15 Mobile/Manufactured Home Park [550.31]

6. A 75-site mobile home park is designed for mobile homes that have a 14,000 VA load per site. The service calculated load for the park is _____.

   (a) 201 kVA
   (b) 222 kVA
   (c) 264 kVA
   (d) 1,200 kVA

Unit 11 — Challenge Questions

11.16 Recreational Vehicle Park [551.73]

7. A recreational vehicle park has 42 sites: nine sites equipped with 50A, 120/240V supply facilities; 30 sites equipped with 20A and 30A, 120/240V supply facilities; and three sites equipped with 20A, 120/240V supply facilities. The minimum feeder calculated load for these sites is _____.

   (a) 83 kVA
   (b) 101 kVA
   (c) 139 kVA
   (d) 158 kVA

PART C—OPTIONAL METHOD—FEEDER/SERVICE LOAD CALCULATIONS

11.17 New Restaurant—Optional Method [220.88]

8. •A new restaurant has a total connected lighting load of 30 kVA. The kitchen equipment includes two gas stoves, one gas grill, three gas ovens, one 75-gallon gas water heater, one 5 kW dishwasher, two 2 kW coffee makers, five 2 kW kitchen appliances on their own circuit, and ten 1.50 kVA small-appliance circuits. Using the optional method, the service calculated load is closest to _____.

   (a) 38 kVA
   (b) 45 kVA
   (c) 50 kVA
   (d) 64 kVA

UNIT 12 Transformer Calculations

INTRODUCTION TO UNIT 12—TRANSFORMER CALCULATIONS

Transformers are essential to electric power distribution. For this reason, you must understand several important concepts about transformers to be successful doing electrical work.

If you remove the cover of a transformer and look inside, you'll see two sets of windings. And you'll immediately understand why they're called that–the manufacturer literally winds wire into a coil. Now, think back to the basic transformer concepts introduced in Chapter 4.

The ratio of the number of turns in the primary side (supply side) to the number on the secondary side (load side) is what determines how the unit *transforms* the power supplied to it. This is a key concept in transformer theory and application.

Typically, the transformation is that of raising or lowering the voltage. Using a transformer, we can step down the 480V facility distribution voltage to the 120V needed at the wall receptacles in the offices or wherever else we want to supply with 120V.

If you look on the nameplate of this particular transformer, you'll see it has a 4:1 ratio and a 120/208V secondary. You get 208V by connecting between two sets of windings (two phases). You get 120V by connecting between one phase and the transformer center tap. Similarly, we can power 277V lighting from a 480/277V transformer.

Transformers are wound into different configurations. Industrial and commercial applications use the delta-wye configuration extensively, so it's important that you understand delta and wye line currents.

We'll look at such concepts as current flow and power loss, and show you how to size the conductors. We'll conclude by discussing the grounding and bonding of transformers. This is an area where mistakes are common. Those are mistakes that, thanks to what you'll learn here, you won't be making.

PART A–GENERAL

Introduction

A transformer is a stationary device used to raise or lower voltage. It transfers electrical energy (power) from one system to another by induction with no physical connection between the two systems.

Some transformers, called isolation transformers, do not raise or lower the voltage. They're simply for the purpose of decoupling the primary from the secondary.

12.1 Transformer Basics

Primary versus Secondary

The transformer winding connected to the voltage source is called the primary. The transformer winding connected to the load is called the secondary. Transformers can be used to either raise or lower voltage. **Figure 12–1**

Mutual Induction

The energy transfer of a transformer is accomplished because the electromagnetic lines of force from the primary winding induce a voltage in the secondary winding. This process is called mutual induction. **Figure 12–2**

Unit 12 | Transformer Calculations

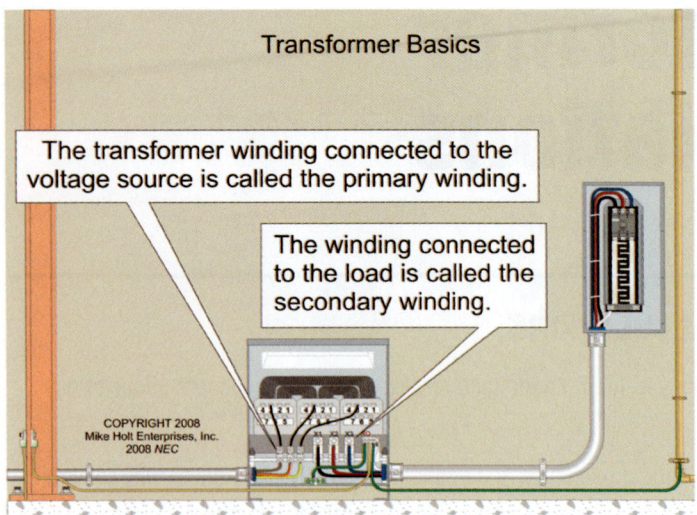

Figure 12–1

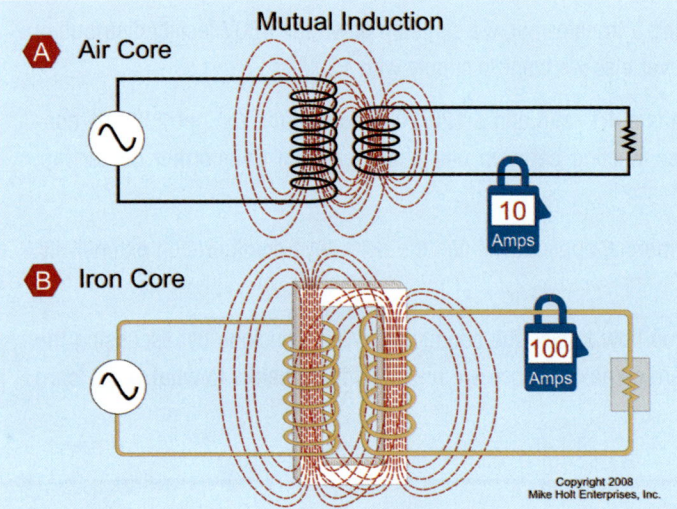

Figure 12–2

For example, if a transformer has three windings on the secondary and each secondary winding has 80V induced, the secondary voltage is 240V. **Figure 12–3**

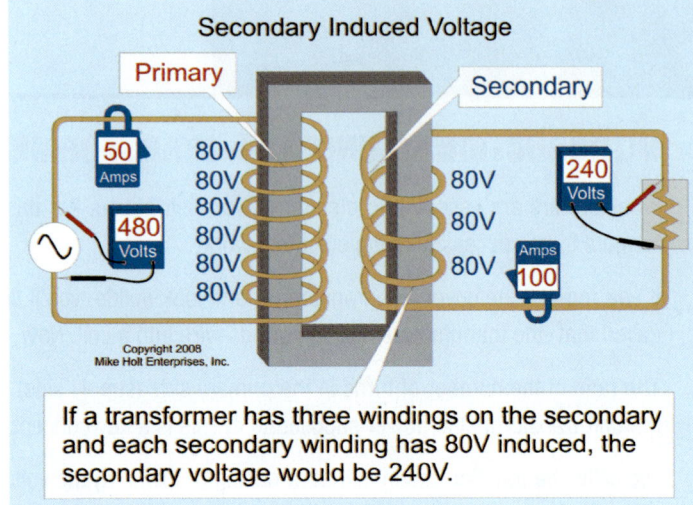

Figure 12–3

The induced voltage in each loop of the secondary winding depends on the number of secondary conductor loops (turns), as compared to the number of primary turns. If both windings have the same number of turns, and if all of the magnetism set up by the primary passes through the secondary, the secondary will deliver the same voltage and power as the primary. This is how an isolation transformer works. **Figure 12–4**

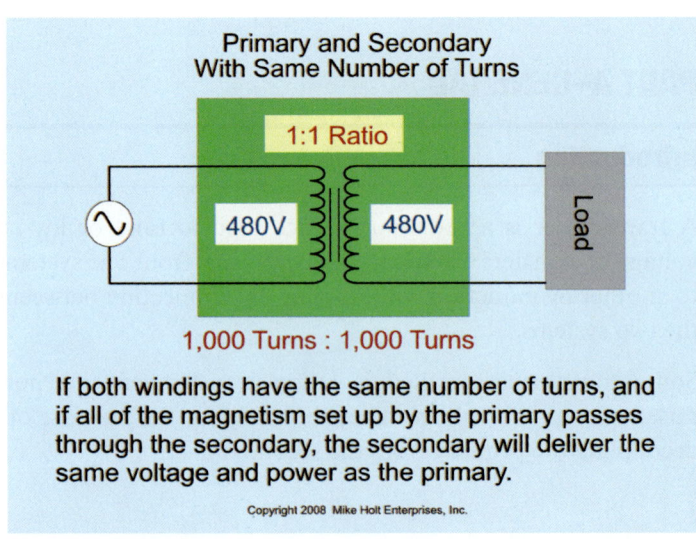

Figure 12–4

The voltage level that can be induced in the secondary winding, from the primary magnetic field, is a function of the number of secondary conductor loops (turns) that are cut by the primary electromagnetic field.

12.2 Secondary Induced Voltage

Voltage induced in the secondary winding of a transformer is equal to the sum of the voltages induced in each loop of the secondary winding.

336 | Mike Holt's Illustrated Guide to NEC Exam Preparation

Transformer Calculations — Unit 12

12.3 Autotransformers

Autotransformers use a single winding for both the primary and secondary.

Autotransformers are often used to step the voltage up from 208V to 240V, or down from 240V to 208V. **Figures 12–5** and **12–6**

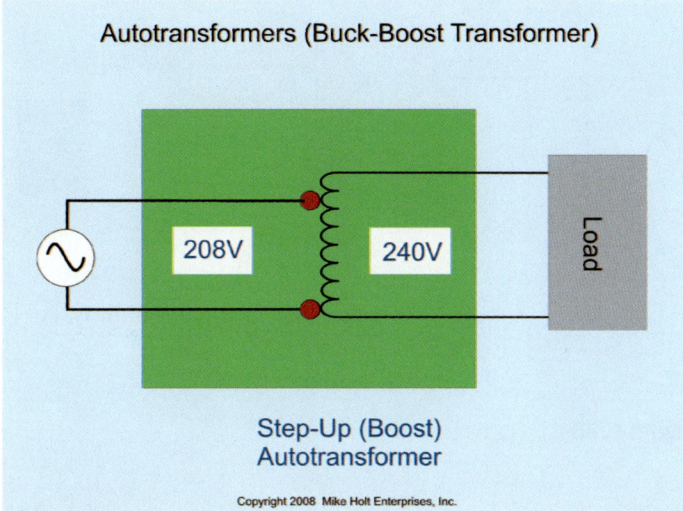

Figure 12–5

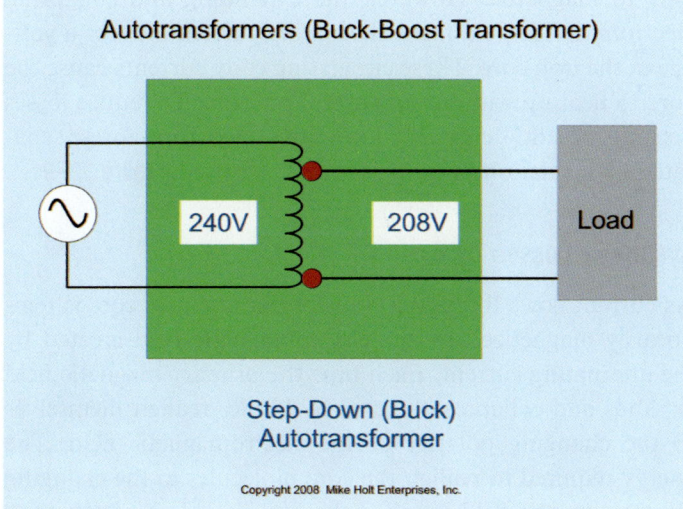

Figure 12–6

Step-Down Transformer

The secondary winding of a step-down transformer has fewer turns than the primary winding, resulting in a lower secondary voltage as compared to the primary.

Step-Up Transformer

The secondary winding of a step-up transformer has more turns than the primary winding, resulting in a higher secondary voltage as compared to the primary.

Author's Comment: "Buck-Boost" transformers used to make small increases or decreases in voltage are autotransformers and are available with multiple voltage taps to meet the needs of various situations.

Lack of Isolation

The disadvantage of an autotransformer is the lack of electrical isolation between the primary and secondary conductors. But they're often used because they're inexpensive.

12.4 Power Losses

In an ideal transformer, all the primary power is transferred from the primary winding to the secondary winding. But real-world transformers have power losses because of conductor resistance, flux leakage, eddy currents, and hysteresis losses.

Conductor Resistance Loss

Transformer primary and secondary windings are generally made of many turns of copper. Conductor resistance is directly proportional to the length of a conductor and inversely proportional to the cross-sectional area of a conductor.

Unit 12 — Transformer Calculations

▶ **Primary Power Loss Example**

Question: What is the primary conductor power loss of a 75 kVA transformer if the primary current rating is 90A and the winding has a resistance of 0.08 ohms? **Figure 12–7A**

(a) 400W   (b) 648W   (c) 1,150W   (d) 1,300W

Answer: (b) 648W

Power = $I^2 \times R$
I = 90A
R = 0.08 ohms
P = $90A^2 \times 0.08$ ohms
P = 648W

Figure 12–7

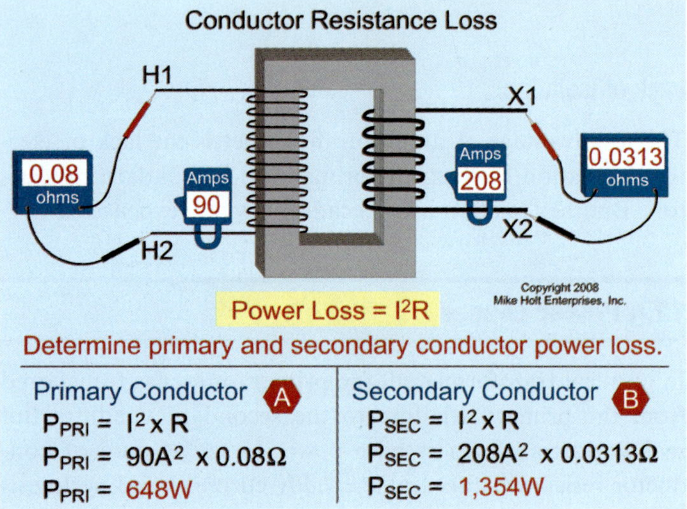

Conductor Resistance Loss

Power Loss = I^2R
Determine primary and secondary conductor power loss.

Primary Conductor Ⓐ
$P_{PRI} = I^2 \times R$
$P_{PRI} = 90A^2 \times 0.08\Omega$
$P_{PRI} = 648W$

Secondary Conductor Ⓑ
$P_{SEC} = I^2 \times R$
$P_{SEC} = 208A^2 \times 0.0313\Omega$
$P_{SEC} = 1,354W$

▶ **Secondary Power Loss Example**

Question: What is the secondary conductor power loss of a 75 kVA transformer if the secondary current rating is 208A and the winding has a resistance of 0.0313 ohms? **Figure 12–7B**

(a) 1,354W   (b) 1,800W   (c) 2,100W   (d) 2,700W

Answer: (a) 1,354W

Power = $I^2 \times R$
I = 208A
R = 0.0313 ohms
P = $208A^2 \times 0.0313$ ohms
P = 1,354W

The total power loss of the transformer is the primary and secondary losses combined: 648W + 1,354W = 2,002W

Flux Leakage Loss

The leakage of the electromagnetic flux lines between the primary and secondary windings of transformers also represents wasted energy. **Figure 12–8**

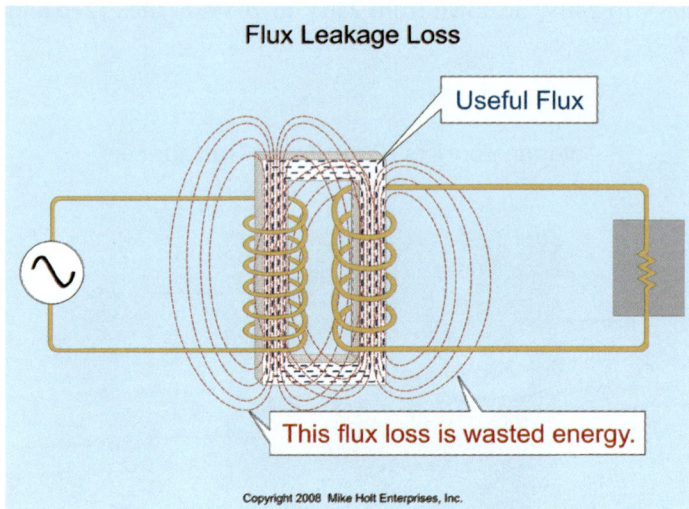

Figure 12–8

Eddy Currents

Iron is the only metal used for transformer cores because it's easy to magnetize. However, the expanding and collapsing electromagnetic field from alternating current induces a voltage in the iron core. These circulating eddy currents cause the core to heat up without any useful purpose. To reduce losses because of eddy currents, long laminated iron cores, separated by insulation (usually lacquer), are used. **Figure 12–9**

Hysteresis Losses

As current flows through a transformer, the iron core is temporarily magnetized by the electromagnetic field created by the alternating current. Each time the primary magnetic field expands and collapses, the core molecules realign themselves to the changing polarity of the electromagnetic field. The energy required to realign the core molecules to the changing electromagnetic field is called the hysteresis loss of the core. **Figure 12–10**

Hysteresis losses are directly proportional to the alternating-current frequency. The higher the frequency, the more times per second the molecules must realign. Hysteresis loss is one of the main reasons why iron-core transformers aren't used in applications involving high frequencies and nonlinear loads.

Transformer Calculations   Unit 12

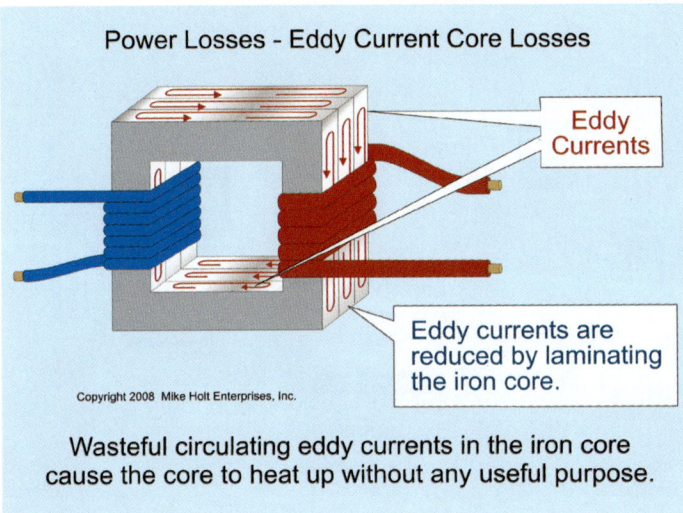

Figure 12–9

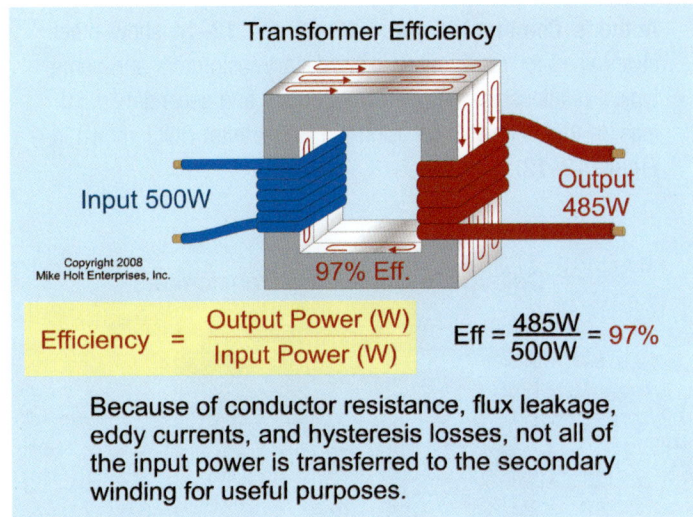

Figure 12–11

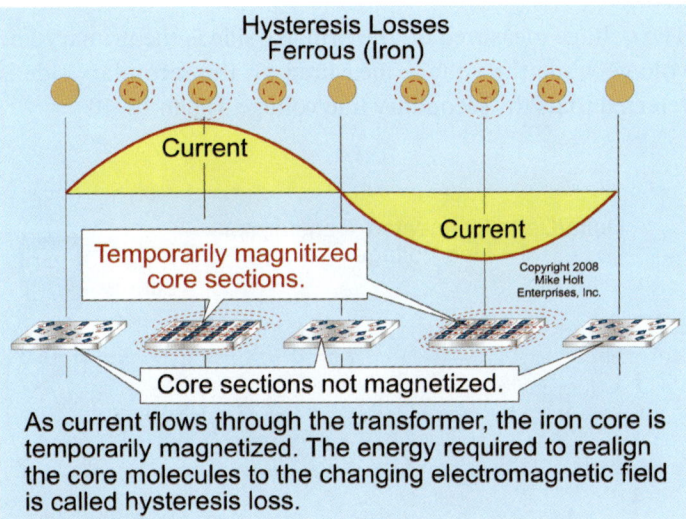

Figure 12–10

12.5 Efficiency

Because of conductor resistance, flux leakage, eddy currents, and hysteresis losses, not all of the input power is transferred to the secondary winding of a transformer for useful purposes. Transformer losses are typically quite small with efficiency approaching 95 to 97 percent. Thus, for most practical purposes, transformer efficiency can be ignored. Figure 12–11

The ratio between the output power and the input power is described as efficiency, and is found with the formula:

Efficiency = Output Power/Input Power

12.6 Delta/Delta Connected Transformers

Delta-Connected Transformer. Delta-connected transformers have three transformer windings connected end-to-end with each other. The line conductors are connected to each point where two windings meet. This system is called "Delta" because when it's drawn out it looks like a triangle (the Greek symbol Delta for the letter D). Figure 12–12

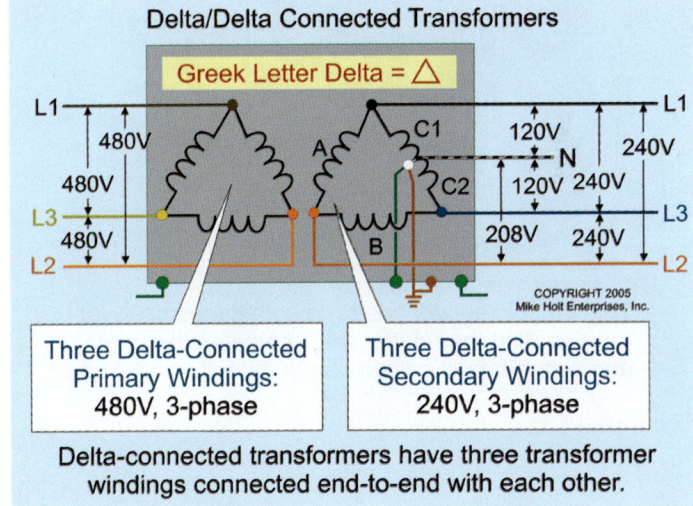

Figure 12–12

Mike Holt Enterprises, Inc. • www.MikeHolt.com • 1.888.NEC.CODE (1.888.632.2633)   339

Unit 12 Transformer Calculations

Author's Comment: Figures 12–13 and **12–14** show other styles used for representing delta/delta transformers and some typical relationships between the primary and secondary windings. In this unit, we'll be using the Greek letter Delta shown in **Figure 12–12**.

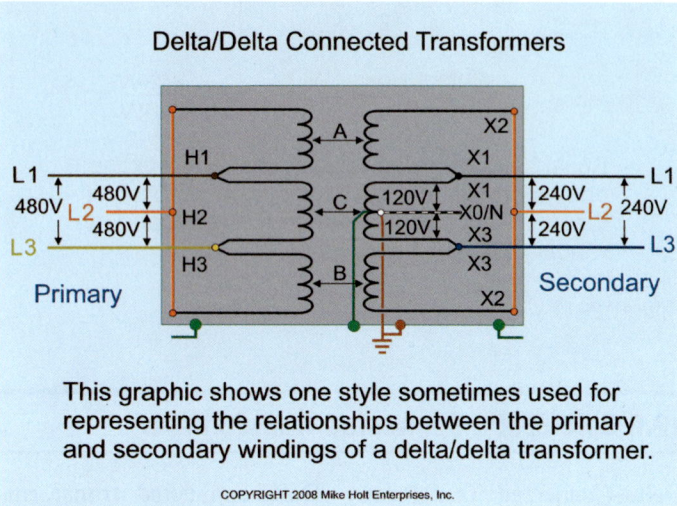

Figure 12–13

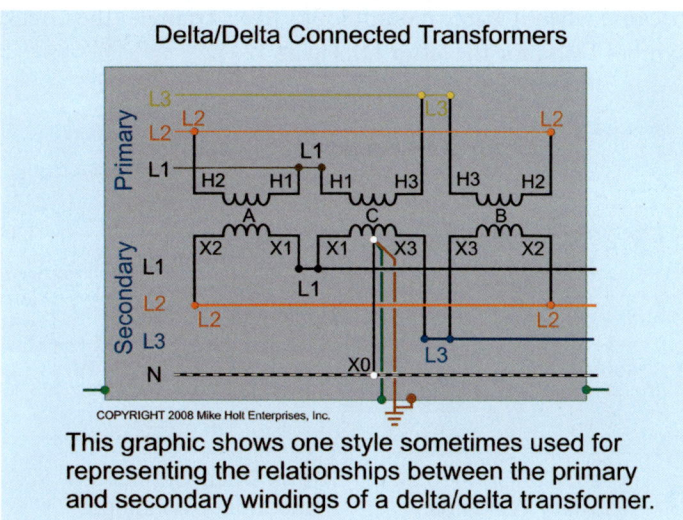

Figure 12–14

The transformer winding connected to the voltage source is called the primary. The transformer winding connected to the load is called the secondary. Transformers can be used to either raise or lower voltage. **Figure 12–15**

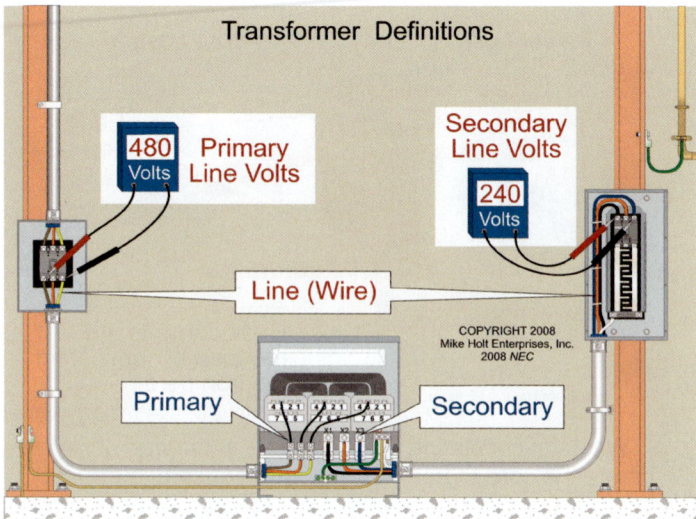

Figure 12–15

The voltage measured on the primary side is the primary line voltage while the voltage measured on the secondary side is referred to as the secondary line voltage. **Figure 12–16**

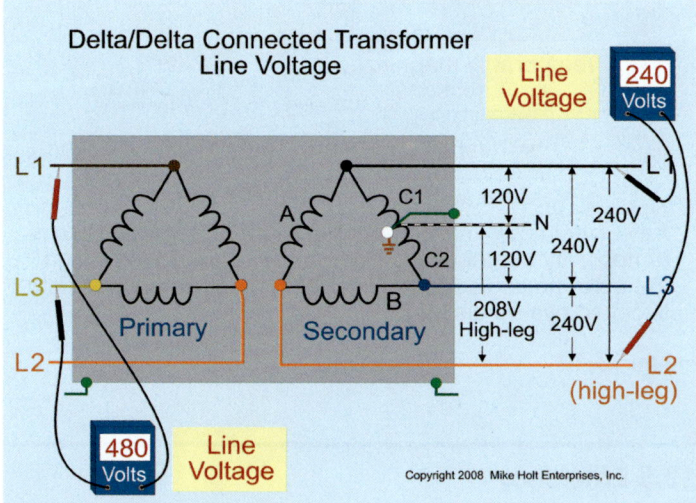

Figure 12–16

The terms "high-leg," "wild-leg," or "bastard-leg" are used to identify the conductor of a delta-connected system that has a voltage rating of 208V to ground. The high-leg voltage is the vector sum of the voltage of transformers A and C1, or transformers B and C2, which equal 120V x 1.732 = 208V for a 120/240V secondary. **Figure 12–17**

340 Mike Holt's Illustrated Guide to NEC Exam Preparation

Transformer Calculations — Unit 12

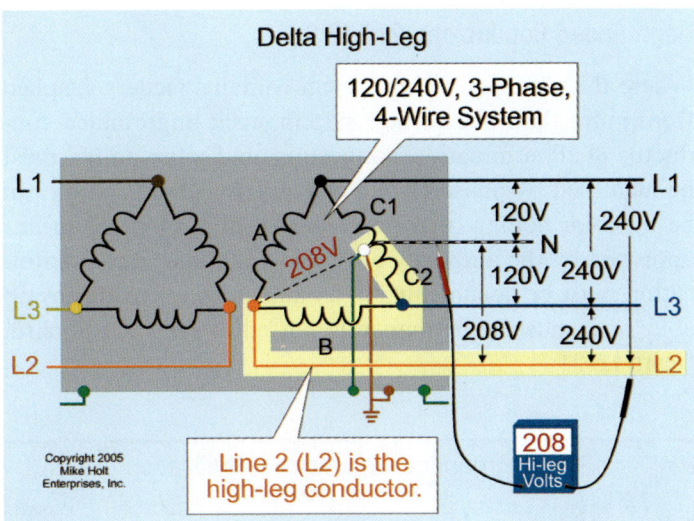

Figure 12–17

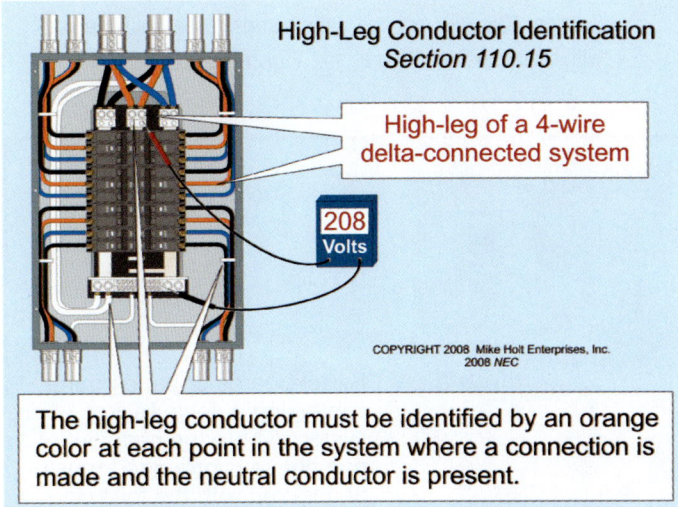

Figure 12–18

Author's Comment: The actual voltage of the high-leg is often less than the nominal system voltage because of voltage drop and transformer loading.

▶ High-leg Voltage Example

Question: What is the actual voltage of the high-leg if the delta-configured secondary is 115/230V, three-phase?

(a) 115V   (b) 199V   (c) 230V   (d) 240V

Answer: (b) 199V

High-leg Voltage = (Line Voltage/2) x 1.732
High-leg Voltage = 115V x 1.732
High-leg Voltage = 199.18V

High-Leg Conductor Identification [110.15]

On a 4-wire, delta-connected, three-phase system, where the midpoint of one phase winding is grounded (high-leg system), the conductor with 208V to ground must be durably and permanently marked by an outer finish orange in color, or other effective means. Such identification must be placed at each point on the system where a connection is made if the neutral conductor is present [110.15, 215.8, and 230.56]. **Figure 12–18**

Author's Comments:

- The high-leg conductor is also called the "wild-leg," "stinger-leg," or "bastard-leg."

- Other important *NEC* rules relating to the high-leg are as follows:

  – Panelboards. Since 1975, panelboards supplied by a 4-wire, delta-connected, three-phase system must have the high-leg conductor terminate to the "B" phase of a panelboard [408.3(E)]. **Figure 12–19**

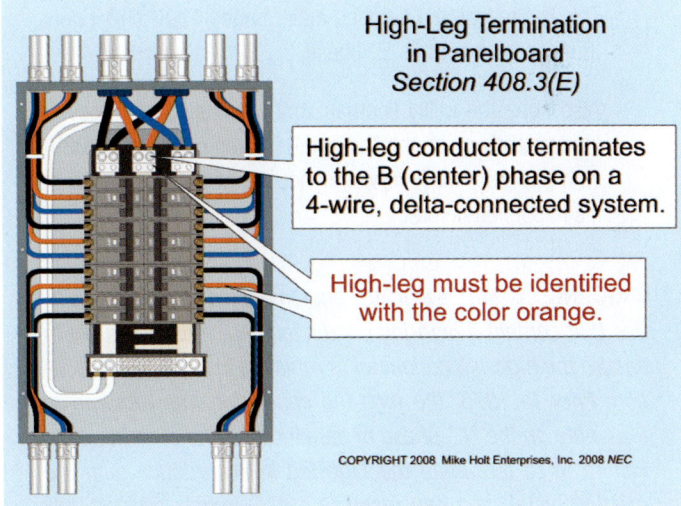

Figure 12–19

Unit 12 Transformer Calculations

– Section 408.3(F) requires panelboards to be field-marked with "Caution 208V to ground." **Figure 12–20**

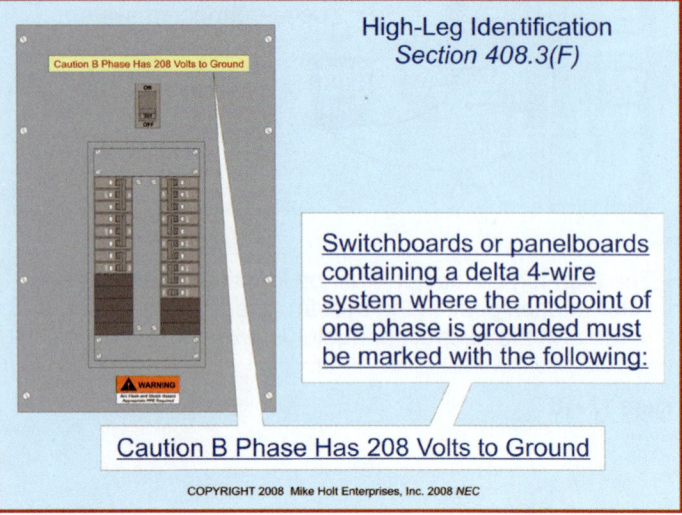

Figure 12–20

– Disconnects. The *NEC* doesn't specify the termination location for the high-leg conductor in switch equipment (Switches Article 404), but the generally accepted practice is to terminate this conductor to the "B" phase.

– Utility Equipment. The ANSI standard for meter equipment requires the high-leg conductor (208V to neutral) to terminate on the "C" (right) phase of the meter socket enclosure. This is because the demand meter needs 120V, and it gets that voltage from the "B" phase.

– Also hope the utility lineman isn't color blind and doesn't inadvertently cross the "orange" high-leg conductor (208V) with the red (120V) service conductor at the weatherhead. It's happened before!

WARNING: *When replacing equipment in existing facilities that contain a high-leg conductor, care must be taken to ensure the high-leg conductor is replaced in the original location. Prior to 1975, the high-leg conductor was required to terminate on the "C" phase of panelboards and switchboards. Failure to re-terminate the high-leg in accordance with the existing installation can result in 120V circuits inadvertently connected to the 208V high-leg, with disastrous results.*

Ungrounded Conductors [215.12(C)]

Where the premises wiring system contains feeders supplied from more than one voltage system, each ungrounded conductor, at all termination, connection, and splice points, must be identified by phase or line and system. Identification can be by color coding, marking tape, tagging, or other means approved by the authority having jurisdiction. Such identification must be documented in a manner that is readily available, or it must be permanently posted at each panelboard. **Figure 12–21**

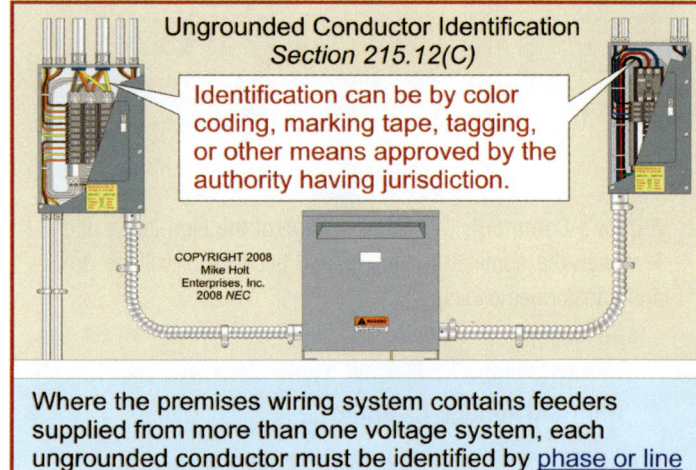

Figure 12–21

Author's Comment: Although the *NEC* doesn't require a specific color code for ungrounded conductors, electricians often use the following color system for power and lighting conductor identification:

- 120/240V, single-phase—black, red, and white

- 120/208V, three-phase—black, red, blue, and white

- 120/240V, three-phase—black, orange, blue, and white

- 277/480V, three-phase—brown, orange, yellow, and gray; or, brown, purple, yellow, and gray

- Conductors with insulation that is green or green with one or more yellow stripes cannot be used for an ungrounded or neutral conductor [250.119].

Transformer Calculations — Unit 12

12.7 Delta/Wye Connected Transformers

Wye-Connected Transformers. Wye-configured transformers have one lead from each of three windings connected to a common point. The other leads from each of the windings are connected to the line conductors. A wye-configured secondary is often represented with a Y-shaped arrangement of the windings. **Figure 12–22**

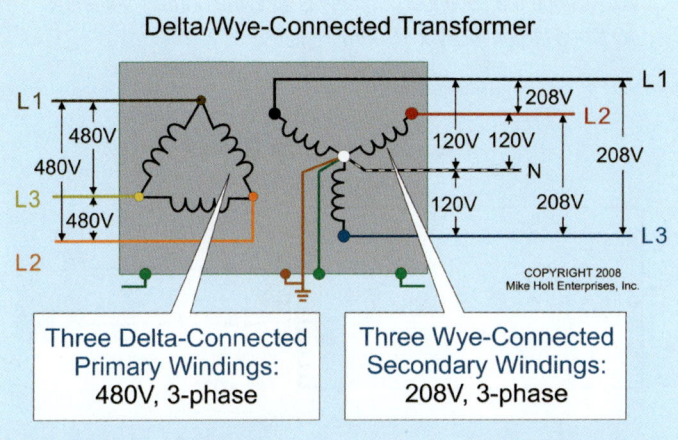

Figure 12–22

Author's Comment: Figures 12–23 and 12–24 show other styles used for representing delta/wye transformers and some typical relationships between the primary and secondary windings.

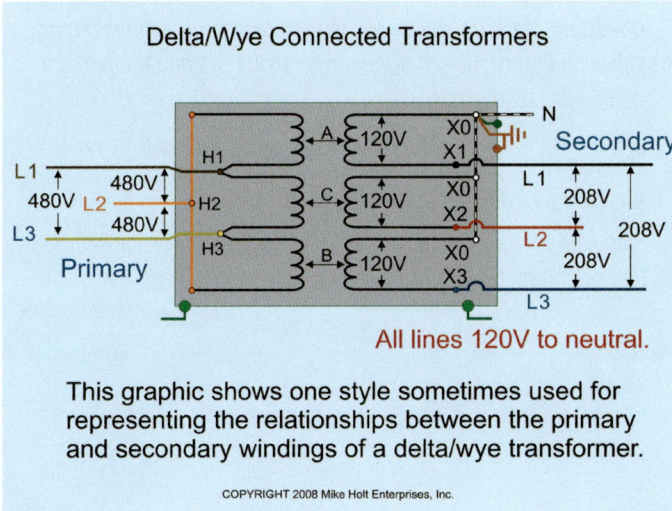

Figure 12–23

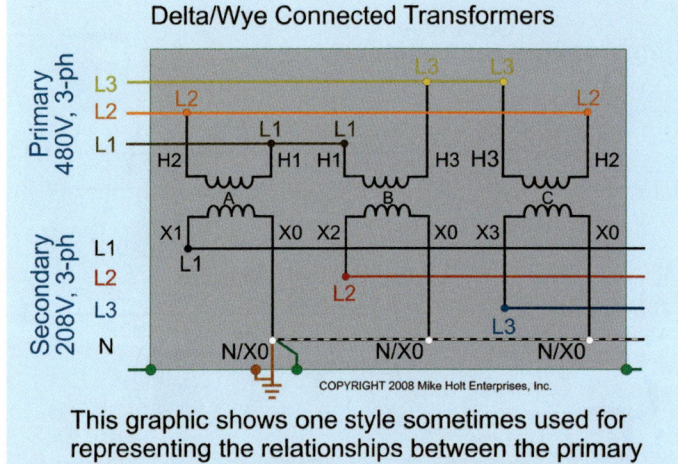

Figure 12–24

The transformer winding connected to the voltage source is called the primary. The transformer winding connected to the load is called the secondary. Transformers can be used to either raise or lower voltage. **Figure 12–25**

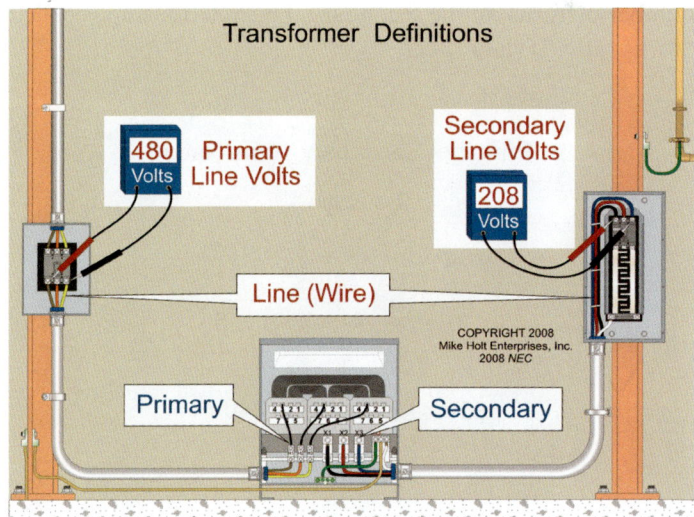

Figure 12–25

The voltage measured on the primary side is the primary line voltage while the voltage measured on the secondary side is referred to as the secondary line voltage. **Figure 12–26**

Mike Holt Enterprises, Inc. • www.MikeHolt.com • 1.888.NEC.CODE (1.888.632.2633) 343

Unit 12 Transformer Calculations

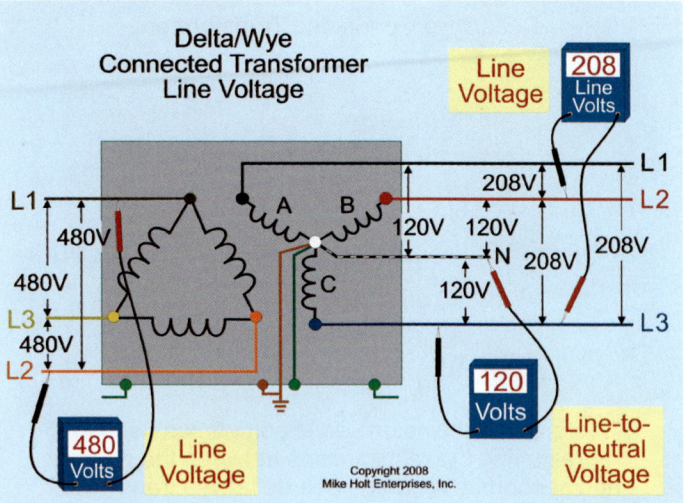

Figure 12–26

12.8 Transformer Turns Ratio

The relationship of the primary winding voltage to the secondary winding voltage is the same as the relationship between the number of turns of wire on the primary and the number of turns on the secondary. This relationship is called "turns ratio." **Figure 12–27**

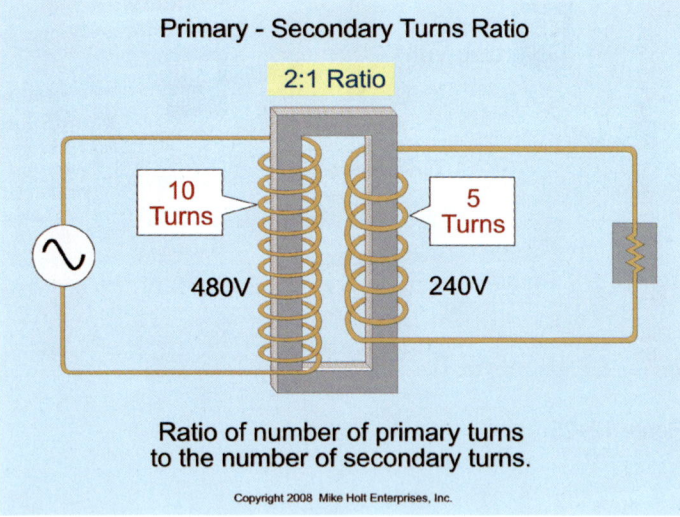

Figure 12–27

▶ **Primary Voltage Example**

Question: What is the primary voltage of the transformer shown in **Figure 12–28** if the turns ratio is 10:1 and the secondary voltage is 12V?

(a) 6V (b) 24V (c) 48V (d) 120V

Answer: (d) 120V

The turns ratio is 10:1. This means the voltage ratio is also 10:1. Since the secondary voltage is 12V, the primary voltage is 10 times larger, 112V x 10 Turns= 120V.

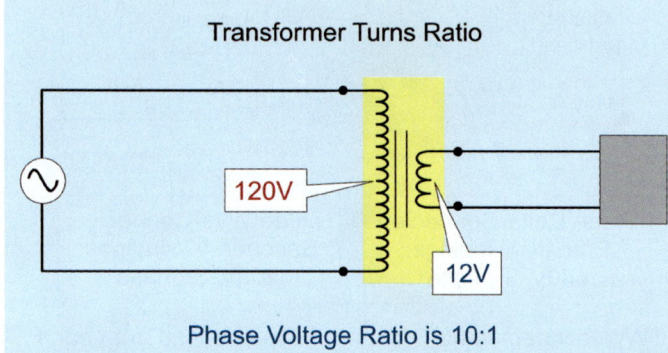

If the ratio is 10 primary turns for 1 secondary turn, and the primary voltage is 120V, then the secondary volts will be 12V. 120V/10 Turns = 12V

Figure 12–28

▶ **Secondary Voltage Example**

Question: What is the secondary voltage of the transformer shown in **Figure 12–29** if the turns ratio is 5:1 and the primary voltage is 120V?

(a) 6V (b) 24V (c) 48V (d) 120V

Answer: (b) 24V

Since the primary voltage is 120V, the secondary voltage is five times smaller, 120V/5 Turns = 24V.

Transformer Calculations   Unit 12

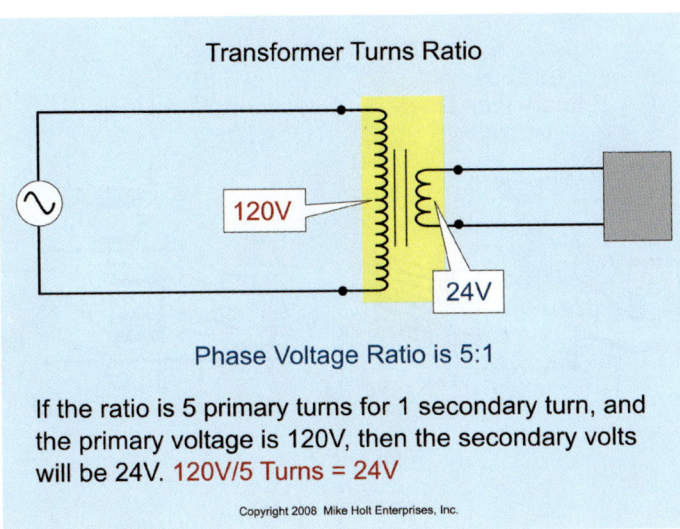

Figure 12–29

▶ **Delta/Wye Example**

Question: What is the turns ratio of the delta/wye transformer shown in **Figure 12–31** if the primary phase voltage is 480V and the secondary is 120V?

(a) 4:1   (b) 1:4   (c) 2:1   (d) 1:2

Answer: (a) 4:1

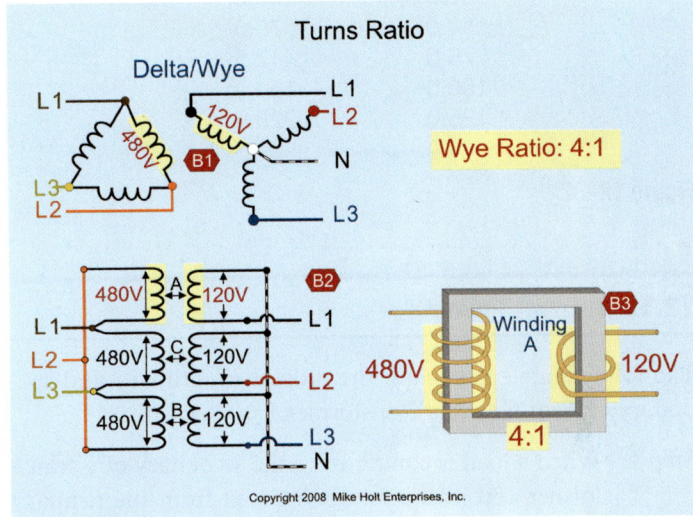

Figure 12–31

▶ **Delta/Delta Example**

Question: What is the turns ratio of the delta/delta transformer shown in **Figure 12–30** if the primary phase voltage is 480V and the secondary is 240V?

(a) 4:1   (b) 1:4   (c) 2:1   (d) 1:2

Answer: (c) 2:1

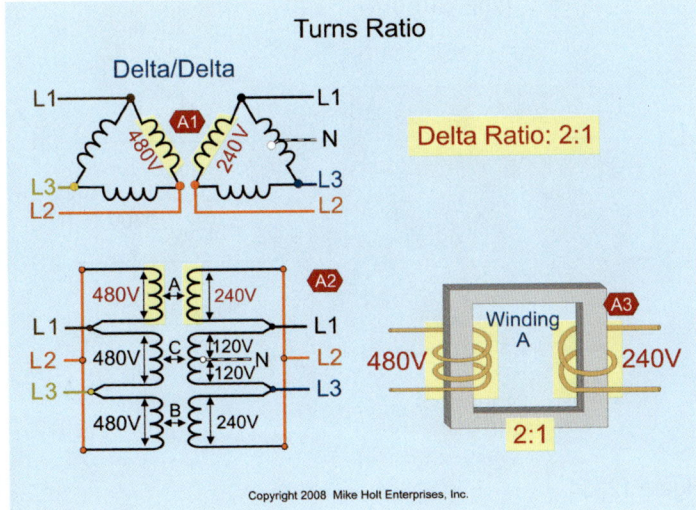

Figure 12–30

Author's Comment: The terms "step-up" and "step-down" are related to the turns ratio of a transformer. If the transformer is connected so that the secondary voltage is higher than the primary voltage, it has increased or "stepped-up" the voltage. If the transformer is connected so the secondary voltage is lower than the primary voltage, it has decreased or "stepped-down" the voltage.

12.9 Transformer kVA Rating

Transformers are rated in kilovolt-amperes (kVA), where 1 kilovolt-ampere = 1,000 volt-amperes (VA). **Figure 12–32** lists common standard transformer sizes used in commercial-industrial applications. However, there are also much larger transformers.

Mike Holt Enterprises, Inc. • www.MikeHolt.com • 1.888.NEC.CODE (1.888.632.2633)   345

Unit 12 — Transformer Calculations

Figure 12–32

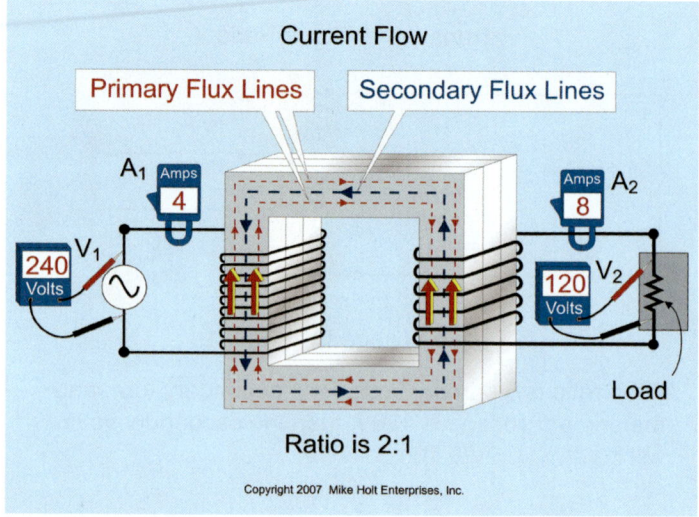

Figure 12–33

12.10 Current Flow

The following steps explain the process of primary and secondary current flow in a transformer.

Step 1: When a load is connected to the secondary of a transformer, secondary voltage induced from the primary magnetic field causes current to flow through the secondary conductor winding.

Step 2: The secondary current flow in the secondary winding creates an electromagnetic field that opposes the primary electromagnetic field.

Step 3: The flux lines from the secondary magnetic field effectively reduce the strength of the primary flux lines, and as a result, less CEMF is generated in the primary winding conductors. With less CEMF to oppose the primary applied voltage, the primary current increases in direct proportion to the secondary current. Figure 12–33

12.11 Line Currents

The line current of a transformer can be calculated using the appropriate formula below for single-phase or three-phase:

Single-phase: I = VA/E
Three-phase: I = VA/(E x 1.732)

Author's Comments:

- Delta/Delta connected three-phase transformers (Figure 12–34) and Delta/Wye connected three-phase transformers (Figure 12–35) have different turns ratios and different internal voltage and current relationships, but the line current can be calculated in each type of transformer using the formula above.

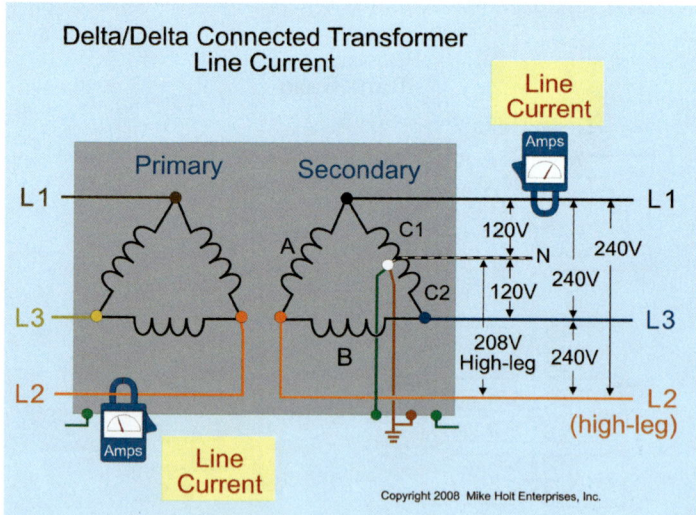

Figure 12–34

Mike Holt's Illustrated Guide to NEC Exam Preparation

Transformer Calculations — Unit 12

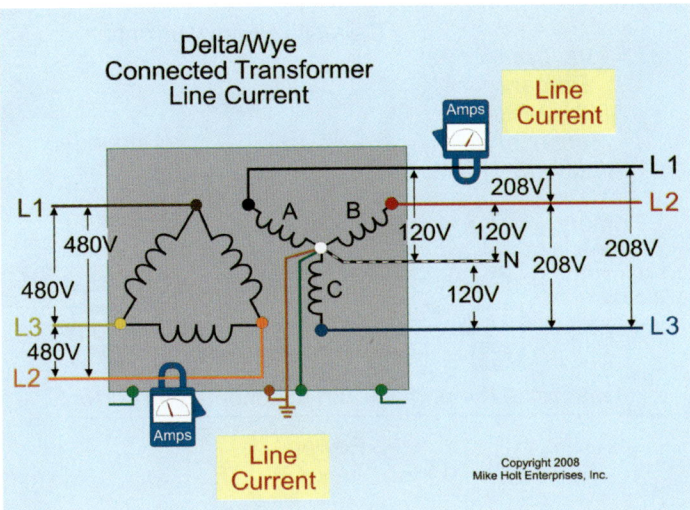

Figure 12–35

- The primary and secondary line currents are inversely proportional to the voltage ratio of the transformer. This means that the winding with the greatest number of turns has a higher voltage and lower current than the winding with the least number of turns, which has a lower voltage and higher current. **Figure 12–36**

- In a typical step-down transformer, the primary has more windings, higher voltage, lower current, and smaller wire. The secondary has fewer windings, lower voltage, higher current, and larger wire. See **Figure 12–36**.

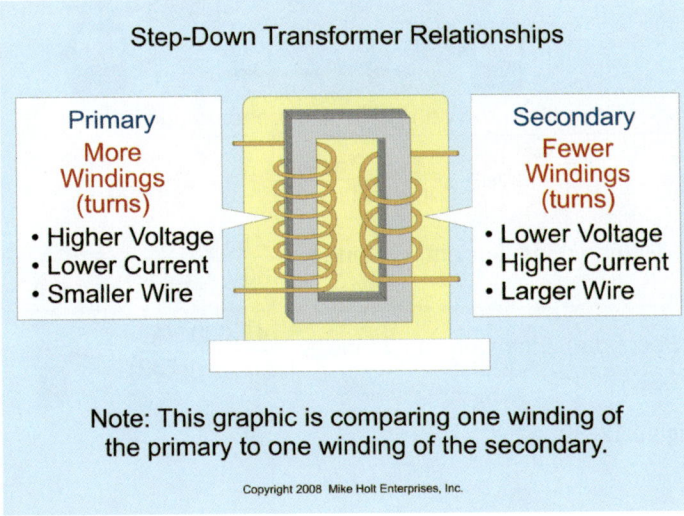

Figure 12–36

Tables 12–1 and 12–2 show the current relationship between kVA and voltage for common size transformers:

Table 12–1. Single-Phase Transformers I = VA/E

kVA Rating	Current at 208V	Current at 240V	Current at 480V
7.50	36A	31A	16A
10.00	48A	42A	21A
15.00	72A	63A	31A
25.00	120A	104A	52A
37.50	180A	156A	78A

Table 12–2. Three-Phase Transformers I = VA/(E x √3)

kVA Rating	Current at 208V	Current at 240V	Current at 480V
15.00	42A	36A	18A
22.50	63A	54A	27A
30.00	83A	72A	36A
37.50	104A	90A	45A
45.00	125A	108A	54A
50.00	139A	120A	60A
75.00	208A	180A	90A
112.50	313A	271A	135A

▶ **Single-Phase Example**

Question: What is the maximum primary and secondary line current at full load for a 480V to 240V, 25 kVA single-phase transformer? **Figure 12–37**

(a) 52/104A (b) 104/52A
(c) 104/208A (d) 208/104A

Answer: (a) 52/104A

I primary = VA/E
I primary = 25,000 VA/480V
I primary = 52A

I secondary = VA/E
I secondary = 25,000 VA/240V
I secondary = 104A

Unit 12 | Transformer Calculations

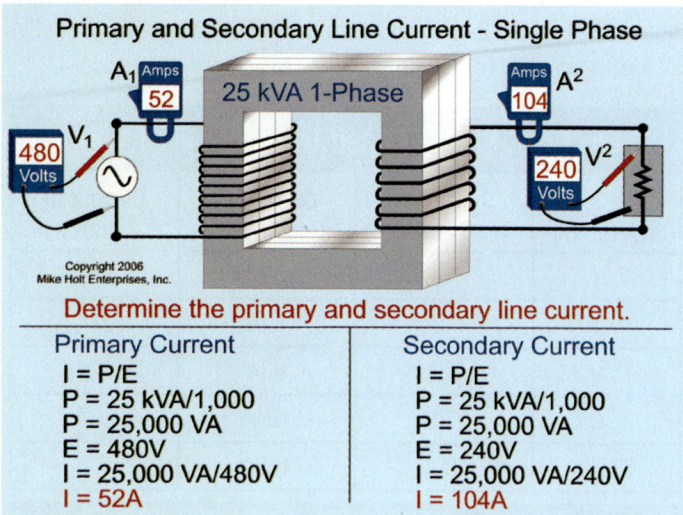

Figure 12–37

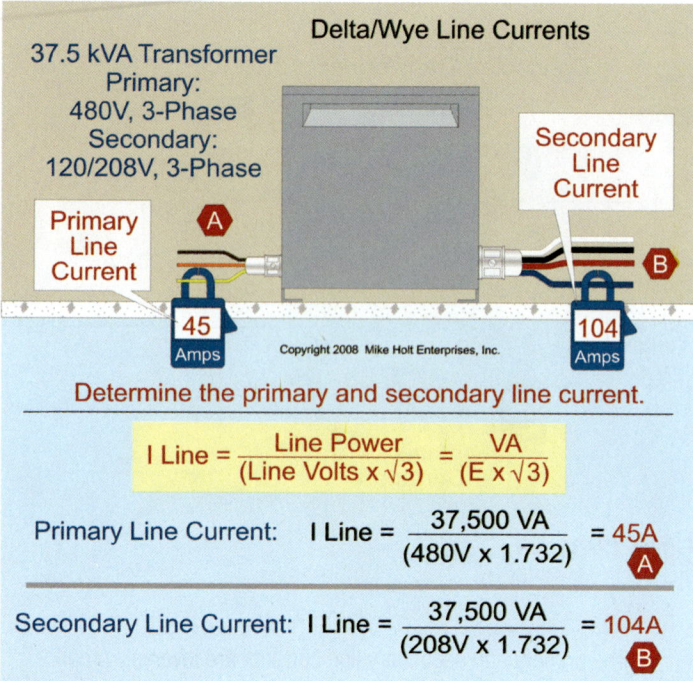

Figure 12–38

▶ **Three-Phase Example 1**

Calculate the primary and secondary current rating for a 37.50 kVA, three-phase, 480V to 120/208V transformer. **Figure 12–38**

Primary Current
$I = VA/(E \times 1.732)$
$I = 37{,}500\ VA/(480V \times 1.732)$
$I = 45A$

Secondary Current
$I = VA/(E \times 1.732)$
$I = 37{,}500\ VA/(208V \times 1.732)$
$I = 104A$

▶ **Three-Phase Example 2**

Calculate the primary and secondary current rating for a 45 kVA, three-phase, 480V to 120/208V transformer. **Figure 12–39**

Primary Current
$I = VA/(E \times 1.732)$
$I = 45{,}000\ VA/(480V \times 1.732)$
$I = 54A$

Secondary Current
$I = VA/(E \times 1.732)$
$I = 45{,}000\ VA/(208V \times 1.732)$
$I = 125A$

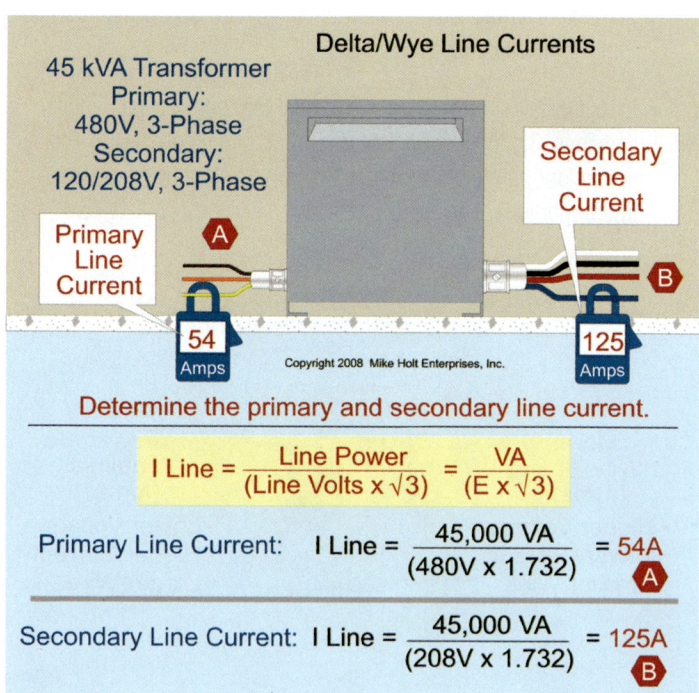

Figure 12–39

348   Mike Holt's Illustrated Guide to NEC Exam Preparation

Transformer Calculations Unit 12

▶ **Three-Phase Example 3**

Calculate the primary and secondary current rating for a 75 kVA, three-phase, 480V to 120/208V transformer. **Figure 12–40**

Primary Current
$I = VA/(E \times 1.732)$
$I = 75,000\ VA/(480V \times 1.732)$
$I = 90A$

Secondary Current
$I = VA/(E \times 1.732)$
$I = 75,000\ VA/(208V \times 1.732)$
$I = 208A$

▶ **Three-Phase Example 4**

Calculate the primary and secondary current rating for a 112.5 kVA, three-phase, 480V to 120/208V transformer. **Figure 12–41**

Primary Current
$I = VA/(E \times 1.732)$
$I = 112,500\ VA/(480V \times 1.732)$
$I = 135A$

Secondary Current
$I = VA/(E \times 1.732)$
$I = 112,500\ VA/(208V \times 1.732)$
$I = 313A$

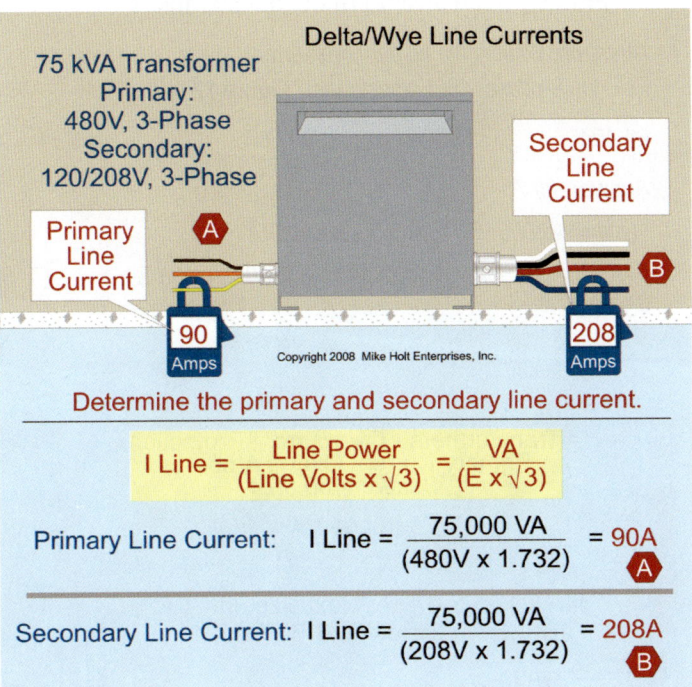

Figure 12–40

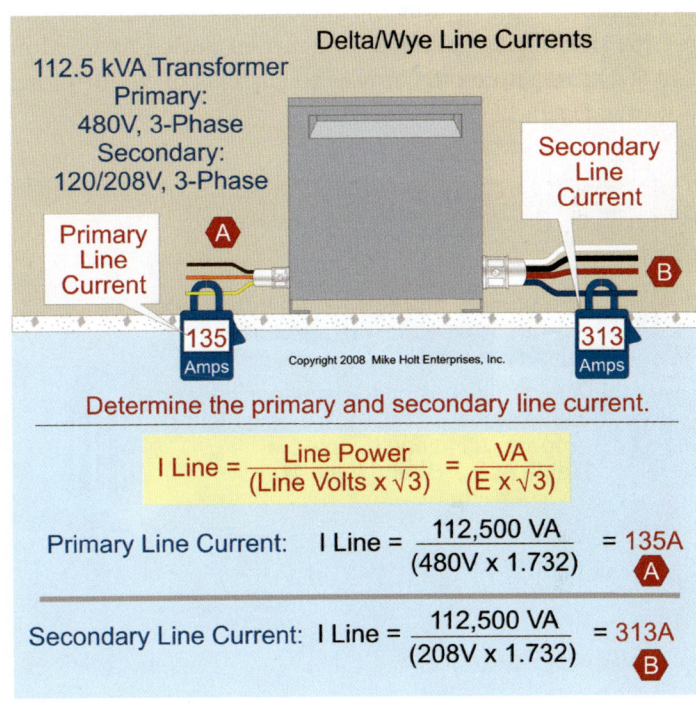

Figure 12–41

Unit 12 — Transformer Calculations

PART B—*NEC* REQUIREMENTS

12.12 Transformer Overcurrent Protection

To protect the primary winding of a transformer against overcurrent, use the percentages listed in Table 450.3(B) and its applicable notes.

Remember that Article 450 is for the protection of the transformer windings, and not the conductors supplying the transformer or leaving the transformer. When using Table 450.3(B), you'll notice that there are two main options: primary protection only, and primary *and* secondary protection. Secondary protection is very seldom required, so we'll concentrate only on primary protection in this discussion.

Where 125 percent of the primary current doesn't correspond to a standard fuse or nonadjustable circuit breaker, you can use the next higher rating of overcurrent device, as listed in 240.6(A). This note only applies to currents of 9A or more, however. **Figure 12–42**

Table 450.3(B) Primary Protection Only

Primary Current Rating	Maximum Protection
9A or More	125%, Note 1
Less Than 9A	167%
Less Than 2A	300%

Note 1. Where 125 percent of the primary current doesn't correspond to a standard rating of a fuse or nonadjustable circuit breaker, the next higher rating is permitted [240.6(A)].

▶ **Primary Overcurrent Protection Example 1**

Question: *What is the primary overcurrent device rating for a 45 kVA, three-phase, 480V transformer?* **Figure 12–43**

(a) 40A (b) 50A (c) 60A (d) 70A

Answer: (d) 70A

I Line = 45,000 VA/(480V x 1.732) = 54A

54A x 1.25 = 68A, next size up 70A, [240.6(A) and Table 450.3(B), Note 1]

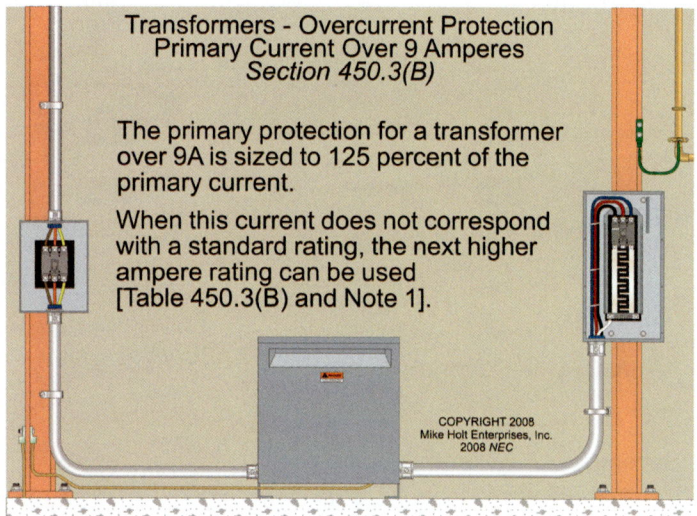

Figure 12–42

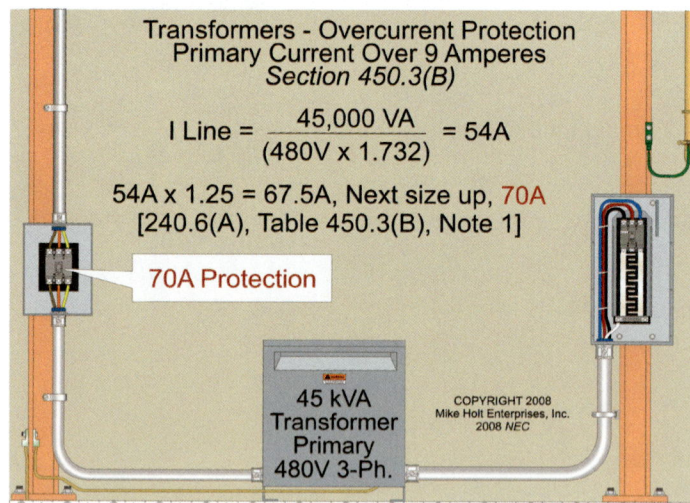

Figure 12–43

Transformer Calculations — Unit 12

▶ **Primary Overcurrent Protection Example 2**

Question: What is the primary overcurrent device rating for a 75 kVA, three-phase, 480V transformer? **Figure 12–44**

(a) 70A (b) 100A (c) 110A (d) 125A

Answer: (d) 125A

I Line = 75,000 VA/(480V x 1.732) = 90A

90A x 1.25 = 113A, next size up 125A, [240.6(A) and Table 450.3(B), Note 1]

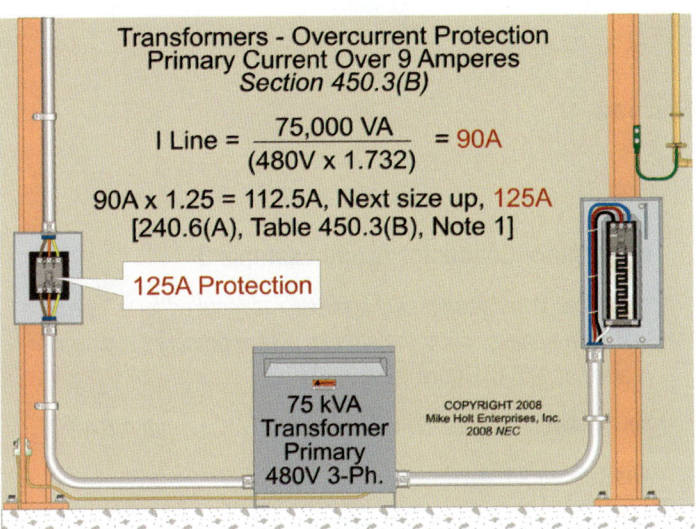

Figure 12–44

▶ **Primary Overcurrent Protection Example 3**

Question: What is the primary overcurrent device rating for a 112.50 kVA, three-phase, 480V transformer? **Figure 12–45**

(a) 70A (b) 100A (c) 110A (d) 125A

Answer: (c) 175A

I Line = 112,500 VA/(480V x 1.732) = 135A

135A x 1.25 = 169A, next size up 175A, [240.6(A) and Table 450.3(B), Note 1]

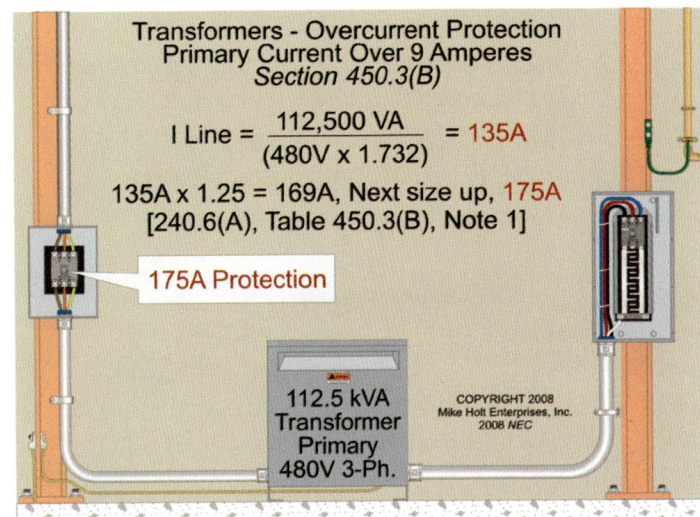

Figure 12–45

12.13 Primary Conductor Sizing

Conductors must be sized no less than 125 percent of the continuous loads, plus 100 percent of the noncontinuous loads, based on the terminal temperature rating ampacities as listed in Table 310.16, before any ampacity adjustment [210.19(A)(1)].

In addition, conductors must be protected against overcurrent in accordance with their ampacity after ampacity adjustment, as specified in 310.15 [240.4], however the next higher standard rating of overcurrent device (above the ampacity of the conductors being protected) is permitted, provided the overcurrent device rating doesn't exceed 800A [240.4(B)].

▶ **Primary Conductor Sizing Example 1**

Question: What size primary conductor is required for a 45 kVA, three-phase, 480V transformer, where primary overcurrent is sized at 70A? **Figure 12–46**

(a) 6 AWG (b) 4 AWG (c) 3 AWG (d) 2 AWG

Answer: (b) 4 AWG

Step 1: Size the primary conductor at 125% of the primary current rating.

I Line = 45,000 VA/(480V x 1.732) = 54A

54A x 1.25 = 68A, 4 AWG rated 85A [110.14(C)(1) and Table 310.16]

(continued in next column)

Unit 12 — Transformer Calculations

Step 2: Verify that the conductors are protected in accordance with their ampacities [240.4].

4 AWG is rated 85A at 75°C, and it's permitted to be protected by the 70A primary overcurrent device.

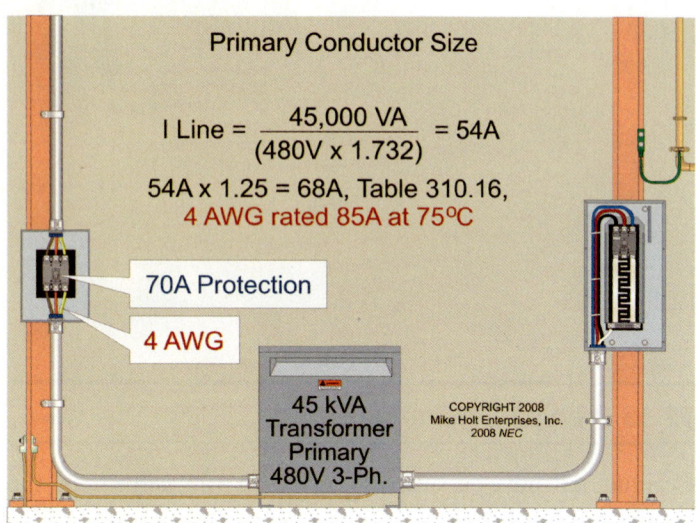

Figure 12–46

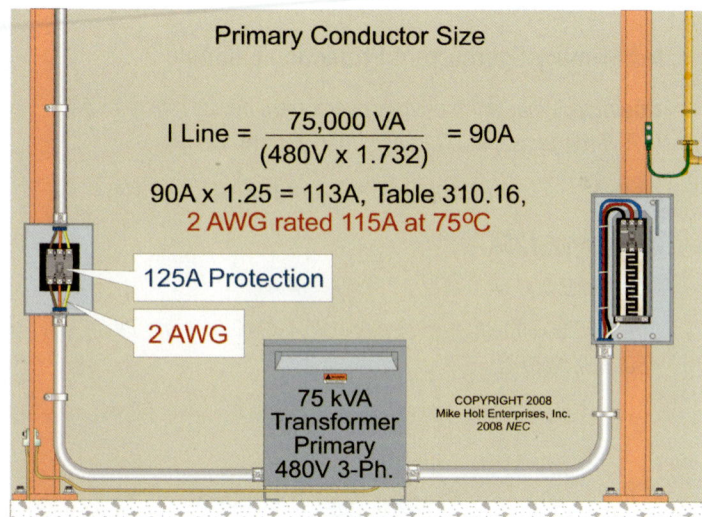

Figure 12–47

▶ **Primary Conductor Sizing Example 2**

Question: What size primary conductor is required for a 75 kVA, three-phase, 480V transformer, where primary overcurrent is sized at 125A? Figure 12–47

(a) 6 AWG   (b) 4 AWG   (c) 3 AWG   (d) 2 AWG

Answer: (d) 2 AWG

Step 1: Size the primary conductor at 125% of the primary current rating.

I Line = 75,000 VA/(480V x 1.732) = 90A

90A x 1.25 = 113A, 2 AWG rated 115A at 75°C [110.14(C)(1) and Table 310.16]

Step 2: Verify that the conductors are protected in accordance with their ampacities [240.4].

2 AWG is rated 115A at 75°C, and it's permitted to be protected by the 125A primary overcurrent device [240.4(B)].

▶ **Primary Conductor Sizing Example 3**

Question: What size primary conductor is required for a 112.50 kVA, three-phase, 480V transformer, where primary overcurrent is sized at 175A? Figure 12–48

(a) 1/0 AWG   (b) 2/0 AWG   (c) 3/0 AWG   (d) 4/0 AWG

Answer: (b) 2/0 AWG

Step 1: Size the primary conductor at 125% of the primary current rating.

I Line = 112,500 VA/(480V x 1.732) = 135A

135A x 1.25 = 169A, 2/0 AWG rated 175A [110.14(C)(1) and Table 310.16]

Step 2: Verify that the conductors are protected in accordance with their ampacities [240.4].

2/0 AWG is rated 175A at 75°C, and it's permitted to be protected by the 175A primary overcurrent device.

12.14 Secondary Conductor Sizing

The ampacity of the secondary conductor must not be less than the rating of the device supplied by the secondary conductors or the overcurrent device at the termination of the secondary conductors [240.21(C)(2)]. Assume the secondary conductors are intended to carry the full capacity of the transformer continuously.

Transformer Calculations | Unit 12

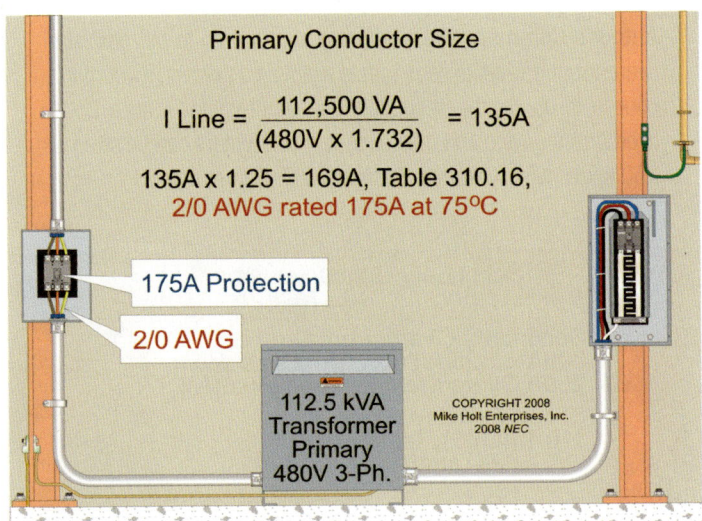

Figure 12–48

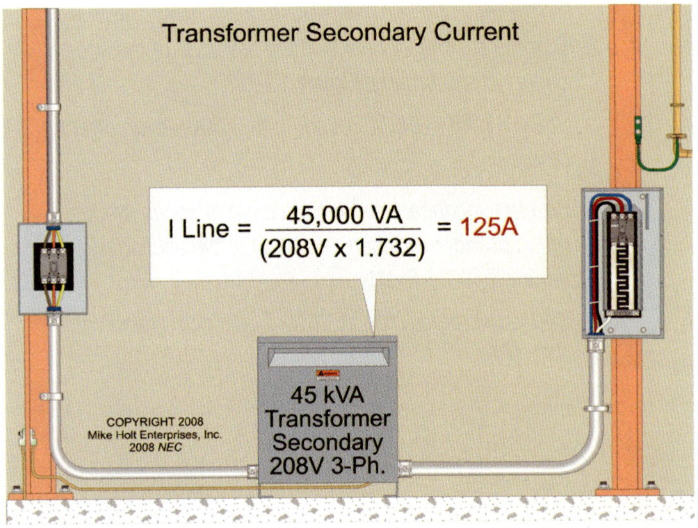

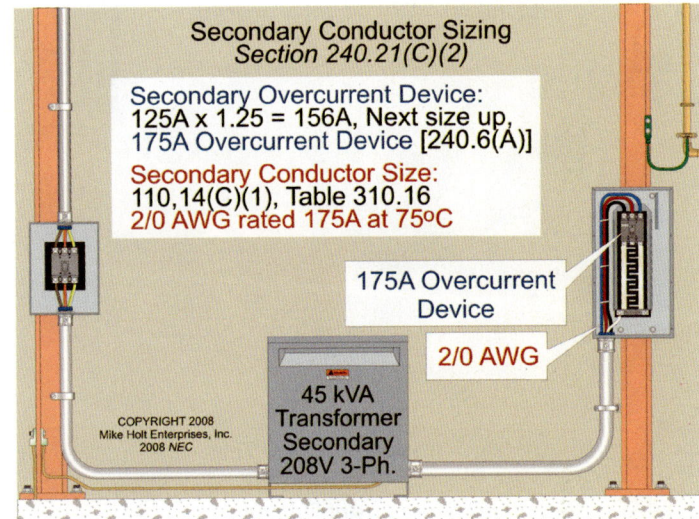

Figure 12–49

Step 1: Determine the rating of the device supplied by the secondary conductors at 125 percent of the secondary rating.

Step 2: Size the secondary conductors so that they have an ampere rating of "not less" than the device rating supplied by the secondary conductors.

▶ **Secondary Conductor Sizing Example 1**

Question: What size secondary conductor is required for a 45 kVA, three-phase, 480-120/208V transformer?

(a) 2 AWG (b) 1 AWG (c) 1/0 AWG (d) 2/0 AWG

Answer: (d) 2/0 AWG

Step 1: Determine the secondary current rating. **Figure 12–49**

$I = VA/(E \times 1.732)$
$I = 45,000 \, VA/(208V \times 1.732)$
$I = 125A$

Step 2: Size the secondary device rating at 125% of the secondary current rating. **Figure 12–50**

$125A \times 1.25 = 156A$, 175A overcurrent device

Step 3: Size the secondary conductor where it has an ampere rating of "not less" than the rating of the secondary device [240.21(C)(2)].

2/0 AWG rated 175A at 75°C 110.14(C)(1) and Table 310.16].

Figure 12–50

▶ **Secondary Conductor Sizing Example 2**

Question: What size secondary conductor is required for a 75 kVA, three-phase, 480-120/208V transformer?

(a) 3/0 AWG (b) 4/0 AWG (c) 300 kcmil (d) 500 kcmil

Answer: (c) 300 kcmil

Step 1: Determine the secondary current rating. **Figure 12–51**

$I = VA/(E \times 1.732)$
$I = 75,000 \, VA/(208V \times 1.732)$
$I = 208A$

(continued in next column)

Unit 12 Transformer Calculations

Step 2: Size the secondary device rating at 125% of the secondary current rating. **Figure 12–52**

208A x 1.25 = 260A, round up to a 300A overcurrent device

Step 3: Size the secondary conductor where it has an ampere rating of "not less" than the rating of the secondary device [240.21(C)(2)].

300 kcmil rated 310A at 75ºC [110.14(C)(1) and Table 310.16].

Author's Comment: We're allowed to round up to the next size overcurrent device as seen in this example. However, if you don't choose to do so, the secondary can be protected with a 250A overcurrent device and the secondary conductors can be sized at 250 kcmil, which is rated 255A at 75ºC. **Figure 12–53**

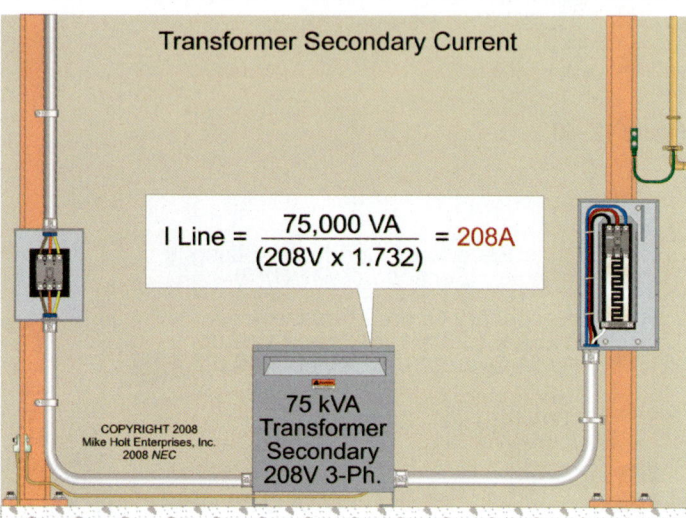

Figure 12–51

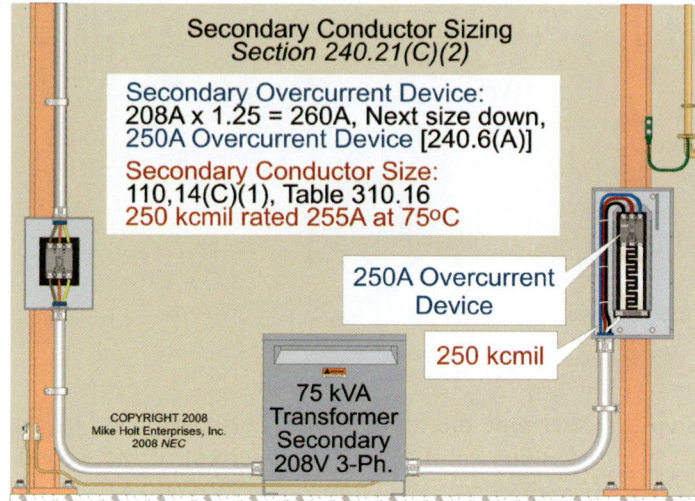

Figure 12–53

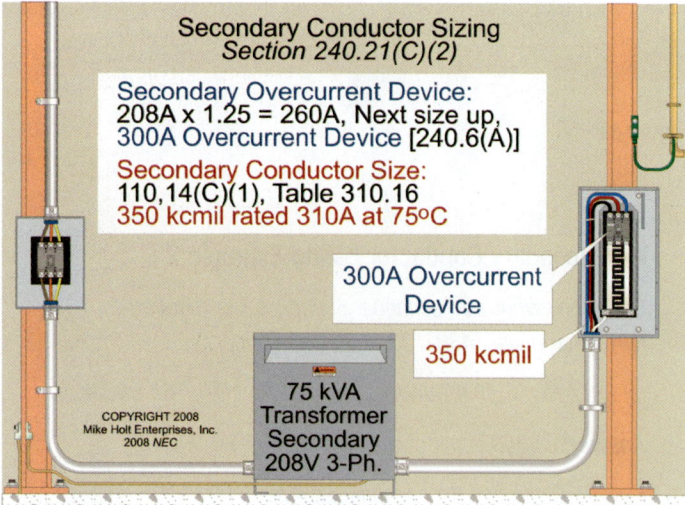

Figure 12–52

▶ **Secondary Conductor Sizing Example 3**

Question: What size secondary conductor is required for a 112.50 kVA, three-phase, 480-120/208V transformer?

(a) Two parallel 1/0 AWG (b) Two parallel 2/0 AWG
(c) Two parallel 3/0 AWG (d) Two parallel 4/0 AWG

Answer: (c) Two parallel 3/0 AWG

Step 1: Determine the secondary current rating. **Figure 12–54**

I = VA/(E x 1.732)
I = 112,500 VA/(208V x 1.732)
I = 313A

Step 2: Size the secondary device rating at 125% of the primary current rating. **Figure 12–55**

313A x 1.25 = 391A, 400A overcurrent device

Step 3: Size the secondary conductors where they have an ampere rating of "not less" than the rating of the secondary device [240.21(C)(2)].

Two sets of parallel 3/0 AWG conductors, each rated 200A at 75ºC [110.14(C)(1) and Table 310.16]

354 — Mike Holt's Illustrated Guide to NEC Exam Preparation

Transformer Calculations — Unit 12

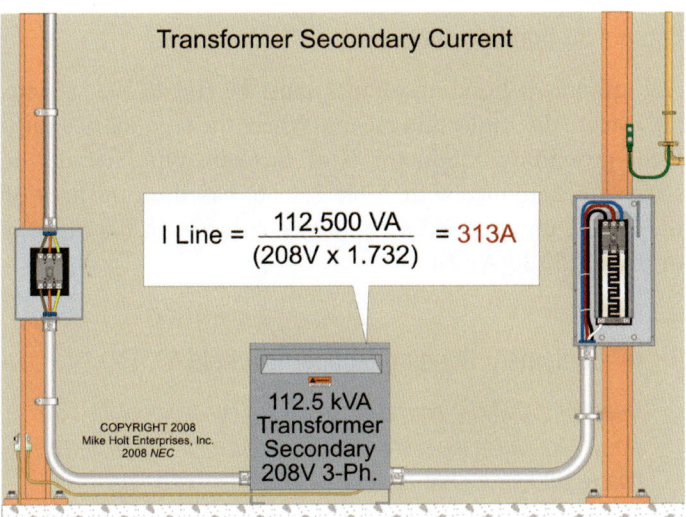

Figure 12–54

▶ **System Bonding Jumper Example 1**

Question: What size system bonding jumper is required for a 45 kVA, three-phase, 480-120/208V transformer, when the secondary conductors are 2/0 AWG? **Figure 12–56**

(a) 6 AWG   (b) 4 AWG   (c) 3 AWG   (d) 2 AWG

Answer: (b) 4 AWG [Table 250.66]

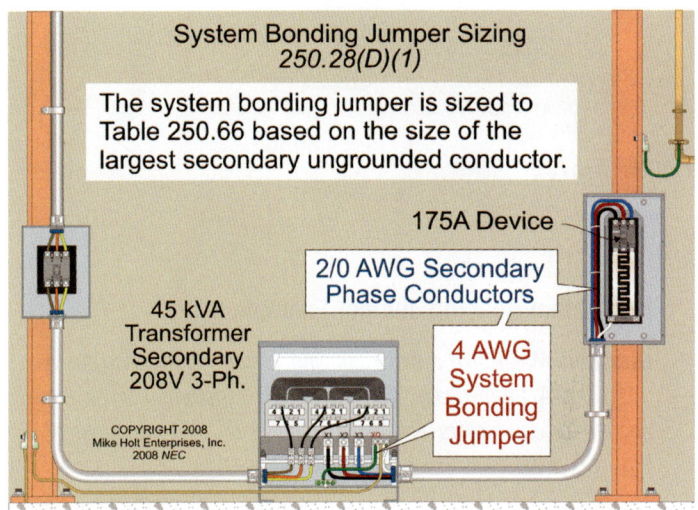

Figure 12–56

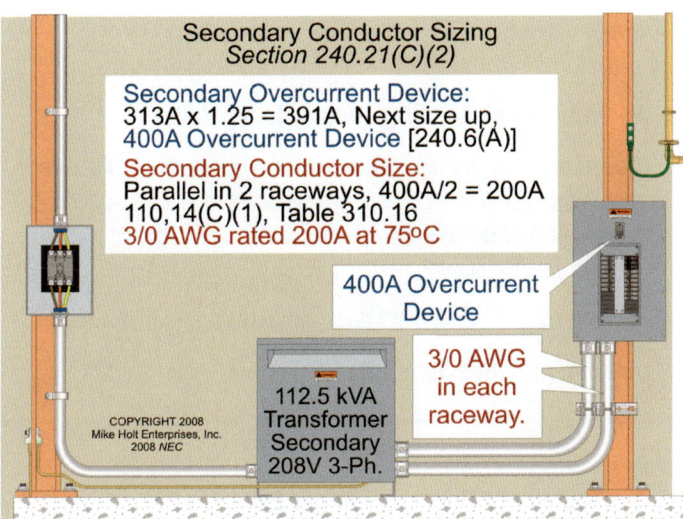

Figure 12–55

▶ **System Bonding Jumper Example 2**

Question: What size system bonding jumper is required for a 75 kVA, three-phase, 480-120/208V transformer, when the secondary conductors are 250 kcmil? **Figure 12–57**

(a) 6 AWG   (b) 4 AWG   (c) 3 AWG   (d) 2 AWG

Answer: (d) 2 AWG [Table 250.66]

12.15 Grounding and Bonding

System Bonding Jumper

A system bonding jumper, sized in accordance with Table 250.66 based on the area of the secondary conductors [250.28(D)], must be installed at the same location where the grounding electrode conductor terminates to the neutral terminal of the separately derived system; either at the separately derived system or the system disconnecting means, but not at both locations [250.30(A)(1) and (A)(3)].

Unit 12 — Transformer Calculations

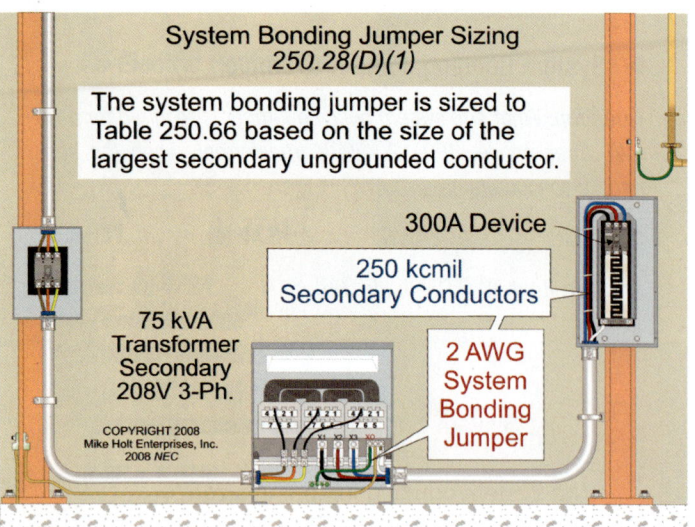

Figure 12–57

▶ System Bonding Jumper Example 3

Question: What size system bonding jumper is required for a 112.50 kVA, three-phase, 480-120/208V transformer, when the secondary conductors are two parallel 3/0 AWG conductors? **Figure 12–58**

(a) 6 AWG (b) 4 AWG (c) 3 AWG (d) 2 AWG

Answer: (c) 2 AWG

Cmil area of 3/0 = 167,800 cmil [Chapter 9, Table 8]

167,000 cmil x 2 = 335,600 cmil

Table 250.66 based on 335,600 cmil = 2 AWG

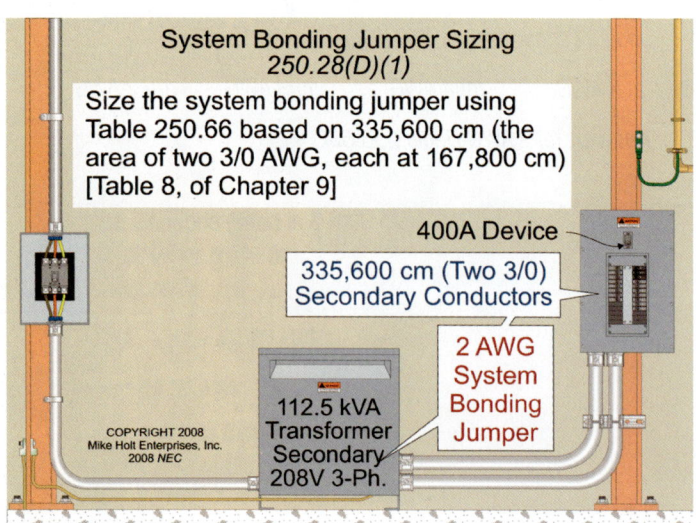

Figure 12–58

Equipment Bonding Jumper

An equipment bonding jumper must be run to the secondary system disconnecting means. Where the secondary equipment grounding conductor is of the wire type, it must be sized in accordance with Table 250.66, based on the area of the largest ungrounded secondary conductor in the raceway or cable [250.30(A)(2)].

▶ Equipment Bonding Jumper Example 1

Question: What size equipment grounding conductor of the wire type is required to the secondary disconnect for a 45 kVA, three-phase, 480-120/208V transformer, when the secondary conductors are 2/0 AWG?

(a) 6 AWG (b) 4 AWG (c) 3 AWG (d) 2 AWG

Answer: (b) 4 AWG [Table 250.66]

▶ Equipment Bonding Jumper Example 2

Question: What size equipment grounding conductor of the wire type is required to the secondary disconnect for a 75 kVA, three-phase, 480-120/208V transformer, when the secondary conductors are 250 kcmil?

(a) 6 AWG (b) 4 AWG (c) 3 AWG (d) 2 AWG

Answer: (d) 2 AWG [Table 250.66]

▶ Equipment Bonding Jumper Example 3

Question: What size equipment grounding conductor of the wire type is required to the secondary disconnect for a 112.50 kVA, three-phase, 480-120/208V transformer, when the secondary conductors are two parallel 3/0 AWG conductors?

(a) 6 AWG (b) 4 AWG (c) 3 AWG (d) 2 AWG

Answer: (c) 2 AWG

Cmil area of 3/0 = 167,800 cmil [Chapter 9, Table 8]

167,000 cmil x 2 = 335,600 cmil

Table 250.66 based on 335,600 cmil = 2 AWG

Transformer Calculations — Unit 12

Grounding Electrode Conductor

A grounding electrode conductor must connect the neutral terminal of a separately derived system to a grounding electrode of a type identified in 250.30(A)(7). The grounding electrode conductor must be sized in accordance with 250.66, based on the area of the ungrounded secondary conductor [250.30(A)(3)].

▶ **Grounding Electrode Conductor Example 1**

Question: What size grounding electrode conductor is required for a 45 kVA, three-phase, 480-120/208V transformer, when the secondary conductors are 2/0 AWG? **Figure 12–59**

(a) 6 AWG   (b) 4 AWG   (c) 3 AWG   (d) 2 AWG

Answer: (b) 4 AWG [Table 250.66]

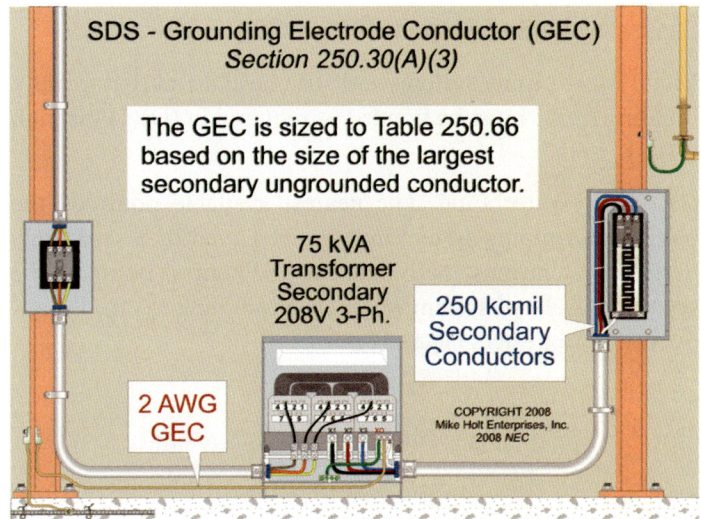

Figure 12–60

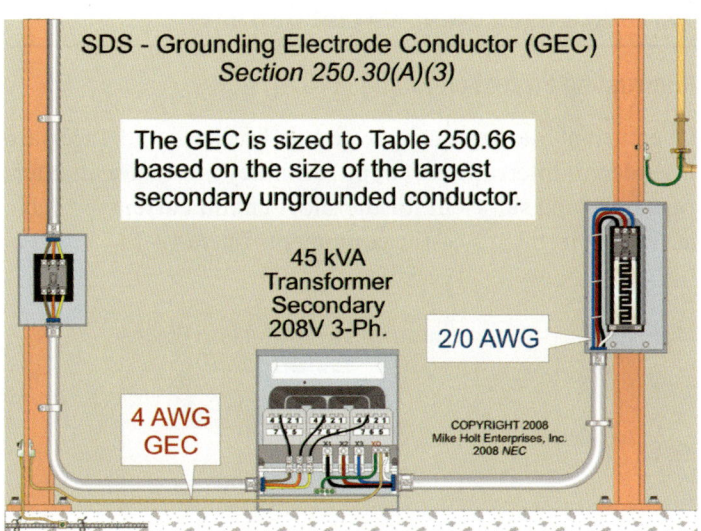

Figure 12–59

▶ **Grounding Electrode Conductor Example 2**

Question: What size grounding electrode conductor is required for a 75 kVA, three-phase, 480-120/208V transformer, when the secondary conductors are 250 kcmil? **Figure 12–60**

(a) 6 AWG   (b) 4 AWG   (c) 3 AWG   (d) 2 AWG

Answer: (d) 2 AWG [Table 250.66]

▶ **Grounding Electrode Conductor Example 3**

Question: What size grounding electrode conductor is required for a 112.50 kVA, three-phase, 480-120/208V transformer, when the secondary conductors are two parallel 3/0 AWG conductors? **Figure 12–61**

(a) 6 AWG   (b) 4 AWG   (c) 3 AWG   (d) 2 AWG

Answer: (d) 2 AWG

Cmil area of 3/0 = 167,800 cmil [Chapter 9, Table 8]

167,000 cmil x 2 = 335,600 cmil

Table 250.66 based on 335,600 cmil = 2 AWG

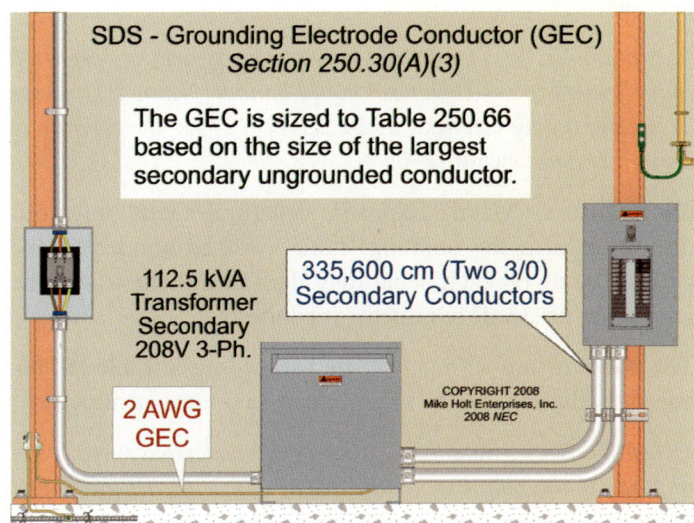

Figure 12–61

Unit 12 — Transformer Calculations

Separately Derived Systems [250.104(D)]

Metal water piping systems and structural metal that forms a building frame must be bonded as required in (D)(1) through (D)(3).

(1) Metal Water Pipe. The nearest available point of the metal water piping system in the area served by a separately derived system must be bonded to the neutral point of the separately derived system where the grounding electrode conductor is connected. **Figure 12–62**

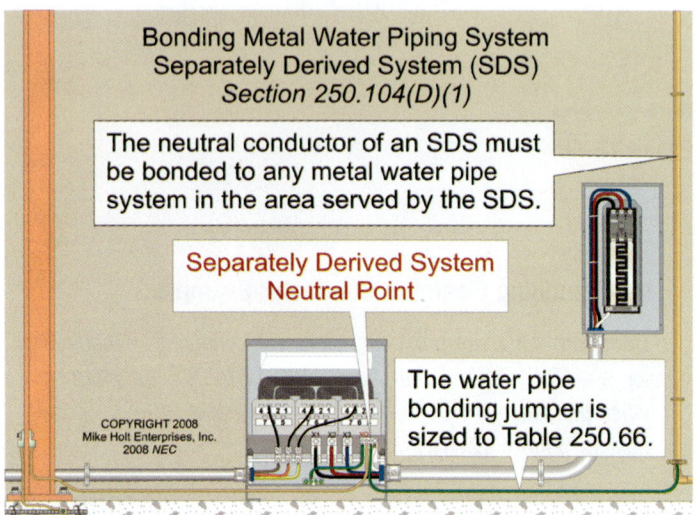

Figure 12–62

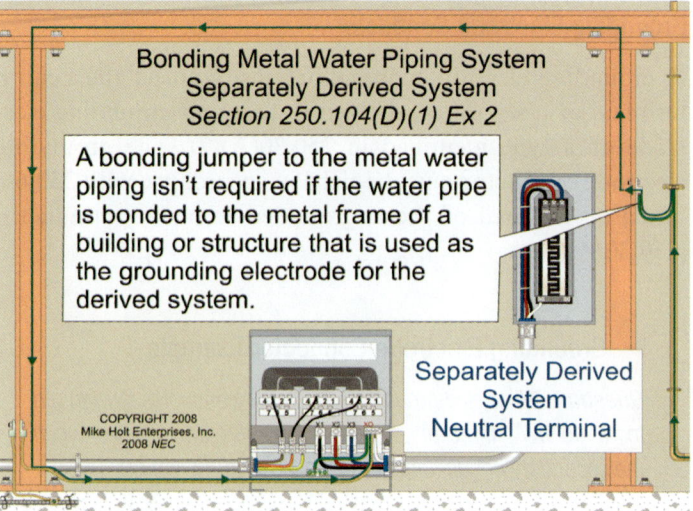

Figure 12–63

The bonding jumper must be sized in accordance with Table 250.66, based on the area of the ungrounded secondary conductors.

Exception No. 2: The metal water piping system is permitted to be bonded to the structural metal building frame if it serves as the grounding electrode [250.52(A)(1)] for the separately derived system. **Figure 12–63**

(2) Structural Metal. Exposed structural metal interconnected to form the building frame must be bonded to the neutral point of each separately derived system where the grounding electrode conductor is connected.

The bonding jumper must be sized according to Table 250.66, based on the area of the ungrounded secondary conductors.

Exception No. 1: Bonding to the separately derived system isn't required if the metal structural frame serves as the grounding electrode [250.52(A)(2)] for the separately derived system.

12.16 Available Short-Circuit Current

Interrupting Protection Rating [110.9]

Overcurrent devices such as circuit breakers and fuses are intended to interrupt the circuit, and they must have an interrupting rating sufficient for the short-circuit current available at the line terminals of the equipment. **Figure 12–64**

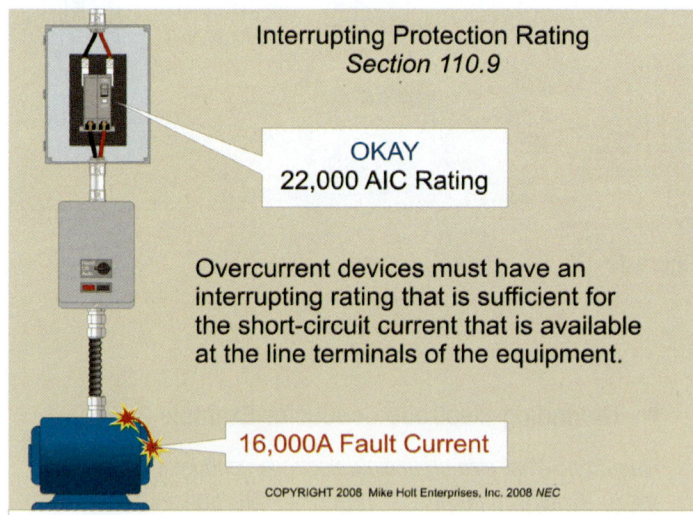

Figure 12–64

Author's Comments:

- See the definition of "Interrupting Rating" in Article 100.

Transformer Calculations Unit 12

- Unless marked otherwise, the ampere interrupting rating for circuit breakers is 5,000A [240.83(C)], and for fuses it's 10,000A [240.60(C)(3)]. **Figure 12–65**

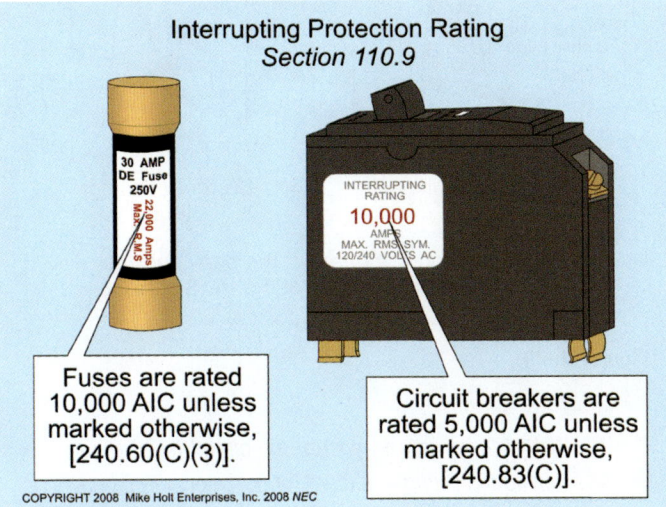

Figure 12–65

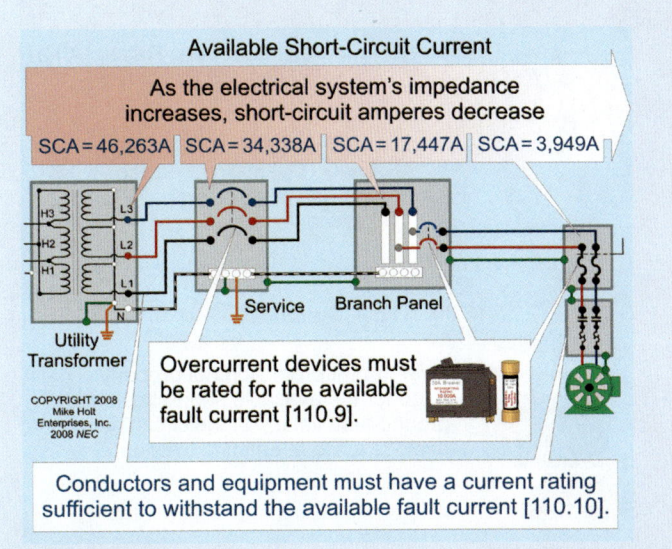

Figure 12–66

AVAILABLE SHORT-CIRCUIT CURRENT

Available short-circuit current is the current, in amperes, available at a given point in the electrical system. This available short-circuit current is first determined at the secondary terminals of the utility transformer. Thereafter, the available short-circuit current is calculated at the terminals of service equipment, then at branch-circuit panelboards and other equipment. The available short-circuit current is different at each point of the electrical system. It's highest at the utility transformer and lowest at the branch-circuit load.

The available short-circuit current depends on the impedance of the circuit, which increases moving downstream from the utility transformer. The greater the circuit impedance (utility transformer and the additive impedances of the circuit conductors), the lower the available short-circuit current. **Figure 12–66**

Factors that affect the available short-circuit current at the utility transformer include the system voltage, the transformer kVA rating, and its impedance (expressed in a percentage on the equipment nameplate). Properties that impact the impedance of the circuit include the conductor material (copper versus aluminum), conductor size, conductor length, and motor-operated equipment supplied by the circuit.

> **Author's Comment:** Many people in the industry describe Amperes Interrupting Rating (AIR) as "Amperes Interrupting Capacity" (AIC).

> **DANGER:** *Extremely high values of current flow (caused by short circuits or ground faults) produce tremendously destructive thermal and magnetic forces. If the circuit overcurrent device isn't rated to interrupt the current at the available fault values at its listed voltage rating, it can explode while attempting to open the circuit overcurrent device resulting from a short circuit or ground fault, which can cause serious injury or death, as well as property damage.* **Figure 12–67**

Short-Circuit Current Rating [110.10]

Electrical equipment must have a short-circuit current rating that permits the circuit overcurrent device to open the circuit upon a short circuit or ground fault without extensive damage to the electrical components of the circuit. For example, a motor controller must have a sufficient short-circuit rating for the available fault current.

> **Author's Comments:**
> - See the definition of "Controller" in Article 100.

Unit 12 — Transformer Calculations

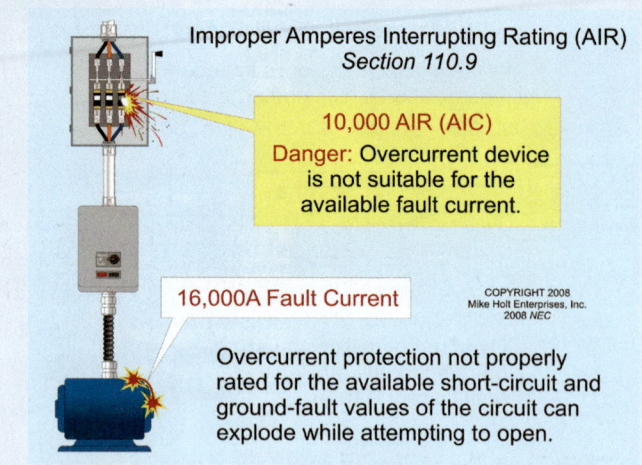

Figure 12–67

- If the fault exceeds the controller's 5,000A short-circuit current rating, the controller can explode, endangering persons and property. **Figure 12–68**

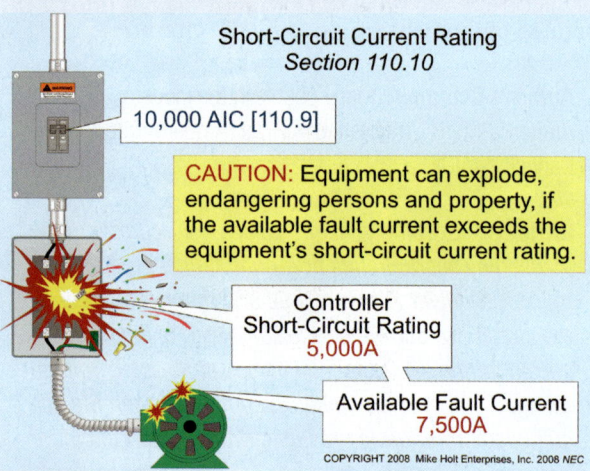

Figure 12–68

To solve this problem, a current-limiting overcurrent device (fast-clearing fuse) can be used to reduce the let-through current to less than 5,000A. **Figure 12–69**

Author's Comment: For more information on the application of current-limiting devices, see 240.2 and 240.60(B).

Interrupting Rating. Circuit breakers have an interrupting rating of 5,000A unless marked otherwise [240.83(C)], and fuses have an interrupting rating of 10,000A unless marked otherwise [240.60].

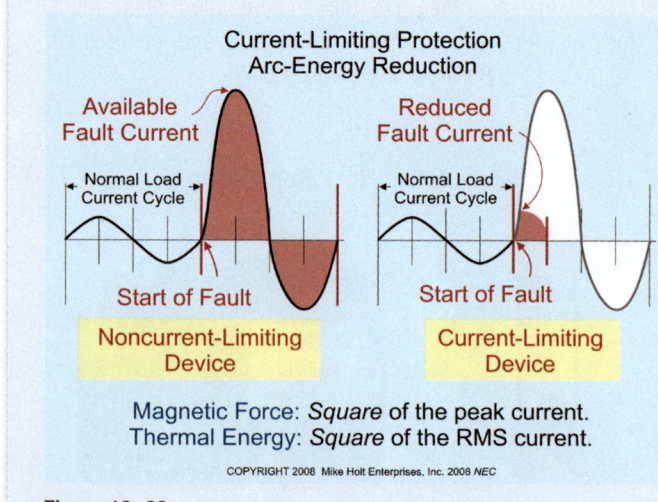

Figure 12–69

WARNING: *Take care to ensure the circuit breaker has an interrupting rating sufficient for the short-circuit current available at the line terminals of the equipment. Using a circuit breaker with inadequate interrupting current rating can cause equipment to be destroyed from a line-to-line or ground fault, and result in death or serious injury. See 110.9 for more details.* **Figure 12–70**

Author's Comment: Refer to Article 450 for information on clearances and ventilation when installing transformers.

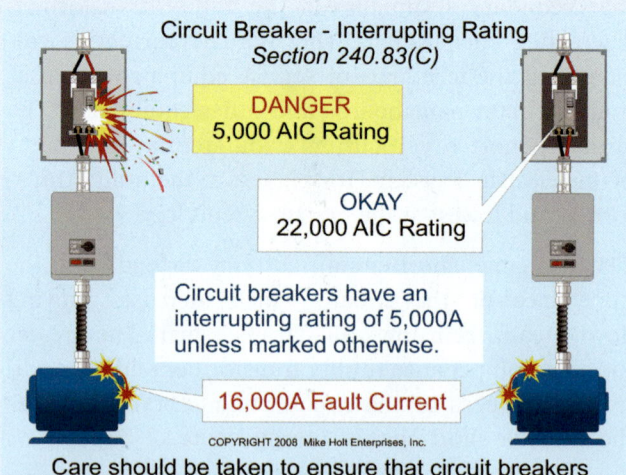

Figure 12–70

360   Mike Holt's Illustrated Guide to NEC Exam Preparation

Unit 12 Conclusion

CONCLUSION TO UNIT 12—TRANSFORMER CALCULATIONS

If you can remember that transformers really are *transforming* the electricity that goes through them, you'll find it easier to apply transformer theory. Except for autotransformers, all transformers consist of two sets of interacting coils. The turns ratio of the primary set to the secondary set determines the degree to which the voltage is transformed—that is, raised or lowered.

To keep from making grounding and bonding errors, remember that there's a difference between grounding and bonding. Grounding is the connection to the earth which is accomplished by means of the grounding electrode and the grounding electrode conductor. The connection to earth won't provide an effective ground fault return path. That's the role of proper bonding. With transformers, a vital link in that path is the system bonding jumper that must be installed on the secondary of the transformer, either in the transformer or in the panel. The system bonding jumper completes the bonding path back to the source (the transformer secondary), to provide an effective fault current path.

Unit 12 Practice Questions

UNIT 12—TRANSFORMER CALCULATIONS PRACTICE QUESTIONS

(•Indicates that 75% or fewer of those who took this exam answered the question correctly.)

PART A–GENERAL

12.6 Delta/Delta Connected Transformers

1. Delta configured means that the windings of three single-phase transformers are connected in _____ with each other. A delta-connected transformer is represented by the Greek letter delta.

    (a) series
    (b) parallel
    (c) series-parallel
    (d) a and b

2. What 4-wire system is called the high-leg system because the voltage from one conductor-to-ground is between 190 volts and 208 volts to ground?

    (a) Delta
    (b) Wye

3. The term _____ is used to identify the ungrounded (hot) conductors in an electrical system.

    (a) load
    (b) line
    (c) system
    (d) grounded

4. _____ voltage is the voltage measured between any two ungrounded (hot) conductors.

    (a) Load
    (b) Line
    (c) Grounded
    (d) any of these

5. In a delta-configured transformer, the line voltage is equal to the phase voltage.

    (a) True
    (b) False

6. If the secondary voltage of a delta/delta transformer is 120/240V, the high-leg voltage-to-ground (or neutral) is approximately _____.

    (a) 191V
    (b) 196V
    (c) 202V
    (d) 208V

7. On a 4-wire, delta-connected system where the midpoint of one phase winding is grounded, the conductor having the higher phase voltage-to-ground shall be durably and permanently marked by an outer finish that is _____ in color.

    (a) black
    (b) red
    (c) blue
    (d) orange

12.7 Delta/Wye Connected Transformers

8. _____-connected means a connection of three single-phase transformer windings to a common point (neutral), with the other end of each winding connected to the line conductors.

    (a) Delta
    (b) Wye
    (c) a or b
    (d) none of these

9. In a wye-configured transformer, the line voltage is the same as the phase (winding) voltage.

    (a) True
    (b) False

Practice Questions Unit 12

12.8 Transformer Turns Ratio

10. The turns ratio is the relationship between the number of primary windings compared to the number of secondary windings. This is the same as the turns ratio between the primary phase voltage and the secondary phase voltage. For under 600V, a typical delta/delta system phase voltage ratio is ____ and a typical delta/wye system phase voltage ratio is ____.

    (a) 1:2, 1:4
    (b) 2:1, 4:1
    (c) 4:1, 2:1
    (d) 10:1, 20:1

11. What is the turns ratio for a 480V to 120/208V, three-phase, delta/wye-configured transformer?

    (a) 4:1
    (b) 1:4
    (c) 1:2
    (d) 2:1

12. What is the voltage ratio of a single-phase 480V to 240V transformer?

    (a) 4:1
    (b) 1:4
    (c) 1:2
    (d) 2:1

12.10 Current Flow

13. When a load is connected to the secondary winding of a transformer, current flows through the secondary winding. The current flowing in the secondary creates an electromagnetic field that opposes the primary electromagnetic field.

    (a) True
    (b) False

12.11 Line Currents

14. In a delta-configured transformer, the ratio of the primary-to-secondary line current is directly proportional (the same ratio) to the voltage turns ratio.

    (a) True
    (b) False

15. What is the primary line current for a fully loaded 45 kVA, 480V to 120/240V, three-phase transformer?

    (a) 43A
    (b) 54A
    (c) 108A
    (d) 124A

16. What is the secondary line current for a fully loaded 45 kVA, 480V to 240V, three-phase transformer?

    (a) 43A
    (b) 54A
    (c) 108A
    (d) 124A

17. The line current for both delta and wye-configured three-phase transformers can be calculated by the formula:

    $I_{Line} = VA_{Line}/(E_{Line} \times 1.732)$

    (a) True
    (b) False

18. What is the primary line current for a fully loaded 22 kVA, 480V to 120/208V, three-phase transformer?

    (a) 22A
    (b) 26A
    (c) 54A
    (d) 61A

19. What is the secondary line current for a fully loaded 22 kVA, 480V to 120/208V, three-phase transformer?

    (a) 22A
    (b) 27A
    (c) 54A
    (d) 61A

PART B—NEC REQUIREMENTS

12.12 Transformer Overcurrent Protection

20. The primary overcurrent protection for a transformer rated 600V, nominal, or less, having a primary current rating of over 9A or more must be set at not more than ____ percent of the primary current.

    (a) 125
    (b) 167
    (c) 200
    (d) 300

Unit 12 — Practice Questions

21. Overcurrent protection for the secondary of a transformer rated 600V or less having a secondary current rating of 9A or more must be set at not more than _____ percent of the secondary current rating.

    (a) 100
    (b) 125
    (c) 150
    (d) 170

12.13 Primary Conductor Sizing

22. A 45 kVA, 480V primary to 120/208V secondary, three-phase transformer with a 70A primary overcurrent device will require at least _____ primary conductors.

    (a) 6 AWG
    (b) 4 AWG
    (c) 2 AWG
    (d) 1 AWG

23. A 75 kVA, 480V primary to 120/208V secondary, three-phase transformer with a 125A primary overcurrent device will require at least _____ primary conductors.

    (a) 6 AWG
    (b) 4 AWG
    (c) 2 AWG
    (d) 1 AWG

12.14 Secondary Conductor Sizing

24. A 45 kVA, 480V primary to 120/208V secondary, three-phase transformer with a 175A secondary overcurrent device will require at least _____ secondary conductors.

    (a) 2/0 AWG
    (b) 4/0 AWG
    (c) 350 kcmil
    (d) 400 kcmil

25. A 75 kVA, 480V primary to 120/208V secondary, three-phase transformer with a 300A secondary overcurrent device will require at least _____ secondary conductors.

    (a) 2/0 AWG
    (b) 4/0 AWG
    (c) 350 kcmil
    (d) 400 kcmil

12.15 Grounding and Bonding

26. The connection of the system bonding jumper for a separately derived system shall be made _____ on the separately derived system from the source to the first system disconnecting means or overcurrent device.

    (a) in at least two locations
    (b) in every location that the grounded conductor is present
    (c) at any single point
    (d) none of these

27. The grounding electrode conductor for a single separately derived system is used to connect the grounded conductor of the derived system to the grounding electrode.

    (a) True
    (b) False

28. In an area served by a separately derived system, the _____ shall be connected to the grounded conductor of the separately derived system.

    (a) structural steel
    (b) metal piping
    (c) metal building skin
    (d) a and b

29. The size of the grounding electrode conductor shall not be smaller than that identified in _____, based on the area of the largest ungrounded supply conductor.

    (a) 250.66
    (b) 250.122
    (c) Table 310.16
    (d) none of these

30. In an ac system, the size of the grounding electrode conductor to a concrete-encased electrode shall not be required to be larger than a(n) _____ copper conductor.

    (a) 10 AWG
    (b) 8 AWG
    (c) 6 AWG
    (d) 4 AWG

12.16 Available Short-Circuit Current

31. Equipment intended to interrupt current at fault levels shall have an interrupting rating sufficient for the nominal circuit voltage and the current that is available at the line terminals of the equipment.

    (a) True
    (b) False

Unit 12 Challenge Questions

UNIT 12—TRANSFORMER CALCULATIONS CHALLENGE QUESTIONS

(•Indicates that 75% or fewer of those who took this exam answered the question correctly.)

PART A–GENERAL

12.11 Line Currents

1. •The primary line current of a three-phase, 37.50 kVA, delta/delta transformer with a 480V primary and a 120/240V secondary is closest to _____.

   (a) 45A
   (b) 50A
   (c) 65A
   (d) 90A

2. •What is the secondary line current for a 45 kVA delta/delta transformer that has a line voltage turns ratio of 2:1, if the primary operates at 460V?

   (a) 36A
   (b) 55A
   (c) 113A
   (d) 240A

3. •A 4,160V to 240V, three-phase transformer is connected delta/delta. The secondary phase current is 300A and the secondary line voltage is 240V. How much current will be flowing in the conductors supplying the transformer (line current)?

   (a) 17A
   (b) 20A
   (c) 30A
   (d) 50A

4. •What is the primary line current for a delta/wye 480V to 120/208V, three-phase transformer that has a secondary line current of 200A?

   (a) 31A
   (b) 53A
   (c) 72A
   (d) 87A

5. •The secondary line current for a wye-connected transformer is approximately _____ if the secondary phase voltage is 277V and the load is 10 kVA, three-phase.

   (a) 7A
   (b) 10A
   (c) 12A
   (d) 32A

6. •The VA load per line for a 15 hp, 208V, three-phase motor can be determined by which of these formulas?

   (a) 1.732 x E x I
   (b) (E x I)/1.732
   (c) E x I
   (d) E x 1.732

PART B–*NEC* REQUIREMENTS

12.12 Transformer Overcurrent Protection

7. •The maximum size overcurrent device for primary only protection of a three-phase, 37.50 kVA, delta/delta transformer with a 480V primary and a 120/240V secondary is _____.

   (a) 40A
   (b) 45A
   (c) 50A
   (d) 60A

Unit 12 — Challenge Questions

8. The maximum size overcurrent device for primary only protection of a three-phase 45 kVA transformer with a 480V primary and a 120/208V secondary is _____ .

   (a) 50A
   (b) 60A
   (c) 70A
   (d) 90A

9. The maximum size overcurrent device for primary only protection of a three-phase 75 kVA transformer with a 480V primary and a 120/208V secondary is _____ .

   (a) 90A
   (b) 100A
   (c) 125A
   (d) 150A

10. The maximum size overcurrent device for primary only protection of a three-phase 112.50 kVA transformer with a 480V primary and a 120/208V secondary is _____ .

    (a) 100A
    (b) 125A
    (c) 150A
    (d) 175A

CHAPTER 4

NEC Practice Quizzes, Challenge Quizzes, and Final NEC Exams

- **NEC Practice Quizzes**
- **NEC Challenge Quizzes**
- **Final NEC Exams**

The first three chapters of this book focused on the theory and calculation topics necessary to pass an exam. This chapter provides a review of *Code* requirements. There are ten sets of Practice Questions in Straight Order and 10 sets of Practice Questions in Random Order that you should complete to begin with. If you are short on time before your exam, complete all of the straight order questions first, then return to the random order questions. After you have done the practice questions to provide a good *Code* review, you should next complete the seven Challenge Quizzes to hone your skills in finding *Code* requirements. To complete your review, there are two 100 question Final Exams that will help you evaluate your exam taking skills.

Answer these questions from the 2008 *NEC* book. Practice using your *Code* book for the straight-order questions, and then as you progress, you may want to see how many of the questions you can answer from memory. Most licensing exams are open book, but check with your testing authority in your region.

Mike Holt's Understanding the National Electrical Code Volume 1 and *Mike Holt's Understanding the National Electrical Code Volume 2* will help if you feel you need more *Code* review.

Chapter 4 Notes

Unit 1: Practice Questions in Straight Order—Article 90.1 through 225.6

Use the 2008 *NEC* to answer the following questions.

1. The *Code* contains provisions considered necessary for safety, which will not necessarily result in ___.

   (a) efficient use
   (b) convenience
   (c) good service or future expansion of electrical use
   (d) all of these

2. The *NEC* applies to the installation of _____.

   (a) electrical conductors and equipment within or on public and private buildings
   (b) outside conductors and equipment on the premises
   (c) optical fiber cables
   (d) all of these

3. Installations of communications equipment that are under the exclusive control of communications utilities, and located outdoors or in building spaces used exclusively for such installations _____ covered by the *NEC*.

   (a) are
   (b) are sometimes
   (c) are not
   (d) may

4. Electric utilities may include entities that install, operate, and maintain _____.

   (a) communications systems (telephone, CATV, Internet, satellite, or data services)
   (b) electric supply systems (generation, transmission, or distribution systems)
   (c) local area network wiring on premises
   (d) a or b

5. Chapters 1 through 4 of the *NEC* apply _____.

   (a) generally to all electrical installations
   (b) only to special occupancies and conditions
   (c) only to special equipment and material
   (d) all of these

6. The authority having jurisdiction has the responsibility _____.

   (a) for making interpretations of rules
   (b) for deciding upon the approval of equipment and materials
   (c) for waiving specific requirements in the *Code* and permitting alternate methods and material if safety is maintained
   (d) all of these

7. If the *NEC* requires new products that are not yet available at the time a new edition is adopted, the _____ may permit the use of the products that comply with the previous edition of the code adopted by that jurisdiction.

   (a) electrical engineer
   (b) master electrician
   (c) authority having jurisdiction
   (d) permit holder

8. Factory-installed _____ wiring of equipment need not be inspected at the time of installation of the equipment, except to detect alterations or damage.

   (a) external
   (b) associated
   (c) internal
   (d) all of these

9. Admitting close approach, not guarded by locked doors, elevation, or other effective means, is referred to as _____.

   (a) accessible (as applied to equipment)
   (b) accessible (as applied to wiring methods)
   (c) accessible, readily
   (d) all of these

Unit 1 — Practice Questions in Straight Order—Article 90.1 through 225.6

10. Capable of being removed or exposed without damaging the building structure or finish, or not permanently closed in by the structure or finish of the building is known as _____.

    (a) accessible (as applied to equipment)
    (b) accessible (as applied to wiring methods)
    (c) accessible, readily
    (d) all of these

11. Capable of being reached quickly for operation, renewal, or inspections without resorting to portable ladders and such is known as _____.

    (a) accessible (as applied to equipment)
    (b) accessible (as applied to wiring methods)
    (c) accessible, readily
    (d) all of these

12. A device that establishes a connection between the conductors of the attached flexible cord and the conductors connected to the receptacle is called a(n) _____.

    (a) attachment plug
    (b) plug cap
    (c) plug
    (d) any of these

13. The conductors between the final overcurrent device protecting the circuit and the outlet(s) are known as _____ conductors.

    (a) feeder
    (b) branch-circuit
    (c) home run
    (d) none of these

14. An enclosure for either surface mounting or flush mounting provided with a frame in which a door can be hung is called a(n) _____.

    (a) enclosure
    (b) outlet box
    (c) cutout box
    (d) cabinet

15. A circuit breaker is a device designed to _____ the circuit automatically on a predetermined overcurrent without damage to itself when properly applied within its rating.

    (a) blow
    (b) disconnect
    (c) connect
    (d) open

16. _____ is a term indicating that there is an intentional delay in the tripping action of the circuit breaker, which decreases as the magnitude of the current increases.

    (a) Adverse time
    (b) Inverse time
    (c) Time delay
    (d) Timed unit

17. Communications equipment includes equipment used for the transmission of _____.

    (a) audio
    (b) video
    (c) data
    (d) all of these

18. A(n) _____ is a device by which the conductors of a circuit can be disconnected from their source of supply.

    (a) feeder
    (b) enclosure
    (c) disconnecting means
    (d) conductor interrupter

19. A fixed, stationary, or portable self-contained, electrically illuminated equipment with words or symbols designed to convey information or attract attention describes _____.

    (a) an electric sign
    (b) equipment
    (c) appliances
    (d) none of these

20. As used in the *NEC*, equipment includes _____.

    (a) fittings
    (b) appliances
    (c) machinery
    (d) all of these

21. A _____ is a building or portion of a building in which one or more self-propelled vehicles can be kept for use, sale, storage, rental, repair, exhibition, or demonstration purposes.

    (a) garage
    (b) residential garage
    (c) service garage
    (d) commercial garage

Practice Questions in Straight Order—Article 90.1 through 225.6 — Unit 1

22. The word "Earth" best describes what *NEC* term?

    (a) Bonded
    (b) Ground
    (c) Effective ground-fault current path
    (d) Guarded

23. "Connected to ground or to a conductive body that extends the ground connection" is called _____.

    (a) equipment grounding
    (b) bonded
    (c) grounded
    (d) all of these

24. A "Class A" GFCI protection device is designed to trip when the ground-fault current to ground is _____ or higher.

    (a) 4 mA
    (b) 5 mA
    (c) 6 mA
    (d) none of these

25. A conducting object through which a direct connection to earth is established is a _____.

    (a) bonding conductor
    (b) grounding conductor
    (c) grounding electrode
    (d) grounded conductor

26. A handhole enclosure is an enclosure for use in underground systems, provided with an open or closed bottom, and sized to allow personnel to _____.

    (a) enter and exit freely
    (b) reach into but not enter
    (c) have full working space
    (d) examine it visually

27. Within sight means visible and not more than _____ distant from the equipment.

    (a) 10 ft
    (b) 20 ft
    (c) 25 ft
    (d) 50 ft

28. A _____ location may be temporarily subject to dampness and wetness.

    (a) dry
    (b) damp
    (c) moist
    (d) wet

29. Conduit installed underground or encased in concrete slabs that are in direct contact with the earth is considered a _____ location.

    (a) dry
    (b) damp
    (c) wet
    (d) moist

30. A(n) _____ is a point on the wiring system at which current is taken to supply utilization equipment.

    (a) box
    (b) receptacle
    (c) outlet
    (d) device

31. An opening in an outlet box where one or more receptacles have been installed is called _____.

    (a) a device
    (b) equipment
    (c) a receptacle
    (d) a receptacle outlet

32. When one electrical circuit controls another circuit through a relay, the first circuit is called a _____.

    (a) control circuit
    (b) remote-control circuit
    (c) signal circuit
    (d) controller

33. The conductors from the electric utility that deliver electric energy to the premises is called _____.

    (a) branch circuit
    (b) feeder
    (c) service
    (d) none of these

34. Conductors shall be _____ unless otherwise provided.

    (a) bare
    (b) stranded
    (c) copper
    (d) aluminum

35. Only wiring methods recognized as _____ are included in this *Code*.

    (a) expensive
    (b) efficient
    (c) suitable
    (d) cost-effective

36. Unused openings other than those intended for the operation of equipment, intended for mounting purposes, or permitted as part of the design for listed equipment shall be _____.

    (a) filled with cable clamps or connectors only
    (b) taped over with electrical tape
    (c) repaired only by welding or brazing in a metal slug
    (d) effectively closed to afford protection substantially equivalent to the wall of the equipment

37. Many terminations and equipment are marked with _____.

    (a) an etching tool
    (b) a removable label
    (c) a tightening torque
    (d) the manufacturer's initials

38. Conductor ampacity shall be determined using the _____ column of Table 310.16 for circuits rated 100A or less or marked for 14 AWG through 1 AWG conductors, unless the equipment terminals are listed for use with conductors that have higher temperature ratings.

    (a) 30°C
    (b) 60°C
    (c) 75°C
    (d) 90°C

39. For circuits rated 100A or less, when the equipment terminals are listed for use with 75°C conductors, the _____ column of Table 310.16 shall be used to determine the ampacity of THHN conductors installed.

    (a) 30°C
    (b) 60°C
    (c) 75°C
    (d) 90°C

40. Conductors shall have their ampacity determined using the _____ column of Table 310.16 for circuits rated over 100A, or marked for conductors larger than 1 AWG, unless the equipment terminals are listed for use with higher temperature rated conductors.

    (a) 30°C
    (b) 60°C
    (c) 75°C
    (d) 90°C

41. On a 4-wire, delta-connected system where the midpoint of one phase winding is grounded, the conductor having the higher phase voltage to ground shall be durably and permanently marked by an outer finish that is _____ in color.

    (a) black
    (b) red
    (c) blue
    (d) orange

42. Electrical equipment such as switchboards, panelboards, industrial control panels, meter socket enclosures, and motor control centers in commercial and industrial occupancies that are likely to require _____ while energized shall be field-marked to warn qualified persons of the danger associated with an arc flash.

    (a) examination
    (b) adjustment
    (c) servicing or maintenance
    (d) any of these

43. The term rainproof is typically used in conjunction with Enclosure-Type Number _____.

    (a) 3
    (b) 3R
    (c) 3RX
    (d) a and b

44. The required working space for access to live parts operating at 300 volts-to-ground, where there are exposed live parts on one side and grounded parts on the other side, is _____.

    (a) 3 ft
    (b) 3½ ft
    (c) 4 ft
    (d) 4½ ft

45. The dimension of working space for access to live parts operating at 300V, nominal-to-ground, where there are exposed live parts on both sides of the workspace is _____ according to Table 110.26(A)(1).

    (a) 3 ft
    (b) 3½ ft
    (c) 4 ft
    (d) 4½ ft

46. Working space distances for enclosed live parts shall be measured from the _____ of equipment or apparatus, if the live parts are enclosed.

    (a) enclosure
    (b) opening
    (c) a or b
    (d) none of these

47. The working space in front of the electric equipment shall not be less than _____ wide, or the width of the equipment, whichever is greater.

    (a) 15 in.
    (b) 30 in.
    (c) 40 in.
    (d) 60 in.

48. Equipment associated with the electrical installation can be located above or below other electrical equipment within their working space when the associated equipment does not extend more than _____ from the front of the electrical equipment.

    (a) 3 in.
    (b) 6 in.
    (c) 12 in.
    (d) 30 in.

49. When normally enclosed live parts are exposed for inspection or servicing, the working space, if in a passageway or general open space, shall be suitably _____.

    (a) accessible
    (b) guarded
    (c) open
    (d) enclosed

50. For equipment rated 1,200A or more and over 6 ft wide that contains overcurrent devices, switching devices, or control devices, there shall be one entrance to and egress from the required working space not less than 24 in. wide and _____ high at each end of the working space.

    (a) 5 ft 6 in.
    (b) 6 ft
    (c) 6 ft 6 in.
    (d) any of these

51. The minimum headroom of working spaces about motor control centers is _____.

    (a) 3 ft
    (b) 5 ft
    (c) 6 ft
    (d) 6½ ft

52. The dedicated equipment space for electrical equipment that is required for panelboards is measured from the floor to a height of _____ above the equipment, or to the structural ceiling, whichever is lower.

    (a) 3 ft
    (b) 6 ft
    (c) 12 ft
    (d) 30 ft

53. In locations where electrical equipment is likely to be exposed to _____, enclosures or guards shall be so arranged and of such strength as to prevent such damage.

    (a) corrosion
    (b) physical damage
    (c) magnetic fields
    (d) weather

54. An insulated grounded conductor of _____ or smaller shall be identified by a continuous white or gray outer finish, or by three continuous white stripes on other than green insulation along its entire length.

    (a) 8 AWG
    (b) 6 AWG
    (c) 4 AWG
    (d) 3 AWG

55. Grounded conductors larger than _____ can be identified by distinctive white or gray markings at their terminations.

    (a) 10 AWG
    (b) 8 AWG
    (c) 6 AWG
    (d) 4 AWG

56. Receptacles shall have the terminal intended for connection to the grounded conductor identified by a metal or metal coating that is substantially _____ in color.

    (a) green
    (b) white
    (c) gray
    (d) b or c

57. The rating of a branch circuit shall be determined by the rating of the _____.

    (a) ampacity of the largest device connected to the circuit
    (b) average of the ampacity of all devices
    (c) branch-circuit overcurrent device
    (d) ampacity of the branch-circuit conductors according to Table 310.16

58. Each multiwire branch circuit shall be provided with a means that will simultaneously disconnect all _____ conductors at the point where the branch circuit originates.

    (a) circuit
    (b) grounded
    (c) grounding
    (d) ungrounded

59. Where two or more branch circuits supply devices or equipment on the same yoke, a means to disconnect simultaneously all ungrounded conductors that supply those devices or equipment shall be provided _____.

    (a) at the point where the branch circuit originates
    (b) at the location of the device or equipment
    (c) at the point where the feeder originates
    (d) none of these

60. GFCI protection shall be provided for all 15A and 20A, 125V receptacles installed in a dwelling unit _____.

    (a) attic
    (b) garage
    (c) laundry room
    (d) all of these

61. Two or more _____ small-appliance branch circuits shall be provided to supply power for receptacle outlets in the dwelling unit kitchen, dining room, breakfast room, pantry, or similar dining areas.

    (a) 15A
    (b) 20A
    (c) 30A
    (d) either 20A or 30A

62. There shall be a minimum of one _____ branch circuit for the laundry outlet(s) required by 210.52(F).

    (a) 15A
    (b) 20A
    (c) 30A
    (d) b and c

63. The recommended maximum total voltage drop on branch-circuit conductors is _____ percent.

    (a) 2
    (b) 3
    (c) 4
    (d) 6

64. When connected to a branch circuit supplying two or more 15A receptacles, each receptacle shall not supply a total cord-and-plug-connected load in excess of _____.

    (a) 12A
    (b) 16A
    (c) 20A
    (d) 24A

65. If a 20A branch circuit supplies multiple receptacles, the receptacles shall have an ampere rating of no less than _____.

    (a) 10A
    (b) 15A
    (c) 20A
    (d) 30A

66. The total rating of utilization equipment fastened in place, shall not exceed _____ percent of the branch-circuit ampere rating where lighting units and cord-and-plug-connected utilization equipment are supplied.

    (a) 50
    (b) 75
    (c) 100
    (d) 125

67. In multi-occupancy buildings, branch circuits for _____ shall not originate from equipment that supplies an individual dwelling unit or tenant space.

    (a) central alarm
    (b) parking lot lighting
    (c) common area receptacles
    (d) all of these

68. When applying the general provisions for receptacle spacing to the rooms of a dwelling unit which require receptacles in the wall space, no point along the floor line in any wall space of a dwelling unit may be more than _____ from an outlet.

    (a) 6 ft
    (b) 8 ft
    (c) 10 ft
    (d) 12 ft

69. In a dwelling unit, each wall space _____ or wider requires a receptacle.

    (a) 2 ft
    (b) 3 ft
    (c) 4 ft
    (d) 5 ft

70. In dwelling units, when determining the spacing of receptacle outlets, _____ on exterior walls shall not be considered wall space.

    (a) fixed panels
    (b) fixed glass
    (c) sliding panels
    (d) all of these

71. A receptacle outlet shall be installed at each dwelling unit kitchen wall countertop space that is 12 in. or wider and receptacle outlets shall be installed so that no point along the wall line is more than _____, measured horizontally from a receptacle outlet in that space.

    (a) 10 in.
    (b) 12 in.
    (c) 16 in.
    (d) 24 in.

72. In dwelling units, at least one receptacle outlet shall be installed at each peninsular countertop having a long dimension of _____ in. or greater, and a short dimension of _____ in. or greater.

    (a) 12, 24
    (b) 24, 12
    (c) 24, 48
    (d) 48, 24

73. Kitchen and dining room countertop receptacle outlets in dwelling units shall be installed above the countertop surface, and not more than ___ above the countertop.

    (a) 12 in.
    (b) 18 in.
    (c) 20 in.
    (d) 24 in.

74. In dwelling units, the required bathroom receptacle outlet can be installed on the side or face of the basin cabinet if no lower than _____ below the countertop.

    (a) 12 in.
    (b) 18 in.
    (c) 24 in.
    (d) 36 in.

75. At least one receptacle outlet not more than _____ above a balcony, deck or porch shall be installed within the perimeter at each balcony, deck or porch of a dwelling unit.

    (a) 3 ft
    (b) 6½ ft
    (c) 8 ft
    (d) 24 in.

76. For a one-family dwelling, at least one receptacle outlet shall be required in each _____.

    (a) basement
    (b) attached garage
    (c) detached garage with electric power
    (d) all of these

77. At least one receptacle outlet shall be installed within 18 in. of the top of a show window for each _____, or major fraction thereof, of show-window area measured horizontally at its maximum width.

    (a) 10 ft
    (b) 12 ft
    (c) 18 ft
    (d) 24 ft

78. A 15A or 20A, 125V receptacle outlet shall be located within 25 ft of heating, air-conditioning, and refrigeration equipment for _____ occupancies.

    (a) dwelling
    (b) commercial
    (c) industrial
    (d) all of these

79. At least one wall switch-controlled lighting outlet shall be installed in every habitable room and bathroom of a guest room or guest suite of hotels, motels, and similar occupancies. A receptacle outlet controlled by a wall switch may be used to meet this requirement in other than _____.

    (a) bathrooms
    (b) kitchens
    (c) sleeping areas
    (d) a and b

UNIT 1 — Practice Questions in Straight Order—Article 90.1 through 225.6

80. For attics and underfoot spaces containing heating, air-conditioning, and refrigeration equipment, at least one lighting outlet containing a switch or controlled by a wall switch shall be installed _____.

    (a) at the equipment requiring servicing
    (b) near the equipment requiring servicing
    (c) a or b
    (d) none of these

81. When a feeder supplies _____ in which equipment grounding conductors are required, the feeder shall include or provide an equipment grounding conductor.

    (a) equipment disconnecting means
    (b) electrical systems
    (c) branch circuits
    (d) electric-discharge lighting equipment

82. On a 4-wire, delta-connected, three-phase system, where the midpoint of one phase winding is grounded, the conductor with the higher phase voltage-to-ground must be durably and permanently marked by an outer finish _____ in color, or other effective means.

    (a) brown
    (b) orange
    (c) yellow
    (d) all of these

83. Where a premises wiring system contains feeders supplied from more than one nominal voltage system, each ungrounded conductor of a feeder shall be identified by phase and system by _____, or other approved means.

    (a) color coding
    (b) marking tape
    (c) tagging
    (d) any of these

84. When calculations in Article 220 result in a fraction of an ampere that is less than _____, such fractions can be dropped.

    (a) 0.49
    (b) 0.50
    (c) 0.51
    (d) 0.80

85. Where fixed multioutlet assemblies are used in other than dwelling units or the guest rooms of hotels or motels, each _____ or fraction thereof of each separate and continuous length shall be considered as 180 VA where appliances are unlikely to be used simultaneously.

    (a) 5 ft
    (b) 5½ ft
    (c) 6 ft
    (d) 6½ ft

86. For other than dwelling occupancies, banks, or office buildings, each receptacle outlet shall be calculated at not less than _____ VA.

    (a) 90
    (b) 180
    (c) 270
    (d) 360

87. The minimum feeder load for show-window lighting is _____ per-linear-foot.

    (a) 180 VA
    (b) 200 VA
    (c) 300 VA
    (d) 400 VA

88. For other than dwelling units or guest rooms of hotels or motels, the feeder and service calculation for track lighting shall be calculated at 150 VA for every _____ or portion thereof of track installed.

    (a) 1 ft
    (b) 2 ft
    (c) 3 ft
    (d) 4 ft

89. Feeder and service loads for fixed electric space heating shall be calculated at _____ percent of the total connected load.

    (a) 80
    (b) 100
    (c) 125
    (d) 200

90. A dwelling unit containing three 120V small-appliance branch circuits has a calculated load of _____ VA for the small appliance circuits.

    (a) 1,500
    (b) 3,000
    (c) 4,500
    (d) 6,000

91. The load for electric clothes dryers in a dwelling unit shall be _____ watts or the nameplate rating, whichever is larger, per dryer.

    (a) 1,500
    (b) 4,500
    (c) 5,000
    (d) 8,000

92. Using the standard load calculation method, the feeder demand factor for five household clothes dryers is _____ percent.

    (a) 50
    (b) 70
    (c) 85
    (d) 100

93. To determine the feeder calculated load for ten 3 kW household cooking appliances, use _____ of Table 220.55.

    (a) Column A
    (b) Column B
    (c) Column C
    (d) none of these

94. The feeder/service calculated load for a multifamily dwelling containing nine 12 kW ranges is _____.

    (a) 13,000W
    (b) 14,700W
    (c) 16,000W
    (d) 24,000W

95. Table 220.56 may be applied to determine the load for thermostatically controlled or intermittently used _____ and other kitchen equipment in a commercial kitchen.

    (a) commercial electric cooking equipment
    (b) dishwasher booster heaters
    (c) water heaters
    (d) all of these

96. The maximum unbalanced feeder load for household electric ranges, wall-mounted ovens, and counter-mounted cooking units shall be considered as _____ percent of the load on the ungrounded conductors.

    (a) 50
    (b) 70
    (c) 85
    (d) 115

97. There shall be no reduction in the size of the neutral or grounded conductor on _____ loads supplied from a 4-wire, wye connected, three-phase system.

    (a) dwelling unit
    (b) hospital
    (c) nonlinear
    (d) motel

98. Under the optional method for calculating a single-family dwelling service, general loads beyond the initial 10 kVA are assessed at a _____ percent demand factor.

    (a) 40
    (b) 50
    (c) 60
    (d) 75

99. A demand factor of _____ percent applies to a multi-family dwelling with ten units if the optional calculation method is used.

    (a) 43
    (b) 50
    (c) 60
    (d) 75

100. Open individual conductors shall not be smaller than _____ AWG copper for spans up to 50 ft in length and _____ AWG copper for a longer span, unless supported by a messenger wire.

    (a) 10, 8
    (b) 8, 8
    (c) 6, 8
    (d) 6, 6

Unit 1: Practice Questions in Random Order—Article 100 through 215.10

Use the 2008 *NEC* to answer the following questions.

1. "Connected to ground without the insertion of any resistor or impedance device" is _____.

    (a) grounded
    (b) solidly grounded
    (c) effectively grounded
    (d) grounding conductor

2. The white conductor within a cable can be used for the _____ conductor where permanently reidentified to indicate its use as an ungrounded conductor at each location where the conductor is visible and accessible.

    (a) grounded
    (b) ungrounded
    (c) a and b
    (d) none of these

3. The screw shell of a luminaire or lampholder shall be connected to the _____.

    (a) grounded conductor
    (b) ungrounded conductor
    (c) equipment grounding conductor
    (d) forming shell terminal

4. No _____ shall be attached to any terminal or lead so as to reverse designated polarity.

    (a) grounded conductor
    (b) grounding conductor
    (c) ungrounded conductor
    (d) grounding connector

5. A system intended to provide protection of equipment from damaging line-to-ground fault currents by opening all ungrounded conductors of the faulted circuit at current levels less than the supply circuit overcurrent device defines the term _____.

    (a) ground-fault protection of equipment
    (b) guarded
    (c) personal protection
    (d) automatic protection

6. A neutral conductor is the conductor connected to the _____ of a system, which is intended to carry neutral current under normal conditions.

    (a) grounding electrode
    (b) neutral point
    (c) intersystem bonding termination
    (d) none of these

7. The installed conductive path that connects normally noncurrent-carrying metal parts of equipment together and to the system grounded conductor or to the grounding electrode conductor, or both, is known as a(n) _____.

    (a) grounding electrode conductor
    (b) grounding conductor
    (c) equipment grounding conductor
    (d) none of these

8. A _____ is an accommodation that combines living, sleeping, sanitary, and storage facilities within a compartment.

    (a) guest room
    (b) guest suite
    (c) dwelling unit
    (d) single-family dwelling

Practice Questions in Random Order—Article 100 through 215.10 — Unit 1

9. A _____ is an accommodation with two or more contiguous rooms comprising a compartment that provides living, sleeping, sanitary, and storage facilities.

   (a) guest room
   (b) guest suite
   (c) dwelling unit
   (d) single-family dwelling

10. All 15A and 20A, 125V receptacles installed in bathrooms of _____ shall have ground-fault circuit-interrupter (GFCI) protection for personnel.

    (a) guest rooms in hotels/motels
    (b) dwelling units
    (c) office buildings
    (d) all of these

11. A clothes closet is defined as a _____ room or space intended primarily for storage of garments and apparel.

    (a) habitable
    (b) nonhabitable
    (c) conditioned
    (d) finished

12. GFCI protection shall be provided for all 15A and 20A, 125V receptacles in dwelling unit accessory buildings that have a floor located at or below grade level not intended as _____ and limited to storage areas, work areas, or similar use.

    (a) habitable rooms
    (b) finished space
    (c) a or b
    (d) none of these

13. Article 200 contains the requirements for _____.

    (a) identification of terminals
    (b) grounded conductors in premises wiring systems
    (c) identification of grounded conductors
    (d) all of these

14. Acceptable to the authority having jurisdiction means _____.

    (a) identified
    (b) listed
    (c) approved
    (d) labeled

15. All 15A and 20A, 125V receptacles _____ of commercial occupancies shall have GFCI protection.

    (a) in bathrooms
    (b) on rooftops
    (c) in kitchens
    (d) all of these

16. Conductor sizes are expressed in American Wire Gage (AWG) or _____.

    (a) inches
    (b) circular mils
    (c) square inches
    (d) AWG

17. Cables are considered _____ if rendered inaccessible by the structure or finish of the building.

    (a) inaccessible
    (b) concealed
    (c) hidden
    (d) enclosed

18. Outline lighting may include an arrangement of _____ to outline or call attention to the shape of a building.

    (a) incandescent lamps
    (b) electric-discharge lighting
    (c) electrically powered light sources
    (d) any of these

19. Concrete, brick, or tile walls are considered _____, as it applies to working space requirements.

    (a) inconsequential
    (b) in the way
    (c) grounded
    (d) none of these

20. A branch circuit that supplies only one utilization equipment is a(n) _____ branch circuit.

    (a) individual
    (b) general-purpose
    (c) isolated
    (d) special-purpose

21. The *NEC* defines a(n) _____ as one who has skills and knowledge related to the construction and operation of the electrical equipment and installations and has received safety training to recognize and avoid the hazards involved.

    (a) inspector
    (b) master electrician
    (c) journeyman electrician
    (d) qualified person

22. The *NEC* does not apply to electric utility-owned wiring and equipment _____.

    (a) installed by an electrical contractor
    (b) installed on public property
    (c) consisting of service drops installed by a utility
    (d) in a utility office building

23. The *NEC* requires that electrical work be _____.

    (a) installed in a neat and workmanlike manner
    (b) installed under the supervision of a qualified person
    (c) completed before being inspected
    (d) all of these

24. GFCI protection shall be provided for all 15A and 20A, 125V receptacles _____ in dwelling unit kitchens.

    (a) installed to serve the countertop surfaces
    (b) within 6 ft of the sink
    (c) for all receptacles
    (d) that are readily accessible

25. Premises wiring includes _____ wiring from the service point or power source to the outlets.

    (a) interior
    (b) exterior
    (c) underground
    (d) all of these

26. The highest current at rated voltage that a device is intended to interrupt under standard test conditions is the _____.

    (a) interrupting rating
    (b) manufacturer's rating
    (c) interrupting capacity
    (d) withstand rating

27. For the purpose of determining the placement of receptacles in a dwelling unit kitchen, a(n) _____ countertop is measured from the connecting edge.

    (a) island
    (b) usable
    (c) peninsular
    (d) cooking

28. Bonded can be described as _____ to establish electrical continuity and conductivity.

    (a) isolated
    (b) guarded
    (c) connected
    (d) separated

29. A separate portion of a raceway system that provides access through a removable cover to the interior of the system, defines the term _____.

    (a) junction box
    (b) accessible raceway
    (c) conduit body
    (d) cutout box

30. "Recognized as suitable for the specific purpose, function, use, environment, and application" is the definition of _____.

    (a) labeled
    (b) identified (as applied to equipment)
    (c) listed
    (d) approved

31. GFCI protection shall be provided for all 15A and 20A, 125V receptacles located within 6 ft of a dwelling unit _____.

    (a) laundry sink
    (b) utility sink
    (c) wet bar sink
    (d) all of these

32. In rooms other than kitchens and bathrooms of dwelling units, one or more receptacles controlled by a wall switch shall be permitted in lieu of _____.

    (a) lighting outlets
    (b) luminaires
    (c) the receptacles required by 210.52(b) and (D)
    (d) all of these

33. A device that provides a means to connect _____ systems grounding and bonding conductors to the building grounding electrode system is an intersystem bonding termination.

    (a) limited energy
    (b) low-voltage
    (c) communications
    (d) power and lighting

34. A kitchen is defined as an area with a sink and _____ facilities for food preparation and cooking.

    (a) listed
    (b) labeled
    (c) temporary
    (d) permanent

35. Equipment or materials to which a symbol or other identifying mark of a product evaluation organization that is acceptable to the authority having jurisdiction has been attached is known as _____.

    (a) listed
    (b) labeled
    (c) approved
    (d) identified

36. Conductor terminal and splicing devices must be _____ for the conductor material and they must be properly installed and used.

    (a) listed
    (b) approved
    (c) identified
    (d) any of these

37. The _____ of any system is the ratio of the maximum demand of a system, or part of a system, to the total connected load of a system.

    (a) load
    (b) demand factor
    (c) minimum load
    (d) calculated factor

38. Ground-fault protection of equipment shall not be required at a feeder disconnect if ground-fault protection of equipment is provided on the _____ side of the feeder and on the load side of any transformer supplying the feeder.

    (a) load
    (b) supply
    (c) service
    (d) none of these

39. Unless identified for use in the operating environment, no conductors or equipment shall be _____ having a deteriorating effect on the conductors or equipment.

    (a) located in damp or wet locations
    (b) exposed to fumes, vapors, or gases
    (c) exposed to excessive temperatures
    (d) all of these

40. The voltage of a circuit is defined by the *Code* as the _____ root-mean-square (effective) difference of potential between any two conductors of the circuit.

    (a) lowest
    (b) greatest
    (c) average
    (d) nominal

41. In dwelling units, the voltage between conductors that supply the terminals of _____ shall not exceed 120V, nominal.

    (a) luminaires
    (b) cord-and-plug-connected loads of 1,440 VA or less
    (c) cord-and-plug-connected loads of more than ¼ hp
    (d) a and b

42. The connection between the grounded conductor and the equipment grounding conductor at the service is accomplished by installing a(n) _____ jumper.

    (a) main bonding
    (b) system bonding
    (c) equipment bonding
    (d) circuit bonding

43. A conductor used to connect the system grounded conductor or equipment to a grounding electrode or to a point on the grounding electrode system is called the _____ conductor.

    (a) main grounding
    (b) common main
    (c) equipment grounding
    (d) grounding electrode

44. In dwelling units, lighting outlets can be controlled by occupancy sensors where equipped with a _____ that will allow the sensor to function as a wall switch.

    (a) manual override
    (b) photo cell
    (c) sensor
    (d) none of these

45. The _____ has the responsibility for deciding on the approval of equipment and materials.

    (a) manufacturer
    (b) authority having jurisdiction
    (c) testing agency
    (d) none of these

46. In judging equipment for approval, considerations such as the following shall be evaluated:

    (a) mechanical strength
    (b) wire-bending space
    (c) arcing effects
    (d) all of these

47. The continuity of a grounded conductor shall not depend on a connection to a _____.

    (a) metallic enclosure
    (b) raceway
    (c) cable armor
    (d) any of these

48. Receptacles installed in a dwelling unit kitchen to serve countertop surfaces shall be supplied by not fewer than _____ small-appliance branch circuits.

    (a) one
    (b) two
    (c) three
    (d) no minimum

49. There shall be a minimum of _____ receptacle(s) installed outdoors at a one-family or two-family dwelling unit.

    (a) one
    (b) two
    (c) three
    (d) four

50. The number of receptacle outlets for guest rooms in hotels and motels shall not be less than that required for a dwelling unit. These receptacles can be located to be convenient for permanent furniture layout, but at least _____ receptacle outlets shall be readily accessible.

    (a) one
    (b) two
    (c) three
    (d) four

Unit 2

Practice Questions in Straight Order—Article 225.16 through 250.102

Use the 2008 *NEC* to answer the following questions.

1. The point of attachment of overhead premises wiring to a building shall in no case be less than _____ above finished grade.

   (a) 8 ft
   (b) 10 ft
   (c) 12 ft
   (d) 15 ft

2. Overhead feeder conductors installed over roofs shall have a vertical clearance of _____ above the roof surface, unless permitted by an exception.

   (a) 3 ft
   (b) 8 ft
   (c) 12 ft
   (d) 15 ft

3. If a set of 120/240V overhead feeder conductors terminates at a through-the-roof raceway or approved support, with less than 6 ft of these conductors passing over the roof overhang, the minimum clearance above the roof for these conductors is _____.

   (a) 12 in.
   (b) 18 in.
   (c) 2 ft
   (d) 5 ft

4. The requirement for maintaining a 3 ft vertical clearance from the edge of the roof shall not apply to the final feeder conductor span where the conductors are attached to _____.

   (a) a building pole
   (b) the side of a building
   (c) an antenna
   (d) the base of a building

5. Overhead feeder conductors to a building shall have a vertical clearance of final spans above, or within _____ measured horizontally from the platforms, projections, or surfaces from which they might be reached.

   (a) 3 ft
   (b) 6 ft
   (c) 8 ft
   (d) 10 ft

6. Raceways on exterior surfaces of buildings or other structures shall be arranged to drain, and in _____ locations shall be raintight.

   (a) damp
   (b) wet
   (c) dry
   (d) all of these

7. Vegetation such as trees shall not be used for support of _____.

   (a) conductor spans
   (b) surface wiring methods
   (c) luminaires
   (d) electric equipment

8. The disconnecting means for a building supplied by a feeder shall be installed at a(n) _____ location.

   (a) accessible
   (b) readily accessible
   (c) outdoor
   (d) indoor

9. In a multiple-occupancy building, each occupant shall have access to his or her own _____.

   (a) disconnecting means
   (b) building drops
   (c) building-entrance assembly
   (d) lateral conductors

Unit 2 — Practice Questions in Straight Order—Article 225.16 through 250.102

10. When the disconnecting means for a building supplied by a feeder is a power-operated switch or circuit breaker, it shall be able to be opened by hand in the event of a _____.

    (a) ground fault
    (b) short circuit
    (c) power surge
    (d) power failure

11. Additional services shall be permitted for different voltages, frequencies, or phases, or for different uses such as for _____.

    (a) gymnasiums
    (b) different rate schedules
    (c) flea markets
    (d) special entertainment events

12. Service conductors supplying a building or other structure shall not _____ of another building or other structure.

    (a) be installed on the exterior walls
    (b) pass through the interior
    (c) a and b
    (d) none of these

13. Service conductors installed as unjacketed multiconductor cable shall have a minimum clearance of _____ from windows that are designed to be opened, doors, porches, stairs, fire escapes, or similar locations.

    (a) 3 ft
    (b) 4 ft
    (c) 6 ft
    (d) 10 ft

14. Overhead service conductors to a building shall maintain a vertical clearance of final spans above, or within, _____ measured horizontally from the platforms, projections, or surfaces from which they might be reached.

    (a) 3 ft
    (b) 6 ft
    (c) 8 ft
    (d) 10 ft

15. The minimum size service-drop conductor permitted is _____ AWG copper or _____ AWG aluminum or copper-clad aluminum.

    (a) 8, 8
    (b) 8, 6
    (c) 6, 8
    (d) 6, 6

16. Service drops installed over roofs shall have a vertical clearance of _____ above the roof surface, unless a lesser distance is permitted by an exception.

    (a) 3 ft
    (b) 8 ft
    (c) 12 ft
    (d) 15 ft

17. If a set of 120/240V service-drop conductors terminates at a through-the-roof raceway or approved support, with less than 6 ft of these conductors passing over the roof overhang, the minimum clearance above the roof for these service conductors shall be _____.

    (a) 12 in.
    (b) 18 in.
    (c) 2 ft
    (d) 5 ft

18. The requirement to maintain a 3 ft vertical clearance from the edge of a roof does not apply to the final conductor span where the service drop is attached to _____.

    (a) a service pole
    (b) the side of a building
    (c) an antenna
    (d) the base of a building

19. Overhead service-drop conductors shall have a horizontal clearance of _____ from a pool.

    (a) 8 ft
    (b) 10 ft
    (c) 12 ft
    (d) 14 ft

20. The minimum point of attachment of the service-drop conductors to a building shall not be less than _____ above finished grade.

    (a) 8 ft
    (b) 10 ft
    (c) 12 ft
    (d) 15 ft

21. Where raceway-type service masts are used, all raceway fittings shall be _____ for use with service masts.

    (a) identified
    (b) approved
    (c) heavy-duty
    (d) listed

22. Service-lateral conductors shall have _____.

    (a) adequate mechanical strength
    (b) sufficient ampacity for the loads calculated
    (c) a and b
    (d) none of these

23. Underground copper service-lateral conductors shall not be smaller than _____ AWG copper.

    (a) 3
    (b) 4
    (c) 6
    (d) 8

24. Service-lateral conductors that supply power to limited loads of a single branch circuit shall not be smaller than _____.

    (a) 14 AWG copper
    (b) 12 AWG aluminum
    (c) 12 AWG copper
    (d) 12 AWG aluminum

25. Service-entrance conductors shall be sized not less than _____ percent of the continuous load, plus 100 percent of the noncontinuous load.

    (a) 100
    (b) 115
    (c) 125
    (d) 150

26. Service cables mounted in contact with a building shall be supported at intervals not exceeding _____.

    (a) 24 in.
    (b) 30 in.
    (c) 3 ft
    (d) 4 ft

27. Service heads shall be located _____, unless impracticable.

    (a) above the point of attachment
    (b) below the point of attachment
    (c) even with the point of attachment
    (d) none of these

28. To prevent moisture from entering service equipment, service-entrance conductors shall _____.

    (a) be connected to service-drop conductors below the level of the service head
    (b) have drip loops formed on the service-entrance conductors
    (c) a or b
    (d) a and b

29. A service disconnecting means shall be installed at a _____ location.

    (a) dry
    (b) readily accessible
    (c) outdoor
    (d) indoor

30. Each service disconnecting means shall be permanently _____ to identify it as a service disconnect.

    (a) identified
    (b) positioned
    (c) marked
    (d) none of these

31. In a multiple-occupancy building where electric service and electrical maintenance are provided by the building management under continuous building management supervision, the service disconnecting means can be accessible to authorized _____ only.

    (a) inspectors
    (b) tenants
    (c) management personnel
    (d) qualified persons

32. When the service disconnecting means is a power-operated switch or circuit breaker, it shall be able to be opened by hand in the event of a _____.

    (a) ground fault
    (b) short circuit
    (c) power surge
    (d) power supply failure

33. For installations that supply only limited loads of a single branch circuit, the service disconnecting means shall have a rating not less than _____.

    (a) 15A
    (b) 20A
    (c) 25A
    (d) 30A

34. For installations consisting of not more than two 2-wire branch circuits, the service disconnecting means shall have a rating of not less than _____.

    (a) 15A
    (b) 20A
    (c) 25A
    (d) 30A

35. Ground-fault protection of equipment shall be provided for solidly grounded wye electrical services of more than 150 volts-to-ground, but not exceeding 600V phase-to-phase for each service disconnecting means rated _____ or more.

    (a) 1,000A
    (b) 1,500A
    (c) 2,000A
    (d) 2,500A

36. Conductor overload protection shall not be required where the interruption of the _____ would create a hazard, such as in a material-handling magnet circuit or fire pump circuit. However, short-circuit protection is required.

    (a) circuit
    (b) line
    (c) phase
    (d) system

37. If the circuit's overcurrent device exceeds _____, the conductor ampacity must have a rating not less than the rating of the overcurrent device.

    (a) 800A
    (b) 1,000A
    (c) 1,200A
    (d) 2,000A

38. Overcurrent protection shall not exceed _____.

    (a) 15A for 14 AWG copper
    (b) 20A for 12 AWG copper
    (c) 30A for 10 AWG copper
    (d) all of these

39. Flexible cords used in listed extension cord sets shall be considered protected against overcurrent when used within/(in) _____.

    (a) indoor installations
    (b) unclassified locations
    (c) the extension cord's listing requirements
    (d) 50 ft of the branch-circuit panelboard

40. Which of the following is not standard size fuses or inverse time circuit breakers?

    (a) 45A
    (b) 70A
    (c) 75A
    (d) 80A

41. The standard ampere ratings for fuses includes _____.

    (a) 1A
    (b) 6A
    (c) 601A
    (d) all of these

42. Ground-fault protection of equipment shall be provided for solidly grounded wye electrical systems of more than 150 volts-to-ground, but not exceeding 600V phase-to-phase for each individual device used as a building or structure main disconnecting means rated _____, or more.

    (a) 1,000A
    (b) 1,500A
    (c) 2,000A
    (d) 2,500A

43. Overcurrent devices shall be _____.

    (a) accessible (as applied to wiring methods)
    (b) accessible (as applied to equipment)
    (c) readily accessible
    (d) inaccessible to unauthorized personnel

44. Overcurrent devices shall be readily accessible and installed so the center of the grip of the operating handle of the switch or circuit breaker, when in its highest position, is not more than _____ above the floor or working platform.

    (a) 2 ft
    (b) 4 ft 6 in.
    (c) 5 ft
    (d) 6 ft 7 in.

45. _____ shall not be located over the steps of a stairway.

    (a) Disconnect switches
    (b) Overcurrent devices
    (c) Knife switches
    (d) Transformers

Practice Questions in Straight Order—Article 225.16 through 250.102 — Unit 2

46. Plug fuses of the Edison-base type shall have a maximum rating of _____.

    (a) 20A
    (b) 30A
    (c) 40A
    (d) 50A

47. Which of the following statements about Type S fuses are true?

    (a) Adapters shall fit Edison-base fuseholders.
    (b) Adapters are designed to be easily removed.
    (c) Type S fuses shall be classified as not over 125V and 30A.
    (d) a and c

48. Type _____ fuse adapters shall be designed so that once inserted in a fuseholder they cannot be removed.

    (a) A
    (b) E
    (c) S
    (d) P

49. Circuit breakers shall be capable of being closed and opened by manual operation. Operation by other means, such as electrical or pneumatic, shall be permitted if means for _____ operation are also provided.

    (a) automated
    (b) timed
    (c) manual
    (d) shunt trip

50. For engineered series combination systems where circuit breakers are installed in series downstream of overcurrent devices with a higher AIC rating, the engineer shall ensure that the _____ device remains passive during the interruption period of the fully rated device.

    (a) downstream
    (b) upstream
    (c) dynamic impedance
    (d) current limiting

51. A ground-fault current path is an electrically conductive path from the point of a ground fault through normally noncurrent-carrying conductors, equipment, or the earth to the _____.

    (a) ground
    (b) earth
    (c) electrical supply source
    (d) none of these

52. Examples of ground-fault current paths include any combination of conductive materials including _____.

    (a) equipment grounding conductors
    (b) metallic raceways
    (c) metal water and gas piping
    (d) all of these

53. For grounded systems, noncurrent-carrying conductive materials enclosing electrical conductors or equipment, or forming part of such equipment, shall be connected together and to the _____ to establishes an effective ground-fault current path.

    (a) ground
    (b) earth
    (c) electrical supply source
    (d) none of these

54. In grounded systems, normally noncurrent-carrying electrically conductive materials that are likely to become energized shall be _____ in a manner that establishes an effective ground-fault current path.

    (a) connected together
    (b) connected to the electrical supply source
    (c) connected to the closest grounded conductor
    (d) a and b

55. For grounded systems, electrical equipment and conductive material likely to become energized, shall be installed in a manner that creates a _____ from any point on the wiring system where a ground fault may occur to the electrical supply source.

    (a) circuit facilitating the operation of the overcurrent device
    (b) low-impedance path
    (c) path capable of safely carrying the ground-fault current likely to be imposed on it
    (d) all of these

56. For grounded systems, the electrical equipment electrically conductive material likely to become energized, shall be installed in a manner that creates a low-impedance circuit capable of safely carrying the maximum ground-fault current likely to be imposed on it from where a ground fault may occur to the _____.

    (a) ground
    (b) earth
    (c) electrical supply source
    (d) none of these

57. For ungrounded systems, noncurrent-carrying conductive materials enclosing electrical conductors or equipment shall be connected to _____ in a manner that will limit the voltage imposed by lightning or unintentional contact with higher-voltage lines.

    (a) ground
    (b) earth
    (c) electrical supply source
    (d) none of these

58. An alternate ac power source such as an on-site generator is not a separately derived system if the _____ is solidly interconnected to a service-supplied system grounded conductor.

    (a) ignition system
    (b) fuel cell
    (c) grounded conductor
    (d) line conductor

59. _____ AC systems operating at 480V shall have ground detectors installed on the system.

    (a) Grounded
    (b) Solidly grounded
    (c) Effectively grounded
    (d) Ungrounded

60. Where the main bonding jumper is installed from the grounded conductor terminal bar to the equipment grounding terminal bar in service equipment, the _____ is permitted to be connected to the equipment grounding terminal bar.

    (a) grounding conductor
    (b) grounded conductor
    (c) grounding electrode conductor
    (d) none of these

61. For a grounded system, an unspliced _____ shall be used to connect the equipment grounding conductor(s) and the service disconnect enclosure to the grounded conductor of the system within the enclosure for each service disconnect.

    (a) grounding electrode
    (b) main bonding jumper
    (c) busbar
    (d) insulated copper conductor

62. Where service-entrance phase conductors are installed in parallel, the size of the grounded conductor in each raceway shall be based on the size of the ungrounded service-entrance conductor in the raceway, but not smaller than _____.

    (a) 1/0 AWG
    (b) 2/0 AWG
    (c) 3/0 AWG
    (d) 4/0 AWG

63. Where a main bonding jumper is a screw only, the screw shall be identified by _____ that shall be visible with the screw installed.

    (a) a silver or white finish
    (b) an etched ground symbol
    (c) hexagonal head
    (d) a green finish

64. Where the supply conductors are larger than 1,100 kcmil copper or 1,750 kcmil aluminum, the main bonding jumper shall have an area that is _____ the area of the largest phase conductor when of the same material.

    (a) at least equal to
    (b) at least 50 percent of
    (c) not less than 12½ percent of
    (d) not more than 12½ percent of

65. The connection of the system bonding jumper for a separately derived system shall be made _____ on the separately derived system from the source to the first system disconnecting means or overcurrent device.

    (a) in at least two locations
    (b) in every location that the grounded conductor is present
    (c) at any single point
    (d) none of these

66. The common grounding electrode conductor installed for multiple separately derived systems shall not be smaller than _____.

    (a) 1/0 AWG
    (b) 2/0 AWG
    (c) 3/0 AWG
    (d) 4/0 AWG copper

67. Each tap conductor to a common grounding electrode conductor for multiple separately derived systems shall be sized in accordance with _____ based on the derived phase conductors of the separately derived system it serves.

    (a) 250.66
    (b) 250.118
    (c) 250.122
    (d) 310.15

68. The size of the grounding electrode conductor for a building or structure supplied by a feeder shall not be smaller than that identified in _____, based on the largest ungrounded supply conductor.

    (a) 250.66
    (b) 250.122
    (c) Table 310.16
    (d) not specified

69. The frame of a portable generator shall not be required to be connected to a _____ if the generator only supplies equipment mounted on the generator or cord-and-plug equipment using receptacles mounted on the generator.

    (a) grounding electrode
    (b) grounded conductor
    (c) ungrounded conductor
    (d) equipment grounding conductor

70. The frame of a vehicle-mounted generator shall not be required to be connected to a _____ if the generator only supplies equipment mounted on the vehicle or cord-and-plug equipment using receptacles mounted on the vehicle.

    (a) grounding electrode
    (b) grounded conductor
    (c) ungrounded conductor
    (d) equipment grounding conductor

71. Concrete-encased electrodes of _____ shall not be required to be part of the grounding electrode system where the steel reinforcing bars or rods aren't accessible for use without disturbing the concrete.

    (a) hazardous (classified) locations
    (b) health care facilities
    (c) existing buildings or structures
    (d) agricultural buildings with equipotential planes

72. In order for a metal underground water pipe to be used as a grounding electrode, it shall be in direct contact with the earth for _____ .

    (a) 5 ft
    (b) 10 ft or more
    (c) less than 10 ft
    (d) 20 ft or more

73. Interior metal water piping located more than _____ from the point of entrance to the building shall not be used as a part of the grounding electrode system, or as a conductor to interconnect electrodes that are part of the grounding electrode system.

    (a) 2 ft
    (b) 4 ft
    (c) 5 ft
    (d) 6 ft

74. A bare 4 AWG copper conductor installed horizontally near the bottom or vertically, and within that portion of a concrete foundation or footing that is in direct contact with the earth can be used as a grounding electrode when the conductor is at least _____ in length.

    (a) 10 ft
    (b) 15 ft
    (c) 20 ft
    (d) 25 ft

75. An electrode encased by at least 2 in. of concrete, located horizontally near the bottom or vertically and within that portion of a concrete foundation or footing that is in direct contact with the earth, shall be permitted as a grounding electrode when it consists of _____.

    (a) at least 20 ft of ½ in. or larger steel reinforcing bars or rods
    (b) at least 20 ft of bare copper conductor of 4 AWG or larger
    (c) a or b
    (d) none of these

76. Grounding electrodes that are driven rods require a minimum of _____ in contact with the soil.

    (a) 6 ft
    (b) 8 ft
    (c) 10 ft
    (d) 12 ft

Unit 2 — Practice Questions in Straight Order—Article 225.16 through 250.102

77. Grounding electrodes of the rod type less than ⅝ in. in diameter shall be listed and shall not be less than _____ in diameter.

    (a) ½ in.
    (b) ¾ in.
    (c) 1 in.
    (d) 1¼ in.

78. Where practicable, rod, pipe, and plate electrodes shall be installed _____.

    (a) directly below the electrical meter
    (b) on the north side of the building
    (c) below permanent moisture level
    (d) all of these

79. Where a metal underground water pipe is used as a grounding electrode, the continuity of the grounding path or the bonding connection to interior piping shall not rely on _____ and similar equipment.

    (a) bonding jumpers
    (b) water meters or filtering devices
    (c) grounding clamps
    (d) all of these

80. Where the supplemental electrode is a rod, that portion of the bonding jumper that is the sole connection to the supplemental grounding electrode shall not be required to be larger than _____ AWG.

    (a) 8
    (b) 6
    (c) 4
    (d) 1

81. When a ground ring is used as a grounding electrode, it shall be buried at a depth below the earth's surface of not less than _____.

    (a) 18 in.
    (b) 24 in.
    (c) 30 in.
    (d) 8 ft

82. Ground rod electrodes shall be installed so that at least _____ of the length is in contact with the soil.

    (a) 5 ft
    (b) 8 ft
    (c) one-half
    (d) 80 percent

83. The upper end of a ground rod electrode shall be _____ ground level unless the aboveground end and the grounding electrode conductor attachment are protected against physical damage.

    (a) above
    (b) flush with
    (c) below
    (d) b or c

84. Where rock bottom is encountered when driving a ground rod at an angle up to 45 degrees, the electrode can be buried in a trench that is at least _____ deep.

    (a) 18 in.
    (b) 30 in.
    (c) 4 ft
    (d) 8 ft

85. Auxiliary grounding electrodes can be connected to the _____.

    (a) equipment grounding conductor
    (b) grounded conductor
    (c) a and b
    (d) none of these

86. Where used outside, aluminum or copper-clad aluminum grounding electrode conductors shall not be terminated within _____ of the earth.

    (a) 6 in.
    (b) 12 in.
    (c) 15 in.
    (d) 18 in.

87. Grounding electrode conductors _____ and larger that are not subject to physical damage can be run exposed along the surface of the building construction.

    (a) 10 AWG
    (b) 8 AWG
    (c) 6 AWG
    (d) 4 AWG

88. Grounding electrode conductors smaller than _____ shall be in rigid metal conduit, IMC, PVC conduit, electrical metallic tubing, or cable armor.

    (a) 10 AWG
    (b) 8 AWG
    (c) 6 AWG
    (d) 4 AWG

89. Grounding electrode conductors shall be installed in one continuous length without a splice or joint, unless spliced _____.

    (a) by connecting to a busbar
    (b) by irreversible compression-type connectors listed as grounding and bonding equipment
    (c) by the exothermic welding process
    (d) any of these

90. A service consisting of 12 AWG service-entrance conductors requires a grounding electrode conductor sized no less than _____.

    (a) 10 AWG
    (b) 8 AWG
    (c) 6 AWG
    (d) 4 AWG

91. The largest size grounding electrode conductor required is _____ copper.

    (a) 6 AWG
    (b) 1/0 AWG
    (c) 3/0 AWG
    (d) 250 kcmil

92. What size copper grounding electrode conductor is required for a service that has three sets of 600 kcmil copper conductors per phase?

    (a) 1 AWG
    (b) 1/0 AWG
    (c) 2/0 AWG
    (d) 3/0 AWG

93. In an ac system, the size of the grounding electrode conductor to a concrete-encased electrode shall not be required to be larger than a _____ copper conductor.

    (a) 10 AWG
    (b) 8 AWG
    (c) 6 AWG
    (d) 4 AWG

94. Bonding jumpers shall be used around _____ knockouts that are punched or otherwise formed so as to impair the electrical connection to ground.

    (a) concentric
    (b) eccentric
    (c) field-punched
    (d) a or b

95. Service raceways threaded into metal service equipment such as bosses (hubs) are considered to be effectively _____ to the service metal enclosure.

    (a) attached
    (b) bonded
    (c) grounded
    (d) none of these

96. A means external to enclosures for connecting intersystem _____ conductors shall be provided at service equipment and disconnecting means of other buildings or structures.

    (a) bonding
    (b) grounding
    (c) secondary
    (d) a and b

97. When bonding enclosures, metal raceways, frames, and fittings, any nonconductive paint, enamel, or similar coating shall be removed at _____.

    (a) contact surfaces
    (b) threads
    (c) contact points
    (d) all of these

98. What is the minimum size copper bonding jumper for a service raceway containing 4/0 THHN aluminum conductors?

    (a) 6 AWG aluminum
    (b) 4 AWG aluminum
    (c) 4 AWG copper
    (d) 3 AWG copper

99. A service is supplied by three metal raceways, each containing 600 kcmil ungrounded conductors. Determine the bonding jumper size for each service raceway.

    (a) 1/0 AWG
    (b) 3/0 AWG
    (c) 250 kcmil
    (d) 500 kcmil

100. What is the minimum size copper equipment bonding jumper for a 40A circuit?

    (a) 14 AWG
    (b) 12 AWG
    (c) 10 AWG
    (d) 8 AWG

Unit 2: Practice Questions in Random Order—Article 90.2 through 250.104

Use the 2008 *NEC* to answer the following questions.

1. The additional service disconnecting means for fire pumps, emergency systems, legally required standby, or optional standby services, shall be installed remote from the one to six service disconnecting means for normal service to minimize the possibility of _____ interruption of supply.

    (a) intentional
    (b) accidental
    (c) simultaneous
    (d) prolonged

2. In ungrounded systems, electrical equipment, wiring, and other electrically conductive material likely to become energized shall be installed in a manner that creates a low-impedance circuit from any point on the wiring system to the electrical supply source to facilitate the operation of overcurrent devices should a(n) _____ from a different phase occur on the wiring system.

    (a) isolated ground
    (b) second ground fault
    (c) arc fault
    (d) high impedance fault

3. Grounded electrical systems shall be connected to earth in a manner that will _____.

    (a) limit voltages due to lightning, line surges, or unintentional contact with higher-voltage lines
    (b) stabilize the voltage-to-ground during normal operation
    (c) facilitate overcurrent device operation in case of ground faults
    (d) a and b

4. The grounding conductor connection to the grounding electrode shall be made by _____.

    (a) listed lugs
    (b) exothermic welding
    (c) listed pressure connectors
    (d) any of these

5. Grounding conductors and bonding jumpers shall be connected by _____.

    (a) listed pressure connectors
    (b) terminal bars
    (c) exothermic welding
    (d) any of these

6. Overhead Type SE cable can be formed in a _____ and taped with self-sealing weather-resistant thermoplastic.

    (a) loop
    (b) circle
    (c) gooseneck
    (d) none of these

7. Where documented safe switching procedures are established and maintained and monitored by _____ individuals, the disconnecting means for a building supplied by a feeder can be located elsewhere on the premises.

    (a) maintenance
    (b) management
    (c) service
    (d) qualified

8. Ground clamps and fittings shall be protected from physical damage by being enclosed in _____ where there may be a possibility of physical damage.

    (a) metal
    (b) wood
    (c) the equivalent of a or b
    (d) none of these

9. For a single separately derived system, the grounding electrode conductor connects the grounding electrode to the grounded conductor of the derived system at the same point on the separately derived system where the _____ is connected.

    (a) metering equipment
    (b) transfer switch
    (c) system bonding jumper
    (d) largest circuit breaker

10. Service metal raceways and metal-clad cables are considered effectively bonded when using threadless couplings and connectors that are _____.

    (a) nonmetallic
    (b) made up tight
    (c) sealed
    (d) classified

11. Flexible cords approved for and used with a specific listed appliance or luminaire shall be considered to be protected by the branch-circuit overcurrent device when _____.

    (a) not more than 6 ft in length
    (b) 20 AWG and larger
    (c) applied within the listing requirements
    (d) 16 AWG and larger

12. A device comprised of _____ or more receptacles shall be calculated at not less than 90 VA per receptacle.

    (a) one
    (b) two
    (c) three
    (d) four

13. A building or structure shall be supplied by a maximum of _____ feeder(s) or branch circuit(s), unless specifically permitted otherwise.

    (a) one
    (b) two
    (c) three
    (d) four

14. A _____ is a single unit that provides independent living facilities for persons, including permanent provisions for living, sleeping, cooking, and sanitation.

    (a) one-family dwelling
    (b) two-family dwelling
    (c) dwelling unit
    (d) multifamily dwelling

15. A building that contains three or more dwelling units is called a _____.

    (a) one-family dwelling
    (b) two-family dwelling
    (c) dwelling unit
    (d) multifamily dwelling

16. A load is considered to be continuous if the maximum current is expected to continue for _____ or more.

    (a) one-half hour
    (b) one hour
    (c) two hours
    (d) three hours

17. On or attached to the surface or behind access panels designed to allow access is known as _____.

    (a) open
    (b) uncovered
    (c) exposed
    (d) bare

18. Circuit breakers shall _____ all ungrounded conductors of the circuit.

    (a) open
    (b) close
    (c) isolate
    (d) inhibit

19. The disconnecting means for a building supplied by a feeder shall plainly indicate whether it is in the _____ position.

    (a) open or closed
    (b) correct
    (c) up or down
    (d) none of these

20. The 3 VA per-square-foot general lighting load for dwelling units does not include _____.

    (a) open porches
    (b) garages
    (c) unused or unfinished spaces not adaptable for future use
    (d) all of these

Unit 2 — Practice Questions in Random Order—Article 90.2 through 250.104

21. This *Code* covers the installation of _____ for public and private premises, including buildings, structures, mobile homes, recreational vehicles, and floating building.

    (a) optical fiber cables
    (b) electrical equipment
    (c) raceways
    (d) all of these

22. Where the resistance-to-ground of 25 ohms or less is not achieved for a single rod electrode, _____.

    (a) other means besides electrodes shall be used in order to provide grounding
    (b) the single rod electrode shall be augmented by one additional electrode
    (c) no additional electrodes are required
    (d) none of these

23. An outlet intended for the direct connection of a lampholder or a luminaire is a(n) _____.

    (a) outlet
    (b) receptacle outlet
    (c) lighting outlet
    (d) general-purpose outlet

24. _____ are designed for surface mounting with a swinging door.

    (a) Outlet boxes
    (b) Cabinets
    (c) Cutout boxes
    (d) none of these

25. Illumination shall be provided for all working spaces about service equipment, switchboards, panelboards, and motor control centers _____.

    (a) over 600V
    (b) located indoors
    (c) rated 1,200A or more
    (d) using automatic means of control

26. The equipment bonding jumper on the supply side of services shall be sized according to the _____.

    (a) overcurrent device rating
    (b) service-entrance conductor size
    (c) service-drop size
    (d) load to be served

27. Outside wiring shall not be installed beneath openings through which materials may be moved, and shall not be installed where they will obstruct entrance to these buildings' openings.

    (a) True
    (b) False

28. Service conductors shall not be installed beneath openings through which materials may be moved and shall not be installed where they will obstruct entrance to these buildings' openings.

    (a) True
    (b) False

29. Where service-entrance conductors are paralleled in two or more raceways or cables, the bonding jumper for each raceway or cable shall be based on the size of the _____ in each raceway or cable.

    (a) overcurrent protection for conductors
    (b) grounded conductors
    (c) service-entrance conductors
    (d) sum of all conductors in the raceway

30. The _____ of a circuit shall be selected and coordinated to permit the circuit protective devices to clear a fault without extensive damage to the electrical components of the circuit.

    (a) overcurrent protective devices
    (b) total circuit impedance
    (c) component short-circuit current ratings
    (d) all of these

31. Each _____ service conductor shall have overload protection.

    (a) overhead
    (b) underground
    (c) ungrounded
    (d) none of these

32. Hazards often occur because of _____.

    (a) overloading of wiring systems by methods or usage not in conformity with the *NEC*
    (b) initial wiring not providing for increases in the use of electricity
    (c) a and b
    (d) none of these

33. _____ on equipment to be grounded shall be removed from contact surfaces to ensure good electrical continuity.

    (a) Paint
    (b) Lacquer
    (c) Enamel
    (d) any of these

34. An accessory, such as a locknut, intended to perform a mechanical function best describes a _____.

    (a) part
    (b) equipment
    (c) device
    (d) fitting

35. The *Code* provides rules for the minimum number of receptacle outlets required in a dwelling unit. The minimum required receptacle outlets shall be in addition to receptacle outlets that are _____.

    (a) part of a luminaire or appliance
    (b) switched by a wall switch for use as required illumination
    (c) located more than 5½ ft above the floor
    (d) all of these

36. Local metal underground systems or structures such as _____ are permitted to serve as grounding electrodes.

    (a) piping systems
    (b) underground tanks
    (c) underground metal well casings
    (d) all of these

37. A solderless pressure connector is a device that _____ between conductors or between conductors and a terminal by means of mechanical pressure.

    (a) provides access
    (b) protects the wiring
    (c) is never needed
    (d) establishes a connection

38. A unit of an electrical system that carries or controls electric energy as its principal function is a(n) _____.

    (a) raceway
    (b) fitting
    (c) device
    (d) enclosure

39. Overhead service cables shall be equipped with a _____.

    (a) raceway
    (b) service head
    (c) cover
    (d) all of these

40. In a dwelling unit, illumination for outdoor entrances that have grade-level access can be controlled by _____.

    (a) remote
    (b) central
    (c) automatic control
    (d) any of these

41. Wiring methods permitted for service-entrance conductors include _____.

    (a) rigid metal conduit
    (b) electrical metallic tubing
    (c) PVC conduit
    (d) all of these

42. Service cables, where subject to physical damage, shall be protected by _____.

    (a) rigid metal conduit
    (b) IMC
    (c) Schedule 80 PVC conduit
    (d) any of these

43. A _____ is a chamber to which one or more ducts are connected and form part of the air distribution system.

    (a) riser
    (b) plenum
    (c) a or b
    (d) none of these

44. A value assigned to a circuit or system for the purpose of conveniently designating its voltage class such as 120/240V is called _____ voltage.

    (a) root-mean-square
    (b) circuit
    (c) nominal
    (d) source

45. A(n) _____ system is a premises wiring system whose power is derived from a source of electric energy that has no direct electrical connection to supply conductors originating in another system.

    (a) separately derived
    (b) classified
    (c) direct
    (d) emergency

46. The *NEC* defines a _____ as: "all circuit conductors between the service equipment, the source of a separately derived system, or other power-supply source and the final branch-circuit overcurrent device."

    (a) service
    (b) feeder
    (c) branch circuit
    (d) all of these

47. In a multiple-occupancy building, each occupant shall have access to the occupant's _____.

    (a) service disconnecting means
    (b) service drops
    (c) distribution transformer
    (d) lateral conductors

48. The _____ is the point of connection between the serving utility and the premises wiring.

    (a) service entrance
    (b) service point
    (c) overcurrent protection
    (d) beginning of the wiring system

49. Where isolated metal water piping systems are installed in a multi-occupancy building, the water pipes can be bonded with bonding jumpers sized per Table 250.122, based on the size of _____.

    (a) service-entrance conductors
    (b) feeder conductors
    (c) rating of the service equipment overcurrent device
    (d) rating of overcurrent device supplying the occupancy.

50. The _____ is the necessary equipment, usually consisting of a circuit breaker(s) or switch(es) and fuse(s) and their accessories, connected to the load end of service conductors, and intended to constitute the main control and cutoff of the supply.

    (a) service equipment
    (b) service
    (c) service disconnect
    (d) service overcurrent device

Unit 3: Practice Questions in Straight Order— Article 250.118 through 334.2

Use the 2008 *NEC* to answer the following questions.

1. Listed FMC can be used as the equipment grounding conductor if the length in any ground return path does not exceed 6 ft and the circuit conductors contained in the conduit are protected by overcurrent devices rated at _____ or less.

    (a) 15A
    (b) 20A
    (c) 30A
    (d) 60A

2. An equipment grounding conductor shall be identified by _____.

    (a) a continuous outer finish that is green
    (b) being bare
    (c) a continuous outer finish that is green with one or more yellow stripes
    (d) any of these

3. Conductors with the color _____ insulation shall not be used for ungrounded or grounded conductors.

    (a) green
    (b) green with one or more yellow stripes
    (c) a or b
    (d) white

4. When ungrounded circuit conductors are increased in size, the equipment grounding conductor must be proportionately increased in size according to _____ of the ungrounded conductors.

    (a) ampacity
    (b) circular mil area
    (c) diameter
    (d) none of these

5. Equipment grounding conductors for motor branch circuits shall be sized in accordance with Table 250.122, based on the rating of the _____ device.

    (a) motor overload
    (b) motor over-temperature
    (c) motor short-circuit and ground-fault protective
    (d) feeder overcurrent protection

6. On the load side of the service disconnecting means, the _____ circuit conductor can be ground meter enclosures if all meter enclosures are located near the service disconnecting means and ground-fault protection is not installed.

    (a) grounding
    (b) bonding
    (c) grounded
    (d) phase

7. A(n) _____ shall be used to connect the grounding terminal of a grounding-type receptacle to a grounded box.

    (a) equipment bonding jumper
    (b) grounded conductor jumper
    (c) a or b
    (d) a and b

8. Receptacle yokes designed and _____ as self-grounding can establish the grounding circuit between the device yoke and a grounded outlet box.

    (a) approved
    (b) advertised
    (c) listed
    (d) installed

Unit 3 — Practice Questions in Straight Order—Article 250.118 through 334.2

9. The receptacle grounding terminal of an isolated ground receptacle shall be connected to a(n) _____ equipment grounding conductor run with the circuit conductors.

   (a) insulated
   (b) covered
   (c) bare
   (d) solid

10. A connection between equipment grounding conductors and a metal box shall be by a(n) _____.

    (a) grounding screw used for no other purpose
    (b) equipment listed for grounding
    (c) a listed grounding device
    (d) any of these

11. The conductors used to connect the surge protective device to ground shall not be any longer than _____ and shall avoid unnecessary bends.

    (a) 6 in.
    (b) 12 in.
    (c) 18 in.
    (d) necessary

12. Conductors of ac and dc circuits, rated 600V or less, can occupy the same _____ provided that all conductors have an insulation rating equal to the maximum voltage applied to any conductor.

    (a) enclosure
    (b) cable
    (c) raceway
    (d) all of these

13. Where cables or nonmetallic raceways are installed through bored holes in joists, rafters, or wood members, holes shall be bored so that the edge of the hole is _____ the nearest edge of the wood member.

    (a) not less than 1¼ in. from
    (b) immediately adjacent to
    (c) not less than 1/16 in. from
    (d) 90° away from

14. Cables laid in wood notches require protection against nails or screws by using a steel plate at least _____ thick, installed before the building finish is applied.

    (a) 1/16 in.
    (b) 1/8 in.
    (c) ¼ in.
    (d) ½ in.

15. Where Type NM cable passes through factory or field openings in metal members, it shall be protected by _____ bushings or _____ grommets that cover metal edges.

    (a) approved
    (b) identified
    (c) listed
    (d) none of these

16. Where Type NM cables pass through cut or drilled slots or holes in metal members, the cable shall be protected by _____ securely covering all metal edges fastened in the opening prior to installation of the cable.

    (a) listed bushings
    (b) listed grommets
    (c) plates
    (d) a or b

17. Where nails or screws are likely to penetrate nonmetallic-sheathed cable or ENT installed through metal framing members, a steel sleeve, steel plate, or steel clip not less than _____ in thickness shall be used to protect the cable or tubing.

    (a) 1/16 in.
    (b) 1/8 in.
    (c) ½ in.
    (d) ¾ in.

18. Cables or raceways installed under metal-corrugated sheet roof decking shall be supported so the nearest outside surface of the cable or raceway is not less than _____ from the nearest surface of the roof decking.

    (a) ½ in.
    (b) 1 in.
    (c) 1½ in.
    (d) 2 in.

19. Rigid metal conduit that is directly buried outdoors shall have at least _____ of cover.

    (a) 6 in.
    (b) 12 in.
    (c) 18 in.
    (d) 24 in.

20. When installing PVC conduit underground without concrete cover, there shall be a minimum of _____ of cover.

    (a) 6 in.
    (b) 12 in.
    (c) 18 in.
    (d) 22 in.

21. What is the minimum cover requirement for Type UF cable supplying power to a 120V, 15A GFCI-protected circuit outdoors under a driveway of a one-family dwelling?

    (a) 6 in.
    (b) 12 in.
    (c) 16 in.
    (d) 24 in.

22. Type UF cable used with a 24V landscape lighting system can be have a minimum cover of _____.

    (a) 6 in.
    (b) 12 in.
    (c) 18 in.
    (d) 24 in.

23. Where direct buried conductors and cables emerge from grade, they shall be protected by enclosures or raceways to a point at least _____ above finished grade.

    (a) 3 ft
    (b) 6 ft
    (c) 8 ft
    (d) 10 ft

24. Direct-buried service conductors that are not encased in concrete and that are buried 18 in. or more below grade shall have their location identified by a warning ribbon placed in the trench at least _____ above the underground installation.

    (a) 6 in.
    (b) 10 in.
    (c) 12 in.
    (d) 18 in.

25. Backfill used for underground wiring shall not _____.

    (a) damage the wiring method
    (b) prevent compaction of the fill
    (c) contribute to the corrosion of the raceway
    (d) all of these

26. All conductors of the same circuit shall be _____, unless otherwise specifically permitted in the *Code*.

    (a) in the same raceway or cable
    (b) in close proximity in the same trench
    (c) the same size
    (d) a or b

27. Cables or raceways installed using directional boring equipment shall be _____ for this purpose.

    (a) marked
    (b) listed
    (c) labeled
    (d) approved

28. Which of the following metal parts shall be protected from corrosion?

    (a) ferrous metal raceways
    (b) ferrous metal elbows
    (c) ferrous boxes
    (d) all of these

29. Aluminum raceways, cable trays, cablebus, auxiliary gutters, cable armor, boxes, cable sheathing, cabinets, elbows, couplings, nipples, fittings, supports, and support hardware _____ shall be provided with supplementary corrosion protection.

    (a) embedded or encased in concrete
    (b) in direct contact with the earth
    (c) likely to become energized
    (d) a or b

30. Nonmetallic raceways, cable trays, boxes, cables with a nonmetallic outer jacket, fittings, and support hardware shall be _____.

    (a) listed as sunlight resistant
    (b) identified as sunlight resistant
    (c) a and b
    (d) a or b

31. Where nonmetallic wiring methods are subject to exposure to chemical solvents or vapors, they shall be inherently resistant to chemicals based upon their being _____.

    (a) listed for the chemical
    (b) identified for the chemical
    (c) a and b
    (d) a or b

Unit 3 — Practice Questions in Straight Order—Article 250.118 through 334.2

32. An exposed wiring system for indoor wet locations where walls are frequently washed shall be mounted so that there is at least a _____ between the mounting surface and the electrical equipment.

    (a) ¼ in. airspace
    (b) separation by insulated bushings
    (c) separation by noncombustible tubing
    (d) none of these

33. Raceways or cable trays containing electric conductors shall not contain any pipe or tube for steam, water, air, gas, drainage, or any service other than _____.

    (a) as permitted by the authority having jurisdiction
    (b) electrical
    (c) pneumatic
    (d) as designed by the engineer

34. Metal raceways, cable armors, and other metal enclosures shall be _____ joined together into a continuous electric conductor so as to provide effective electrical continuity.

    (a) electrically
    (b) permanently
    (c) metallically
    (d) none of these

35. The independent support wires for wiring in a fire-rated ceiling assembly shall be distinguishable from fire-rated suspended-ceiling framing support wires by _____.

    (a) color
    (b) tagging
    (c) other effective means
    (d) any of these

36. Metal or nonmetallic raceways, cable armors, and cable sheaths _____ between cabinets, boxes, fittings or other enclosures or outlets.

    (a) can be attached with electrical tape
    (b) are allowed gaps for expansion
    (c) shall be continuous
    (d) none of these

37. Raceways and cables installed into the _____ of open-bottom equipment shall not be required to be mechanically secured to the equipment.

    (a) bottom
    (b) sides
    (c) top
    (d) any of these

38. Conductors in raceways shall be _____ between outlets, boxes, devices, and so forth.

    (a) continuous
    (b) installed
    (c) copper
    (d) in conduit

39. When the opening to an outlet, junction, or switch point is less than 8 in. in any dimension, each conductor shall be long enough to extend at least _____ outside the opening of the enclosure.

    (a) 0 in.
    (b) 3 in.
    (c) 6 in.
    (d) 12 in.

40. Raceways shall be _____ between outlet, junction, or splicing points prior to the installation of conductors.

    (a) installed complete
    (b) tested for ground faults
    (c) a minimum of 80 percent completed
    (d) none of these

41. Short sections of raceways used for _____ shall not be required to be installed complete between outlet, junction, or splicing points.

    (a) meter to service enclosure connection
    (b) protection of cables from physical damage
    (c) nipples
    (d) separately derived systems

42. A vertical run of 4/0 AWG copper shall be supported at intervals not exceeding _____.

    (a) 40 ft
    (b) 80 ft
    (c) 100 ft
    (d) 120 ft

43. Openings around electrical penetrations through fire-resistant-rated walls, partitions, floors, or ceilings shall _____ to maintain the fire-resistance rating.

    (a) be documented
    (b) not be permitted
    (c) be firestopped using approved methods
    (d) be enlarged

44. No wiring of any type shall be installed in ducts used to transport _____.

    (a) dust
    (b) flammable vapors
    (c) loose stock
    (d) all of these

45. Equipment and devices shall only be permitted within ducts or plenum chambers used to transport environmental air if necessary for their direct action upon, or sensing of, the _____.

    (a) contained air
    (b) air quality
    (c) air temperature
    (d) none of these

46. The space above a hung ceiling used for environmental air-handling purposes is an example of _____, and the wiring limitations of _____ apply.

    (a) a plenum used for environmental air, 300.22(B)
    (b) other space used for environmental air, 300.22(C)
    (c) a duct used for environmental air, 300.22(B)
    (d) none of these

47. Wiring methods permitted in the ceiling areas used for environmental air include _____.

    (a) electrical metallic tubing
    (b) FMC of any length
    (c) RMC without an overall nonmetallic covering
    (d) all of these

48. Where installed in raceways, conductors _____ AWG and larger shall be stranded.

    (a) 10
    (b) 8
    (c) 6
    (d) 4

49. In general, the minimum size conductor permitted for use in parallel installations is _____ AWG.

    (a) 10
    (b) 4
    (c) 1
    (d) 1/0

50. Parallel conductors shall have the same _____.

    (a) length
    (b) material
    (c) cross-sectional area
    (d) all of these

51. Parallel conductors shall _____.

    (a) be the same length and conductor material
    (b) have the same circular mil area and insulation type
    (c) be terminated in the same manner
    (d) all of these

52. The minimum size conductor permitted for branch circuits under 600V is _____ AWG.

    (a) 14
    (b) 12
    (c) 10
    (d) 8

53. There are four principal determinants of conductor operating temperature, one of which is _____ generated internally in the conductor as the result of load current flow.

    (a) friction
    (b) magnetism
    (c) heat
    (d) none of these

54. THWN insulated conductors are rated _____.

    (a) 75°C
    (b) for wet locations
    (c) a and b
    (d) not enough information

55. The ampacities listed in the Tables of Article 310.16 do not take _____ into consideration.

    (a) continuous loads
    (b) voltage drop
    (c) insulation
    (d) wet locations

56. The ampacity of a conductor can be different along the length of the conductor. The higher ampacity can be used beyond the point of transition for a distance of no more than _____ ft, or no more than _____ percent of the circuit length figured at the higher ampacity, whichever is less.

    (a) 10, 10
    (b) 10, 20
    (c) 15, 15
    (d) 20, 10

57. Where six current-carrying conductors are run in the same conduit or cable, the ampacity of each conductor shall be adjusted by a factor of _____ percent.

    (a) 40
    (b) 60
    (c) 80
    (d) 90

58. Conductor derating factors shall not apply to conductors in nipples having a length not exceeding _____

    (a) 12 in.
    (b) 24 in.
    (c) 36 in.
    (d) 48 in.

59. The ampacity adjustment factors of Table 310.15(B)(2)(a) does not apply to Type AC or Type MC cable without an overall outer jacket, if which of the following conditions are met?

    (a) Each cable has not more than three current-carrying conductors.
    (b) The conductors are 12 AWG copper.
    (c) No more than 20 current-carrying conductors are bundled or stacked.
    (d) all of these

60. Where conductors or cables are installed in conduits exposed to direct sunlight on or above rooftops, the ambient temperature shall be increased by _____ where the conduits are less than ½ in. from the rooftop.

    (a) 30°F
    (b) 40°F
    (c) 50°F
    (d) 60°F

61. A neutral conductor that carries only the unbalanced current from other conductors of the same circuit shall not be required to be counted when applying the provisions of 310.15(B)(2)(a).

    (a) neutral
    (b) grounded
    (c) grounding
    (d) none of these

62. Surface-type cabinets, cutout boxes, and meter socket enclosures in damp or wet locations shall be mounted so there is at least _____ airspace between the enclosure and the wall or supporting surface.

    (a) ¹⁄₁₆ in.
    (b) ¼ in.
    (c) 1¼ in.
    (d) 6 in.

63. In walls constructed of wood or other _____ material, electrical cabinets shall be flush with the finished surface or project therefrom.

    (a) nonconductive
    (b) porous
    (c) fibrous
    (d) combustible

64. Noncombustible surfaces that are broken or incomplete shall be repaired so there will be no gaps or open spaces greater than _____ at the edge of a cabinet or cutout box employing a flush-type cover.

    (a) ¹⁄₃₂ in.
    (b) ¹⁄₁₆ in.
    (c) ⅛ in.
    (d) ¼ in.

65. Openings in cabinets, cutout boxes, and meter socket enclosures through which conductors enter shall be _____.

    (a) adequately closed
    (b) made using concentric knockouts only
    (c) centered in the cabinet wall
    (d) identified

66. Nonmetallic cables can enter the top of surface-mounted cabinets, cutout boxes, and meter socket enclosures through nonflexible raceways not less than 18 in. or more than _____ ft in length if all of the required conditions are met.

    (a) 3
    (b) 10
    (c) 25
    (d) 100

Practice Questions in Straight Order—Article 250.118 through 334.2 — Unit 3

67. Enclosures for switches or overcurrent devices used for conductors feeding through shall not fill all the wiring space at any cross section to more than _____ percent of the cross-sectional area of the space.

    (a) 20
    (b) 30
    (c) 40
    (d) 60

68. Nonmetallic boxes can be used with _____.

    (a) nonmetallic cables
    (b) nonmetallic raceways
    (c) flexible cords
    (d) all of these

69. The total volume occupied by two internal cable clamps, six 12 AWG conductors, and a single-pole switch is _____.

    (a) 2.0 cu in.
    (b) 4.50 cu in.
    (c) 14.50 cu in.
    (d) 20.25 cu in.

70. According to the *NEC*, the volume of a 3 x 2 x 2 in. device box is _____.

    (a) 8 cu in.
    (b) 10 cu in.
    (c) 12 cu in.
    (d) 14 cu in.

71. When Type NM cable is used with nonmetallic boxes not larger than 2¼ x 4 in., securing the cable to the box shall not be required if the cable is fastened within _____ of that box.

    (a) 6 in.
    (b) 8 in.
    (c) 10 in.
    (d) 12 in.

72. In noncombustible walls or ceilings, the front edge of a box, plaster ring, extension ring, or listed extender employing a flush-type cover, shall be set back not more than _____ from the finished surface.

    (a) ⅛ in.
    (b) ¼ in.
    (c) ⅜ in.
    (d) ½ in.

73. In walls or ceilings constructed of wood or other combustible surface material, boxes, plaster rings, extension rings, or listed extenders shall _____.

    (a) be flush with the surface
    (b) project from the surface
    (c) a or b
    (d) be set back no more than ¼ in.

74. _____ can be used to fasten boxes to structural members of a building using brackets on the outside of the enclosure.

    (a) Nails
    (b) Screws
    (c) Bolts
    (d) a and b

75. A wood brace used for supporting a box for structural mounting shall have a cross-section not less than nominal _____.

    (a) 1 x 2 in.
    (b) 2 x 2 in.
    (c) 2 x 3 in.
    (d) 2 x 4 in.

76. Outlet boxes can be secured to suspended-ceiling framing members by mechanical means such as _____, or by other means identified for the suspended-ceiling framing member(s).

    (a) bolts
    (b) screws
    (c) rivets
    (d) all of these

77. Two intermediate metal or rigid metal conduits threaded wrenchtight into the enclosure can be used to support an outlet box containing devices or luminaires, if each raceway is supported within _____ of the box.

    (a) 12 in.
    (b) 18 in.
    (c) 24 in.
    (d) 36 in.

78. Boxes used at luminaire or lampholder outlets in a ceiling shall be designed for the purpose and shall be required to support a luminaire weighing a minimum of _____.

    (a) 20 lb
    (b) 30 lb
    (c) 40 lb
    (d) 50 lb

Unit 3 — Practice Questions in Straight Order—Article 250.118 through 334.2

79. A wall-mounted luminaire weighing not more than _____ can be supported to a device box with no fewer than two No. 6 or larger screws.

    (a) 4 lb
    (b) 6 lb
    (c) 8 lb
    (d) 10 lb

80. A luminaire that weighs more than _____ can be supported by an outlet box that is listed and marked for the weight of the luminaire.

    (a) 20 lb
    (b) 30 lb
    (c) 40 lb
    (d) 50 lb

81. Floor boxes shall be _____ specifically for this application.

    (a) identified
    (b) listed
    (c) marked
    (d) none of these

82. Listed outlet boxes to support ceiling-suspended fans that weigh more than _____ lb shall have their allowable weight marked on the box.

    (a) 35 lb
    (b) 50 lb
    (c) 60 lb
    (d) 70 lb

83. Equipment weighing less than 6 lb can be supported to any box or plaster ring secured to a box, provided the equipment is secured with at least two _____ or larger screws.

    (a) No. 6
    (b) No. 8
    (c) No. 10
    (d) self tapping

84. Pull boxes or junction boxes with any dimension over _____ shall have all conductors cabled or racked in an approved manner.

    (a) 3 ft
    (b) 6 ft
    (c) 9 ft
    (d) 12 ft

85. Listed boxes and handhole enclosures designed for underground installation can be directly buried when covered by _____, if their location is effectively identified and accessible.

    (a) concrete
    (b) gravel
    (c) noncohesive granulated soil
    (d) b or c

86. Handhole enclosures shall be designed and installed to withstand _____.

    (a) 600 lb
    (b) 3,000 lb
    (c) 6,000 lb
    (d) all loads likely to be imposed

87. Underground raceways and cable assemblies entering a handhole enclosure shall extend into the enclosure, but they are not required to be _____.

    (a) bonded
    (b) insulated
    (c) mechanically connected to the handhole enclosure
    (d) below minimum cover requirements after leaving the handhole

88. Handhole enclosure covers shall have an identifying _____ that prominently identifies the function of the enclosure, such as "electric."

    (a) mark
    (b) logo
    (c) a or b
    (d) manual

89. Handhole enclosure covers shall require the use of tools to open, or they shall weigh over _____.

    (a) 45 lb
    (b) 70 lb
    (c) 100 lb
    (d) 200 lb

90. When Type AC cable is run across the top of a floor joist in an attic without permanent ladders or stairs, substantial guard strips within _____ of the scuttle hole, or attic entrance, shall protect the cable.

    (a) 3 ft
    (b) 4 ft
    (c) 5 ft
    (d) 6 ft

91. Type AC cable shall be secured at intervals not exceeding 4½ ft and within _____ of every outlet box, cabinet, conduit body, or fitting.

    (a) 6 in.
    (b) 8 in.
    (c) 10 in.
    (d) 12 in.

92. Type AC cable installed horizontally through wooden or metal framing members are considered supported where support doesn't exceed _____.

    (a) 2 ft
    (b) 3 ft
    (c) 4½ ft
    (d) 6 ft

93. Armored cable used to connect recessed luminaires or equipment within an accessible ceiling can be unsecured for lengths up to _____.

    (a) 2 ft
    (b) 3 ft
    (c) 4½ ft
    (d) 6 ft

94. At Type AC cable terminations, a(n) _____ shall be provided.

    (a) fitting (or box design) that protects the conductors from abrasion
    (b) insulating bushing between the conductors and the cable armor
    (c) a and b
    (d) none of these

95. Type MC cable shall not be _____ unless the metallic sheath or armor is resistant to the conditions, or is protected by material resistant to the conditions.

    (a) used for direct burial in the earth
    (b) embedded in concrete
    (c) exposed to cinder fill
    (d) all of these

96. Bends made in interlocked or corrugated sheath Type MC cable shall have a radius of at least _____ times the external diameter of the metallic sheath.

    (a) 5
    (b) 7
    (c) 10
    (d) 12

97. Type MC cable containing four or fewer conductors, sized no larger than 10 AWG, shall be secured within _____ of every box, cabinet, fitting, or other cable termination.

    (a) 8 in.
    (b) 12 in.
    (c) 18 in.
    (d) 24 in.

98. Type MC cable shall be secured at intervals not exceeding _____.

    (a) 3 ft
    (b) 4 ft
    (c) 6 ft
    (d) 8 ft

99. Fittings used for connecting Type MC cable to boxes, cabinets, or other equipment shall _____.

    (a) be nonmetallic only
    (b) be listed and identified for such use
    (c) be listed and identified as weatherproof
    (d) include anti-shorting bushings

100. Type _____ cable is a wiring method that encloses two or more insulated conductors within a nonmetallic jacket.

    (a) AC
    (b) MC
    (c) NM
    (d) b and c

Unit 3 — Practice Questions in Random Order—Article 90.2 through 320.30

Use the 2008 *NEC* to answer the following questions.

1. Where cables and nonmetallic raceways are installed parallel to framing members, the nearest outside surface of the cable or raceway shall be _____ the nearest edge of the framing member where nails or screws are likely to penetrate.

   (a) not less than 1¼ in. from
   (b) immediately adjacent to
   (c) not less than 1/16 in. from
   (d) 90° away from

2. The terminal of a wiring device for the connection of the equipment grounding conductor shall be identified by a green-colored, _____.

   (a) not readily removable terminal screw with a hexagonal head
   (b) hexagonal, not readily removable terminal nut
   (c) pressure wire connector
   (d) any of these

3. _____ is defined as the area between the top of direct-burial cable and the top surface of the finished grade.

   (a) Notch
   (b) Cover
   (c) Gap
   (d) none of these

4. Where the box is mounted on the surface, direct metal-to-metal contact between the device yoke and the box shall be permitted to ground the receptacle to the box if at least _____ of the insulating washers of the receptacle is removed.

   (a) one
   (b) two
   (c) three
   (d) none of these

5. A listed exposed work cover can be the grounding and bonding means when the device is attached to the cover with at least _____ fasteners and the cover mounting holes are located on a non-raised portion of the cover.

   (a) one
   (b) two
   (c) three
   (d) none of these

6. Where independent support wires of a ceiling assembly are used to support raceways, cable assemblies, or boxes above a ceiling, they shall be secured at _____ ends.

   (a) one
   (b) both
   (c) a or b
   (d) none of these

7. When counting the number of conductors in a box, a conductor running through the box with an unbroken loop or coil not less than twice the minimum length required for free conductors shall be counted as _____ conductor(s).

   (a) one
   (b) two
   (c) three
   (d) four

8. Equipment grounding conductor(s), and not more than _____ fixture wires smaller than 14 AWG can be omitted from the calculations where they enter the box from a domed luminaire or similar canopy and terminate within that box.

   (a) one
   (b) two
   (c) three
   (d) four

9. Enclosures not over 100 cu in. having threaded entries and not containing a device shall be considered to be adequately supported where _____ or more conduits are threaded wrenchtight into the enclosure and each conduit is secured within 3 ft of the enclosure.

    (a) one
    (b) two
    (c) three
    (d) none of these

10. For individual dwelling units of _____ dwellings, Table 310.15(B)(6) can be used to size 3-wire, single-phase, 120/240V service or feeder conductors that serve as the main power feeder.

    (a) one-family
    (b) two-family
    (c) multifamily
    (d) any of these

11. Cable wiring methods shall not be used as a means of support for _____.

    (a) other cables
    (b) raceways
    (c) nonelectrical equipment
    (d) all of these

12. The number and size of conductors permitted in a raceway is limited to _____.

    (a) permit heat to dissipate
    (b) prevent damage to insulation during installation
    (c) prevent damage to insulation during removal of conductors
    (d) all of these

13. Listed FMC and LFMC shall contain an equipment grounding conductor if the raceway is installed for the reason of _____.

    (a) physical protection
    (b) flexibility after installation
    (c) protection from moisture
    (d) communications systems

14. All conductors of a circuit, including the grounded and equipment grounding conductors, shall be contained within the same _____, unless permitted elsewhere in the Code.

    (a) raceway
    (b) cable
    (c) trench
    (d) all of these

15. Where portions of a cable raceway or sleeve are subjected to different temperatures and condensation is known to be a problem, the _____ shall be filled with an approved material to prevent the circulation of warm air to a colder section of the raceway or sleeve.

    (a) raceways
    (b) sleeve
    (c) a or b
    (d) none of these

16. Which conductor type has an insulation temperature rating of 90°C?

    (a) RH
    (b) RHW
    (c) THHN
    (d) TW

17. Surface-mounted enclosures shall be _____.

    (a) rigidly and securely fastened in place
    (b) supported by cables that protrude from the box
    (c) supported by cable entries from the top and permitted to rest against the supporting surface
    (d) none of these

18. Conduits or raceways through which moisture may contact live parts shall be _____ at either or both ends.

    (a) sealed
    (b) plugged
    (c) bushed
    (d) a or b

19. Exposed structural metal interconnected to form a metal building frame that is not intentionally grounded and is likely to become energized, shall be bonded to the _____.

    (a) service equipment enclosure
    (b) grounded conductor at the service
    (c) grounding electrode conductor where of sufficient size
    (d) any of these

Unit 3 — Practice Questions in Random Order—Article 90.2 through 320.30

20. Conductors other than service conductors shall not be installed in the same _____.
    (a) service raceway
    (b) service cable
    (c) enclosure
    (d) a or b

21. The noncurrent-carrying metal parts of service equipment, such as _____, shall be bonded together.
    (a) service raceways or service cable armor
    (b) service equipment enclosures containing service conductors, including meter fittings, boxes, or the like, interposed in the service raceway or armor
    (c) service cable trays
    (d) all of these

22. Overhead service conductors from the last pole or other aerial support to and including the splices, if any, are called _____ conductors.
    (a) service-entrance
    (b) service-drop
    (c) service-lateral
    (d) overhead service

23. _____ in dwelling units shall supply only loads within that dwelling unit or loads associated only with that dwelling unit.
    (a) Service-entrance conductors
    (b) Ground-fault protection
    (c) Branch circuits
    (d) none of these

24. In the *NEC*, the words _____ indicate a mandatory requirement.
    (a) shall
    (b) shall not
    (c) shall be permitted
    (d) a or b

25. When the *Code* uses _____, it means the identified actions are allowed but not required, and the authority having jurisdiction is not to restrict an installation from being done in that manner.
    (a) shall
    (b) shall not
    (c) shall be permitted
    (d) a or b

26. Lightning protection system ground terminals _____ be bonded to the building grounding electrode system.
    (a) shall
    (b) shall not
    (c) shall be permitted to
    (d) none of these

27. Each cable entering a cutout box _____.
    (a) shall be secured to the cutout box
    (b) can be sleeved through a chase
    (c) shall have a maximum of two cables per connector
    (d) all of these

28. An 800A fuse rated at 600V _____ on a 250V system.
    (a) shall not be used
    (b) shall be used
    (c) can be used
    (d) none of these

29. Supplementary overcurrent protection _____.
    (a) shall not be used in luminaires
    (b) may be used as a substitute for a branch-circuit overcurrent device
    (c) may be used to protect internal circuits of equipment
    (d) shall be readily accessible

30. For equipment rated 1,200A or more and over 6 ft wide that contains overcurrent devices, switching devices, or control devices; and where the entrance to the working space has a personnel door less than 25 ft from the working space, the door _____.
    (a) shall open either in or out with simple pressure and shall not have any lock
    (b) shall open in the direction of egress and be equipped with panic hardware or other devices so the door can open under simple pressure
    (c) shall be equipped with a locking means
    (d) shall be equipped with an electronic opener

31. Wiring shall be installed so that the completed system will be free from _____, other than as required or permitted elsewhere in the *Code*.
    (a) short circuits
    (b) ground faults
    (c) connections to the earth
    (d) all of these

32. A device that, when interrupting currents in its current-limiting range, reduces the current flowing in the faulted circuit to a magnitude substantially less than that obtainable in the same circuit if the device were replaced with a solid conductor having comparable impedance, is a(n) _____ protective device.

    (a) short-circuit
    (b) overload
    (c) ground-fault
    (d) current-limiting

33. The prospective symmetrical fault current at a nominal voltage to which an apparatus or system is able to be connected without sustaining damage exceeding defined acceptance criteria is known as _____.

    (a) short-circuit current rating
    (b) withstand rating
    (c) interrupting rating
    (d) available fault current

34. The current in amperes a conductor can carry continuously, where the temperature will not be raised in excess of the conductor's insulation temperature rating is called its _____. See 310.10 and 310.15 for details and examples. Figure 100–4

    (a) short-circuit rating
    (b) ground-fault rating
    (c) ampacity
    (d) all of these

35. Direct-buried conductors, cables, or raceways, which are subject to movement by settlement or frost, shall be arranged to prevent damage to the _____ or to equipment connected to the raceways.

    (a) siding of the building mounted on
    (b) landscaping around the cable or raceway
    (c) the enclosed conductors
    (d) expansion fitting

36. Receptacle outlets installed for a specific appliance in a dwelling unit, such as laundry equipment, shall be located within _____ of the intended location of the appliance.

    (a) sight
    (b) 3 ft
    (c) 6 ft
    (d) readily accessible, no maximum distance

37. The following systems shall be installed in accordance with the *NEC* requirements:

    (a) signaling
    (b) communications
    (c) electrical conductors, equipment, and raceways
    (d) all of these

38. Where a luminaire stud or hickey is present in the box, a _____ volume allowance in accordance with Table 314.16(b) shall be made for each type of fitting, based on the largest conductor present in the box.

    (a) single
    (b) double
    (c) single allowance for each gang
    (d) none of these

39. Where one or more equipment grounding conductors enter a box, a _____ volume allowance in accordance with Table 314.16(b) shall be made, based on the largest equipment grounding conductor.

    (a) single
    (b) double
    (c) triple
    (d) none of these

40. In straight pulls, the length of the box shall not be less than _____ times the trade size of the largest raceway.

    (a) six
    (b) eight
    (c) twelve
    (d) none of these

41. Where angle or U pulls are made, the distance between each raceway entry inside the box and the opposite wall of the box shall not be less than _____ times the trade size of the largest raceway in a row.

    (a) six
    (b) eight
    (c) twelve
    (d) none of these

42. Where it is unlikely that two or more noncoincident loads will be in use simultaneously, only the _____ loads used at one time is required to be used in computing the total load to a feeder.

    (a) smaller of the
    (b) largest of the
    (c) difference between the
    (d) none of these

43. Connection of conductors to terminal parts shall ensure a thoroughly good connection without damaging the conductors and shall be made by means of _____.

    (a) solder lugs
    (b) pressure connectors
    (c) splices to flexible leads
    (d) any of these

44. Grounding electrode conductors shall be made of:

    (a) solid
    (b) stranded
    (c) insulated or bare
    (d) any of these

45. A _____ wiring method can be used for a surface extension from a cover, and the wiring method shall include an equipment grounding conductor.

    (a) solid
    (b) flexible
    (c) rigid
    (d) cord

46. When installing direct-buried cables, a _____ shall be used at the end of a conduit that terminates underground.

    (a) splice kit
    (b) terminal fitting
    (c) bushing
    (d) b or c

47. Short-radius conduit bodies such as capped elbows, and service-entrance elbows that enclose conductors 6 AWG and smaller shall not contain _____.

    (a) splices
    (b) taps
    (c) devices
    (d) any of these

48. Type AC cable can be supported and secured by _____.

    (a) staples
    (b) cable ties
    (c) straps
    (d) all of these

49. Working space shall not be used for _____.

    (a) storage
    (b) raceways
    (c) lighting
    (d) accessibility

50. A circuit breaker with a _____ voltage rating, such as 240V or 480V, can be used where the nominal voltage between any two conductors does not exceed the circuit breaker's voltage rating.

    (a) straight
    (b) slash
    (c) high
    (d) low

UNIT 4 — Practice Questions in Straight Order—Article 100 through 378.10

Use the 2008 *NEC* to answer the following questions.

1. A panel, including buses and automatic overcurrent devices, designed to be placed in a cabinet or cutout box and accessible only from the front is called a _____.

   (a) switchboard
   (b) disconnect
   (c) panelboard
   (d) switch

2. Special permission would be the written consent from the _____.

   (a) testing laboratory
   (b) manufacture
   (c) owner
   (d) AHJ

3. A(n) _____ is intended to provide limited overcurrent protection for specific applications and utilization equipment, such as luminaires and appliances. This limited protection is in addition to the protection provided by the required branch-circuit overcurrent protective device.

   (a) supplementary overcurrent protective device
   (b) surge protection device
   (c) arc-fault circuit interrupter
   (d) Class A GFCI

4. The temperature rating associated with the ampacity of a _____ shall be so selected and coordinated so as not to exceed the lowest temperature rating of any connected termination, conductor, or device.

   (a) terminal
   (b) conductor
   (c) device
   (d) all of these

5. Each disconnecting means shall be legibly marked to indicate its purpose unless located and arranged so _____.

   (a) that it can be locked out and tagged
   (b) it is not readily accessible
   (c) the purpose is evident
   (d) that it operates at less than 300 volts-to-ground

6. Multiwire branch circuits shall _____.

   (a) supply only line-to-neutral loads
   (b) not be permitted in dwelling units
   (c) have their conductors originate from different panelboards
   (d) none of these

7. In a dwelling unit, at least one lighting outlet _____ shall be located at the point of entry to the attic, under-floor space, utility room, or basement where these spaces are used for storage or contain equipment requiring servicing.

   (a) that is unswitched
   (b) containing a switch
   (c) controlled by a wall switch
   (d) b or c

8. Service-drop conductors shall have _____.

   (a) sufficient ampacity to carry the load
   (b) adequate mechanical strength
   (c) a or b
   (d) a and b

9. Circuit breakers used to switch high-intensity discharge lighting circuits shall be listed and marked as _____.

   (a) SWD
   (b) HID
   (c) a or b
   (d) a and b

Mike Holt Enterprises, Inc. • www.MikeHolt.com • 1.888.NEC.CODE (1.888.632.2633) 411

Unit 4 — Practice Questions in Straight Order—Article 100 through 378.10

10. A circuit breaker with a _____ rating, such as 120/240V or 277/480V can be used on a solidly grounded circuit where the nominal voltage of any conductor to ground does not exceed the lower of the two values, and the nominal voltage between any two conductors does not exceed the higher value.

    (a) straight
    (b) slash
    (c) high
    (d) low

11. Main bonding jumpers and system bonding jumpers shall not be smaller than the sizes shown in _____.

    (a) Table 250.66
    (b) Table 250.122
    (c) Table 310.16
    (d) Chapter 9, Table 8

12. A grounded conductor shall not be connected to normally noncurrent-carrying metal parts of equipment on the _____ of the point of grounding of the separately derived system except as otherwise permitted in Article 250.

    (a) supply
    (b) grounded side
    (c) high-voltage side
    (d) load side

13. An unspliced _____ that is sized based on the derived phase conductors shall be used to connect the equipment grounding conductors of a separately derived system to the grounded conductor.

    (a) system bonding jumper
    (b) equipment grounding conductor
    (c) grounded conductor
    (d) grounding electrode conductor

14. In an area served by a separately derived system, the _____ shall be connected to the grounded conductor of the separately derived system.

    (a) structural steel
    (b) metal piping
    (c) metal building skin
    (d) a and b

15. Equipment bonding jumpers on the supply side of the service shall be no smaller than the sizes shown in _____.

    (a) Table 250.66
    (b) Table 250.122
    (c) Table 310.16
    (d) Table 310.15(B)(6)

16. The bonding jumper used to bond the metal water piping system shall be sized in accordance with _____.

    (a) Table 250.66
    (b) Table 250.122
    (c) Table 310.16
    (d) Table 310.15(B)(6)

17. A building or structure that is supplied by a feeder shall have the interior metal water piping system bonded with a conductor sized in accordance with _____.

    (a) Table 250.66
    (b) Table 250.122
    (c) Table 310.16
    (d) none of these

18. Wiring methods installed behind panels that allow access shall be _____ according to their applicable articles.

    (a) supported
    (b) painted
    (c) in a metal raceway
    (d) all of these

19. Metal raceways shall not be _____ by welding to the raceway.

    (a) supported
    (b) terminated
    (c) connected
    (d) all of these

20. Electrical installations in hollow spaces, vertical shafts, and ventilation or air-handling ducts shall be made so that the possible spread of fire or products of combustion is not _____.

    (a) substantially increased
    (b) allowed
    (c) inherent
    (d) possible

21. When mounting an enclosure in a finished surface, the enclosure shall be _____ secured to the surface by clamps, anchors, or fittings identified for the application.

    (a) temporarily
    (b) partially
    (c) never
    (d) rigidly

22. Conductors, splices or terminations in a handhole enclosure shall be listed as _____.

    (a) suitable for wet locations
    (b) suitable for damp locations
    (c) suitable for direct burial in the earth
    (d) none of these

23. Type NM cable shall be _____.

    (a) marked
    (b) approved
    (c) identified
    (d) listed

24. Type NM cable shall not be used _____.

    (a) in other than dwelling units
    (b) in the air void of masonry block not subject to excessive moisture
    (c) for exposed work
    (d) embedded in poured cement, concrete, or aggregate

25. Type NM cable shall be protected from physical damage by _____.

    (a) EMT
    (b) PVC conduit
    (c) RMC without an overall nonmetallic covering
    (d) any of these

26. Type NM cable on a wall of an unfinished basement installed in a listed raceway shall have a(n) _____ installed at the point where the cable enters the raceway.

    (a) insulating bushing or adapter
    (b) sealing fitting
    (c) bonding bushing
    (d) junction box

27. Grommets or bushings for the protection of Type NM cable shall be _____ for the purpose.

    (a) marked
    (b) approved
    (c) identified
    (d) listed

28. Type NM cable can be supported and secured by _____.

    (a) staples
    (b) cable ties
    (c) straps
    (d) all of these

29. Where more than two Type NM cables are installed through the same opening in wood framing that is to be fire- or draft-stopped, the allowable ampacity of each conductor shall be _____.

    (a) no more than 20A
    (b) adjusted in accordance with Table 310.15(B)(2)(a)
    (c) limited to 30A
    (d) calculated by an engineer

30. The insulation temperature rating of conductors in Type NM cable shall be _____.

    (a) 60°C
    (b) 75°C
    (c) 90°C
    (d) 105°C

31. Type TC cable can be used _____.

    (a) for power and lighting circuits
    (b) in cable trays
    (c) in Class 1 control circuits
    (d) all of these

32. Type _____ cable is an assembly primarily for services not over 600V.

    (a) NM
    (b) TC
    (c) SE
    (d) none of these

33. Type SE cable can be used as _____ in wiring systems where all of the circuit conductors of the cable are of the thermoset or thermoplastic type.

    (a) branch circuits
    (b) feeders
    (c) a or b
    (d) neither a or b

34. Type SE cables can be used for branch circuits or feeders where the insulated conductors are used for circuit wiring and the uninsulated conductor is used only for _____ purposes.

    (a) grounded connection
    (b) equipment grounding
    (c) remote control and signaling
    (d) none of these

35. Type _____ cable is a factory assembly of conductors with an overall covering of nonmetallic material suitable for direct burial in the earth.

    (a) NM
    (b) UF
    (c) SE
    (d) TC

36. Type UF cable shall not be used in _____.

    (a) motion picture studios
    (b) storage battery rooms
    (c) hoistways
    (d) all of these

37. Type UF cable shall not be used _____.

    (a) in any hazardous (classified) location except as otherwise permitted in this Code
    (b) embedded in poured cement, concrete, or aggregate
    (c) where exposed to direct rays of the sun, unless identified as sunlight resistant
    (d) all of these

38. The ampacity of Type UF cable shall be that of _____ in accordance with 310.15.

    (a) 60°C conductors
    (b) 75°C conductors
    (c) 90°C conductors
    (d) 105°C

39. Where practicable, contact of dissimilar metals shall be avoided in an IMC raceway installation to prevent _____.

    (a) corrosion
    (b) galvanic action
    (c) shorts
    (d) none of these

40. A run of IMC shall not contain more than the equivalent of _____ quarter bends between pull points such as conduit bodies and boxes.

    (a) one
    (b) two
    (c) three
    (d) four

41. Trade size 1 IMC shall be supported at intervals not exceeding _____.

    (a) 8 ft
    (b) 10 ft
    (c) 12 ft
    (d) 14 ft

42. Where IMC enters a box, fitting, or other enclosure, _____ shall be provided to protect the wire from abrasion, unless the design of the box, fitting, or enclosure affords equivalent protection.

    (a) a bushing
    (b) duct seal
    (c) electrical tape
    (d) seal fittings

43. Galvanized steel, stainless steel and red brass RMC can be installed in concrete, in direct contact with the earth, or in areas subject to severe corrosive influences when protected by _____ and judged suitable for the condition.

    (a) ceramic
    (b) corrosion protection
    (c) backfill
    (d) a natural barrier

44. Aluminum fittings and enclosures can be used with _____ conduit where not subject to severe corrosive influences.

    (a) steel rigid metal
    (b) aluminum rigid metal
    (c) PVC-coated rigid conduit only
    (d) a and b

45. The minimum radius of a field bend on trade size 1¼ RMC is _____.

    (a) 7 in.
    (b) 8 in.
    (c) 10 in.
    (d) 14 in.

46. A run of RMC shall not contain more than the equivalent of _____ quarter bends between pull points such as conduit bodies and boxes.

    (a) one
    (b) two
    (c) three
    (d) four

47. RMC can be unsupported between enclosures when the raceway contains no _____, oversized knockouts are not encountered, and the raceway does not exceed 18" in length.

    (a) conductors
    (b) couplings
    (c) connectors
    (d) any of these

48. Running threads shall not be used on RMC for connection at _____.

    (a) boxes
    (b) cabinets
    (c) couplings
    (d) meter sockets

49. Where RMC enters a box, fitting, or other enclosure, _____ shall be provided to protect the wire from abrasion, unless the design of the box, fitting, or enclosure affords equivalent protection.

    (a) a bushing
    (b) duct seal
    (c) electrical tape
    (d) seal fittings

50. FMC shall not be installed _____.

    (a) in wet locations
    (b) embedded in poured concrete
    (c) where subject to physical damage
    (d) all of these

51. Bends in FMC shall be made so that the conduit is not damaged and the internal diameter of the conduit is _____.

    (a) larger than 3/8 in.
    (b) not effectively reduced
    (c) increased
    (d) larger than 1 in.

52. Bends in FMC _____ between pull points.

    (a) shall not be made
    (b) need not be limited (in degrees)
    (c) shall not exceed 360 degrees
    (d) shall not exceed 180 degrees

53. Cut ends of FMC shall be trimmed or otherwise finished to remove rough edges, except where fittings _____.

    (a) are the crimp-on type
    (b) thread into the convolutions
    (c) contain insulated throats
    (d) are listed for grounding

54. FMC shall be supported and secured _____.

    (a) at intervals not exceeding 4½ ft
    (b) within 8 in. on each side of a box where fished
    (c) where fished
    (d) at intervals not exceeding 6 ft

55. Flexible metal conduit shall not be required to be _____ where fished between access points through concealed spaces in finished buildings or structures and supporting is impractical.

    (a) secured
    (b) supported
    (c) complete
    (d) (a) and (b)

56. In a concealed FMC installation, _____ connectors shall not be used.

    (a) straight
    (b) angle
    (c) grounding-type
    (d) none of these

57. The use of LFMC shall be permitted for _____.

    (a) direct burial where listed and marked for the purpose
    (b) exposed work
    (c) concealed work
    (d) all of these

58. LFMC shall not be required to be secured or supported where fished between access points through _____ spaces in finished buildings or structures and supporting is impractical.

    (a) concealed
    (b) exposed
    (c) hazardous
    (d) completed

59. _____ connectors shall not be used for concealed installations of LFMC.

    (a) Straight
    (b) Angle
    (c) Grounding-type
    (d) none of these

60. When LFMC is used to connect equipment requiring flexibility after installation, a(n) _____ conductor shall be installed.

    (a) main bonding
    (b) grounded
    (c) equipment grounding
    (d) none of these

61. Extreme _____ may cause PVC conduit to become brittle, and therefore more susceptible to damage from physical contact.

    (a) sunlight
    (b) corrosive conditions
    (c) heat
    (d) cold

62. PVC conduit shall not be used _____, unless specifically permitted.

    (a) in hazardous (classified) locations
    (b) for the support of luminaires or other equipment
    (c) where subject to physical damage unless identified for such use
    (d) all of these

63. Bends in PVC conduit shall _____ between pull points.

    (a) not be made
    (b) not be limited in degrees
    (c) be limited to 360 degrees
    (d) be limited to 180 degrees

64. PVC conduit shall be securely fastened within _____ of each box.

    (a) 6 in.
    (b) 12 in.
    (c) 24 in.
    (d) 36 in.

65. _____ is a nonmetallic raceway of circular cross section, with couplings, connectors, and fittings for the installation of electrical conductors.

    (a) ENT
    (b) PVC conduit
    (c) LFNMC
    (d) HDPE

66. HDPE shall be permitted _____.

    (a) in discrete lengths
    (b) in continuous lengths from a reel
    (c) in 20 ft lengths
    (d) a or b

67. HDPE conduit shall not be subjected to ambient temperatures in excess of _____, unless listed otherwise.

    (a) 50°C
    (b) 60°C
    (c) 75°C
    (d) 90°C

68. Bends shall be made in HDPE conduit _____.

    (a) in a manner that will not damage the raceway
    (b) so as not to significantly reduce the internal diameter of the raceway
    (c) with mechanical bending tools
    (d) a and b

69. Joints between HDPE conduit shall be made using _____.

    (a) expansion fittings
    (b) an approved method
    (c) a listed method
    (d) none of these

70. HDPE conduit can be joined using _____.

    (a) heat fusion
    (b) electrofusion
    (c) mechanical fittings
    (d) any of these

71. Where equipment grounding is required for an installation of HDPE, a separate equipment grounding conductor shall be _____.

    (a) an insulated copper conductor
    (b) installed within the conduit
    (c) a stranded bare copper conductor
    (d) a solid bare copper conductor

72. ___ is a factory assembly of conductors or cables inside a nonmetallic, smooth-wall conduit with a circular cross section.

    (a) HDPE
    (b) ENT
    (c) NUCC
    (d) RTRC

73. NUCC and its associated fittings shall be _____.

    (a) listed
    (b) metallic
    (c) identified
    (d) none of these

74. The use of NUCC shall be permitted _____.

    (a) for direct burial underground installations
    (b) to be encased or embedded in concrete
    (c) in cinder fill
    (d) all of these

75. NUCC shall not be used _____.

    (a) in exposed locations
    (b) inside buildings
    (c) in hazardous (classified) locations
    (d) all of these

76. Bends in NUCC can be _____ so that the conduit will not be damaged and the internal diameter of the conduit will not be effectively reduced.

    (a) manually made
    (b) made only with approved benders
    (c) made with RMC bending shoes
    (d) made using an open flame torch

77. Bends in NUCC shall _____ between termination points.

    (a) not be made
    (b) not be limited in degrees
    (c) be limited to 360 degrees
    (d) be limited to 180 degrees

78. In order to _____ NUCC, the conduit shall be trimmed away from the conductors or cables using an approved method that will not damage the conductor or cable insulation or jacket.

    (a) facilitate installing
    (b) enhance the appearance of the installation of
    (c) terminate
    (d) provide safety to the persons installing

79. Joints between NUCC, fittings, and boxes shall be made by _____.

    (a) a qualified person
    (b) set screw fittings
    (c) an approved method
    (d) exothermic welding

80. RTRC conduit can be described as a rigid nonmetallic conduit of circular cross section, with integral or associated _____ for the installation of electrical conductors and cables.

    (a) couplings
    (b) connectors
    (c) fittings
    (d) all of these

81. ___ is a listed raceway of circular cross section having an outer liquidtight, nonmetallic, sunlight-resistant jacket over a flexible inner core.

    (a) LFNC
    (b) ENT
    (c) NUCC
    (d) RTRC

82. The number of conductors permitted in LFNC shall not exceed the percentage fill specified in _____.

    (a) Chapter 9, Table 1
    (b) Table 250.66
    (c) Table 310.16
    (d) 240.6

83. Bends in LFNC shall be made so that the conduit will not be damaged and the internal diameter of the conduit will not be effectively reduced. Bends can be made _____.

    (a) manually without auxiliary equipment
    (b) with bending equipment identified for the purpose
    (c) with any kind of conduit bending tool that will work
    (d) by use of an open flame torch

84. Bends in LFNC shall _____ between pull points.

    (a) not be made
    (b) not be limited in degrees
    (c) be limited to 360 degrees
    (d) be limited to 180 degrees

85. _____ is a listed thinwall, metallic tubing of circular cross section used for the installation and physical protection of electrical conductors when joined together with listed fittings.

    (a) LFNC
    (b) EMT
    (c) NUCC
    (d) RTRC

86. When EMT is installed in wet locations, all supports, bolts, straps, and screws shall be _____.

    (a) of corrosion-resistant materials
    (b) protected against corrosion
    (c) a or b
    (d) of nonmetallic materials only

87. EMT shall not be used where _____.

    (a) subject to severe physical damage
    (b) protected from corrosion only by enamel
    (c) used for the support of luminaires
    (d) any of these

88. ENT is composed of a material resistant to moisture and chemical atmospheres, and is _____.

    (a) flexible
    (b) flame retardant
    (c) fireproof
    (d) flammable

89. When a building is supplied with a(n) _____ fire sprinkler system, ENT can be installed exposed or concealed in buildings of any height.

    (a) listed
    (b) identified
    (c) NFPA 13-2007
    (d) none of these

90. ENT and fittings can be_____, provided fittings identified for this purpose are used.

    (a) encased in poured concrete
    (b) embedded in a concrete slab on grade where the tubing is placed on sand or approved screenings
    (c) a or b
    (d) none of these

91. ENT shall not be used where exposed to the direct rays of the sun, unless identified as _____.

    (a) high-temperature rated
    (b) sunlight resistant
    (c) Schedule 80
    (d) never can be

92. The number of conductors permitted in ENT shall not exceed the percentage fill specified in _____.

    (a) Chapter 9, Table 1
    (b) Table 250.66
    (c) Table 310.16
    (d) 240.6

93. Cut ends of ENT shall be trimmed inside and _____ to remove rough edges.

    (a) outside
    (b) tapered
    (c) filed
    (d) beveled

94. Electric nonmetallic tubing shall not be required to be secured where fished between access points for _____ work in finished buildings or structures and supporting is impractical.

    (a) concealed
    (b) exposed
    (c) hazardous
    (d) completed

95. Wireways shall be permitted for _____.

    (a) exposed work
    (b) concealed work
    (c) wet locations if listed for the purpose
    (d) a and c

96. Wireways can pass transversely through a wall _____.

    (a) if the length passing through the wall is unbroken
    (b) if the wall is of fire-rated construction
    (c) in hazardous (classified) locations
    (d) if the wall is not of fire-rated construction

97. Where insulated conductors are deflected within a metal wireway, the wireway shall be sized to meet the bending requirements corresponding to _____ wire per terminal in Table 312.6(A).

    (a) one
    (b) two
    (c) three
    (d) none of these

98. Power distribution blocks installed in metal wireways shall _____.

   (a) allow for sufficient wire-bending space at terminals
   (b) not have uninsulated exposed live parts
   (c) a or b
   (d) a and b

99. _____ is a flame-retardant raceway with hinged or removable covers for housing and protecting electric conductors and cable, and in which conductors are placed after its been installed as a complete system.

   (a) Nonmetallic cable
   (b) Nonmetallic raceway
   (c) Nonmetallic wireway
   (d) none of these

100. Nonmetallic wireways shall be permitted for _____.

   (a) exposed work
   (b) concealed work
   (c) wet locations if listed for the purpose
   (d) a and c

Unit 4: Practice Questions in Random Order—Article 90.1 through 358.12

Use the 2008 *NEC* to answer the following questions.

1. A circuit breaker with a _____ rating, such as 120/240V or 277/480V can be used on a solidly grounded circuit where the nominal voltage of any conductor to ground does not exceed the lower of the two values, and the nominal voltage between any two conductors does not exceed the higher value.

   (a) straight
   (b) slash
   (c) high
   (d) low

2. In a concealed FMC installation, _____ connectors shall not be used.

   (a) straight
   (b) angle
   (c) grounding-type
   (d) none of these

3. _____ connectors shall not be used for concealed installations of LFMC.

   (a) Straight
   (b) Angle
   (c) Grounding-type
   (d) none of these

4. In an area served by a separately derived system, the _____ shall be connected to the grounded conductor of the separately derived system.

   (a) structural steel
   (b) metal piping
   (c) metal building skin
   (d) a and b

5. EMT shall not be used where _____.

   (a) subject to severe physical damage
   (b) protected from corrosion only by enamel
   (c) used for the support of luminaires
   (d) any of these

6. Electrical installations in hollow spaces, vertical shafts, and ventilation or air-handling ducts shall be made so that the possible spread of fire or products of combustion is not _____.

   (a) substantially increased
   (b) allowed
   (c) inherent
   (d) possible

7. Service-drop conductors shall have _____.

   (a) sufficient ampacity to carry the load
   (b) adequate mechanical strength
   (c) a or b
   (d) a and b

8. Conductors, splices or terminations in a handhole enclosure shall be listed as _____.

   (a) suitable for wet locations
   (b) suitable for damp locations
   (c) suitable for direct burial in the earth
   (d) none of these

9. Extreme _____ may cause PVC conduit to become brittle, and therefore more susceptible to damage from physical contact.

   (a) sunlight
   (b) corrosive conditions
   (c) heat
   (d) cold

10. A(n) _____ is intended to provide limited overcurrent protection for specific applications and utilization equipment, such as luminaires and appliances. This limited protection is in addition to the protection provided by the required branch-circuit overcurrent protective device.

    (a) supplementary overcurrent protective device
    (b) surge protection device
    (c) arc-fault circuit interrupter
    (d) Class A GFCI

11. A grounded conductor shall not be connected to normally noncurrent-carrying metal parts of equipment on the _____ of the point of grounding of the separately derived system except as otherwise permitted in Article 250.

    (a) supply
    (b) grounded side
    (c) high voltage side
    (d) load side

12. Multiwire branch circuits shall _____.

    (a) supply only line-to-neutral loads
    (b) not be permitted in dwelling units
    (c) have their conductors originate from different panelboards
    (d) none of these

13. Wiring methods installed behind panels that allow access shall be _____ according to their applicable articles.

    (a) supported
    (b) painted
    (c) in a metal raceway
    (d) all of these

14. Metal raceways shall not be _____ by welding to the raceway.

    (a) supported
    (b) terminated
    (c) connected
    (d) all of these

15. Circuit breakers used to switch high-intensity discharge lighting circuits shall be listed and marked as _____.

    (a) SWD
    (b) HID
    (c) a or b
    (d) a and b

16. A panel, including buses and automatic overcurrent devices, designed to be placed in a cabinet or cutout box and accessible only from the front.

    (a) switchboard
    (b) disconnect
    (c) panelboard
    (d) switch

17. An unspliced _____ that is sized based on the derived phase conductors shall be used to connect the equipment grounding conductors of a separately derived system to the grounded conductor.

    (a) system bonding jumper
    (b) equipment grounding conductor
    (c) grounded conductor
    (d) grounding electrode conductor

18. Main bonding jumpers and system bonding jumpers shall not be smaller than the sizes shown in _____.

    (a) Table 250.66
    (b) Table 250.122
    (c) Table 310.16
    (d) Chapter 9, Table 8

19. Equipment bonding jumpers on the supply side of the service shall be no smaller than the sizes shown in _____.

    (a) Table 250.66
    (b) Table 250.122
    (c) Table 310.16
    (d) Table 310.15(B)(6)

20. The bonding jumper used to bond the metal water piping system shall be sized in accordance with _____.

    (a) Table 250.66
    (b) Table 250.122
    (c) Table 310.16
    (d) Table 310.15(B)(6)

21. A building or structure that is supplied by a feeder shall have the interior metal water piping system bonded with a conductor sized in accordance with _____.

    (a) Table 250.66
    (b) Table 250.122
    (c) Table 310.16
    (d) none of these

22. When mounting an enclosure in a finished surface, the enclosure shall be _____ secured to the surface by clamps, anchors, or fittings identified for the application.

    (a) temporarily
    (b) partially
    (c) never
    (d) rigidly

23. The temperature rating associated with the ampacity of a _____ shall be so selected and coordinated so as not to exceed the lowest temperature rating of any connected termination, conductor, or device.

    (a) terminal
    (b) conductor
    (c) device
    (d) all of these

24. Special permission would be the written consent from the _____.

    (a) testing laboratory
    (b) manufacturer
    (c) owner
    (d) AHJ

25. In a dwelling unit, at least one lighting outlet _____ shall be located at the point of entry to the attic, under-floor space, utility room, or basement where these spaces are used for storage or contain equipment requiring servicing.

    (a) that is unswitched
    (b) containing a switch
    (c) controlled by a wall switch
    (d) b or c

26. Each disconnecting means shall be legibly marked to indicate its purpose unless located and arranged so _____.

    (a) that it can be locked out and tagged
    (b) it is not readily accessible
    (c) the purpose is evident
    (d) that it operates at less than 300 volts-to-ground

27. Type AC cable installed in thermal insulation shall have conductors that are rated at 90°C. The ampacity of the cable in this application shall be based on _____.

    (a) that of 60°C conductors
    (b) that of 75°C conductors
    (c) that of 90°C conductors
    (d) none of these

28. An arc-fault circuit interrupter can be located at the first outlet to provide protection for the remaining portion of the branch circuit if _____.

    (a) the arc-fault circuit interrupter is installed within 6 ft of the branch-circuit overcurrent device
    (b) the circuit conductors up to the arc-fault circuit interrupter are in a metal raceway or Type AC cable
    (c) a and b
    (d) the branch circuit serves only lighting loads

29. Joints between PVC conduit, couplings, fittings, and boxes shall be made by _____.

    (a) the authority having jurisdiction
    (b) set screw fittings
    (c) an approved method
    (d) expansion fittings

30. When a single equipment grounding conductor is used for multiple circuits in the same raceway, cable or cable tray, the single equipment grounding conductor shall be sized according to _____.

    (a) the combined rating of all the overcurrent devices
    (b) the largest overcurrent device of the multiple circuits
    (c) the combined rating of all the loads
    (d) any of these

31. High-impedance grounded neutral systems shall be permitted for three-phase ac systems of 480 volts to 1,000 volts where _____.

    (a) the conditions of maintenance ensure that only qualified persons service the installation
    (b) ground detectors are installed on the system
    (c) line-to-neutral loads are not served
    (d) all of these

32. Outside secondary conductors can be of unlimited length without overcurrent protection at the point they receive their supply if:

    (a) The conductors are suitably protected from physical damage.
    (b) The conductors terminate at a single circuit breaker or a single set of fuses that limits the load to the ampacity of the conductors.
    (c) a and b
    (d) none of these

33. The next higher standard rating overcurrent device above the ampacity of the ungrounded conductors being protected is permitted, provided all of the following conditions are met:

    (a) The conductors do not supply multioutlet receptacle branch circuits.
    (b) The ampacity of a conductor, after ampacity adjustment and/or correction, doesn't correspond with the standard rating of a fuse or circuit breaker.
    (c) The protection device rating doesn't exceed 800A.
    (d) all of these

34. Ground-fault protection of equipment shall be required for a feeder disconnect if _____.

    (a) the disconnect is rated 800A, 208V
    (b) the disconnect is rated 800A, 480V
    (c) the disconnect is rated 1,000A, 208V
    (d) the disconnect is rated 1,000A, 480V

35. The grounding electrode conductor shall be connected to the grounded service conductor at _____.

    (a) the load end of the service drop
    (b) the meter equipment
    (c) the service disconnect
    (d) any of these

36. The two to six disconnects for a disconnecting means for a building supplied by a feeder shall be _____.

    (a) the same size
    (b) grouped
    (c) in the same enclosure
    (d) none of these

37. When the service contains two to six service disconnecting means, they shall be _____.

    (a) the same size
    (b) grouped
    (c) in the same enclosure
    (d) none of these

38. Where an equipment bonding jumper of the wire type is run with the derived phase conductors from the source of a separately derived system to the first disconnecting means, it shall be sized in accordance with 250.102(C), based on _____.

    (a) the size of the primary conductors
    (b) the size of the secondary overcurrent protection
    (c) the size of the derived phase conductors
    (d) one third the size of the primary grounded conductor

39. Outside feeder tap conductors can be of unlimited length without overcurrent protection at the point they receive their supply if:

    (a) The tap conductors are suitably protected from physical damage.
    (b) The tap conductors terminate at a single circuit breaker or a single set of fuses that limits the load to the ampacity of the conductors.
    (c) a and b
    (d) none of these

40. When applying the demand factors of Table 220.56, the feeder or service demand load shall not be less than the sum of _____.

    (a) the total number of receptacles at 180 VA per receptacle outlet
    (b) the VA rating of all of the small-appliance branch circuits combined
    (c) the largest two kitchen equipment loads
    (d) the kitchen heating and air-conditioning loads

41. Where required by the *Code*, markings on all electrical equipment shall contain voltage, current, wattage, or other ratings with durability to withstand _____.

    (a) the voltages encountered
    (b) painting and other finishes applied
    (c) the environment involved
    (d) lack of planning by the installer

42. Conductors smaller than 1/0 AWG can be connected in parallel to supply control power, provided _____.

    (a) they are all contained within the same raceway or cable
    (b) each parallel conductor has an ampacity sufficient to carry the entire load
    (c) the circuit overcurrent device rating does not exceed the ampacity of any individual parallel conductor
    (d) all of these

43. Cut ends of RMC shall be _____ or otherwise finished to remove rough edges.

    (a) threaded
    (b) reamed
    (c) painted
    (d) galvanized

44. For circuits over 250 volts-to-ground (277/480V), electrical continuity can be maintained between a box or enclosure where no oversized, concentric, or eccentric knockouts are encountered, and a metal conduit by _____.

    (a) threadless fittings for cables with metal sheath
    (b) double locknuts on threaded conduit (one inside and one outside the box or enclosure)
    (c) fittings that have shoulders that seat firmly against the box with a locknut on the inside or listed fittings
    (d) all of these

45. Service equipment, service raceways, and service conductor enclosures shall be bonded _____.

    (a) to the grounded service conductor
    (b) by threaded raceways into enclosures, couplings, hubs, conduit bodies, etc.
    (c) by listed bonding devices with bonding jumpers
    (d) any of these

46. NUCC larger than _____ shall not be used.

    (a) trade size 1
    (b) trade size 2
    (c) trade size 3
    (d) trade size 4

47. Where a service raceway enters a building or structure from a(n) _____, it shall be sealed in accordance with 300.5(G).

    (a) transformer vault
    (b) underground distribution system
    (c) cable tray
    (d) overhead rack

48. Any current in excess of the rated current of equipment or the ampacity of a conductor is called _____.

    (a) trip current
    (b) faulted
    (c) overcurrent
    (d) shorted

49. The *Code* isn't a design specification standard or instruction manual for the untrained and unqualified.

    (a) True
    (b) False

50. The *NEC* does not cover electrical installations in ships, watercraft, railway rolling stock, aircraft, or automotive vehicles.

    (a) True
    (b) False

Unit 5

Practice Questions in Straight Order—Article 378.10 through 410.151

Use the 2008 *NEC* to answer the following questions.

1. Nonmetallic wireways can pass transversely through a wall _____.

   (a) if the length through the wall is unbroken
   (b) if the wall is not of fire-rated construction
   (c) in hazardous (classified) locations
   (d) if the wall is of fire-rated construction

2. The sum of the cross-sectional areas of all contained conductors at any cross section of a nonmetallic wireway shall not exceed _____.

   (a) 20 percent
   (b) 30 percent
   (c) 40 percent
   (d) 50 percent

3. Expansion fittings for nonmetallic wireways shall be provided to compensate for thermal expansion and contraction, where the length change is expected to be _____ or greater in a straight run.

   (a) 1/16 in.
   (b) 1/4 in.
   (c) 1/2 in.
   (d) 6 in.

4. Conductors, including splices and taps, shall not fill the nonmetallic wireway to more than _____ percent of its area at that point.

   (a) 25
   (b) 75
   (c) 80
   (d) 125

5. A multioutlet assembly can be installed in _____.

   (a) dry locations
   (b) wet locations
   (c) a and b
   (d) damp locations

6. A multioutlet assembly shall not be installed _____.

   (a) in hoistways
   (b) where subject to severe physical damage
   (c) where subject to corrosive vapors
   (d) all of these

7. A strut-type channel raceway shall not be installed _____.

   (a) in concealed locations
   (b) where subject to corrosive vapors if protected solely by enamel
   (c) a or b
   (d) none of these

8. A surface mount strut-type channel raceway shall be secured to the mounting surface with retention straps external to the channel at intervals not exceeding _____ and within 3 ft of each outlet box, cabinet, junction box, or other channel raceway termination.

   (a) 3 ft
   (b) 5 ft
   (c) 6 ft
   (d) 10 ft

9. Conductors, including splices and taps, shall not fill a strut-type channel raceway to more than _____ percent of its area at that point.

   (a) 25
   (b) 75
   (c) 80
   (d) 125

Unit 5 — Practice Questions in Straight Order—Article 378.10 through 410.151

10. The maximum number of conductors permitted in any surface raceway shall be _____.

    (a) no more than 30 percent of the inside diameter
    (b) no greater than the number for which it was designed
    (c) no more than 75 percent of the cross-sectional area
    (d) that which is permitted in Table 312.6(A)

11. Surface metal raceways shall be secured and supported at intervals _____.

    (a) in accordance with the manufacturer's installation instructions
    (b) appropriate for the building design
    (c) not exceeding 4 ft
    (d) not exceeding 8 ft

12. The conductors, including splices and taps, in a metal surface raceway having a removable cover shall not fill the raceway to more than _____ percent of its cross-sectional area at that point.

    (a) 38
    (b) 40
    (c) 53
    (d) 75

13. Surface metal raceway enclosures providing a transition from other wiring methods shall have a means for connecting a(n) _____.

    (a) grounded conductor
    (b) ungrounded conductor
    (c) equipment grounding conductor
    (d) all of these

14. Surface nonmetallic raceways shall be permitted _____.

    (a) in dry locations
    (b) where concealed
    (c) in hoistways
    (d) all of these

15. The maximum number of conductors permitted in any surface nonmetallic raceway shall be _____.

    (a) no more than 30 percent of the inside diameter
    (b) no greater than the number for which it was designed
    (c) no more than 75 percent of the cross-sectional area
    (d) that which is permitted in Table 312.6(A)

16. The conductors, including splices and taps, in a surface nonmetallic surface having a cover capable of being opened in place, shall not fill the raceway to more than _____ percent of its cross-sectional area at that point.

    (a) 38
    (b) 40
    (c) 53
    (d) 75

17. Cable trays can be used as a support system for _____.

    (a) service conductors, feeders, and branch circuits
    (b) communications circuits
    (c) control and signaling circuits
    (d) all of these

18. Cable trays and their associated fittings shall be _____ for the intended use.

    (a) listed
    (b) approved
    (c) identified
    (d) none of these

19. Any of the following wiring methods can be installed in a cable tray:

    (a) metal raceways
    (b) nonmetallic raceways
    (c) cables
    (d) all of these

20. Cable tray systems shall not be used _____.

    (a) in hoistways
    (b) where subject to severe physical damage
    (c) in hazardous (classified) locations
    (d) a and b

21. Supports for cable trays shall be provided in accordance with _____.

    (a) installation instructions
    (b) the *NEC*
    (c) listing requirements
    (d) all of these

22. Cable trays shall be _____ except as permitted by 392.6(G).

    (a) exposed
    (b) accessible
    (c) concealed
    (d) a and b

23. In industrial facilities where conditions of maintenance and supervision ensure that only qualified persons will service the installation, cable tray systems can be used to support _____.

    (a) raceways
    (b) cables
    (c) boxes and conduit bodies
    (d) all of these

24. For raceways terminating at a cable tray, a(n) _____ cable tray clamp or adapter shall be used to securely fasten the raceway to the cable tray system.

    (a) listed
    (b) approved
    (c) identified
    (d) none of these

25. Steel or aluminum cable tray systems can be used as an equipment grounding conductor, provided the cable tray sections and fittings are identified for _____ purposes, among other requirements.

    (a) equipment grounding
    (b) special
    (c) industrial
    (d) all

26. Cable _____ made and insulated by approved methods can be located within a cable tray provided they are accessible, and do not project above the side rails where the splices are subject to physical damage.

    (a) connections
    (b) jumpers
    (c) splices
    (d) conductors

27. A box shall not be required where cables or conductors from cable trays are installed in bushed conduit and tubing used as support or for protection against _____.

    (a) abuse
    (b) unauthorized access
    (c) physical damage
    (d) tampering

28. Where single conductor cables comprising each phase, neutral, or grounded conductor of a circuit are connected in parallel in a cable tray, the conductors shall be installed _____, to prevent current unbalance in the paralleled conductors due to inductive reactance.

    (a) in groups consisting of not more than three conductors per phase or neutral
    (b) in groups consisting of not more than one conductor per phase or neutral
    (c) as individual conductors securely bound to the cable tray
    (d) in separate groups

29. HPD cord shall be permitted for _____.

    (a) not hard usage
    (b) hard usage
    (c) extra-hard usage
    (d) all of these

30. Where flexible cords are used in ambient temperatures exceeding _____ the temperature correction factors from Table 310.16 shall be applied to the ampacity from Table 400.5.

    (a) 30°C
    (b) 60°C
    (c) 75°C
    (d) 90°C

31. Flexible cords and cables shall not be used where _____.

    (a) run through holes in walls, ceilings, or floors
    (b) run through doorways, windows, or similar openings
    (c) attached to building surfaces, unless permitted by 368.56(B)
    (d) all of these

32. Flexible cords and cables shall be connected to fittings so that tension will not be transmitted to joints or terminal screws by _____.

    (a) knotting the cord
    (b) winding the cord with tape
    (c) fittings designed for the purpose
    (d) any of these

33. Flexible cords and cables shall be protected by _____ where passing through holes in covers, outlet boxes, or similar enclosures.

    (a) bushings
    (b) fittings
    (c) a or b
    (d) none of these

34. In industrial establishments where conditions of maintenance and supervision ensure that only qualified persons service the installation, flexible cords and cables can be installed in aboveground raceways that are no longer than _____, to protect the flexible cord or cable from physical damage.

    (a) 25 ft
    (b) 50 ft
    (c) 100 ft
    (d) no limit

35. One conductor of flexible cords intended to be used as a(n) _____ conductor shall have a continuous marker readily distinguishing it from the other conductors.

    (a) grounded
    (b) equipment grounding
    (c) ungrounded
    (d) all of these

36. The ampacity of 18 TFFN is _____.

    (a) 6A
    (b) 8A
    (c) 10A
    (d) 14A

37. The smallest size fixture wire permitted by the *NEC* is _____ AWG.

    (a) 22
    (b) 20
    (c) 18
    (d) 16

38. The number of fixture wires in a single conduit or tubing shall not exceed the percentage fill specified in _____.

    (a) Chapter 9, Table 1
    (b) Table 250.66
    (c) Table 310.16
    (d) 240.6

39. Fixture wires are used to connect luminaires to the _____ conductors supplying the luminaires.

    (a) service
    (b) branch-circuit
    (c) feeder
    (d) none of these

40. Which of the following switches must indicate whether they are in the open (off) or closed (on) position?

    (a) General-use switches
    (b) Motor-circuit switches
    (c) Circuit breakers
    (d) all of these

41. Switches and circuit breakers used as switches can be mounted _____ if they are installed adjacent to motors, appliances, or other equipment that they supply and are accessible by portable means.

    (a) not higher than 6 ft 7 in.
    (b) higher than 6 ft 7 in.
    (c) in the mechanical equipment room
    (d) up to 8 ft high

42. Snap switches shall not be grouped or ganged in enclosures unless the voltage between adjacent devices does not exceed _____.

    (a) 100V
    (b) 200V
    (c) 300V
    (d) 400V

43. A multipole, general-use snap switch shall not be fed from more than a single circuit unless it is listed and marked as a 2-circuit or 3-circuit switch, or unless its voltage rating is not less than the nominal _____ voltage of the system supplying the circuits.

    (a) line-to-ground
    (b) line-to-neutral
    (c) line-to-line
    (d) phase-to-phase

44. A snap switch that does not have means for connection to an equipment grounding conductor shall be permitted for replacement purposes only where the wiring method does not include an equipment grounding conductor and the switch is _____.

    (a) provided with a faceplate of nonconducting, noncombustible material
    (b) GFCI-protected
    (c) a or b
    (d) none of these

45. Nonmetallic enclosures for switches and circuit breakers shall be installed with a wiring method that provides or includes _____.

    (a) a grounded conductor
    (b) an equipment grounding conductor
    (c) an inductive balance
    (d) none of these

46. Alternating-current general-use snap switches are permitted to control _____.

    (a) resistive and inductive loads that do not exceed the ampere and voltage rating of the switch
    (b) tungsten-filament lamp loads that do not exceed the ampere rating of the switch at 120V
    (c) motor loads that do not exceed 80 percent of the ampere and voltage rating of the switch
    (d) all of these

47. Snap switches rated _____ or less directly connected to aluminum conductors shall be listed and marked CO/ALR.

    (a) 15A
    (b) 20A
    (c) 25A
    (d) 30A

48. General-use _____ switches shall be used only to control permanently installed incandescent luminaires unless listed for control of other loads.

    (a) dimmer
    (b) fan speed control
    (c) timer
    (d) all of these

49. Switches shall be marked with _____.

    (a) current
    (b) voltage
    (c) maximum horsepower, if horsepower rated
    (d) all of these

50. A switching device with a marked OFF position shall completely disconnect all _____ conductors of the load it controls.

    (a) grounded
    (b) ungrounded
    (c) grounding
    (d) all of these

51. Receptacles and cord connectors shall be rated not less than _____ at 125V, or at 250V, and shall be of a type not suitable for use as lampholders.

    (a) 10A
    (b) 15A
    (c) 20A
    (d) 30A

52. Receptacles and cord connectors having equipment grounding conductor contacts shall have those contacts connected to a(n) _____ conductor.

    (a) grounded
    (b) ungrounded
    (c) equipment grounding
    (d) neutral

53. Where a grounding means exists in the receptacle enclosure a(n) _____-type receptacle shall be used.

    (a) isolated ground
    (b) grounding
    (c) GFCI
    (d) dedicated

54. When replacing receptacles in locations that would require GFCI protection under the current *NEC*, _____ receptacles shall be installed.

    (a) dedicated
    (b) isolated ground
    (c) GFCI-protected
    (d) grounding

55. When replacing a nongrounding-type receptacle where attachment to an equipment grounding conductor does not exist in the receptacle enclosure, the receptacle can use a _____.

    (a) nongrounding-type receptacle
    (b) grounding receptacle
    (c) GFCI-type receptacle
    (d) a or c

56. Receptacles mounted in boxes set back of the wall surface shall be installed so that the mounting _____ of the receptacle is held rigidly at the surface.

    (a) screws or nails
    (b) yoke or strap
    (c) faceplate
    (d) none of these

57. Receptacles mounted in boxes flush with the finished surface or projecting beyond it shall be installed so that the mounting yoke or strap of the receptacle is _____.

    (a) held rigidly against the box or box cover
    (b) mounted behind the wall surface
    (c) held rigidly at the finished surface
    (d) none of these

58. Receptacles in countertops and similar work surfaces in dwelling units shall not be installed _____.

    (a) in the sides of cabinets
    (b) in a face-up position
    (c) on GFCI circuits
    (d) on the kitchen small-appliance circuit

59. Receptacles shall not be grouped or ganged in enclosures unless the voltage between adjacent devices does not exceed _____.

    (a) 100V
    (b) 200V
    (c) 300V
    (d) 400V

60. An outdoor receptacle in a location protected from the weather, or in another damp location, shall be installed in an enclosure that is weatherproof when the receptacle is _____.

    (a) covered
    (b) enclosed
    (c) protected
    (d) none of these

61. Receptacles installed outdoors, in a location protected from the weather or other damp locations, shall be in an enclosure that is _____ when the receptacle is covered.

    (a) raintight
    (b) weatherproof
    (c) rainproof
    (d) weathertight

62. Nonlocking 15A and 20A, 125V and 250V receptacles installed in damp locations shall be listed as _____.

    (a) raintight
    (b) watertight
    (c) weatherproof
    (d) weather resistant

63. A receptacle installed in an outlet box flush-mounted in a finished surface in a damp or wet location shall be made weatherproof by means of a weatherproof faceplate assembly that provides a _____ connection between the plate and the finished surface.

    (a) sealed
    (b) weathertight
    (c) sealed and protected
    (d) watertight

64. Grounding-type attachment plugs shall be used only with a cord having a(n) _____ conductor.

    (a) equipment grounding
    (b) isolated
    (c) computer circuit
    (d) insulated

65. In dwelling units, 125V, 15A and 20A receptacles installed _____ shall be listed as tamper resistant.

    (a) in bedrooms
    (b) outdoors, above 6 ft 6 in
    (c) above counter tops
    (d) all areas

66. Each switchboard or panelboard used as service equipment shall be provided with a main bonding jumper within the panelboard, or within one of the sections of the switchboard, for connecting the grounded service conductor on its _____ side to the switchboard or panelboard frame.

    (a) load
    (b) supply
    (c) phase
    (d) high-leg

67. In switchboards and panelboards, load terminals for field wiring shall be so located that it is not necessary to reach across or beyond a(n) _____ ungrounded line bus in order to make connections.

    (a) insulated
    (b) uninsulated
    (c) grounded
    (d) ungrounded

68. Panelboards supplied by a three-phase, 4-wire, delta-connected system shall have the phase with higher voltage-to-ground (high-leg) connected to the _____ phase.

(a) A
(b) B
(c) C
(d) any of these

69. Conduits and raceways, including end fittings, shall not rise more than _____ above the bottom of a switchboard enclosure.

(a) 3 in.
(b) 4 in.
(c) 5 in.
(d) 6 in.

70. Unused openings for circuit breakers and switches in switchboards and panelboards shall be closed using _____ or other approved means that provide protection substantially equivalent to the wall of the enclosure.

(a) duct seal and tape
(b) identified closures
(c) exothermic welding
(d) sheet metal

71. A panelboard shall be protected by an overcurrent protective device within the panelboard, or at any point on the _____ side of the panelboard.

(a) load
(b) supply
(c) a or b
(d) none of these

72. When a panelboard is supplied from a transformer, the overcurrent protection shall be located on the _____, unless otherwise permitted by 240.21(C)(1).

(a) secondary side of the transformer
(b) primary side of the transformer
(c) none is required
(d) a or b

73. Each _____ conductor shall terminate within the panelboard at an individual terminal that is not used for another conductor.

(a) grounded
(b) ungrounded
(c) grounding
(d) all of these

74. A panelboard shall be provided with physical means to prevent the installation of more _____ devices than that number for which the panelboard was designed, rated, and listed.

(a) overcurrent
(b) equipment
(c) circuit breaker
(d) all of these

75. The minimum spacing between busbars of opposite polarity held in free air inside a panelboard is _____ when operating at not over 125V.

(a) ½ in.
(b) 1 in.
(c) 2 in.
(d) 4 in.

76. Article 410 covers luminaires, portable luminaires, lampholders, pendants, incandescent filament lamps, arc lamps, electric discharge lamps, and _____, and the wiring and equipment forming part of such products and lighting installations.

(a) decorative lighting products
(b) lighting accessories for temporary seasonal and holiday use
(c) portable flexible lighting products
(d) all of these

77. Closet storage space is defined as a volume bounded by the sides and back closet walls extending from the closet floor vertically to a height of _____ or the highest clothes-hanging rod at a horizontal distance of 2 ft from the sides and back of the closet walls.

(a) 6 ft
(b) 7 ft
(c) 8 ft
(d) 9 ft

78. Lighting track is a manufactured assembly designed to support and _____ luminaires that are capable of being readily repositioned on the track.

(a) connect
(b) protect
(c) energize
(d) all of these

79. A luminaire marked "Suitable for Damp Locations" _____ be permitted to be used in a wet location.

    (a) shall
    (b) shall not
    (c) a or b
    (d) none of these

80. Luminaires can be installed in a commercial cooking hood if the luminaire is identified for use within a _____ cooking hood.

    (a) nonresidential
    (b) commercial
    (c) multifamily
    (d) all of these

81. No part of cord-connected luminaires, chain-, cable-, or cord-suspended luminaires, lighting track, pendants, or paddle fans shall be located within a zone measured 3 ft horizontally and _____ vertically from the top of the bathtub rim or shower stall threshold.

    (a) 4 ft
    (b) 6 ft
    (c) 8 ft
    (d) 10 ft

82. Luminaires located within the actual outside dimension of a bathtub and shower shall be marked for damp locations, or marked for wet locations where they are _____.

    (a) below 7 ft in. height
    (b) below 6 ft 7 in. in height
    (c) subject to shower spray
    (d) not GFCI-protected

83. Luminaires using a _____ lamp, that is subject to physical damage and installed in playing and spectator seating areas of indoor sports, mixed-use, or all-purpose facilities, shall be of the type that protects the lamp with a glass or plastic lens.

    (a) mercury vapor
    (b) metal halide
    (c) fluorescent
    (d) a or b

84. The following types of luminaires can be installed in a clothes closet:

    (a) A surface or recessed incandescent luminaire with enclosed lamp.
    (b) A surface or recessed fluorescent luminaire.
    (c) A surface-mounted or recessed LED luminaire with a completely enclosed light source.
    (d) all of these

85. Coves for luminaires shall be located so that the lamps and equipment can be properly installed and _____.

    (a) maintained
    (b) protected from physical damage
    (c) tested
    (d) inspected

86. Electric-discharge luminaires supported independently of the outlet box shall be connected to the branch circuit through _____.

    (a) raceways
    (b) Type MC, AC, MI, or NM cable
    (c) flexible cords
    (d) any of these

87. The maximum weight of a luminaire permitted to be supported by the screw shell of a lampholder is _____.

    (a) 2 lb
    (b) 3 lb
    (c) 4 lb
    (d) 6 lb

88. Handholes in poles supporting luminaires shall not be required for poles _____ or less in height above finished grade, if the pole is provided with a hinged base.

    (a) 5 ft
    (b) 10 ft
    (c) 15 ft
    (d) 20 ft

89. Metal raceways shall be bonded to the metal pole with a(n) _____.

    (a) grounding electrode
    (b) grounded conductor
    (c) equipment grounding conductor
    (d) any of these

90. Luminaires attached to the framing of a suspended ceiling shall be secured to the framing member(s) by mechanical means such as _____.

   (a) bolts
   (b) screws
   (c) rivets
   (d) any of these

91. _____ clips identified for use with the type of ceiling-framing member(s) and luminaires shall be permitted to be used to secure luminaires to the ceiling-framing members.

   (a) marked
   (b) labeled
   (c) identified
   (d) listed

92. Exposed conductive parts of luminaires shall be _____.

   (a) connected to an equipment grounding conductor
   (b) painted
   (c) removed
   (d) a and b

93. Recessed incandescent luminaires shall have _____ protection and shall be identified as thermally protected.

   (a) physical
   (b) corrosion
   (c) thermal
   (d) all of these

94. A recessed luminaire not identified for contact with insulation shall have all recessed parts spaced not less than _____ from combustible materials, except for points of support.

   (a) ¼ in.
   (b) ½ in.
   (c) 1¼ in.
   (d) 6 in.

95. Type IC recessed luminaires are permitted to make contact with combustible material at _____.

   (a) recessed parts
   (b) points of support
   (c) portions passing through or finishing off the opening in the building structure
   (d) all of these

96. Thermal insulation shall not be installed above or within _____ of the recessed luminaire's enclosure, wiring compartment, or ballast unless it is a Type IC luminaire.

   (a) ½ in.
   (b) 3 in.
   (c) 6 in.
   (d) 12 in.

97. The minimum distance that an outlet box containing tap supply conductors is permitted to be placed from a recessed luminaire is _____.

   (a) 1 ft
   (b) 2 ft
   (c) 3 ft
   (d) 4 ft

98. The raceway or cable for tap conductors to recessed luminaires shall have a minimum length of _____

   (a) 6 in.
   (b) 12 in.
   (c) 18 in.
   (d) 24 in.

99. Surface-mounted luminaires with a ballast shall have a minimum clearance of _____ from combustible low-density cellulose fiberboard, unless the luminaire is marked for surface mounting on combustible low-density cellulose fiberboard.

   (a) ½ in.
   (b) 1 in.
   (c) 1½ in.
   (d) 2 in.

100. Lighting track shall not be installed less than _____ above the finished floor except where protected from physical damage or where the track operates at less than 30V rms, open-circuit voltage.

   (a) 4 ft
   (b) 5 ft
   (c) 5½ ft
   (d) 6 ft

Unit 5: Practice Questions in Random Order—Article 90.4 through 410.90

Use the 2008 *NEC* to answer the following questions.

1. 20A, 125V receptacles subject to _____ can have an enclosure that is weatherproof when the attachment plug is removed.

   (a) shower spray
   (b) beating rain
   (c) corrosive conditions
   (d) routine high-pressure spray

2. A receptacle is considered to be in a location protected from the weather when located under roofed open porches, canopies, marquees, and the like, where it will not be subjected to _____.

   (a) spray from a hose
   (b) a direct lightning hit
   (c) beating rain or water runoff
   (d) falling or wind-blown debris

3. Where combination surface metal raceways are used for both signaling conductors and lighting and power circuits, the different systems shall be run in separate compartments identified by _____ of the interior finish.

   (a) stamping
   (b) imprinting
   (c) color coding
   (d) any of these

4. Where combination surface nonmetallic raceways are used for both signaling conductors and lighting and power circuits, the different systems shall be run in separate compartments identified by _____ of the interior finish.

   (a) stamping
   (b) imprinting
   (c) color coding
   (d) any of these

5. Luminaires shall be wired so that the _____ of each lampholder is connected to the same luminaire or circuit conductor or terminal.

   (a) stem
   (b) arm
   (c) supplemental protection
   (d) screw shell

6. A cable tray is a unit or assembly of units or sections and associated fittings forming a _____ system used to securely fasten or support cables and raceways.

   (a) structural
   (b) flexible
   (c) movable
   (d) secure

7. Flexible cords and cables shall not be concealed behind building _____, or run through doorways, windows, or similar openings.

   (a) structural ceilings
   (b) suspended or dropped ceilings
   (c) floors or walls
   (d) all of these

8. The allowable ampacity of flexible cords and cables is found in _____.

   (a) Table 310.16
   (b) Tables 400.5(a) and (B)
   (c) Chapter 9, Table 1
   (d) Table 430.52

9. Flexible cords shall not be used as a substitute for _____ wiring.

   (a) temporary
   (b) fixed
   (c) overhead
   (d) none of these

Practice Questions in Random Order—Article 90.4 through 410.90 — Unit 5

10. The ampacity adjustment factors of 310.15(B)(2)(a) shall not apply to conductors installed in surface metal raceways where _____.

    (a) the cross-sectional area exceeds 4 sq in.
    (b) the current-carrying conductors do not exceed 30 in number
    (c) the total cross-sectional area of all conductors does not exceed 20 percent of the interior cross-sectional area of the raceway
    (d) all of these

11. The ampacity adjustment factors of 310.15(B)(2)(a) shall not apply to conductors installed in strut-type channel raceways where _____.

    (a) the cross-sectional area of the raceway is at least 4 sq in.
    (b) the number of current-carrying conductors does not exceed 30
    (c) the sum of the cross-sectional areas of all contained conductors does not exceed 20 percent of the interior cross-sectional area of the strut-type channel raceway
    (d) all of these

12. Poles over 20 ft in height above grade, without a hinged base, that support luminaires shall meet which of the following requirements?

    (a) They shall have an accessible handhole (sized 2 x 4 in.) with a suitable cover.
    (b) The equipment grounding terminal shall be accessible from the handhole.
    (c) a and b
    (d) none of these

13. A receptacle connected to one of the dwelling unit small-appliance branch circuits can be used to supply an electric clock.

    (a) True
    (b) False

14. Surge protective devices shall be marked with their short-circuit current rating and shall not be installed where the available fault current is in excess of that rating.

    (a) True
    (b) False

15. Exposed runs of Type AC cable can be installed on the underside of joists where supported at each joist and located so they are not subject to physical damage.

    (a) True
    (b) False

16. Running threads shall not be used on IMC for connection at couplings.

    (a) True
    (b) False

17. Single-throw knife switches shall be installed so that gravity will tend to close the switch.

    (a) True
    (b) False

18. Isolated ground receptacles installed in nonmetallic boxes shall be covered with a nonmetallic faceplate, unless the box contains a feature or accessory that permits the effective grounding of the faceplate.

    (a) True
    (b) False

19. Lampholders of the screw-shell type shall be installed for use as lampholders only.

    (a) True
    (b) False

20. By special permission, the authority having jurisdiction may waive specific requirements in this *Code* where it is assured that equivalent objectives can be achieved by establishing and maintaining effective safety.

    (a) True
    (b) False

21. Branch-circuit neutral conductors that supply a continuous load shall have an ampacity of not less than 125 percent of the continuous load.

    (a) True
    (b) False

22. A receptacle connected to a dwelling unit small-appliance circuit can supply gas-fired ranges, ovens, or counter-mounted cooking units.

    (a) True
    (b) False

Unit 5 — Practice Questions in Random Order—Article 90.4 through 410.90

23. Receptacles installed behind a bed in the guest rooms in hotels and motels shall be located to prevent the bed from contacting an attachment plug, or the receptacle shall be provided with a suitable guard.

    (a) True
    (b) False

24. Where a building or structure is supplied by more than one service, a permanent plaque or directory shall be installed at each service disconnect location denoting all other services supplying that building or structure and the area served by each.

    (a) True
    (b) False

25. Service disconnecting means shall not be installed in bathrooms.

    (a) True
    (b) False

26. The grounding of electrical systems, circuit conductors, surge arresters, surge protective devices, and conductive normally non–current-carrying metal parts of equipment shall be installed and arranged in a manner that will prevent objectionable current.

    (a) True
    (b) False

27. A grounding electrode conductor, sized in accordance with 250.66, shall be used to connect the equipment grounding conductors, the service-equipment enclosures, and, where the system is grounded, the grounded service conductor to a grounding electrode.

    (a) True
    (b) False

28. The grounding electrode for a separately derived system shall be as near as practicable to, and preferably in the same area as, the grounding electrode conductor connection to the system.

    (a) True
    (b) False

29. The structural metal frame of a building can be used as the required circuit equipment grounding conductor.

    (a) True
    (b) False

30. Surge protective devices shall only be located outdoors.

    (a) True
    (b) False

31. Conduit bodies that are durably and legibly marked by the manufacturer with their volume can contain splices, taps, or devices

    (a) True
    (b) False

32. Type AC cable installed through, or parallel to, framing members shall be protected against physical damage from penetration by screws or nails.

    (a) True
    (b) False

33. The cut ends of PVC conduit must be trimmed to remove the burrs and rough edges.

    (a) True
    (b) False

34. A 30A, 208V receptacle installed in a wet location, where the product intended to be plugged into it is not attended while in use, shall have an enclosure that is weatherproof with the attachment plug cap inserted or removed.

    (a) True
    (b) False

35. The term "Luminaire" includes an individual lampholder.

    (a) True
    (b) False

36. Where a remote-control device actuates the service disconnecting means, the service disconnecting means must still be at a readily accessible location either outside the building or structure, or nearest the point of entry of the service conductors.

    (a) True
    (b) False

37. Temporary currents resulting from accidental conditions, such as ground faults, are not considered to be objectionable currents.

    (a) True
    (b) False

Practice Questions in Random Order—Article 90.4 through 410.90 — Unit 5

38. A grounding electrode shall be required if a building or structure is supplied by a feeder.

    (a) True
    (b) False

39. Metal parts of cord-and-plug-connected equipment shall be connected to an equipment grounding conductor that terminates to a grounding-type attachment plug.

    (a) True
    (b) False

40. Type UF cable shall not be used where subject to physical damage.

    (a) True
    (b) False

41. Conductors larger than that for which the wireway is designed can be installed in any wireway.

    (a) True
    (b) False

42. Switches and circuit breakers used as switches shall be installed so that they may be operated from a readily accessible place.

    (a) True
    (b) False

43. A receptacle shall not be installed within, or directly over, a bathtub or shower space.

    (a) True
    (b) False

44. When an electric-discharge luminaire is surface mounted over a concealed outlet box, which is not its sole means of support, the luminaire shall provide access to the conductor wiring within the outlet box by means of suitable openings in the back of the luminaire.

    (a) True
    (b) False

45. Listed or labeled equipment shall be installed and used in accordance with any instructions included in the listing or labeling.

    (a) True
    (b) False

46. Where a branch circuit supplies continuous loads, or any combination of continuous and noncontinuous loads, the rating of the overcurrent device shall not be less than the noncontinuous load plus 125 percent of the continuous load.

    (a) True
    (b) False

47. A grounding electrode at a separate building or structure shall be required where one multiwire branch circuit serves the building or structure.

    (a) True
    (b) False

48. Metal gas piping shall be considered bonded by the equipment grounding conductor of the circuit that is likely to energize the piping.

    (a) True
    (b) False

49. A grounded circuit conductor can be ground noncurrent-carrying metal parts of equipment, raceways, and other enclosures on the supply side or within the enclosure of the ac service-disconnecting means.

    (a) True
    (b) False

50. A Type 2 surge protective device is permitted to be connected on either the line or load side of the service equipment.

    (a) True
    (b) False

Unit 6: Practice Questions in Straight Order—Article 410.151 through 502.1

Use the 2008 *NEC* to answer the following questions.

1. Lighting track shall not be installed within the zone measured 3 ft horizontally and _____ vertically from the top of the bathtub rim or shower stall threshold.

   (a) 2 ft
   (b) 3 ft
   (c) 4 ft
   (d) 8 ft

2. A lighting system covered by Article 411 shall have a power supply rated not more than _____ and not more than 30V.

   (a) 15A
   (b) 20A
   (c) 25A
   (d) 30A

3. Lighting systems operating at 30V or less shall not be installed within _____ ft of pools, spas, fountains, or similar locations.

   (a) 5 ft
   (b) 6 ft
   (c) 10 ft
   (d) 20 ft

4. Exposed secondary circuits of lighting systems operating at 30V or less shall be _____.

   (a) installed in a Chapter 3 wiring method
   (b) supplied by a Class 2 power source shall use Class 2 cable in accordance with Article 725
   (c) at least 7 ft above the finished floor unless listed for a lower installation height
   (d) any of these

5. Branch-circuit conductors to individual appliances shall not be sized _____ than required by the appliance markings.

   (a) larger
   (b) smaller

6. Individual circuits for nonmotor-operated appliances that are continuously loaded shall have the branch-circuit rating sized not less than _____ percent of the appliance marked ampere rating, unless otherwise listed.

   (a) 80
   (b) 100
   (c) 125
   (d) 150

7. If a protective device rating is marked on an appliance, the branch-circuit overcurrent device rating shall not exceed _____ percent of the protective device rating marked on the appliance.

   (a) 50
   (b) 80
   (c) 100
   (d) 115

8. If a branch circuit supplies a single nonmotor-operated appliance, the rating of overcurrent protection shall not exceed _____ if the overcurrent protection rating is not marked and the appliance is rated 13.30A or less.

   (a) 15A
   (b) 20A
   (c) 25A
   (d) 30A

438   Mike Holt's Illustrated Guide to NEC Exam Preparation

9. If a branch circuit supplies a single nonmotor-operated appliance, the rating of overcurrent protection shall not exceed _____ percent of the appliance rated current if the overcurrent protection rating is not marked and the appliance is rated over 13.30A.

   (a) 100
   (b) 125
   (c) 150
   (d) 160

10. Central heating equipment, other than fixed electric space-heating equipment, shall be supplied by a(n) _____ branch circuit.

    (a) multiwire
    (b) individual
    (c) multipurpose
    (d) small-appliance

11. Fixed storage-type water heater having a capacity of _____ gallons or less shall be considered a continuous load for the purposes of sizing branch circuits.

    (a) 60
    (b) 75
    (c) 90
    (d) 120

12. A waste disposal can be cord-and-plug-connected, but the cord shall not be less than 18 in. or more than _____ in length.

    (a) 30 in.
    (b) 36 in.
    (c) 42 in.
    (d) 48 in.

13. The length of the cord for a dishwasher or trash compactor shall not be longer than _____, measured from the rear of the appliance.

    (a) 2 ft
    (b) 4 ft
    (c) 6 ft
    (d) 8 ft

14. Wall-mounted ovens and counter-mounted cooking units complete shall be permitted to be _____.

    (a) permanently connected
    (b) cord-and-plug-connected
    (c) a or b
    (d) none of these

15. For cord-and-plug-connected household electric ranges, an attachment plug and receptacle connection at the rear base of the range can serve as the disconnecting means, if it is _____.

    (a) less than 40A
    (b) a flush-mounted receptacle
    (c) GFCI-protected
    (d) accessible by the removal of a drawer

16. Appliances that have a unit switch with a marked _____ position that disconnects all the ungrounded conductors can serve as the appliance disconnecting means.

    (a) on
    (b) off
    (c) on/off
    (d) all of these

17. The term ⊠vending machine⊠ means any self-service device that dispenses products or merchandise without the necessity of replenishing the device between each vending operation and designed to require insertion of a _____.

    (a) coin or paper currency
    (b) token or card
    (c) key or payment by other means
    (d) all of these

18. Electric drinking fountains shall be _____ -protected.

    (a) GFCI
    (b) AFCI
    (c) a or b
    (d) none of these

19. Fixed electric space-heating equipment shall be considered a _____ load.

    (a) noncontinuous
    (b) intermittent
    (c) continuous
    (d) none of these

20. Means shall be provided to simultaneously disconnect the _____ of all fixed electric space-heating equipment from all ungrounded conductors.

    (a) heater
    (b) motor controller(s)
    (c) supplementary overcurrent protective device(s)
    (d) all of these

Unit 6 — Practice Questions in Straight Order—Article 410.151 through 502.1

21. A unit switch with a marked _____ position that is part of a fixed space heater, and disconnects all ungrounded conductors, shall be permitted to serve as the required disconnecting means.

    (a) on
    (b) closed
    (c) off
    (d) none of these

22. GFCI protection shall be provided for electrically heated floors in _____ locations.

    (a) bathroom
    (b) hydromassage bathtub
    (c) kitchen
    (d) a and b

23. Duct heater controller equipment shall have a disconnecting means installed within _____ the controller.

    (a) 25 ft of
    (b) sight from
    (c) the side of
    (d) none of these

24. For general motor applications, the motor branch-circuit short-circuit and ground-fault protection device shall be sized based on the _____ values.

    (a) motor nameplate
    (b) NEMA standard
    (c) *NEC* Table
    (d) Factory Mutual

25. The motor _____ currents listed in Tables 430.247 through 430.250 shall be used to determine the ampacity of motor circuit conductors and short-circuit and ground-fault protection devices.

    (a) nameplate
    (b) full-load
    (c) power factor
    (d) service factor

26. Motors shall be located so that adequate _____ is provided and so that maintenance, such as lubrication of bearings and replacing of brushes, can be readily accomplished.

    (a) space
    (b) ventilation
    (c) protection
    (d) all of these

27. In determining the highest rated motor for purposes of 430.24, the highest rated motor shall be based on the rated full-load current as selected _____.

    (a) from the motor nameplate
    (b) from Tables 430.247, 430.248, 430.249, and 430.250
    (c) from taking the horsepower times 746 watts
    (d) using the largest horsepower motor

28. Feeder tap conductors supplying motor circuits, with an ampacity at least one-third that of the feeder, shall not exceed _____ in length.

    (a) 10 ft
    (b) 15 ft
    (c) 20 ft
    (d) 25 ft

29. Overload devices are intended to protect motors, motor control apparatus, and motor branch-circuit conductors against _____.

    (a) excessive heating due to motor overloads
    (b) excessive heating due to failure to start
    (c) short circuits and ground faults
    (d) a and b

30. Motor overload protection shall not be required where _____.

    (a) conductors are oversized by 125 percent
    (b) conductors are part of a limited-energy circuit
    (c) it might introduce additional or increased hazards
    (d) short-circuit protection is provided

31. A separate overload device used to protect continuous-duty motors rated more than 1 hp shall be selected to trip at no more than _____ percent of the motor nameplate full-load current rating if marked with a service factor of 1.16.

    (a) 110
    (b) 115
    (c) 120
    (d) 125

32. The ultimate trip current of a thermally protected motor with a full-load current not exceeding 9A shall not exceed _____ percent of the motor full-load current.

    (a) 140
    (b) 156
    (c) 170
    (d) 175

Practice Questions in Straight Order—Article 410.151 through 502.1

33. Unless overload protection is provided by other approved means, the minimum number of overload units required for a three-phase ac motor is _____.

    (a) one
    (b) two
    (c) three
    (d) any of these

34. The maximum rating or setting of an inverse time breaker used as the motor branch-circuit short-circuit and ground-fault protective device for a single-phase motor is _____ percent of the full-load current given in Table 430.248.

    (a) 125
    (b) 175
    (c) 250
    (d) 300

35. A motor can be provided with combined overcurrent protection using a single protective device to provide branch-circuit _____ where the rating of the device provides the overload protection specified in 430.32.

    (a) short-circuit protection
    (b) ground-fault protection
    (c) motor overload protection
    (d) all of these

36. A feeder supplying fixed motor load(s) shall have a protective device with a rating or setting _____ branch-circuit short-circuit and ground-fault protective device for any motor in the group, plus the sum of the full-load currents of the other motors of the group.

    (a) not greater than the largest rating or setting of the
    (b) 125 percent of the largest rating of any
    (c) equal to the largest rating of any
    (d) none of these

37. Overcurrent protection for motor control circuits shall not exceed _____ percent if the conductor is based on Table 310.17 and does not extend beyond the motor control equipment enclosure.

    (a) 100
    (b) 150
    (c) 300
    (d) 400

38. Motor control circuit conductors that extend beyond the motor control equipment enclosure shall have short-circuit and ground-fault protection sized not greater than _____ percent of value specified in Table 310.16 for 60°C conductors.

    (a) 100
    (b) 150
    (c) 300
    (d) 400

39. The motor controller shall have horsepower ratings at the application voltage not _____ than the horsepower rating of the motor.

    (a) lower than
    (b) higher than
    (c) equal to
    (d) none of these

40. A _____ shall be located in sight from the motor location and the driven machinery location.

    (a) controller
    (b) protection device
    (c) disconnecting means
    (d) all of these

41. The motor disconnecting means shall not be required to be in sight from the motor and the driven machinery location, provided _____.

    (a) the controller disconnecting means is capable of being individually locked in the open position
    (b) the provisions for locking are permanently installed on, or at, the switch or circuit breaker used as the controller disconnecting means
    (c) locating the motor disconnecting means within sight of the motor is impractical or introduces additional or increased hazards to people or property
    (d) all of these

42. The disconnecting means for a motor controller shall be designed so that it does not _____ automatically.

    (a) open
    (b) close
    (c) restart
    (d) shut down

Unit 6 — Practice Questions in Straight Order—Article 410.151 through 502.1

43. The motor disconnecting means for a motor shall _____ whether it is in the open (off) or closed (on) position.

    (a) plainly indicate
    (b) provide current
    (c) be in the upper position
    (d) none of these

44. A motor disconnecting means can be a _____.

    (a) listed molded case circuit breaker
    (b) listed motor-circuit switch rated in horsepower
    (c) listed molded case switch
    (d) any of these

45. If a motor disconnecting means is a motor-circuit switch, it shall be rated in _____.

    (a) horsepower
    (b) watts
    (c) amperes
    (d) locked-rotor current

46. A switch or circuit breaker can be used as both the controller and disconnecting means if it _____.

    (a) opens all ungrounded conductors
    (b) is protected by an overcurrent device in each ungrounded conductor
    (c) is manually operable, or both power and manually operable
    (d) all of these

47. The _____ current for a hermetic refrigerant motor-compressor is the current resulting when the motor-compressor is operated at the rated load, rated voltage, and rated frequency of the equipment it serves.

    (a) full-load current
    (b) nameplate rating
    (c) selection current
    (d) rated-load current

48. The rules of _____ shall apply to air-conditioning and refrigerating equipment that does not incorporate a hermetic refrigerant motor-compressor.

    (a) Article 422
    (b) Article 424
    (c) Article 430
    (d) all of these

49. Equipment such as _____ shall be considered appliances, and the provisions of Article 422 apply in addition to Article 440.

    (a) room air conditioners
    (b) household refrigerators and freezers
    (c) drinking water coolers and beverage dispensers
    (d) all of these

50. Short-circuit and ground-fault protection for an individual a/c motor-compressor shall not exceed _____ percent of the motor-compressor rated-load current or branch-circuit protection current, whichever is greater.

    (a) 80
    (b) 125
    (c) 175
    (d) 250

51. Branch-circuit conductors supplying a single a/c motor-compressor shall have an ampacity not less than _____ percent of either the motor-compressor rated-load current or the branch-circuit selection current, whichever is greater.

    (a) 100
    (b) 125
    (c) 150
    (d) 200

52. Conductors supplying more than one a/c motor-compressor shall have an ampacity not less than the sum of the rated load or branch-circuit current ratings, whichever is larger, of all the a/c motor-compressors plus the full-load currents of any other motors, plus _____ percent of the highest motor or motor compressor rating in the group.

    (a) 25
    (b) 50
    (c) 80
    (d) 100

53. The total rating of a cord-and-plug-connected room air conditioner, connected to the same branch circuit which supplies lighting units, other appliances, or general-use receptacles, shall not exceed _____ percent of the branch-circuit rating.

    (a) 40
    (b) 50
    (c) 70
    (d) 80

54. When supplying a room air conditioner rated 120V, the length of a flexible supply cord shall not exceed _____.

    (a) 4 ft
    (b) 6 ft
    (c) 8 ft
    (d) 10 ft

55. Each generator shall be provided with a _____ listing the manufacturer's name, the rated frequency, power factor, number of phases (if of alternating current), and its rating in kilowatts or kilovolt-amperes.

    (a) list
    (b) faceplate
    (c) nameplate
    (d) sticker

56. The ampacity of ungrounded conductors from the generator terminals to the first distribution device containing overcurrent protection shall not be less than _____ percent of the nameplate rating of the generator.

    (a) 75
    (b) 115
    (c) 125
    (d) 140

57. The primary overcurrent protection for a transformer rated 600V, nominal, or less, having a primary current rating of over 9A or more must be set at not more than _____.

    (a) 125
    (b) 167
    (c) 200
    (d) 300

58. Overcurrent protection for the secondary of a transformer rated 600V or less having a primary current rating of 9A or more must be set at not more than _____ percent of the primary current rating.

    (a) 100
    (b) 125
    (c) 150
    (d) 170

59. Transformers with ventilating openings shall be installed so that the ventilating openings _____.

    (a) are a minimum 18 in. above the floor
    (b) are not blocked by walls or obstructions
    (c) are aesthetically located
    (d) are vented to the exterior of the building

60. Transformers and transformer vaults shall be readily accessible to qualified personnel for inspection and maintenance, except _____.

    (a) dry-type transformers 600V or less, located in the open on walls, columns, or structures
    (b) dry-type transformers rated not more than 50 kVA/600V installed in hollow spaces of buildings not permanently closed in by structure
    (c) a or b
    (d) none of these

61. Indoor transformers of greater than _____ rating shall be installed in a transformer room of fire-resistant construction.

    (a) 75 kVA
    (b) 87½ kVA
    (c) 112½ kVA
    (d) 35,000 kVA

62. Personnel doors for transformer vaults shall _____ and be equipped with panic bars, pressure plates, or other devices that are normally latched but open under simple pressure.

    (a) be clearly identified
    (b) swing out
    (c) a and b
    (d) a or b

63. Ventilation openings for transformer vaults must be as far as possible from _____.

    (a) doors
    (b) windows
    (c) combustible material
    (d) any of these

64. Capacitors shall be _____ so that persons cannot come into accidental contact or bring conducting materials into accidental contact with exposed energized parts, terminals, or buses associated with them.

    (a) enclosed
    (b) located
    (c) guarded
    (d) any of these

Unit 6 — Practice Questions in Straight Order—Article 410.151 through 502.1

65. The ampacity of capacitor circuit conductors shall not be less than _____ percent of the rated current of the capacitor.

    (a) 100
    (b) 115
    (c) 125
    (d) 135

66. An overcurrent device shall be provided in each ungrounded conductor for each capacitor bank. The rating or setting of the overcurrent device shall be _____.

    (a) as low as practicable
    (b) not to exceed 20A
    (c) not to exceed 40A
    (d) not to exceed 100A

67. A disconnecting means shall be provided in each ungrounded conductor for each capacitor bank, and shall _____.

    (a) open all ungrounded conductors simultaneously
    (b) be permitted to disconnect the capacitor from the line as a regular operating procedure
    (c) be rated no less than 135 percent of the rated current of the capacitor
    (d) all of these

68. _____ equipment excludes dust and doesn't permit arcs, sparks, or heat inside the enclosure to ignite accumulations or atmospheric suspensions of a specified dust on or in the vicinity outside of the enclosure.

    (a) Dust-ignitionproof
    (b) Dusttight
    (c) a or b
    (d) a and

69. _____ enclosures are constructed so that dust will not enter under specific test conditions.

    (a) Dust-ignitionproof
    (b) Dusttight
    (c) a or b
    (d) a and

70. Except as modified in Articles 500 through 504, all installation requirements contained in _____ of the NEC apply to electrical equipment and wiring installed in hazardous (classified) locations.

    (a) Chapter 1 and 2
    (b) Chapter 3 and 4
    (c) a or b
    (d) a and b

71. Hazardous (classified) locations shall be classified depending on the properties of the _____ that may be present, and the likelihood that a flammable or combustible concentration or quantity is present.

    (a) flammable liquid produced vapors
    (b) flammable gases
    (c) combustible dusts
    (d) all of these

72. Class I, Division 1 locations are those in which ignitible concentrations of _____ can exist under normal operating conditions.

    (a) combustible dust
    (b) easily ignitible fibers or flyings
    (c) flammable gases or flammable liquid produced vapors
    (d) pyrotechnics

73. Class II locations are those that are hazardous because of the presence of _____.

    (a) combustible dust
    (b) easily ignitible fibers/flyings
    (c) flammable gases or vapors
    (d) flammable liquids or gases

74. Locations in which combustible dust is in the air under normal operating conditions in quantities sufficient to produce explosive or ignitible mixtures are classified as _____.

    (a) Class I, Division 2
    (b) Class II, Division 1
    (c) Class II, Division 2
    (d) Class III, Division 1

Practice Questions in Straight Order—Article 410.151 through 502.1 — Unit 6

75. Class III locations are those that are hazardous because of the presence of _____.
    (a) combustible dust
    (b) easily ignitible fibers or materials producing combustible flyings
    (c) flammable gases or vapors
    (d) flammable liquids or gases

76. Class III, Division _____ locations include areas where easily ignitible fibers/flyings are handled, manufactured, or used.
    (a) 1
    (b) 2
    (c) 3
    (d) all of these

77. A Class III, Division_____ location is where easily ignitible fibers/flyings are stored or handled but not manufactured.
    (a) 1
    (b) 2
    (c) 3
    (d) all of these

78. Acceptable protection techniques for electrical and electronic equipment in hazardous (classified) locations includes:
    (a) Explosionproof Apparatus
    (b) Dust Ignitionproof
    (c) Dusttight
    (d) all of these

79. Suitability of identified equipment for use in a hazardous (classified) location shall be determined by:
    (a) Equipment listing or labeling.
    (b) Evidence of equipment evaluation from a qualified testing laboratory or inspection agency concerned with product evaluation.
    (c) Evidence acceptable to the authority having jurisdiction, such as a manufacturer's self-evaluation or an owner's engineering judgment.
    (d) any of these

80. Equipment shall be identified not only for the class of hazardous (classified) location, but also for the explosive, combustible, or ignitible properties of the specific _____ present.
    (a) gas or vapor
    (b) dust
    (c) fibers/flyings
    (d) all of these

81. Equipment installed in hazardous (classified) locations shall be marked to show the _____.
    (a) class
    (b) group
    (c) temperature class (T Code) or operating temperature at a 40°C ambient temperature
    (d) all of these

82. Threaded conduits or fittings referred to in hazardous (classified) locations shall be made wrenchtight to _____.
    (a) prevent sparking when a fault current flows
    (b) ensure the explosionproof integrity of the conduit system
    (c) a and b
    (d) none of these

83. For listed explosionproof equipment, factory threaded entries shall be made up with at least _____ threads fully engaged.
    (a) 4
    (b) 4½
    (c) 5
    (d) 6

84. Article 501 covers the requirements for electrical and electronic equipment and wiring for all voltages in Class I locations where fire or explosion hazards may exist due to _____.
    (a) flammable gases
    (b) flammable vapors
    (c) flammable liquids
    (d) any of these

85. A sealing fitting shall be installed within _____ of either side of the boundary where a conduit leaves a Class I, Division 1 location.

    (a) 5 ft
    (b) 6 ft
    (c) 8 ft
    (d) 10 ft

86. Where the Class I, Division 1 location boundary is below grade, a sealing fitting shall be installed _____.

    (a) after the conduit emerges from grade
    (b) before the conduit emerges from grade
    (c) within 10 ft of where the conduit emerges from grade
    (d) none of these

87. A conduit seal fitting shall be installed in each conduit that passes from a Class I, Division 2 location into an unclassified location and it is not required to be _____.

    (a) listed
    (b) installed
    (c) explosionproof
    (d) accessible

88. A sealing fitting shall not be required if a conduit passes completely through a Class I, Division 2 location if the termination points of the unbroken conduit are in unclassified locations and it has no fittings less than _____ beyond each boundary of the hazardous (classified) location.

    (a) 6 in.
    (b) 12 in.
    (c) 18 in.
    (d) 24 in.

89. The minimum thickness of sealing compound in Class I shall not be less than the trade size of the conduit or sealing fitting and, in no case, less than _____.

    (a) 1/8 in.
    (b) 1/4 in.
    (c) 3/8 in.
    (d) 5/8 in.

90. The cross-sectional area of the conductors permitted in a sealing fitting shall not exceed _____ percent of the cross-sectional area of RMC of the same trade size.

    (a) 25
    (b) 50
    (c) 100
    (d) 125

91. Meters, instruments and relays installed in Class I, Division 2 locations can have switches, circuit breakers, and make-and-break contacts installed in general-purpose enclosures if current-interrupting contacts are _____.

    (a) immersed in oil
    (b) enclosed within a hermetically-sealed chamber
    (c) a or b
    (d) a and b

92. In Class I, Division 1 locations, control transformers, impedance coils, and resistors, along with any switching mechanism associated with them, shall be provided with enclosures identified for _____.

    (a) Class I, Division 1 locations
    (b) control transformer duty
    (c) general duty
    (d) NEMA 3R

93. Totally-enclosed motors shall have no external surface with an operating temperature that exceeds _____ when operating in Class I, Division 1 locations.

    (a) absolute zero
    (b) 80 percent of ignition temperature of the gas or liquid involved
    (c) 100°C
    (d) 40°C

94. Luminaires installed in Class I, Division 1 locations shall be identified as a complete assembly for the Class I, Division 1 location and shall be clearly marked to indicate the _____.

    (a) maximum wattage of lamps
    (b) minimum conductor size
    (c) maximum overcurrent protection permitted
    (d) all of these

95. Pendant luminaires shall be suspended by and supplied through threaded rigid metal conduit stems or threaded steel intermediate conduit stems, and threaded joints shall be provided with set-screws or other effective means to prevent loosening.

    (a) True
    (b) False

96. Boxes, box assemblies, or fittings used for the support of luminaires in Class I locations shall be identified for Class I locations.

    (a) True
    (b) False

97. Portable lighting equipment must be _____ for use in a Class I, Division 1 location, unless the luminaire is mounted on movable stands and connected by a flexible cord.

    (a) identified
    (b) approved
    (c) marked
    (c) listed

98. In Class I, Division 1 locations, all utilization equipment must be _____ for use in a Class I, Division 1 location.

    (a) identified
    (b) approved
    (c) marked
    (c) listed

99. Flexible cords in hazardous (classified) locations must _____.

    (a) be listed as extra-hard usage
    (b) connect to terminals in an approved manner
    (c) have suitable seals
    (d) all of these

100. Article 502 covers the requirements for electrical and electronic equipment and wiring in Class II, Division 1 and 2 locations where fire or explosion hazards may exist due to _____.

    (a) gases or vapors
    (b) fibers/flyings
    (c) dust
    (d) all of these

Unit 6: Practice Questions in Random Order—Article 90.2 through 502.10

Use the 2008 *NEC* to answer the following questions.

1. Generators must have a disconnecting means that is lockable in the open position, except where:

   (a) The driving means for the generator can be readily shut down.
   (b) The generator isn't arranged to operate in parallel with another generator or other source of voltage.
   (c) a and b
   (d) a or b

2. An attachment plug and receptacle can serve as the disconnecting means for a single-phase room air conditioner rated 250 volts or less if _____.

   (a) the manual controls on the room air conditioner are readily accessible and located within 1.8 m (6 ft) of the floor
   (b) an approved manually operable disconnecting means is installed in a readily accessible location within sight from the room air conditioner
   (c) a or b
   (d) a and b

3. When determining a Class I, Division 2 location, _____ is a factor that should be considered in determining the classification and extent of the location.

   (a) the quantity of flammable material that might escape in case of an accident
   (b) the adequacy of ventilating equipment
   (c) the record of the industry or business with respect to explosions or fires
   (d) all of these

4. When sealing fittings are required for Class I locations, they shall comply with the following rule(s):

   (a) They shall be listed for Class I locations and shall be accessible.
   (b) The minimum thickness of the sealing compound shall not be less than the trade size of the sealing fitting and, in no case, less than ⅝ in.
   (c) Splices and taps shall not be made in the conduit seal.
   (d) all of these

5. Wiring methods permitted in Class I, Division 1 locations include _____.

   (a) threaded rigid metal or threaded steel IMC
   (b) flexible fittings listed for Class I, Division 1 locations
   (c) boxes approved for Class I, Division 1 locations
   (d) all of these

6. Raceways permitted as a wiring method in Class II, Division 1 locations include _____.

   (a) threaded RMC and steel IMC
   (b) PVC conduit
   (c) electrical metallic tubing
   (d) any of these

7. A motor controller shall open all conductors to the motor where not used as a disconnecting means.

   (a) True
   (b) False

8. Boxes and fittings in a Class I, Division 2 location for switches and circuit breakers shall be installed in an explosionproof enclosure meeting the requirements for Class I, Division 1 locations.

   (a) True
   (b) False

9. Lighting systems operating at 30V or less shall be listed or assembled with listed components.

   (a) True
   (b) False

10. Each motor shall be provided with an individual controller.

    (a) True
    (b) False

11. Lighting systems operating at 30V or less can be concealed or extended through a building wall, floor or ceiling without regard to the wiring method used.

    (a) True
    (b) False

12. Motors, generators, or other rotating electric machinery identified for Class I, Division 2 locations can be used in Class I, Division 1 locations.

    (a) True
    (b) False

13. When armored cable is run parallel to the sides of rafters, studs, or floor joists in an accessible attic, the cable shall be protected with running boards.

    (a) True
    (b) False

14. Type NM cables shall not be used in one- and two-family dwellings exceeding three floors above grade.

    (a) True
    (b) False

15. The common point on a wye-connection in a polyphase system describes a neutral point.

    (a) True
    (b) False

16. Service conductors originate at the service point and terminate at service equipment.

    (a) True
    (b) False

17. A single receptacle installed on an individual branch circuit shall have an ampere rating not less than that of the branch circuit.

    (a) True
    (b) False

18. When supplying a grounded system at a separate building or structure, an equipment grounding conductor shall be run with the supply conductors and connected to the building disconnecting means.

    (a) True
    (b) False

19. Conductors shall be installed within a raceway, cable, or enclosure.

    (a) True
    (b) False

20. Type NM cable can be installed as open runs in dropped or suspended ceilings in other than one- and two-family and multifamily dwellings.

    (a) True
    (b) False

21. FMC can be installed exposed or concealed where not subject to physical damage.

    (a) True
    (b) False

22. When a building is supplied with a fire sprinkler system, ENT can be installed above any suspended ceiling.

    (a) True
    (b) False

23. Metal multioutlet assemblies can pass through a dry partition, provided no receptacle is concealed in the partition and the cover of the exposed portion of the system can be removed.

    (a) True
    (b) False

24. The disconnecting means for the controller and motor shall open all ungrounded supply conductors.

    (a) True
    (b) False

25. Totally enclosed motors shall have a device to de-energize the motor or provide an alarm if there is an increase in temperature of the motor beyond designed limits when operating in Class I, Division 1 locations.

    (a) True
    (b) False

Unit 6 — Practice Questions in Random Order—Article 90.2 through 502.10

26. All 15A and 20A, 125V receptacles installed in crawl spaces at or below grade level of dwelling units shall have GFCI protection.

    (a) True
    (b) False

27. At least one wall switch-controlled lighting outlet must be installed in every habitable room and bathroom of a dwelling unit.

    (a) True
    (b) False

28. The demand factors of Table 220.56 apply to space heating, ventilating, or air-conditioning equipment.

    (a) True
    (b) False

29. Two or more grounding electrodes bonded together are considered a single grounding electrode system.

    (a) True
    (b) False

30. Ferrous metal enclosures for grounding electrode conductors shall be electrically continuous, from the point of attachment to cabinets or equipment, to the grounding electrode.

    (a) True
    (b) False

31. In general, areas where acids and alkali chemicals are handled and stored may present corrosive conditions, particularly when wet or damp.

    (a) True
    (b) False

32. Each current-carrying conductor of a paralleled set of conductors shall be counted as a current-carrying conductor for the purpose of applying the adjustment factors of 310.15(B)(2)(a).

    (a) True
    (b) False

33. A strut-type channel raceway is a metallic raceway intended to be mounted to the surface of, or suspended from, a structure with associated accessories for the installation of electrical conductors and cables.

    (a) True
    (b) False

34. Lighting track is a manufactured assembly and its length may not be altered by the addition or subtraction of sections of track.

    (a) True
    (b) False

35. If a permanently installed electric baseboard heater has factory-installed receptacle outlets, the receptacle is permitted to be connected to the heater circuits.

    (a) True
    (b) False

36. Explanatory material, such as references to other standards, references to related sections of the *NEC*, or information related to a *Code* rule, are included in the form of Fine Print Notes (FPNs).

    (a) True
    (b) False

37. The white conductor within a cable can be used for switch loops if the white conductor is permanently reidentified to indicate its use at all terminations where the conductor is visible and accessible.

    (a) True
    (b) False

38. AC circuits of less than 50V shall be grounded if supplied by a transformer whose supply system exceeds 150 volts-to-ground.

    (a) True
    (b) False

39. Fittings and connectors shall be used only with the specific wiring methods for which they are designed and listed.

    (a) True
    (b) False

40. Wiring methods and equipment installed behind suspended-ceiling panels shall be arranged and secured to allow access to the electrical equipment.

    (a) True
    (b) False

41. Surface extensions shall be made by mounting and mechanically securing an extension ring over the box, unless otherwise permitted.

    (a) True
    (b) False

42. Type NM cable shall closely follow the surface of the building finish or running boards when run exposed.

    (a) True
    (b) False

43. Where PVC conduit enters a box, fitting, or other enclosure, a bushing or adapter shall be provided to protect the conductor from abrasion unless the design of the box, fitting, or enclosure affords equivalent protection.

    (a) True
    (b) False

44. ENT is not permitted in hazardous (classified) locations, unless permitted in other articles of the *Code*.

    (a) True
    (b) False

45. Nonmetallic surface metal raceway is a nonmetallic raceway that is intended to be mounted to the surface of a structure, with associated couplings, connectors, boxes, and stings for the installation of electrical conductors.

    (a) True
    (b) False

46. Snap switches, including dimmer and similar control switches, shall be connected to an equipment grounding conductor and shall provide a means to connect metal faceplates to the equipment grounding conductor, whether or not a metal faceplate is installed.

    (a) True
    (b) False

47. A separate overcurrent device shall not be required for a capacitor connected on the load side of a motor overload protective device.

    (a) True
    (b) False

48. The *Code* covers underground mine installations and self-propelled mobile surface mining machinery and its attendant electrical trailing cable.

    (a) True
    (b) False

49. When breaks occur in dwelling unit kitchen countertop spaces for rangetops, refrigerators or sinks, each countertop surface shall be considered a separate counter space for determining receptacle placement.

    (a) True
    (b) False

50. AC systems of 50V to 1,000V that supply premises wiring systems shall be grounded where the system is three-phase, 4-wire, wye connected with the neutral conductor used as a circuit conductor.

    (a) True
    (b) False

Unit 7: Practice Questions in Straight Order—Article 502.10 through 550.2

Use the 2008 *NEC* to answer the following questions.

1. Boxes and fittings used for taps, joints, or terminal connections shall be _____ where installed in Class II, Division 1 locations.

   (a) explosionproof
   (b) identified for Class II locations
   (c) dusttight
   (d) weatherproof

2. Where flexibility is required, the following wiring methods are permitted in a Class II, Division 1 location:

   (a) dusttight flexible connectors
   (b) liquidtight flexible metal conduit
   (c) flexible cords listed for extra-hard usage
   (d) any of these

3. Raceways permitted as a wiring method in Class II, Division 2 locations include _____.

   (a) RMC and IMC
   (b) electrical metallic tubing
   (c) PVC conduit
   (d) a or b

4. Rigid metal conduit and IMC shall not be required to be threaded when used in Class II, Division 2 locations.

   (a) True
   (b) False

5. Where flexibility is required, the following wiring methods are permitted in a Class II, Division 2 location:

   (a) dusttight flexible connectors
   (b) liquidtight flexible metal conduit
   (c) flexible cords listed for extra-hard usage
   (d) any of these

6. In Class II locations, dust must be prevented from entering the required dust-ignitionproof enclosure from a raceway by one of the following methods:

   (a) permanent effective seal
   (b) horizontal raceway not less than 10 ft long
   (c) vertical raceway that extends downward for not less than 5 ft
   (d) any of these

7. In Class II locations, a permitted method of bonding is the use of bonding jumpers with proper fittings.

   (a) True
   (b) False

8. In Class II, Division 1 locations, switches, circuit breakers, motor controllers, and fuses, including pushbuttons, relays, and similar devices shall be provided with identified _____ enclosures.

   (a) explosionproof
   (b) dust-ignitionproof
   (c) dusttight
   (d) weatherproof

9. In Class II, Division 2 locations, enclosures for fuses, switches, circuit breakers, and motor controllers, including pushbuttons, relays, and similar devices shall be _____.

   (a) dusttight
   (b) raintight
   (c) rated as Class I, Division 1 explosionproof
   (d) general duty

10. In Class II, Division 1 locations, control transformers, solenoids, impedance coils and resistors, and any overcurrent devices or switching mechanisms associated with them, shall have dust-ignitionproof enclosures identified for _____.

    (a) Class I, Division 1 locations
    (b) control transformer duty
    (c) general duty
    (d) Class II locations

11. In Class II, Division 1 locations, motors, generators, or other rotating electric machinery shall be _____.

    (a) identified for Class II, Division 1 locations
    (b) totally enclosed pipe-ventilated, and meet the temperature limitations of 502.5
    (c) general duty
    (d) a or b

12. Pendant luminaires installed in Class II, Division 1 locations shall be suspended by threaded conduit stems, by chains with approved fittings, or by other approved means. Stems _____.

    (a) shall not be longer than 12 in.
    (b) over 12 in. shall be provided with lateral bracing
    (c) shall be provided with a fitting or flexible connector listed for the location
    (d) any of these

13. Luminaires for fixed lighting in Class II, Division 2 locations shall be identified for use in such locations, or shall be equipped with _____ enclosures.

    (a) dust-ignitionproof
    (b) dusttight
    (c) dustproof
    (d) explosionproof

14. Flexible cords used in Class II, Division 1 or 2 locations _____.

    (a) shall be listed for hard usage
    (b) shall be listed for extra-hard usage
    (c) shall not be permitted
    (d) none of these

15. In Class II, Division 1 locations, receptacles and attachment plugs shall be identified _____.

    (a) as explosionproof
    (b) for Class II locations
    (c) with laminated tags
    (d) for general duty

16. Signaling, alarm, remote-control and communications system circuits containing contacts installed in Class II, Division 2 locations shall be in _____ enclosures effective January 1, 2011.

    (a) dust-ignitionproof
    (b) dusttight
    (c) dustproof
    (d) explosionproof

17. Article 503 covers the requirements for electrical and electronic equipment and wiring in Class III locations where fire or explosion hazards may exist due to ignitible _____.

    (a) gases or vapors
    (b) fibers/flyings
    (c) dust
    (d) all of these

18. Raceways permitted in a Class III location include _____.

    (a) RMC and IMC
    (b) Type MC cable with listed termination fittings
    (c) electrical metallic tubing
    (d) any of these

19. Boxes and fittings shall be _____ where installed in a Class III location.

    (a) explosionproof
    (b) dust-ignitionproof
    (c) dusttight
    (d) weatherproof

20. In Class III locations, switches, circuit breakers, motor controllers, and fuses, including pushbuttons, relays, and similar devices, shall be provided with _____.

    (a) Class I enclosures
    (b) general duty enclosures
    (c) dusttight enclosures
    (d) seals at each enclosure

21. In Class III locations, _____ used as or in conjunction with control equipment for motors, generators, and appliances shall be provided with dusttight enclosures.

    (a) transformers
    (b) impedance coils
    (c) resistors
    (d) all of these

22. Luminaires in Class III locations exposed to physical damage shall be protected by a(n) _____ guard.

    (a) plastic
    (b) metal
    (c) suitable
    (d) explosionproof

23. In a Class III location, pendant luminaires suspended by stems longer than _____ ft shall be provided with a fitting or flexible connector identified for the location, or it shall be provided with effective bracing.

    (a) 1
    (b) 2
    (c) 3
    (d) 4

24. In Class III, portable lighting equipment shall be equipped with handles and protected with substantial guards. Lampholders shall be of the unswitched type with no provisions for _____.

    (a) receiving attachment plugs
    (b) grounding connections
    (c) lamp installation
    (d) hooks or hangers

25. In Class III, flexible cords shall _____.

    (a) be listed as extra-hard usage
    (b) contain an equipment grounding conductor
    (c) terminate in an approved manner
    (d) all of these

26. In Class III, receptacles and attachment plugs shall be of the grounding type, shall be designed so as to minimize the accumulation or the entry of _____, and shall prevent the escape of sparks or molten particles.

    (a) gases or vapors
    (b) particles of combustion
    (c) fibers/flyings
    (d) none of these

27. Article 504 covers the installation of intrinsically safe apparatus, wiring, and systems for _____ locations.

    (a) Class I
    (b) Class II
    (c) Class III
    (d) all of these

28. An assembly of interconnected intrinsically safe apparatus, associated apparatus, and interconnecting cables designed so that they are intrinsically safe circuits is a(n) _____.

    (a) intrinsically safe system
    (b) safe location
    (c) reclassified location
    (d) associated system

29. Intrinsically safe apparatus, associated apparatus, and other equipment must be installed in accordance with the _____ drawing(s).

    (a) approved
    (b) control
    (c) identified
    (d) any of these

30. Intrinsically safe and associated apparatus are permitted to be installed in _____.

    (a) any hazardous (classified) location for which they have been identified
    (b) Class I locations
    (c) Class II locations
    (d) Class III locations

31. Conductors of intrinsically safe circuits shall not be placed in any _____ with conductors of any nonintrinsically safe system.

    (a) raceway
    (b) cable tray
    (c) cable
    (d) any of these

32. Conductors of intrinsically safe circuits shall be separated at least _____ from conductors of any nonintrinsically safe circuits within the same enclosures.

    (a) 2 in.
    (b) 6 in.
    (c) 12 in.
    (d) 18 in.

33. Conductors and cables of intrinsically safe circuits run in other than raceway or cable tray systems shall be separated by at least _____ and secured from conductors and cables of any nonintrinsically safe circuits.

    (a) 2 in.
    (b) 6 in.
    (c) 12 in.
    (d) 18 in.

34. In hazardous (classified) locations, intrinsically safe apparatus shall be _____ in accordance with 250.100.

    (a) secured
    (b) bonded
    (c) painted
    (d) not be used

35. Conduits, cable trays, and other wiring methods used for intrinsically safe systems shall be identified by permanently affixed labels with the wording "Intrinsic Safety Wiring." The labels shall be visible after installation and the spacing between labels shall not exceed _____ ft.

    (a) 3
    (b) 10
    (c) 25
    (d) 50

36. Color coding shall be permitted to identify intrinsically safe conductors where they are colored light _____ and where no other conductors colored light _____ are used.

    (a) red
    (b) blue
    (c) yellow
    (d) any of these

37. Article _____ contains the wiring requirements for service and repair operations in connection with self-propelled vehicles in which volatile flammable liquids or flammable gases are used for fuel or power.

    (a) 500
    (b) 501
    (c) 511
    (d) 514

38. A building or portions of a building where engine overhauls, painting, body and fender work, and repairs that require _____ of the motor vehicle fuel tank are performed defines the term major repair garage.

    (a) removal
    (b) replacement
    (c) filling
    (d) draining

39. For each floor area inside a major repair garage where ventilation is not provided and flammable liquids having a flash point below 100°F are transferred, the entire area up to a level of _____ above the floor is considered to be a Class I, Division 2 location.

    (a) 6 in.
    (b) 12 in.
    (c) 18 in.
    (d) 24 in.

40. In major repair garages where CNG (compressed natural gas) vehicles are repaired or stored, the area within _____ in. of the ceiling is classified as Class I, Division 2.

    (a) 6
    (b) 12
    (c) 18
    (d) 24

41. Any pit below a major repair garage floor level of a lubrication or service room is considered to be a Class I, _____ location up to 18 in. above floor level.

    (a) Division 1
    (b) Division 2
    (c) a or b
    (d) a and b

42. Any pit below a minor repair garage floor level of a lubrication or service room is considered to be a Class I, _____ location up to 18 in. above floor level.

    (a) Division 1
    (b) Division 2
    (c) a or b
    (d) a and b

43. The area used for _____ of alcohol-based windshield washer fluid into motor vehicles in repair garages shall be unclassified.

    (a) storage
    (b) handling
    (c) dispensing into motor vehicles
    (d) any of these

44. Portable lighting equipment used in commercial garages shall be identified for _____ locations, unless the lamp and its cord are supported or arranged in such a manner that they cannot be used in the locations classified in 511.3.

    (a) hard usage
    (b) Class I, Division 2
    (c) Class I, Division 1
    (d) general duty

45. Flexible cords for pendants installed above Class I locations in a commercial garage shall be _____.

    (a) suitable for the type of service
    (b) listed for hard usage
    (c) a and b
    (d) shall not be used

46. A permanently mounted luminaire in a commercial garage, located over lanes on which vehicles are commonly driven, shall be located not less than _____ above floor level.

    (a) 10 ft
    (b) 12 ft
    (c) 14 ft
    (d) 16 ft

47. Article 513 applies to buildings or structures in which service, repairs, or alterations are performed on _____.

    (a) boats
    (b) aircraft
    (c) automobiles
    (d) any of these

48. Any pit or depression below the level of an aircraft hangar floor shall be classified as a _____ location that extends up to said floor level.

    (a) Class I, Division 1
    (b) Class I, Division 2
    (c) Class II, Division 1
    (d) Class III

49. The entire area of an aircraft hangar, including any adjacent and communicating areas not suitably cut off from the hangar, shall be classified as a Class I, Division 2 or Zone 2 location up to a level of _____ above the floor.

    (a) 6 in.
    (b) 12 in.
    (c) 18 in.
    (d) 24 in.

50. Within the vicinity of aircraft in an aircraft hangar, the area within 5 ft horizontally from aircraft power plants or aircraft fuel tanks shall be classified as a Class I, Division 2 location that extends upward from the floor to a level of _____ above the upper surface of wings and of engine enclosures.

    (a) 2 ft
    (b) 3 ft
    (c) 5 ft
    (d) 6 ft

51. Stock rooms and similar areas adjacent to classified locations of aircraft hangars, but effectively isolated and adequately ventilated, shall be designated as _____ locations.

    (a) Class I, Division 2
    (b) Class II, Division 1
    (c) Class II, Division 2
    (d) unclassified

52. Attachment plugs and receptacles in Class I locations of aircraft hangars shall be _____.

    (a) identified for use in Class I locations
    (b) designed so that they cannot be energized while the connections are being made or broken
    (c) a or b
    (d) none of these

53. Fixed wiring in an aircraft hangar not in a Class I location shall be _____.

    (a) metal raceways
    (b) Type MI, TC, or MC cable
    (c) nonmetallic raceways
    (d) a or b

54. For pendants in an aircraft hangar, not installed in Class I locations, flexible cords suitable for the type of service and identified for _____ shall be used.

    (a) hard usage
    (b) extra-hard usage
    (c) general duty
    (d) a or b

55. In aircraft hangars, equipment less than _____ above wings and engine enclosures of aircraft and may produce arcs, sparks, or particles of hot metal shall be of the totally enclosed type or constructed so as to prevent the escape of sparks or hot metal particles.

    (a) 2 ft
    (b) 3 ft
    (c) 10 ft
    (d) 20 ft

56. Wiring in or under the aircraft hangar floor shall comply with the requirements for _____ locations.

    (a) Class I, Division 1
    (b) Class I, Division 2
    (c) a or b
    (d) none of these

57. All 15A and 20A, 125V receptacles installed in aircraft hangars in areas where _____ is(are) used shall have GFCI protection.

    (a) electrical diagnostic equipment
    (b) electrical hand tools
    (c) portable lighting equipment
    (d) any of these

58. Article 514 applies to _____ and fleet vehicle motor fuel dispensing facilities.

    (a) motor fuel dispensing facilities
    (b) marine fuel dispensing facilities
    (c) commercial gas stations for motor vehicles only
    (d) a and b

59. Locations where flammable liquids having a flash point _____ can be unclassified.

    (a) above 100°F
    (b) below 100°F
    (c) 100°F
    (d) none of these

60. Underground wiring to motor fuel dispensers shall be installed in _____.

    (a) threaded rigid metal conduit
    (b) threaded steel IMC
    (c) PVC conduit when buried under not less than 2 ft of cover
    (d) any of these

61. A listed seal for motor fuel dispensers shall be _____.

    (a) provided in each conduit run entering a dispenser
    (b) provided in each conduit run leaving a dispenser
    (c) the first fitting after the conduit emerges from the earth or concrete at a dispenser
    (d) all of these

62. The emergency controls for attended self-service stations shall be located no more than _____ from motor fuel dispensers.

    (a) 20 ft
    (b) 50 ft
    (c) 75 ft
    (d) 100 ft

63. The emergency controls for unattended self-service stations shall be located not less than _____ or more than _____ from motor fuel dispensers.

    (a) 10 ft, 25 ft
    (b) 20 ft, 50 ft
    (c) 20 ft, 100 ft
    (d) 50 ft, 100 ft

64. In motor fuel dispensing facilities, all metal raceways, the metal armor or metallic sheath on cables, and all noncurrent-carrying metal parts of fixed portable electrical equipment _____ shall be grounded and bonded.

    (a) operating at under 600V
    (b) regardless of voltage
    (c) over 300V
    (d) under 50V

65. Article 517 applies to electrical wiring in health care facilities such as _____.

    (a) dental offices
    (b) veterinary facilities
    (c) a and b
    (d) none of these

66. The requirements of Article 517 apply to buildings or portions of buildings in which medical, _____, or surgical care is provided.

    (a) psychiatric
    (b) nursing
    (c) obstetrical
    (d) any of these

Unit 7 — Practice Questions in Straight Order—Article 502.10 through 550.2

67. A limited care facility is an area used on a(n) _____ basis for the housing of four or more persons who are incapable of self-preservation because of age, physical limitation due to accident or illness, or mental limitations, mental illness, or chemical dependency.

    (a) occasional
    (b) 10-hour or less per day
    (c) 24-hour
    (d) temporary

68. A _____ is the location of an inpatient sleeping bed; or the bed or procedure table used in a critical care area.

    (a) patient bed location
    (b) patient care area
    (c) a or b
    (d) none of these

69. The patient care area is any portion of a health care facility where patients are intended to be _____.

    (a) examined
    (b) treated
    (c) registered
    (d) a or b

70. The patient care of a health care facility does not include business offices, corridors, lounges, day rooms, dining rooms, or similar areas.

    (a) True
    (b) False

71. In patient care areas, the grounding terminals of all receptacles and all noncurrent-carrying conductive surfaces of fixed electric equipment _____ shall be connected to an insulated copper equipment grounding conductor.

    (a) operating at over 100V
    (b) likely to become energized
    (c) subject to personal contact
    (d) all of these

72. Each general care area patient bed location shall be provided with a minimum of _____ receptacle(s), which can be single, duplex, or a combination with each duplex counting as two receptacles.

    (a) one
    (b) two
    (c) three
    (d) four

73. Receptacles located within the rooms, bathrooms, playrooms, activity rooms, and patient care areas of pediatric wards shall be listed as _____, or shall employ a listed _____ cover.

    (a) tamper resistant
    (b) isolated
    (c) GFCI-protected
    (d) specification grade

74. Luminaires located more than 7½ ft above the floor can be connected to the equipment grounding return path complying with 517.13(A), without being connected to an insulated equipment grounding conductor.

    (a) True
    (b) False

75. Examples of assembly occupancies include _____.

    (a) restaurants
    (b) conference rooms
    (c) pool rooms
    (d) all of these

76. Temporary wiring in exhibition halls for display booths in assembly occupancies is permitted to be in accordance with Article 590, except _____.

    (a) GFCI protection is not required
    (b) hard or extra-hard usage cords and can be laid on floors where protected from the general public
    (c) GFCI protection is required
    (d) a and b

77. Which of the following wiring methods are permitted in an assembly occupancy of fire-rated construction?

    (a) metal raceways
    (b) Type MC cable
    (c) Type AC cable
    (d) all of these

78. In assembly occupancies of fire-rated construction, nonmetallic raceways encased in not less than _____ of concrete shall be permitted.

    (a) 1 in.
    (b) 2 in.
    (c) 3 in.
    (d) 4 in.

79. In assembly occupancies, Type NM cable, AC cable, ENT, and PVC conduit can be installed in those portions of the building not required to be of _____ construction by the applicable building code.

    (a) Class I, Division 1
    (b) fire-rated
    (c) occupancy-rated
    (d) aboveground

80. Article _____ covers the installation of portable wiring and equipment for carnivals, circuses, exhibitions, fairs, traveling attractions, and similar functions.

    (a) 518
    (b) 525
    (c) 590
    (d) all of these

81. The individual responsible for starting, stopping, and controlling an amusement ride or supervising a concession is known as the _____.

    (a) operator
    (b) director
    (c) manager
    (d) facilitator

82. Overhead wiring outside of tents and concession areas of carnivals and circuses which are accessible to pedestrians shall maintain a vertical clearance of _____ above finished grade, sidewalks, or from platforms, projections, or surfaces from which the wiring might be reached.

    (a) 3 ft
    (b) 6 ft
    (c) 8 ft
    (d) 10 ft

83. Portable carnival or circus structures shall be maintained not less than _____ in any direction from overhead conductors operating at 600 volts or less.

    (a) 7½ ft
    (b) 10 ft
    (c) 12½ ft
    (d) 15 ft

84. Electrical equipment and wiring methods for carnival or circuses on portable structures shall have mechanical protection where subject to _____.

    (a) public access
    (b) physical damage
    (c) exposure to the weather
    (d) operator access

85. Electrical service equipment at carnivals and circuses shall not be accessible to unqualified persons, unless _____.

    (a) lockable
    (b) readily accessible
    (c) accessible
    (d) none of these

86. Cord connectors for carnivals, circuses, and fairs can be laid on the ground when the connectors are _____ for a wet location.

    (a) listed
    (b) labeled
    (c) approved
    (d) identified

87. Each portable structure at a carnival, circus, or fair shall be provided with a disconnect switch within sight of and within _____ of the operator's station.

    (a) 3 ft
    (b) 6 ft
    (c) 8 ft
    (d) 10 ft

88. Portable distribution and termination boxes installed outdoors at carnivals, circuses, or fairs shall be weatherproof and mounted so the bottom of the enclosure is not less than _____ above the ground.

    (a) 6 in.
    (b) 8 in.
    (c) 10 in.
    (d) 12 in.

89. GFCI protection is not permitted at carnivals, circuses, and fairs for _____.

    (a) portable structures
    (b) egress lighting
    (c) a and b
    (d) none of these

90. The _____ of the circuit supplying metal parts of portable structures at a carnival, circuit or fair can serve as the required bonding means.

   (a) grounded conductor
   (b) equipment grounding conductor
   (c) grounding electrode conductor
   (d) ungrounded conductor

91. At circuses and carnivals, all equipment to be grounded shall be connected to a(n) _____ conductor of a type recognized by 250.118.

   (a) equipment grounding
   (b) grounded
   (c) grounding electrode
   (d) ungrounded

92. Agricultural buildings where excessive dust and dust with water may accumulate, including all areas of _____ confinement systems where litter dust or feed dust may accumulate shall comply with Article 547.

   (a) poultry
   (b) livestock
   (c) fish
   (d) all of these

93. Agricultural buildings where corrosive atmospheres exist include areas where the following exist:

   (a) poultry and animal excrement
   (b) corrosive particles which may combine with water
   (c) area of periodic washing water and cleansing agents
   (d) all of these

94. An equipotential plane is an area where wire mesh or other conductive elements are embedded in or placed under concrete bonded to _____.

   (a) all metal structures
   (b) fixed nonelectrical equipment that may become energized
   (c) the electrical grounding system
   (d) all of these

95. Cables installed in agricultural buildings shall be secured within _____ of each cabinet, box, or fitting.

   (a) 8 in.
   (b) 12 in.
   (c) 18 in.
   (d) 24 in.

96. In damp or wet locations of agricultural buildings, equipment enclosures and fittings shall be located or equipped to prevent moisture from _____ within the enclosure, box, conduit body, or fitting.

   (a) entering
   (b) accumulating
   (c) a or b
   (d) none of these

97. An equipment grounding conductor installed within an agricultural building shall be a(n) _____ conductor.

   (a) aluminum
   (b) copper
   (c) copper-clad
   (d) any of these

98. All 15A and 20A, 125V, single-phase general-purpose receptacles installed _____ of agricultural buildings shall be GFCI-protected.

   (a) in areas having an equipotential plane
   (b) outdoors
   (c) in dirt confinement areas for livestock
   (d) any of these

99. The equipotential plane in an agricultural building shall be connected to the electrical grounding system with a copper, insulated, covered, or bare conductor and not smaller than _____.

   (a) 10 AWG
   (b) 8 AWG
   (c) 6 AWG
   (d) 4 AWG

100. A _____ is a structure transportable in one or more sections that is built on a permanent chassis and designed to be used as a dwelling, with or without a permanent foundation.

   (a) manufactured home
   (b) mobile home
   (c) dwelling unit
   (d) all of these

Unit 7: Practice Questions in Random Order—Article 90.2 through 547.10

Use the 2008 *NEC* to answer the following questions.

1. Single-pole breakers utilizing identified handle ties are not permitted to be used for the required disconnecting means for motor fuel dispensing equipment.

    (a) True
    (b) False

2. When installed indoors for carnivals, circuses, and fairs, flexible cords and flexible cables shall be listed for wet locations and shall be sunlight resistant.

    (a) True
    (b) False

3. Enclosures and fittings installed in areas of agricultural buildings where excessive dust may be present shall be designed to minimize the entrance of dust and shall have no openings through which dust can enter the enclosure.

    (a) True
    (b) False

4. Circuits of transformer-powered communications or signaling systems shall not be required to be enclosed in raceways unless required by Chapters 7 or 8.

    (a) True
    (b) False

5. Wiring for an amusement ride, attraction, tent, or similar structure at a carnival, circus, or fairs shall not be supported by a ride or structure unless specifically designed for the purpose.

    (a) True
    (b) False

6. Each motor fuel dispensing device shall be provided with a means to remove all external voltage sources, including feedback, during periods of maintenance and service of the fuel dispensing equipment.

    (a) True
    (b) False

7. In agricultural building locations where surfaces are periodically washed or sprayed with water, boxes, conduit bodies, and fittings shall be weatherproof.

    (a) True
    (b) False

8. In Class II, Division 2 locations, flexible cord can serve as the supporting means for a pendant luminaire.

    (a) True
    (b) False

9. Parking garages used for parking or storage can be considered as an unclassified location.

    (a) True
    (b) False

10. Wiring for lighting located inside tents and concession areas at carnivals, circuses, and fairs, where subject to physical damage shall be provided with mechanical protection.

    (a) True
    (b) False

11. GFCI protection shall be provided at carnivals, circuses, and fairs for all 15A and 20A, 125V nonlocking-type receptacle that are readily accessible to the general public.

    (a) True
    (b) False

Unit 7 — Practice Questions in Random Order—Article 90.2 through 547.10

12. Raceways shall be provided with expansion fittings where necessary to compensate for thermal expansion and contraction.

    (a) True
    (b) False

13. A box or conduit body shall not be required where cables enter or exit from conduit or tubing that is used to provide cable support or protection against physical damage.

    (a) True
    (b) False

14. Conductors must be insulated and installed as part of a recognized wiring method.

    (a) True
    (b) False

15. ENT shall be permitted for direct earth burial when used with fittings listed for this purpose.

    (a) True
    (b) False

16. Receptacles mounted to and supported by a cover shall be secured by more than one screw unless listed and identified for securing by a single screw.

    (a) True
    (b) False

17. Where more than one motor disconnecting means is provided in the same motor branch circuit, at least one of the disconnecting means shall be readily accessible.

    (a) True
    (b) False

18. GFCI protection is not required for nonlocking-type receptacles that only facilitate the quick disconnecting and reconnecting of electrical equipment.

    (a) True
    (b) False

19. An equipotential plane shall be installed in all concrete floor confinement areas of livestock buildings, and all outdoor confinement areas with a concrete slab that contain metallic equipment accessible to livestock and that may become energized.

    (a) True
    (b) False

20. Compliance with either the metric or the inch-pound unit of measurement system shall be permitted.

    (a) True
    (b) False

21. An overload is the same as a short circuit or ground fault.

    (a) True
    (b) False

22. Equipment intended to interrupt current at fault levels shall have an interrupting rating sufficient for the nominal circuit voltage and the current that is available at the line terminals of the equipment.

    (a) True
    (b) False

23. In a dwelling unit, illumination from a lighting outlet shall be provided at the exterior side of each outdoor entrance or exit that has grade-level access.

    (a) True
    (b) False

24. Underground service conductors shall be protected against physical damage.

    (a) True
    (b) False

25. For grounded systems, normally noncurrent-carrying conductive materials enclosing electrical conductors or equipment shall be connected to earth so as to limit the voltage-to-ground on these materials.

    (a) True
    (b) False

26. AC systems of 50V to 1,000V that supply premises wiring systems shall be grounded where supplied by a three-phase, 4-wire, delta-connected system in which the midpoint of one phase winding is used as a circuit conductor.

    (a) True
    (b) False

27. Equipment bonding jumpers shall be of copper or other corrosion-resistant material.

    (a) True
    (b) False

28. Direct-buried conductors or cables can be spliced or tapped without the use of splice boxes when the splice or tap is made in accordance with 110.14(B).

    (a) True
    (b) False

29. A box or conduit body shall not be required for splices and taps in direct-buried conductors and cables as long as the splice is made with a splicing device that is identified for the purpose.

    (a) True
    (b) False

30. Boxes, conduit bodies, and fittings installed in wet locations shall be required to be listed for use in wet locations.

    (a) True
    (b) False

31. In indoor locations, other than dwellings, fluorescent luminaires that utilize double-ended lamps and contain ballast(s) shall have a disconnecting means either internal or external to each luminaire.

    (a) True
    (b) False

32. Where the motor short-circuit and ground-fault protection devices determined by Table 430.52 do not correspond to the standard sizes or ratings, a higher rating that does not exceed the next higher standard ampere rating shall be permitted.

    (a) True
    (b) False

33. A separate disconnecting means shall be installed where a capacitor is connected on the load side of a motor controller.

    (a) True
    (b) False

34. The removal of shielding material or the separation of the twisted pairs isn't required within the conduit seal fitting in a Class I, Division 1 hazardous (classified) location.

    (a) True
    (b) False

35. Boxes and fittings used to support luminaires in Class I, Division 1 locations shall be identified for Class I locations.

    (a) True
    (b) False

36. Equipotential planes shall be installed in outdoor concrete slabs where metallic equipment is located that may be energized and is accessible to livestock, other than poultry.

    (a) True
    (b) False

37. All 15A and 20A, 125V receptacles installed in dwelling unit boathouses shall have GFCI protection.

    (a) True
    (b) False

38. Receptacle outlets can be installed below the countertop surface in dwelling units when necessary for the physically impaired, or if there is no means available to mount a receptacle above an island or peninsular countertop.

    (a) True
    (b) False

39. When locating lighting outlets in dwelling units, a vehicle door in a garage shall be considered an outdoor entrance.

    (a) True
    (b) False

40. Overhead service conductors spans can be supported to hardwood trees.

    (a) True
    (b) False

41. A box or conduit body shall not be required for conductors in handhole enclosures.

    (a) True
    (b) False

42. When IMC is cut in the field, reaming is required to remove the burrs and rough edges.

    (a) True
    (b) False

43. Power distribution blocks installed in metal wireways shall be listed.

    (a) True
    (b) False

44. The maximum size conductors permitted in a nonmetallic surface raceway shall not be larger than that for which the wireway is designed.

    (a) True
    (b) False

45. Outdoor luminaires and associated equipment can be supported by trees.

    (a) True
    (b) False

46. Where a motor installation includes a capacitor connected on the load side of the motor overload device, the rating or setting of the motor overload device shall be based on the improved power factor of the motor circuit.

    (a) True
    (b) False

47. The removal of shielding material or the separation of the twisted pairs isn't required within the conduit seal fitting in a Class I, Division 2 hazardous (classified) location.

    (a) True
    (b) False

48. Utilities may be subject to compliance with codes and standards covering their regulated activities as adopted under governmental law or regulation.

    (a) True
    (b) False

49. A signaling circuit is any electric circuit that energizes signaling equipment.

    (a) True
    (b) False

50. Where a lighting outlet(s) is installed for interior stairways, there shall be a wall switch at each floor level and each landing level that includes an entryway where the stairway between floor levels has four risers or more.

    (a) True
    (b) False

Unit 8 — Practice Questions in Straight Order—Article 550.13 through 680.22

Use the 2008 *NEC* to answer the following questions.

1. GFCI protection in a mobile home shall be provided for _____.

    (a) receptacle outlets installed outdoors and in compartments accessible from outside
    (b) receptacles within 6 ft of a wet bar sink
    (c) all receptacles in bathrooms including receptacles in luminaires
    (d) all of these

2. A receptacle outlet for mobile and manufactured home heat tape that is used to protect cold water inlet piping shall be _____-protected and it shall be connected to an interior branch circuit other than a small-appliance branch circuit where all of the outlets of the circuit are on the load side of the GFCI.

    (a) AFCI
    (b) GFCI
    (c) a or b
    (d) none of these

3. Branch circuits that supply 15A and 20A, 120V outlets in bedrooms of mobile homes and manufactured homes shall be _____-protected.

    (a) GFCI
    (b) weatherproof-in-use covers
    (c) AFCI
    (d) none of these

4. Mobile home service equipment shall be located adjacent to the mobile home, located in sight from but not more than _____ from the exterior wall of the mobile home it serves.

    (a) 15 ft
    (b) 20 ft
    (c) 30 ft
    (d) 40 ft

5. Mobile home service equipment shall be rated not less than _____ at 120/240V.

    (a) 50A
    (b) 60A
    (c) 100A
    (d) 200A

6. An outdoor disconnecting means for a mobile home shall be installed so the bottom of the enclosure is not less than _____ above the finished grade or working platform.

    (a) 1 ft
    (b) 2 ft
    (c) 3 ft
    (d) 6 ft

7. Feeder conductors to a mobile home shall consist of _____.

    (a) a listed cord
    (b) a permanently installed feeder consisting of four color-coded, insulated conductors
    (c) a or b
    (d) none of these

8. Mobile home and manufactured home feeder-circuit conductors shall have a capacity not less _____.

    (a) 50A
    (b) 60A
    (c) 100A
    (d) 200A

9. A minimum of 20 percent of all recreational vehicle sites with electrical supply shall be equipped with a _____,125/250V receptacle.

(a) 15A
(b) 20A
(c) 30A
(d) 50A

10. A minimum of 70 percent of all recreational vehicle sites with electrical supply shall be equipped with a _____,125V receptacle.

(a) 15A
(b) 20A
(c) 30A
(d) 50A

11. Electrical service and feeder loads for a recreational vehicle park shall be calculated at a minimum of _____ per site equipped with only 20A supply facilities (not including tent sites).

(a) 1,200 VA
(b) 2,400 VA
(c) 3,500 VA
(d) 9,600 VA

12. Private, noncommercial docking facilities constructed or occupied for the use of the owner or residents of the associated _____ are not covered by Article 555.

(a) single-family dwelling
(b) multifamily dwelling
(c) a or b
(d) none of these

13. A _____ is an enclosed assembly that can include receptacles, circuit breakers, fused switches, fuses, watt-hour meter(s), and monitoring means approved for marine use.

(a) marine power receptacle
(b) marine outlet
(c) marine power outlet
(d) any of these

14. At marinas and boatyards, the electrical datum plane, in land areas subject to tidal fluctuation, is a horizontal plane _____ above the highest high tide under normal circumstances.

(a) 1 ft
(b) 2 ft
(c) 3 ft
(d) 4 ft

15. In marinas and boatyards, the bottom of enclosures for transformers shall not be located below _____.

(a) 2 ft above the dock
(b) 18 in. above the electrical datum plane
(c) the electrical datum plane
(d) a dock

16. Service equipment for floating docks or marinas shall be located _____ the floating structure.

(a) adjacent to
(b) on
(c) in
(d) any of these

17. Electrical connections in marinas and boatyards shall be located _____ unless the conductor splices are contained within sealed wire connector systems listed and identified for submersion.

(a) at least 12 in. above the deck of a floating pier
(b) not less than 12 in. above the deck of a fixed pier
(c) not below the electrical datum plane
(d) all of these

18. Where shore power accommodations provide two receptacles specifically for an individual boat slip, and these receptacles have different voltages, only the receptacle with the _____ shall be required to be calculated.

(a) smaller kW demand
(b) larger kW demand
(c) higher voltage
(d) none of these

19. A _____ must be used to serve as the required shore power receptacle disconnecting means and it must be identified as to which receptacle it controls.

(a) circuit breaker
(b) switch
(c) a and b
(d) a or b

20. In marinas or boatyards, the *NEC* requires a(n) _____ disconnecting means, which allows individual boats to be isolated from their supply connection.

    (a) accessible
    (b) readily accessible
    (c) remote
    (d) any of these

21. The disconnecting means for a shore power connection shall be not more than _____ from the receptacle it controls.

    (a) 12 in.
    (b) 24 in.
    (c) 30 in.
    (d) 36 in.

22. Receptacles that provide shore power for boats shall be rated not less than _____.

    (a) 15A
    (b) 20A
    (c) 30A
    (d) 60A

23. Fifteen and 20A, single-phase, 125V receptacles used for "other than shore power" in marinas shall be provided with _____.

    (a) lockouts
    (b) GFCI protection
    (c) warning labels
    (d) all of these

24. Electrical wiring for_____ shall be installed on the side of the wharf, pier, or dock opposite from the liquid fuel piping system.

    (a) power
    (b) lighting
    (c) fuel dispensing equipment
    (d) all of these

25. Temporary wiring methods shall be acceptable only if _____, based on the conditions of use and any special requirements of the temporary installation.

    (a) listed
    (b) identified
    (c) approved
    (d) any of these

26. Temporary electrical power and lighting installations shall be permitted for a period not to exceed 90 days for _____ decorative lighting and similar purposes.

    (a) Christmas
    (b) New Year's
    (c) July 4th
    (d) holiday

27. Temporary electrical power and lighting shall be permitted during emergencies and for _____.

    (a) tests
    (b) experiments
    (c) developmental work
    (d) all of these

28. Temporary wiring shall be _____ immediately upon the completion of construction or purpose for which the wiring was installed.

    (a) disconnected
    (b) removed
    (c) de-energized
    (d) any of these

29. Receptacles on construction sites shall not be installed on the _____ as temporary lighting.

    (a) same branch circuit
    (b) the same feeders
    (c) a or b
    (d) none of these

30. For temporary installations, lamps shall be protected from accidental contact by a suitable _____ or by the use of a lampholder with a guard.

    (a) globe
    (b) luminaire
    (c) fitting
    (d) porcelain fitting

31. At construction sites, boxes are not required for temporary wiring splices of _____.

    (a) multiconductor cords
    (b) multiconductor cables
    (c) a or b
    (d) none of these

32. Flexible cords and flexible cables used for temporary wiring shall _____.

    (a) be protected from accidental damage
    (b) be protected where passing through doorways
    (c) avoid sharp corners and projections
    (d) all of these

33. For temporary installations, cable assemblies, flexible cords, and flexible cables shall be supported by _____ or similar fittings so as not to damage the wiring.

    (a) staples
    (b) cable ties
    (c) straps
    (d) any of these

34. Vegetation shall not be used to support overhead conductor spans of _____ for temporary installations.

    (a) branch circuits
    (b) feeders
    (c) a or b
    (d) none of these

35. Ground-fault protection for personnel is required for all temporary wiring used for construction, remodeling, maintenance, repair, or demolition of buildings, structures, or equipment, from power derived from an _____.

    (a) electric utility company
    (b) on-site generated power source
    (c) a or b
    (d) none of these

36. All _____, 125V receptacle outlets that are not part of the permanent wiring of the building or structure and are used by personnel for temporary power shall be GFCI-protected.

    (a) 15A
    (b) 20A
    (c) 30A
    (d) all of these

37. For temporary power, GFCI protection can be incorporated into a _____.

    (a) circuit breaker
    (b) receptacle
    (c) cord set
    (d) any of these

38. Receptacles rated other than 125V, single-phase, 15, 20, and 30A for temporary installations shall be protected by a _____.

    (a) GFCI device
    (b) written assured equipment grounding conductor program
    (c) AFCI device
    (d) a or b

39. Article 600 covers the installation of conductors and equipment for _____.

    (a) outline lighting
    (b) skeleton tubing
    (c) neon tubing
    (d) all of these

40. A section sign is a sign or _____, shipped as subassemblies that require field-installed wiring between the subassemblies to complete the overall sign.

    (a) outline lighting system
    (b) skeleton tubing system
    (c) neon system
    (d) sign body

41. Electric signs and outline lighting shall be _____.

    (a) approved by special permission
    (b) listed
    (c) identified for the location
    (d) a or b

42. Field-installed skeleton outline lighting consisting of listed luminaires are not required to be _____ as an assembly.

    (a) listed
    (b) identified for the purpose
    (c) marked
    (d) none of these

43. Each commercial occupancy accessible to pedestrians shall have at least one outside sign outlet in an accessible location at each entrance supplied by a branch circuit rated at least _____.

    (a) 15A
    (b) 20A
    (c) 30A
    (d) 40A

44. Where the disconnecting means is out of the line of sight from any section of the sign or outline lighting able to be energized, the disconnecting means must be capable of being _____ in the open position.

    (a) secured
    (b) bolted
    (c) locked
    (d) none of these

45. The bottom of sign and outline lighting system equipment shall be at least _____ above areas accessible to vehicles unless protected from physical damage.

    (a) 12 ft
    (b) 14 ft
    (c) 16 ft
    (d) 18 ft

46. Neon tubing, other than _____ readily accessible to pedestrians, shall be protected from physical damage.

    (a) Class I, Division 1 locations
    (b) dry-location portable signs
    (c) fixed equipment
    (d) wet-location portable signs

47. Signs and outline lighting systems shall be installed so that adjacent combustible materials are not subjected to temperatures in excess of _____.

    (a) 60°C
    (b) 75°C
    (c) 90°C
    (d) 105°C

48. A portable or mobile electric sign in a wet or damp location shall have GFCI _____.

    (a) located on the sign
    (b) located in the power-supply cord within 12 in. of the attachment plug
    (c) as an integral part of the attachment plug of the supply cord
    (d) b or c

49. A working space not less than 3 ft high by 3 ft wide by _____ deep is required for each ballast, transformer, and electronic power supply where not installed in a sign.

    (a) 2 ft
    (b) 3 ft
    (c) 4 ft
    (d) 6 ft

50. Ballasts, transformers, and electronic power supplies can be above a suspended ceiling, provided the enclosures are _____.

    (a) securely fastened in place
    (b) don't use the suspended-ceiling grid for support.
    (c) a or b
    (d) a and b

51. A manufactured wiring system is a system assembled by a manufacturer, which cannot be inspected at the building site without _____.

    (a) a permit
    (b) a manufacturer's representative present
    (c) damage or destruction to the assembly
    (d) an engineer's supervision

52. Manufactured wiring systems shall be permitted in _____ locations.

    (a) accessible
    (b) dry
    (c) wet
    (d) a and b

53. Manufactured wiring systems constructed with Type MC cable shall be supported and secured at intervals not exceeding _____.

    (a) 3 ft
    (b) 4 ft
    (c) 6 ft
    (d) 8 ft

54. At least _____ 125V, single-phase, 15A or 20A, duplex receptacle(s) shall be provided in each elevator machine room and elevator machinery space.

    (a) one
    (b) two
    (c) three
    (d) four

55. A separate _____ shall supply the elevator hoistway pit lighting and receptacle(s). The required lighting shall not be connected to the load side of a ground-fault circuit interrupter.

    (a) feeder
    (b) subpanel
    (c) emergency system
    (d) branch circuit

Unit 8 — Practice Questions in Straight Order—Article 550.13 through 680.22

56. The lighting switch for hoistway pits shall be readily accessible from the _____.

    (a) pit access door
    (b) elevator car
    (c) floor of the pit
    (d) machinery room

57. At least _____ 125V, single-phase, 15A or 20A, duplex receptacle(s) shall be provided in the elevator hoistway pit.

    (a) one
    (b) two
    (c) three
    (d) four

58. Where multiple driving machines are connected to a single elevator, escalator, moving walk, or pumping unit, there shall be one disconnecting means to disconnect the _____.

    (a) motors
    (b) control valve operating magnets
    (c) a and b
    (d) none of these

59. The disconnecting means for an elevator or escalator shall be an enclosed externally operable circuit breaker or fused motor-circuit switch capable of _____.

    (a) interrupting six times the locked-rotor current
    (b) being locked in the open position
    (c) including overload protection
    (d) serving as a transfer switch

60. The location of the disconnecting means for an elevator shall be _____ to qualified persons.

    (a) accessible
    (b) readily accessible
    (c) disclosed only
    (d) accessible only with a key

61. Where there is more than one driving machine in an elevator machine room, the disconnecting means shall be numbered to correspond to the identifying number of the _____.

    (a) driving machine they control
    (b) circuit feeding it
    (c) panel it is fed from
    (d) all of these

62. All 15A and 20A, 125V receptacles installed in machine rooms and machinery spaces for elevators, escalators, moving walks, and lifts shall have GFCI protection by a _____.

    (a) GFCI receptacle
    (b) GFCI circuit breaker
    (c) a or b
    (d) none of these

63. Each 15A and 20A, 125V receptacle installed in pits, in hoistways, on elevator car tops, and in escalator and moving walk wellways shall be _____.

    (a) on a GFCI-protected circuit
    (b) of the GFCI type
    (c) a or b
    (d) none of these

64. An arc welder shall have overcurrent protection rated or set at not more than _____ percent of the rated primary current of the welder.

    (a) 100
    (b) 125
    (c) 150
    (d) 200

65. A disconnecting means shall be provided in the supply circuit for each arc welder not equipped with _____.

    (a) a governor
    (b) a shunt-trip device
    (c) an integral disconnect
    (d) GFCI protection

66. The required ampacity for the supply conductors for a resistance welder with a duty cycle of 15 percent and a primary current of 21A is _____.

    (a) 5.67A
    (b) 6.72A
    (c) 8.19A
    (d) 9.45A

67. A _____ shall be provided to disconnect each resistance welder and its control equipment from the supply circuit.

    (a) switch
    (b) circuit breaker
    (c) magnetic starter
    (d) a or b

68. Amplifiers, loudspeakers, and other equipment shall be located or protected so as to guard against environmental exposure or physical damage that might cause _____.

    (a) a fire
    (b) shock
    (c) personal hazard
    (d) all of these

69. Exposed audio cables shall be secured by _____ or similar fittings designed and installed so as not to damage the cable.

    (a) straps
    (b) staples
    (c) hangers
    (d) any of these

70. Audio cables identified for future use shall be marked with a tag of sufficient durability to withstand _____.

    (a) moisture
    (b) humidity
    (c) the environment involved
    (d) none of these

71. Audio system equipment supplied by branch-circuit power shall not be located within _____ of the inside wall of a pool, spa, hot tub, fountain, or tidal high-water mark.

    (a) 18 in.
    (b) 2 ft
    (c) 5 ft
    (d) 10 ft

72. Loudspeakers of permanent audio systems which are installed in fire-resistance-rated partitions, walls, or ceilings shall be listed for the purpose or installed in an enclosure or recess that _____.

    (a) maintains the fire-resistance rating
    (b) is no more than 4 in. deep
    (c) is no more than 6 ft 6 in. high
    (d) all of these

73. For Information Technology Equipment, supply circuits and interconnecting cables not terminated to equipment and not identified for future use with a tag are considered to be _____.

    (a) abandoned
    (b) unused
    (c) future use
    (d) none of these

74. Article 645 does not apply unless an information technology equipment room contains _____.

    (a) a disconnecting means complying with 645.10
    (b) a separate heating/ventilating/air-conditioning (HVAC) system
    (c) separation by fire-resistance-rated walls, floors, and ceiling
    (d) all of these

75. Branch-circuit conductors for data processing equipment in information technology equipment rooms shall have an ampacity not less than _____ of the total connected load.

    (a) 80 percent
    (b) 100 percent
    (c) 125 percent
    (d) the sum

76. Where power cables associated with the information technology are installed under a raised floor, they shall not be longer than _____ ft.

    (a) 10
    (b) 12
    (c) 15
    (d) 16

77. Interconnecting cables under raised floors that support information technology equipment shall be Type _____.

    (a) RF
    (b) Type UF
    (c) LS
    (d) DP

78. _____ conductor cables 4 AWG and marked "for use in cable trays" shall be permitted for equipment grounding within a raised floor of an information technology equipment room.

    (a) Green
    (b) Insulated
    (c) Single
    (d) all of these

79. The accessible portion of abandoned interconnecting cables under an information technology room raised floor shall be removed, unless the cables are contained within a _____ raceway.

    (a) metallic
    (b) nonmetallic
    (c) a or b
    (d) none of these

80. In information technology equipment rooms, a single disconnecting means can be used to control _____.

    (a) the HVAC systems to the room
    (b) the power to electronic computer/data processing equipment
    (c) the electronic computer/data processing equipment and the building supply
    (d) the HVAC systems to the room and power to electronic computer/data processing equipment

81. Where a pushbutton is used to disconnect power in the information technology equipment room, pushing the button _____ shall disconnect the power.

    (a) in
    (b) out
    (c) a or b
    (d) none of these

82. Exposed noncurrent-carrying metal parts of an information technology system shall be _____.

    (a) bonded to an equipment grounding conductor
    (b) double insulated
    (c) GFCI-protected
    (d) a or b

83. According to the *NEC*, sensitive electronic equipment is a separately derived system that operates at _____ volts-to-ground and _____ volts line-to-line.

    (a) 30, 60
    (b) 60, 120
    (c) 120, 120
    (d) 120, 240

84. The purpose of sensitive electronic equipment is to reduce objectionable noise in sensitive electronic equipment locations.

    (a) commercial
    (b) industrial
    (c) a and b
    (d) institutional

85. The voltage drop on technical power systems for sensitive electronic equipment shall not exceed _____ percent for branch circuits.

    (a) 1.50
    (b) 2
    (c) 2.50
    (d) 2.80

86. The voltage drop on technical power systems for sensitive electronic equipment shall not exceed _____ percent for feeder and branch-circuit conductors combined.

    (a) 1.50
    (b) 2
    (c) 2.50
    (d) 3

87. Permanently installed swimming pools include those constructed in the ground or partially in the ground, and capable of holding water in a depth greater than _____ in.

    (a) 36
    (b) 42
    (c) 48
    (d) 54

88. Flexible cords for fixed or stationary pool equipment shall _____.

    (a) not exceed 3 ft
    (b) have a copper equipment grounding conductor not smaller than 12 AWG
    (c) terminate in a grounding-type attachment plug
    (d) all of these

89. Insulated overhead utility service conductors that are cabled together with a bare messenger and operate at not over 750 volts-to-ground shall maintain a _____ clearance in any direction to the water level.

    (a) 14 ft
    (b) 16 ft
    (c) 20 ft
    (d) 22½ ft

90. Cables for communications systems such as _____ shall be located no less than 10 ft from the water's edge of swimming and wading pools, diving structures, observation stands, towers, or platforms.

    (a) telephone
    (b) radio
    (c) CATV
    (d) all of these

91. Overhead network-powered broadband communications systems conductors shall be located no less than _____ from the water's edge of swimming and wading pools, or the base of diving structures.

    (a) 10 ft
    (b) 12 ft
    (c) 18 ft
    (d) none of these

92. All electric pool water heaters shall have the heating elements not exceed _____ amperes.

    (a) 20A
    (b) 35A
    (c) 48A
    (d) 60A

93. Underground PVC conduit shall be not less than _____ from the inside wall of the pool, unless space limitations prevent otherwise.

    (a) 5 ft
    (b) 8 ft
    (c) 10 ft
    (d) 25 ft

94. Underground PVC cannot be located less than 5 ft from the inside wall of a pool and shall be buried not less than _____.

    (a) 6 in.
    (b) 10 in.
    (c) 12 in.
    (d) 18 in.

95. The maintenance disconnecting means required for swimming pool equipment shall be _____ and at least 5 ft from the water's edge, unless separated by a permanently installed barrier.

    (a) readily accessible
    (b) within sight of its equipment
    (c) capable of being locked in the open position
    (d) a and b

96. Duplex receptacles that provide power for pump motors related to the circulation and sanitation system of pools shall be located at least _____ from the inside walls of the pool.

    (a) 3 ft
    (b) 5 ft
    (c) 10 ft
    (d) 12 ft

97. Grounding-type GFCI-protected receptacles of the single and locking type for motors related to the circulation and sanitation system of a pool can be located not less than _____ from the inside walls of the pool.

    (a) 3 ft
    (b) 6 ft
    (c) 8 ft
    (d) 12 ft

98. All 15A and 20A, 125V receptacles located within _____ of the inside walls of a pool shall be GFCI-protected.

    (a) 8 ft
    (b) 10 ft
    (c) 15 ft
    (d) 20 ft

99. Circuits rated 15A or 20A, 125V or 250V supplying pool pump motors for a permanently installed pool shall be _____ whether the pump is cord-and-plug-connected or direct connected.

    (a) AFCI-protected
    (b) GFCI-protected
    (c) a or b
    (d) a and b

100. Luminaires installed above the pool or the area extending _____ ft horizontally from the inside walls of the pool shall be installed at a height not less than 12 ft above the maximum water level of the pool.

    (a) 3 ft
    (b) 5 ft
    (c) 10 ft
    (d) 12 ft

Unit 8 — Practice Questions in Random Order—Article 90.2 through 680.22

Use the 2008 *NEC* to answer the following questions.

1. Junction boxes used in sensitive electronic equipment systems shall be clearly marked to indicate _____.

   (a) the system frequency
   (b) the system voltage
   (c) the distribution panel
   (d) b and c

2. An insulated copper equipment grounding conductor is required for all circuits at a marina.

   (a) True
   (b) False

3. Electronic organs or other electronic musical instruments are included in the scope of Article 640.

   (a) True
   (b) False

4. Ventilation in the underfloor area of an information equipment room shall be used in that room only, unless provided with fire/smoke dampers at the point of penetration of the room boundary.

   (a) True
   (b) False

5. Audio distribution cable not terminated at equipment and not identified for future use with a tag is considered abandoned.

   (a) True
   (b) False

6. Only wiring, raceways, and cables used directly in connection with an elevator shall be permitted inside the hoistway and the machine room.

   (a) True
   (b) False

7. General purpose Type CL2, CM, or CATV cables are permitted within the raised floor area of an information technology equipment room.

   (a) True
   (b) False

8. Supplying 120/240V 3-wire shore power receptacles from a 120/208V 3-wire supply may cause overheating or malfunctioning of connected equipment.

   (a) True
   (b) False

9. The accessible portion of abandoned audio distribution cables shall be removed.

   (a) True
   (b) False

10. Electric swimming pool equipments can be installed in pits where drainage adequately prevents water accumulation during abnormal operation.

    (a) True
    (b) False

11. Flexible cords and flexible cables are permitted for the electrical connection of permanently installed equipment racks of audio systems to facilitate access to equipment.

    (a) True
    (b) False

12. Except as specifically modified by Article 590, all other *Code* requirements for permanent wiring apply to temporary wiring installations.

    (a) True
    (b) False

474 Mike Holt's Illustrated Guide to NEC Exam Preparation

13. Metal equipment racks and enclosures for permanent audio system installations shall be grounded.

    (a) True
    (b) False

14. The 3 VA per-square-foot general lighting load for dwelling units includes general-use receptacles and lighting outlets.

    (a) True
    (b) False

15. Supplementary overcurrent devices used in luminaires or appliances are not required to be readily accessible.

    (a) True
    (b) False

16. Mechanical elements used to terminate a grounding electrode conductor to a grounding electrode shall be accessible.

    (a) True
    (b) False

17. Where circuit conductors are spliced or terminated on equipment within a box, any equipment grounding conductors associated with those circuit conductors shall be connected to the box with devices suitable for the use.

    (a) True
    (b) False

18. Metal faceplates for receptacles shall be grounded.

    (a) True
    (b) False

19. Circuit directories can include labels that depend on transient conditions of occupancy.

    (a) True
    (b) False

20. Lighting track fittings can be equipped with general-purpose receptacles.

    (a) True
    (b) False

21. In Class I locations, the locknut-bushing and double-locknut types of contacts shall not be depended on for bonding purposes.

    (a) True
    (b) False

22. In Class II, Division 1 locations, multiwire branch circuits are not permitted unless a disconnect opens all of the ungrounded circuit conductors simultaneously.

    (a) True
    (b) False

23. The number of conductors in a raceway for permanent audio system installations shall not be limited by the percentage fill specified in Chapter 9, Table 1.

    (a) True
    (b) False

24. Utilities may include entities that are designated or recognized by governmental law or regulation by public service/utility commissions.

    (a) True
    (b) False

25. Some cleaning and lubricating compounds can cause severe deterioration of many _____ materials used for insulating and structural applications in equipment.

    (a) True
    (b) False

26. In other than dwelling units, GFCI protection shall be provided for all outdoor 15A and 20A, 125V receptacles.

    (a) True
    (b) False

27. The grounding electrode conductor for a single separately derived system is used to connect the grounded conductor of the derived system to the grounding electrode.

    (a) True
    (b) False

28. Grounding electrode conductor connections to a concrete-encased or buried grounding electrode shall be accessible.

    (a) True
    (b) False

29. The equipment grounding conductor shall not be required to be larger than the circuit conductors.

    (a) True
    (b) False

30. The arrangement of grounding connections shall be such that the disconnection or the removal of a receptacle, luminaire, or other device does not interrupt the grounding continuity.

    (a) True
    (b) False

31. Raceways, cable assemblies, boxes, cabinets, and fittings shall be securely fastened in place.

    (a) True
    (b) False

32. On a three-phase, 4-wire, wye circuit, where the major portion of the load consists of nonlinear loads, the neutral conductor shall be counted when applying 310.15(B)(2) adjustment factors.

    (a) True
    (b) False

33. Type MC cable installed through, or parallel to, framing members shall be protected against physical damage from penetration by screws or nails by 1¼ in. separation or protected by a suitable metal plate.

    (a) True
    (b) False

34. Type SE cable can be used for interior wiring as long as it complies with the installation requirements of Part II of Article 334.

    (a) True
    (b) False

35. Where NUCC enters a box, fitting, or other enclosure, a bushing or adapter shall be provided to protect the conductor or cable from abrasion unless the design of the box, fitting, or enclosure provides equivalent protection.

    (a) True
    (b) False

36. EMT shall not be threaded.

    (a) True
    (b) False

37. Strut-type channel raceway shall be permitted as an equipment grounding conductor in accordance with 250.118(13).

    (a) True
    (b) False

38. The conductor adjustment factors only apply to the number of current-carrying conductors in the cable and not to the number of conductors in the cable tray.

    (a) True
    (b) False

39. Where an equipment grounding conductor isn't present in the outlet box for a luminaire, the luminaire must be made of insulating material and must not have any exposed conductive parts.

    (a) True
    (b) False

40. The connected load on lighting track is permitted to exceed the rating of the track under some conditions.

    (a) True
    (b) False

41. Article 440 applies to electric motor-driven air-conditioning and refrigerating equipment that has a hermetic refrigerant motor-compressor.

    (a) True
    (b) False

42. When FMC or LFMC is used as permitted in Class I, Division 2 locations, it shall not be used as the sole ground-fault current path.

    (a) True
    (b) False

43. In Class III locations, locknut-bushing and double-locknut types of fittings may be depended on for bonding purposes.

    (a) True
    (b) False

44. Type NM cable can be installed above a Class I location in a commercial garage.

    (a) True
    (b) False

45. The continuity of the grounding conductor system used to reduce electrical shock hazards at carnivals, fairs, and similar locations shall be verified each time that portable electrical equipment is connected.

    (a) True
    (b) False

46. In the, the term mobile home includes manufactured homes, unless otherwise indicated.

    (a) True
    (b) False

47. The ampacity of the supply conductors to an individual electric arc welder shall not be less than the effective current value on the rating plate.

    (a) True
    (b) False.

48. In dwelling units, a 125V receptacle shall be installed a minimum of 6 ft and a maximum of 20 ft from the inside wall of a pool.

    (a) True
    (b) False

49. A structure is that which is built or constructed.

    (a) True
    (b) False

50. The minimum headroom for working spaces about service equipment, switchboards, panelboards, or motor control centers shall be 6½ ft, except for service equipment or panelboards in existing dwelling units that do not exceed 200A.

    (a) True
    (b) False

UNIT 9 — Practice Questions in Straight Order—Article 680.23 through 760.1

Use the 2008 *NEC* to answer the following questions.

1. Branch circuit supplying luminaires operating at more than _____ volts shall be GFCI-protected.

   (a) 6V
   (b) 12V
   (c) 15V
   (d) 18V

2. Wet-niche luminaires shall be installed with the top of the luminaire lens not less than _____ below the normal water level of the pool.

   (a) 6 in.
   (b) 12 in.
   (c) 18 in.
   (d) 24 in.

3. When PVC conduit extends from the pool light forming shell to a pool junction box, an 8 AWG _____ conductor shall be installed in the raceway.

   (a) solid bare
   (b) solid insulated
   (c) stranded insulated
   (d) b or c

4. Branch-circuit wiring for underwater luminaires shall be installed in _____.

   (a) RMC or IMC
   (b) LFNC and PVC conduit
   (c) any wiring method
   (d) a or b

5. EMT where installed on buildings, is permitted to contain branch-circuit wiring for underwater pool luminaires.

   (a) True
   (b) False

6. Wet-niche luminaires shall be connected to an equipment grounding conductor not smaller than _____ AWG.

   (a) 12
   (b) 10
   (c) 8
   (d) 6

7. A pool light junction box connected to a conduit that extends directly to a forming shell shall be _____ for this use.

   (a) listed
   (b) identified
   (c) marked
   (d) a and b

8. The junction box connected to a conduit that extends to the forming shell of the luminaire that operates at over 15V shall be located not less than _____ above the ground level or pool deck.

   (a) 4 in.
   (b) 6 in.
   (c) 8 in.
   (d) 12 in.

9. A pool light junction box that has a raceway that extends directly to an underwater pool light forming shell shall be located not less than _____ from the outdoor pool or spa.

   (a) 2 ft
   (b) 3 ft
   (c) 4 ft
   (d) 6 ft

Mike Holt's Illustrated Guide to NEC Exam Preparation

10. Junction boxes for pool lighting shall be located not less than _____ from the inside wall of a pool unless separated by a fence or wall.

    (a) 3 ft
    (b) 4 ft
    (c) 6 ft
    (d) 8 ft

11. The enclosure for a transformer or ground-fault circuit interrupter connected to a conduit that extends directly to a pool light forming shell shall be _____ for this purpose.

    (a) labeled
    (b) listed
    (c) approved
    (d) a and b

12. Junction boxes connected to conduits that extend directly to forming shells of swimming pool no-niche luminaires shall be provided with a number of grounding terminals that is at least _____ the number of conduit entries.

    (a) one more than
    (b) two more than
    (c) the same as
    (d) none of these

13. The _____ conductor terminals of a junction box (pool deck box), transformer enclosure, or other enclosure in the supply circuit to a wet-niche luminaire shall be connected to the equipment grounding terminal of the panelboard.

    (a) equipment grounding
    (b) grounded
    (c) grounding electrode
    (d) ungrounded

14. The feeder to a swimming pool panelboard at a separate building or structure can be supplied with any Chapter 3 wiring method provided the feeder has a separate insulated copper equipment grounding conductor.

    (a) True
    (b) False

15. The pool structure, including the reinforcing metal of the pool shell and deck, shall be bonded together.

    (a) True
    (b) False

16. Which of the following shall be bonded?

    (a) Metal parts of electrical equipment associated with the pool water circulating system.
    (b) Pool structural metal.
    (c) Metal fittings within or attached to the pool.
    (d) all of these

17. Metal conduit and metal piping within _____ horizontally of the inside walls of the pool shall be bonded.

    (a) 4 ft
    (b) 5 ft
    (c) 8 ft
    (d) 10 ft

18. The electric motors, controllers, and wiring for an electrically operated pool cover shall be _____.

    (a) located at least 5 ft from the inside wall of the pool
    (b) separated from the pool by a permanent barrier
    (c) a and b
    (d) a or b

19. The electric motor and controller for an electrically operated pool cover shall be _____.

    (a) GFCI-protected
    (b) AFCI-protected
    (c) a and b
    (d) a or b

20. All 15A and 20A, 125V receptacles located within _____ of the inside walls of a storable pool shall be GFCI-protected.

    (a) 8 ft
    (b) 10 ft
    (c) 15 ft
    (d) 20 ft

21. Electric equipment, including power-supply cords, used with storable pools shall be _____.

    (a) AFCI-protected
    (b) GFCI-protected
    (c) a or b
    (d) none of these

22. Receptacles shall be located not less than _____ from the inside walls of a storable pool.

    (a) 5 ft
    (b) 6 ft
    (c) 15 ft
    (d) 20 ft

Unit 9 — Practice Questions in Straight Order—Article 680.23 through 760.1

23. In spas or hot tubs, a clearly labeled emergency shutoff or control switch shall be _____, not less than 5 ft away, and within sight of the spa or hot tub.

    (a) accessible
    (b) readily accessible
    (c) available
    (d) none of these

24. Listed packaged spa or hot tub equipment assemblies, or self-contained spas or hot tubs installed outdoors, are permitted to have flexible connections using _____.

    (a) LFMC or LFNC in lengths of not more than 6 ft
    (b) cords not longer than 15 ft, where GFCI-protected
    (c) a or b
    (d) none of these

25. Listed spa and hot tub packaged units installed indoors, rated 20A or less, can be cord-and-plug-connected.

    (a) True
    (b) False

26. At least one 15A or 20A, 125V receptacle on a general-purpose branch circuit shall be located a minimum of _____ and a maximum of 10 ft from the inside wall of a spa or hot tub installed indoors.

    (a) 18 in.
    (b) 2 ft
    (c) 6 ft
    (d) no minimum

27. Receptacles rated 30A or less, 125V within 10 ft of the inside walls of an indoor spa or hot tub, shall be _____.

    (a) GFCI-protected
    (b) AFCI-protected
    (c) a or b
    (d) none of these

28. Luminaires and ceiling fans located over or within 5 ft, measured horizontally, from the inside walls of an indoor spa or hot tub shall have a mounting height of not less than _____ above the maximum water level when not GFCI-protected.

    (a) 4.7 ft
    (b) 5 ft
    (c) 7 ft 6 in.
    (d) 12 ft

29. Luminaires and ceiling fans located over or within 5 ft, measured horizontally, from the inside walls of an indoor spa or hot tub shall have a mounting height of not less than _____ above the maximum water level when GFCI-protected.

    (a) 4.7 ft
    (b) 5 ft
    (c) 7 ft 6 in.
    (d) 12 ft

30. Metal parts of electric equipment associated with an indoor spa or hot tub water circulating system shall be bonded.

    (a) True
    (b) False

31. Metal raceways and metal piping within _____ of the inside walls of an indoor spa or hot tub, and not separated from the indoor spa or hot tub by a permanent barrier, shall be bonded.

    (a) 4 ft
    (b) 5 ft
    (c) 7 ft
    (d) 12 ft

32. Metal parts associated with an indoor spa or hot tub shall be bonded by _____.

    (a) the interconnection of threaded metal piping and fittings
    (b) metal-to-metal mounting on a common frame or base
    (c) a solid copper bonding jumper not smaller than 8 AWG solid
    (d) any of these

33. Receptacle outlet(s) for a _____ shall be GFCI-protected.

    (a) self-contained spa or hot tub
    (b) packaged spa or hot tub equipment assembly
    (c) field-assembled spa or hot tub with a heater load of 50A or less
    (d) all of these

34. The branch circuit supplying submersible fountain equipment shall be _____, unless the equipment is listed for operation at not more than 15V.

    (a) 240V
    (b) GFCI-protected
    (c) a and b
    (d) none of these

35. Luminaires installed in fountains shall be _____.

    (a) installed with the top of the luminaire lens below the normal water level
    (b) listed for above-water use
    (c) have the lens guarded or be listed for use without a guard
    (d) any of these

36. Luminaires installed in a fountain shall _____.

    (a) be capable of being removed from the water for relamping or normal maintenance
    (b) not be permanently embedded into the fountain structure
    (c) a and b
    (d) a or b

37. Metal piping systems associated with a fountain shall be bonded to the equipment grounding conductor of the _____.

    (a) branch circuit supplying the fountain
    (b) bonding grid
    (c) equipotential plane
    (d) grounding electrode system

38. Fountain equipment supplied by a flexible cord shall have all exposed noncurrent-carrying metal parts grounded by an insulated copper equipment grounding conductor that is an integral part of the cord.

    (a) True
    (b) False

39. Cord-and-plug-connected equipment in fountains _____ shall have GFCI protection.

    (a) except pumps
    (b) including power-supply cords
    (c) less than 6 ft high
    (d) except power-supply cords

40. Flexible cords immersed in or exposed to water in a fountain shall be _____.

    (a) extra-hard usage type
    (b) listed with a "W" suffix
    (c) encased in not less than 2 in. of concrete
    (d) a and b

41. Each circuit supplying a sign within or adjacent to a fountain shall _____.

    (a) have GFCI protection
    (b) be capable of being locked in the open position
    (c) operate at less than 15V
    (d) be an intrinsically safe circuit

42. A portable electric sign shall not be placed in or within _____ from the inside walls of a fountain.

    (a) 5 ft
    (b) 10 ft
    (c) 15 ft
    (d) 20 ft

43. All 15A and 20A, single-phase, 125V through 250V receptacles located within _____ of a fountain edge shall have GFCI protection.

    (a) 8 ft
    (b) 10 ft
    (c) 15 ft
    (d) 20 ft

44. Metal piping systems and all grounded metal parts in contact with the circulating water of a hydromassage bathtub shall be bonded together using a(n) _____ solid copper bonding jumper not smaller than 8 AWG.

    (a) insulated
    (b) covered
    (c) bare
    (d) any of these

45. The 8 AWG solid bonding jumper required for hydromassage bathtubs shall not be required to be extended to any _____.

    (a) remote panelboard
    (b) service equipment
    (c) electrode
    (d) all of these

Unit 9 — Practice Questions in Straight Order—Article 680.23 through 760.1

46. Conductors from a generator to supply a fire pump shall terminate to a generator disconnecting means dedicated for serving the fire pump located in _____ enclosure from any other generator disconnecting means.

    (a) the same
    (b) a dedicated
    (c) a separate
    (d) an explosionproof

47. The _____ shall be selected to carry indefinitely the sum of the locked-rotor current of the fire pump motor(s), pressure maintenance pump motor(s), and full-load current of associated fire pump accessory equipment.

    (a) overcurrent protective device(s)
    (b) pump motor conductors
    (c) a and b
    (d) none of these

48. Transformers that supply a fire pump motor shall be sized no less than _____ percent of the sum of the fire pump motor(s) and pressure maintenance pump motors, and 100 percent of associated fire pump accessory equipment.

    (a) 100
    (b) 125
    (c) 250
    (d) 300

49. Primary overcurrent protective device for a transformer supplying a fire pump shall carry the sum of the locked-rotor current of the fire pump motor(s), pressure maintenance pump motor(s), and the full-load current of associated fire pump accessory equipment _____.

    (a) for 15 minutes
    (b) for 45 minutes
    (c) for 3 hours
    (d) indefinitely

50. Fire pump supply conductors on the load side of the final disconnecting means and overcurrent device(s) can be routed through a building(s) using _____.

    (a) 2 in. of concrete encasement
    (b) a listed 2-hour fire-rated assembly dedicated to the fire pump circuit
    (c) a listed electrical circuit protective system with a minimum 2-hour fire rating
    (d) any of these

51. Feeder conductors supplying fire pump motors and accessory equipment shall be sized no less than _____ percent of the sum of the motor full-load currents, plus 100 percent of the ampere rating of the fire pump accessory equipment.

    (a) 100
    (b) 125
    (c) 250
    (d) 600

52. Ground-fault protection of equipment _____ for fire pumps.

    (a) shall not be permitted
    (b) shall be provided
    (c) shall be permitted
    (d) shall be listed

53. The voltage at the line terminals of a fire pump motor controller shall not drop more than _____ percent below the controller's normal rated voltage under motor-starting conditions.

    (a) 5
    (b) 10
    (c) 15
    (d) any of these

54. When a fire pump motor operates at 115 percent of its full-load current rating, the supply voltage at the motor terminals shall not drop more than _____ percent below the voltage rating of the motor.

    (a) 5
    (b) 10
    (c) 15
    (d) any of these

55. Emergency systems are generally installed where artificial illumination is required for safe exiting and for panic control in buildings occupied by large numbers of persons, such as _____ and similar institutions.

    (a) hotels
    (b) theaters and sports arenas
    (c) health care facilities
    (d) all of these

Practice Questions in Straight Order—Article 680.23 through 760.1 — Unit 9

56. Emergency systems may provide power for ventilation where essential to maintain life, fire detection and alarm systems, elevators, fire pumps, public safety communications systems, industrial processes where current interruption would produce serious _____.

    (a) production slowdowns
    (b) life safety or health hazards
    (c) a and b
    (d) a or b

57. Except as modified by Article 700, all requirements contained in *NEC* Chapters 1 through 4 apply to installations of emergency systems.

    (a) True
    (b) False

58. A written record shall be kept of required tests and maintenance on emergency systems.

    (a) True
    (b) False

59. An emergency system shall have adequate capacity and rating for _____ to be operated simultaneously.

    (a) 80% of the loads
    (b) 100% of the load
    (c) 125% of the load
    (d) none of these

60. A portable or temporary alternate source _____ whenever the emergency generator is out of service for major maintenance or repair.

    (a) shall not be required
    (b) is recommended
    (c) shall be available
    (d) shall be avoided

61. Emergency transfer equipment, including transfer switches, shall be _____.

    (a) automatic
    (b) identified for emergency use
    (c) approved by the authority having jurisdiction
    (d) all of these

62. Automatic transfer switches shall be _____.

    (a) electrically operated
    (b) mechanically held
    (c) listed for emergency system use
    (d) all of these

63. An emergency transfer switch shall supply _____.

    (a) emergency loads
    (b) computer equipment
    (c) UPS equipment
    (d) all of these

64. When practicable, audible and visual signal devices for an emergency system shall be provided to indicate _____.

    (a) that the battery is carrying load
    (b) derangement of the emergency source
    (c) that the battery charger is not functioning
    (d) all of these

65. Wiring from an emergency source or emergency source distribution overcurrent protection to emergency loads shall be kept independent of all other wiring and equipment except in _____.

    (a) transfer equipment enclosures
    (b) exit or emergency luminaires supplied from two sources
    (c) a common junction box attached to exit or emergency luminaires supplied from two sources
    (d) all of these

66. Wiring from an emergency source can supply any combination of emergency, legally required, or optional loads, provided the conductors _____.

    (a) terminate in separate vertical sections of a switchboard
    (b) terminate in the same vertical section of a switchboard
    (c) terminate in junction box identified for emergency use
    (d) are identified as emergency conductors

67. Emergency circuit wiring shall be designed and located to minimize the hazards that might cause failure because of _____.

    (a) flooding
    (b) fire
    (c) icing
    (d) all of these

Unit 9 — Practice Questions in Straight Order—Article 680.23 through 760.1

68. Emergency lighting and/or emergency power in a building or group of buildings shall be available within _____ seconds.

    (a) 5
    (b) 10
    (c) 30
    (d) 60

69. A storage battery supplying emergency lighting and power shall maintain not less than 87½ percent of normal voltage at total load for a period of at least _____ hour(s).

    (a) 1
    (b) 1½
    (c) 2
    (d) 2½

70. Where an internal combustion engine is used as the prime mover for an emergency system, an on-site fuel supply shall be provided for not less than _____ hours of full-demand operation of the system.

    (a) 2
    (b) 3
    (c) 4
    (d) 5

71. Branch-circuit overcurrent devices in emergency circuits shall be accessible to _____ only.

    (a) the authority having jurisdiction
    (b) authorized persons
    (c) the general public
    (d) qualified persons

72. Overcurrent devices for emergency power systems _____ all supply-side overcurrent protective devices.

    (a) shall be selectively coordinated with
    (b) can be selectively coordinated with
    (c) shall be the same amperage as
    (d) shall be a higher amperage than

73. A legally required standby system is intended to automatically supply power to _____ in the event of failure of the normal source.

    (a) those systems classed as emergency systems
    (b) selected loads
    (c) a and b
    (d) none of these

74. The _____ shall conduct or witness a test of the complete legally required standby system upon installation.

    (a) electrical engineer
    (b) authority having jurisdiction
    (c) qualified person
    (d) manufacturer's representative

75. Where batteries are used for _____ of prime movers of legally required standby systems, the authority having jurisdiction shall require periodic maintenance.

    (a) control
    (b) starting or ignition
    (c) a and b
    (d) none of these

76. A legally required standby system shall have adequate capacity and rating for _____ that are expected to operate simultaneously on the standby system.

    (a) all of the loads
    (b) 80% of the loads
    (c) 125% of the loads
    (d) none of these

77. Legally required standby system equipment shall be suitable for _____ at its terminals.

    (a) the maximum available fault current
    (b) the maximum overload current only
    (c) the minimum fault current
    (d) a one-hour rating

78. Transfer equipment for legally required systems, including automatic transfer switches, shall be _____.

    (a) automatic
    (b) identified for standby use
    (c) approved by the authority having jurisdiction
    (d) all of these

79. Automatic transfer switches on legally required standby systems shall be electrically operated and _____ held.

    (a) electrically
    (b) mechanically
    (c) gravity
    (d) any of these

80. Where acceptable to the authority having jurisdiction, connections ahead of and not within the same cabinet, enclosure, or vertical switchboard section as the service disconnecting means shall be permitted for _____ standby service.

    (a) emergency
    (b) legally required
    (c) optional
    (d) all of these

81. The branch-circuit overcurrent devices for legally required standby systems shall be accessible only to _____.

    (a) the authority having jurisdiction
    (b) authorized persons
    (c) the general public
    (d) qualified persons

82. Overcurrent devices for legally required power systems shall be _____ with all supply-side overcurrent protective devices.

    (a) series rated
    (b) selectively coordinated
    (c) installed in parallel
    (d) any of these

83. Article 702 applies to _____ generators used for backup power to public or private facilities or property where life safety does not depend on the performance of the system.

    (a) permanently installed
    (b) portable
    (c) a and b
    (d) none of these

84. Optional standby systems are typically installed to provide an alternate source of power for _____.

    (a) data processing and communication systems
    (b) emergency systems for health care facilities
    (c) emergency systems for hospitals
    (d) none of these

85. Optional standby systems utilizing automatic transfer equipment shall have adequate capacity and rating for the supply of _____.

    (a) all emergency lighting and power loads
    (b) the load, as calculated in Article 220
    (c) 100 percent of the appliance loads and 50 percent of the lighting loads
    (d) 100 percent of the lighting loads and 75 percent of the appliance loads

86. For optional standby systems, the temporary connection of a portable generator without transfer equipment shall be permitted where conditions of maintenance and supervision ensure that only qualified persons will service the installation, and where the normal supply is physically isolated by _____.

    (a) a lockable disconnecting means
    (b) the disconnection of the normal supply conductors
    (c) an extended power outage
    (d) a or b

87. Since Class 3 control circuits permit higher allowable levels of voltage and current than Class 2 control circuits, additional _____ are specified to provide protection against the electric shock hazard.

    (a) circuits
    (b) safeguards
    (c) conditions
    (d) requirements

88. Class 2 and Class 3 circuits installed _____ on the surface of ceilings and walls shall be supported by the building structure in such a manner that the cable will not be damaged by normal building use.

    (a) exposed
    (b) concealed
    (c) hidden
    (d) a and b

89. Class 2 cables identified for future use shall be marked with a tag of sufficient durability to withstand _____.

    (a) moisture
    (b) humidity
    (c) the environment involved
    (d) none of these

90. Remote-control circuits to safety-control equipment shall be classified as _____ if the failure of the equipment to operate introduces a direct fire or life hazard.

    (a) Class 1
    (b) Class 2
    (c) Class 3
    (d) Class I, Division 1

91. The power source for a Class 2 circuit shall be _____.

    (a) a listed Class 2 transformer
    (b) a listed Class 2 power supply
    (c) other listed equipment marked to identify the Class 2 power source
    (d) any of these

92. Equipment supplying Class 2 or Class 3 circuits shall be durably marked where plainly visible to indicate _____.

    (a) each circuit that is a Class 2 or Class 3 circuit
    (b) the circuit VA rating
    (c) the size of conductors serving each circuit
    (d) all of these

93. Cables and conductors of Class 2 and Class 3 circuits _____ be placed with conductors of electric light, power, Class 1, nonpower-limited fire alarm circuits, and medium power network-powered broadband communications circuits.

    (a) shall be permitted to
    (b) shall not
    (c) shall
    (d) none of these

94. Conductors of Class 2 and Class 3 circuits shall not be placed in any enclosure, raceway, cable, or similar fittings with conductors of Class 1 or electric light or power conductors, unless _____.

    (a) insulated for the maximum voltage present
    (b) totally comprised of aluminum conductors
    (c) separated by a barrier
    (d) all of these

95. Audio system circuits using Class 2 or Class 3 wiring methods, are not permitted in the same cable or raceway with _____.

    (a) other audio system circuits
    (b) Class 2 conductors or cables
    (c) Class 3 conductors or cables
    (d) b or c

96. Class 2 or Class 3 cables, installed in vertical runs penetrating more than one floor or installed in a shaft, shall be Type _____.

    (a) CL2R
    (b) CL3R
    (c) CL2P
    (d) any of these

97. Class 2 and Class 3 plenum cables listed as suitable for use in other spaces used for environmental air shall be Type _____.

    (a) CL2P and CL3P
    (b) CL2R and CL3R
    (c) CL2 and CL3
    (d) PLTC

98. Class 2 and Class 3 riser cables listed as suitable for use in a vertical run in a shaft, or from floor to floor shall be Type _____.

    (a) CL2P and CL3P
    (b) CL2R and CL3R
    (c) CL2 and CL3
    (d) PLTC

99. Article 760 covers the requirements for the installation of wiring and equipment of _____.

    (a) communications systems
    (b) antennas
    (c) fire alarm systems
    (d) fiber optics

100. Fire alarm systems include _____.

    (a) fire detection and alarm notification
    (b) guard's tour
    (c) sprinkler water flow
    (d) all of these

UNIT 9 — Practice Questions in Random Order—Article 90.3 through 760.2

Use the 2008 *NEC* to answer the following questions.

1. Emergency lighting systems shall be designed and installed so that the failure of any individual lighting element, such as the burning out of a lamp, will not leave in total darkness any space that requires emergency illumination.

   (a) True
   (b) False

2. Exposed Class 2 and Class 3 cables shall be supported by straps, staples, hangers, or similar fittings designed and installed so as not to damage the cable.

   (a) True
   (b) False

3. Means for testing all emergency lighting and power systems during maximum anticipated load conditions shall be provided.

   (a) True
   (b) False

4. Accessible portions of abandoned Class 2 and Class 3 cables shall be removed.

   (a) True
   (b) False

5. Fire alarm cables that are not terminated at equipment and not identified for future use with a tag are considered abandoned.

   (a) True
   (b) False

6. The alternate source for emergency systems shall have ground-fault protection of equipment.

   (a) True
   (b) False

7. The alternate source for legally required standby systems shall not be required to have ground-fault protection of equipment.

   (a) True
   (b) False

8. Legally required standby systems that are tested upon installation and found to be acceptable to the authority having jurisdiction shall not be required to undergo any future tests unless the equipment is modified.

   (a) True
   (b) False

9. A transfer switch shall be required for all fixed or portable optional standby power systems for buildings or structures at which an electric utility is either the normal or standby source, unless permitted by an exception.

   (a) True
   (b) False

10. A sign shall be placed at the service-entrance equipment indicating the type and location of on-site emergency power sources.

    (a) True
    (b) False

11. A written record shall be kept of required tests and maintenance on legally required standby systems.

    (a) True
    (b) False

12. Testing legally required standby system lighting and power systems during maximum anticipated load conditions shall be avoided so as not to tax the standby system unnecessarily.

    (a) True
    (b) False

UNIT 9 — Practice Questions in Random Order—Article 90.3 through 760.2

13. Optional standby system wiring can occupy the same raceways, cables, boxes, and cabinets with other general wiring.

    (a) True
    (b) False

14. Raceways shall not be used as a means of support for Class 2 or Class 3 cables.

    (a) True
    (b) False

15. Class 2 and Class 3 cable not terminated at equipment and not identified for future use with a tag is considered abandoned.

    (a) True
    (b) False

16. Ground-fault circuit-interrupter protection shall be provided for outlets not exceeding 240V that supply boat hoists installed in dwelling unit locations.

    (a) True
    (b) False

17. A meter disconnect switch located ahead of service equipment must have a short-circuit current rating equal to or greater than the available short-circuit current and be capable of interrupting the load served.

    (a) True
    (b) False

18. Grounding electrode conductor taps from a separately derived system to a common grounding electrode conductor are permitted when a building or structure has multiple separately derived systems, provided that the taps terminate at the same point as the system bonding jumper.

    (a) True
    (b) False

19. Exothermic or irreversible compression connections, together with the mechanical means used to attach to fire-proofed structural metal, shall not be required to be accessible.

    (a) True
    (b) False

20. When determining the number of current-carrying conductors, a grounding or bonding conductor shall not be counted when applying the provisions of 310.15(B)(2)(a) _____.

    (a) True
    (b) False

21. PVC conduit shall be permitted for exposed work where subject to physical damage if identified for such use.

    (a) True
    (b) False

22. Surface metal raceway is a metallic raceway that is intended to be mounted to the surface of a structure, with associated couplings, connectors, boxes, and stings for the installation of electrical conductors.

    (a) True
    (b) False

23. Fixture wires shall not be used for branch-circuit wiring, except as permitted in other articles of the *Code*.

    (a) True
    (b) False

24. The *NEC* requires a lighting outlet on the wall in clothes closets.

    (a) True
    (b) False

25. Motor controllers and terminals of control circuit devices shall be connected with copper conductors unless identified for use with a different conductor.

    (a) True
    (b) False

26. Portable or transportable equipment with a self-contained power supply, such as battery-operated equipment, could potentially become an ignition source in hazardous (classified) locations.

    (a) True
    (b) False

27. In Class I, Division 1 locations, a multiwire branch circuit can be protected using single-pole breakers.

    (a) True
    (b) False

28. Due to its power limitations, a Class 2 circuit is considered safe from a fire initiation standpoint and provides acceptable protection from electric shock.

    (a) True
    (b) False

29. NFPA 70E-2004, *Standard for Electrical Safety in the Workplace*, provides information to help determine the electrical safety training requirements expected of a "qualified person."

    (a) True
    (b) False

30. Accepted industry workmanship practices are described in ANSI/NEC A 1-2006, *Standard Practices for Good Workmanship in Electrical Contracting*, and other ANSI approved installation standards.

    (a) True
    (b) False

31. The *NEC* requires tested series-rated installations of circuit breakers or fuses to be field-marked to indicate the current rating for which the system has been installed.

    (a) True
    (b) False

32. Heating, cooling, or ventilating equipment shall not be installed in the dedicated space above a panelboard or switchboard.

    (a) True
    (b) False

33. A laundry receptacle outlet shall not be required in each dwelling unit of a multifamily building, even if laundry facilities are provided on the premises for all building occupants.

    (a) True
    (b) False

34. Single-pole breakers with identified handle ties can be used to protect each ungrounded conductor for line-to-line connected loads.

    (a) True
    (b) False

35. For grounded systems, the earth is considered an effective ground-fault current path.

    (a) True
    (b) False

36. The auxiliary electrode shall not be used as an effective ground-fault current path.

    (a) True
    (b) False

37. When an underground metal water piping system is used as a grounding electrode, bonding shall be provided around insulated joints and around any equipment that is likely to be disconnected for repairs or replacement.

    (a) True
    (b) False

38. Article 285 covers surge protective devices rated over 1 kV.

    (a) True
    (b) False

39. Electrical wiring within the cavity of a fire-rated floor-ceiling or roof-ceiling assembly shall not be supported by the ceiling assembly or ceiling support wires.

    (a) True
    (b) False

40. Where conductors in parallel are run in separate raceways, the raceways shall have the same electrical characteristics.

    (a) True
    (b) False

41. Where one or more internal cable clamps are present in the box, a single volume allowance in accordance with Table 314.16(b) shall be made, based on the largest conductor present in the box.

    (a) True
    (b) False

42. Horizontal runs of IMC supported by openings through framing members at intervals not exceeding 10 ft and securely fastened within 3 ft of terminations shall be permitted.

    (a) True
    (b) False

43. PVC conduit can support nonmetallic conduit bodies not larger than the largest raceway, but the conduit bodies shall not contain devices, luminaires, or other equipment.

    (a) True
    (b) False

Unit 9 — Practice Questions in Random Order—Article 90.3 through 760.2

44. Luminaires that require adjustment or aiming after installation can be cord-connected without an attachment plug, provided the exposed cord is of the hard-usage type and is not longer than that required for maximum adjustment.

    (a) True
    (b) False

45. Motor control circuits shall be arranged so they will be disconnected from all sources of supply when the disconnecting means is in the open position.

    (a) True
    (b) False

46. Transformers and capacitors installed in Class I, Division 1 locations containing a liquid that will burn shall be installed in vaults.

    (a) True
    (b) False

47. Flexible cords are not permitted in Class I locations.

    (a) True
    (b) False

48. Intrinsically safe apparatus, metal enclosures, and metal raceways shall be connected to the equipment grounding conductor.

    (a) True
    (b) False

49. Branch circuits serving patient care areas shall be installed in a metal raceway or cable listed in 250.118 as an effective ground-fault current path.

    (a) True
    (b) False

50. Communications wiring such as telephone, antenna, and CATV wiring within a building shall not be required to comply with the installation requirements of Chapters 1 through 7, except where it is specifically referenced in Chapter 8.

    (a) True
    (b) False

Unit 10: Practice Questions in Straight Order—Article 760.24 through Chapter 9

Use the 2008 *NEC* to answer the following questions.

1. Exposed fire alarm circuit cables shall be supported by the building structure using straps, staples, hangers, cable ties or similar fittings designed and installed so as not to damage the cable.

    (a) True
    (b) False

2. Accessible portions of abandoned fire alarm cable shall be removed.

    (a) True
    (b) False

3. Fire alarm cables identified for future use shall be marked with a tag of sufficient durability to withstand _____.

    (a) moisture
    (b) humidity
    (c) the environment involved
    (d) none of these

4. Fire alarm circuits shall be identified at all terminal and junction locations in a manner that helps prevent unintentional signals on fire alarm system circuits during _____.

    (a) installation
    (b) testing and servicing
    (c) renovations
    (d) all of these

5. The power source for a nonpower-limited fire alarm circuit shall not operate at more than _____.

    (a) 120V
    (b) 300V
    (c) 600V
    (d) 20A

6. The power source for a nonpower-limited fire alarm circuit shall not be supplied through a(n) _____.

    (a) ground-fault circuit interrupter
    (b) arc-fault circuit interrupter
    (c) inverse time circuit breaker
    (d) a or b

7. For nonpower-limited fire alarm circuits, an 18 AWG conductor shall be considered protected if the overcurrent device protecting the circuit is not over _____.

    (a) 7A
    (b) 10A
    (c) 15A
    (d) 20A

8. Nonpower-limited fire alarm circuit conductors are permitted to be in the same cable, enclosure, or raceway with power-supply circuits where connected to the same equipment.

    (a) True
    (b) False

9. The number of nonpower-limited fire alarm conductors in a raceway shall not be required to comply with the fill requirements contained in 300.17.

    (a) True
    (b) False

10. The power source for a power-limited fire alarm circuit can be supplied through a ground-fault circuit interrupter or an arc-fault circuit interrupter.

    (a) True
    (b) False

11. Fire alarm equipment supplying power-limited fire alarm circuits shall be durably marked where plainly visible to indicate each circuit that is _____.

    (a) supplied by a nonpower-limited fire alarm circuit
    (b) a power-limited fire alarm circuit
    (c) a fire alarm circuit
    (d) none of these

12. Cable splices or terminations in power-limited fire alarm systems shall be made in listed _____ or utilization equipment.

    (a) fittings
    (b) boxes or enclosures
    (c) fire alarm devices
    (d) all of these

13. Generally speaking, conductors for lighting or power may occupy the same enclosure or raceway with conductors of power-limited fire alarm circuits.

    (a) True
    (b) False

14. Audio system circuits using Class 2 or Class 3 wiring methods shall not be installed in the same cable or raceway with _____.

    (a) other audio system circuits
    (b) power-limited fire alarm conductors or cables
    (c) a or b
    (d) none of these

15. Power-limited fire alarm cables can be supported by strapping, taping, or attaching to the exterior of a conduit or raceway.

    (a) True
    (b) False

16. Power-limited fire alarm cable used in a _____ location shall be listed for use in _____ locations or have a moisture-impervious metal sheath.

    (a) dry
    (b) damp
    (c) wet
    (d) hazardous

17. Power-limited fire alarm cables installed within buildings shall be _____ as being resistant to the spread of fire.

    (a) marked FR
    (b) listed
    (c) identified
    (d) color-coded

18. Coaxial cables used in power-limited fire alarm systems shall have an insulation rating of not less than _____.

    (a) 100V
    (b) 300V
    (c) 600V
    (d) 1,000V

19. An optical fiber cable is a factory assembly of one or more optical fibers having a(n) _____ covering.

    (a) conductive
    (b) nonconductive
    (c) overall
    (d) metallic

20. Composite optical fiber cables contain optical fibers and _____.

    (a) strength members
    (b) vapor barriers
    (c) current-carrying electrical conductors
    (d) none of these

21. Conductive optical fiber cables contain noncurrent-carrying conductive members such as metallic _____.

    (a) strength members
    (b) vapor barriers
    (c) armor or sheath
    (d) any of these

22. Nonconductive optical fiber cable contains no metallic members and no other _____ materials.

    (a) electrically conductive
    (b) inductive
    (c) synthetic
    (d) insulating

23. Optical fiber cables not terminated at equipment, and not identified for future use with a tag are considered abandoned.

    (a) True
    (b) False

24. Access to electrical equipment shall not be denied by an accumulation of optical fiber cables that _____ removal of panels, including suspended-ceiling panels.

    (a) prevents
    (b) hinders
    (c) blocks
    (d) require

25. Accepted industry practices for optical fiber installations are described in _____.

    (a) ANSI/NEC A/BICSI 568, *Standard for Installing Commercial Building Telecommunications Cabling*
    (b) ANSI/NEC A/FOA 301, *Standard for Installing and Testing Fiber Optic Cables*
    (c) other ANSI-approved installation standards
    (d) any of these

26. Exposed optical fiber cables shall be supported by the building structure using hardware including straps, staples, cable ties, hangers, or similar fittings designed and installed so as not to damage the cable.

    (a) True
    (b) False

27. Optical fiber cables installed _____ on the surface of ceilings and walls shall be supported by the building structure in such a manner that the cable will not be damaged by normal building use.

    (a) exposed
    (b) concealed
    (c) hidden
    (d) a and b

28. Accessible portions of abandoned optical fiber cable shall be removed.

    (a) True
    (b) False

29. Openings around penetrations of optical fiber cables and raceways through fire-resistant–rated walls, partitions, floors, or ceilings shall be _____ using approved methods to maintain the fire-resistance rating.

    (a) closed
    (b) opened
    (c) draft stopped
    (d) firestopped

30. Where exposed to contact with electric light or power conductors, the non–current-carrying metallic members of optical fiber cables entering buildings shall be _____.

    (a) grounded as specified in 770.100 as close to the point of entrance as practicable
    (b) interrupted as close to the point of entrance as practicable by an insulating joint or equivalent device
    (c) a or b
    (d) a and b

31. When optical fiber cable is installed in a Chapter 3 raceway, the raceway shall be installed in accordance with Chapter 3 requirements.

    (a) True
    (b) False

32. Conductive optical fiber cables can occupy the same cable tray or raceway with conductors for electric light, power, and Class 1 circuits.

    (a) True
    (b) False

33. Optical fibers shall be permitted within the same composite cable as electric light, power, and Class 1 circuits operating at 600V, or less where the functions of the optical fibers and the electrical conductors are associated.

    (a) True
    (b) False

34. Optical fiber cables shall not be _____ to the exterior of any raceway as a means of support.

    (a) strapped
    (b) taped
    (c) attached
    (d) all of these

35. Type OFNG, OFN, OFCG, and OFC optical fiber cables can be installed in vertical riser runs when _____.

    (a) encased in a metal raceway
    (b) located in a fireproof shaft having firestops at each floor
    (c) a or b
    (d) none of these

36. Optical fiber plenum cables listed as suitable for use in other space used for environmental air includes Types _____.

    (a) OFNP and OFCP
    (b) OFNR and OFCR
    (c) OFNG and OFCG
    (d) OFN and OFC

37. Optical fiber riser cables listed as suitable for use in a vertical run in a shaft or from floor to floor includes Types _____.

    (a) OFNP and OFCP
    (b) OFNR and OFCR
    (c) OFNG and OFCG
    (d) OFN and OFC

UNIT 10 — Practice Questions in Straight Order—Article 760.24 through Chapter 9

38. General-purpose optical fiber cables listed as suitable for general-purpose use, with the exception of risers and plenums, includes Types _____.

    (a) OFNP and OFCP
    (b) OFNR and OFCR
    (c) OFNG and OFCG
    (d) OFN and OFC

39. Optical fiber cables listed as suitable for general-purpose use, with the exception of risers, plenums, and other spaces used for environmental air are Types _____.

    (a) OFNP and OFCP
    (b) OFNR and OFCR
    (c) OFNG and OFCG
    (d) OFN and OFC

40. Communications cables not terminated at both ends with a connector or other equipment and not identified for future use with a tag are considered abandoned.

    (a) True
    (b) False

41. A communications circuit that is in such a position that, in case of failure of supports or _____, contact with another circuit may result, and is considered to be exposed to accidental contact.

    (a) insulation
    (b) shield
    (c) fittings
    (d) grounding conductor

42. The point of entrance of a communications circuit is the point _____ at which the communications cable emerges from an external wall, from a concrete floor slab, or from a RMC or an IMC connected by a grounding conductor to an electrode.

    (a) outside a building
    (b) within a building
    (c) on the building
    (d) none of these

43. Communications circuits and equipment installed in a location that is _____ in accordance with 500.5 shall comply with the applicable requirements of Chapter 5.

    (a) designed
    (b) classified
    (c) located
    (d) approved

44. Equipment intended to be permanently electrically connected to a telecommunications network shall be listed.

    (a) True
    (b) False

45. Exposed communications cables shall be secured by hardware including straps, staples, cable ties, hangers, or similar fittings designed and installed so as not to damage the cable.

    (a) True
    (b) False

46. Accessible portions of abandoned communications cable shall be removed.

    (a) True
    (b) False

47. Openings around penetrations of communications cables and raceways through fire-resistant–rated walls, partitions, floors, or ceilings shall be _____ using approved methods to maintain the fire-resistance rating.

    (a) closed
    (b) opened
    (c) draft stopped
    (d) firestopped

48. Where overhead communications cables enter buildings, they shall _____.

    (a) be located below the electric light or power conductors, where practicable
    (b) not be attached to a cross-arm that carries electric light or power conductors
    (c) have a vertical clearance of not less than 8 ft from all points of roofs above which they pass
    (d) all of these

49. Communications cables shall not be required to be listed where the length of the cable within the building, measured from its point of entrance, does not exceed _____ and the cable enters the building from the outside and is terminated in an enclosure or on a listed primary protector.

    (a) 25 ft
    (b) 30 ft
    (c) 50 ft
    (d) 100 ft

Practice Questions in Straight Order—Article 760.24 through Chapter 9 — Unit 10

50. When practicable, a separation of at least _____ shall be maintained between communications cables on buildings and lightning conductors.

    (a) 6 ft
    (b) 8 ft
    (c) 10 ft
    (d) 12 ft

51. The metallic sheath members of communications cable entering or attached to buildings shall be _____.

    (a) grounded at the point of emergence through an exterior wall
    (b) grounded at the point of emergence through a concrete floor slab
    (c) interrupted as close to the point of entrance as practicable by an insulating joint
    (d) any of these

52. In one- and two-family dwellings, the primary protector grounding conductor for communications systems shall be as short as practicable, not to exceed _____ in length.

    (a) 5 ft
    (b) 8 ft
    (c) 10 ft
    (d) 20 ft

53. In one- and two-family dwellings where it is not practicable to achieve an overall maximum primary protector grounding conductor length of 20 ft, a separate ground rod not less than _____ shall be driven and it shall be connected to the power grounding electrode system with a 6 AWG conductor.

    (a) 5 ft
    (b) 8 ft
    (c) 10 ft
    (d) 20 ft

54. Limiting the length of the primary protector grounding conductors for communications circuits helps to reduce voltage between the building's _____ and communications systems during lightning events.

    (a) power
    (b) fire alarm
    (c) lighting
    (d) lightning protection

55. For buildings with grounding means but without an intersystem bonding termination, the grounding conductor for communications circuits shall terminate to the nearest _____.

    (a) the building or structure grounding electrode system
    (b) interior metal water piping system, within 5 ft from its point of entrance
    (c) the service equipment enclosure
    (d) any of these

56. Communications electrodes must be bonded to the power grounding electrode system using a minimum _____ copper bonding jumper.

    (a) 10 AWG
    (b) 8 AWG
    (c) 6 AWG
    (d) 4 AWG

57. Where communications wires and cables are installed in a Chapter 3 raceway, the raceway shall be installed in accordance with Chapter 3 requirements.

    (a) True
    (b) False

58. Communications cables shall be listed.

    (a) True
    (b) False

59. Communications wires and cables shall be separated by at least 2 in. from conductors of _____ circuits, unless permitted otherwise.

    (a) power
    (b) lighting
    (c) Class 1
    (d) any of these

60. Type CMX communications cables that are less than ¼ in. in diameter can be installed in _____.

    (a) one- or two-family dwellings
    (b) multifamily dwellings in nonconcealed spaces
    (c) a or b
    (d) none of these

61. Communications plenum cable shall be _____ as being suitable for use in other spaces used for environmental air.

    (a) marked
    (b) identified
    (c) approved
    (d) listed

62. Communications _____ cables shall be listed as being suitable for use in a vertical run in a shaft, or from floor to floor, and shall also be listed as having fire-resistant characteristics capable of preventing the carrying of fire from floor to floor.

    (a) plenum
    (b) riser
    (c) general-purpose
    (d) none of these

63. Outdoor antennas and lead-in conductors shall be securely supported and the lead-in conductors shall be securely attached to the antenna, but they shall not be attached to the electric service mast.

    (a) True
    (b) False

64. Underground antenna conductors for radio and television receiving equipment shall be separated at least _____ from any light, power, or Class 1 circuit conductors.

    (a) 12 in.
    (b) 18 in.
    (c) 5 ft
    (d) 6 ft

65. Indoor antenna and lead-in conductors for radio and television receiving equipment shall be separated by at least _____ from conductors of any electric light, power, or Class 1 circuit conductors, unless otherwise permitted.

    (a) 2 in.
    (b) 12 in.
    (c) 18 in.
    (d) 6 ft

66. Indoor antenna lead-in conductors for radio and television receiving equipment can be in the same enclosure with conductors of other wiring systems where separated by an effective, permanently installed barrier.

    (a) True
    (b) False

67. Each lead-in conductor from an outdoor antenna shall be provided with a listed antenna discharge unit, unless enclosed in a grounded metallic shield.

    (a) True
    (b) False

68. The grounding conductor for an antenna mast shall be _____ protected where subject to physical damage.

    (a) electrically
    (b) mechanically
    (c) arc fault
    (d) none of these

69. If a separate grounding electrode is installed for the radio and television equipment, it shall be bonded to the building's electrical power grounding electrode system with a bonding jumper not smaller than _____ AWG.

    (a) 10
    (b) 8
    (c) 6
    (d) 1/0

70. Unshielded lead-in antenna conductors for amateur transmitting stations attached to buildings shall be firmly mounted at least _____ clear of the surface of the building on nonabsorbent insulating supports.

    (a) 1 in.
    (b) 2 in.
    (c) 3 in.
    (d) 4 in.

71. Coaxial cable is a round assembly composed of a conductor inside a metallic tube or shield, separated by _____ material covered by an insulating jacket.

    (a) insulating
    (b) conductive
    (c) isolating
    (d) dielectric

72. Access to electrical equipment shall not be denied by an accumulation of CATV coaxial cables that _____ removal of suspended-ceiling panels.

    (a) prevents
    (b) hinders
    (c) blocks
    (d) require

73. CATV coaxial cables installed _____ on the surface of ceilings and walls shall be supported by the building structure in such a manner that the cables will not be damaged by normal building use.

    (a) exposed
    (b) concealed
    (c) hidden
    (d) a and b

74. Openings around penetrations of CATV coaxial cables and raceways through fire-resistant–rated walls, partitions, floors, or ceilings shall be _____ using approved methods to maintain the fire-resistance rating.

(a) closed
(b) opened
(c) draft stopped
(d) firestopped

75. Overhead coaxial cables for a CATV system shall be separated by at least _____ from lightning conductors, where practicable.

(a) 3 in.
(b) 6 in.
(c) 2 ft
(d) 6 ft

76. Community antenna television and radio system coaxial cables shall not be required to be listed and marked where the length of the cable within the building, measured from its point of entrance, does not exceed _____, the cable enters the building from the outside, and the cable is terminated at a grounding block.

(a) 25 ft
(b) 30 ft
(c) 50 ft
(d) 100 ft

77. The conductor used to ground the outer cover of a CATV coaxial cable shall be _____.

(a) insulated
(b) 14 AWG minimum
(c) bare
(d) a and b

78. In one- and two-family dwellings, the grounding conductor for CATV shall be as short as practicable, not to exceed _____ in length.

(a) 5 ft
(b) 8 ft
(c) 10 ft
(d) 20 ft

79. In one- and two-family dwellings where it is not practicable to achieve an overall maximum grounding conductor length of _____ for CATV, a separate grounding electrode as specified in 250.52(A)(5), (6), or (7) shall be used.

(a) 5 ft
(b) 8 ft
(c) 10 ft
(d) 20 ft

80. Limiting the length of the primary protector grounding conductors for community antenna television and radio systems reduces voltages between the building's _____ and communications systems during lightning events.

(a) power
(b) fire alarm
(c) lighting
(d) lightning protection

81. The grounding conductor for CATV coaxial cable shall be _____ protected where subject to physical damage.

(a) electrically
(b) arc fault
(c) physically
(d) none of these

82. A bonding jumper not smaller than _____ copper or equivalent shall be connected between the CATV system's grounding electrode and the power grounding electrode system at the building or structure served where separate electrodes are used.

(a) 12 AWG
(b) 8 AWG
(c) 6 AWG
(d) 4 AWG

83. CATV coaxial cable can be placed in a raceway, compartment, outlet box, or junction box with the conductors of light or power circuits, or Class 1 circuits when _____.

(a) installed in rigid metal conduit
(b) separated by a permanent barrier
(c) insulated
(d) none of these

84. Coaxial cables for community antenna television systems, installed in vertical runs and penetrating more than one floor, or cables installed in vertical runs in a shaft, shall be Type _____.

    (a) CATV
    (b) CATVX
    (c) CATVR
    (d) any of these

85. The permitted conduit fill for five conductors is _____ percent.

    (a) 35
    (b) 40
    (c) 55
    (d) 60

86. The permitted raceway fill for two conductors is _____ percent.

    (a) 30
    (b) 31
    (c) 40
    (d) 53

87. When calculating raceway conductor fill, equipment grounding conductors shall _____.

    (a) not be required to be counted
    (b) have the actual dimensions used
    (c) not be counted if in a nipple
    (d) not be counted if for a wye three-phase balanced load

88. When the calculated number of conductors, all of the same size, installed in a conduit or in tubing includes a decimal, the next higher whole number shall be used when this decimal is _____ or larger.

    (a) 0.40
    (b) 0.60
    (c) 0.70
    (d) 0.80

89. For the purposes of conduit fill, a multiconductor cable or flexible cord of two or more conductors is considered _____ conductor(s).

    (a) one
    (b) two
    (c) three
    (d) four

90. The inside diameter of trade size 1 IMC is _____.

    (a) 0.314 in.
    (b) 0.826 in.
    (c) 1.000 in.
    (d) 1.105 in.

91. The cross-sectional area of trade size 1 IMC is approximately _____.

    (a) 0.62 sq in.
    (b) 0.96 sq in.
    (c) 1.22 sq in.
    (d) 2.13 sq in.

92. Trade size 1½ RMC nipple containing three conductors can be filled to an area of _____.

    (a) 0.882 sq in.
    (b) 1.072 sq in.
    (c) 1.243 sq in.
    (d) 1.343 sq in.

93. The cross-sectional area (sq in.) of a single 12 THHN conductor is _____.

    (a) 0.0133
    (b) 0.0147
    (c) 0.0233
    (d) 0.0321

94. A 10 AWG stranded copper conductor has a cross-sectional area of _____.

    (a) 0.007 sq in.
    (b) 0.011 sq in.
    (c) 0.012 sq in.
    (d) 0.106 sq in.

95. A 250 kcmil bare conductor has a diameter of _____.

    (a) 0.557 in.
    (b) 0.575 in.
    (c) 0.690 in.
    (d) 0.755 in.

96. The area in sq in. for a 1/0 AWG bare aluminum conductor is _____.

    (a) 0.087 sq in.
    (b) 0.109 sq in.
    (c) 0.137 sq in.
    (d) 0.173 sq in.

97. The circular mil area of a 12 AWG conductor is _____.

    (a) 6,100
    (b) 6,530
    (c) 10,380
    (d) 26,240

98. The ac ohms-to-neutral impedance per 1,000 ft of 4/0 AWG aluminum conductor in a steel raceway is _____.

    (a) 0.06 ohms
    (b) 0.10 ohms
    (c) 0.11 ohms
    (d) 0.22 ohms

99. The ac ohms-to-neutral impedance per 1,000 ft of 2/0 AWG copper conductor in a steel raceway is _____.

    (a) 0.06 ohms
    (b) 0.10 ohms
    (c) 0.11 ohms
    (d) 0.22 ohms

100. The ac ohms-to-neutral impedance per 1,000 ft of 4/0 AWG aluminum in a steel raceway is the same as the ohms-to-neutral impedance of 1,000 ft of _____ copper installed in a steel raceway.

    (a) 1 AWG
    (b) 1/0 AWG
    (c) 2/0 AWG
    (d) 250 kcmil

Unit 10: Practice Questions in Random Order—Article 90.3 through Annex C

Use the 2008 *NEC* to answer the following questions.

1. CATV cables installed in other spaces used for environmental air shall be Type CATVP.

   (a) True
   (b) False

2. Exposed CATV cables shall be secured by hardware such as straps, staples, cable ties, hangers, or similar fittings designed and installed so as not to damage the cables.

   (a) True
   (b) False

3. Network-powered broadband cable not terminated at equipment other than a connector, and not identified for future use with a tag, is considered abandoned.

   (a) True
   (b) False

4. Accessible portions of abandoned CATV cable shall be removed.

   (a) True
   (b) False

5. Antenna discharge units shall be located outside the building only.

   (a) True
   (b) False

6. The outer conductive shield of a CATV coaxial cable entering a building shall be grounded as close to the point of entrance as practicable.

   (a) True
   (b) False

7. The grounding conductor for an antenna mast or antenna discharge unit shall be run to the grounding electrode in as straight a line as practicable.

   (a) True
   (b) False

8. If the building or structure served has an intersystem bonding termination, the grounding conductor for an antenna mast shall be connected to the intersystem bonding termination.

   (a) True
   (b) False

9. The grounding conductor for an antenna mast or antenna discharge unit, if copper, shall not be smaller than 10 AWG.

   (a) True
   (b) False

10. CATV cable not terminated at equipment other than a coaxial cable connector and not identified for future use with a tag is considered abandoned.

    (a) True
    (b) False

11. Coaxial cables for CATV shall be listed.

    (a) True
    (b) False

12. A raceway is an enclosure designed for the installation of conductors, cables, or busbars.

    (a) True
    (b) False

Practice Questions in Random Order—Article 90.3 through Annex C — Unit 10

13. Where a mast is used for overhead conductor support, it shall be of adequate strength or be supported by braces or guys to withstand safely the strain imposed by the overhead drop.
    (a) True
    (b) False

14. Conductors supplied under the tap rules shall not supply another conductor using the tap rules.
    (a) True
    (b) False

15. Reinforcing bars for use as a concrete-encased electrode can be bonded together by the usual steel tie wires or other effective means.
    (a) True
    (b) False

16. Transient voltage surge suppressors (TVSSs) are also known as surge protective devices.
    (a) True
    (b) False

17. Type UF cable is permitted to be used for inside wiring.
    (a) True
    (b) False

18. Bushings or adapters shall be provided at ENT terminations to protect the conductors from abrasion, unless the box, fitting, or enclosure design provides equivalent protection.
    (a) True
    (b) False

19. When grouping conductors of switch loops in the same raceway, it is not required to include a grounded conductor in every switch loop.
    (a) True
    (b) False

20. Incandescent luminaires that have open lamps, and pendant-type luminaires, shall be permitted in clothes closets where proper clearance is maintained from combustible products.
    (a) True
    (b) False

21. Transformers and capacitors installed in Class I, Division 1 locations that do not contain flammable liquids shall not be required to be installed in vaults if they are identified for Class I locations.
    (a) True
    (b) False

22. Electrical equipment enclosures at marinas must be securely and substantially supported by structural members, independent of any raceway connected to them.
    (a) True
    (b) False

23. Feeders for temporary installations can be cable assemblies or multiconductor cords or cables identified for hard usage or extra-hard usage.
    (a) True
    (b) False

24. Wired partitions for office furnishings that are secured to building surfaces shall be connected to the building electrical system by a Chapter 3 wiring method.
    (a) True
    (b) False

25. Class 1, 2, and 3 circuits installed through fire-resistant-rated walls, partitions, floors, or ceilings must be firestopped to limit the possible spread of fire or products of combustion.
    (a) True
    (b) False

26. When conduit nipples having a maximum length not exceeding 24 in. are installed between boxes, the nipple can be filled to 60 percent.
    (a) True
    (b) False

27. Annex C contains tables that list the number of conductors or fixture wires permitted in a raceway when the conductors are all of the same size and type.
    (a) True
    (b) False

28. Each doorway leading into a transformer vault from the building interior shall be provided with a tight-fitting door having a minimum fire rating of _____ hours.

    (a) two
    (b) three
    (c) four
    (d) six

29. There shall be no more than _____ disconnects installed for each service or for each set of service-entrance conductors as permitted in 230.2 and 230.40.

    (a) two
    (b) four
    (c) six
    (d) eight

30. A hospital is a building or portion thereof used for the medical, psychiatric, obstetrical, or surgical care, on a 24-hour basis, of _____ or more inpatients.

    (a) two
    (b) three
    (c) four
    (d) five

31. A nursing home is an area used for the housing and nursing care, on a 24-hour basis, of _____ or more persons who, because of mental or physical incapacity, might be unable to provide for their own needs and safety without the assistance of another person.

    (a) two
    (b) three
    (c) four
    (d) five

32. There shall be no more than _____ switches or circuit breakers to serve as the disconnecting means for a building supplied by a feeder.

    (a) two
    (b) four
    (c) six
    (d) eight

33. For a circuit to be considered a multiwire branch circuit, it shall have _____.

    (a) two or more ungrounded conductors with a voltage potential between them
    (b) a grounded conductor having equal voltage potential between it and each ungrounded conductor of the circuit
    (c) a grounded conductor connected to the neutral or grounded terminal of the system
    (d) all of these

34. Where Type NM cable is run at angles with joists in unfinished basements and crawl spaces, it is permissible to secure cables not smaller than _____ conductors directly to the lower edges of the joist.

    (a) two, 6 AWG
    (b) three, 8 AWG
    (c) three, 10 AWG
    (d) a or b

35. A sign shall be placed at the service-entrance equipment indicating the _____ of on-site optional standby power sources.

    (a) type
    (b) location
    (c) manufacturer
    (d) a and b

36. A sign shall be placed at the service equipment indicating the _____ of on-site legally required standby power sources.

    (a) type
    (b) location
    (c) manufacturer
    (d) a and b

37. Luminaires containing a metal halide lamp shall be provided with a containment barrier that encloses the lamp, or shall be provided with a physical means that only allows the use of a(n) _____.

    (a) Type "O" lamp
    (b) Type PAR lamp
    (c) a or b
    (d) inert gas

38. Manufactured wiring systems shall be constructed with _____.

    (a) Type AC
    (b) Type MC cable
    (c) a or b
    (d) none of these

39. Listed plenum signaling raceways enclosing _____ cable can be installed in other spaces used for environmental air.

    (a) Type CL2P
    (b) Type CL3P
    (c) a or b
    (d) none of these

40. A surge protective device (SPD) intended for installation on the load side of the service disconnect overcurrent device, including SPDs located at the branch panel, is a _____ SPD.

    (a) Type I
    (b) Type II
    (c) Type III
    (d) Type IV

41. ___ is a raceway of circular cross section made of a helically wound, formed, interlocked metal strip of either steel or aluminum.

    (a) Type MC cable
    (b) Type AC Cable
    (c) LFMC
    (d) FMC

42. Circuit breakers used to switch 120V or 277V fluorescent-lighting circuits shall be listed and marked _____.

    (a) UL
    (b) SWD or HID
    (c) Amps
    (d) VA

43. Conductors are considered outside a building when they are installed:

    (a) Under not less than 2 in. of concrete beneath a building or structure.
    (b) Within a building or structure in a raceway encased in not less than a 2 in. thickness of concrete or brick.
    (c) Installed in a vault that meets the construction requirements of Article 450.
    (d) all of these

44. Multiwire branch circuits for temporary wiring shall be provided with a means to disconnect simultaneously all _____ conductors where the branch circuit originates.

    (a) underground
    (b) overhead
    (c) ungrounded
    (d) grounded

45. Type USE cable is not permitted for _____ wiring.

    (a) underground
    (b) interior
    (c) a or b
    (d) a and b

46. _____ shall not be used as grounding electrodes.

    (a) Underground gas piping systems
    (b) Aluminum
    (c) Metal well casings
    (d) a and b

47. All 15A and 20A, 125V receptacles installed in _____ of dwelling units shall have GFCI protection.

    (a) unfinished attics
    (b) finished attics
    (c) unfinished basements
    (d) finished basements

48. Each sign and outline lighting system shall be controlled by an externally operable switch or circuit breaker that opens all _____ conductors.

    (a) ungrounded
    (b) grounded
    (c) equipment grounding
    (d) all of these

49. In multiwire branch circuits, the continuity of the _____ conductor shall not be dependent upon the device connections.

    (a) ungrounded
    (b) grounded
    (c) grounding electrode
    (d) a and b

Unit 10 — Practice Questions in Random Order—Article 90.3 through Annex C

50. A flexible cord conductor intended to be used as a(n) _____ conductor shall have a continuous identifying marker readily distinguishing it from the other conductor or conductors.

    (a) ungrounded
    (b) equipment grounding
    (c) service
    (d) high-leg

51. Three-way and four-way switches shall be wired so that all switching is done only in the _____ circuit conductor.

    (a) ungrounded
    (b) grounded
    (c) equipment ground
    (d) neutral

52. Where an ac system operating at less than 1,000V is grounded at any point, the _____ conductor(s) shall be run to each service disconnecting means and shall be connected to each disconnecting means grounded conductor(s) terminal or bus.

    (a) ungrounded
    (b) grounded
    (c) grounding
    (d) none of these

53. The NEC defines a(n) _____ as a structure that stands alone or that is cut off from adjoining structures by firewalls, with all openings therein protected by approved fire doors.

    (a) unit
    (b) apartment
    (c) building
    (d) utility

54. Where LFMC is used in a Class II Division 1 location as permitted in 502.10, it shall not be _____.

    (a) unsupported
    (b) permitted in lengths exceeding 6 ft
    (c) used as the sole ground-fault current path
    (d) used unless listed for use in Class I locations

55. Where LFMC is used in a Class III location as permitted in 503.10, it shall not be _____.

    (a) unsupported
    (b) permitted in lengths greater than 6 ft
    (c) used as the sole ground-fault current path
    (d) permitted unless listed for use in a Class I hazardous (classified) location

56. Underground service conductors between the street main and the first point of connection to the service entrance conductors are known as the _____.

    (a) utility service
    (b) service lateral
    (c) service drop
    (d) main service conductors

57. The motor branch-circuit short-circuit and ground-fault protective device shall be capable of carrying the _____ current of the motor.

    (a) varying
    (b) starting
    (c) running
    (d) continuous

58. Sufficient access and _____ shall be provided and maintained about all electrical equipment to permit ready and safe operation and maintenance of such equipment.

    (a) ventilation
    (b) cleanliness
    (c) circulation
    (d) working space

59. Enclosures for overcurrent devices shall be mounted in a _____ position unless impracticable.

    (a) vertical
    (b) horizontal
    (c) vertical or horizontal
    (d) there are no requirements

60. GFCI protection shall be required for all 125V receptacles not exceeding 30A and located within 6 ft measured _____ from the inside walls of a hydromassage bathtub.

    (a) vertically
    (b) horizontally
    (c) across
    (d) none of these

61. Cartridge fuses and fuseholders shall be classified according to _____ ranges.

    (a) voltage
    (b) amperage
    (c) a or b
    (d) a and b

62. Fuseholders for cartridge fuses shall be so designed that it is difficult to put a fuse of any given class into a fuseholder that is designed for a _____ lower or a _____ higher than that of the class to which the fuse belongs.

    (a) voltage, wattage
    (b) wattage, voltage
    (c) voltage, current
    (d) current, voltage

63. Separately installed pressure connectors shall be used with conductors at the _____ not exceeding the ampacity at the listed and identified temperature rating of the connector.

    (a) voltages
    (b) temperatures
    (c) listings
    (d) ampacities

64. Unbroken lengths of surface nonmetallic raceways can be run through dry _____.

    (a) walls
    (b) partitions
    (c) floors
    (d) all of these

65. Horizontal runs of RMC supported by openings through _____ at intervals not exceeding 10 ft and securely fastened within 3 ft of termination points shall be permitted.

    (a) walls
    (b) trusses
    (c) rafters
    (d) framing members

66. Unbroken lengths of surface metal raceways can be run through dry _____.

    (a) walls
    (b) partitions
    (c) floors
    (d) all of these

67. Luminaires installed in Class II, Division 1 locations shall be protected from physical damage by a suitable _____.

    (a) warning label
    (b) pendant
    (c) guard or location
    (d) all of these

68. Luminaires installed in Class II, Division 2 locations shall be protected from physical damage by a suitable _____.

    (a) warning label
    (b) pendant
    (c) guard or by location
    (d) all of these

69. Luminaires installed in Class I, Division 1 locations shall be protected from physical damage by a suitable _____.

    (a) warning label
    (b) pendant
    (c) guard or location
    (d) all of these

70. Luminaires installed in Class I, Division 2 locations shall be protected from physical damage by a suitable _____.

    (a) warning label
    (b) pendant
    (c) guard or location
    (d) all of these

71. An intrinsically safe circuit is a circuit in which any spark or thermal effect is incapable of causing ignition of a mixture of flammable or combustible material in air under _____.

    (a) water
    (b) prescribed test conditions
    (c) supervision
    (d) duress

72. Materials such as straps, bolts, etc., associated with the installation of RMC in wet locations shall be _____.

    (a) weatherproof
    (b) weathertight
    (c) corrosion resistant
    (d) none of these

73. Materials such as straps, bolts, screws, etc., that are associated with the installation of IMC in wet locations shall be _____.

    (a) weatherproof
    (b) weathertight
    (c) corrosion resistant
    (d) none of these

74. Fixed electrical equipment installed above a Class I location in a commercial garage shall be _____.

    (a) well ventilated
    (b) located above the level of any defined Class I location
    (c) identified for the location
    (d) b or c

75. The interior of underground raceways shall be considered a _____ location.

    (a) wet
    (b) dry
    (c) damp
    (d) corrosive

76. Where raceways are installed in wet locations above grade, the interior of these raceways shall be considered a _____ location.

    (a) wet
    (b) dry
    (c) damp
    (d) corrosive

77. Where _____ may be present in an agricultural building, enclosures and fittings shall have corrosion-resistance properties suitable for the conditions.

    (a) wet dust
    (b) corrosive gases or vapors
    (c) other corrosive conditions
    (d) any of these

78. Power-supply conductors and Class 1 circuit conductors can occupy the same cable, enclosure, or raceway _____.

    (a) where both are functionally associated with the equipment powered
    (b) where the circuits involved are not a mixture of ac and dc
    (c) under no circumstances
    (d) none of these

79. A strut-type channel raceway can be installed _____.

    (a) where exposed
    (b) as a power pole
    (c) unbroken through walls, partitions, and floors
    (d) all of these

80. Overcurrent devices shall not be located _____.

    (a) where exposed to physical damage
    (b) near easily ignitible materials, such as in clothes closets
    (c) in bathrooms of dwelling units
    (d) all of these

81. Type TC cable shall not be installed _____.

    (a) where it will be exposed to physical damage
    (b) outside of a raceway or cable tray system, unless permitted in 336.10(7)
    (c) direct buried unless identified for such use
    (d) all of these

82. Track lighting shall not be installed _____.

    (a) where likely to be subjected to physical damage
    (b) in wet or damp locations
    (c) a and b
    (d) none of these

83. Plug fuses of the Edison-base type shall be used _____.

    (a) where overfusing is necessary
    (b) as a replacement in existing installations
    (c) as a replacement for Type S fuses
    (d) 50A and above

84. HDPE conduit can be installed _____.

    (a) where subject to chemicals for which the conduit is listed
    (b) in direct burial installations in earth or concrete
    (c) for conductors operating at a temperature above the rating of the raceway
    (d) all of these

85. Surface metal raceways shall not be used _____.

    (a) where subject to severe physical damage
    (b) where subject to corrosive vapors
    (c) in hoistways
    (d) all of these

86. Wet-niche luminaires installed in swimming pools shall be removable from the water for inspection, relamping or other maintenance. The luminaire maintenance location shall be accessible _____.

    (a) while the pool is drained
    (b) without entering the pool water
    (c) during construction
    (d) all of these

87. Guest rooms or guest suites provided with permanent provisions for _____ shall have receptacle outlets installed in accordance with all of the applicable requirements for a dwelling unit as specified in 210.52(a) and 210.52(D).

    (a) whirlpool tubs
    (b) bathing
    (c) cooking
    (d) internet access

88. A receptacle outlet shall be installed in dwelling units for every kitchen and dining area countertop space _____, and no point along the wall line shall be more than 2 ft, measured horizontally, from a receptacle outlet in that space.

    (a) wider than 10 in.
    (b) wider than 3 ft
    (c) 18 in. or wider
    (d) 12 in. or wider

89. A main bonding jumper shall be a _____ or similar suitable conductor.

    (a) wire
    (b) bus
    (c) screw
    (d) any of these

90. The identification of _____ to which a grounded conductor is to be connected shall be substantially white in color.

    (a) wire connectors
    (b) circuit breakers
    (c) terminals
    (d) ground rods

91. Access to equipment shall not be denied by an accumulation of communications _____ that prevents the removal of suspended-ceiling panels.

    (a) wires
    (b) cables
    (c) ductwork
    (d) a and b

92. Where the air conditioner disconnecting means is not within sight from the equipment, the provision for locking or adding a lock to the disconnecting means shall be on the switch or circuit breaker and remain in place _____ the lock installed.

    (a) with
    (b) without
    (c) with or without
    (d) none of these

93. A luminaire or lighting assembly can be cord-connected if located _____ the outlet, the cord is visible for its entire length outside the luminaire, and the cord is not subject to strain or physical damage.

    (a) within
    (b) directly below
    (c) directly above
    (d) adjacent to

94. Exposed noncurrent-carrying metal parts of fixed equipment likely to become energized shall be connected to the equipment conductor where located _____.

    (a) within 8 ft vertically or 5 ft horizontally of ground or grounded metal objects and subject to contact by persons
    (b) in wet or damp locations and not isolated
    (c) in electrical contact with metal
    (d) any of these

95. Signs or outline lighting systems operated by electronic or electromechanical controllers can have a disconnecting means located _____ when capable of being locked in the open position.

    (a) within sight of the controller
    (b) in the same enclosure with the controller
    (c) a or b
    (d) none of these

96. A switchboard or panelboard containing a 4-wire, _____ system where the midpoint of one phase winding is grounded, shall be legibly and permanently field-marked to caution that one phase has a higher voltage to ground.

    (a) wye-connected
    (b) delta-connected
    (c) solidly grounded
    (d) ungrounded

97. A three-phase, 4-wire, _____ power system used to supply power to nonlinear loads may necessitate that the power system design allow for the possibility of high harmonic current on the neutral conductor.

(a) wye-connected
(b) delta-connected
(c) wye-delta-connected
(d) none of these

98. Plaster, drywall, or plasterboard surfaces that are broken or incomplete around boxes employing a flush-type cover or faceplate shall be repaired so there will be no gaps or open spaces larger than _____ at the edge of the box.

(a) 1/16 in.
(b) 1/8 in.
(c) 1/4 in.
(d) 1/2 in.

99. A 10 AWG single strand (solid) copper conductor has a cross-sectional area of _____.

(a) 0.008 sq in.
(b) 0.012 sq in.
(c) 0.101 sq in.
(d) 0.106 sq in.

100. Where the circuit breaker handles are operated vertically, the "up" position of the handle shall be the _____.

(a) "on" position
(b) "off" position
(c) tripped position
(d) any of these

Article 90 through Chapter 9

CHALLENGE QUIZ 1

Use the 2008 *NEC* to answer the following questions.

1. Where overcurrent protection is provided as part of an industrial machine, the machine shall be marked _____.

    (a) "Overcurrent Protection Provided At Machine Supply Terminals"
    (b) "This Unit Contains Overcurrent Protection"
    (c) "Fuses Or Circuit Breaker Enclosed"
    (d) "Overcurrent Protected"

2. Type IGS cable is defined as a factory assembly of one or more conductors, each individually insulated and enclosed in a loose-fit, nonmetallic flexible conduit as an integrated gas spacer cable rated _____ volts.

    (a) 0 through 6,000
    (b) 600 through 6,000
    (c) 0 through 600
    (d) 3,000 through 6,000

3. Bare conductors in auxiliary gutters shall be securely and rigidly supported so that the minimum clearance between bare current-carrying metal parts of different potential mounted on the same surface will not be less than 2 in., nor less than _____, for parts in free air.

    (a) ¼ in.
    (b) ½ in.
    (c) 1 in.
    (d) 2 in.

4. A motor terminal housing enclosing rigidly mounted motor terminals shall have a minimum of _____ between line terminals for a 230V motor.

    (a) ¼ in.
    (b) ⅜ in.
    (c) ½ in.
    (d) ⅝ in.

5. The radius of the inner edge of any bend in Type MI cable shall not be less than five times the external diameter of the metallic sheath for cable not more than _____ in external diameter.

    (a) ½ in.
    (b) ⅝ in.
    (c) ¾ in.
    (d) 1½ in.

6. Flat-top underfloor raceways over 4 in. but not over 8 in. wide, with a minimum 1 in. spacing between raceways, shall be covered with concrete to a depth of not less than 1 in. Raceways spaced less than 1 in. apart shall be covered with concrete to a depth of _____.

    (a) 1 in.
    (b) 1½ in.
    (c) 2 in.
    (d) 4 in.

7. Dry-type transformers installed indoors rated over _____ shall be installed in a vault.

    (a) 1,000V
    (b) 20,000V
    (c) 35,000V
    (d) 50,000V

8. In a cablebus, the size and number of conductors shall be that for which the cablebus is designed, and in no case smaller than _____ AWG.

    (a) 1/0
    (b) 2/0
    (c) 3/0
    (d) 4/0

Challenge Quiz 1 — Article 90 through Chapter 9

9. Except by special permission, no conductor larger than _____ AWG shall be installed in a cellular metal floor raceway.

    (a) 1
    (b) 1/0
    (c) 4/0
    (d) no restriction

10. The branch-circuit protective device can serve as the controller for a stationary motor rated at _____ or less that is normally left running and cannot be damaged by overload or failure to start.

    (a) ⅛ hp
    (b) ¼ hp
    (c) ⅜ hp
    (d) ½ hp

11. Each length of HDPE conduit shall be clearly and durably marked not less than every _____ ft.

    (a) 3
    (b) 5
    (c) 10
    (d) 20

12. A vented alkaline-type battery, as it relates to storage batteries, operating at less than 250V shall be installed with not more than _____ cells in the series circuit of any one tray.

    (a) 10
    (b) 12
    (c) 18
    (d) 20

13. A receiving station outdoor wire-strung antenna conductor with a span of 75 ft shall be at least _____ if a copper-clad steel conductor is used.

    (a) 17 AWG
    (b) 14 AWG
    (c) 12 AWG
    (d) 10 AWG

14. An outdoor wire-strung antenna conductor of a receiving station with a 75 ft span using a hard-drawn copper conductor shall not be less than _____.

    (a) 17 AWG
    (b) 14 AWG
    (c) 12 AWG
    (d) 10 AWG

15. In network-powered broadband communications systems, all separate electrodes shall be bonded together using a minimum _____ copper bonding jumper.

    (a) 10 AWG
    (b) 8 AWG
    (c) 6 AWG
    (d) 4 AWG

16. Busways shall be securely supported, unless otherwise designed and marked as such, at intervals not to exceed _____.

    (a) 3 ft
    (b) 5 ft
    (c) 8 ft
    (d) 10 ft

17. Capacitors containing more than _____ of flammable liquid shall be enclosed in vaults or outdoor fenced enclosures.

    (a) 3 gallons
    (b) 5 gallons
    (c) 10 gallons
    (d) 11 gallons

18. When an overload device sized in accordance with 430.32(A)(1) and (B)(1) is not sufficient to start the motor, higher-size sensing elements shall be permitted if the trip current does not exceed _____ percent of the motor nameplate full-load current rating, when the motor is marked with a service factor of 1.12.

    (a) 100
    (b) 110
    (c) 120
    (d) 130

19. The disconnecting means for a torque motor shall have an ampere rating of at least _____ percent of the motor nameplate current.

    (a) 100
    (b) 115
    (c) 125
    (d) 175

20. Transformer vaults containing more than _____ total kVA transformer capacity shall be provided with a drain or other means that will carry off any accumulation of oil or water in the vault, unless impracticable.

    (a) 100
    (b) 150
    (c) 200
    (d) 250

21. Circuit breakers rated at _____ amperes or less and _____ volts or less shall have the ampere rating molded, stamped, etched, or similarly marked into their handles or escutcheon areas.

    (a) 100, 600
    (b) 600, 100
    (c) 1,000, 6,000
    (d) 6,000, 1,000

22. The pilot light provided within a portable stage switchboard enclosure shall have overcurrent protection rated or set at not more than _____.

    (a) 10A
    (b) 15A
    (c) 20A
    (d) 30A

23. Circuits and equipment operating at less than 50V shall have receptacles that are rated at not less than _____.

    (a) 10A
    (b) 15A
    (c) 20A
    (d) 30A

24. External surfaces of pipeline and vessel heating equipment that operate at temperatures exceeding _____ shall be physically guarded, isolated, or thermally insulated to protect against contact by personnel in the area.

    (a) 110°F
    (b) 120°F
    (c) 130°F
    (d) 140°F

25. For straight pulls, the length of a pull box shall not be less than _____ times the outside diameter, over the sheath, of the largest shielded or lead-covered conductor or cable entering the box on systems over 600V.

    (a) 12
    (b) 16
    (c) 24
    (d) 48

26. For angle or U-pulls, the distance between the shielded conductor entry (for systems over 600V, nominal) and the opposite wall of the box shall not be less than _____ times the outside diameter of the largest cable or conductor.

    (a) 12
    (b) 16
    (c) 24
    (d) 36

27. The distance between a shielded cable or conductor entry and its exit from the box shall not be less than _____ times the outside diameter of that cable or conductor on a system of over 600V.

    (a) 12
    (b) 16
    (c) 24
    (d) 36

28. Connection by means of wire-binding screws, studs, or nuts having upturned lugs or the equivalent shall be permitted for _____ or smaller conductors.

    (a) 12 AWG
    (b) 10 AWG
    (c) 8 AWG
    (d) 6 AWG

29. Conductors supplying outlets for arc and xenon projectors of the professional type shall not be smaller than _____ and shall have an ampacity not less than the projector rating.

    (a) 12 AWG
    (b) 10 AWG
    (c) 8 AWG
    (d) 6 AWG

Challenge Quiz 1 — Article 90 through Chapter 9

30. For switchboards that are not totally enclosed, a space of _____ or more shall be provided between the top of the switchboard and any combustible ceiling.

    (a) 12 in.
    (b) 18 in.
    (c) 2 ft
    (d) 3 ft

31. Electric-discharge luminaires having an open-circuit voltage exceeding _____ shall not be installed in or on dwelling occupancies.

    (a) 120V
    (b) 250V
    (c) 600V
    (d) 1,000V

32. On electric space-heating cables, blue leads indicate a cable rated for use on a nominal circuit voltage of _____.

    (a) 120V
    (b) 208V
    (c) 240V
    (d) 277V

33. A disconnecting means that serves a hermetic refrigerant motor-compressor shall have an ampere rating of at least _____ percent of the nameplate rated-load current or branch-circuit selection current, whichever is greater.

    (a) 80
    (b) 100
    (c) 115
    (d) 125

34. Flat cable assemblies shall have conductors of _____ AWG special stranded copper conductors.

    (a) 14
    (b) 12
    (c) 10
    (d) all of these

35. The maximum size of conductors in Type UF cable shall be _____ AWG.

    (a) 14
    (b) 10
    (c) 1/0
    (d) 4/0

36. The minimum size parallel conductors permitted for elevator lighting shall be _____, provided the combined ampacity is equivalent to at least that of a 14 AWG conductor.

    (a) 20 AWG
    (b) 16 AWG
    (c) 14 AWG
    (d) 1/0 AWG

37. An open span length of 200 ft for antenna conductors of hard-drawn copper located at an amateur transmitting and receiving station requires a minimum conductor size of _____.

    (a) 14 AWG
    (b) 12 AWG
    (c) 10 AWG
    (d) 8 AWG

38. The residual voltage of a capacitor, rated not over 600V, shall be reduced to 50V or less within _____ after the capacitor is disconnected from the source of supply.

    (a) 15 seconds
    (b) 45 seconds
    (c) one minute
    (d) two minutes

39. Motor control circuit transformers, with a primary current rating of less than 2A, can have the primary protection device set at no more than _____ percent of the rated primary current rating.

    (a) 150
    (b) 200
    (c) 400
    (d) 500

40. A building disconnecting means that supplies only limited loads of a single branch circuit shall have a rating of not less than _____.

    (a) 15A
    (b) 20A
    (c) 25A
    (d) 30A

41. For installations consisting of not more than two 2-wire branch circuits, the building disconnecting means shall have a rating of not less than _____.

    (a) 15A
    (b) 20A
    (c) 25A
    (d) 30A

42. General-purpose branch circuits wired with flat conductor cable shall not be rated more than _____.

    (a) 15A
    (b) 20A
    (c) 30A
    (d) 40A

43. The rating of the attachment plug and receptacle shall not exceed _____ at 250V for a cord-and-plug-connected a/c motor-compressor.

    (a) 15A
    (b) 20A
    (c) 30A
    (d) 40A

44. Fixture wires of 18 AWG or 16 AWG, shall be permitted for the control and operating circuits of X-ray equipment when protected by an overcurrent device not larger than _____.

    (a) 15A
    (b) 20A
    (c) 25A
    (d) 30A

45. Pilot lights, instruments, potential transformers, current transformers, and other switchboard devices with potential coils shall be supplied by a circuit that is protected by overcurrent devices rated _____.

    (a) 10A or less
    (b) 15A or less
    (c) 15A or more
    (d) 20A or less

46. Conductors for an appliance circuit supplying more than one appliance or appliance receptacle in an installation operating at less than 50V shall not be smaller than _____ AWG copper or equivalent.

    (a) 18
    (b) 14
    (c) 12
    (d) 10

47. The minimum size copper conductor permitted for Type MC cable is _____.

    (a) 18 AWG
    (b) 16 AWG
    (c) 14 AWG
    (d) 12 AWG

48. The conductors contained within Type ITC cable shall be rated 300V, and not smaller than _____ AWG nor larger than _____ AWG.

    (a) 22, 12
    (b) 18, 10
    (c) 16, 8
    (d) 14, 1/0

49. The maximum spacing of receptacle outlets over countertops in the kitchen in a mobile home is _____.

    (a) 2 ft
    (b) 4 ft
    (c) 6 ft
    (d) one on each side of the sink

50. Galvanized steel, stainless steel, and red brass RMC shall be permitted in or under cinder fill subject to permanent moisture, when protected on all sides by a layer of noncinder concrete not less than _____ thick.

    (a) 2 in.
    (b) 4 in.
    (c) 6 in.
    (d) 18 in.

CHALLENGE QUIZ 2

Article 90 through Chapter 9

Use the 2008 *NEC* to answer the following questions.

1. Where knob-and-tube conductors pass through wood cross members in plastered partitions, conductors shall be protected by noncombustible, nonabsorbent, insulating tubes extending not less than _____ beyond the wood member.

    (a) 2 in.
    (b) 3 in.
    (c) 4 in.
    (d) 6 in.

2. For installations of resistors and reactors, a thermal barrier shall be required if the space between them and any combustible material is less than _____

    (a) 2 in.
    (b) 3 in.
    (c) 6 in.
    (d) 12 in.

3. A buried iron or steel plate used as a grounding electrode shall expose not less than _____ of surface area to exterior soil.

    (a) 2 sq ft
    (b) 4 sq ft
    (c) 9 sq ft
    (d) 10 sq ft

4. The minimum size conductor for operating control and signaling circuits in an elevator shall be _____.

    (a) 20 AWG
    (b) 16 AWG
    (c) 14 AWG
    (d) 12 AWG

5. Steel cable trays shall not be used as equipment grounding conductors for circuits with ground-fault protection above _____.

    (a) 200A
    (b) 300A
    (c) 600A
    (d) 800A

6. Tap devices used in Type FC assemblies shall be rated not less than _____ or more than 300 volts-to-ground.

    (a) 15A
    (b) 20A
    (c) 30A
    (d) 40A

7. Live parts of electrical equipment operating at _____ or more shall be guarded against accidental contact by approved enclosures or by suitable permanent, substantial partitions, or screens arranged so that only qualified persons have access to the space within reach of the live parts.

    (a) 20V
    (b) 30V
    (c) 50V
    (d) 100V

8. Fixture wires used as pendant conductors for incandescent luminaires with intermediate or candelabra-base lampholders shall not be smaller than _____ AWG.

    (a) 22
    (b) 18
    (c) 16
    (d) 14

9. The working clearance for a distribution panelboard in a recreational vehicle shall be not less than _____.

   (a) 24 in. wide
   (b) 30 in. deep
   (c) 30 in. wide
   (d) a and b

10. Electric space-heating equipment employing resistance-type heating elements rated more than _____ shall have the heating elements subdivided.

    (a) 24A
    (b) 36A
    (c) 48A
    (d) 60A

11. Resistance-type heating elements in electric space-heating equipment shall be protected at not more than _____.

    (a) 24A
    (b) 36A
    (c) 48A
    (d) 60A

12. Exposed live parts of motors and controllers operating at _____ or more between terminals shall be guarded against accidental contact by enclosure or by location.

    (a) 24V
    (b) 50V
    (c) 60V
    (d) 150V

13. Electrical equipment permanently mounted on skids, and the skids themselves, shall be connected to the equipment grounding conductor sized as required by _____.

    (a) 250.50
    (b) 250.66
    (c) 250.122
    (d) 310.15

14. The ampacity 1/0 THHN copper conductor in free air is _____.

    (a) 185A
    (b) 215A
    (c) 260A
    (d) 300A

15. Electric pipe organ circuits shall be arranged so that 26 AWG and 28 AWG conductors are protected from overcurrent by an overcurrent device rated not more than _____.

    (a) 2A
    (b) 4A
    (c) 6A
    (d) 8A

16. Type MI cable shall be supported and secured at intervals not exceeding _____.

    (a) 3 ft
    (b) 3½ ft
    (c) 5 ft
    (d) 6 ft

17. Each length of RMC shall be clearly and durably identified in every _____.

    (a) 3 ft
    (b) 5 ft
    (c) 10 ft
    (d) 20 ft

18. Cleat-type lampholders located at least _____ above the floor can have exposed terminals.

    (a) 3 ft
    (b) 6 ft
    (c) 8 ft
    (d) 10 ft

19. Unless an individual switch is provided for each luminaire located over combustible material, lampholders shall be located at least _____ above the floor, or shall be located or guarded so that the lamps cannot be readily removed or damaged.

    (a) 3 ft
    (b) 6 ft
    (c) 8 ft
    (d) 10 ft

20. Individual unit equipment for legally required standby illumination shall be permanently fixed in place. Flexible cord-and-plug connection shall be permitted, provided the cord does not exceed _____ in length.

    (a) 12 in.
    (b) 18 in.
    (c) 3 ft
    (d) 6 ft

Challenge Quiz 2 — Article 90 through Chapter 9

21. Type FCC cable consists of _____ copper conductors placed edge-to-edge, separated and enclosed within an insulating assembly.

    (a) 3 or more square
    (b) 2 or more round
    (c) 3 or more flat
    (d) 2 or more flat

22. Supports for concealed knob-and-tube wiring shall be installed within _____ in. of each side of each tap or splice and at intervals not exceeding _____ ft.

    (a) 2, 3½
    (b) 3, 2½
    (c) 4, 6½
    (d) 6, 4½

23. The maximum trade size FMT is _____.

    (a) ⅜
    (b) ½
    (c) ¾
    (d) 1

24. Type ITC cable is used for instrumentation and control circuits operating at _____.

    (a) 150V or less and 5A or less
    (b) 300V or less and 0.50A or less
    (c) 480V or less and 10A or less
    (d) over 600V

25. Panelboards equipped with snap switches rated at 30A or less shall have overcurrent protection not exceeding _____.

    (a) 30A
    (b) 50A
    (c) 100A
    (d) 200A

26. Infrared lamps for commercial and industrial heating appliances shall have overcurrent protection not exceeding _____.

    (a) 30A
    (b) 40A
    (c) 50A
    (d) 60A

27. A mobile home that is factory-equipped with gas or oil-fired central heating equipment and cooking appliances can be provided with a listed mobile home power-supply cord rated _____.

    (a) 30A
    (b) 35A
    (c) 40A
    (d) 50A

28. Soft-drawn or medium-drawn copper lead-in conductors for receiving antenna systems shall be permitted where the maximum span between points of support is less than _____.

    (a) 10 ft
    (b) 20 ft
    (c) 30 ft
    (d) 35 ft

29. A professional motion picture projector is one that uses _____ film and has on each edge 212 perforations per meter, or a type using carbon arc, xenon, or other light source equipment that develops hazardous gases, dust, or radiation.

    (a) 35 mm
    (b) 70 mm
    (c) 105 mm
    (d) a or b

30. Floor-mounted flat conductor cable and fittings shall be covered with carpet squares no larger than _____.

    (a) 24 inches square
    (b) 30 sq in. area
    (c) 36 inches square
    (d) 36 sq in. area

31. Overcurrent protection for Type ITC cable shall not exceed _____ for 20 AWG and larger conductors.

    (a) 3A
    (b) 5A
    (c) 10A
    (d) 15A

32. Open wiring on insulators within _____ from the floor shall be considered exposed to physical damage.

    (a) 2 ft
    (b) 4 ft
    (c) 7 ft
    (d) 8 ft

33. Smooth-sheath Type MC cable with an external diameter greater than ¾ in. but not greater than 1½ in. shall have a bending radius not more than _____ times the cable external diameter.

    (a) five
    (b) 10
    (c) 12
    (d) 13

34. The radius of the curve of the inner edge of any bend, during or after installation, shall not be less than _____ times the diameter of Type USE or SE cable.

    (a) five
    (b) seven
    (c) 10
    (d) 12

35. In one- and two-family and multifamily dwellings, the grounding conductor for network-powered broadband communications systems shall be as short as permissible, not to exceed _____ in length.

    (a) 5 ft
    (b) 6 ft
    (c) 10 ft
    (d) 20 ft

36. In one- and two-family dwellings where it is not practicable to achieve an overall maximum primary protector grounding conductor length of 20 ft or less for network-powered broadband communications systems, and a grounding means is not present, a separate _____ communications ground rod shall be driven and be bonded to the power grounding electrode system with a 6 AWG conductor.

    (a) 5 ft
    (b) 8 ft
    (c) 10 ft
    (d) 20 ft

37. Auxiliary gutters shall not be used to extend a distance greater than _____ beyond the equipment that it supplements.

    (a) 10 ft
    (b) 25 ft
    (c) 30 ft
    (d) 50 ft

38. The ampacity of supply branch-circuit conductors and the rating of the circuit overcurrent devices for X-ray equipment shall not be less than _____.

    (a) 50 percent of the momentary rating
    (b) 100 percent of the long-time rating
    (c) the larger of a or b
    (d) the smaller of a or b

39. A disconnecting means for X-ray equipment shall have adequate capacity for at least _____ percent of the input required for the momentary rating of the equipment or _____ percent of the input required for the long-time rating of the equipment, whichever is greater

    (a) 50, 100
    (b) 50, 125
    (c) 100, 100
    (d) 125, 125

40. The continuous current-carrying capacity of 1½ sq in. copper busbar mounted in an unventilated sheet metal auxiliary gutter is _____.

    (a) 500A
    (b) 650A
    (c) 750A
    (d) 1,500A

41. The maximum ampere rating of a 4 in. x ½ in. busbar 4 ft long and installed in an auxiliary gutter is _____.

    (a) 500A
    (b) 650A
    (c) 750A
    (d) 2,000A

42. Busways can be extended vertically through dry floors if totally enclosed (unventilated) where passing through, and for a minimum distance of _____ above the floor to provide adequate protection from physical damage.

    (a) 6 ft
    (b) 6½ ft
    (c) 8 ft
    (d) 10 ft

43. A separation of at least _____ shall be maintained between communications wires and cables on buildings and lightning conductors, where practicable.

    (a) 6 ft
    (b) 8 ft
    (c) 10 ft
    (d) 12 ft

44. Nonmetallic surface extensions shall be secured in place by approved means at intervals not exceeding _____.

    (a) 6 in.
    (b) 8 in.
    (c) 10 in.
    (d) 16 in.

45. Conductors smaller than 8 AWG for open wiring on insulators shall be supported within _____ of a tap or splice.

    (a) 6 in.
    (b) 8 in.
    (c) 10 in.
    (d) 12 in.

46. Noninsulated busbars shall have a minimum space of _____ between the bottom of a switchboard enclosure and busbar or other obstructions.

    (a) 6 in.
    (b) 8 in.
    (c) 10 in.
    (d) 12 in.

47. Electric space-heating cables shall be furnished complete with factory-assembled nonheating leads at least _____ in length.

    (a) 6 in.
    (b) 18 in.
    (c) 3 ft
    (d) 7 ft

48. THW insulation has a _____ rating when installed within electric-discharge lighting equipment, such as through fluorescent luminaires.

    (a) 60°C
    (b) 75°C
    (c) 90°C
    (d) 105°C

49. When bare conductors are permitted, their ampacities shall be limited to _____.

    (a) 60°C
    (b) 75°C
    (c) 90°C
    (d) those permitted for the insulated conductors of the same size

50. The maximum voltage permitted between ungrounded conductors of flat conductor cable systems is _____.

    (a) 150V
    (b) 250V
    (c) 300V
    (d) 600V

CHALLENGE QUIZ 3

Article 90 through Chapter 9

Use the 2008 *NEC* to answer the following questions.

1. A secondary tie of a transformer is a circuit operating at _____, nominal, or less, between phases that connects two power sources or power-supply points.

   (a) 600V
   (b) 1,000V
   (c) 12,000V
   (d) 35,000V

2. Type MV cable is defined as a single or multiconductor solid dielectric insulated cable rated _____ volts or higher.

   (a) 601
   (b) 1,001
   (c) 2,001
   (d) 6,001

3. Nominal Battery Voltage, as it relates to storage batteries, is defined as the voltage calculated on the basis of _____ per cell for the lead-acid type and _____ per cell for the alkali type.

   (a) 1.2V, 2V
   (b) 2V, 1.2V
   (c) 6V, 12V
   (d) 12V, 24V

4. Torque requirements for motor control circuit device terminals shall be a minimum of _____ lb-in. (unless otherwise identified) for screw-type pressure terminals used for 14 AWG and smaller copper conductors.

   (a) 7
   (b) 9
   (c) 10
   (d) 15

5. The combined area of all conductors installed at any point of an underfloor raceway shall not exceed _____ percent of the internal cross-sectional area.

   (a) 0
   (b) 40
   (c) 60
   (d) 75

6. AC or DC general-use snap switches may be used for control of inductive loads not exceeding _____ percent of the ampere rating of the switch at the applied voltage.

   (a) 50
   (b) 75
   (c) 90
   (d) 100

7. A recessed luminaire shall be installed so that adjacent combustible material will not be subjected to temperatures in excess of _____°C.

   (a) 75
   (b) 90
   (c) 125
   (d) 150

8. For phase converters serving variable loads, the ampacity of the single-phase supply conductors shall not be less than _____ percent of the single-phase input full-load amperes listed on the phase converter nameplate.

   (a) 75
   (b) 100
   (c) 125
   (d) 150

Challenge Quiz 3 — Article 90 through Chapter 9

9. Fixed electric space-heating equipment requiring supply conductors with insulation rated over _____ shall be clearly and permanently marked.

    (a) 60°C
    (b) 75°C
    (c) 90°C
    (d) all of these

10. On fixed stage equipment, portable strip lights and connector strips shall be wired with conductors having insulation rated suitable for the temperature but not less than _____.

    (a) 75°C
    (b) 90°C
    (c) 125°C
    (d) 200°C

11. Individual open conductors shall not be installed within _____ of grade level or where exposed to physical damage.

    (a) 8 ft
    (b) 10 ft
    (c) 12 ft
    (d) 15 ft

12. TPT and TST cords shall be permitted in lengths not exceeding _____ when attached directly to a portable appliance rated 50W or less.

    (a) 8 ft
    (b) 10 ft
    (c) 15 ft
    (d) 20 ft

13. The minimum clearance between an electric space-heating cable and an outlet box used for surface luminaires shall not be less than _____.

    (a) 6 in.
    (b) 8 in.
    (c) 14 in.
    (d) 18 in.

14. The disconnecting means for motor circuits rated 600 volts, nominal, or less shall have an ampere rating not less than _____ percent of the full-load current rating of the motor.

    (a) 80
    (b) 100
    (c) 115
    (d) 125

15. The maximum setting for ground-fault protection in a service disconnecting means is _____.

    (a) 800A
    (b) 1,000A
    (c) 1,200A
    (d) 2,000A

16. Conductors located above an electric space heated ceiling are considered to be operating in an ambient temperature of _____.

    (a) 20°C
    (b) 30°C
    (c) 50°C
    (d) 86°C

17. In a health care facility, receptacles and attachment plugs in a hazardous (classified) location within an anesthetizing area shall be listed for use in Class I, Group _____ locations.

    (a) A
    (b) B
    (c) C
    (d) D

18. The ground-fault protection system for service equipment shall be _____ when first installed on-site.

    (a) a Class A device
    (b) identified
    (c) turned on
    (d) performance tested

19. The essential electrical systems in a health care facility shall be supplied from _____.

    (a) a normal source generally supplying the entire electrical system
    (b) one or more alternate sources for use when the normal source is interrupted
    (c) a or b
    (d) a and b

20. Type _____ cable is a fabricated assembly of conductors in a flexible metal sheath with an internal bonding strip in intimate contact with the armor for its entire length.

    (a) AC
    (b) MC
    (c) NM
    (d) b and c

21. Type _____ is a factory assembly of insulated circuit conductors within an armor of interlocking metal tape, or a smooth or corrugated metallic sheath.

    (a) AC
    (b) MC
    (c) NM
    (d) b and c

22. Luminaires shall be marked with the maximum lamp wattage, manufacturer's _____, trademark, or other suitable means of identification.

    (a) address
    (b) name
    (c) phone number
    (d) none of these

23. Ballasts for fluorescent lighting installed indoors shall have _____ protection.

    (a) AFCI
    (b) supplementary
    (c) integral thermal
    (d) none of these

24. Service-entrance conductors for a manufactured building shall be installed _____.

    (a) after erection at the building site
    (b) before erection only where the point of attachment is known prior to manufacture
    (c) before erection at the building site
    (d) a or b

25. The outer sheath of Type MI cable is made of _____.

    (a) aluminum
    (b) steel alloy
    (c) copper
    (d) b or c

26. Where livestock is housed, that portion of the equipment grounding conductor run underground to the agricultural building or structure from a distribution point shall be insulated or covered _____.

    (a) aluminum
    (b) copper
    (c) copper-clad aluminum
    (d) none of these

27. Fuses shall be marked with their _____.

    (a) ampere and voltage rating
    (b) interrupting rating where other than 10,000A
    (c) name or trademark of the manufacturer
    (d) all of these

28. Resistors and reactors rated over 600V shall be isolated by _____ to protect personnel from accidental contact with energized parts.

    (a) an enclosure
    (b) elevation
    (c) a or b
    (d) a and b

29. For cord-and-plug-connected electric vehicle supply equipment, the listed system of personnel protection can be _____.

    (a) an integral part of the attachment plug
    (b) in the power-supply cable not more than 12 in. from the attachment plug
    (c) a or b
    (d) none of these

30. The wiring methods permitted in theaters, audience areas of motion picture and television studios, performance areas, and similar locations are _____.

    (a) any metal raceway
    (b) nonmetallic raceways encased in 2 in. of concrete
    (c) Type MC or AC cable with an insulated equipment grounding conductor
    (d) any of these

31. The fuel cell system shall be evaluated and _____ for its intended application.

    (a) approved
    (b) identified
    (c) listed
    (d) marked

32. Splices and taps shall not be located within luminaire _____.

    (a) arms or stems
    (b) bases or screw shells
    (c) a and b
    (d) a or b

33. Electric vehicle supply equipment identified for and intended to be interconnected to a vehicle, and also serve _____, shall be listed as suitable for that purpose.

(a) as an optional standby system
(b) as an electric power production source
(c) to provide bidirectional power feed
(d) all of these

34. The standard length of RMC shall _____.

(a) be in lengths of 10 ft
(b) include a coupling on each length
(c) be threaded on each end
(d) all of these

35. The disconnecting means for a photovoltaic system shall _____.

(a) be installed at a readily accessible location
(b) outside the building or inside nearest the point of entrance
(c) consist of not more than six switches or six circuit breakers
(d) all of these

36. Where network-powered broadband communications system aerial cables are installed outside and entering buildings, they shall _____.

(a) be located below the electric light or power conductors, where practicable
(b) not be attached to a cross-arm that carries electric light or power conductors
(c) have a vertical clearance of not less than 8 ft from all points of roofs above which they pass
(d) all of these

37. Bare Type FCC cable ends shall _____.

(a) be sealed
(b) be insulated
(c) use listed insulating ends
(d) all of these

38. Pipeline resistance heating elements include heating _____.

(a) blankets
(b) tape
(c) cables
(d) all of these

39. Cablebus framework that is _____ shall be permitted as the equipment grounding conductor for branch circuits and feeders.

(a) bonded
(b) welded
(c) protected
(d) galvanized

40. _____ shall be installed so that the wiring contained can be rendered accessible without removing any part of the building or, in underground circuits, without excavating sidewalks, paving, or earth.

(a) Boxes
(b) Conduit bodies
(c) Handhole enclosures
(d) all of these

41. A separate _____ shall be provided for the elevator car lights, receptacle(s), auxiliary lighting power source, and ventilation on each elevator car.

(a) branch circuit
(b) disconnecting means
(c) connection
(d) none of these

42. Nonmetallic surface extensions shall be permitted in _____ when occupied for residential or office purposes.

(a) buildings not over three stories high
(b) buildings over four stories high
(c) all buildings
(d) none of these

43. Barriers shall be placed in all service switchboards such that no uninsulated, ungrounded service _____ or service terminal is exposed to inadvertent contact by persons or maintenance equipment while servicing load terminations.

(a) busbar
(b) conductors
(c) cables
(d) none of these

44. Where Type MI cable terminates, a(n) _____ shall be installed immediately after stripping to prevent the entrance of moisture into the insulation.

(a) bushing
(b) connector
(c) flexible fitting
(d) end seal fitting

45. Resistance heating elements of embedded deicing and snow-melting _____ shall not be installed where they bridge expansion joints unless provision is made for expansion and contraction.

    (a) cables
    (b) units
    (c) panels
    (d) all of these

46. Switches or circuit breakers shall not disconnect the grounded conductor of a circuit unless the switch or circuit breaker _____.

    (a) can be opened and closed by hand levers only
    (b) simultaneously disconnects all conductors of the circuit
    (c) opens the grounded conductor before it disconnects the ungrounded conductors
    (d) none of these

47. Knife switches rated for more than 1,200A at 250V or less _____.

    (a) can be used only as isolating switch
    (b) shall not be opened under load
    (c) shall be placed so that gravity tends to close them
    (d) a and b

48. When the term exposed, as it applies to live parts, is used in the *Code*, it refers to _____.

    (a) capable of being inadvertently touched or approached nearer than a safe distance by a person
    (b) parts that are not suitably guarded, isolated, or insulated
    (c) wiring on, or attached to, the surface or behind panels designed to allow access
    (d) a and b

49. The distribution point is also known as the _____.

    (a) center yard pole
    (b) meterpole
    (c) common distribution point
    (d) all of these

50. According to the *Code*, automatic is self-acting when actuated by some impersonal influence, such as _____.

    (a) change in current
    (b) temperature
    (c) mechanical configuration
    (d) all of these

Article 90 through Chapter 9

CHALLENGE QUIZ 4

Use the 2008 *NEC* to answer the following questions.

1. In cellular concrete floor raceways, a cell is defined as a single, enclosed _____ space in a floor made of precast cellular concrete slabs, the direction of the cell being parallel to the direction of the floor member.

   (a) circular
   (b) oval
   (c) tubular
   (d) hexagonal

2. Aboveground bulk storage tanks shall be classified as _____ locations for the space between 5 ft and 10 ft from the open end of a vent.

   (a) Class I, Division 1
   (b) Class I, Division 2
   (c) Class II, Division 1
   (d) Class II, Division 2

3. Locations where flammable paints are dried, with the ventilating equipment interlocked with the electrical equipment, may be designated _____ locations by the authority having jurisdiction.

   (a) Class I, Division 2
   (b) unclassified
   (c) Class II, Division 2
   (d) Class II, Division 1

4. Inserts for cellular metal floor raceways shall be leveled to the floor grade and sealed against the entrance of _____.

   (a) concrete
   (b) water
   (c) moisture
   (d) all of these

5. An auxiliary gutter is designed to contain _____.

   (a) conductors
   (b) overcurrent devices
   (c) busways
   (d) none of these

6. A motor for general use shall be marked with a time rating of _____.

   (a) continuous
   (b) 5 or 15 minutes
   (c) 30 or 60 minutes
   (d) any of these

7. A hermetic motor-compressor controller shall have a _____ current rating not less than the respective nameplate rating(s) on the compressor.

   (a) continuous-duty full-load
   (b) locked-rotor
   (c) a or b
   (d) a and b

8. The difference in construction between Type NM cable and Type NMC cable is that Type NMC cable is _____, which Type NM is not.

   (a) corrosion resistant
   (b) flame-retardant
   (c) fungus-resistant
   (d) a and c

9. Exposed elements of impedance heating systems shall be physically guarded, isolated, or thermally insulated with a _____ jacket to protect against contact by personnel in the area.

   (a) corrosion resistant
   (b) waterproof
   (c) weatherproof
   (d) flame-retardant

10. Provisions shall be made for sufficient diffusion and ventilation of the gases from a storage battery to prevent the accumulation of a(n) _____ mixture.

    (a) corrosive
    (b) explosive
    (c) toxic
    (d) all of these

11. The top shield installed over all floor-mounted Type FCC cable shall completely _____ all cable runs, corners, connectors, and ends.

    (a) cover
    (b) encase
    (c) protect
    (d) none of these

12. The conductor insulation of Type MI cable shall be a highly compressed refractory mineral that provides proper _____ for all conductors.

    (a) covering
    (b) spacing
    (c) resistance
    (d) none of these

13. A(n) _____ shall be considered equivalent to an overcurrent trip unit for the purpose of providing overcurrent protection of conductors.

    (a) current transformer
    (b) overcurrent relay
    (c) a and b
    (d) a or b

14. Letters used to designate the number of conductors within a cable are _____.

    (a) D—Two insulated conductors laid parallel
    (b) M—Two or more insulated conductors twisted spirally
    (c) T—Two or more insulated conductors twisted in parallel
    (d) a and b

15. Bends in Type MI cable shall be made so that the cable will not be _____.

    (a) damaged
    (b) shortened
    (c) a and b
    (d) none of these

16. Audible and visual signal devices shall be provided on optional standby systems to indicate _____, where practicable.

    (a) derangement of the optional standby source
    (b) that the optional standby source is carrying load
    (c) that the battery charger is not functioning
    (d) a and b

17. Audible and visual signal devices shall be provided on legally required standby systems, where practicable, to indicate _____.

    (a) derangement of the standby source
    (b) that the standby source is carrying load
    (c) that the battery charger is not functioning
    (d) all of these

18. Each run of nonmetallic extension shall terminate in a fitting that covers the _____.

    (a) device
    (b) box
    (c) end of the extension
    (d) end of the assembly

19. When an outlet is _____ from a cellular concrete floor raceway, the sections of circuit conductors supplying the outlet shall be removed from the raceway.

    (a) discontinued
    (b) abandoned
    (c) removed
    (d) any of these

20. An oil switch can be used as both the motor controller and disconnecting means on a circuit whose rating _____ or 100A.

    (a) does not exceed 240V
    (b) does not exceed 300V
    (c) does not exceed 600V
    (d) exceeds 600V

21. Switchboards that have any exposed live parts shall be installed in permanently _____ locations, and then only where under competent supervision and accessible only to qualified persons.

    (a) dry
    (b) mounted
    (c) supported
    (d) all of these

22. An enclosure or piece of equipment constructed so that dust will not enter the enclosure under specified test conditions is known as _____.

    (a) dusttight
    (b) dustproof
    (c) dust rated
    (d) all of these

23. Edison-base fuseholders shall be used only if they are made to accept _____ fuses by the use of adapters.

    (a) Edison-base
    (b) medium-base
    (c) heavy-duty base
    (d) Type S

24. Bonding shall be provided where necessary to ensure _____ and the capacity to conduct safely any fault current likely to be imposed.

    (a) electrical continuity
    (b) fiduciary responsibility
    (c) listing requirements
    (d) electrical demand

25. The power supply to contact conductors of a crane in Class III locations shall be _____.

    (a) electrically isolated from all other systems
    (b) equipped with an acceptable ground detector
    (c) have an alarm in the case of a ground fault
    (d) all of these

26. Utilization equipment is equipment that utilizes electricity for _____ purposes.

    (a) electromechanical
    (b) heating
    (c) lighting
    (d) any of these

27. If a(n) _____ shutdown is necessary to reduce hazards to persons, the overload sensing devices can be connected to a supervised alarm instead of causing immediate interruption of the motor circuit.

    (a) emergency
    (b) normal
    (c) orderly
    (d) none of these

28. Extensions from flat cable assemblies shall be made by approved wiring methods, within the _____, installed at either end of the flat cable assembly runs.

    (a) end-caps
    (b) junction boxes
    (c) surface metal raceway
    (d) underfloor metal raceway

29. Each elevator shall have a means for disconnecting all ungrounded main power-supply conductors for each unit _____.

    (a) excluding the emergency power system
    (b) including the emergency or standby power system
    (c) excluding the emergency power system if it is automatic
    (d) and the power supply may not be an emergency power system

30. Which of the following shall not be connected to the life safety branch in a hospital?

    (a) exit signs
    (b) elevators
    (c) general illumination
    (d) communications systems

31. Network-powered broadband communications cables installed _____ on the surface of ceilings and walls shall be supported by the structural components of the building in such a manner that the network-powered broadband communications cables will not be damaged by normal building use.

    (a) exposed
    (b) concealed
    (c) hidden
    (d) a and b

32. Snap switches installed in boxes shall have the _____ seated against the finished wall surface.

    (a) extension plaster ears
    (b) body
    (c) toggle
    (d) all of these

33. Feeders to floating buildings can be installed with _____ where flexibility is required.

    (a) extra-hard usage portable power cable listed for both wet locations and sunlight resistance
    (b) LFMC with approved fittings
    (c) LFNC with approved fittings
    (d) all of these

34. Each Type FCC transition assembly shall incorporate means for _____.

    (a) facilitating the entry of the Type FCC cable into the assembly
    (b) connecting the Type FCC cable to the grounded conductors
    (c) electrically connecting the assembly to the metal cable shields and equipment grounding conductors
    (d) all of these

35. Nonmetallic cable trays shall be made of _____ material.

    (a) fire-resistant
    (b) waterproof
    (c) corrosive
    (d) flame-retardant

36. _____ identified for use on lighting track shall be designed specifically for the track on which they are to be installed.

    (a) Fittings
    (b) Receptacles
    (c) Devices
    (d) all of these

37. The radius of the curve of the inner edge of any bend shall not be less than _____ for Type AC cable.

    (a) five times the largest conductor within the cable
    (b) three times the diameter of the cable
    (c) five times the diameter of the cable
    (d) six times the outside diameter of the conductors

38. PVC conduit and fittings for use above ground shall have the following characteristics _____.

    (a) flame retardance
    (b) resistance to low temperatures and sunlight
    (c) resistance to distortion from heat
    (d) all of these

39. The overall covering of Type UF cable is _____.

    (a) flame retardant
    (b) moisture, fungus, and corrosion resistant
    (c) suitable for direct burial in the earth
    (d) all of these

40. The metallic sheath of metal-clad cable shall be continuous and _____.

    (a) flame-retardant
    (b) weatherproof
    (c) close fitting
    (d) all of these

41. Portable luminaires shall be wired with _____ recognized by 400.4, and have an attachment plug of the polarized or grounding type.

    (a) flexible cable
    (b) flexible cord
    (c) nonmetallic flexible cable
    (d) nonmetallic flexible cord

42. Metal shields for flat conductor cable shall be electrically continuous to the _____.

    (a) floor
    (b) cable
    (c) equipment grounding conductor
    (d) none of these

43. Duty on escalator and moving walk driving machine motors shall be rated as _____.

    (a) full time
    (b) continuous
    (c) various
    (d) long term

44. A(n) _____ branch circuit supplies energy to one or more outlets to which appliances are to be connected.

    (a) general purpose
    (b) multiwire
    (c) individual
    (d) appliance

45. An effective ground-fault current path is an intentionally constructed, permanent, low-impedance path designed and intended to carry fault current from the point of a ground fault on a wiring system to _____.

    (a) ground
    (b) earth
    (c) the electrical supply source
    (d) none of these

Challenge Quiz 4 — Article 90 through Chapter 9

46. The system bonding jumper is the connection between the _____ conductor and the equipment grounding conductor at a separately derived system.

    (a) grounded
    (b) ungrounded
    (c) grounding electrode
    (d) equipment bonding jumper

47. Metal boxes shall be _____ in accordance with Article 250.

    (a) grounded
    (b) bonded
    (c) a and b
    (d) none of these

48. Fixed outdoor deicing and snow-melting equipment shall be provided with a means for disconnection from all _____ conductors.

    (a) grounded
    (b) grounding
    (c) ungrounded
    (d) all of these

49. A(n) _____ is an unintentional, electrically conducting connection between an ungrounded normally current-carrying conductor of an electrical circuit, and the normally non–current-carrying conductors, metallic enclosures, metallic raceways, metallic equipment, or earth.

    (a) grounded conductor
    (b) ground fault
    (c) equipment ground
    (d) bonding jumper

50. Surrounded by a case, housing, fence, or wall(s) that prevents persons from accidentally contacting energized parts is called _____.

    (a) guarded
    (b) covered
    (c) protection
    (d) enclosed

CHALLENGE QUIZ 5: Article 90 through Chapter 9

Use the 2008 *NEC* to answer the following questions.

1. Equipment enclosed with a means of sealing or locking so that live parts cannot be made accessible without opening the enclosure is said to be _____.

    (a) guarded
    (b) protected
    (c) sealable
    (d) lockable

2. Handles or levers of circuit breakers, and similar parts that may move suddenly in such a way that persons in the vicinity are likely to be injured by being struck by them, shall be _____.

    (a) guarded
    (b) isolated
    (c) a and b
    (d) a or b

3. A hoistway is any _____ in which an elevator or dumbwaiter is designed to operate.

    (a) hatchway or well hole
    (b) vertical opening or space
    (c) shaftway
    (d) all of these

4. Each service disconnecting means shall be suitable for _____.

    (a) hazardous (classified) locations
    (b) wet locations
    (c) dry locations
    (d) the prevailing conditions

5. Electric space-heating cables shall not be installed over cabinets whose clearance from the ceiling is less than the minimum _____ dimension of the cabinet to the nearest cabinet edge that is open to the room or area.

    (a) horizontal
    (b) vertical
    (c) overall
    (d) depth

6. An atmosphere containing ignitible concentrations of _____ is classified as Group C hazardous (classified) location.

    (a) hydrogen
    (b) ethylene
    (c) gasoline
    (d) all of these

7. In Class _____ locations for Groups A, B, C, and D, the classification involves determinations of maximum explosion pressure and maximum safe clearance between parts of a clamped joint in an enclosure.

    (a) I
    (b) II
    (c) III
    (d) all of these

8. The header on a cellular concrete floor raceway shall be installed _____ to the cells.

    (a) in a straight line
    (b) at right angles
    (c) a and b
    (d) none of these

9. FMT can be used _____.

   (a) in dry and damp locations
   (b) for direct burial
   (c) in lengths over 6 ft
   (d) for a maximum of 1,000V

10. Cable trays shall _____.

    (a) include fittings for changes in direction and elevation
    (b) have side rails or equivalent structural members
    (c) be made of corrosion-resistant material or protected from corrosion as required by 300.6
    (d) all of these

11. Lampholders installed over highly combustible material shall be of the _____ type.

    (a) industrial
    (b) switched
    (c) unswitched
    (d) residential

12. Constant-voltage generators, except ac generator exciters, shall be protected from overloads by _____ or other acceptable overcurrent protective means suitable for the conditions of use.

    (a) inherent design
    (b) circuit breakers
    (c) fuses
    (d) any of these

13. Where a generator for an optional standby system is installed outdoors and equipped with a readily accessible disconnecting means located _____, an additional disconnecting means is not required where ungrounded conductors serve or pass through the building or structure.

    (a) inside the building or structure
    (b) within sight of the building or structure
    (c) inside the generator enclosure
    (d) a or c

14. Type S fuses, fuseholders, and adapters shall be designed so that _____ would be difficult.

    (a) installation
    (b) tampering
    (c) shunting
    (d) b or c

15. Varying duty is defined as _____.

    (a) intermittent operation in which the load conditions are regularly recurrent
    (b) operation at a substantially constant load for an indefinite length of time
    (c) operation for alternate intervals of load and rest, or load, no load, and rest
    (d) operation at loads, and for intervals of time, both of which may be subject to wide variations

16. An electrically driven or controlled machine with motors, used to transport and distribute water for agricultural purposes, is called a(n) _____.

    (a) irrigation machine
    (b) electric water distribution system
    (c) center pivot irrigation machine
    (d) automatic water distribution system

17. _____ means that an object is not readily accessible to persons unless special means for access are used.

    (a) Isolated
    (b) Secluded
    (c) Protected
    (d) Locked

18. Which of the following statements about Type MI cable is correct?

    (a) It may be used in any hazardous (classified) location.
    (b) It shall not be permitted for branch circuits.
    (c) It shall be securely supported at intervals not exceeding 10 ft.
    (d) none of these

19. Where single-conductor Type MI cables are used, all ungrounded conductors and, when used, the _____ conductor, shall be grouped together to minimize induced voltage on the metal sheath.

    (a) larger
    (b) neutral
    (c) grounding
    (d) largest

20. For cellular concrete floor raceways, junction boxes shall be _____ the floor grade and sealed against the free entrance of water or concrete.

    (a) leveled to
    (b) above
    (c) below
    (d) perpendicular to

21. Flat cable assemblies can supply tap devices for _____ loads.

    (a) lighting
    (b) small appliance
    (c) small power
    (d) all of these

22. If the motor control circuit transformer is located in the controller enclosure, the transformer shall be connected to the _____ side of the control circuit disconnecting means.

    (a) line
    (b) load
    (c) adjacent
    (d) none of these

23. Indoor transformers rated over 35,000V, and insulated with nonflammable dielectric fluid identified as nonflammable, shall be installed in a vault furnished with a _____.

    (a) liquid confinement area
    (b) pressure-relief vent
    (c) means for absorbing or venting any gases generated by arcing
    (d) all of these

24. Where installed to reduce electrical noise for electronic equipment, electrical continuity of the metal raceway shall not be required and the metal raceway can terminate to a(n) _____ nonmetallic fitting(s) or spacer on the electronic equipment.

    (a) listed
    (b) labeled
    (c) identified
    (d) marked

25. Insulated conductors used within a switchboard shall be _____.

    (a) listed
    (b) flame-retardant
    (c) rated for the highest voltage it may contact
    (d) all of these

26. Flexible conductors used to supply portable stage equipment shall be _____ cords or cables.

    (a) listed
    (b) extra-hard usage
    (c) hard usage
    (d) a and b

27. Connections from cellular concrete floor raceway headers to cabinets shall be made by means of _____.

    (a) listed metal raceways
    (b) PVC raceways
    (c) listed fittings
    (d) a and c

28. To guard live parts over 50V but less than 600V, equipment is permitted to be _____.

    (a) located in a room accessible only to qualified persons
    (b) located on a balcony accessible only to qualified persons
    (c) elevated 8 ft or more above the floor or other working surface
    (d) any of these

29. Manual switches controlling emergency circuits shall be convenient to authorized persons responsible for their _____.

    (a) maintenance
    (b) actuation
    (c) inspection
    (d) evaluation

30. Plans, specifications, and other building details for construction of manufactured buildings are included in the _____ details.

    (a) manufactured building
    (b) building component
    (c) building structure
    (d) building system

31. The purpose or use of panelboard circuits and circuit _____, including spare positions, shall be legibly identified on a circuit directory located on the face or inside of the door of a panelboard, and at each switch on a switchboard.

    (a) manufacturers
    (b) conductors
    (c) feeders
    (d) modifications

32. Attachment plugs and cord connectors shall be listed and marked with the _____.

    (a) manufacturer's name or identification
    (b) voltage rating
    (c) amperage rating
    (d) all of these

Challenge Quiz 5 — Article 90 through Chapter 9

33. Wooden plugs driven into holes in _____ or similar materials shall not be used for securing electrical equipment.

    (a) masonry
    (b) concrete
    (c) plaster
    (d) all of these

34. Bare aluminum or copper-clad aluminum grounding electrode conductors shall not be used where in direct contact with _____ or where subject to corrosive conditions.

    (a) masonry or the earth
    (b) bare copper conductors
    (c) wooden framing members
    (d) all of these

35. When an outlet from an underfloor raceway is discontinued, the circuit conductors supplying the outlet _____.

    (a) may be spliced
    (b) may be reinsulated
    (c) may be cut and capped off
    (d) shall be removed from the raceway

36. Auxiliary gutters shall be constructed and installed so that adequate _____ continuity of the complete system is secured.

    (a) mechanical
    (b) electrical
    (c) a or b
    (d) a and b

37. Dimensions of Type S fuses, fuseholders, and adapters shall be standardized to permit interchangeability regardless of the _____.

    (a) model
    (b) manufacturer
    (c) amperage
    (d) voltage

38. HDPE conduit shall be resistant to _____.

    (a) moisture
    (b) chemical atmospheres
    (c) impact and crushing
    (d) all of these

39. Insulated conductors used in wet locations shall be _____.

    (a) moisture-impervious metal sheathed
    (b) RHW, TW, THW, THHW, THWN, or XHHW
    (c) listed for wet locations
    (d) any of these

40. Receptacles of dc plugging boxes shall be rated at not _____ when used on a stage or set of a motion picture studio.

    (a) more than 30A
    (b) less than 20A
    (c) less than 30A
    (d) more than 20A

41. _____ can be installed in messenger-supported wiring.

    (a) Multiconductor service-entrance cable
    (b) Mineral Insulated (Type MI) cable
    (c) Multiconductor underground feeder cable
    (d) all of these

42. In cellular concrete floor raceways, a grounding conductor shall connect the insert receptacle to a _____.

    (a) negative ground connection provided in the raceway
    (b) negative ground connection provided on the header
    (c) positive ground connection provided on the header
    (d) grounded terminal located within the insert

43. Type _____ cable is a factory assembly of insulated conductors under a nonmetallic sheath for installation in cable trays or raceways.

    (a) NM
    (b) TC
    (c) SE
    (d) UF

44. Fixed electric heating equipment for pipelines and vessels shall be considered a(n) _____ load.

    (a) noncontinuous
    (b) insignificant
    (c) continuous
    (d) proprietary

45. A conductor encased within material of composition or thickness that is not recognized by the *NEC* as electrical insulation is considered _____.

    (a) noninsulating
    (b) bare
    (c) covered
    (d) protected

46. Motor overload protection shall not be shunted or cut out during the starting period if the motor is _____.

    (a) not automatically started
    (b) automatically started
    (c) manually started
    (d) none of these

47. Use of Type FCC systems in damp locations is _____.

    (a) not permitted
    (b) permitted
    (c) permitted provided the system is encased in concrete
    (d) permitted by special permission only

48. An isolating switch is one that is _____.

    (a) not readily accessible to persons unless special means for access are used
    (b) capable of interrupting the maximum operating overload current of a motor
    (c) intended for use in general distribution and branch circuits
    (d) intended for isolating an electrical circuit from the source of power

49. Conductors and busbars on a switchboard or panelboard shall be located so as to be free from _____ and shall be held firmly in place.

    (a) obstructions
    (b) physical damage
    (c) a and b
    (d) none of these

50. Plug fuses of 15A or less shall be identified by a(n) _____ configuration of the window, cap, or other prominent part to distinguish them from fuses of higher ampere ratings.

    (a) octagonal
    (b) rectangular
    (c) hexagonal
    (d) triangular

CHALLENGE QUIZ 6: Article 90 through Chapter 9

Use the 2008 *NEC* to answer the following questions.

1. Type FCC cable shall be clearly and durably marked _____.

   (a) on the top side at intervals not exceeding 30 in.
   (b) on both sides at intervals not exceeding 24 in.
   (c) with conductor material, maximum temperature, and ampacity
   (d) b and c

2. Open motors having commutators or collector rings shall be located or protected so that sparks cannot reach adjacent combustible material. This shall not prohibit the installation of these motors _____.

   (a) on wooden floors
   (b) over combustible fiber
   (c) under combustible material
   (d) none of these

3. No more than _____ layers of flat conductor cable can cross at any one point.

   (a) one
   (b) two
   (c) three
   (d) four

4. A capacitor operating at over 600V shall be provided with means to reduce the residual voltage to 50V or less within _____ after it is disconnected from the source of supply.

   (a) one minute
   (b) three minutes
   (c) five minutes
   (d) seven minutes

5. The service disconnecting means shall plainly indicate whether it is in the _____ position.

   (a) open or closed
   (b) tripped
   (c) up or down
   (d) correct

6. An exposed wiring support system using a messenger wire to support insulated conductors is known as _____.

   (a) open wiring
   (b) messenger-supported wiring
   (c) field wiring
   (d) none of these

7. Overload relays and other devices for motor overload protection that are not capable of _____ shall be protected by fuses, circuit breakers, or motor short-circuit protector.

   (a) opening short circuits
   (b) clearing overloads
   (c) opening ground faults
   (d) a or c

8. In anesthetizing locations of health care facilities, low-voltage equipment in frequent contact with persons shall _____.

   (a) operate on an electrical potential of 10V or less
   (b) be moisture resistant
   (c) be intrinsically safe or double-insulated
   (d) any of these

9. Conductors within flexible cords and cables shall not be associated together in such a way that the _____ temperature of the conductors is exceeded.

   (a) operating
   (b) governing
   (c) ambient
   (d) limiting

10. An apparatus that is capable of withstanding an explosion of a specified gas or vapor that may occur within it, and of preventing the ignition of a specified gas or vapor surrounding the enclosure by sparks, flashes, or explosion of the gas or vapor within, and that operates at such an external temperature that a surrounding flammable atmosphere will not be ignited thereby defines the term _____.

    (a) overcurrent device
    (b) thermal apparatus
    (c) explosionproof apparatus
    (d) bomb casing

11. Each electric appliance shall be provided with a(n) _____ giving the identifying name and the rating in volts and amperes, or in volts and watts.

    (a) pamphlet
    (b) nameplate
    (c) auxiliary statement
    (d) owner's manual

12. According to Article 625, automotive-type vehicles for on-road use include _____.

    (a) passenger automobiles
    (b) trucks
    (c) neighborhood electric vehicles
    (d) all of these

13. If the transfer switch for a portable generator switches the _____ conductor, then it is being used as a separately derived system and the portable generator shall be grounded in accordance with 250.30.

    (a) phase
    (b) equipment grounding
    (c) grounded
    (d) all of these

14. Cablebus is ordinarily assembled at the _____.

    (a) point of installation
    (b) manufacturer's location
    (c) distributor's location
    (d) none of these

15. X-ray equipment mounted on a permanent base equipped with wheels for moving while completely assembled describes _____.

    (a) portable
    (b) mobile
    (c) movable
    (d) room

16. A fuel cell is an electrochemical system that consumes _____ to produce an electric current.

    (a) power
    (b) water
    (c) heat
    (d) fuel

17. Network-powered broadband communications system cables shall be separated at least 2 in. from conductors of _____ circuits.

    (a) power
    (b) electric light
    (c) Class 1
    (d) any of these

18. Grounding and bonding connection devices shall not be connected by _____.

    (a) pressure connections
    (b) solder
    (c) lugs
    (d) approved clamps

19. The service conductors shall be connected to the service disconnecting means by _____ or other approved means.

    (a) pressure connectors
    (b) clamps
    (c) solder
    (d) a or b

Challenge Quiz 6 — Article 90 through Chapter 9

20. Each vented cell of a battery, as it relates to storage batteries, shall be equipped with _____ that is(are) designed to prevent destruction of the cell due to ignition of gases within the cell by an external spark or flame under normal operating conditions.

    (a) pressure relief
    (b) a flame arrester
    (c) fluid level indicators
    (d) none of these

21. Sealing compound shall be used in Type MI cable termination stings to _____.

    (a) preventing the passage of gas or vapor
    (b) excluding moisture and other fluids from the cable insulation
    (c) limiting a possible explosion
    (d) preventing the escape of powder

22. The phase converter disconnecting means shall be _____ and located in sight from the phase converter.

    (a) protected from physical damage
    (b) readily accessible
    (c) easily visible
    (d) clearly identified

23. Electric space-heating cables shall not extend beyond the room or area in which they _____.

    (a) provide heat
    (b) originate
    (c) terminate
    (d) are connected

24. Constructed, protected, or treated so as to prevent rain from interfering with the successful operation of the apparatus under specified test conditions defines the term _____.

    (a) raintight
    (b) waterproof
    (c) weathertight
    (d) rainproof

25. Receptacles, receptacle housings, and self-contained devices used with flat conductor cable systems shall be _____.

    (a) rated a minimum of 20A
    (b) rated a minimum of 15A
    (c) identified for this use
    (d) none of these

26. Wiring on luminaire chains and other movable parts shall be _____.

    (a) rated for 110°C
    (b) stranded
    (c) hard-usage rated
    (d) none of these

27. When service equipment has ground-fault protection installed, it may be necessary to review the overall wiring system for proper selective overcurrent protection _____.

    (a) rating
    (b) coordination
    (c) devices
    (d) none of these

28. In mobile/manufactured homes, portable appliances could be _____, if these appliances can be moved from one place to another in normal use.

    (a) refrigerators
    (b) range equipment
    (c) clothes washers
    (d) all of these

29. Fixed electric space-heating equipment shall be installed to provide the _____ spacing between the equipment and adjacent combustible material, unless it is listed for direct contact with combustible material.

    (a) required
    (b) minimum
    (c) maximum
    (d) safest

30. Flat conductor cable shall not be installed in _____.

    (a) residential buildings
    (b) schools
    (c) hospitals
    (d) any of these

31. An attachment plug and receptacle can serve as the disconnecting means for cord-connected _____.

    (a) room air conditioners
    (b) household refrigerators and freezers
    (c) drinking water coolers and beverage dispensers
    (d) all of these

32. When devices or plug-in connections for tapping off feeders or branch circuits from busway include an externally operable fusible switch that is out of reach, _____ shall be provided for operation of the disconnecting means from the floor.

    (a) ropes
    (b) chains
    (c) hook sticks
    (d) any of these

33. Type _____ is a multiconductor cable identified for use as underground service-entrance cable.

    (a) SE
    (b) NM
    (c) UF
    (d) USE

34. A conductor used for open wiring on insulators that penetrates a wall, floor, or other framing member shall be carried through a _____.

    (a) separate sleeve or tube
    (b) weatherproof tube
    (c) tube of absorbent material
    (d) grounded metallic tube

35. The cablebus assembly is designed to carry _____ current and to withstand the magnetic forces of such current.

    (a) service
    (b) load
    (c) fault
    (d) grounded

36. A branch-circuit overcurrent device is capable of providing protection of _____.

    (a) service conductors
    (b) feeders
    (c) branch circuits
    (d) all of these

37. The distribution point is an electrical supply point from which _____ to agricultural buildings or structures under single management are supplied.

    (a) service drops or service conductors
    (b) feeders or branch circuits
    (c) a or b
    (d) none of these

38. Loop wiring in an underfloor raceway _____ to be a splice or tap.

    (a) shall be considered
    (b) shall not be considered
    (c) shall not be permitted
    (d) none of these

39. Radiant heating cables embedded in or below a pool deck _____.

    (a) shall not be installed within 5 ft horizontally from the inside walls of the pool
    (b) shall be mounted at least 12 ft vertically above the pool deck
    (c) shall not be permitted
    (d) none of these

40. Loop wiring _____ in a cellular metal raceway.

    (a) shall not be permitted
    (b) shall not be considered a splice or tap
    (c) shall be considered a splice or tap when used
    (d) none of these

41. A motor _____ device that can restart a motor automatically after overload tripping shall not be installed if automatic restarting of the motor can result in injury to persons.

    (a) short-circuit
    (b) ground-fault
    (c) overcurrent
    (d) overload

42. The radius of the inner edge of any bend in Type MI cable shall not be less than _____ times the external diameter of the metallic sheath for any cable having a diameter greater than ¾ in., but not more than 1 in.

    (a) three
    (b) six
    (c) eight
    (d) 10

43. When nails are used to mount knobs for the support of open wiring on insulators, they shall not be smaller than _____-penny.

    (a) six
    (b) eight
    (c) ten
    (d) twelve

Challenge Quiz 6 — Article 90 through Chapter 9

44. Bends in Type ITC cable shall be made _____.

    (a) so as not to exceed 45 degrees
    (b) so as not to damage the cable
    (c) not less than five times the diameter of the cable
    (d) using listed bending tools

45. The combination of all components and subsystems that convert solar energy into electrical energy is called a _____ system.

    (a) solar
    (b) solar voltaic
    (c) separately derived source
    (d) solar photovoltaic

46. Type MI cable conductors shall be made of _____, nickel, or nickel-coated copper with a resistance corresponding to standard AWG and kcmil sizes.

    (a) solid copper
    (b) solid or stranded copper
    (c) stranded copper
    (d) solid copper or aluminum

47. Conductors for nonpower-limited fire alarm circuits shall be _____.

    (a) solid copper
    (b) stranded copper
    (c) copper or aluminum
    (d) a or b

48. Switchboards shall be placed so as to reduce to a minimum the probability of communicating _____ to adjacent combustible materials.

    (a) sparks
    (b) backfeed
    (c) fire
    (d) all of these

49. Flat cable assemblies shall not be installed outdoors or in wet or damp locations unless _____ for the use.

    (a) special permission is granted
    (b) approved
    (c) identified
    (d) none of these

50. The messenger of messenger-supported wiring shall be supported at dead ends and at intermediate locations so as to eliminate _____ on the circuit conductors.

    (a) static
    (b) magnetism
    (c) tension
    (d) induction

Article 90 through Chapter 9

CHALLENGE QUIZ 7

Use the 2008 *NEC* to answer the following questions.

1. Where motors are provided with terminal housings, the housings shall be of _____ and of substantial construction.

   (a) steel
   (b) iron
   (c) metal
   (d) copper

2. Type FCC cable systems are permitted on wall surfaces in _____.

   (a) surface metal raceways
   (b) cable trays
   (c) busways
   (d) any of these

3. A large single panel, frame, or assembly of panels on which switches, overcurrent and other protective devices, buses, and instruments are mounted is a _____. They are generally accessible from the rear as well as from the front and are not intended to be installed in cabinets.

   (a) switchboard
   (b) panel box
   (c) switch box
   (d) panelboard

4. Conductors with thermoplastic and fibrous outer braid, such as TBS insulation, are used for wiring _____.

   (a) switchboards only
   (b) in a dry location
   (c) in a wet location
   (d) luminaires

5. Luminaires shall be wired with conductors having insulation suitable for the environmental conditions and _____ to which the conductors will be subjected.

   (a) temperature
   (b) voltage
   (c) current
   (d) all of these

6. Open wiring on insulators is a(n) _____ wiring method using cleats, knobs, tubes, and flexible tubing for the protection and support of single insulated conductors run in or on buildings.

   (a) temporary
   (b) acceptable
   (c) enclosed
   (d) exposed

7. Performance area includes the stage and audience seating area associated with a _____ stage structure, whether indoors or outdoors.

   (a) temporary
   (b) permanent
   (c) a or b
   (d) a and b

8. Overcurrent devices for nonpower-limited fire alarm circuits shall be located at the point where the conductor to be protected _____.

   (a) terminates at the load
   (b) is spliced to any other conductor
   (c) receives its supply
   (d) none of these

Challenge Quiz 7 — Article 90 through Chapter 9

9. Overcurrent devices for Class 1 circuits shall be located at the point where the conductor to be protected _____.

    (a) terminates to the load
    (b) is spliced to any other conductor
    (c) receives its supply
    (d) none of these

10. Underfloor raceways shall be laid so that a straight line from the center of one _____ to the center of the next _____ coincides with the centerline of the raceway system.

    (a) termination point, termination point
    (b) junction box, junction box
    (c) receptacle, receptacle
    (d) panelboard, panelboard

11. Each run of cable tray shall be _____ before the installation of cables.

    (a) tested for 25 ohms resistance
    (b) insulated
    (c) completed
    (d) all of these

12. Type ITC cable shall be installed in industrial establishments where the conditions of maintenance and supervision ensure that only _____ will service the installation.

    (a) the authority having jurisdiction
    (b) authorized persons
    (c) the general public
    (d) qualified persons

13. Service-entrance conductors can be spliced or tapped by clamped or bolted connections at any time as long as _____.

    (a) the free ends of conductors are covered with an insulation that is equivalent to that of the conductors or with an insulating device identified for the purpose
    (b) wire connectors or other splicing means installed on conductors that are buried in the earth are listed for direct burial
    (c) no splice is made in a raceway
    (d) all of these

14. Ferrous metal raceways, boxes, fittings, supports, and support hardware can be installed in concrete or in direct contact with the earth or other areas subject to severe corrosive influences, where _____ approved for the conditions.

    (a) the soil is
    (b) made of material
    (c) the qualified installer is
    (d) none of these

15. Where solid knobs are used for concealed knob and tube wiring, conductors shall be securely tied to them by _____ equivalent to that of the conductor.

    (a) tie wires having insulation
    (b) conductors having an AWG
    (c) nonconductive material
    (d) none of these

16. Which of the following appliances installed in residential occupancies need not be connected to an equipment grounding conductor?

    (a) toaster
    (b) aquarium
    (c) dishwasher
    (d) refrigerator

17. Circuit breakers shall be marked with their ampere rating in a manner that is durable and visible after installation. Such marking can be made visible by removal of a _____.

    (a) trim
    (b) cover
    (c) box
    (d) a or b

18. A raintight enclosure is constructed or protected so that exposure to a beating rain will not result in the entrance of water under specified test conditions.

    (a) True
    (b) False

19. A thermal protector may consist of one or more sensing elements integral with the motor or motor-compressor and an external control device.

    (a) True
    (b) False

20. Soldered splices shall first be spliced or joined so as to be mechanically and electrically secure without solder and then be soldered.

    (a) True
    (b) False

21. Currents that introduce noise or data errors in electronic equipment are considered objectionable currents.

    (a) True
    (b) False

22. Metal enclosures and raceways containing service conductors shall be connected to the grounded system conductor if the electrical system is grounded.

    (a) True
    (b) False

23. A bushing shall be permitted in lieu of a box or terminal where the conductors emerge from a raceway and enter or terminate at equipment such as open switchboards, unenclosed control equipment, or similar equipment.

    (a) True
    (b) False

24. Prewired raceway assemblies shall be used only where specifically permitted in the *NEC* for the applicable wiring method.

    (a) True
    (b) False

25. Conduit bodies containing conductors larger than 6 AWG shall have a cross-sectional area at least twice that of the largest conduit to which they can be attached.

    (a) True
    (b) False

26. Boxes and conduit bodies, covers, extension rings, plaster rings, and the like shall be durably and legibly marked with the manufacturer's name or trademark.

    (a) True
    (b) False

27. Type FC cable is an assembly of parallel conductors with an insulating material specifically designed for field installation in surface metal raceways.

    (a) True
    (b) False

28. Type FCC systems shall be permitted for individual branch circuits.

    (a) True
    (b) False

29. Where the outer sheath of Type MI cable is made of copper, it shall provide an adequate path to serve as an equipment grounding conductor.

    (a) True
    (b) False

30. A cablebus system shall include approved fittings for dead ends.

    (a) True
    (b) False

31. When installing duct heaters, sufficient clearance shall be maintained to permit replacement and adjustment of controls and heating elements.

    (a) True
    (b) False

32. Ground-fault protection of equipment shall be provided for electric heat tracing and heating panels installed on pipelines or vessels except in certain industrial installations where there is alarm indication of ground faults.

    (a) True
    (b) False

33. For a skin-effect heating installation complying with Article 427, the provisions of 300.20 shall apply to the installation of a single conductor in a ferromagnetic envelope (metal enclosure).

    (a) True
    (b) False

34. A branch-circuit overcurrent device can serve as the disconnecting means for a stationary motor of $\frac{1}{8}$ hp or less.

    (a) True
    (b) False

35. Equipment in an area containing ignitible concentrations of acetylene under normal operating conditions shall be classified as a Class I, Group A location.

    (a) True
    (b) False

36. An atmosphere classified as hazardous Group E contains combustible metal dusts.

    (a) True
    (b) False

37. An atmosphere containing carbon black, charcoal, coal, or coke dusts that have been sensitized by other materials so they present an explosion hazard is classified as Group F.

    (a) True
    (b) False

38. An atmosphere classified as hazardous Group G contains combustible dusts such as flour, grain, wood, plastic, and chemicals.

    (a) True
    (b) False

39. In Class I, Division 2 locations, fused or unfused disconnect and isolating switches for transformers or capacitor banks that are not intended to interrupt current in normal performance can be installed in general-purpose enclosures.

    (a) True
    (b) False

40. Seals in Class II hazardous (classified) locations shall be explosionproof.

    (a) True
    (b) False

41. Multiwire branch circuits shall be permitted in a Class I, Zone 1 location if all ungrounded conductors of the circuit are opened simultaneously.

    (a) True
    (b) False

42. Receptacles or their cover plates, supplied from the emergency system in hospitals shall have a distinctive color or marking so as to be readily identifiable.

    (a) True
    (b) False

43. Receptacles or their cover plates, supplied from the emergency system in nursing homes shall have a distinctive color or marking so as to be readily identifiable.

    (a) True
    (b) False

44. Theater fixed stage switchboards that are not completely enclosed, dead-front and dead-rear or recessed into a wall, shall be provided with a metal hood extending the full length of the board to protect all equipment on the board from falling objects.

    (a) True
    (b) False

45. Lights and receptacles installed in theater dressing rooms adjacent to the mirrors and above the dressing table counter, shall be controlled by wall switches in the dressing rooms.

    (a) True
    (b) False

46. Switches not required for projectors, flood or other special effect lamps shall not be installed in projection rooms.

    (a) True
    (b) False

47. The *NEC* specifies wiring methods for manufactured buildings.

    (a) True
    (b) False

48. Aluminum conductors and copper-clad aluminum conductors shall be permitted for branch-circuit wiring in mobile homes.

    (a) True
    (b) False

49. Wiring systems for office furnishings shall be identified as suitable for providing power for lighting accessories and appliances in wired partitions.

    (a) True
    (b) False

50. An electric vehicle connector is a device that, by insertion into an electric vehicle inlet, establishes an electrical connection to the electric vehicle for the purpose of charging and information exchange.

    (a) True
    (b) False

FINAL EXAM 1

Practice Questions in Random Order—Article 90 through Annex C

Use the 2008 *NEC* to answer the following questions.

1. Where the circuit breaker handles are operated vertically, the "up" position of the handle shall be the _____.

   (a) "on" position
   (b) "off" position
   (c) tripped position
   (d) any of these

2. A dwelling unit containing two 120V laundry branch circuits has a calculated load of _____VA for the laundry circuits.

   (a) 1,500
   (b) 3,000
   (c) 4,500
   (d) 6,000

3. Overhead feeder conductors shall have a minimum _____ vertical clearance over residential property and driveways, as well as those commercial areas not subject to truck traffic, where the voltage is limited to 300 volts-to-ground.

   (a) 10 ft
   (b) 12 ft
   (c) 15 ft
   (d) 18 ft

4. Service-drop conductors shall have a minimum of _____ vertical clearance from final grade over residential property and driveways, as well as over commercial areas not subject to truck traffic where the voltage is limited to 300 volts-to-ground.

   (a) 10 ft
   (b) 12 ft
   (c) 15 ft
   (d) 18 ft

5. The feeder conductor ampacity shall not be less than that of the service-entrance conductors where the feeder conductors carry the total load supplied by service conductors with an ampacity of _____ or less.

   (a) 30A
   (b) 55A
   (c) 60A
   (d) 100A

6. An equipment bonding jumper can be installed on the outside of a raceway, providing the length of the equipment bonding jumper is not more than _____ and the equipment bonding jumper is routed with the raceway.

   (a) 12 in.
   (b) 24 in.
   (c) 36 in.
   (d) 72 in.

7. The mobile home park secondary electrical distribution system to mobile home lots shall be _____.

   (a) 120/208V, single-phase
   (b) 120/240V single-phase
   (c) a or b
   (d) none of these

8. Branch-circuit conductors supplying a single continuous-duty motor shall have an ampacity not less than _____ rating.

   (a) 125 percent of the motor's nameplate current
   (b) 125 percent of the motor's full-load current as determined by 430.6(A)(1)
   (c) 125 percent of the motor's full locked-rotor
   (d) 80 percent of the motor's full-load current

9. A 3-conductor SJE cable (one conductor is used for grounding) has a maximum ampacity of _____ for each 16 AWG conductor.

(a) 9A
(b) 11A
(c) 13A
(d) 15A

10. Receptacles rated _____ or less directly connected to aluminum conductors shall be listed and marked CO/ALR.

(a) 15A
(b) 20A
(c) 25A
(d) 30A

11. _____, 125 and 250V receptacles installed in a wet location shall have an enclosure that is weatherproof whether or not the attachment plug cap is inserted.

(a) 15A
(b) 20A
(c) a and b
(d) none of these

12. EMT can be unsupported between enclosures when the raceways contain no couplings, oversized knockouts are not encountered, and the raceway does not exceed _____ in length.

(a) 18 in
(b) 24 in
(c) 36 in
(d) 42 in

13. A minimum depth of _____ working space to live parts operating at 277 volts-to-ground is required where there are exposed live parts on one side and no live or grounded parts on the other side.

(a) 2 ft
(b) 3 ft
(c) 4 ft
(d) 6 ft

14. For permanently connected appliances rated over _____ or 1/8 hp, the branch-circuit circuit breaker can serve as the disconnecting means where the circuit breaker is within sight from the appliance, or is capable of being locked in the open position with a permanently installed locking provision.

(a) 200 VA
(b) 300 VA
(c) 400 VA
(d) 500 VA

15. Optical fiber cables are not required to be listed and marked where the length of the cable within the building, measured from its point of entrance, does not exceed _____ and the cable enters the building from the outside and is terminated in an enclosure.

(a) 25 ft
(b) 30 ft
(c) 50 ft
(d) 100 ft

16. The grounded conductor brought to service equipment shall be routed with the phase conductors and shall not be smaller than specified in Table _____ when the service-entrance conductors are not larger than 1,100 kcmil copper.

(a) 250.66
(b) 250.122
(c) 310.16
(d) 430.52

17. Type MC cable installed horizontally through wooden or metal framing members are considered secured and supported where such support doesn't exceed _____ intervals.

(a) 3 ft
(b) 4 ft
(c) 6 ft
(d) 8 ft

18. Switching devices shall be at least _____ ft horizontally from the inside walls of a pool unless the switch is listed as being acceptable for use within 5 ft.

(a) 3 ft
(b) 5 ft
(c) 10 ft
(d) 12 ft

19. Surface-mounted fluorescent luminaires in clothes closets shall be permitted on the wall above the door, or on the ceiling, provided there is a minimum clearance of _____ between the luminaire and the nearest point of a storage space.

    (a) 3 in.
    (b) 6 in.
    (c) 9 in.
    (d) 12 in.

20. The voltage between conductors in a surface metal raceway shall not exceed _____ unless the metal has a thickness of not less than 0.040 in. nominal.

    (a) 150V
    (b) 300V
    (c) 600V
    (d) 1,000V

21. Lighting track shall have two supports for a single section of _____ or shorter in length and each individual section of not more than 4 ft attached to it shall have one additional support.

    (a) 2 ft
    (b) 4 ft
    (c) 6 ft
    (d) 8 ft

22. Where run horizontally, nonmetallic wireways shall be supported at each end, at intervals not to exceed _____, and at each joint, unless listed for other support intervals.

    (a) 3 ft
    (b) 6 ft
    (c) 8 ft
    (d) 10 ft

23. Receptacle outlets in floors shall not be counted as part of the required number of receptacle outlets to service dwelling unit wall spaces unless they are located within _____ of the wall.

    (a) 6 in.
    (b) 12 in.
    (c) 18 in.
    (d) 24 in.

24. CATV coaxial cable can deliver power to equipment that is directly associated with the radio frequency distribution system if voltage is not over _____ volts and if the current supply is from a transformer or other power-limiting device.

    (a) 60
    (b) 120
    (c) 180
    (d) 270

25. Where raceways contain insulated circuit conductors _____ AWG and larger, the conductors shall be protected from abrasion during and after installation by a fitting that provides a smooth, rounded insulating surface.

    (a) 8
    (b) 6
    (c) 4
    (d) 2

26. Each resistance welder shall have an overcurrent device rated or set at not more than _____ percent of the rated primary current of the welder.

    (a) 80
    (b) 100
    (c) 125
    (d) 300

27. Separate data processing units can be interconnected by means of listed cables and cable assemblies. Where exposed to physical damage, the installation shall be protected by _____.

    (a) a GFCI circuit breaker
    (b) approved means
    (c) permanent floor coverings
    (d) metal raceways

28. Tap connections to the common grounding electrode conductor for multiple separately derived systems shall be made at an accessible location by _____.

    (a) a listed connector
    (b) listed connections to aluminum or copper busbars
    (c) by the exothermic welding process
    (d) any of these

29. A separate branch circuit shall supply elevator machine room/machinery space lighting and receptacle(s). The required lighting shall not be connected to the load side of _____.

    (a) a local subpanel
    (b) an SWD-type circuit breaker
    (c) an HID-type circuit breaker
    (d) a GFCI

30. Joints between lengths of ENT, couplings, fittings, and boxes shall be made by _____.

    (a) a qualified person
    (b) set screw fittings
    (c) an approved method
    (d) exothermic welding

31. Tap conductors not over 25 ft shall be permitted, providing the _____.

    (a) ampacity of the tap conductors is not less than one-third the rating of the overcurrent device protecting the feeder conductors being tapped
    (b) tap conductors terminate in a single circuit breaker or set of fuses that limit the load to the ampacity of the tap conductors
    (c) tap conductors are suitably protected from physical damage
    (d) all of these

32. When FMC is used to install equipment where flexibility is required after installation, _____ shall be installed.

    (a) an equipment grounding conductor
    (b) an expansion fitting
    (c) flexible nonmetallic connectors
    (d) a grounded conductor one size larger

33. A contact device installed at an outlet for the connection of an attachment plug is known as a(n) _____.

    (a) attachment point
    (b) tap
    (c) receptacle
    (d) wall plug

34. Listed equipment or materials is included in a _____ published by a testing laboratory acceptable to the authority having jurisdiction.

    (a) book
    (b) digest
    (c) manifest
    (d) list

35. The ungrounded and grounded conductors of each _____ shall be grouped by wire ties or similar means in at least one location within the panelboard or other point of origination.

    (a) branch circuit
    (b) multiwire branch circuit
    (c) feeder circuit
    (d) service-entrance conductor

36. A _____ rated in amperes shall be permitted as a controller for all motors.

    (a) branch-circuit inverse time circuit breaker
    (b) molded case switch
    (c) a and b
    (d) none of these

37. Internal parts of electrical equipment, including _____, shall not be damaged or contaminated by foreign materials such as paint, plaster, cleaners, abrasives, or corrosive residues.

    (a) busbars
    (b) wiring terminals
    (c) insulators
    (d) all of these

38. The number of conductors permitted in PVC conduit shall not exceed the percentage fill specified in _____.

    (a) Chapter 9, Table 1
    (b) Table 250.66
    (c) Table 310.16
    (d) 240.6

39. Class 2 and Class 3 cables listed as suitable for general-purpose use with the exception of risers, ducts, plenums, and other spaces used for environmental air, shall be Type _____.

    (a) CL2P and CL3P
    (b) CL2R and CL3R
    (c) CL2 and CL3
    (d) PLTC

40. Areas designated as hazardous (classified) shall be properly _____ and shall be available to those authorized to design, install, inspect, maintain, or operate electrical equipment at these locations.

    (a) cleaned
    (b) documented
    (c) maintained
    (d) all of these

41. Coaxial cables used for CATV systems shall not be strapped, taped, or attached by any means to the exterior of any _____ as a means of support.

    (a) conduit
    (b) raceway
    (c) raceway-type mast intended for overhead spans of such cables
    (d) a or b

42. Raceways, cable trays, cablebus, auxiliary gutters, cable armor, boxes, cable sheathing, cabinets, elbows, couplings, fittings, supports, and support hardware shall be of materials suitable for _____.

    (a) corrosive locations
    (b) wet locations
    (c) the environment in which they are to be installed
    (d) none of these

43. Type NM cable protected from physical damage by a raceway shall not be required to be _____ within the raceway.

    (a) covered
    (b) insulated
    (c) secured
    (d) unspliced

44. Service-drop conductors and service-entrance conductors shall be arranged so that _____ will not enter the service raceway or equipment.

    (a) dust
    (b) vapor
    (c) water
    (d) none of these

45. The _____ shall conduct or witness an acceptance test of the complete emergency system upon installation and periodically afterward.

    (a) electrical engineer
    (b) authority having jurisdiction
    (c) qualified person
    (d) manufacturer's representative

46. When LFNC is used to connect equipment requiring flexibility after installation, a separate _____ shall be installed.

    (a) equipment grounding conductor
    (b) expansion fitting
    (c) flexible nonmetallic connector
    (d) none of these

47. Intrinsically safe conduit or cable runs that leave a Class I or II location shall be sealed. The seal shall be _____.

    (a) explosionproof or flameproof
    (b) flameproof
    (c) a and b
    (d) none of these

48. Hydromassage bathtubs and their associated electrical components shall be on an individual branch circuit and protected by a(n) _____ GFCI.

    (a) exposed
    (b) accessible
    (c) readily accessible
    (d) concealed

49. Communications cables installed _____ on the surface of ceilings and walls shall be supported by the building structure in such a manner that the cable will not be damaged by normal building use.

    (a) exposed
    (b) concealed
    (c) hidden
    (d) a and b

50. The cut ends of HDPE conduit shall be _____ to avoid rough edges.

    (a) filed on the inside
    (b) trimmed inside and outside
    (c) cut only with a hacksaw
    (d) all of these

51. As defined by 230.95, the rating of the service disconnect shall be considered to be the rating of the largest _____ that can be installed or the highest continuous current trip setting for which the actual overcurrent device installed in a circuit breaker is rated or can be adjusted.

    (a) fuse
    (b) circuit
    (c) conductor
    (d) all of these

52. Plug-in-type circuit breakers that are back-fed shall be _____ by an additional fastener that requires more than a pull to release.

    (a) grounded
    (b) secured in place
    (c) shunt tripped
    (d) none of these

53. A circuit conductor that is intentionally grounded is called a _____.

    (a) grounding conductor
    (b) unidentified conductor
    (c) grounded conductor
    (d) grounding electrode conductor

54. The *NEC* is _____.

    (a) intended to be a design manual
    (b) meant to be used as an instruction guide for untrained persons
    (c) for the practical safeguarding of persons and property
    (d) published by the Bureau of Standards

55. Threadless couplings and connectors used with RMC in wet locations shall be _____.

    (a) listed for wet locations
    (b) listed for damp locations
    (c) nonabsorbent
    (d) weatherproof

56. In Class 1 Division 2 locations, PVC conduit shall be permitted where _____.

    (a) metal conduit does not provide sufficient corrosion resistance
    (b) only qualified persons service the installation
    (c) installed in industrial locations
    (d) all of these

57. A building or structure shall be supplied by a maximum of _____ service(s), unless specifically permitted otherwise.

    (a) one
    (b) two
    (c) three
    (d) as many as desired

58. Dwelling units shall contain at least _____ communications outlet(s).

    (a) one
    (b) two
    (c) three
    (d) four

59. Transformer vaults shall be located where they can be ventilated to the outside air without using flues or ducts, where _____.

    (a) permitted
    (b) practicable
    (c) required
    (d) all of these

60. Conductors in ferrous metal raceways and enclosures shall be arranged so as to avoid heating the surrounding ferrous metal by alternating-current induction. To accomplish this, the _____ conductor(s) shall be grouped together.

    (a) phase
    (b) grounded
    (c) equipment grounding
    (d) all of these

61. The disconnecting means for air-conditioning and refrigerating equipment shall be _____ from the air-conditioning or refrigerating equipment.

    (a) readily accessible
    (b) within sight
    (c) a or b
    (d) a and b

62. A _____ is a device that governs, in some predetermined manner, the electric power delivered to the apparatus to which it is connected.

    (a) relay
    (b) breaker
    (c) transformer
    (d) controller

63. In Class I locations, receptacles and attachment plugs shall be of the type providing for _____ a flexible cord and shall be identified for the location.

    (a) sealing compound around
    (b) quick connection to
    (c) connection to the equipment grounding conductor of
    (d) none of these

64. Buildings or structures supplied by multiple services or feeders must use the same _____ to ground enclosures and equipment in or on that building.

    (a) service
    (b) disconnect
    (c) electrode system
    (d) any of these

Practice Questions in Random Order—Article 90 through Annex C — Final Exam 1

65. Each strap containing one or more devices shall count as a _____ volume allowance in accordance with Table 314.16(B), based on the largest conductor connected to a device(s) or equipment supported by the strap.

    (a) single
    (b) double
    (c) triple
    (d) none of these

66. Boxes can be supported from a multiconductor cord or cable, provided the conductors are protected from _____.

    (a) strain
    (b) temperature
    (c) sunlight
    (d) abrasion

67. Signs and outline lighting systems shall be marked with _____.

    (a) the manufacturer's name, trademark, or other means of identification
    (b) input voltage
    (c) current rating
    (d) all of these

68. For ungrounded systems, noncurrent-carrying conductive materials enclosing electrical conductors or equipment, or forming part of such equipment, shall be connected together and to the supply system equipment in a manner that creates a low-impedance path for ground-fault current that is capable of carrying _____.

    (a) the maximum branch-circuit current
    (b) at least twice the maximum ground-fault current
    (c) the maximum fault current likely to be imposed on it
    (d) the equivalent to the main service rating

69. In Class III locations, motors, generators, and other rotating machinery shall be _____.

    (a) totally enclosed nonventilated
    (b) totally enclosed pipe-ventilated
    (c) totally enclosed fan-cooled
    (d) any of these

70. In Class II, Division 2 locations, motors, generators, or other rotating electrical equipment shall be _____.

    (a) totally enclosed nonventilated or pipe-ventilated
    (b) totally enclosed water-air-cooled or fan-cooled
    (c) dust-ignitionproof
    (d) any of these

71. A switch constructed so that it can be installed in device boxes or on box covers, or otherwise used in conjunction with wiring systems recognized by the *NEC* is called a _____ switch.

    (a) transfer
    (b) motor-circuit
    (c) general-use snap
    (d) bypass isolation

72. The material located in the *NEC* Annexes are part of the *Code* and shall be complied with.

    (a) True
    (b) False

73. Where no statutory requirement exists, the authority having jurisdiction could be a property owner or his/her agent, such as an architect or engineer.

    (a) True
    (b) False

74. The dedicated space above a panelboard extends to a dropped or suspended ceiling, which is considered a structural ceiling.

    (a) True
    (b) False

75. An individual 20A branch circuit can supply a single dwelling unit bathroom for receptacle outlet(s) and other equipment within the same bathroom.

    (a) True
    (b) False

76. Where a portion of the dwelling unit basement is finished into one or more habitable rooms, each separate unfinished portion shall have a receptacle outlet installed.

    (a) True
    (b) False

77. Where more than one concrete encased electrode is present at a building or structure, it shall be permitted to connect to only one of them.

    (a) True
    (b) False

78. Metal enclosures and raceways for other than service conductors shall be connected to the grounded conductor.

    (a) True
    (b) False

79. Where conductors are run in parallel, the equipment grounding conductor must be installed with each parallel conductor set and it is not required to be larger than the circuit conductors.

(a) True
(b) False

80. Where used, the surge protective device shall be connected to the grounded conductor of the circuit.

(a) True
(b) False

81. Ceiling-support wires used for the support of electrical raceways and cables within nonfire-rated assemblies shall be distinguishable from the suspended-ceiling framing support wires.

(a) True
(b) False

82. No conductor shall be used where its operating temperature exceeds that designated for the type of insulated conductor involved.

(a) True
(b) False

83. Type UF cable can be used for service conductors.

(a) True
(b) False

84. Threadless couplings and connectors used on threaded IMC ends shall be listed for the purpose.

(a) True
(b) False

85. Article 392 limits the use of cable trays to industrial establishments only.

(a) True
(b) False

86. Luminaires can be used as a raceway for circuit conductors when listed and marked for use as a raceway.

(a) True
(b) False

87. Meters, instruments, instrument transformers, resistors, rectifiers, and thermionic tubes in Class I, Division 1 locations shall be installed in enclosures identified for use in Class I, Division 1 locations.

(a) True
(b) False

88. Battery chargers and the batteries being charged can be located n any area of a commercial garage.

(a) True
(b) False

89. Each circuit leading to motor fuel dispensing equipment shall be provided with a clearly identified and readily accessible switch or other acceptable means to disconnect all conductors of the circuit.

(a) True
(b) False

90. Metal faceplates for switches and receptacles can be connected to the equipment grounding conductor by means of a metal mounting screw(s) securing the faceplate to a grounded outlet box or grounded wiring device in patient care areas.

(a) True
(b) False

91. Service equipment shall be mounted on solid backing and be installed so as to be protected from the weather, unless of weatherproof construction.

(a) True
(b) False

92. The purpose of the equipotential plane is to prevent a difference in voltage from developing within the plane area.

(a) True
(b) False

93. Equipment enclosures on piers must be located so as not to interfere with mooring lines.

(a) True
(b) False

94. Type NM cables can be used for feeder temporary installations without height limitation and without concealment.

(a) True
(b) False

95. A spa or hot tub is a hydromassage pool or tub and is not designed to have the contents drained or discharged after each use.

(a) True
(b) False

96. An 8 AWG or larger solid copper equipotential bonding conductor shall be extended to service equipment to eliminate voltage gradients in the pool area.

    (a) True
    (b) False

97. Legally required standby system wiring can occupy the same raceways, cables, boxes, and cabinets with other general-purpose wiring.

    (a) True
    (b) False

98. Access to electrical equipment shall not be denied by an accumulation of remote-control, signaling, or power-limited cables that prevent removal of panels, including suspended-ceiling panels.

    (a) True
    (b) False

99. Cables and conductors of two or more power-limited fire alarm circuits can be installed in the same cable, enclosure, cable tray or raceway.

    (a) True
    (b) False

100. Cabinets, cutout boxes, and meter socket enclosures installed in wet locations shall be _____.

    (a) waterproof
    (b) raintight
    (c) weatherproof
    (d) watertight

FINAL EXAM 2: Practice Questions in Random Order—Article 90 through Annex C

Use the 2008 *NEC* to answer the following questions.

1. The minimum working space on a circuit that is 120 volts-to-ground, with exposed live parts on one side and no live or grounded parts on the other side of the working space, is _____.

   (a) 1 ft
   (b) 3 ft
   (c) 4 ft
   (d) 6 ft

2. Branch-circuit conductors within _____ of a ballast shall have an insulation temperature rating not lower than 90°C (194°F).

   (a) 1 in.
   (b) 3 in.
   (c) 6 in.
   (d) 8 in.

3. The minimum clearance for overhead feeder conductors not exceeding 600V that pass over commercial areas subject to truck traffic is _____.

   (a) 10 ft
   (b) 12 ft
   (c) 15 ft
   (d) 18 ft

4. The minimum clearance for service-drop conductors not exceeding 600V that pass over commercial areas subject to truck traffic is _____.

   (a) 10 ft
   (b) 12 ft
   (c) 15 ft
   (d) 18 ft

5. Hallways in dwelling units that are _____ long or longer require a receptacle outlet.

   (a) 6 ft
   (b) 8 ft
   (c) 10 ft
   (d) 12 ft

6. Where installed in a wet location, all _____ receptacle(s) shall be listed as weather-resistant.

   (a) 125V, 30A nonlocking
   (b) 250V, 15A nonlocking
   (c) 125V, 30A locking
   (d) 250V, 15A locking

7. When service-entrance conductors exceed 1,100 kcmil for copper, the required grounded conductor for the service shall be sized not less than _____ percent of the area of the largest ungrounded service-entrance conductor.

   (a) 9
   (b) 11
   (c) 12½
   (d) 15

8. The maximum length of a feeder tap conductor in a high-bay manufacturing building over 35 ft high shall be _____.

   (a) 15 ft
   (b) 20 ft
   (c) 50 ft
   (d) 100 ft

Practice Questions in Random Order—Article 90 through Annex C — Final Exam 2

9. In clothes closets, recessed incandescent or LED luminaires with a completely enclosed lamp can be installed in the wall or the ceiling, provided there is a minimum clearance of _____ between the luminaire and the nearest point of a storage space.

   (a) 3 in.
   (b) 6 in.
   (c) 9 in.
   (d) 12 in.

10. What is the park electrical wiring system load, after applying the demand factors, for a mobile home park having six mobile homes?

    (a) 4,640 VA
    (b) 27,840 VA
    (c) 96,000 VA
    (d) 144,000 VA

11. Switches shall be located at least _____, measured horizontally, from the inside walls of an indoor spa or hot tub.

    (a) 4 ft
    (b) 5 ft
    (c) 7 ft 6 in.
    (d) 12 ft

12. The demand factor for marina load calculations of 45 shore power receptacles is _____ percent.

    (a) 40
    (b) 50
    (c) 60
    (d) 70

13. Circuit breakers having an interrupting current rating of other than _____ shall have their interrupting rating marked on the circuit breaker.

    (a) 5,000A
    (b) 10,000A
    (c) 22,000A
    (d) 50,000A

14. Where run vertically, nonmetallic wireways shall be securely supported at intervals not exceeding _____, unless listed otherwise, with no more than one joint between supports.

    (a) 4 ft
    (b) 6 ft
    (c) 8 ft
    (d) 10 ft

15. Where multiple services or separately derived systems supply portable structures of carnival, circuses, or fairs, all structures separated by less than _____ shall be bonded together at the portable structures.

    (a) 6 ft
    (b) 8 ft
    (c) 10 ft
    (d) 12 ft

16. What is the minimum cover requirement for direct burial Type UF cable installed outdoors that supplies a 120V, 30A circuit?

    (a) 6 in.
    (b) 12 in.
    (c) 18 in.
    (d) 24 in.

17. The _____ pool bonding conductor shall be connected to the equipotential bonding grid either by exothermic welding or by pressure connectors in accordance with 250.8.

    (a) 8 AWG
    (b) insulated or bare
    (c) copper
    (d) all of these

18. For stationary motors of 2 hp or less and 300V or less on ac circuits, the controller can be an ac-rated only general-use snap switch where the motor full-load current rating is not more than _____ percent of the rating of the switch.

    (a) 50
    (b) 60
    (c) 70
    (d) 80

19. Conductors that supply one or more resistance welders shall be protected by an overcurrent device rated or set at not more than _____ percent of the conductor rating.

    (a) 80
    (b) 100
    (c) 125
    (d) 300

20. Conductors supplying several motors shall not be sized smaller than _____ percent of the full-load current rating of the highest rated motor plus the sum of the full-load current ratings of all other motors in the group, plus the ampacity or other loads.

    (a) 80
    (b) 100
    (c) 125
    (d) 150

21. For cord-and-plug-connected appliances, _____ plug and receptacle is permitted to serve as the disconnecting means.

    (a) a labeled
    (b) an accessible
    (c) a metal enclosed
    (d) none of these

22. In Class I, Division 1 locations, all apparatus and equipment of signaling, alarm, remote-control, and communications systems _____ shall be identified for Class I, Division 1 locations.

    (a) above 50V
    (b) above 100 volts-to-ground
    (c) regardless of voltage
    (d) except under 24V

23. The _____ rating of a conductor is the maximum temperature, at any location along its length, which the conductor can withstand over a prolonged period of time without serious degradation.

    (a) ambient
    (b) temperature
    (c) maximum withstand
    (d) short-circuit

24. Receptacles incorporating an isolated grounding conductor connection intended for the reduction of electrical noise shall be identified by _____ on the face of the receptacle.

    (a) an orange triangle
    (b) a green triangle
    (c) the color orange
    (d) the engraved word "ISOLATED"

25. Each circuit leading to or through motor fuel dispensing equipment, including equipment for remote pumping systems, shall be provided with a switch or other acceptable means to disconnect _____ from the source of supply all conductors of the circuit, including the grounded conductor, if any.

    (a) automatically
    (b) simultaneously
    (c) manually
    (d) individually

26. A pool transformer used for the supply of underwater luminaires, together with the transformer enclosure, shall _____.

    (a) be of the isolated-winding type with an ungrounded secondary
    (b) have a grounded metal barrier between the primary and secondary windings
    (c) be listed for the purpose
    (d) all of these

27. Electrically conductive materials that are likely to _____ in ungrounded systems shall be connected together and to the supply system grounded equipment in a manner that creates a low-impedance path for ground-fault current that is capable of carrying the maximum fault current likely to be imposed on it.

    (a) become energized
    (b) require service
    (c) be removed
    (d) be coated with paint or nonconductive materials

28. When HDPE conduit enters a box, fitting, or other enclosure, a(n) _____ shall be provided to protect the conductor from abrasion, unless the box design provides such protection.

    (a) bushing
    (b) adapter
    (c) a or b
    (d) reducing bushing

29. Bends in PVC conduit shall be made only _____.

    (a) by hand forming the bend
    (b) with bending equipment identified for the purpose
    (c) with a truck exhaust pipe
    (d) by use of an open flame torch

30. A single receptacle is a single contact device with no other contact device on the same _____.

    (a) circuit
    (b) yoke
    (c) run
    (d) equipment

31. Short sections of metal enclosures or raceways used to provide support or protection of _____ from physical damage shall not be required to be connected to the equipment grounding conductor.

    (a) conduit
    (b) under 600V feeders
    (c) cable assemblies
    (d) none of these

32. In completed installations, each outlet box shall have a _____.

    (a) cover
    (b) faceplate
    (c) canopy
    (d) any of these

33. LFNC shall be permitted for _____.

    (a) direct burial where listed and marked for the purpose
    (b) exposed work
    (c) outdoors where listed and marked for this purpose
    (d) all of these

34. A(n) _____ shall be located at the point of entry to an elevator machine room.

    (a) directory
    (b) lighting switch
    (c) control circuit disconnecting means
    (d) emergency exit map

35. A _____ location would be protected from weather and not subject to saturation with water or other liquids; such locations include partially protected locations under canopies, marquees, and roofed open porches.

    (a) dry
    (b) damp
    (c) wet
    (d) moist

36. A device intended for the protection of personnel that functions to de-energize a circuit or portion thereof within an established period of time when a current to ground exceeds the values established for a Class A device, is a(n) _____.

    (a) dual-element fuse
    (b) inverse time breaker
    (c) ground-fault circuit interrupter
    (d) safety switch

37. In commercial garages, GFCI protection shall be provided on all 15A and 20A, 125V receptacles installed where _____ is(are) to be used.

    (a) electrical diagnostic equipment
    (b) electrical hand tools
    (c) portable lighting equipment
    (d) any of these

38. In emergency systems, only appliances and lamps required for emergency use, shall be supplied by _____.

    (a) emergency circuits
    (b) multiwire branch circuits
    (c) HID-rated circuit breakers
    (d) a and b

39. Type MC cable can be unsupported where _____.

    (a) fished between concealed access points in finished buildings or structures and support is impracticable
    (b) not more than 2 ft in length at terminals where flexibility is necessary
    (c) not more than 6 ft from the last point of support within an accessible ceiling for the connection of luminaires or other electrical equipment
    (d) a or c

40. Where raceways or cables enter above the level of uninsulated live parts of cabinets, cutout boxes, and meter socket enclosures in a wet location, a(n) _____ shall be used.

    (a) fitting listed for wet locations
    (b) explosionproof seal
    (c) fitting listed for damp locations
    (d) insulated fitting

Final Exam 2 — Practice Questions in Random Order—Article 90 through Annex C

41. _____ is a raceway of circular cross section having an outer liquidtight, nonmetallic, sunlight-resistant jacket over an inner flexible metal core.

    (a) FMC
    (b) LFNMC
    (c) LFMC
    (d) none of these

42. Compliance with the provisions of the *NEC* will result in _____.

    (a) good electrical service
    (b) an efficient electrical system
    (c) an electrical system essentially free from hazard
    (d) all of these

43. Where more than one nominal voltage system exists in a building, each _____ conductor of a branch circuit shall be identified by phase and system at all termination, connection, and splice points.

    (a) grounded
    (b) ungrounded
    (c) grounding
    (d) all of these

44. When equipment grounding conductors are installed in panelboards, a _____ shall be secured inside the cabinet.

    (a) grounded conductor
    (b) terminal lug
    (c) terminal bar
    (d) none of these

45. Metal water piping system(s) shall be bonded to the _____.

    (a) grounded conductor at the service
    (b) service equipment enclosure
    (c) equipment grounding bar or bus at any panelboard within a single occupancy building
    (d) a or b

46. A generator set for a legally required standby system shall _____.

    (a) have means for automatically starting the prime mover
    (b) have two hours of fuel supply for full demand operation available on-site if the prime mover is an internal combustion engine
    (c) not be solely dependent on a public utility gas system
    (d) all of these

47. Armored cable shall not be installed _____.

    (a) in damp or wet locations
    (b) where subject to physical damage
    (c) where exposed to corrosive fumes or vapors
    (d) all of these

48. Where exposed to direct rays of the sun, insulated conductors and jacketed cables installed in cable trays shall be _____ as being sunlight resistant.

    (a) listed
    (b) approved
    (c) identified
    (d) none of these

49. Section signs shall be marked to indicate that field wiring and _____ are required.

    (a) listing
    (b) labeling
    (c) inspection
    (d) installation instructions

50. Where corrosion protection is necessary for ferrous metal equipment and the conduit is threaded in the field, the threads shall be coated with a(n) _____, electrically conductive, corrosion-resistant compound.

    (a) marked
    (b) listed
    (c) labeled
    (d) approved

51. Decorative lighting and similar accessories used for holiday lighting and similar purposes shall be _____.

    (a) marked
    (b) listed
    (c) arc-fault protected
    (d) GFCI-protected

52. Luminaires installed in Class II, Division 1 locations shall be identified for Class II locations and shall be clearly marked to indicate the _____.

    (a) maximum wattage of lamps for which designed
    (b) minimum conductor size
    (c) maximum overcurrent protection permitted
    (d) all of these

53. Tap connections to a common grounding electrode conductor for multiple separately derived systems may be made to a copper or aluminum busbar that is _____.

    (a) not over ½ in. x 4 in.
    (b) not over ¼ in. x 2 in.
    (c) at least ¼ in. x 2 in.
    (d) a and c

54. EMT couplings and connectors shall be made up _____.

    (a) of metal
    (b) in accordance with industry standards
    (c) tight
    (d) none of these

55. Disconnecting means for air-conditioning or refrigerating equipment can be installed _____ the air-conditioning or refrigerating equipment, but not on panels that are designed to allow access to the equipment, and not over the equipment nameplate.

    (a) on
    (b) within
    (c) a or b
    (d) none of these

56. In health care facilities, receptacles with insulated grounding terminals shall be identified; such identification shall be visible _____.

    (a) on rough-in inspection
    (b) by removal of faceplates
    (c) after installation
    (d) on the blueprints only

57. Overcurrent protection for conductors and equipment is designed to _____ the circuit if the current reaches a value that will cause an excessive or dangerous temperature in conductors or conductor insulation.

    (a) open
    (b) close
    (c) monitor
    (d) record

58. On a three-phase, 4-wire, delta-connected service where the midpoint of one phase winding is grounded, the service conductor having the higher phase voltage-to-ground shall be durably and permanently marked by an outer finish that is _____ in color at each termination or junction point.

    (a) orange
    (b) red
    (c) blue
    (d) any of these

59. Electrical equipment that depends on _____ for cooling of exposed surfaces shall be installed so that airflow over such surfaces is not prevented by walls or by adjacent installed equipment.

    (a) outdoor air
    (b) natural circulation of air and convection principles
    (c) artificial cooling and circulation
    (d) magnetic induction

60. The choice of overcurrent protective devices so that an overcurrent condition will be localized and restrict outages to the circuit or equipment affected, is called _____.

    (a) overcurrent protection
    (b) interrupting capacity
    (c) selective coordination
    (d) overload protection

61. Threadless couplings approved for use with IMC in wet locations shall be _____.

    (a) rainproof
    (b) listed for wet locations
    (c) moistureproof
    (d) concrete-tight

62. Threadless couplings and connectors used with RMC buried in masonry or concrete shall be the _____ type.

    (a) raintight
    (b) wet and damp location
    (c) nonabsorbent
    (d) concrete-tight

63. Electrical equipment rooms or enclosures housing electrical apparatus that are controlled by a lock(s) shall be considered _____ to qualified persons.

    (a) readily accessible
    (b) accessible
    (c) available
    (d) none of these

64. Hydromassage bathtub electrical equipment shall be _____ without damaging the building structure or building finish.

    (a) readily accessible
    (b) accessible
    (c) within sight
    (d) none of these

65. Metal wireways are sheet metal troughs with _____ for housing and protecting electric conductors and cable.

    (a) removable covers
    (b) hinged covers
    (c) a or b
    (d) none of these

66. A device or utilization equipment wider than a single 2 in. device box shall have _____ volume allowances provided for each gang required for mounting.

    (a) single
    (b) double
    (c) triple
    (d) none of these

67. Luminaires for fixed lighting in Class III shall have enclosures for lamps and lampholders that are designed to prevent the escape of _____.

    (a) sparks
    (b) burning material
    (c) hot metal
    (d) all of these

68. Where batteries are used for _____ in auxiliary engines of emergency systems, the authority having jurisdiction shall require periodic maintenance.

    (a) starting
    (b) control or ignition
    (c) a and b
    (d) none of these

69. _____ is a nonferrous, nonmagnetic metal that has no heating due to hysteresis heating.

    (a) Steel
    (b) Iron
    (c) Aluminum
    (d) all of these

70. Intrinsically safe circuits must be identified at _____ locations to prevent unintentional interference during testing and servicing.

    (a) terminal
    (b) junction
    (c) a or b
    (d) a and b

71. Additional services shall be permitted for a single building or other structure sufficiently large to make two or more services necessary if permitted by _____.

    (a) the registered design professional
    (b) special permission
    (c) the engineer of record
    (d) master electricians

72. A ground ring encircling the building or structure can be used as a grounding electrode when _____.

    (a) the ring is in direct contact with the earth
    (b) the ring consists of at least 20 ft of bare conductor
    (c) the bare copper conductor is not smaller than 2 AWG
    (d) all of these

73. A wet-niche luminaire is intended to be installed in a _____.

    (a) transformer
    (b) forming shell
    (c) hydromassage bathtub
    (d) all of these

74. The authority having jurisdiction shall not be required to enforce any requirements of Chapter 7 (Signaling Conditions) or Chapter 8 (Communications Systems).

    (a) True
    (b) False

75. Ungrounded describes "not connected to ground or a conductive body that extends the ground connection."

    (a) True
    (b) False

Practice Questions in Random Order—Article 90 through Annex C — Final Exam 2

76. All 15A or 20A, 120V branch circuits that supply outlets in dwelling unit family rooms, dining rooms, living rooms, parlors, libraries, dens, bedrooms, sunrooms, recreation rooms, closets, hallways, or similar rooms or areas shall be AFCI-protected by a listed arc-fault circuit interrupter of the combination type.

    (a) True
    (b) False

77. In dwelling units, outdoor receptacles can be connected to one of the 20A small-appliance branch circuits.

    (a) True
    (b) False

78. Dwelling unit or mobile home feeder conductors need not be larger than the service conductors and can be sized according to 310.15(B)(6).

    (a) True
    (b) False

79. The grounding electrode for a lightning protection system can be used as a grounding electrode system for the buildings or structures.

    (a) True
    (b) False

80. Equipment grounding conductors for feeder taps are not required to be larger than the tap conductors.

    (a) True
    (b) False

81. Surge protective devices shall be listed.

    (a) True
    (b) False

82. Raceways can be used as a means of support of Class 2 circuit conductors or cables that connect to the same equipment.

    (a) True
    (b) False

83. Flat Type NM cables shall not be stapled on edge.

    (a) True
    (b) False

84. Type UF cable can be used in commercial garages.

    (a) True
    (b) False

85. The maximum size conductors permitted in a metal surface raceway shall not be larger than that for which the wireway is designed.

    (a) True
    (b) False

86. Switches shall not be installed within wet locations in tub or shower spaces unless installed as part of a listed tub or shower assembly.

    (a) True
    (b) False

87. It is frequently possible to locate much of the equipment in less hazardous or unclassified locations and thus reduce the amount of special equipment required.

    (a) True
    (b) False

88. When provisions for limited flexibility are necessary in Class I, Division 2 locations, FMC with listed fittings can be used.

    (a) True
    (b) False

89. Circuit breakers in Class I, Division 2 locations that are not hermetically sealed or oil-immersed shall be installed in a Class I, Division 1 enclosure.

    (a) True
    (b) False

90. Type NM cable shall be permitted in agricultural buildings.

    (a) True
    (b) False

91. Type NM cables can be used for branch-circuit temporary installations without height limitation and without concealment.

    (a) True
    (b) False

92. Nonmetallic raceways can be installed within the raised floor area of an information technology equipment room.

    (a) True
    (b) False

Final Exam 2 — Practice Questions in Random Order—Article 90 through Annex C

93. When sizing a feeder for the fixed appliance loads in dwelling units, a demand factor of 75 percent of the total nameplate ratings can be applied if there are _____ or more appliances fastened in place on the same feeder.

    (a) two
    (b) three
    (c) four
    (d) five

94. The walls and roofs of transformer vaults shall be constructed of materials that have adequate structural strength for the conditions with a minimum fire-resistance rating of _____ hours.

    (a) two
    (b) three
    (c) four
    (d) six

95. Surface-mounted switches or circuit breakers in a damp or wet location shall be enclosed in a _____ enclosure or cabinet that complies with 312.2.

    (a) weatherproof
    (b) rainproof
    (c) watertight
    (d) raintight

96. Type AC cable is permitted in _____ installations.

    (a) wet
    (b) cable tray
    (c) exposed
    (d) b and c

97. Flexible cords and cables can be used for _____.

    (a) wiring of luminaires
    (b) connection of portable luminaires or appliances
    (c) connection of utilization equipment to facilitate frequent interchange
    (d) all of these

98. Unit equipment for emergency systems shall be on the same branch circuit that serves the normal lighting in the area and connected _____ any local switches.

    (a) with
    (b) ahead of
    (c) after
    (d) downstream of

99. Surface-mounted luminaires _____ located over or within 5 ft, measured horizontally, from the inside walls of an indoor spa or hot tub can be installed at less than 7 ft 6 in. above the maximum water level when GFCI-protected.

    (a) with a glass or plastic globe
    (b) with a nonmetallic body or a metallic body isolated from contact
    (c) suitable for use in a damp location
    (d) all of these

100. A _____ is an area that includes a basin with a toilet, tub, or shower.

    (a) bath area
    (b) bathroom
    (c) rest area
    (d) none of these

Did you find this book helpful? Want more? Order our 2008 Editions of *Master/Contractor* or *Journeyman Simulated Exams* which cover Theory, Code and Calculation questions in a realistic exam format.

Visit www.MikeHolt.com.

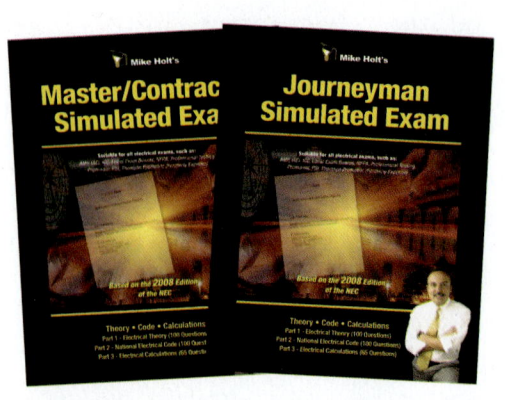

Discount 25% off

Supreme Estimating Library

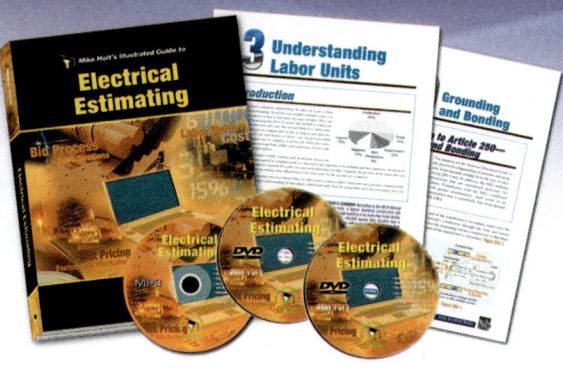

The Supreme Estimating Libraries include a textbook, instructor's guide, two DVDs and an MP3 audio CD:

- **Electrical Estimating textbook/workbook and instructor's guide**
- *Electrical Estimating two 6-hour DVDs*
- *MP3 audio CD containing 12 hours of audio on one disk*

Please contact the office directly to place your order and be sure to mention the coupon code DC08EXB.

Discount 25% off

National Electrical Code Library

The NEC Library includes three textbooks and ten DVDs

- **Understanding the National Electrical Code—Volume 1 textbook**
- **Understanding the National Electrical Code—Volume 2 textbook**
- **NEC Exam Practice Questions book**
- *General Requirements DVD*
- *Grounding versus Bonding DVD*
- *Wiring Methods and materials DVD*
- *Equipment for General Use DVD*
- *Special Occupancies DVD*
- *Special Equipment DVD*
- *Limited Energy and Communication Systems DVD*

Please contact the office directly to place your order and be sure to mention the coupon code DC08EXB.

Discount 25% off

Grounding versus Bonding Library

The Grounding versus Bonding Library includes a textbook/workbook, MP3 Audio CD, 16-Hour Online Program, and two DVDs.

- **Grounding versus Bonding textbook/workbook**
- *Grounding versus Bonding two 4.5 hour DVDs*
- *16-Hour Online Grounding versus Bonding Program*
- *Grounding versus Bonding MP3 Audio CD*

Please contact the office directly to place your order and be sure to mention the coupon code DC08EXB.

Order today to take advantage of these special offers! Call 1.888.NEC.CODE
For access to lots of great free resources visit www.MikeHolt.com

Mike Holt Enterprises, Inc.
3604 Parkway Blvd. Suite 3, Leesburg, FL 34748

Chapter 4 Notes

CHAPTER 4 Notes

Chapter 4 Notes

CHAPTER 4 Notes

Chapter 4 Notes

Chapter 4 Notes

CHAPTER 4 Notes